TRIGONOMETRY

Definition of the Six Trigonometric Functions

Right triangle definitions, where $0 < \theta < \pi/2$.

$$\sin \theta = \frac{\text{opp}}{\text{hyp}} \quad \csc \theta = \frac{\text{hyp}}{\text{opp}}$$

$$\cos \theta = \frac{\text{adj}}{\text{hyp}} \quad \sec \theta = \frac{\text{hyp}}{\text{adj}}$$

$$\tan \theta = \frac{\text{opp}}{\text{adj}} \quad \cot \theta = \frac{\text{adj}}{\text{opp}}$$

Circular function definitions, where θ is any angle.

$r = \sqrt{x^2 + y^2}$

$$\sin \theta = \frac{y}{r} \quad \csc \theta = \frac{r}{y}$$

$$\cos \theta = \frac{x}{r} \quad \sec \theta = \frac{r}{x}$$

$$\tan \theta = \frac{y}{x} \quad \cot \theta = \frac{x}{y}$$

Reciprocal Identities

$$\sin x = \frac{1}{\csc x} \quad \sec x = \frac{1}{\cos x} \quad \tan x = \frac{1}{\cot x}$$

$$\csc x = \frac{1}{\sin x} \quad \cos x = \frac{1}{\sec x} \quad \cot x = \frac{1}{\tan x}$$

Tangent and Cotangent Identities

$$\tan x = \frac{\sin x}{\cos x} \quad \cot x = \frac{\cos x}{\sin x}$$

Pythagorean Identities

$\sin^2 x + \cos^2 x = 1$

$1 + \tan^2 x = \sec^2 x \qquad 1 + \cot^2 x = \csc^2 x$

Cofunction Identities

$$\sin\left(\frac{\pi}{2} - x\right) = \cos x \quad \cos\left(\frac{\pi}{2} - x\right) = \sin x$$

$$\csc\left(\frac{\pi}{2} - x\right) = \sec x \quad \tan\left(\frac{\pi}{2} - x\right) = \cot x$$

$$\sec\left(\frac{\pi}{2} - x\right) = \csc x \quad \cot\left(\frac{\pi}{2} - x\right) = \tan x$$

Reduction Formulas

$\sin(-x) = -\sin x \quad \cos(-x) = \cos x$

$\csc(-x) = -\csc x \quad \tan(-x) = -\tan x$

$\sec(-x) = \sec x \qquad \cot(-x) = -\cot x$

Sum and Difference Formulas

$\sin(u \pm v) = \sin u \cos v \pm \cos u \sin v$

$\cos(u \pm v) = \cos u \cos v \mp \sin u \sin v$

$$\tan(u \pm v) = \frac{\tan u \pm \tan v}{1 \mp \tan u \tan v}$$

Double-Angle Formulas

$\sin 2u = 2 \sin u \cos u$

$\cos 2u = \cos^2 u - \sin^2 u = 2 \cos^2 u - 1 = 1 - 2 \sin^2 u$

$$\tan 2u = \frac{2 \tan u}{1 - \tan^2 u}$$

Power-Reducing Formulas

$$\sin^2 u = \frac{1 - \cos 2u}{2}$$

$$\cos^2 u = \frac{1 + \cos 2u}{2}$$

$$\tan^2 u = \frac{1 - \cos 2u}{1 + \cos 2u}$$

Sum-to-Product Formulas

$$\sin u + \sin v = 2 \sin\left(\frac{u + v}{2}\right) \cos\left(\frac{u - v}{2}\right)$$

$$\sin u - \sin v = 2 \cos\left(\frac{u + v}{2}\right) \sin\left(\frac{u - v}{2}\right)$$

$$\cos u + \cos v = 2 \cos\left(\frac{u + v}{2}\right) \cos\left(\frac{u - v}{2}\right)$$

$$\cos u - \cos v = -2 \sin\left(\frac{u + v}{2}\right) \sin\left(\frac{u - v}{2}\right)$$

Product-to-Sum Formulas

$$\sin u \sin v = \frac{1}{2}[\cos(u - v) - \cos(u + v)]$$

$$\cos u \cos v = \frac{1}{2}[\cos(u - v) + \cos(u + v)]$$

$$\sin u \cos v = \frac{1}{2}[\sin(u + v) + \sin(u - v)]$$

$$\cos u \sin v = \frac{1}{2}[\sin(u + v) - \sin(u - v)]$$

Calculus of a Single Variable

Early Transcendental Functions

Fourth Edition

Calculus of a Single Variable
Early Transcendental Functions

Fourth Edition

Ron Larson
The Pennsylvania State University
The Behrend College

Robert Hostetler
The Pennsylvania State University
The Behrend College

Bruce H. Edwards
University of Florida

Houghton Mifflin Company   Boston   New York

Publisher: Richard Stratton
Sponsoring Editor: Cathy Cantin
Development Manager: Maureen Ross
Associate Editor: Yen Tieu
Editorial Associate: Elizabeth Kassab
Supervising Editor: Karen Carter
Senior Project Editor: Patty Bergin
Editorial Assistant: Julia Keller
Art and Design Manager: Gary Crespo
Executive Marketing Manager: Brenda Bravener-Greville
Senior Marketing Manager: Danielle Curran
Director of Manufacturing: Priscilla Manchester
Cover Design Manager: Tony Saizon

We have included examples and exercises that use real-life data as well as technology output from a variety of software. This would not have been possible without the help of many people and organizations. Our wholehearted thanks goes to all for their time and effort.

Cover photograph: "Music of the Spheres" by English sculptor John Robinson is a three-foot-tall sculpture in bronze that has one continuous edge. You can trace its edge three times around before returning to the starting point. To learn more about this and other works by John Robinson, see the Centre for the Popularisation of Mathematics, University of Wales, at *http://www.popmath.org.uk/sculpture/gallery2.html*.

Trademark Acknowledgments: TI is a registered trademark of Texas Instruments, Inc. Mathcad is a registered trademark of MathSoft, Inc. Windows, Microsoft, and MS-DOS are registered trademarks of Microsoft, Inc. Mathematica is a registered trademark of Wolfram Research, Inc. DERIVE is a registered trademark of Texas Instruments, Inc. IBM is a registered trademark of International Business Machines Corporation. Maple is a registered trademark of Waterloo Maple, Inc. HM ClassPrep is a trademark of Houghton Mifflin Company. Diploma is a registered trademark of Brownstone Research Group.

Printed in the U.S.A.

Library of Congress Control Number: 2005933919

Instructor's exam copy:
ISBN 13: 978-0-618-73070-4
ISBN 10: 0-618-73070-2
For orders, use student ISBNs:
ISBN 13: 978-0-618-60625-2
ISBN 10: 0-618-60625-4

4 5 6 7 8 9-DOW-10-09 08

Contents

Chapter 4   Applications of Differentiation   203

Chapter 5   Integration   283

Additional Appendices   The following appendices are available at the textbook website at
college.hmco.com/pic/larsoncalculussvetf4e, on the HM mathSpace® Student CD-ROM,
and the HM ClassPrep™ with HM Testing CD-ROM.

Welcome to *Calculus of a Single Variable: Early Transcendental Functions*, Fourth Edition. With each edition, we have listened to you, our users, and incorporated many of your suggestions for improvement.

1st

2nd

3rd

4th

A Text Formed by Its Users

Through your support and suggestions, the text has evolved over four editions to include these extensive enhancements:

- Comprehensive exercise sets containing a wide variety of problems such as skill-building exercises, applications, explorations, writing exercises, critical thinking exercises, and theoretical problems

- Abundant real-life applications that accurately represent the diverse uses of calculus

- Many open-ended activities and investigations

- Clear, uncluttered text presentation with full annotations and labels and a carefully planned page layout

- Comprehensive, four-color art program

- Comprehensive and mathematically rigorous text

- Technology used throughout as both a problem-solving tool and an investigative tool

- A comprehensive program of additional resources available in print, on CD-ROM, and online

- With 5 different volumes of the text available, you can choose the sequence, amount of content, and teaching approach that is best for you and your students (see pages xii–xiii)

- References to the history of calculus and to the mathematicians who developed it, including over 50 biographical sketches available on the HM mathSpace® Student CD-ROM

- References to over 50 articles from mathematical journals are available at *www.MathArticles.com*

What's New and Different in the Fourth Edition

In the Fourth Edition, we continue to offer instructors and students a text that is pedagogically sound, mathematically precise, and still comprehensible. There are many changes in the mathematics, prose, art, and design; the more significant changes are noted here.

- *New Chapter Openers* Each Chapter Opener has two parts: a description of the concepts that are covered in the chapter and a thought-provoking question about a real-life application from the chapter.

- *New Introduction to Differential Equations* The topic of differential equations is now introduced in Chapter 6 in the first semester of calculus, to better prepare students for their courses in disciplines such as engineering, physics, and chemistry. The chapter contains six sections: 6.1 *Slope Fields and Euler's Method*, 6.2 *Differential Equations: Growth and Decay*, 6.3 *Differential Equations: Separation of Variables*, 6.4 *The Logistic Equation*, 6.5 *First-Order Linear Differential Equations*, and 6.6 *Predator-Prey Differential Equations*.

- **Revised Exercise Sets** The exercise sets have been carefully and extensively examined to ensure they are rigorous and cover all topics suggested by our users. Many new skill-building and challenging exercises have been added.

- **Updated Data** All data in the examples and exercise sets have been updated.

- Eduspace® combines numerous dynamic resources with online home-work and testing materials to create a comprehensive online learning system. Students benefit from having immediate access to algorithmic tutorial practice, videos, and resources such as a color graphing calculator. Instructors benefit from time-saving grading resources, as well as dynamic instructional tools such as animations, explorations, and Computer Algebra System Labs.

- *Study and Solutions Guides* The worked-out solutions to the odd-numbered text exercises are now provided on a CD-ROM, in Eduspace®, and at *www.CalcChat.com*.

Although we carefully and thoroughly revised the text by enhancing the usefulness of some features and topics and by adding others, we did not change many of the things that our colleagues and the over two million students who have used this book have told us work for them. The *Calculus: Early Transcendental Functions*, Fourth Edition, program offers comprehensive coverage of the material required by students in calculus courses, including carefully stated theories and proofs.

We hope you will enjoy the Fourth Edition. We welcome any comments, as well as suggestions for continued improvement.

Ron Larson   Robert Hostetler   Bruce H. Edwards

Over 25 Years of Success, Leadership, and Innovation

The bestselling authors Larson, Hostetler, and Edwards continue to offer instructors and students more flexible teaching and learning options for the calculus course.

Calculus Textbook Options

CALCULUS: Early Transcendental Functions

The early transcendental functions calculus course is available in a variety of textbook configurations to address the different ways instructors teach—and students take—their classes.

Designed for the three-semester course

Designed for the two-semester course

Designed for single-semester course

Designed for third semester of Calculus

Also available for the *Calculus: Early Transcendental Functions*, Fourth Edition, program by Larson, Hostetler, and Edwards

- Eduspace® online learning system
- HM mathSpace® Student CD-ROMs
- Instructional DVDs and videos

For more information on these—and more—electronic course materials, please turn to pages xv-xvii.

CALCULUS

For instructors who prefer the traditional calculus course
sequence, the following textbook sequences are available.

- Calculus I, II, and III
- Calculus I and II and Calculus III
- Calculus I, Calculus II, and Calculus III

CALCULUS WITH PRECALCULUS

To give more students access to calculus by easing
the transition from precalculus, the following textbook
sequence is available.

- Precalculus and Calculus I, Calculus II, and Calculus III

CALCULUS WITH LATE TRIGONOMETRY

For instructors who introduce the trigonometric functions
in the second semester, the following textbook is available.

- Calculus I, II, and III

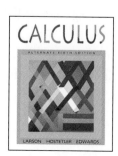

Comprehensive Calculus Resources

The Integrated Learning System for the *Calculus: Early Transcendental Functions*, Fourth Edition, program addresses the changing needs of today's instructors and students. Recognizing that the calculus course is presented in a variety of teaching and learning environments, we offer extensive resources that support the textbook program in print, CD-ROM, and online formats.

- Online homework practice
- Testing
- Tutoring
- Graded homework
- Classroom management
- Online course
- Interactive resources

ONE
INTEGRATED LEARNING SYSTEM

The teaching and learning resources you need in the format you prefer

The Integrated Learning System for the *Calculus: Early Transcendental Functions*, Fourth Edition, program offers dynamic teaching tools for instructors and interactive learning resources for students in the following flexible course delivery formats.

- Eduspace® online learning system
- HM mathSpace® Student CD-ROM
- Instructional DVDs and videos
- HM ClassPrep™ with HM Testing CD-ROM
- Companion Textbook Websites
- *Study and Solutions Guide* in two volumes available in print and electronically
- *Complete Solutions Guide* in three volumes (for instructors only) available only electronically

Enhanced! Eduspace® Online Calculus

Eduspace®, powered by Blackboard®, is ready to use and easy to integrate into the calculus course. It provides comprehensive homework exercises, tutorials, and testing keyed to the textbook by section.

Features

- Algorithmically generated tutorial exercises for unlimited practice
- Comprehensive problem sets for graded homework
- Interactive (multimedia) textbook pages with video lectures, animations, and much more.
- SMARTHINKING® live, online tutoring for students
- Color graphing calculator
- Ample prerequisite skills review with customized student self-study plan
- Chapter tests
- Link to CalcChat
- Electronic version of all textbook exercises
- Links to detailed, stepped-out solutions to odd-numbered textbook exercises

Enhanced! HM mathSpace® Student CD-ROM

For the student, HM mathSpace® CD-ROM offers a wealth of learning resources keyed to the textbook by section.

Features

- Algorithmically generated tutorial questions for unlimited practice of prerequisite skills
- Point-of-use links to additional tools, animations, and simulations
- Link to CalcChat
- Color graphing calculator
- Chapter tests

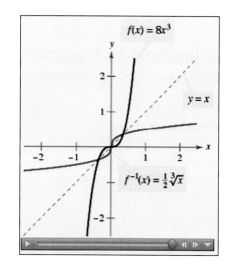

For additional information about the Larson, Hostetler, and Edwards Calculus program, go to *college.hmco.com/info/larsoncalculus*.

New! HM ClassPrep™ with HM Testing Instructor CD-ROM

This valuable CD-ROM contains an array of useful **instructor resources** keyed to the textbook.

Features

- **Complete Solutions Guide** by Bruce Edwards
  This resource contains worked-out solutions to all text-book exercises in electronic format. It is available in three volumes: Volume I covers Chapters 1–6, Volume II covers Chapters 7–11, and Volume III covers Chapters 11–15.

- **Instructor's Resource Guide** by Ann Rutledge Kraus
  This resource contains an abundance of resources keyed to the textbook by chapter and section, including chapter summaries, teaching strategies, multiple versions of chapter tests, final exams, and gateway tests, and suggested solutions to the Chapter Openers, Explorations, Section Projects, and Technology features in the text in electronic format.

- **Test Item File** The *Test Item File* contains a sample question for every algorithm in HM Testing in electronic format.

- HM Testing test generator
- Digital textbook art
- Textbook Appendices D–F, containing additional presentations with exercises covering precalculus review, rotation and the general second degree equation, and complex numbers.
- Downloadable graphing calculator programs

New! HM Testing (powered by Diploma™)

For the instructor, HM Testing is a robust test-generating system.

Features

- Comprehensive set of algorithmic test items
- Can produce chapter tests, cumulative tests, and final exams
- Online testing
- Gradebook function

Enhanced! Instructional DVDs and Videos

These comprehensive DVD and video presentations complement the textbook topic coverage and have a variety of uses, including supplementing an online or hybrid course, giving students the opportunity to catch up if they miss a class, and providing substantial course material for self-study and review.

Features

- Comprehensive topic coverage from Calculus I, II, and III
- Additional explanations of calculus concepts, sample problems, and applications

Enhanced! Companion Textbook Website

The free Houghton Mifflin website at *college.hmco.com/pic/larsoncalculussvetf4e* contains an abundance of instructor and student resources.

Features

- Downloadable graphing calculator programs
- Textbook Appendices D–F, containing additional presentations with exercises covering precalculus review, rotation and the general second-degree equation, and complex numbers
- Algebra Review Summary
- Calculus Labs
- 3-D rotatable graphs

Printed Resources

For the convenience of students, the Study and Solutions Guides are available as printed supplements, but are also available in electronic format.

Study and Solutions Guide by Bruce Edwards
    This student resource contains detailed, worked-out solutions to all odd-numbered textbook exercises. It is available in two volumes: Volume I covers Chapters 1–10 and Volume II covers Chapters 11–15.

For additional information about the Larson, Hostetler, and Edwards Calculus program, go to *college.hmco.com/info/larsoncalculus*.

Features

Chapter Openers

Each chapter opens with a real-life application of the concepts presented in the chapter, illustrated by a photograph. Open-ended and thought-provoking questions about the application encourage the student to consider how calculus concepts relate to real-life situations. A brief summary with a graphical component highlights the primary mathematical concepts presented in the chapter, and explains why they are important.

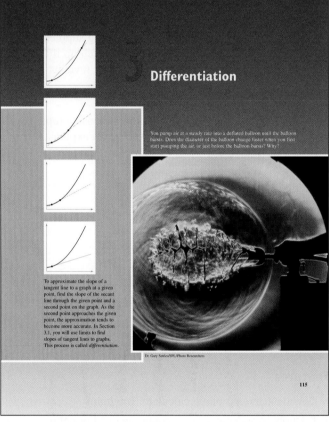

Section Openers

Every section begins with an outline of the key concepts covered in the section. This serves as a class planning resource for the instructor and a study and review guide for the student.

Explorations

For selected topics, Explorations offer the opportunity to discover calculus concepts before they are formally introduced in the text, thus enhancing student understanding. This optional feature can be omitted at the discretion of the instructor with no loss of continuity in the coverage of the material.

Historical Notes

Integrated throughout the text, Historical Notes help students grasp the basic mathematical foundations of calculus.

Theorems

All Theorems and Definitions are highlighted for emphasis and easy reference. Proofs are shown for selected theorems to enhance student understanding.

Study Tip

Located at point of use throughout the text, Study Tips advise students on how to avoid common errors, address special cases, and expand upon theoretical concepts.

Graphics

Numerous graphics throughout the text enhance student understanding of complex calculus concepts (especially in three-dimensional representations), as well as real-life applications.

Example

To enhance the usefulness of the text as a study and learning tool, the Fourth Edition contains numerous Examples. The detailed, worked-out Solutions (many with side comments to clarify the steps or the method) are presented graphically, analytically, and/or numerically to provide students with opportunities for practice and further insight into calculus concepts. Many Examples incorporate real-data analysis.

Open Exploration

Eduspace® contains Open Explorations, which investigate selected Examples using computer algebra systems (*Maple*, *Mathematica*, *Derive*, and *Mathcad*). The icon identifies these Examples.

Notes

Instructional Notes accompany many of the Theorems, Definitions, and Examples to offer additional insights or describe generalizations.

Exercises

The core of every calculus text, Exercises provide opportunities for exploration, practice, and comprehension. The Fourth Edition contains over 8000 Section and Chapter Review Exercises, carefully graded in each set from skill-building to challenging. The extensive range of problem types includes true/false, writing, conceptual, real-data modeling, and graphical analysis.

In Exercises 63–66, find k such that the line is tangent to the graph of the function.

Function	Line
63. $f(x) = x^2 - kx$	$y = 4x - 9$
64. $f(x) = k - x^2$	$y = -4x + 7$
65. $f(x) = \dfrac{k}{x}$	$y = -\dfrac{3}{4}x + 3$

In Exercises 81–86, describe the x-values at which f is differentiable.

81. $f(x) = \dfrac{1}{x+1}$ **82.** $f(x) = |x^2 - 9|$

83. $f(x) = (x-3)^{2/3}$ **84.** $f(x) = \dfrac{x^2}{x^2 - 4}$

Putnam Exam Challenge

186. Let $f(x) = a_1 \sin x + a_2 \sin 2x + \cdots + a_n \sin nx$, where $a_1, a_2, \ldots, a_n$ are real numbers and where n is a positive integer. Given that $|f(x)| \le |\sin x|$ for all real x, prove that $|a_1 + 2a_2 + \cdots + na_n| \le 1$.

187. Let k be a fixed positive integer. The nth derivative of $\dfrac{1}{x^k - 1}$ has the form

$$\frac{P_n(x)}{(x^k - 1)^{n+1}}$$

where $P_n(x)$ is a polynomial. Find $P_n(1)$.

These problems were composed by the Committee on the Putnam Prize Competition.
© The Mathematical Association of America. All rights reserved.

True or False?   In Exercises 183–185, determine whether the statement is true or false. If it is false, explain why or give an example that shows it is false.

183. If $y = (1 - x)^{1/2}$, then $y' = \frac{1}{2}(1 - x)^{-1/2}$.

184. If $f(x) = \sin^2(2x)$, then $f'(x) = 2(\sin 2x)(\cos 2x)$.

185. If y is a differentiable function of u, u is a differentiable function of v, and v is a differentiable function of x, then

167. *Modeling Data*   The table shows the temperature T (°F) at which water boils at selected pressures p (pounds per square inch). (*Source: Standard Handbook of Mechanical Engineers*)

p	5	10	14.696 (1 atm)	20
T	162.24°	193.21°	212.00°	227.96°

p	30	40	60	80	100
T	250.33°	267.25°	292.71°	312.03°	327.81°

A model that approximates the data is

$$T = 87.97 + 34.96 \ln p + 7.91 \sqrt{p}.$$

(a) Use a graphing utility to plot the data and graph the model.

(b) Find the rate of change of T with respect to p when $p = 10$ and $p = 70$.

P.S.   Problem Solving   201

P.S.   Problem Solving

See www.CalcChat.com for worked-out solutions to odd-numbered exercises.

1. Consider the graph of the parabola $y = x^2$.

(a) Find the radius r of the largest possible circle centered on the y-axis that is tangent to the parabola at the origin, as indicated in the figure. This circle is called the **circle of curvature** (see Section 12.5). Use a graphing utility to graph the circle and parabola in the same viewing window.

(b) Find the center $(0, b)$ of the circle of radius 1 centered on the y-axis that is tangent to the parabola at two points, as indicated in the figure. Use a graphing utility to graph the circle and parabola in the same viewing window.

Figure for 1(a)   **Figure for 1(b)**

2. Graph the two parabolas $y = x^2$ and $y = -x^2 + 2x - 5$ in the same coordinate plane. Find equations of the two lines simultaneously tangent to both parabolas.

3. (a) Find the polynomial $P_1(x) = a_0 + a_1 x$ whose value and slope agree with the value and slope of $f(x) = \cos x$ at the point $x = 0$.

(b) Find the polynomial $P_2(x) = a_0 + a_1 x + a_2 x^2$ whose value and first two derivatives agree with the value and first two derivatives of $f(x) = \cos x$ at the point $x = 0$. This polynomial is called the second-degree **Taylor polynomial** of $f(x) = \cos x$ at $x = 0$.

(c) Complete the table comparing the values of f and P_2. What do you observe?

x	-1.0	-0.1	-0.001	0	0.001	0.1	1.0
$\cos x$							
$P_2(x)$							

(d) Find the third-degree Taylor polynomial of $f(x) = \sin x$ at $x = 0$.

4. (a) Find an equation of the tangent line to the parabola $y = x^2$ at the point $(2, 4)$.

(b) Find an equation of the normal line to $y = x^2$ at the point $(2, 4)$. (The normal line is perpendicular to the tangent line.) Where does this line intersect the parabola a second time?

(c) Find equations of the tangent line and normal line to $y = x^2$ at the point $(0, 0)$.

(d) Prove that for any point $(a, b) \ne (0, 0)$ on the parabola $y = x^2$, the normal line intersects the graph a second time.

5. Find a third-degree polynomial $p(x)$ that is tangent to the line $y = 14x - 13$ at the point $(1, 1)$, and tangent to the line $y = -2x - 5$ at the point $(-1, -3)$.

6. Find a function of the form $f(x) = a + b \cos cx$ that is tangent to the line $y = 1$ at the point $(0, 1)$, and tangent to the line

$$y = x + \frac{3}{2} - \frac{\pi}{4}$$

at the point $\left(\dfrac{\pi}{4}, \dfrac{3}{2}\right)$.

7. The graph of the **eight curve**,

$$x^4 = a^2(x^2 - y^2), a \ne 0,$$

is shown below.

(a) Explain how you could use a graphing utility to obtain the graph of this curve.

(b) Use a graphing utility to graph the curve for various values of the constant a. Describe how a affects the shape of the curve.

(c) Determine the points on the curve where the tangent line is horizontal.

8. The graph of the **pear-shaped quartic**,

$$b^2 y^2 = x^3(a - x), a, b > 0,$$

is shown below.

(a) Explain how you could use a graphing utility to obtain the graph of this curve.

(b) Use a graphing utility to graph the curve for various values of the constants a and b. Describe how a and b affect the shape of the curve.

(c) Determine the points on the curve where the tangent line is horizontal.

P.S. Problem Solving

Each chapter concludes with a set of thought-provoking and challenging exercises that provide opportunities for the student to explore the concepts in the chapter further.

Technology

Throughout the text, the use of a graphing utility or computer algebra system is suggested as appropriate for problem-solving as well as exploration and discovery. For example, students may choose to use a graphing utility to execute complicated computations, to visualize theoretical concepts, to discover alternative approaches, or to verify the results of other solution methods. However, students are not required to have access to a graphing utility to use this text effectively. In addition to describing the benefits of using technology to learn calculus, the text also addresses its possible misuse or misinterpretation.

Additional Features

Additional teaching and learning resources are integrated throughout the textbook, including Section Projects, journal references, and Writing About Concepts Exercises.

Acknowledgments

We would like to thank the many people who have helped us at various stages of this project over the years. Their encouragement, criticisms, and suggestions have been invaluable to us.

For the Fourth Edition

Andre Adler
Illinois Institute of Technology

Evelyn Bailey
Oxford College of Emory University

Katherine Barringer
Central Virginia Community College

Robert Bass
Gardner-Webb University

Joy Becker
University of Wisconsin Stout

Michael Bezusko
Pima Community College

Bob Bradshaw
Ohlone College

Robert Brown
The Community College of Baltimore County (Essex Campus)

Joanne Brunner
DePaul University

Minh Bui
Fullerton College

Fang Chen
Oxford College of Emory University

Alex Clark
University of North Texas

Jeff Dodd
Jacksonville State University

Daniel Drucker
Wayne State University

Pablo Echeverria
Camden County College

Angela Hare
Messiah College

Karl Havlak
Angelo State University

James Herman
Cecil Community College

Xuezhang Hou
Towson University

Gene Majors
Fullerton College

Suzanne Molnar
College of St. Catherine

Karen Murany
Oakland Community College

Keith Nabb
Moraine Valley Community College

Stephen Nicoloff
Paradise Valley Community College

James Pommersheim
Reed College

James Ralston
Hawkeye Community College

Chip Rupnow
Martin Luther College

Mark Snavely
Carthage College

Ben Zandy
Fullerton College

For the Fourth Edition Technology Program

Jim Ball
Indiana State University

Arek Goetz
San Francisco State University

Marcelle Bessman
Jacksonville University

John Gosselin
University of Georgia

Tim Chappell
Penn Valley Community College

Shahryar Heydari
Piedmont College

Oiyin Pauline Chow
Harrisburg Area Community College

Douglas B. Meade
University of South Carolina

Julie M. Clark
Hollins University

Teri Murphy
University of Oklahoma

Jim Dotzler
Nassau Community College

Howard Speier
Chandler-Gilbert Community College

Murray Eisenberg
University of Massachusetts at Amherst

Reviewers of Previous Editions

Raymond Badalian
Los Angeles City College

Kathy Hoke
University of Richmond

Norman A. Beirnes
University of Regina

Howard E. Holcomb
Monroe Community College

Christopher Butler
Case Western Reserve University

Gus Huige
University of New Brunswick

Dane R. Camp
New Trier High School, IL

E. Sharon Jones
Towson State University

Jon Chollet
Towson State University

Robert Kowalczyk
University of Massachusetts–Dartmouth

Barbara Cortzen
DePaul University

Anne F. Landry
Dutchess Community College

Patricia Dalton
Montgomery College

Robert F. Lax
Louisiana State University

Luz M. DeAlba
Drake University

Beth Long
Pellissippi State Technical College

Dewey Furness
Ricks College

Gordon Melrose
Old Dominion University

Javier Garza
Tarleton State University

Bryan Moran
Radford University

Claire Gates
Vanier College

David C. Morency
University of Vermont

Lionel Geller
Dawson College

Guntram Mueller
University of Massachusetts–Lowell

Carollyne Guidera
University College of Fraser Valley

Donna E. Nordstrom
Pasadena City College

Irvin Roy Hentzel
Iowa State University

Larry Norris
North Carolina State University

Mikhail Ostrovskii
Catholic University of America

Lynn Smith
Gloucester County College

Jim Paige
Wayne State College

Linda Sundbye
Metropolitan State College of Denver

Eleanor Palais
Belmont High School, MA

Anthony Thomas
University of Wisconsin–Platteville

James V. Rauff
Millikin University

Robert J. Vojack
Ridgewood High School, NJ

Lila Roberts
Georgia Southern University

Michael B. Ward
Bucknell University

David Salusbury
John Abbott College

Charles Wheeler
Montgomery College

John Santomas
Villanova University

During the past four years, several users of the Third Edition wrote to us with suggestions. We considered each and every one of them when preparing the manuscript for the Fourth Edition. A special note of thanks goes to the instructors and to the students who have used earlier editions of the text.

We would like to thank the staff at Larson Texts, Inc., who assisted with proofreading the manuscript, preparing and proofreading the art package, and checking and typesetting the supplements.

On a personal level, we are grateful to our wives, Deanna Gilbert Larson, Eloise Hostetler, and Consuelo Edwards, for their love, patience, and support. Also, a special note of thanks goes to R. Scott O'Neil.

If you have suggestions for improving this text, please feel free to write to us. Over the years we have received many useful comments from both instructors and students, and we value these very much.

Ron Larson   Robert Hostetler   Bruce H. Edwards

Preparation for Calculus

Two types of racecars designed and built by NASCAR teams are short track cars and super-speedway (long track) cars. Super-speedway racecars are subjected to extensive testing in wind tunnels like the one shown in the photo. Short track racecars and super-speedway racecars are designed either to allow for as much downforce as possible or to reduce the amount of drag on the racecar. Which design do you think is used for each type of racecar? Why?

Mathematical models are commonly used to describe data sets. These models can be represented by many different types of functions such as linear, quadratic, cubic, rational, and trigonometric functions. In Chapter 1, you will review how to find, graph, and compare mathematical models for different data sets.

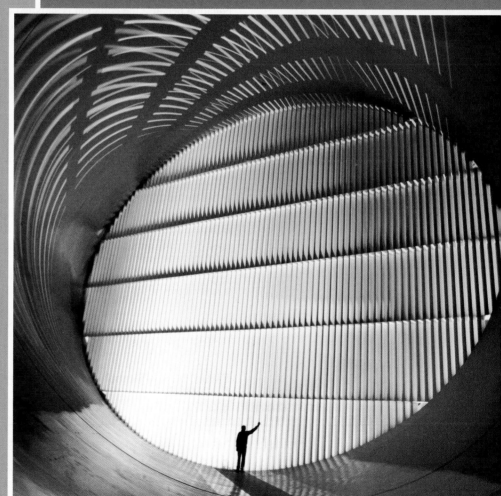

Graphs and Models

- Sketch the graph of an equation.
- Find the intercepts of a graph.
- Test a graph for symmetry with respect to an axis and the origin.
- Find the points of intersection of two graphs.
- Interpret mathematical models for real-life data.

The Graph of an Equation

In 1637, the French mathematician René Descartes revolutionized the study of mathematics by joining its two major fields—algebra and geometry. With Descartes's coordinate plane, geometric concepts could be formulated analytically and algebraic concepts could be viewed graphically. The power of this approach is such that within a century, much of calculus had been developed.

The same approach can be followed in your study of calculus. That is, by viewing calculus from multiple perspectives—*graphically*, *analytically*, and *numerically*—you will increase your understanding of core concepts.

Consider the equation $3x + y = 7$. The point $(2, 1)$ is a **solution point** of the equation because the equation is satisfied (is true) when 2 is substituted for x and 1 is substituted for y. This equation has many other solutions, such as $(1, 4)$ and $(0, 7)$. To systematically find other solutions, solve the original equation for y.

$$y = 7 - 3x \qquad \text{Analytic approach}$$

Then construct a table of values by substituting several values of x.

x	0	1	2	3	4
y	7	4	1	-2	-5

Numerical approach

From the table, you can see that $(0, 7)$, $(1, 4)$, $(2, 1)$, $(3, -2)$, and $(4, -5)$ are solutions of the original equation $3x + y = 7$. Like many equations, this equation has an infinite number of solutions. The set of all solution points is the **graph** of the equation, as shown in Figure 1.1.

NOTE   Even though we refer to the sketch shown in Figure 1.1 as the graph of $3x + y = 7$, it really represents only a *portion* of the graph. The entire graph would extend beyond the page.

In this course, you will study many sketching techniques. The simplest is point plotting—that is, you plot points until the basic shape of the graph seems apparent.

EXAMPLE 1   Sketching a Graph by Point Plotting

Sketch the graph of $y = x^2 - 2$.

Solution   First construct a table of values. Then plot the points shown in the table.

x	-2	-1	0	1	2	3
y	2	-1	-2	-1	2	7

Finally, connect the points with a *smooth curve*, as shown in Figure 1.2. This graph is a **parabola.** It is one of the conics you will study in Chapter 10.

RENÉ DESCARTES (1596–1650)

Descartes made many contributions to philosophy, science, and mathematics. The idea of representing points in the plane by pairs of real numbers and representing curves in the plane by equations was described by Descartes in his book *La Géométrie*, published in 1637.

Graphical approach: $3x + y = 7$
Figure 1.1

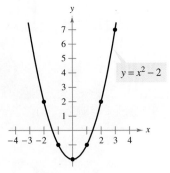

The parabola $y = x^2 - 2$
Figure 1.2

One disadvantage of point plotting is that to get a good idea about the shape of a graph, you may need to plot many points. With only a few points, you could misrepresent the graph. For instance, suppose that to sketch the graph of

$$y = \tfrac{1}{30}x(39 - 10x^2 + x^4)$$

you plotted only five points:

$$(-3, -3), (-1, -1), (0, 0), (1, 1), \text{ and } (3, 3)$$

as shown in Figure 1.3(a). From these five points, you might conclude that the graph is a line. This, however, is not correct. By plotting several more points, you can see that the graph is more complicated, as shown in Figure 1.3(b).

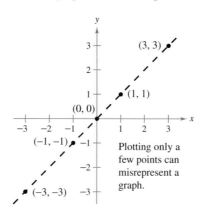

Plotting only a few points can misrepresent a graph.

(a)

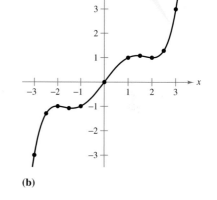

(b)

Figure 1.3

Comparing Graphical and Analytic Approaches Use a graphing utility to graph each of the following. In each case, find a viewing window that shows the important characteristics of the graph.

a. $y = x^3 - 3x^2 + 2x + 5$

b. $y = x^3 - 3x^2 + 2x + 25$

c. $y = -x^3 - 3x^2 + 20x + 5$

d. $y = 3x^3 - 40x^2 + 50x - 45$

e. $y = -(x + 12)^3$

f. $y = (x - 2)(x - 4)(x - 6)$

A purely graphical approach to this problem would involve a simple "guess, check, and revise" strategy. What types of things do you think an analytic approach might involve? For instance, does the graph have symmetry? Does the graph have turns? If so, where are they?

As you proceed through Chapters 2, 3, and 4 of this text, you will study many new analytic tools that will help you analyze graphs of equations such as these.

TECHNOLOGY   Technology has made sketching of graphs easier. Even with technology, however, it is possible to misrepresent a graph badly. For instance, each of the graphing utility screens in Figure 1.4 shows a portion of the graph of

$$y = x^3 - x^2 - 25.$$

From the screen on the left, you might assume that the graph is a line. From the screen on the right, however, you can see that the graph is not a line. Thus, whether you are sketching a graph by hand or using a graphing utility, you must realize that different "viewing windows" can produce very different views of a graph. In choosing a viewing window, your goal is to show a view of the graph that fits well in the context of the problem.

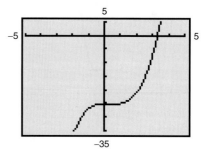

Graphing utility screens of $y = x^3 - x^2 - 25$

Figure 1.4

NOTE   In this text, we use the term *graphing utility* to mean either a graphing calculator or computer graphing software such as *Maple, Mathematica, Derive, Mathcad,* or the *TI-89.*

Intercepts of a Graph

Two types of solution points that are especially useful when graphing an equation are those having zero as their x- or y-coordinate. Such points are called **intercepts** because they are the points at which the graph intersects the x- or y-axis. The point $(a, 0)$ is an **x-intercept** of the graph of an equation if it is a solution point of the equation. To find the x-intercepts of a graph, let y be zero and solve the equation for x. The point $(0, b)$ is a **y-intercept** of the graph of an equation if it is a solution point of the equation. To find the y-intercepts of a graph, let x be zero and solve the equation for y.

NOTE   Some texts denote the x-intercept as the x-coordinate of the point $(a, 0)$ rather than the point itself. Unless it is necessary to make a distinction, we will use the term *intercept* to mean either the point or the coordinate.

It is possible for a graph to have no intercepts, or it might have several. For instance, consider the four graphs shown in Figure 1.5.

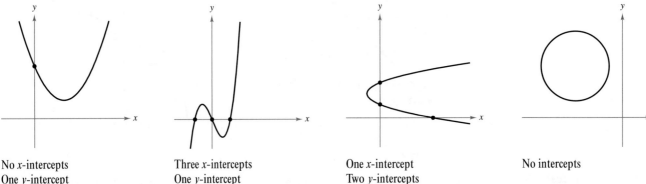

No x-intercepts
One y-intercept
Figure 1.5

Three x-intercepts
One y-intercept

One x-intercept
Two y-intercepts

No intercepts

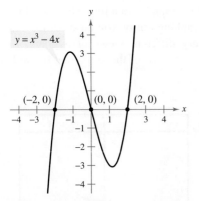

Intercepts of a graph
Figure 1.6

EXAMPLE 2   Finding x- and y-intercepts

Find the x- and y-intercepts of the graph of $y = x^3 - 4x$.

Solution   To find the x-intercepts, let y be zero and solve for x.

$$x^3 - 4x = 0 \qquad \text{Let } y \text{ be zero.}$$
$$x(x - 2)(x + 2) = 0 \qquad \text{Factor.}$$
$$x = 0, 2, \text{ or } -2 \qquad \text{Solve for } x.$$

Because this equation has three solutions, you can conclude that the graph has three x-intercepts:

$(0, 0), (2, 0), \text{ and} (-2, 0).$ x-intercepts

To find the y-intercepts, let x be zero. Doing so produces $y = 0$. So, the y-intercept is

$(0, 0).$ y-intercept

(See Figure 1.6.)

TECHNOLOGY   Example 2 uses an analytic approach to finding intercepts. When an analytic approach is not possible, you can use a graphical approach by finding the points where the graph intersects the axes. Use a graphing utility to approximate the intercepts.

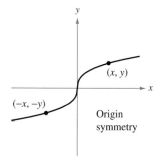

Figure 1.7

Symmetry of a Graph

Knowing the symmetry of a graph *before* attempting to sketch it is useful because you need only half as many points to sketch the graph. The following three types of symmetry can be used to help sketch the graph of an equation (see Figure 1.7).

1. A graph is **symmetric with respect to the *y*-axis** if, whenever (x, y) is a point on the graph, $(-x, y)$ is also a point on the graph. This means that the portion of the graph to the left of the *y*-axis is a mirror image of the portion to the right of the *y*-axis.

2. A graph is **symmetric with respect to the *x*-axis** if, whenever (x, y) is a point on the graph, $(x, -y)$ is also a point on the graph. This means that the portion of the graph above the *x*-axis is a mirror image of the portion below the *x*-axis.

3. A graph is **symmetric with respect to the origin** if, whenever (x, y) is a point on the graph, $(-x, -y)$ is also a point on the graph. This means that the graph is unchanged by a rotation of $180°$ about the origin.

Tests for Symmetry

1. The graph of an equation in *x* and *y* is symmetric with respect to the *y*-axis if replacing *x* by $-x$ yields an equivalent equation.

2. The graph of an equation in *x* and *y* is symmetric with respect to the *x*-axis if replacing *y* by $-y$ yields an equivalent equation.

3. The graph of an equation in *x* and *y* is symmetric with respect to the origin if replacing *x* by $-x$ and *y* by $-y$ yields an equivalent equation.

The graph of a polynomial has symmetry with respect to the *y*-axis if each term has an even exponent (or is a constant). For instance, the graph of

$$y = 2x^4 - x^2 + 2 \qquad \text{\textit{y}-axis symmetry}$$

has symmetry with respect to the *y*-axis. Similarly, the graph of a polynomial has symmetry with respect to the origin if each term has an odd exponent, as illustrated in Example 3.

EXAMPLE 3   Testing for Origin Symmetry

Show that the graph of

$$y = 2x^3 - x$$

is symmetric with respect to the origin.

Solution

$$
\begin{aligned}
y &= 2x^3 - x & &\text{Write original equation.} \\
-y &= 2(-x)^3 - (-x) & &\text{Replace } x \text{ by } -x \text{ and } y \text{ by } -y. \\
-y &= -2x^3 + x & &\text{Simplify.} \\
y &= 2x^3 - x & &\text{Equivalent equation}
\end{aligned}
$$

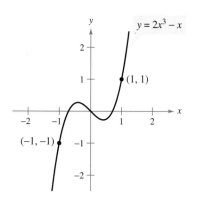

Origin symmetry
Figure 1.8

Because the replacement produces an equivalent equation, you can conclude that the graph of $y = 2x^3 - x$ is symmetric with respect to the origin, as shown in Figure 1.8.

Figure 1.9

 EXAMPLE 4   **Using Intercepts and Symmetry to Sketch a Graph**

Sketch the graph of $x - y^2 = 1$.

Solution   The graph is symmetric with respect to the x-axis because replacing y by $-y$ yields an equivalent equation.

$$x - y^2 = 1 \qquad \text{Write original equation.}$$
$$x - (-y)^2 = 1 \qquad \text{Replace } y \text{ by } -y.$$
$$x - y^2 = 1 \qquad \text{Equivalent equation}$$

This means that the portion of the graph below the x-axis is a mirror image of the portion above the x-axis. To sketch the graph, first sketch the portion above the x-axis. Then reflect in the x-axis to obtain the entire graph, as shown in Figure 1.9.

TECHNOLOGY   Graphing utilities are designed so that they most easily graph equations in which y is a function of x (see Section 1.3 for a definition of **function**). To graph other types of equations, you need to split the graph into two or more parts *or* you need to use a different graphing mode. For instance, to graph the equation in Example 4, you can split it into two parts.

$$y_1 = \sqrt{x - 1} \qquad \text{Top portion of graph}$$
$$y_2 = -\sqrt{x - 1} \qquad \text{Bottom portion of graph}$$

Points of Intersection

A **point of intersection** of the graphs of two equations is a point that satisfies both equations. You can find the points of intersection of two graphs by solving their equations simultaneously.

EXAMPLE 5   **Finding Points of Intersection**

Find all points of intersection of the graphs of $x^2 - y = 3$ and $x - y = 1$.

Solution   Begin by sketching the graphs of both equations on the *same* rectangular coordinate system, as shown in Figure 1.10. Having done this, it appears that the graphs have two points of intersection. To find these two points, you can use the following steps.

$$y = x^2 - 3 \qquad \text{Solve first equation for } y.$$
$$y = x - 1 \qquad \text{Solve second equation for } y.$$
$$x^2 - 3 = x - 1 \qquad \text{Equate } y\text{-values.}$$
$$x^2 - x - 2 = 0 \qquad \text{Write in general form.}$$
$$(x - 2)(x + 1) = 0 \qquad \text{Factor.}$$
$$x = 2 \text{ or } -1 \qquad \text{Solve for } x.$$

The corresponding values of y are obtained by substituting $x = 2$ and $x = -1$ into either of the original equations. Doing this produces two points of intersection:

$$(2, 1) \quad \text{and} \quad (-1, -2). \qquad \text{Points of intersection}$$

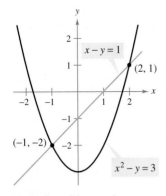

Two points of intersection
Figure 1.10

STUDY TIP   You can check the points of intersection from Example 5 by substituting into *both* of the original equations or by using the *intersect* feature of a graphing utility.

 indicates that in the HM mathSpace® *CD-ROM and the online* Eduspace® *system for this text, you will find an Open Exploration, which further explores this example using the computer algebra systems* Maple, Mathcad, Mathematica, *and* Derive.

Mathematical Models

Real-life applications of mathematics often use equations as **mathematical models.** In developing a mathematical model to represent actual data, you should strive for two (often conflicting) goals—accuracy and simplicity. That is, you want the model to be simple enough to be workable, yet accurate enough to produce meaningful results. Section 1.4 explores these goals more completely.

EXAMPLE 6   Comparing Two Mathematical Models

The Mauna Loa Observatory in Hawaii has been measuring the increasing concentration of carbon dioxide in Earth's atmosphere since 1958.

The Mauna Loa Observatory in Hawaii records the carbon dioxide concentration y (in parts per million) in Earth's atmosphere. The January readings for various years are shown in Figure 1.11. In the July 1990 issue of *Scientific American*, these data were used to predict the carbon dioxide level in Earth's atmosphere in the year 2035. The article used the quadratic model

$$y = 316.2 + 0.70t + 0.018t^2 \qquad \text{Quadratic model for 1960–1990 data}$$

where $t = 0$ represents 1960, as shown in Figure 1.11(a).

The data shown in Figure 1.11(b) represent the years 1980 through 2002 and can be modeled by

$$y = 306.3 + 1.56t \qquad \text{Linear model for 1980–2002 data}$$

where $t = 0$ represents 1960. What was the prediction given in the *Scientific American* article in 1990? Given the new data for 1990 through 2002, does this prediction for the year 2035 seem accurate?

(a)

(b)

Figure 1.11

Solution   To answer the first question, substitute $t = 75$ (for 2035) into the quadratic model.

$$y = 316.2 + 0.70(75) + 0.018(75)^2 = 469.95 \qquad \text{Quadratic model}$$

So, the prediction in the *Scientific American* article was that the carbon dioxide concentration in Earth's atmosphere would reach about 470 parts per million in the year 2035. Using the linear model for the 1980–2002 data, the prediction for the year 2035 is

$$y = 306.3 + 1.56(75) = 423.3. \qquad \text{Linear model}$$

So, based on the linear model for 1980–2002, it appears that the 1990 prediction was too high.

NOTE   The models in Example 6 were developed using a procedure called *least squares regression* (see Section 13.9). The quadratic and linear models have correlations given by $r^2 = 0.997$ and $r^2 = 0.996$, respectively. The closer r^2 is to 1, the "better" the model.

In Exercises 1–4, match the equation with its graph. [Graphs are labeled (a), (b), (c), and (d).]

(a)

(b)

(c)

(d)

1. $y = -\frac{1}{2}x + 2$

2. $y = \sqrt{9 - x^2}$

3. $y = 4 - x^2$

4. $y = x^3 - x$

In Exercises 5–14, sketch the graph of the equation by point plotting.

5. $y = \frac{3}{2}x + 1$

6. $y = 6 - 2x$

7. $y = 4 - x^2$

8. $y = (x - 3)^2$

9. $y = |x + 2|$

10. $y = |x| - 1$

11. $y = \sqrt{x} - 4$

12. $y = \sqrt{x + 2}$

13. $y = \frac{2}{x}$

14. $y = \frac{1}{x - 1}$

 In Exercises 15 and 16, describe the viewing window that yields the figure.

15. $y = x^3 - 3x^2 + 4$

16. $y = |x| + |x - 10|$

 In Exercises 17 and 18, use a graphing utility to graph the equation. Move the cursor along the curve to approximate the unknown coordinate of each solution point accurate to two decimal places.

17. $y = \sqrt{5 - x}$ (a) $(2, y)$ (b) $(x, 3)$

18. $y = x^5 - 5x$ (a) $(-0.5, y)$ (b) $(x, -4)$

In Exercises 19–26, find any intercepts.

19. $y = x^2 + x - 2$

20. $y^2 = x^3 - 4x$

21. $y = x^2\sqrt{25 - x^2}$

22. $y = (x - 1)\sqrt{x^2 + 1}$

23. $y = \dfrac{3(2 - \sqrt{x})}{x}$

24. $y = \dfrac{x^2 + 3x}{(3x + 1)^2}$

25. $x^2 y - x^2 + 4y = 0$

26. $y = 2x - \sqrt{x^2 + 1}$

In Exercises 27–38, test for symmetry with respect to each axis and to the origin.

27. $y = x^2 - 2$

28. $y = x^2 - x$

29. $y^2 = x^3 - 4x$

30. $y = x^3 + x$

31. $xy = 4$

32. $xy^2 = -10$

33. $y = 4 - \sqrt{x + 3}$

34. $xy - \sqrt{4 - x^2} = 0$

35. $y = \dfrac{x}{x^2 + 1}$

36. $y = \dfrac{x^2}{x^2 + 1}$

37. $y = |x^3 + x|$

38. $|y| - x = 3$

In Exercises 39–56, sketch the graph of the equation. Identify any intercepts and test for symmetry.

39. $y = -3x + 2$

40. $y = -\frac{1}{2}x + 2$

41. $y = \frac{1}{2}x - 4$

42. $y = \frac{2}{3}x + 1$

43. $y = 1 - x^2$

44. $y = x^2 + 3$

45. $y = (x + 3)^2$

46. $y = 2x^2 + x$

47. $y = x^3 + 2$

48. $y = x^3 - 4x$

49. $y = x\sqrt{x + 2}$

50. $y = \sqrt{9 - x^2}$

51. $x = y^3$

52. $x = y^2 - 4$

53. $y = \dfrac{1}{x}$

54. $y = \dfrac{10}{x^2 + 1}$

55. $y = 6 - |x|$

56. $y = |6 - x|$

 In Exercises 57–60, use a graphing utility to graph the equation. Identify any intercepts and test for symmetry.

57. $y^2 - x = 9$

58. $x^2 + 4y^2 = 4$

59. $x + 3y^2 = 6$

60. $3x - 4y^2 = 8$

In Exercises 61–68, find the point(s) of intersection of the graphs of the equations.

61. $x + y = 2$
   $2x - y = 1$

62. $2x - 3y = 13$
   $5x + 3y = 1$

63. $x^2 + y = 6$
   $x + y = 4$

64. $x = 3 - y^2$
   $y = x - 1$

The symbol *indicates an exercise in which you are instructed to use graphing technology or a symbolic computer algebra system. The solutions of other exercises may also be facilitated by use of appropriate technology.*

65. $x^2 + y^2 = 5$

$x - y = 1$

66. $x^2 + y^2 = 25$

$2x + y = 10$

67. $y = x^3$

$y = x$

68. $y = x^3 - 4x$

$y = -(x + 2)$

 In Exercises 69–72, use a graphing utility to find the point(s) of intersection of the graphs. Check your results analytically.

69. $y = x^3 - 2x^2 + x - 1$

$y = -x^2 + 3x - 1$

70. $y = x^4 - 2x^2 + 1$

$y = 1 - x^2$

71. $y = \sqrt{x + 6}$

$y = \sqrt{-x^2 - 4x}$

72. $y = -|2x - 3| + 6$

$y = 6 - x$

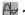 **73.** *Modeling Data*   The table shows the Consumer Price Index (CPI) for selected years. *(Source: Bureau of Labor Statistics)*

Year	1970	1975	1980	1985	1990	1995	2000
CPI	38.8	53.8	82.4	107.6	130.7	152.4	172.2

(a) Use the regression capabilities of a graphing utility to find a mathematical model of the form $y = at^2 + bt + c$ for the data. In the model, y represents the CPI and t represents the year, with $t = 0$ corresponding to 1970.

(b) Use a graphing utility to plot the data and graph the model. Compare the data with the model.

(c) Use the model to predict the CPI for the year 2010.

74. *Modeling Data*   The table shows the average number of acres per farm in the United States for selected years. *(Source: U.S. Department of Agriculture)*

Year	1950	1960	1970	1980	1990	2000
Acreage	213	297	374	426	460	434

(a) Use the regression capabilities of a graphing utility to find a mathematical model of the form $y = at^2 + bt + c$ for the data. In the model, y represents the average acreage and t represents the year, with $t = 0$ corresponding to 1950.

(b) Use a graphing utility to plot the data and graph the model. Compare the data with the model.

(c) Use the model to predict the average number of acres per farm in the United States in the year 2010.

75. *Break-Even Point*   Find the sales necessary to break even $(R = C)$ if the cost C of producing x units is

$$C = 5.5\sqrt{x} + 10{,}000 \qquad \text{Cost equation}$$

and the revenue R for selling x units is

$$R = 3.29x. \qquad \text{Revenue equation}$$

 76. *Copper Wire*   The resistance y in ohms of 1000 feet of solid copper wire at 77°F can be approximated by the model

$$y = \frac{10{,}770}{x^2} - 0.37, \qquad 5 \le x \le 100$$

where x is the diameter of the wire in mils (0.001 inch). Use a graphing utility to graph the model. If the diameter of the wire is doubled, the resistance is changed by about what factor?

Writing About Concepts

In Exercises 77 and 78, write an equation whose graph has the indicated property. (There may be more than one correct answer.)

77. The graph has intercepts at $x = -2$, $x = 4$, and $x = 6$.

78. The graph has intercepts at $x = -\frac{5}{2}$, $x = 2$, and $x = \frac{3}{2}$.

79. Each table shows solution points for one of the following equations.

(i) $y = kx + 5$

(ii) $y = x^2 + k$

(iii) $y = kx^{3/2}$

(iv) $xy = k$

Match each equation with the correct table and find k. Explain your reasoning.

(a)

x	1	4	9
y	3	24	81

(b)

x	1	4	9
y	7	13	23

(c)

x	1	4	9
y	36	9	4

(d)

x	1	4	9
y	-9	6	71

80. (a) Prove that if a graph is symmetric with respect to the x-axis and to the y-axis, then it is symmetric with respect to the origin. Give an example to show that the converse is not true.

(b) Prove that if a graph is symmetric with respect to one axis and to the origin, then it is symmetric with respect to the other axis.

True or False?   In Exercises 81–84, determine whether the statement is true or false. If it is false, explain why or give an example that shows it is false.

81. If $(1, -2)$ is a point on a graph that is symmetric with respect to the x-axis, then $(-1, -2)$ is also a point on the graph.

82. If $(1, -2)$ is a point on a graph that is symmetric with respect to the y-axis, then $(-1, -2)$ is also a point on the graph.

83. If $b^2 - 4ac > 0$ and $a \ne 0$, then the graph of $y = ax^2 + bx + c$ has two x-intercepts.

84. If $b^2 - 4ac = 0$ and $a \ne 0$, then the graph of $y = ax^2 + bx + c$ has only one x-intercept.

In Exercises 85 and 86, find an equation of the graph that consists of all points (x, y) having the given distance from the origin. (For a review of the Distance Formula, see Appendix D.)

85. The distance from the origin is twice the distance from $(0, 3)$.

86. The distance from the origin is $K (K \ne 1)$ times the distance from $(2, 0)$.

Section 1.2   Linear Models and Rates of Change

- Find the slope of a line passing through two points.
- Write the equation of a line given a point and the slope.
- Interpret slope as a ratio or as a rate in a real-life application.
- Sketch the graph of a linear equation in slope-intercept form.
- Write equations of lines that are parallel or perpendicular to a given line.

$\Delta y = y_2 - y_1 = $ change in y
$\Delta x = x_2 - x_1 = $ change in x
Figure 1.12

The Slope of a Line

The **slope** of a nonvertical line is a measure of the number of units the line rises (or falls) vertically for each unit of horizontal change from left to right. Consider the two points (x_1, y_1) and (x_2, y_2) on the line in Figure 1.12. As you move from left to right along this line, a vertical change of

$$\Delta y = y_2 - y_1 \qquad \text{Change in } y$$

units corresponds to a horizontal change of

$$\Delta x = x_2 - x_1 \qquad \text{Change in } x$$

units. (Δ is the Greek uppercase letter *delta*, and the symbols Δy and Δx are read "delta y" and "delta x.")

Definition of the Slope of a Line

The **slope** m of the nonvertical line passing through (x_1, y_1) and (x_2, y_2) is

$$m = \frac{\Delta y}{\Delta x} = \frac{y_2 - y_1}{x_2 - x_1}, \qquad x_1 \neq x_2.$$

Slope is not defined for vertical lines.

NOTE   When using the formula for slope, note that

$$\frac{y_2 - y_1}{x_2 - x_1} = \frac{-(y_1 - y_2)}{-(x_1 - x_2)} = \frac{y_1 - y_2}{x_1 - x_2}.$$

So, it does not matter in which order you subtract *as long as* you are consistent and both "subtracted coordinates" come from the same point.

Figure 1.13 shows four lines: one has a positive slope, one has a slope of zero, one has a negative slope, and one has an "undefined" slope. In general, the greater the absolute value of the slope of a line, the steeper the line is. For instance, in Figure 1.13, the line with a slope of -5 is steeper than the line with a slope of $\frac{1}{5}$.

If m is positive, then the line rises from left to right.
Figure 1.13

If m is zero, then the line is horizontal.

If m is negative, then the line falls from left to right.

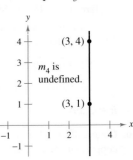

If m is undefined, then the line is vertical.

Equations of Lines

Any two points on a nonvertical line can be used to calculate its slope. This can be verified from the similar triangles shown in Figure 1.14. (Recall that the ratios of corresponding sides of similar triangles are equal.)

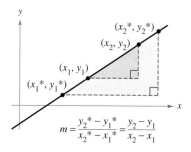

$$m = \frac{y_2{}^* - y_1{}^*}{x_2{}^* - x_1{}^*} = \frac{y_2 - y_1}{x_2 - x_1}$$

Any two points on a nonvertical line can be used to determine its slope.
Figure 1.14

You can write an equation of a nonvertical line if you know the slope of the line and the coordinates of one point on the line. Suppose the slope is m and the point is (x_1, y_1). If (x, y) is any other point on the line, then

$$\frac{y - y_1}{x - x_1} = m.$$

This equation, involving the two variables x and y, can be rewritten in the form

$$y - y_1 = m(x - x_1)$$

which is called the **point-slope equation of a line.**

Point-Slope Equation of a Line

An equation of the line with slope m passing through the point (x_1, y_1) is given by $y - y_1 = m(x - x_1)$.

EXAMPLE 1   Finding an Equation of a Line

Find an equation of the line that has a slope of 3 and passes through the point $(1, -2)$.

Solution

$$
\begin{aligned}
y - y_1 &= m(x - x_1) && \text{Point-slope form} \\
y - (-2) &= 3(x - 1) && \text{Substitute } -2 \text{ for } y_1, 1 \text{ for } x_1, \text{ and } 3 \text{ for } m. \\
y + 2 &= 3x - 3 && \text{Simplify.} \\
y &= 3x - 5 && \text{Solve for } y.
\end{aligned}
$$

(See Figure 1.15.)

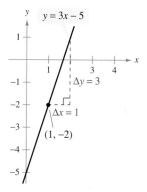

The line with a slope of 3 passing through the point $(1, -2)$
Figure 1.15

NOTE   Remember that only nonvertical lines have a slope. Vertical lines, on the other hand, cannot be written in point-slope form. For instance, the equation of the vertical line passing through the point $(1, -2)$ is $x = 1$.

Ratios and Rates of Change

The slope of a line can be interpreted as either a *ratio* or a *rate*. If the *x*- and *y*-axes have the same unit of measure, the slope has no units and is a **ratio.** If the *x*- and *y*-axes have different units of measure, the slope is a rate or **rate of change.** In your study of calculus, you will encounter applications involving both interpretations of slope.

EXAMPLE 2 Population Growth and Engineering Design

a. The population of Kentucky was 3,687,000 in 1990 and 4,042,000 in 2000. Over this 10-year period, the average rate of change of the population was

$$\text{Rate of change} = \frac{\text{change in population}}{\text{change in years}}$$

$$= \frac{4{,}042{,}000 - 3{,}687{,}000}{2000 - 1990}$$

$$= 35{,}500 \text{ people per year.}$$

If Kentucky's population continues to increase at this rate for the next 10 years, it will have a population of 4,397,000 in 2010 (see Figure 1.16).   *(Source: U.S. Census Bureau)*

b. In tournament water-ski jumping, the ramp rises to a height of 6 feet on a raft that is 21 feet long, as shown in Figure 1.17. The slope of the ski ramp is the ratio of its height (the rise) to the length of its base (the run).

$$\text{Slope of ramp} = \frac{\text{rise}}{\text{run}}$$ Rise is vertical change, run is horizontal change.

$$= \frac{6 \text{ feet}}{21 \text{ feet}}$$

$$= \frac{2}{7}$$

In this case, note that the slope is a ratio and has no units.

Population of Kentucky in census years
Figure 1.16

Dimensions of a water-ski ramp
Figure 1.17

The rate of change found in Example 2(a) is an **average rate of change.** An average rate of change is always calculated over an interval. In this case, the interval is [1990, 2000]. In Chapter 3 you will study another type of rate of change called an *instantaneous rate of change.*

Graphing Linear Models

Many problems in analytic geometry can be classified into two basic categories: (1) Given a graph, what is its equation? and (2) Given an equation, what is its graph? The point-slope equation of a line can be used to solve problems in the first category. However, this form is not especially useful for solving problems in the second category. The form that is better suited to sketching the graph of a line is the **slope-intercept** form of the equation of a line.

The Slope-Intercept Equation of a Line

The graph of the linear equation

$$y = mx + b$$

is a line having a *slope* of m and a *y-intercept* at $(0, b)$.

EXAMPLE 3   Sketching Lines in the Plane

Sketch the graph of each equation.

a. $y = 2x + 1$ **b.** $y = 2$ **c.** $3y + x - 6 = 0$

Solution

a. Because $b = 1$, the y-intercept is $(0, 1)$. Because the slope is $m = 2$, you know that the line rises two units for each unit it moves to the right, as shown in Figure 1.18(a).

b. Because $b = 2$, the y intercept is $(0, 2)$. Because the slope is $m = 0$, you know that the line is horizontal, as shown in Figure 1.18(b).

c. Begin by writing the equation in slope-intercept form.

$$3y + x - 6 = 0 \qquad \text{Write original equation.}$$
$$3y = -x + 6 \qquad \text{Isolate } y\text{-term on the left.}$$
$$y = -\frac{1}{3}x + 2 \qquad \text{Slope-intercept form}$$

In this form, you can see that the y-intercept is $(0, 2)$ and the slope is $m = -\frac{1}{3}$. This means that the line falls one unit for every three units it moves to the right, as shown in Figure 1.18(c).

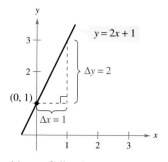

(a) $m = 2$; line rises

Figure 1.18

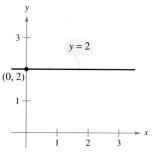

(b) $m = 0$; line is horizontal

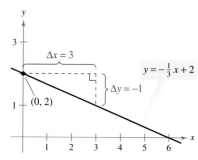

(c) $m = -\frac{1}{3}$; line falls

Because the slope of a vertical line is not defined, its equation cannot be written in the slope-intercept form. However, the equation of *any* line can be written in the **general form**

$$Ax + By + C = 0$$ General form of the equation of a line

where A and B are not *both* zero. For instance, the vertical line given by $x = a$ can be represented by the general form $x - a = 0$.

Summary of Equations of Lines

1. General form: $Ax + By + C = 0$
2. Vertical line: $x = a$
3. Horizontal line: $y = b$
4. Point-slope form: $y - y_1 = m(x - x_1)$
5. Slope-intercept form: $y = mx + b$

Parallel and Perpendicular Lines

The slope of a line is a convenient tool for determining whether two lines are parallel or perpendicular, as shown in Figure 1.19. Specifically, nonvertical lines with the same slope are parallel and nonvertical lines whose slopes are negative reciprocals are perpendicular.

Parallel lines
Figure 1.19

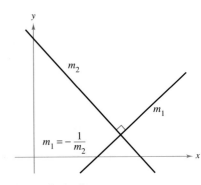

Perpendicular lines

STUDY TIP   In mathematics, the phrase "if and only if" is a way of stating two implications in one statement. For instance, the first statement at the right could be rewritten as the following two implications.

a. If two distinct nonvertical lines are parallel, then their slopes are equal.

b. If two distinct nonvertical lines have equal slopes, then they are parallel.

Parallel and Perpendicular Lines

1. Two distinct nonvertical lines are **parallel** if and only if their slopes are equal—that is, if and only if

   $$m_1 = m_2.$$

2. Two nonvertical lines are **perpendicular** if and only if their slopes are negative reciprocals of each other—that is, if and only if

   $$m_1 = -\frac{1}{m_2}.$$

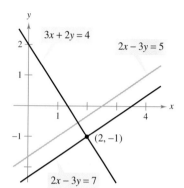

$3x + 2y = 4$

$2x - 3y = 5$

$(2, -1)$

$2x - 3y = 7$

Lines parallel and perpendicular to
$2x - 3y = 5$

Figure 1.20

EXAMPLE 4   **Finding Parallel and Perpendicular Lines**

Find the general form of the equation of the line that passes through the point $(2, -1)$ and is

a. parallel to the line $2x - 3y = 5$.   **b.** perpendicular to the line $2x - 3y = 5$.

(See Figure 1.20.)

Solution   By writing the linear equation $2x - 3y = 5$ in slope-intercept form, $y = \frac{2}{3}x - \frac{5}{3}$, you can see that the given line has a slope of $m = \frac{2}{3}$.

a. The line through $(2, -1)$ that is parallel to the given line also has a slope of $\frac{2}{3}$.

$$y - y_1 = m(x - x_1) \qquad \text{Point-slope form}$$
$$y - (-1) = \tfrac{2}{3}(x - 2) \qquad \text{Substitute.}$$
$$3(y + 1) = 2(x - 2) \qquad \text{Simplify.}$$
$$2x - 3y - 7 = 0 \qquad \text{General form}$$

Note the similarity to the original equation.

b. Using the negative reciprocal of the slope of the given line, you can determine that the slope of a line perpendicular to the given line is $-\frac{3}{2}$. So, the line through the point $(2, -1)$ that is perpendicular to the given line has the following equation.

$$y - y_1 = m(x - x_1) \qquad \text{Point-slope form}$$
$$y - (-1) = -\tfrac{3}{2}(x - 2) \qquad \text{Substitute.}$$
$$2(y + 1) = -3(x - 2) \qquad \text{Simplify.}$$
$$3x + 2y - 4 = 0 \qquad \text{General form}$$

TECHNOLOGY PITFALL   The slope of a line will appear distorted if you use different tick-mark spacing on the x- and y-axes. For instance, the graphing calculator screens in Figures 1.21(a) and 1.21(b) both show the lines given by

$$y = 2x \qquad \text{and} \qquad y = -\tfrac{1}{2}x + 3.$$

Because these lines have slopes that are negative reciprocals, they must be perpendicular. In Figure 1.21(a), however, the lines don't appear to be perpendicular because the tick-mark spacing on the x-axis is not the same as that on the y-axis. In Figure 1.21(b), the lines appear perpendicular because the tick-mark spacing on the x-axis is the same as that on the y-axis. This type of viewing window is said to have a *square setting*.

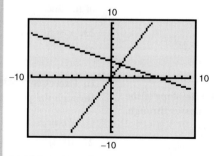

(a) Tick-mark spacing on the x-axis is not the
same as tick-mark spacing on the y-axis.

(b) Tick-mark spacing on the x-axis is the
same as tick-mark spacing on the y-axis.

Figure 1.21

Exercises for Section 1.2

See www.CalcChat.com for worked-out solutions to odd-numbered exercises.

In Exercises 1–6, estimate the slope of the line from its graph. To print an enlarged copy of the graph, go to the website *www.mathgraphs.com*.

1.

2.

3.

4.

5.

6.

In Exercises 7 and 8, sketch the lines through the given point with the indicated slopes. Make the sketches on the same set of coordinate axes.

	Point	Slopes			
7.	$(2, 3)$	(a) 1	(b) -2	(c) $-\frac{3}{2}$	(d) Undefined
8.	$(-4, 1)$	(a) 3	(b) -3	(c) $\frac{1}{3}$	(d) 0

In Exercises 9–14, plot the points and find the slope of the line passing through them.

9. $(3, -4), (5, 2)$

10. $(1, 2), (-2, 4)$

11. $(2, 1), (2, 5)$

12. $(3, -2), (4, -2)$

13. $\left(-\frac{1}{2}, \frac{2}{3}\right), \left(-\frac{3}{4}, \frac{1}{6}\right)$

14. $\left(\frac{7}{8}, \frac{3}{4}\right), \left(\frac{5}{4}, -\frac{1}{4}\right)$

In Exercises 15–18, use the point on the line and the slope of the line to find three additional points that the line passes through. (There is more than one correct answer.)

	Point	Slope		Point	Slope
15.	$(2, 1)$	$m = 0$	**16.**	$(-3, 4)$	m undefined
17.	$(1, 7)$	$m = -3$	**18.**	$(-2, -2)$	$m = 2$

19. *Conveyor Design* A moving conveyor is built to rise 1 meter for each 3 meters of horizontal change.

(a) Find the slope of the conveyor.

(b) Suppose the conveyor runs between two floors in a factory. Find the length of the conveyor if the vertical distance between floors is 10 feet.

20. *Rate of Change* Each of the following is the slope of a line representing daily revenue y in terms of time x in days. Use the slope to interpret any change in daily revenue for a one-day increase in time.

(a) $m = 400$ (b) $m = 100$ (c) $m = 0$

21. *Modeling Data* The table shows the population y (in millions) of the United States for 1996–2001. The variable t represents the time in years, with $t = 6$ corresponding to 1996. *(Source: U.S. Bureau of the Census)*

t	6	7	8	9	10	11
y	269.7	272.9	276.1	279.3	282.3	285.0

(a) Plot the data by hand and connect adjacent points with a line segment.

(b) Use the slope of each line segment to determine the year when the population increased least rapidly.

22. *Modeling Data* The table shows the rate r (in miles per hour) that a vehicle is traveling after t seconds.

t	5	10	15	20	25	30
r	57	74	85	84	61	43

(a) Plot the data by hand and connect adjacent points with a line segment.

(b) Use the slope of each line segment to determine the interval when the vehicle's rate changed most rapidly. How did the rate change?

In Exercises 23–26, find the slope and the *y*-intercept (if possible) of the line.

23. $x + 5y = 20$

24. $6x - 5y = 15$

25. $x = 4$

26. $y = -1$

In Exercises 27–32, find an equation of the line that passes through the point and has the indicated slope. Sketch the line.

	Point	Slope		Point	Slope
27.	$(0, 3)$	$m = \frac{3}{4}$	**28.**	$(-1, 2)$	m undefined
29.	$(0, 0)$	$m = \frac{2}{3}$	**30.**	$(0, 4)$	$m = 0$
31.	$(3, -2)$	$m = 3$	**32.**	$(-2, 4)$	$m = -\frac{3}{5}$

In Exercises 33–42, find an equation of the line that passes through the points, and sketch the line.

33. $(0, 0), (2, 6)$ **34.** $(0, 0), (-1, 3)$

35. $(2, 1), (0, -3)$ **36.** $(-3, -4), (1, 4)$

37. $(2, 8), (5, 0)$ **38.** $(-3, 6), (1, 2)$

39. $(5, 1), (5, 8)$ **40.** $(1, -2), (3, -2)$

41. $\left(\frac{1}{2}, \frac{7}{2}\right), \left(0, \frac{3}{4}\right)$ **42.** $\left(\frac{7}{8}, \frac{3}{4}\right), \left(\frac{5}{4}, -\frac{1}{4}\right)$

43. Find an equation of the vertical line with x-intercept 3.

44. Show that the line with intercepts $(a, 0)$ and $(0, b)$ has the following equation.

$$\frac{x}{a} + \frac{y}{b} = 1, \qquad a \neq 0, b \neq 0$$

In Exercises 45–48, use the result of Exercise 44 to write an equation of the line.

45. x-intercept: $(2, 0)$
    y-intercept: $(0, 3)$

46. x-intercept: $\left(-\frac{2}{3}, 0\right)$
    y-intercept: $(0, -2)$

47. Point on line: $(1, 2)$
    x-intercept: $(a, 0)$
    y-intercept: $(0, a)$
    $(a \neq 0)$

48. Point on line: $(-3, 4)$
    x-intercept: $(a, 0)$
    y-intercept: $(0, a)$
    $(a \neq 0)$

In Exercises 49–56, sketch a graph of the equation.

49. $y = -3$ **50.** $x = 4$

51. $y = -2x + 1$ **52.** $y = \frac{1}{3}x - 1$

53. $y - 2 = \frac{3}{2}(x - 1)$ **54.** $y - 1 = 3(x + 4)$

55. $2x - y - 3 = 0$ **56.** $x + 2y + 6 = 0$

 Square Setting   **In Exercises 57 and 58, use a graphing utility to graph both lines in each viewing window. Compare the graphs. Do the lines appear perpendicular? Are the lines perpendicular? Explain.**

57. $y = x + 6$, $y = -x + 2$

(a)
```
Xmin = -10
Xmax = 10
Xscl = 1
Ymin = -10
Ymax = 10
Yscl = 1
```

(b)
```
Xmin = -15
Xmax = 15
Xscl = 1
Ymin = -10
Ymax = 10
Yscl = 1
```

58. $y = 2x - 3$, $y = -\frac{1}{2}x + 1$

(a)
```
Xmin = -5
Xmax = 5
Xscl = 1
Ymin = -5
Ymax = 5
Yscl = 1
```

(b)
```
Xmin = -6
Xmax = 6
Xscl = 1
Ymin = -4
Ymax = 4
Yscl = 1
```

In Exercises 59–64, write an equation of the line through the point (a) parallel to the given line and (b) perpendicular to the given line.

	Point	*Line*		*Point*	*Line*
59.	$(2, 1)$	$4x - 2y = 3$	**60.**	$(-3, 2)$	$x + y = 7$
61.	$\left(\frac{3}{4}, \frac{7}{8}\right)$	$5x - 3y = 0$	**62.**	$(-6, 4)$	$3x + 4y = 7$
63.	$(2, 5)$	$x = 4$	**64.**	$(-1, 0)$	$y = -3$

Rate of Change   **In Exercises 65–68, you are given the dollar value of a product in 2004 *and* the rate at which the value of the product is expected to change during the next 5 years. Write a linear equation that gives the dollar value V of the product in terms of the year t. (Let $t = 0$ represent 2000.)**

	2004 Value	*Rate*
65.	$2540	$125 increase per year
66.	$156	$4.50 increase per year
67.	$20,400	$2000 decrease per year
68.	$245,000	$5600 decrease per year

 In Exercises 69 and 70, use a graphing utility to graph the parabolas and find their points of intersection. Find an equation of the line through the points of intersection and graph the line in the same viewing window.

69. $y = x^2$
    $y = 4x - x^2$

70. $y = x^2 - 4x + 3$
    $y = -x^2 + 2x + 3$

In Exercises 71 and 72, determine whether the points are collinear. (Three points are *collinear* if they lie on the same line.)

71. $(-2, 1), (-1, 0), (2, -2)$ **72.** $(0, 4), (7, -6), (-5, 11)$

Writing About Concepts

In Exercises 73–75, find the coordinates of the point of intersection of the given segments. Explain your reasoning.

73.

(b, c)
$(-a, 0)$ $(a, 0)$

Perpendicular bisectors

74.

(b, c)
$(-a, 0)$ $(a, 0)$

Medians

75.

(b, c)
$(-a, 0)$ $(a, 0)$

Altitudes

76. Show that the points of intersection in Exercises 73, 74, and 75 are collinear.

77. *Temperature Conversion*   Find a linear equation that expresses the relationship between the temperature in degrees Celsius C and the temperature in degrees Fahrenheit F. Use the fact that water freezes at $0°C$ ($32°F$) and boils at $100°C$ ($212°F$). Use the equation to convert $72°F$ to degrees Celsius.

78. *Reimbursed Expenses*   A company reimburses its sales representatives $150 per day for lodging and meals plus 34¢ per mile driven. Write a linear equation giving the daily cost C to the company in terms of x, the number of miles driven. How much does it cost the company if a sales representative drives 137 miles on a given day?

79. *Career Choice*   An employee has two options for positions in a large corporation. One position pays $12.50 per hour *plus* an additional unit rate of $0.75 per unit produced. The other pays $9.20 per hour *plus* a unit rate of $1.30.

(a) Find linear equations for the hourly wages W in terms of x, the number of units produced per hour, for each option.

(b) Use a graphing utility to graph the linear equations and find the point of intersection.

(c) Interpret the meaning of the point of intersection of the graphs in part (b). How would you use this information to select the correct option if the goal were to obtain the highest hourly wage?

80. *Straight-Line Depreciation*   A small business purchases a piece of equipment for $875. After 5 years, the equipment will be outdated, having no value.

(a) Write a linear equation giving the value y of the equipment in terms of the time x in years, $0 \le x \le 5$.

(b) Find the value of the equipment when $x = 2$.

(c) Estimate (to two-decimal-place accuracy) the time when the value of the equipment is $200.

81. *Apartment Rental*   A real estate office handles an apartment complex with 50 units. When the rent is $580 per month, all 50 units are occupied. However, when the rent is $625, the average number of occupied units drops to 47. Assume that the relationship between the monthly rent p and the demand x is linear. (*Note:* The term *demand* refers to the number of occupied units.)

(a) Write a linear equation giving the demand x in terms of the rent p.

(b) *Linear extrapolation* Use a graphing utility to graph the demand equation and use the *trace* feature to predict the number of units occupied if the rent is raised to $655.

(c) *Linear interpolation* Predict the number of units occupied if the rent is lowered to $595. Verify graphically.

82. *Modeling Data*   An instructor gives regular 20-point quizzes and 100-point exams in a mathematics course. Average scores for six students, given as ordered pairs (x, y) where x is the average quiz score and y is the average test score, are $(18, 87)$, $(10, 55)$, $(19, 96)$, $(16, 79)$, $(13, 76)$, and $(15, 82)$.

(a) Use the regression capabilities of a graphing utility to find the least-squares regression line for the data.

(b) Use a graphing utility to plot the points and graph the regression line in the same viewing window.

(c) Use the regression line to predict the average exam score for a student with an average quiz score of 17.

(d) Interpret the meaning of the slope of the regression line.

(e) The instructor adds 4 points to the average test score of everyone in the class. Describe the changes in the positions of the plotted points and the change in the equation of the line.

83. *Tangent Line*   Find an equation of the line tangent to the circle $x^2 + y^2 = 169$ at the point $(5, 12)$.

84. *Tangent Line*   Find an equation of the line tangent to the circle $(x - 1)^2 + (y - 1)^2 = 25$ at the point $(4, -3)$.

Distance   **In Exercises 85–90, find the distance between the point and the line, or between the lines, using the formula for the distance between the point (x_1, y_1) and the line $Ax + By + C = 0$:**

$$\text{Distance} = \frac{|Ax_1 + By_1 + C|}{\sqrt{A^2 + B^2}}.$$

85. Point: $(0, 0)$
Line: $4x + 3y = 10$

86. Point: $(2, 3)$
Line: $4x + 3y = 10$

87. Point: $(-2, 1)$
Line: $x - y - 2 = 0$

88. Point: $(6, 2)$
Line: $x = -1$

89. Line: $x + y = 1$
Line: $x + y = 5$

90. Line: $3x - 4y = 1$
Line: $3x - 4y = 10$

91. Show that the distance between the point (x_1, y_1) and the line $Ax + By + C = 0$ is

$$\text{Distance} = \frac{|Ax_1 + By_1 + C|}{\sqrt{A^2 + B^2}}.$$

92. Write the distance d between the point $(3, 1)$ and the line $y = mx + 4$ in terms of m. Use a graphing utility to graph the equation. When is the distance 0? Explain the result geometrically.

93. Prove that the diagonals of a rhombus intersect at right angles. (A rhombus is a quadrilateral with sides of equal lengths.)

94. Prove that the figure formed by connecting consecutive midpoints of the sides of any quadrilateral is a parallelogram.

95. Prove that if the points (x_1, y_1) and (x_2, y_2) lie on the same line as (x_1^*, y_1^*) and (x_2^*, y_2^*), then

$$\frac{y_2^* - y_1^*}{x_2^* - x_1^*} = \frac{y_2 - y_1}{x_2 - x_1}.$$

Assume $x_1 \ne x_2$ and $x_1^* \ne x_2^*$.

96. Prove that if the slopes of two nonvertical lines are negative reciprocals of each other, then the lines are perpendicular.

True or False?   **In Exercises 97 and 98, determine whether the statement is true or false. If it is false, explain why or give an example that shows it is false.**

97. The lines represented by $ax + by = c_1$ and $bx - ay = c_2$ are perpendicular. Assume $a \ne 0$ and $b \ne 0$.

98. It is possible for two lines with positive slopes to be perpendicular to each other.

Functions and Their Graphs

- **Use function notation to represent and evaluate a function.**
- **Find the domain and range of a function.**
- **Sketch the graph of a function.**
- **Identify different types of transformations of functions.**
- **Classify functions and recognize combinations of functions.**

Functions and Function Notation

A **relation** between two sets X and Y is a set of ordered pairs, each of the form (x, y), where x is a member of X and y is a member of Y. A **function** from X to Y is a relation between X and Y having the property that any two ordered pairs with the same x-value also have the same y-value. The variable x is the **independent variable,** and the variable y is the **dependent variable.**

Many real-life situations can be modeled by functions. For instance, the area A of a circle is a function of the circle's radius r.

$$A = \pi r^2 \qquad \text{\small{A is a function of r.}}$$

In this case r is the independent variable and A is the dependent variable.

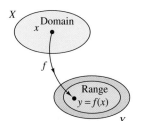

X — Domain, x

f

Range, $y = f(x)$

Y

A real-valued function f of a real variable x
Figure 1.22

Definition of a Real-Valued Function of a Real Variable

Let X and Y be sets of real numbers. A **real-valued function f of a real variable x** from X to Y is a correspondence that assigns to each number x in X exactly one number y in Y.

The **domain** of f is the set X. The number y is the **image** of x under f and is denoted by $f(x)$, which is called the **value of f at x.** The **range** of f is a subset of Y and consists of all images of numbers in X (see Figure 1.22).

Functions can be specified in a variety of ways. In this text, however, we will concentrate primarily on functions that are given by equations involving the dependent and independent variables. For instance, the equation

$$x^2 + 2y = 1 \qquad \text{\small{Equation in implicit form}}$$

defines y, the dependent variable, as a function of x, the independent variable. To **evaluate** this function (that is, to find the y-value that corresponds to a given x-value), it is convenient to isolate y on the left side of the equation.

$$y = \frac{1}{2}(1 - x^2) \qquad \text{\small{Equation in explicit form}}$$

Using f as the name of the function, you can write this equation as

$$f(x) = \frac{1}{2}(1 - x^2). \qquad \text{\small{Function notation}}$$

The original equation, $x^2 + 2y = 1$, **implicitly** defines y as a function of x. When you solve the equation for y, you are writing the equation in **explicit** form.

Function notation has the advantage of clearly identifying the dependent variable as $f(x)$ while at the same time telling you that x is the independent variable and that the function itself is "f." The symbol $f(x)$ is read "f of x." Function notation allows you to be less wordy. Instead of asking "What is the value of y that corresponds to $x = 3$?" you can ask, "What is $f(3)$?"

FUNCTION NOTATION

The word *function* was first used by Gottfried Wilhelm Leibniz in 1694 as a term to denote any quantity connected with a curve, such as the coordinates of a point on a curve or the slope of a curve. Forty years later, Leonhard Euler used the word *function* to describe any expression made up of a variable and some constants. He introduced the notation $y = f(x)$.

In an equation that defines a function, the role of the variable x is simply that of a placeholder. For instance, the function given by

$$f(x) = 2x^2 - 4x + 1$$

can be described by the form

$$f(\;\rule{0.6cm}{0.25cm}\;) = 2(\;\rule{0.6cm}{0.25cm}\;)^2 - 4(\;\rule{0.6cm}{0.25cm}\;) + 1$$

where parentheses are used instead of x. To evaluate $f(-2)$, simply place -2 in each set of parentheses.

$$f(-2) = 2(-2)^2 - 4(-2) + 1 \qquad \text{Substitute } -2 \text{ for } x.$$
$$= 2(4) + 8 + 1 \qquad \text{Simplify.}$$
$$= 17 \qquad \text{Simplify.}$$

NOTE Although f is often used as a convenient function name and x as the independent variable, you can use other symbols. For instance, the following equations all define the same function.

$$f(x) = x^2 - 4x + 7 \qquad \text{Function name is } f, \text{ independent variable is } x.$$
$$f(t) = t^2 - 4t + 7 \qquad \text{Function name is } f, \text{ independent variable is } t.$$
$$g(s) = s^2 - 4s + 7 \qquad \text{Function name is } g, \text{ independent variable is } s.$$

EXAMPLE 1 Evaluating a Function

For the function f defined by

$$f(x) = x^2 + 7,$$

evaluate each of the following.

a. $f(3a)$ **b.** $f(b - 1)$ **c.** $\dfrac{f(x + \Delta x) - f(x)}{\Delta x}, \quad \Delta x \neq 0$

Solution

a. $f(3a) = (3a)^2 + 7 \qquad \text{Substitute } 3a \text{ for } x.$
   $\qquad\quad = 9a^2 + 7 \qquad \text{Simplify.}$

b. $f(b - 1) = (b - 1)^2 + 7 \qquad \text{Substitute } b - 1 \text{ for } x.$
   $\qquad\qquad\;\; = b^2 - 2b + 1 + 7 \qquad \text{Expand binomial.}$
   $\qquad\qquad\;\; = b^2 - 2b + 8 \qquad \text{Simplify.}$

c. $\dfrac{f(x + \Delta x) - f(x)}{\Delta x} = \dfrac{[(x + \Delta x)^2 + 7] - (x^2 + 7)}{\Delta x}$

$$= \frac{x^2 + 2x\Delta x + (\Delta x)^2 + 7 - x^2 - 7}{\Delta x}$$

$$= \frac{2x\Delta x + (\Delta x)^2}{\Delta x}$$

$$= \frac{\Delta x(2x + \Delta x)}{\Delta x}$$

$$= 2x + \Delta x, \qquad \Delta x \neq 0$$

STUDY TIP In calculus, it is important to clearly communicate the domain of a function or expression. For instance, in Example 1(c), the two expressions

$$\frac{f(x + \Delta x) - f(x)}{\Delta x} \quad \text{and} \quad 2x + \Delta x,$$
$$\Delta x \neq 0$$

are equivalent because $\Delta x = 0$ is excluded from the domain of each expression. Without a stated domain restriction, the two expressions would not be equivalent.

NOTE The expression in Example 1(c) is called a *difference quotient* and has a special significance in calculus. You will learn more about this in Chapter 3.

The Domain and Range of a Function

The domain of a function may be described explicitly, or it may be described *implicitly* by an equation used to define the function. The implied domain is the set of all real numbers for which the equation is defined, whereas an explicitly defined domain is one that is given along with the function. For example, the function given by

$$f(x) = \frac{1}{x^2 - 4}, \qquad 4 \le x \le 5$$

has an explicitly-defined domain given by $\{x \colon 4 \le x \le 5\}$. On the other hand, the function given by

$$g(x) = \frac{1}{x^2 - 4}$$

has an implied domain that is the set $\{x \colon x \ne \pm 2\}$.

EXAMPLE 2   Finding the Domain and Range of a Function

a. The domain of the function

$$f(x) = \sqrt{x - 1}$$

is the set of all x-values for which $x - 1 \ge 0$, which is the interval $[1, \infty)$. To find the range, observe that $f(x) = \sqrt{x - 1}$ is never negative. So, the range is the interval $[0, \infty)$, as indicated in Figure 1.23(a).

b. The domain of the tangent function, shown in Figure 1.23(b),

$$f(x) = \tan x$$

is the set of all x-values such that

$$x \ne \frac{\pi}{2} + n\pi, \qquad n \text{ is an integer.} \qquad \text{Domain of tangent function}$$

The range of this function is the set of all real numbers. For a review of the characteristics of this and other trigonometric functions, see Appendix D.

EXAMPLE 3   A Function Defined by More than One Equation

Determine the domain and range of the function.

$$f(x) = \begin{cases} 1 - x, & \text{if } x < 1 \\ \sqrt{x - 1}, & \text{if } x \ge 1 \end{cases}$$

Solution   Because f is defined for $x < 1$ and $x \ge 1$, the domain is the entire set of real numbers. On the portion of the domain for which $x \ge 1$, the function behaves as in Example 2(a). For $x < 1$, the values of $1 - x$ are positive. So, the range of the function is the interval $[0, \infty)$. (See Figure 1.24.)

A function from X to Y is **one-to-one** if to each y-value in the range there corresponds exactly one x-value in the domain. For instance, the function given in Example 2(a) is one-to-one, whereas the functions given in Examples 2(b) and 3 are not one-to-one. A function from X to Y is **onto** if its range consists of all of Y.

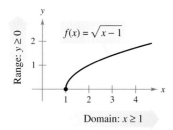

(a) The domain of f is $[1, \infty)$ and the range is $[0, \infty)$.

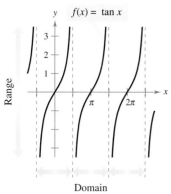

(b) The domain of f is all x-values such that $x \ne \dfrac{\pi}{2} + n\pi$ and the range is $(-\infty, \infty)$.

Figure 1.23

The domain of f is $(-\infty, \infty)$ and the range is $[0, \infty)$.

Figure 1.24

The graph of a function
Figure 1.25

The Graph of a Function

The graph of the function $y = f(x)$ consists of all points $(x, f(x))$, where x is in the domain of f. In Figure 1.25, note that

$$x = \text{the directed distance from the } y\text{-axis}$$
$$f(x) = \text{the directed distance from the } x\text{-axis.}$$

A vertical line can intersect the graph of a function of x at most *once*. This observation provides a convenient visual test, called the **Vertical Line Test,** for functions of x. That is, a graph in the coordinate plane is the graph of a function of x if and only if no vertical line intersects the graph at more than one point. For example, in Figure 1.26(a), you can see that the graph does not define y as a function of x because a vertical line intersects the graph twice. In Figures 1.26(b) and (c), the graphs do define y as a function of x.

(a) Not a function of x

(b) A function of x

(c) A function of x

Figure 1.26

Figure 1.27 shows the graphs of eight basic functions. You should be able to recognize these graphs. (Graphs of the other four basic trigonometric functions are shown in Appendix D.)

Identity function

Squaring function

Cubing function

Square root function

Absolute value function

Rational function

Sine function

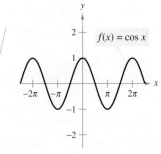

Cosine function

The graphs of eight basic functions
Figure 1.27

Transformations of Functions

Some families of graphs have the same basic shape. For example, compare the graph of $y = x^2$ with the graphs of the four other quadratic functions shown in Figure 1.28.

a.

b.

c.

d.

(a) Vertical shift upward

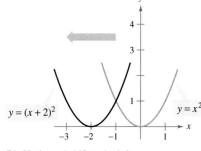

(b) Horizontal shift to the left

(c) Reflection

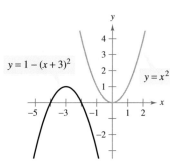

(d) Shift left, reflect, and shift upward

Figure 1.28

Each of the graphs in Figure 1.28 is a **transformation** of the graph of $y = x^2$. The three basic types of transformations illustrated by these graphs are vertical shifts, horizontal shifts, and reflections. Function notation lends itself well to describing transformations of graphs in the plane. For instance, if $f(x) = x^2$ is considered to be the original function in Figure 1.28, the transformations shown can be represented by the following equations.

$y = f(x) + 2$	Vertical shift up two units
$y = f(x + 2)$	Horizontal shift to the left two units
$y = -f(x)$	Reflection about the x-axis
$y = -f(x + 3) + 1$	Shift left three units, reflect about the x-axis, and shift up one unit

---

Basic Types of Transformations $(c > 0)$

Original graph:	$y = f(x)$
Horizontal shift c units to the **right**:	$y = f(x - c)$
Horizontal shift c units to the **left**:	$y = f(x + c)$
Vertical shift c units **downward**:	$y = f(x) - c$
Vertical shift c units **upward**:	$y = f(x) + c$
Reflection (about the x-axis):	$y = -f(x)$
Reflection (about the y-axis):	$y = f(-x)$
Reflection (about the origin):	$y = -f(-x)$

LEONHARD EULER (1707–1783)

In addition to making major contributions to almost every branch of mathematics, Euler was one of the first to apply calculus to real-life problems in physics. His extensive published writings include such topics as shipbuilding, acoustics, optics, astronomy, mechanics, and magnetism.

FOR FURTHER INFORMATION For more on the history of the concept of a function, see the article "Evolution of the Function Concept: A Brief Survey" by Israel Kleiner in *The College Mathematics Journal.* To view this article, go to the website *www.matharticles.com.*

Classifications and Combinations of Functions

The modern notion of a function is derived from the efforts of many seventeenth- and eighteenth-century mathematicians. Of particular note was Leonhard Euler, to whom we are indebted for the function notation $y = f(x)$. By the end of the eighteenth century, mathematicians and scientists had concluded that many real-world phenomena could be represented by mathematical models taken from a collection of functions called **elementary functions.** Elementary functions fall into three categories.

1. Algebraic functions (polynomial, radical, rational)
2. Trigonometric functions (sine, cosine, tangent, and so on)
3. Exponential and logarithmic functions

You can review the trigonometric functions in Appendix D. The other nonalgebraic functions, such as the inverse trigonometric functions and the exponential and logarithmic functions, are introduced in Sections 1.5 and 1.6.

The most common type of algebraic function is a **polynomial function**

$$f(x) = a_n x^n + a_{n-1} x^{n-1} + \cdots + a_2 x^2 + a_1 x + a_0$$

where n is a nonnegative integer. The numbers a_i are **coefficients,** with a_n the **leading coefficient** and a_0 the **constant term** of the polynomial function. If $a_n \neq 0$, then n is the degree of the polynominal function. The zero polynomial $f(x) = 0$ is not assigned a degree. It is common practice to use subscript notation for coefficients of general polynomial functions, but for polynomial functions of low degree, the following simpler forms are often used. (Note that $a \neq 0$.)

Zeroth degree:	$f(x) = a$	Constant function
First degree:	$f(x) = ax + b$	Linear function
Second degree:	$f(x) = ax^2 + bx + c$	Quadratic function
Third degree:	$f(x) = ax^3 + bx^2 + cx + d$	Cubic function

Although the graph of a polynomial function can have several turns, eventually the graph will rise or fall without bound as x moves to the right or left. Whether the graph of

$$f(x) = a_n x^n + a_{n-1} x^{n-1} + \cdots + a_2 x^2 + a_1 x + a_0$$

eventually rises or falls can be determined by the function's degree (odd or even) and by the leading coefficient a_n, as indicated in Figure 1.29. Note that the dashed portions of the graphs indicate that the **Leading Coefficient Test** determines *only* the right and left behavior of the graph.

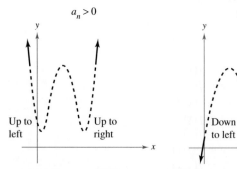

Graphs of polynomial functions of even degree $(n \geq 2)$

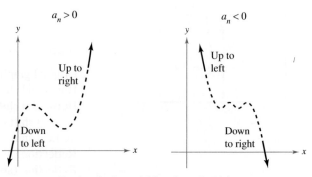

Graphs of polynomial functions of odd degree

The Leading Coefficient Test for polynomial functions
Figure 1.29

Just as a rational number can be written as the quotient of two integers, a **rational function** can be written as the quotient of two polynomials. Specifically, a function f is rational if it has the form

$$f(x) = \frac{p(x)}{q(x)}, \quad q(x) \neq 0$$

where $p(x)$ and $q(x)$ are polynomials.

Polynomial functions and rational functions are examples of **algebraic functions.** An algebraic function of x is one that can be expressed as a finite number of sums, differences, multiples, quotients, and radicals involving x^n. For example, $f(x) = \sqrt{x+1}$ is algebraic. Functions that are not algebraic are **transcendental.** For instance, the trigonometric functions are transcendental.

Two functions can be combined in various ways to create new functions. For example, given $f(x) = 2x - 3$ and $g(x) = x^2 + 1$, you can form the following functions.

$$(f + g)(x) = f(x) + g(x) = (2x - 3) + (x^2 + 1) \qquad \text{Sum}$$
$$(f - g)(x) = f(x) - g(x) = (2x - 3) - (x^2 + 1) \qquad \text{Difference}$$
$$(fg)(x) = f(x)g(x) = (2x - 3)(x^2 + 1) \qquad \text{Product}$$
$$(f/g)(x) = \frac{f(x)}{g(x)} = \frac{2x - 3}{x^2 + 1} \qquad \text{Quotient}$$

You can combine two functions in yet another way, called **composition.** The resulting function is called a **composite function.**

Domain of g

x

$g(x)$

g

f

$f(g(x))$

$f \circ g$

Domain of f

The domain of the composite function $f \circ g$
Figure 1.30

Definition of Composite Function

Let f and g be functions. The function given by $(f \circ g)(x) = f(g(x))$ is called the **composite** of f with g. The domain of $f \circ g$ is the set of all x in the domain of g such that $g(x)$ is in the domain of f (see Figure 1.30).

The composite of f with g is not generally equal to the composite of g with f.

 EXAMPLE 4   **Finding Composites of Functions**

Given $f(x) = 2x - 3$ and $g(x) = \cos x$, find the following.

a. $f \circ g$ **b.** $g \circ f$

Solution

a. $(f \circ g)(x) = f(g(x))$ Definition of $f \circ g$

$\qquad\qquad = f(\cos x)$ Substitute $\cos x$ for $g(x)$.

$\qquad\qquad = 2(\cos x) - 3$ Definition of $f(x)$

$\qquad\qquad = 2 \cos x - 3$ Simplify.

b. $(g \circ f)(x) = g(f(x))$ Definition of $g \circ f$

$\qquad\qquad = g(2x - 3)$ Substitute $2x - 3$ for $f(x)$.

$\qquad\qquad = \cos(2x - 3)$ Definition of $g(x)$

Note that $(f \circ g)(x) \neq (g \circ f)(x)$.

EXPLORATION

Graph each of the following functions with a graphing utility. Determine whether the function is *even*, *odd*, or *neither*.

$$f(x) = x^2 - x^4$$
$$g(x) = 2x^3 + 1$$
$$h(x) = x^5 - 2x^3 + x$$
$$j(x) = 2 - x^6 - x^8$$
$$k(x) = x^5 - 2x^4 + x - 2$$
$$p(x) = x^9 + 3x^5 - x^3 + x$$

Describe a way to identify a function as odd or even by inspecting its equation.

In Section 1.1, an x-intercept of a graph was defined to be a point $(a, 0)$ at which the graph crosses the x-axis. If the graph represents a function f, the number a is a **zero** of f. In other words, *the zeros of a function f are the solutions of the equation $f(x) = 0$.* For example, the function $f(x) = x - 4$ has a zero at $x = 4$ because $f(4) = 0$.

In Section 1.1 you also studied different types of symmetry. In the terminology of functions, a function is **even** if its graph is symmetric with respect to the y-axis, and is **odd** if its graph is symmetric with respect to the origin. The symmetry tests in Section 1.1 yield the following test for even and odd functions.

Test for Even and Odd Functions

The function $y = f(x)$ is **even** if $f(-x) = f(x)$.

The function $y = f(x)$ is **odd** if $f(-x) = -f(x)$.

NOTE Except for the constant function $f(x) = 0$, the graph of a function of x cannot have symmetry with respect to the x-axis because it then would fail the Vertical Line Test for the graph of the function.

EXAMPLE 5   Even and Odd Functions and Zeros of Functions

Determine whether each function is even, odd, or neither. Then find the zeros of the function.

a. $f(x) = x^3 - x$ **b.** $g(x) = 1 + \cos x$

Solution

a. This function is odd because

$$f(-x) = (-x)^3 - (-x) = -x^3 + x = -(x^3 - x) = -f(x).$$

The zeros of f are found as shown.

$$x^3 - x = 0 \qquad \text{Let } f(x) = 0.$$
$$x(x^2 - 1) = 0 \qquad \text{Factor.}$$
$$x(x - 1)(x + 1) = 0 \qquad \text{Factor.}$$
$$x = 0, 1, -1$$

See Figure 1.31(a).

b. This function is even because

$$g(-x) = 1 + \cos(-x) = 1 + \cos x = g(x). \qquad \cos(-x) = \cos(x)$$

The zeros of g are found as shown.

$$1 + \cos x = 0 \qquad \text{Let } g(x) = 0.$$
$$\cos x = -1 \qquad \text{Subtract 1 from each side.}$$
$$x = (2n + 1)\pi, \ n \text{ is an integer} \qquad \text{Zeros of } g$$

See Figure 1.31(b).

NOTE Each of the functions in Example 5 is either even or odd. However, some functions, such as $f(x) = x^2 + x + 1$, are neither even nor odd.

(a) Odd function

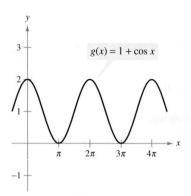

(b) Even function
Figure 1.31

Exercises for Section 1.3

In Exercises 1 and 2, use the graphs of f and g to answer the following.

(a) Identify the domains and ranges of f and g.

(b) Identify $f(-2)$ and $g(3)$.

(c) For what value(s) of x is $f(x) = g(x)$?

(d) Estimate the solution(s) of $f(x) = 2$.

(e) Estimate the solutions of $g(x) = 0$.

1.

2.

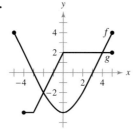

In Exercises 3–12, evaluate (if possible) the function at the given value(s) of the independent variable. Simplify the results.

3. $f(x) = 2x - 3$

  (a) $f(0)$

  (b) $f(-3)$

  (c) $f(b)$

  (d) $f(x - 1)$

4. $f(x) = \sqrt{x + 3}$

  (a) $f(-2)$

  (b) $f(6)$

  (c) $f(-5)$

  (d) $f(x + \Delta x)$

5. $g(x) = 3 - x^2$

  (a) $g(0)$

  (b) $g(\sqrt{3})$

  (c) $g(-2)$

  (d) $g(t - 1)$

6. $g(x) = x^2(x - 4)$

  (a) $g(4)$

  (b) $g(\frac{3}{2})$

  (c) $g(c)$

  (d) $g(t + 4)$

7. $f(x) = \cos 2x$

  (a) $f(0)$

  (b) $f(-\pi/4)$

  (c) $f(\pi/3)$

8. $f(x) = \sin x$

  (a) $f(\pi)$

  (b) $f(5\pi/4)$

  (c) $f(2\pi/3)$

9. $f(x) = x^3$

$$\frac{f(x + \Delta x) - f(x)}{\Delta x}$$

10. $f(x) = 3x - 1$

$$\frac{f(x) - f(1)}{x - 1}$$

11. $f(x) = \dfrac{1}{\sqrt{x - 1}}$

$$\frac{f(x) - f(2)}{x - 2}$$

12. $f(x) = x^3 - x$

$$\frac{f(x) - f(1)}{x - 1}$$

In Exercises 13–18, find the domain and range of the function.

13. $h(x) = -\sqrt{x + 3}$

14. $g(x) = x^2 - 5$

15. $f(t) = \sec \dfrac{\pi t}{4}$

16. $h(t) = \cot t$

17. $f(x) = \dfrac{1}{x}$

18. $g(x) = \dfrac{2}{x - 1}$

In Exercises 19-24, find the domain of the function.

19. $f(x) = \sqrt{x} + \sqrt{1 - x}$

20. $f(x) = \sqrt{x^2 - 3x + 2}$

21. $g(x) = \dfrac{2}{1 - \cos x}$

22. $h(x) = \dfrac{1}{\sin x - \dfrac{1}{2}}$

23. $f(x) = \dfrac{1}{|x + 3|}$

24. $g(x) = \dfrac{1}{|x^2 - 4|}$

In Exercises 25–28, evaluate the function as indicated. Determine its domain and range.

25. $f(x) = \begin{cases} 2x + 1, & x < 0 \\ 2x + 2, & x \geq 0 \end{cases}$

  (a) $f(-1)$ (b) $f(0)$ (c) $f(2)$ (d) $f(t^2 + 1)$

26. $f(x) = \begin{cases} x^2 + 2, & x \leq 1 \\ 2x^2 + 2, & x > 1 \end{cases}$

  (a) $f(-2)$ (b) $f(0)$ (c) $f(1)$ (d) $f(s^2 + 2)$

27. $f(x) = \begin{cases} |x| + 1, & x < 1 \\ -x + 1, & x \geq 1 \end{cases}$

  (a) $f(-3)$ (b) $f(1)$ (c) $f(3)$ (d) $f(b^2 + 1)$

28. $f(x) = \begin{cases} \sqrt{x + 4}, & x \leq 5 \\ (x - 5)^2, & x > 5 \end{cases}$

  (a) $f(-3)$ (b) $f(0)$ (c) $f(5)$ (d) $f(10)$

In Exercises 29 and 30, write the function whose graph is given in the figure.

29.

30.

In Exercises 31–38, sketch a graph of the function and find its domain and range. Use a graphing utility to verify your graph.

31. $f(x) = 4 - x$

32. $g(x) = \dfrac{4}{x}$

33. $h(x) = \sqrt{x - 1}$

34. $f(x) = \frac{1}{2}x^3 + 2$

35. $f(x) = \sqrt{9 - x^2}$

36. $f(x) = x + \sqrt{4 - x^2}$

37. $g(t) = 2 \sin \pi t$

38. $h(\theta) = -5 \cos \dfrac{\theta}{2}$

Writing About Concepts

39. The graph of the distance that a student drives in a 10-minute trip to school is shown in the figure. Give a verbal description of characteristics of the student's drive to school.

40. A student who commutes 27 miles to attend college remembers, after driving for a few minutes, that a term paper that is due has been forgotten. Driving faster than usual, the student returns home, picks up the paper, and once again starts toward school. Sketch a possible graph of the student's distance from home as a function of time.

In Exercises 41–44, use the Vertical Line Test to determine whether *y* is a function of *x*. To print an enlarged copy of the graph, go to the website *www.mathgraphs.com*.

41. $x - y^2 = 0$

42. $\sqrt{x^2 - 4} - y = 0$

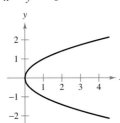

43. $y = \begin{cases} x + 1, & x \le 0 \\ -x + 2, & x > 0 \end{cases}$

44. $x^2 + y^2 = 4$

In Exercises 45–48, determine whether *y* is a function of *x*.

45. $x^2 + y^2 = 4$

46. $x^2 + y = 4$

47. $y^2 = x^2 - 1$

48. $x^2 y - x^2 + 4y = 0$

In Exercises 49–54, use the graph of $y = f(x)$ to match the function with its graph.

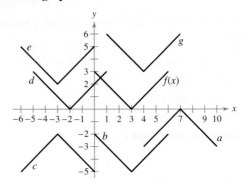

49. $y = f(x + 5)$

50. $y = f(x) - 5$

51. $y = -f(-x) - 2$

52. $y = -f(x - 4)$

53. $y = f(x + 6) + 2$

54. $y = f(x - 1) + 3$

55. Use the graph of *f* shown in the figure to sketch the graph of each function. To print an enlarged copy of the graph, go to the website *www.mathgraphs.com*.

(a) $f(x + 3)$ (b) $f(x - 1)$

(c) $f(x) + 2$ (d) $f(x) - 4$

(e) $3f(x)$ (f) $\frac{1}{4}f(x)$

56. Use the graph of *f* shown in the figure to sketch the graph of each function. To print an enlarged copy of the graph, go to the website *www.mathgraphs.com*.

(a) $f(x - 4)$ (b) $f(x + 2)$

(c) $f(x) + 4$ (d) $f(x) - 1$

(e) $2f(x)$ (f) $\frac{1}{2}f(x)$

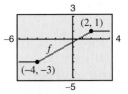

57. Use the graph of $f(x) = \sqrt{x}$ to sketch the graph of each function. In each case, describe the transformation.

(a) $y = \sqrt{x} + 2$ (b) $y = -\sqrt{x}$ (c) $y = \sqrt{x - 2}$

58. Specify a sequence of transformations that will yield each graph of *h* from the graph of the function $f(x) = \sin x$.

(a) $h(x) = \sin\left(x + \dfrac{\pi}{2}\right) + 1$ (b) $h(x) = -\sin(x - 1)$

59. Given $f(x) = \sqrt{x}$ and $g(x) = x^2 - 1$, evaluate each expression.

(a) $f(g(1))$ (b) $g(f(1))$ (c) $g(f(0))$

(d) $f(g(-4))$ (e) $f(g(x))$ (f) $g(f(x))$

60. Given $f(x) = \sin x$ and $g(x) = \pi x$, evaluate each expression.

(a) $f(g(2))$ (b) $f\left(g\left(\dfrac{1}{2}\right)\right)$ (c) $g(f(0))$

(d) $g\left(f\left(\dfrac{\pi}{4}\right)\right)$ (e) $f(g(x))$ (f) $g(f(x))$

In Exercises 61–64, find the composite functions $(f \circ g)$ and $(g \circ f)$. What is the domain of each composite function? Are the two composite functions equal?

61. $f(x) = x^2$

$g(x) = \sqrt{x}$

62. $f(x) = x^2 - 1$

$g(x) = \cos x$

63. $f(x) = \dfrac{3}{x}$

$g(x) = x^2 - 1$

64. $f(x) = \dfrac{1}{x}$

$g(x) = \sqrt{x + 2}$

65. Use the graphs of f and g to evaluate each expression. If the result is undefined, explain why.

(a) $(f \circ g)(3)$
(b) $g(f(2))$
(c) $g(f(5))$
(d) $(f \circ g)(-3)$
(e) $(g \circ f)(-1)$
(f) $f(g(-1))$

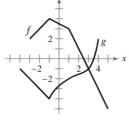

66. *Ripples*   A pebble is dropped into a calm pond, causing ripples in the form of concentric circles. The radius (in feet) of the outer ripple is given by $r(t) = 0.6t$, where t is the time in seconds after the pebble strikes the water. The area of the circle is given by the function $A(r) = \pi r^2$. Find and interpret $(A \circ r)(t)$.

Think About It   **In Exercises 67 and 68, $F(x) = f \circ g \circ h$. Identify functions for f, g, and h. (There are many correct answers.)**

67. $F(x) = \sqrt{2x - 2}$

68. $F(x) = -4 \sin(1 - x)$

In Exercises 69–72, determine whether the function is even, odd, or neither. Use a graphing utility to verify your result.

69. $f(x) = x^2(4 - x^2)$

70. $f(x) = \sqrt[3]{x}$

71. $f(x) = x \cos x$

72. $f(x) = \sin^2 x$

Think About It   **In Exercises 73 and 74, find the coordinates of a second point on the graph of the function f if the given point is on the graph and the function is (a) even and (b) odd.**

73. $\left(-\frac{3}{2}, 4\right)$

74. $(4, 9)$

75. The graphs of f, g, and h are shown in the figure. Decide whether each function is even, odd, or neither.

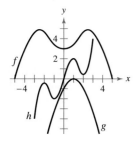

Figure for 75   **Figure for 76**

76. The domain of the function f shown in the figure is $-6 \le x \le 6$.

(a) Complete the graph of f given that f is even.

(b) Complete the graph of f given that f is odd.

Writing Functions   **In Exercises 77–80, write an equation for a function that has the given graph.**

77. Line segment connecting $(-4, 3)$ and $(0, -5)$

78. Line segment connecting $(1, 2)$ and $(5, 5)$

79. The bottom half of the parabola $x + y^2 = 0$

80. The bottom half of the circle $x^2 + y^2 = 4$

Modeling Data   **In Exercises 81–84, match the data with a function from the following list.**

(i) $f(x) = cx$

(ii) $g(x) = cx^2$

(iii) $h(x) = c\sqrt{|x|}$

(iv) $r(x) = c/x$

Determine the value of the constant c for each function such that the function fits the data shown in the table.

81.

x	-4	-1	0	1	4
y	-32	-2	0	-2	-32

82.

x	-4	-1	0	1	4
y	-1	$-\frac{1}{4}$	0	$\frac{1}{4}$	1

83.

x	-4	-1	0	1	4
y	-8	-32	Undef.	32	8

84.

x	-4	-1	0	1	4
y	6	3	0	3	6

85. *Graphical Reasoning*   A thermostat is programmed to lower the temperature during the night automatically (see figure). The temperature T in degrees Celsius is given in terms of t, the time in hours on a 24-hour clock.

(a) Approximate $T(4)$ and $T(15)$.

(b) The thermostat is reprogrammed to produce a temperature $H(t) = T(t - 1)$. How does this change the temperature? Explain.

(c) The thermostat is reprogrammed to produce a temperature $H(t) = T(t) - 1$. How does this change the temperature? Explain.

86. Water runs into a vase of height 30 centimeters at a constant rate. The vase is full after 5 seconds. Use this information and the shape of the vase shown to answer the questions if d is the depth of the water in centimeters and t is the time in seconds (see figure).

(a) Explain why d is a function of t.

(b) Determine the domain and range of the function.

(c) Sketch a possible graph of the function.

30 cm

d

87. *Modeling Data*   The table shows the average number of acres per farm in the United States for selected years. *(Source: U.S. Department of Agriculture)*

Year	1950	1960	1970	1980	1990	2000
Acreage	213	297	374	426	460	434

(a) Plot the data, where A is the acreage and t is the time in years, with $t = 0$ corresponding to 1950. Sketch a freehand curve that approximates the data.

(b) Use the curve in part (a) to approximate $A(15)$.

88. *Automobile Aerodynamics*   The horsepower H required to overcome wind drag on a certain automobile is approximated by

$$H(x) = 0.002x^2 + 0.005x - 0.029, \qquad 10 \le x \le 100$$

where x is the speed of the car in miles per hour.

 (a) Use a graphing utility to graph H.

(b) Rewrite the power function so that x represents the speed in kilometers per hour. [Find $H(x/1.6)$.]

89. *Think About It*   Write the function

$$f(x) = |x| + |x - 2|$$

without using absolute value signs. (For a review of absolute value, see Appendix D.)

 90. *Writing*   Use a graphing utility to graph the polynomial functions $p_1(x) = x^3 - x + 1$ and $p_2(x) = x^3 - x$. How many zeros does each function have? Is there a cubic polynomial that has no zeros? Explain.

91. Prove that the function is odd.

$$f(x) = a_{2n+1}x^{2n+1} + \cdots + a_3 x^3 + a_1 x$$

92. Prove that the function is even.

$$f(x) = a_{2n}x^{2n} + a_{2n-2}x^{2n-2} + \cdots + a_2 x^2 + a_0$$

93. Prove that the product of two even (or two odd) functions is even.

94. Prove that the product of an odd function and an even function is odd.

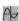 **95.** *Volume*   An open box of maximum volume is to be made from a square piece of material 24 centimeters on a side by cutting equal squares from the corners and turning up the sides (see figure).

x

$24 - 2x$

$\vdash x \dashv \longleftarrow 24 - 2x \longrightarrow \vdash x \dashv$

(a) Write the volume V as a function of x, the length of the corner squares. What is the domain of the function?

(b) Use a graphing utility to graph the volume function and approximate the dimensions of the box that yield a maximum volume.

(c) Use the *table* feature of a graphing utility to verify your answer in part (b). (The first two rows of the table are shown.)

Height, x	Length and Width	Volume, V
1	$24 - 2(1)$	$1[24 - 2(1)]^2 = 484$
2	$24 - 2(2)$	$2[24 - 2(2)]^2 = 800$

96. *Length*   A right triangle is formed in the first quadrant by the x- and y-axes and a line through the point $(3, 2)$. Write the length L of the hypotenuse as a function of x.

True or False?   **In Exercises 97–100, determine whether the statement is true or false. If it is false, explain why or give an example that shows it is false.**

97. If $f(a) = f(b)$, then $a = b$.

98. A vertical line can intersect the graph of a function at most once.

99. If $f(x) = f(-x)$ for all x in the domain of f, then the graph of f is symmetric with respect to the y-axis.

100. If f is a function, then $f(ax) = af(x)$.

Putnam Exam Challenge

101. Let R be the region consisting of the points (x, y) of the Cartesian plane satisfying both $|x| - |y| \le 1$ and $|y| \le 1$. Sketch the region R and find its area.

102. Consider a polynomial $f(x)$ with real coefficients having the property $f(g(x)) = g(f(x))$ for every polynomial $g(x)$ with real coefficients. Determine and prove the nature of $f(x)$.

These problems were composed by the Committee on the Putnam Prize Competition.
© The Mathematical Association of America. All rights reserved.

Section 1.4

Fitting Models to Data

- Fit a linear model to a real-life data set.
- Fit a quadratic model to a real-life data set.
- Fit a trigonometric model to a real-life data set.

Fitting a Linear Model to Data

A basic premise of science is that much of the physical world can be described mathematically and that many physical phenomena are predictable. This scientific outlook was part of the scientific revolution that took place in Europe during the late 1500s. Two early publications connected with this revolution were *On the Revolutions of the Heavenly Spheres* by the Polish astronomer Nicolaus Copernicus, and *On the Structure of the Human Body* by the Belgian anatomist Andreas Vesalius. Each of these books was published in 1543 and each broke with prior tradition by suggesting the use of a scientific method rather than unquestioned reliance on authority.

One method of modern science is gathering data and then describing the data with a mathematical model. For instance, the data given in Example 1 are inspired by Leonardo da Vinci's famous drawing that indicates that a person's height and arm span are equal.

A computer graphics drawing based on the pen and ink drawing of Leonardo da Vinci's famous study of human proportions, called *Vitruvian Man*

 EXAMPLE 1   **Fitting a Linear Model to Data**

A class of 28 people collected the following data, which represent their heights x and arm spans y (rounded to the nearest inch).

(60, 61), (65, 65), (68, 67), (72, 73), (61, 62), (63, 63), (70, 71),
(75, 74), (71, 72), (62, 60), (65, 65), (66, 68), (62, 62), (72, 73),
(70, 70), (69, 68), (69, 70), (60, 61), (63, 63), (64, 64), (71, 71),
(68, 67), (69, 70), (70, 72), (65, 65), (64, 63), (71, 70), (67, 67)

Find a linear model to represent these data.

Solution   There are different ways to model these data with an equation. The simplest would be to observe that x and y are about the same and list the model as simply $y = x$. A more careful analysis would be to use a procedure from statistics called linear regression. (You will study this procedure in Section 13.9.) The least squares regression line for these data is

$$y = 1.006x - 0.23.$$ Least squares regression line

The graph of the model and the data are shown in Figure 1.32. From this model, you can see that a person's arm span tends to be about the same as his or her height.

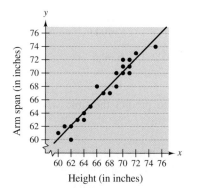

Linear model and data
Figure 1.32

TECHNOLOGY   Many scientific and graphing calculators have built-in least squares regression programs. Typically, you enter the data into the calculator and then run the linear regression program. The program usually displays the slope and y-intercept of the best-fitting line and the *correlation coefficient* r. The correlation coefficient gives a measure of how well the model fits the data. The closer $|r|$ is to 1, the better the model fits the data. For instance, the correlation coefficient for the model in Example 1 is $r \approx 0.97$, which indicates that the model is a good fit for the data. If the r-value is positive, the variables have a positive correlation, as in Example 1. If the r-value is negative, the variables have a negative correlation.

Fitting a Quadratic Model to Data

A function that gives the height s of a falling object in terms of the time t is called a position function. If air resistance is not considered, the position of a falling object can be modeled by

$$s(t) = \tfrac{1}{2}gt^2 + v_0 t + s_0$$

where g is the acceleration due to gravity, v_0 is the initial velocity, and s_0 is the initial height. The value of g depends on where the object is dropped. On Earth, g is approximately -32 feet per second per second, or -9.8 meters per second per second.

To discover the value of g experimentally, you could record the heights of a falling object at several increments, as shown in Example 2.

EXAMPLE 2   Fitting a Quadratic Model to Data

A basketball is dropped from a height of about $5\tfrac{1}{4}$ feet. The height of the basketball is recorded 23 times at intervals of about 0.02 second.* The results are shown in the table.

Time	0.0	0.02	0.04	0.06	0.08	0.099996
Height	5.23594	5.20353	5.16031	5.0991	5.02707	4.95146

Time	0.119996	0.139992	0.159988	0.179988	0.199984	0.219984
Height	4.85062	4.74979	4.63096	4.50132	4.35728	4.19523

Time	0.23998	0.25993	0.27998	0.299976	0.319972	0.339961
Height	4.02958	3.84593	3.65507	3.44981	3.23375	3.01048

Time	0.359961	0.379951	0.399941	0.419941	0.439941
Height	2.76921	2.52074	2.25786	1.98058	1.63488

Find a model to fit these data. Then use the model to predict the time when the basketball will hit the ground.

Solution   Draw a scatter plot of the data, as shown in Figure 1.33. From the scatter plot, you can see that the data do not appear to be linear. It does appear, however, that they might be quadratic. To find a quadratic model, enter the data into a calculator or computer that has a quadratic regression program. You should obtain the model

$$s = -15.45t^2 - 1.302t + 5.2340. \qquad \text{Least squares regression quadratic}$$

Using this model, you can predict the time when the basketball hits the ground by substituting 0 for s and solving the resulting equation for t.

$$0 = -15.45t^2 - 1.302t + 5.2340 \qquad \text{Let } s = 0.$$

$$t = \frac{1.302 \pm \sqrt{(-1.302)^2 - 4(-15.45)(5.2340)}}{2(-15.45)} \qquad \text{Quadratic Formula}$$

$$t \approx 0.54 \qquad \text{Choose positive solution.}$$

The solution is about 0.54 second. In other words, the basketball will continue to fall for about 0.1 second more before hitting the ground.

Scatter plot of data
Figure 1.33

---

* *Data were collected with a Texas Instruments CBL (Calculator-Based Laboratory) System.*

The plane of Earth's orbit about the sun and its axis of rotation are not perpendicular. Instead, Earth's axis is tilted with respect to its orbit. The result is that the amount of daylight received by locations on Earth varies with the time of year. That is, it varies with the position of Earth in its orbit.

Fitting a Trigonometric Model to Data

What is mathematical modeling? This is one of the questions that is asked in the book *Guide to Mathematical Modeling*. Here is part of the answer.*

1. Mathematical modeling consists of applying your mathematical skills to obtain useful answers to real problems.

2. Learning to apply mathematical skills is very different from learning mathematics itself.

3. Models are used in a very wide range of applications, some of which do not appear initially to be mathematical in nature.

4. Models often allow quick and cheap evaluation of alternatives, leading to optimal solutions that are not otherwise obvious.

5. There are no precise rules in mathematical modeling and no "correct" answers.

6. Modeling can be learned only by *doing*.

EXAMPLE 3   Fitting a Trigonometric Model to Data

The number of hours of daylight on Earth depends on the latitude and the time of year. Here are the numbers of minutes of daylight at a location of 20° N latitude on the longest and shortest days of the year: June 21, 801 minutes; December 22, 655 minutes. Use these data to write a model for the amount of daylight d (in minutes) on each day of the year at a location of 20° N latitude. How could you check the accuracy of your model?

Solution   Here is one way to create a model. You can hypothesize that the model is a sine function whose period is 365 days. Using the given data, you can conclude that the amplitude of the graph is $(801 - 655)/2$, or 73. So, one possible model is

$$d = 728 - 73 \sin\left(\frac{2\pi t}{365} + \frac{\pi}{2}\right).$$

In this model, t represents the number of the day of the year, with December 22 represented by $t = 0$. A graph of this model is shown in Figure 1.34. To check the accuracy of this model, we used a weather almanac to find the numbers of minutes of daylight on different days of the year at the location of 20° N latitude.

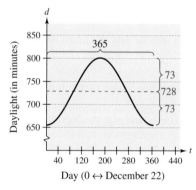

Graph of model
Figure 1.34

Date	Value of t	Actual Daylight	Daylight Given by Model
Dec 22	0	655 min	655 min
Jan 1	10	657 min	656 min
Feb 1	41	676 min	672 min
Mar 1	69	705 min	701 min
Apr 1	100	740 min	739 min
May 1	130	772 min	773 min
Jun 1	161	796 min	796 min
Jun 21	181	801 min	801 min
Jul 1	191	799 min	800 min
Aug 1	222	782 min	785 min
Sep 1	253	752 min	754 min
Oct 1	283	718 min	716 min
Nov 1	314	685 min	681 min
Dec 1	344	661 min	660 min

You can see that the model is fairly accurate.

* Text from Dilwyn Edwards and Mike Hamson, Guide to Mathematical Modelling (Boca Raton: CRC Press, 1990). Used by permission of the authors.

In Exercises 1–4, a scatter plot of data is given. Determine whether the data can be modeled by a linear function, a quadratic function, or a trigonometric function, or that there appears to be no relationship between x and y. To print an enlarged copy of the graph, go to the website *www.mathgraphs.com*.

1.

2.

3.

4.
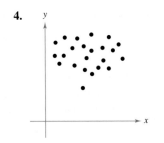

5. *Carcinogens*   Each ordered pair gives the exposure index x of a carcinogenic substance and the cancer mortality y per 100,000 people in the population.

(3.50, 150.1), (3.58, 133.1), (4.42, 132.9),

(2.26, 116.7), (2.63, 140.7), (4.85, 165.5),

(12.65, 210.7), (7.42, 181.0), (9.35, 213.4)

(a) Plot the data. From the graph, do the data appear to be approximately linear?

(b) Visually find a linear model for the data. Graph the model.

(c) Use the model to approximate y if $x = 3$.

6. *Quiz Scores*   The ordered pairs represent the scores on two consecutive 15-point quizzes for a class of 18 students.

(7, 13), (9, 7), (14, 14), (15, 15), (10, 15), (9, 7),

(14, 11), (14, 15), (8, 10), (15, 9), (10, 11), (9, 10),

(11, 14), (7, 14), (11, 10), (14, 11), (10, 15), (9, 6)

(a) Plot the data. From the graph, does the relationship between consecutive scores appear to be approximately linear?

(b) If the data appear to be approximately linear, find a linear model for the data. If not, give some possible explanations.

7. *Hooke's Law*   Hooke's Law states that the force F required to compress or stretch a spring (within its elastic limits) is proportional to the distance d that the spring is compressed or stretched from its original length. That is, $F = kd$, where k is a measure of the stiffness of the spring and is called the *spring constant*. The table shows the elongation d in centimeters of a spring when a force of F newtons is applied.

F	20	40	60	80	100
d	1.4	2.5	4.0	5.3	6.6

Table for 7

(a) Use the regression capabilities of a graphing utility to find a linear model for the data.

(b) Use a graphing utility to plot the data and graph the model. How well does the model fit the data? Explain.

(c) Use the model to estimate the elongation of the spring when a force of 55 newtons is applied.

8. *Falling Object*   In an experiment, students measured the speed s (in meters per second) of a falling object t seconds after it was released. The results are shown in the table.

t	0	1	2	3	4
s	0	11.0	19.4	29.2	39.4

(a) Use the regression capabilities of a graphing utility to find a linear model for the data.

(b) Use a graphing utility to plot the data and graph the model. How well does the model fit the data? Explain your reasoning.

(c) Use the model to estimate the speed of the object after 2.5 seconds.

9. *Energy Consumption and Gross National Product*   The data show the per capita electricity consumption (in millions of Btu) and the per capita gross national product (in thousands of U.S. dollars) for several countries in 2000. *(Source: U.S. Census Bureau)*

Argentina	(73, 12.05)	Bangladesh	(4, 1.59)
Chile	(68, 9.1)	Egypt	(32, 3.67)
Greece	(126, 16.86)	Hong Kong	(118, 25.59)
Hungary	(105, 11.99)	India	(13, 2.34)
Mexico	(63, 8.79)	Poland	(95, 9)
Portugal	(108, 16.99)	South Korea	(167, 17.3)
Spain	(137, 19.26)	Turkey	(47, 7.03)
United Kingdom	(166, 23.55)	Venezuela	(113, 5.74)

(a) Use the regression capabilities of a graphing utility to find a linear model for the data. What is the correlation coefficient?

(b) Use a graphing utility to plot the data and graph the model.

(c) Interpret the graph in part (b). Use the graph to identify the three countries whose data points differ most from the linear model.

(d) Delete the data for the three countries identified in part (c). Fit a linear model to the remaining data and give the correlation coefficient.

10. *Brinell Hardness* The data in the table show the Brinell hardness H of 0.35 carbon steel when hardened and tempered at various temperatures t (degrees Fahrenheit). *(Source: Standard Handbook for Mechanical Engineers)*

t	200	400	600	800	1000	1200
H	534	495	415	352	269	217

(a) Use the regression capabilities of a graphing utility to find a linear model for the data.

(b) Use a graphing utility to plot the data and graph the model. How well does the model fit the data? Explain your reasoning.

(c) Use the model to estimate the hardness when t is 500°F.

11. *Automobile Costs* The data in the table show the variable costs for operating an automobile in the United States for several recent years. The functions y_1, y_2, and y_3 represent the costs in cents per mile for gas and oil, maintenance, and tires, respectively. *(Source: American Automobile Manufacturers Association)*

Year	y_1	y_2	y_3
0	5.40	2.10	0.90
1	6.70	2.20	0.90
2	6.00	2.20	0.90
3	6.00	2.40	0.90
4	5.60	2.50	1.10
5	6.00	2.60	1.40
6	5.90	2.80	1.40
7	6.60	2.80	1.40

(a) Use the regression capabilities of a graphing utility to find a cubic model for y_1 and linear models for y_2 and y_3.

(b) Use a graphing utility to graph y_1, y_2, y_3, and $y_1 + y_2 + y_3$ in the same viewing window. Use the model to estimate the total variable cost per mile in year 12.

12. *Beam Strength* Students in a lab measured the breaking strength S (in pounds) of wood 2 inches thick, x inches high, and 12 inches long. The results are shown in the table.

x	4	6	8	10	12
S	2370	5460	10,310	16,250	23,860

(a) Use the regression capabilities of a graphing utility to fit a quadratic model to the data.

(b) Use a graphing utility to plot the data and graph the model.

(c) Use the model to approximate the breaking strength when $x = 2$.

13. *Health Maintenance Organizations* The bar graph shows the number of people N (in millions) receiving care in HMOs for the years 1990 through 2002. *(Source: Centers for Disease Control)*

(a) Let t be the time in years, with $t = 0$ corresponding to 1990. Use the regression capabilities of a graphing utility to find linear and cubic models for the data.

(b) Use a graphing utility to graph the data and the linear and cubic models.

(c) Use the graphs in part (b) to determine which is the better model.

(d) Use a graphing utility to find and graph a quadratic model for the data.

(e) Use the linear and cubic models to estimate the number of people receiving care in HMOs in the year 2004.

(f) Use a graphing utility to find other models for the data. Which models do you think best represent the data? Explain.

14. *Car Performance* The time t (in seconds) required to attain a speed of s miles per hour from a standing start for a Dodge Avenger is shown in the table. *(Source: Road & Track)*

s	30	40	50	60	70	80	90
t	3.4	5.0	7.0	9.3	12.0	15.8	20.0

(a) Use the regression capabilities of a graphing utility to find a quadratic model for the data.

(b) Use a graphing utility to plot the data and graph the model.

(c) Use the graph in part (b) to state why the model is not appropriate for determining the times required to attain speeds less than 20 miles per hour.

(d) Because the test began from a standing start, add the point (0, 0) to the data. Fit a quadratic model to the revised data and graph the new model.

(e) Does the model from part (d) more accurately model the behavior of the car for low speeds? Explain.

15. *Car Performance* A V8 car engine is coupled to a dynamometer and the horsepower y is measured at different engine speeds x (in thousands of revolutions per minute). The results are shown in the table.

x	1	2	3	4	5	6
y	40	85	140	200	225	245

(a) Use the regression capabilities of a graphing utility to find a cubic model for the data.

(b) Use a graphing utility to plot the data and graph the model.

(c) Use the model to approximate the horsepower when the engine is running at 4500 revolutions per minute.

16. *Boiling Temperature* The table shows the temperatures T (in degrees Fahrenheit) at which water boils at selected pressures p (pounds per square inch). *(Source: Standard Handbook for Mechanical Engineers)*

p	5	10	14.696 (1 atmosphere)	20
T	162.24°	193.21°	212.00°	227.96°

p	30	40	60	80	100
T	250.33°	267.25°	292.71°	312.03°	327.81°

(a) Use the regression capabilities of a graphing utility to find a cubic model for the data.

(b) Use a graphing utility to plot the data and graph the model.

(c) Use the graph to estimate the pressure required for the boiling point of water to exceed 300°F.

(d) Explain why the model would not be correct for pressures exceeding 100 pounds per square inch.

17. *Harmonic Motion* The motion of an oscillating weight suspended by a spring was measured by a motion detector. The data collected and the approximate maximum (positive and negative) displacements from equilibrium are shown in the figure. The displacement y is measured in centimeters and the time t is measured in seconds.

(a) Is y a function of t? Explain.

(b) Approximate the amplitude and period of the oscillations.

(c) Find a model for the data.

(d) Use a graphing utility to graph the model in part (c). Compare the result with the data in the figure.

18. *Temperature* The table shows the normal daily high temperatures for Honolulu H and Chicago C (in degrees Fahrenheit) for month t, with $t = 1$ corresponding to January. *(Source: NOAA)*

t	1	2	3	4	5	6
H	80.1	80.5	81.6	82.8	84.7	86.5
C	29.0	33.5	45.8	58.6	70.1	79.6

t	7	8	9	10	11	12
H	87.5	88.7	88.5	86.9	84.1	81.2
C	83.7	81.8	74.8	63.3	48.4	34.0

(a) A model for Honolulu is

$$H(t) = 84.40 + 4.28 \sin\left(\frac{\pi t}{6} + 3.86\right).$$

Find a model for Chicago.

(b) Use a graphing utility to graph the data and the model for the temperatures in Honolulu. How well does the model fit the data?

(c) Use a graphing utility to graph the data and the model for the temperatures in Chicago. How well does the model fit the data?

(d) Use the models to estimate the average annual temperature in each city. What term of the model did you use? Explain.

(e) What is the period of each model? Is it what you expected? Explain.

(f) Which city has a greater variability of temperatures throughout the year? Which factor of the models determines this variability? Explain.

Writing About Concepts

19. Search for real-life data in a newspaper or magazine. Fit the data to a model. What does your model imply about the data?

20. Describe a real-life situation for each data set. Then describe how a model could be used in the real-life setting.

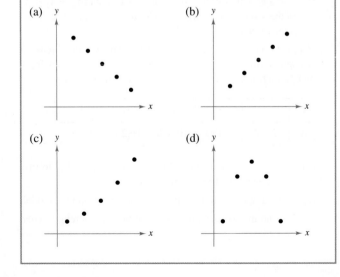

Section 1.5

Inverse Functions

- Verify that one function is the inverse function of another function.
- Determine whether a function has an inverse function.
- Develop properties of the six inverse trigonometric functions.

Inverse Functions

Recall from Section 1.3 that a function can be represented by a set of ordered pairs. For instance, the function $f(x) = x + 3$ from $A = \{1, 2, 3, 4\}$ to $B = \{4, 5, 6, 7\}$ can be written as

$$f: \{(1, 4), (2, 5), (3, 6), (4, 7)\}.$$

By interchanging the first and second coordinates of each ordered pair, you can form the **inverse function** of f. This function is denoted by f^{-1}. It is a function from B to A, and can be written as

$$f^{-1}: \{(4, 1), (5, 2), (6, 3), (7, 4)\}.$$

Note that the domain of f is equal to the range of f^{-1} and vice versa, as shown in Figure 1.35. The functions f and f^{-1} have the effect of "undoing" each other. That is, when you form the composition of f with f^{-1} or the composition of f^{-1} with f, you obtain the identity function.

$$f(f^{-1}(x)) = x \qquad \text{and} \qquad f^{-1}(f(x)) = x$$

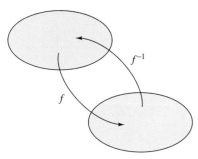

Domain of f = range of f^{-1}
Domain of f^{-1} = range of f
Figure 1.35

EXPLORATION

Finding Inverse Functions   Explain how to "undo" each of the following functions. Then use your explanation to write the inverse function of f.

a. $f(x) = x - 5$

b. $f(x) = 6x$

c. $f(x) = \dfrac{x}{2}$

d. $f(x) = 3x + 2$

e. $f(x) = x^3$

f. $f(x) = 4(x - 2)$

Use a graphing utility to graph each function and its inverse function in the same "square" viewing window. What observation can you make about each pair of graphs?

Definition of Inverse Function

A function g is the **inverse function** of the function f if

$$f(g(x)) = x \text{ for each } x \text{ in the domain of } g$$

and

$$g(f(x)) = x \text{ for each } x \text{ in the domain of } f.$$

The function g is denoted by f^{-1} (read "f inverse").

NOTE   Although the notation used to denote an inverse function resembles *exponential notation*, it is a different use of -1 as a superscript. That is, in general, $f^{-1}(x) \neq 1/f(x)$.

Here are some important observations about inverse functions.

1. If g is the inverse function of f, then f is the inverse function of g.
2. The domain of f^{-1} is equal to the range of f, and the range of f^{-1} is equal to the domain of f.
3. A function need not have an inverse function, but if it does, the inverse function is unique (see Exercise 143).

You can think of f^{-1} as undoing what has been done by f. For example, subtraction can be used to undo addition, and division can be used to undo multiplication. Use the definition of an inverse function to check the following.

$$f(x) = x + c \qquad \text{and} \qquad f^{-1}(x) = x - c \quad \text{are inverse functions of each other.}$$

$$f(x) = cx \qquad \text{and} \qquad f^{-1}(x) = \frac{x}{c}, \ c \neq 0, \quad \text{are inverse functions of each other.}$$

<u>EXAMPLE 1</u>   **Verifying Inverse Functions**

Show that the functions are inverse functions of each other.

$$f(x) = 2x^3 - 1 \quad \text{and} \quad g(x) = \sqrt[3]{\frac{x + 1}{2}}$$

Solution   Because the domains and ranges of both f and g consist of all real numbers, you can conclude that both composite functions exist for all x. The composite of f with g is given by

$$f(g(x)) = 2\left(\sqrt[3]{\frac{x + 1}{2}}\right)^3 - 1$$
$$= 2\left(\frac{x + 1}{2}\right) - 1$$
$$= x + 1 - 1$$
$$= x.$$

The composite of g with f is given by

$$g(f(x)) = \sqrt[3]{\frac{(2x^3 - 1) + 1}{2}}$$
$$= \sqrt[3]{\frac{2x^3}{2}}$$
$$= \sqrt[3]{x^3}$$
$$= x.$$

Because $f(g(x)) = x$ and $g(f(x)) = x$, you can conclude that f and g are inverse functions of each other (see Figure 1.36).

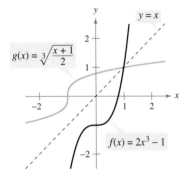

f and g are inverse functions of each other.
Figure 1.36

STUDY TIP   In Example 1, try comparing the functions f and g verbally.

For f: First cube x, then multiply by 2, then subtract 1.

For g: First add 1, then divide by 2, then take the cube root.

Do you see the "undoing pattern"?

In Figure 1.36, the graphs of f and $g = f^{-1}$ appear to be mirror images of each other with respect to the line $y = x$. The graph of f^{-1} is a **reflection** of the graph of f in the line $y = x$. This idea is generalized as follows.

The graph of f^{-1} is a reflection of the graph of f in the line $y = x$.
Figure 1.37

> **Reflective Property of Inverse Functions**
>
> The graph of f contains the point (a, b) if and only if the graph of f^{-1} contains the point (b, a).

To see this, suppose (a, b) is on the graph of f. Then $f(a) = b$ and you can write

$$f^{-1}(b) = f^{-1}(f(a)) = a.$$

So, (b, a) is on the graph of f^{-1}, as shown in Figure 1.37. A similar argument will verify this result in the other direction.

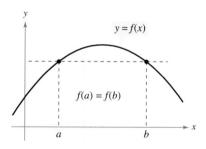

If a horizontal line intersects the graph of f twice, then f is not one-to-one.
Figure 1.38

Existence of an Inverse Function

Not every function has an inverse, and the Reflective Property of Inverse Functions suggests a graphical test for those that do—the **Horizontal Line Test** for an inverse function. This test states that a function f has an inverse function if and only if every horizontal line intersects the graph of f at most once (see Figure 1.38). The following formally states why the Horizontal Line Test is valid.

The Existence of an Inverse Function

A function has an inverse function if and only if it is one-to-one.

EXAMPLE 2   The Existence of an Inverse Function

Which of the functions has an inverse function?

a. $f(x) = x^3 - 1$ **b.** $f(x) = x^3 - x + 1$

Solution

a. From the graph of f given in Figure 1.39(a), it appears that f is one-to-one over its entire domain. To verify this, suppose that there exist x_1 and x_2 such that $f(x_1) = f(x_2)$. By showing that $x_1 = x_2$, it follows that f is one-to-one.

$$f(x_1) = f(x_2)$$
$$x_1^3 - 1 = x_2^3 - 1$$
$$x_1^3 = x_2^3$$
$$\sqrt[3]{x_1^3} = \sqrt[3]{x_2^3}$$
$$x_1 = x_2$$

Because f is one-to-one, you can conclude that f must have an inverse function.

b. From the graph in Figure 1.39(b), you can see that the function does not pass the Horizontal Line Test. In other words, it is not one-to-one. For instance, f has the same value when $x = -1, 0$, and 1.

$$f(-1) = f(1) = f(0) = 1 \qquad \text{Not one-to-one}$$

Therefore, f does not have an inverse function.   ▬▬▬

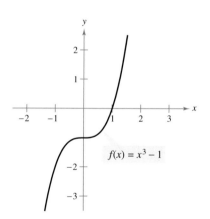

(a) Because f is one-to-one over its entire domain, it has an inverse function.

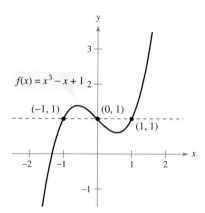

(b) Because f is not one-to-one, it does not have an inverse function.
Figure 1.39

NOTE   Often it is easier to prove that a function has an inverse function than to find the inverse function. For instance, by sketching the graph of $f(x) = x^3 + x - 1$, you can see that it is one-to-one. Yet it would be difficult to determine the inverse of this function algebraically.

Guidelines for Finding an Inverse of a Function

1. Determine whether the function given by $y = f(x)$ has an inverse function.

2. Solve for x as a function of y: $x = g(y) = f^{-1}(y)$.

3. Interchange x and y. The resulting equation is $y = f^{-1}(x)$.

4. Define the domain of f^{-1} to be the range of f.

5. Verify that $f(f^{-1}(x)) = x$ and $f^{-1}(f(x)) = x$.

The domain of f^{-1}, $[0, \infty)$, is the range of f.
Figure 1.40

EXAMPLE 3 Finding an Inverse Function

Find the inverse function of

$$f(x) = \sqrt{2x - 3}.$$

Solution   The function has an inverse function because it is one-to-one on its entire domain (see Figure 1.40). To find an equation for the inverse function, let $y = f(x)$ and solve for x in terms of y.

$\sqrt{2x - 3} = y$	Let $y = f(x)$.
$2x - 3 = y^2$	Square both sides.
$x = \dfrac{y^2 + 3}{2}$	Solve for x.
$y = \dfrac{x^2 + 3}{2}$	Interchange x and y.
$f^{-1}(x) = \dfrac{x^2 + 3}{2}$	Replace y by $f^{-1}(x)$.

The domain of f^{-1} is the range of f, which is $[0, \infty)$. You can verify this result as follows.

$$f(f^{-1}(x)) = \sqrt{2\left(\frac{x^2 + 3}{2}\right) - 3} = \sqrt{x^2} = x, \quad x \geq 0$$

$$f^{-1}(f(x)) = \frac{\left(\sqrt{2x - 3}\right)^2 + 3}{2} = \frac{2x - 3 + 3}{2} = x, \quad x \geq \frac{3}{2}$$

NOTE   Remember that any letter can be used to represent the independent variable. So,

$$f^{-1}(y) = \frac{y^2 + 3}{2}, \qquad f^{-1}(x) = \frac{x^2 + 3}{2}, \qquad \text{and} \qquad f^{-1}(s) = \frac{s^2 + 3}{2}$$

all represent the same function.

Suppose you are given a function that is *not* one-to-one on its entire domain. By restricting the domain to an interval on which the function *is* one-to-one, you can conclude that the new function has an inverse function on the restricted domain.

EXAMPLE 4 Testing Whether a Function Is One-to-One

Show that the sine function

$$f(x) = \sin x$$

is not one-to-one on the entire real line. Then show that f is one-to-one on the closed interval $[-\pi/2, \pi/2]$.

Solution   It is clear that f is not one-to-one, because many different x-values yield the same y-value. For instance,

$$\sin(0) = 0 = \sin(\pi).$$

Moreover, from the graph of $f(x) = \sin x$ in Figure 1.41, you can see that when f is restricted to the interval $[-\pi/2, \pi/2]$, then the restricted function *is* one-to-one.

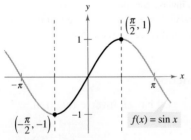

f is one-to-one on the interval $[-\pi/2, \pi/2]$.
Figure 1.41

Inverse Trigonometric Functions

From the graphs of the six basic trigonometric functions, you can see that they do not have inverse functions. (Graphs of the six basic trigonometric functions are shown in Appendix D.) The functions that are called "inverse trigonometric functions" are actually inverses of trigonometric functions whose domains have been restricted.

For instance, in Example 4, you saw that the sine function is one-to-one on the interval $[-\pi/2, \pi/2]$ (see Figure 1.42). On this interval, you can define the inverse of the *restricted* sine function to be

$$y = \arcsin x \qquad \text{if and only if} \qquad \sin y = x$$

where $-1 \le x \le 1$ and $-\pi/2 \le \arcsin x \le \pi/2$. From Figures 1.42 (a) and (b), you can see that you can obtain the graph of $y = \arcsin x$ by reflecting the graph of $y = \sin x$ in the line $y = x$ on the interval $[-\pi/2, \pi/2]$.

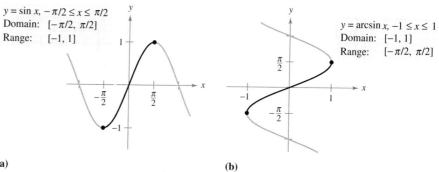

$y = \sin x,\ -\pi/2 \le x \le \pi/2$
Domain: $[-\pi/2, \pi/2]$
Range: $[-1, 1]$

$y = \arcsin x,\ -1 \le x \le 1$
Domain: $[-1, 1]$
Range: $[-\pi/2, \pi/2]$

(a)   **(b)**

Figure 1.42

Under suitable restrictions, each of the six trigonometric functions is one-to-one and so has an inverse function, as indicated in the following definition. (The term "iff" is used to represent the phrase "if and only if.")

EXPLORATION

Inverse Secant Function   In the definition at the right, the inverse secant function is defined by restricting the domain of the secant function to the intervals

$$\left[0, \frac{\pi}{2}\right) \cup \left(\frac{\pi}{2}, \pi\right].$$

Most other texts and reference books agree with this, but some disagree. What other domains might make sense? Explain your reasoning graphically. Most calculators do not have a key for the inverse secant function. How can you use a calculator to evaluate the inverse secant function?

Definition of Inverse Trigonometric Functions

Function	Domain	Range
$y = \arcsin x$ iff $\sin y = x$	$-1 \le x \le 1$	$-\dfrac{\pi}{2} \le y \le \dfrac{\pi}{2}$
$y = \arccos x$ iff $\cos y = x$	$-1 \le x \le 1$	$0 \le y \le \pi$
$y = \arctan x$ iff $\tan y = x$	$-\infty < x < \infty$	$-\dfrac{\pi}{2} < y < \dfrac{\pi}{2}$
$y = \text{arccot } x$ iff $\cot y = x$	$-\infty < x < \infty$	$0 < y < \pi$
$y = \text{arcsec } x$ iff $\sec y = x$	$\lvert x \rvert \ge 1$	$0 \le y \le \pi,\ \ y \ne \dfrac{\pi}{2}$
$y = \text{arccsc } x$ iff $\csc y = x$	$\lvert x \rvert \ge 1$	$-\dfrac{\pi}{2} \le y \le \dfrac{\pi}{2},\ \ y \ne 0$

NOTE   The term arcsin x is read as "the arcsine of x" or sometimes "the angle whose sine is x." An alternative notation for the inverse sine function is $\sin^{-1} x$.

39. Think About It What is the minimum degree of the polynomial function whose graph approximates the given graph? What sign must the leading coefficient have?

(a)

(b)

(c)

(d)

40. Writing The following graphs give the profits *P* for two small companies over a period *p* of 2 years. Create a story to describe the behavior of each profit function for some hypothetical product the company produces.

(a)

(b)
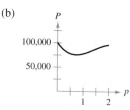

41. Harmonic Motion The motion of an oscillating weight suspended by a spring was measured by a motion detector. The data collected and the approximate maximum (positive and negative) displacements from equilibrium are shown in the figure. The displacement *y* is measured in feet and the time *t* is measured in seconds.

(a) Is *y* a function of *t*? Explain.

(b) Approximate the amplitude and period of the oscillations.

(c) Find a model for the data.

(d) Use a graphing utility to graph the model in part (c). Compare the result with the data in the figure.

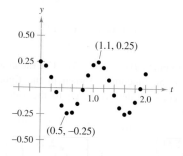

42. Stress Test A machine part was tested by bending it *x* centimeters 10 times per minute until the time *y* (in hours) of failure. The results are recorded in the table.

x	3	6	9	12	15	18	21	24	27	30
y	61	56	53	55	48	35	36	33	44	23

Table for 42

(a) Use the regression capabilities of a graphing utility to find a linear model for the data.

(b) Use a graphing utility to plot the data and graph the model.

(c) Use the graph to determine whether there may have been an error made in conducting one of the tests or in recording the results. If so, eliminate the erroneous point and find a model for the revised data.

 In Exercises 43–48, (a) find the inverse of the function, (b) use a graphing utility to graph *f* **and** f^{-1} **in the same viewing window, and (c) verify that** $f^{-1}(f(x)) = x$ **and** $f(f^{-1}(x)) = x$.

43. $f(x) = \frac{1}{2}x - 3$ **44.** $f(x) = 5x - 7$

45. $f(x) = \sqrt{x + 1}$ **46.** $f(x) = x^3 + 2$

47. $f(x) = \sqrt[3]{x + 1}$ **48.** $f(x) = x^2 - 5, \quad x \geq 0$

In Exercises 49 and 50, sketch the graph of the function by hand.

49. $f(x) = 2 \arctan(x + 3)$ **50.** $h(x) = -3 \arcsin 2x$

In Exercises 51 and 52, evaluate the expression without using a calculator. (*Hint:* Make a sketch of a right triangle.)

51. $\sin\left(\arcsin \frac{1}{2}\right)$ **52.** $\tan(\text{arccot } 2)$

In Exercises 53 and 54, sketch the graph of the function by hand.

53. $f(x) = \ln x + 3$ **54.** $f(x) = \ln(x - 3)$

In Exercises 55 and 56, use the properties of logarithms to expand the logarithmic function.

55. $\ln \sqrt[5]{\dfrac{4x^2 - 1}{4x^2 + 1}}$ **56.** $\ln[(x^2 + 1)(x - 1)]$

In Exercises 57 and 58, write the expression as the logarithm of a single quantity.

57. $\ln 3 + \frac{1}{3} \ln(4 - x^2) - \ln x$

58. $3[\ln x - 2 \ln(x^2 + 1)] + 2 \ln 5$

In Exercises 59 and 60, solve the equation for *x*.

59. $\ln \sqrt{x + 1} = 2$ **60.** $\ln x + \ln(x - 3) = 0$

 In Exercises 61 and 62, (a) find the inverse function of *f*, **(b) use a graphing utility to graph** *f* **and** f^{-1} **in the same viewing window, and (c) verify that** $f^{-1}(f(x)) = x$ **and** $f(f^{-1}(x)) = x$.

61. $f(x) = \ln \sqrt{x}$ **62.** $f(x) = e^{1-x}$

In Exercises 63 and 64, sketch the graph of the function by hand.

63. $y = e^{-x/2}$ **64.** $y = 4e^{-x^2}$

1. Consider the circle $x^2 + y^2 - 6x - 8y = 0$, as shown in the figure.

   (a) Find the center and radius of the circle.

   (b) Find an equation of the tangent line to the circle at the point $(0, 0)$.

   (c) Find an equation of the tangent line to the circle at the point $(6, 0)$.

   (d) Where do the two tangent lines intersect?

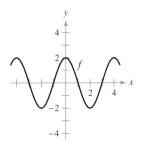

   Figure for 1   Figure for 2

2. There are two tangent lines from the point $(0, 1)$ to the circle $x^2 + (y + 1)^2 = 1$ (see figure). Find equations of these two lines by using the fact that each tangent line intersects the circle in *exactly* one point.

3. The Heaviside function $H(x)$ is widely used in engineering applications.

$$H(x) = \begin{cases} 1, & x \geq 0 \\ 0, & x < 0 \end{cases}$$

Sketch the graph of the Heaviside function and the graphs of the following functions by hand.

   (a) $H(x) - 2$ (b) $H(x - 2)$ (c) $-H(x)$

   (d) $H(-x)$ (e) $\frac{1}{2}H(x)$ (f) $-H(x - 2) + 2$

OLIVER HEAVISIDE (1850–1925)

Heaviside was a British mathematician and physicist who contributed to the field of applied mathematics, especially applications of mathematics to electrical engineering. The *Heaviside function* is a classic type of "on-off" function that has applications to electricity and computer science.

Institute of Electrical Engineers, London

4. Consider the graph of the function f shown below. Use this graph to sketch the graphs of the following functions. To print an enlarged copy of the graph, go to the website *www.mathgraphs.com*.

   (a) $f(x + 1)$ (b) $f(x) + 1$ (c) $2f(x)$ (d) $f(-x)$

   (e) $-f(x)$ (f) $|f(x)|$ (g) $f(|x|)$

5. A rancher plans to fence a rectangular pasture adjacent to a river. The rancher has 100 meters of fence, and no fencing is needed along the river (see figure).

   (a) Write the area A of the pasture as a function of x, the length of the side parallel to the river. What is the domain of A?

   (b) Graph the area function $A(x)$ and estimate the dimensions that yield the maximum amount of area for the pasture.

   (c) Find the dimensions that yield the maximum amount of area for the pasture by completing the square.

   Figure for 5   Figure for 6

6. A rancher has 300 feet of fence to enclose two adjacent pastures.

   (a) Write the total area A of the two pastures as a function of x (see figure). What is the domain of A?

   (b) Graph the area function and estimate the dimensions that yield the maximum amount of area for the pastures.

   (c) Find the dimensions that yield the maximum amount of area for the pastures by completing the square.

7. You are in a boat 2 miles from the nearest point on the coast. You are to go to a point Q located 3 miles down the coast and 1 mile inland (see figure). You can row at 2 miles per hour and walk at 4 miles per hour. Write the total time T of the trip as a function of x.

8. Graph the function $f(x) = e^x - e^{-x}$. From the graph the function appears to be one-to-one. Assuming that the function has an inverse, find $f^{-1}(x)$.

9. One of the fundamental themes of calculus is to find the slope of the tangent line to a curve at a point. To see how this can be done, consider the point $(2, 4)$ on the graph of $f(x) = x^2$.

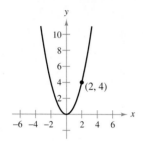

(a) Find the slope of the line joining $(2, 4)$ and $(3, 9)$. Is the slope of the tangent line at $(2, 4)$ greater than or less than this number?

(b) Find the slope of the line joining $(2, 4)$ and $(1, 1)$. Is the slope of the tangent line at $(2, 4)$ greater than or less than this number?

(c) Find the slope of the line joining $(2, 4)$ and $(2.1, 4.41)$. Is the slope of the tangent line at $(2, 4)$ greater than or less than this number?

(d) Find the slope of the line joining $(2, 4)$ and $(2 + h, f(2 + h))$ in terms of the nonzero number h. Verify that $h = 1, -1$, and 0.1 yield the solutions to parts (a)–(c) above.

(e) What is the slope of the tangent line at $(2, 4)$? Explain how you arrived at your answer.

10. Sketch the graph of the function $f(x) = \sqrt{x}$ and label the point $(4, 2)$ on the graph.

(a) Find the slope of the line joining $(4, 2)$ and $(9, 3)$. Is the slope of the tangent line at $(4, 2)$ greater than or less than this number?

(b) Find the slope of the line joining $(4, 2)$ and $(1, 1)$. Is the slope of the tangent line at $(4, 2)$ greater than or less than this number?

(c) Find the slope of the line joining $(4, 2)$ and $(4.41, 2.1)$. Is the slope of the tangent line at $(4, 2)$ greater than or less than this number?

(d) Find the slope of the line joining $(4, 2)$ and $(4 + h, f(4 + h))$ in terms of the nonzero number h.

(e) What is the slope of the tangent line at the point $(4, 2)$? Explain how you arrived at your answer.

11. A large room contains two speakers that are 3 meters apart. The sound intensity I of one speaker is twice that of the other, as shown in the figure. (To print an enlarged copy of the graph, go to the website *www.mathgraphs.com*.) Suppose the listener is free to move about the room to find those positions that receive equal amounts of sound from both speakers. Such a

location satisfies two conditions: (1) the sound intensity at the listener's position is directly proportional to the sound level of a source, and (2) the sound intensity is inversely proportional to the square of the distance from the source.

(a) Find the points on the x-axis that receive equal amounts of sound from both speakers.

(b) Find and graph the equation of all locations (x, y) where one could stand and receive equal amounts of sound from both speakers.

 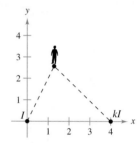

Figure for 11   **Figure for 12**

12. Suppose the speakers in Exercise 11 are 4 meters apart and the sound intensity of one speaker is k times that of the other, as shown in the figure. To print an enlarged copy of the graph, go to the website *www.mathgraphs.com*.

(a) Find the equation of all locations (x, y) where one could stand and receive equal amounts of sound from both speakers.

(b) Graph the equation for the case $k = 3$.

(c) Describe the set of locations of equal sound as k becomes very large.

13. Let d_1 and d_2 be the distances from the point (x, y) to the points $(-1, 0)$ and $(1, 0)$, respectively, as shown in the figure. Show that the equation of the graph of all points (x, y) satisfying $d_1 d_2 = 1$ is $(x^2 + y^2)^2 = 2(x^2 - y^2)$. This curve is called a **lemniscate.** Graph the lemniscate and identify three points on the graph.

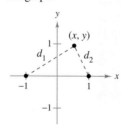

14. Let $f(x) = \dfrac{1}{1 - x}$.

(a) What are the domain and range of f?

(b) Find the composition $f(f(x))$. What is the domain of this function?

(c) Find $f(f(f(x)))$. What is the domain of this function?

(d) Graph $f(f(f(x)))$. Is the graph a line? Why or why not?

2 Limits and Their Properties

Percent removed

Percent removed

Percent removed

The limit process is a fundamental concept of calculus. One technique you can use to estimate a limit is to graph the function and then determine the behavior of the graph as the independent variable approaches a specific value. In Chapter 2, you will learn how to find limits of functions analytically, graphically, and numerically.

Utility companies use a platinum catalyst pollution scrubber to remove pollutants from smokestack emissions. Which do you think is more costly for a utility company to accomplish—removing the first 90 percent of the pollutants, or removing the last 10 percent? Why?

Jeremy Walker/Getty Images

61

Limits That Fail to Exist

In the next three examples you will examine some limits that fail to exist.

EXAMPLE 3 Behavior That Differs from the Right and Left

Show that the limit does not exist.

$$\lim_{x \to 0} \frac{|x|}{x}$$

Solution Consider the graph of the function $f(x) = |x|/x$. From Figure 2.8, you can see that for positive x-values

$$\frac{|x|}{x} = 1, \quad x > 0$$

and for negative x-values

$$\frac{|x|}{x} = -1, \quad x < 0.$$

This means that no matter how close x gets to 0, there will be both positive and negative x-values that yield $f(x) = 1$ and $f(x) = -1$. Specifically, if δ (the lowercase Greek letter *delta*) is a positive number, then for x-values satisfying the inequality $0 < |x| < \delta$, you can classify the values of $|x|/x$ as shown.

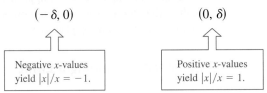

$(-\delta, 0)$ $(0, \delta)$

| Negative x-values yield $\|x\|/x = -1$. | Positive x-values yield $\|x\|/x = 1$. |

This implies that the limit does not exist.

EXAMPLE 4 Unbounded Behavior

Discuss the existence of the limit

$$\lim_{x \to 0} \frac{1}{x^2}.$$

Solution Let $f(x) = 1/x^2$. In Figure 2.9, you can see that as x approaches 0 from either the right or the left, $f(x)$ increases without bound. This means that by choosing x close enough to 0, you can force $f(x)$ to be as large as you want. For instance, $f(x)$ will be larger than 100 if you choose x that is within $\frac{1}{10}$ of 0. That is,

$$0 < |x| < \frac{1}{10} \quad \Longrightarrow \quad f(x) = \frac{1}{x^2} > 100.$$

Similarly, you can force $f(x)$ to be larger than 1,000,000, as follows.

$$0 < |x| < \frac{1}{1000} \quad \Longrightarrow \quad f(x) = \frac{1}{x^2} > 1,000,000$$

Because $f(x)$ is not approaching a real number L as x approaches 0, you can conclude that the limit does not exist.

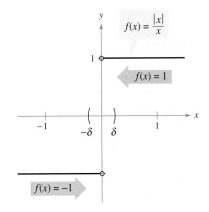

$\displaystyle \lim_{x \to 0} f(x)$ does not exist.

Figure 2.8

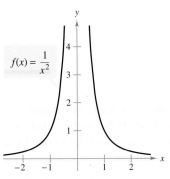

$\displaystyle \lim_{x \to 0} f(x)$ does not exist.

Figure 2.9

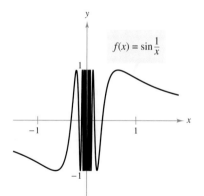

$$f(x) = \sin \frac{1}{x}$$

$\lim_{x \to 0} f(x)$ does not exist.

Figure 2.10

www *EXAMPLE 5*   **Oscillating Behavior**

Discuss the existence of the limit $\lim_{x \to 0} \sin \dfrac{1}{x}$.

Solution   Let $f(x) = \sin(1/x)$. In Figure 2.10, you can see that as x approaches 0, $f(x)$ oscillates between -1 and 1. So, the limit does not exist because no matter how small you choose δ, it is possible to choose x_1 and x_2 within δ units of 0 such that $\sin(1/x_1) = 1$ and $\sin(1/x_2) = -1$, as shown in the table.

x	$2/\pi$	$2/3\pi$	$2/5\pi$	$2/7\pi$	$2/9\pi$	$2/11\pi$	$x \to 0$
$\sin(1/x)$	1	-1	1	-1	1	-1	Limit does not exist.

Common Types of Behavior Associated with Nonexistence of a Limit

1. $f(x)$ approaches a different number from the right side of c than it approaches from the left side.

2. $f(x)$ increases or decreases without bound as x approaches c.

3. $f(x)$ oscillates between two fixed values as x approaches c.

There are many other interesting functions that have unusual limit behavior. An often cited one is the *Dirichlet function*

$$f(x) = \begin{cases} 0, & \text{if } x \text{ is rational.} \\ 1, & \text{if } x \text{ is irrational.} \end{cases}$$

Because this function has *no limit* at any real number c, it is *not continuous* at any real number c. You will study continuity more closely in Section 2.4.

TECHNOLOGY PITFALL   When you use a graphing utility to investigate the behavior of a function near the x-value at which you are trying to evaluate a limit, remember that you can't always trust the pictures that graphing utilities draw. For instance, if you use a graphing utility to graph the function in Example 5 over an interval containing 0, you will most likely obtain an incorrect graph such as that shown in Figure 2.11. The reason that a graphing utility can't show the correct graph is that the graph has infinitely many oscillations over any interval that contains 0.

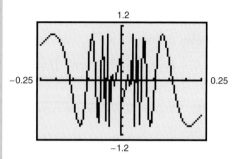

Incorrect graph of $f(x) = \sin(1/x)$
Figure 2.11

PETER GUSTAV DIRICHLET (1805–1859)

In the early development of calculus, the definition of a function was much more restricted than it is today, and "functions" such as the Dirichlet function would not have been considered. The modern definition of function was given by the German mathematician Peter Gustav Dirichlet.

www *indicates that in the* HM mathSpace® *CD-ROM and the online* Eduspace® *system for this text, you will find an Open Exploration, which further explores this example using the computer algebra systems* Maple, Mathcad, Mathematica, *and* Derive.

A Formal Definition of Limit

Let's take another look at the informal description of a limit. If $f(x)$ becomes arbitrarily close to a single number L as x approaches c from either side, then the limit of $f(x)$ as x approaches c is L, written as

$$\lim_{x \to c} f(x) = L.$$

At first glance, this description looks fairly technical. Even so, it is informal because exact meanings have not yet been given to the two phrases

"$f(x)$ becomes arbitrarily close to L"

and

"x approaches c."

The first person to assign mathematically rigorous meanings to these two phrases was Augustin-Louis Cauchy. His ε-δ **definition of limit** is the standard used today.

In Figure 2.12, let ε (the lowercase Greek letter *epsilon*) represent a (small) positive number. Then the phrase "$f(x)$ becomes arbitrarily close to L" means that $f(x)$ lies in the interval $(L - \varepsilon, L + \varepsilon)$. Using absolute value, you can write this as

$$\left| f(x) - L \right| < \varepsilon.$$

Similarly, the phrase "x approaches c" means that there exists a positive number δ such that x lies in either the interval $(c - \delta, c)$ or the interval $(c, c + \delta)$. This fact can be concisely expressed by the double inequality

$$0 < |x - c| < \delta.$$

The first inequality

$$0 < |x - c| \qquad \text{The distance between } x \text{ and } c \text{ is more than 0.}$$

expresses the fact that $x \neq c$. The second inequality

$$|x - c| < \delta \qquad x \text{ is within } \delta \text{ units of } c.$$

states that x is within a distance δ of c.

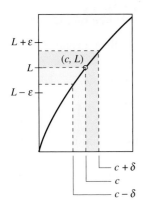

The ε-δ definition of the limit of $f(x)$ as x approaches c

Figure 2.12

Definition of Limit

Let f be a function defined on an open interval containing c (except possibly at c) and let L be a real number. The statement

$$\lim_{x \to c} f(x) = L$$

means that for each $\varepsilon > 0$ there exists a $\delta > 0$ such that if

$$0 < |x - c| < \delta, \quad \text{then} \quad \left| f(x) - L \right| < \varepsilon.$$

FOR FURTHER INFORMATION For more on the introduction of rigor to calculus, see "Who Gave You the Epsilon? Cauchy and the Origins of Rigorous Calculus" by Judith V. Grabiner in *The American Mathematical Monthly*. To view this article, go to the website *www.matharticles.com*.

NOTE Throughout this text, the expression

$$\lim_{x \to c} f(x) = L$$

implies two statements—the limit exists *and* the limit is L.

Some functions do not have limits as $x \to c$, but those that do cannot have two different limits as $x \to c$. That is, *if the limit of a function exists, it is unique* (see Exercise 71).

The next three examples should help you develop a better understanding of the ε-δ definition of a limit.

EXAMPLE 6   Finding a δ for a Given ε

Given the limit

$$\lim_{x \to 3} (2x - 5) = 1$$

find δ such that $|(2x - 5) - 1| < 0.01$ whenever $0 < |x - 3| < \delta$.

Solution   In this problem, you are working with a given value of ε—namely, $\varepsilon = 0.01$. To find an appropriate δ, notice that

$$|(2x - 5) - 1| = |2x - 6| = 2|x - 3|.$$

Because the inequality $|(2x - 5) - 1| < 0.01$ is equivalent to $2|x - 3| < 0.01$, you can choose $\delta = \frac{1}{2}(0.01) = 0.005$. This choice works because

$$0 < |x - 3| < 0.005$$

implies that

$$|(2x - 5) - 1| = 2|x - 3| < 2(0.005) = 0.01$$

as shown in Figure 2.13.

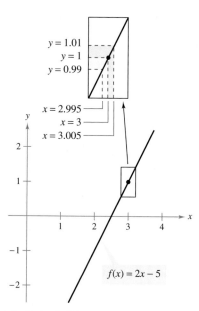

$y = 1.01$
$y = 1$
$y = 0.99$

$x = 2.995$
$x = 3$
$x = 3.005$

$f(x) = 2x - 5$

The limit of $f(x)$ as x approaches 3 is 1.
Figure 2.13

NOTE   In Example 6, note that 0.005 is the *largest* value of δ that will guarantee $|(2x - 5) - 1| < 0.01$ whenever $0 < |x - 3| < \delta$. Any *smaller* positive value of δ would, of course, also work.

In Example 6, you found a δ-value for a *given* ε. This does not prove the existence of the limit. To do that, you must prove that you can find a δ for any ε, as shown in the next example.

EXAMPLE 7   Using the ε-δ Definition of a Limit

Use the ε-δ definition of a limit to prove that

$$\lim_{x \to 2} (3x - 2) = 4.$$

Solution   You must show that for each $\varepsilon > 0$, there exists a $\delta > 0$ such that $|(3x - 2) - 4| < \varepsilon$ whenever $0 < |x - 2| < \delta$. Because your choice of δ depends on ε, you need to establish a connection between the absolute values $|(3x - 2) - 4|$ and $|x - 2|$.

$$|(3x - 2) - 4| = |3x - 6| = 3|x - 2|$$

So, for a given $\varepsilon > 0$, you can choose $\delta = \varepsilon/3$. This choice works because

$$0 < |x - 2| < \delta = \frac{\varepsilon}{3}$$

implies that

$$|(3x - 2) - 4| = 3|x - 2| < 3\left(\frac{\varepsilon}{3}\right) = \varepsilon$$

as shown in Figure 2.14.

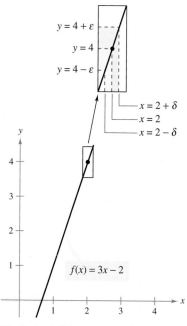

$y = 4 + \varepsilon$
$y = 4$
$y = 4 - \varepsilon$

$x = 2 + \delta$
$x = 2$
$x = 2 - \delta$

$f(x) = 3x - 2$

The limit of $f(x)$ as x approaches 2 is 4.
Figure 2.14

EXAMPLE 8   Using the ε-δ Definition of a Limit

Use the ε-δ definition of a limit to prove that

$$\lim_{x \to 2} x^2 = 4.$$

Solution   You must show that for each $\varepsilon > 0$, there exists a $\delta > 0$ such that

$$|x^2 - 4| < \varepsilon \quad \text{whenever} \quad 0 < |x - 2| < \delta.$$

To find an appropriate δ, begin by writing $|x^2 - 4| = |x - 2||x + 2|$. For all x in the interval $(1, 3)$, you know that $|x + 2| < 5$. So, letting δ be the minimum of $\varepsilon/5$ and 1, it follows that, whenever $0 < |x - 2| < \delta$, you have

$$|x^2 - 4| = |x - 2||x + 2| < \left(\frac{\varepsilon}{5}\right)(5) = \varepsilon$$

as shown in Figure 2.15.

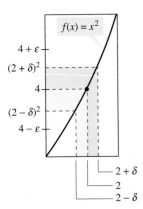

$f(x) = x^2$

The limit of $f(x)$ as x approaches 2 is 4.
Figure 2.15

Throughout this chapter you will use the ε-δ definition of a limit primarily to prove theorems about limits and to establish the existence or nonexistence of particular types of limits. For *finding* limits, you will learn techniques that are easier to use than the ε-δ definition of a limit.

Exercises for Section 2.2

See www.CalcChat.com for worked-out solutions to odd-numbered exercises.

In Exercises 1–10, complete the table and use the result to estimate the limit. Use a graphing utility to graph the function to confirm your result.

1. $\lim\limits_{x \to 2} \dfrac{x - 2}{x^2 - x - 2}$

x	1.9	1.99	1.999	2.001	2.01	2.1
$f(x)$						

2. $\lim\limits_{x \to 2} \dfrac{x - 2}{x^2 - 4}$

x	1.9	1.99	1.999	2.001	2.01	2.1
$f(x)$						

3. $\lim\limits_{x \to 3} \dfrac{[1/(x + 1)] - (1/4)}{x - 3}$

x	2.9	2.99	2.999	3.001	3.01	3.1
$f(x)$						

4. $\lim\limits_{x \to -3} \dfrac{\sqrt{1 - x} - 2}{x + 3}$

x	-3.1	-3.01	-3.001	-2.999	-2.99	-2.9
$f(x)$						

5. $\lim\limits_{x \to 0} \dfrac{\sin x}{x}$

x	-0.1	-0.01	-0.001	0.001	0.01	0.1
$f(x)$						

6. $\lim\limits_{x \to 0} \dfrac{\cos x - 1}{x}$

x	-0.1	-0.01	-0.001	0.001	0.01	0.1
$f(x)$						

7. $\lim\limits_{x \to 0} \dfrac{e^x - 1}{x}$

x	-0.1	-0.01	-0.001	0.001	0.01	0.1
$f(x)$						

8. $\lim\limits_{x \to 0} \dfrac{4}{1 + e^{1/x}}$

x	-0.1	-0.01	-0.001	0.001	0.01	0.1
$f(x)$						

9. $\lim\limits_{x \to 0} \dfrac{\ln(x + 1)}{x}$

x	−0.1	−0.01	−0.001	0.001	0.01	0.1
f(x)						

10. $\lim\limits_{x \to 2} \dfrac{\ln x - \ln 2}{x - 2}$

x	1.9	1.99	1.999	2.001	2.01	2.1
f(x)						

In Exercises 11–20, use the graph to find the limit (if it exists). If the limit does not exist, explain why.

11. $\lim\limits_{x \to 3} (4 - x)$

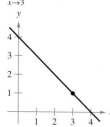

12. $\lim\limits_{x \to 1} (x^2 + 2)$

13. $\lim\limits_{x \to 3} \dfrac{|x - 3|}{x - 3}$

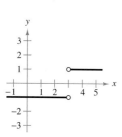

14. $\lim\limits_{x \to 1} f(x)$

$$f(x) = \begin{cases} x^2 + 3, & x \neq 1 \\ 1, & x = 1 \end{cases}$$

15. $\lim\limits_{x \to 1} \sqrt[3]{x} \ln|x - 2|$

16. $\lim\limits_{x \to 0} \dfrac{4}{2 + e^{1/x}}$

17. $\lim\limits_{x \to \pi/2} \tan x$

18. $\lim\limits_{x \to 0} 2 \cos \dfrac{1}{x}$

19. $\lim\limits_{x \to 0} \sec x$

20. $\lim\limits_{x \to 2} \dfrac{1}{x - 2}$

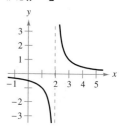

In Exercises 21 and 22, use the graph of the function f to decide whether the value of the given quantity exists. If it does, find it. If not, explain why.

21. (a) $f(1)$

(b) $\lim\limits_{x \to 1} f(x)$

(c) $f(4)$

(d) $\lim\limits_{x \to 4} f(x)$

22. (a) $f(-2)$

(b) $\lim\limits_{x \to -2} f(x)$

(c) $f(0)$

(d) $\lim\limits_{x \to 0} f(x)$

(e) $f(2)$

(f) $\lim\limits_{x \to 2} f(x)$

(g) $f(4)$

(h) $\lim\limits_{x \to 4} f(x)$

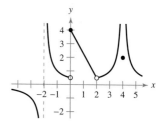

In Exercises 23 and 24, use the graph of f to identify the values of c for which $\lim\limits_{x \to c} f(x)$ exists.

23.

24.

In Exercises 25 and 26, sketch the graph of f. Then identify the values of c for which $\lim\limits_{x \to c} f(x)$ exists.

25. $f(x) = \begin{cases} x^2, & x \le 2 \\ 8 - 2x, & 2 < x < 4 \\ 4, & x \ge 4 \end{cases}$

26. $f(x) = \begin{cases} \sin x, & x < 0 \\ 1 - \cos x, & 0 \le x \le \pi \\ \cos x, & x > \pi \end{cases}$

In Exercises 27 and 28, sketch a graph of a function f that satisfies the given values. (There are many correct answers.)

27. $f(0)$ is undefined.

$\lim\limits_{x \to 0} f(x) = 4$

$f(2) = 6$

$\lim\limits_{x \to 2} f(x) = 3$

28. $f(-2) = 0$

$f(2) = 0$

$\lim\limits_{x \to -2} f(x) = 0$

$\lim\limits_{x \to 2} f(x)$ does not exist.

29. Modeling Data The cost of a telephone call between two cities is $0.75 for the first minute and $0.50 for each additional minute or fraction thereof. A formula for the cost is given by

$$C(t) = 0.75 - 0.50[\![-(t - 1)]\!]$$

where t is the time in minutes.

(*Note:* $[\![x]\!]$ = greatest integer n such that $n \le x$. For example, $[\![3.2]\!] = 3$ and $[\![-1.6]\!] = -2$.)

(a) Use a graphing utility to graph the cost function for $0 < t \le 5$.

(b) Use the graph to complete the table and observe the behavior of the function as t approaches 3.5. Use the graph and the table to find

$$\lim\limits_{t \to 3.5} C(t).$$

t	3	3.3	3.4	3.5	3.6	3.7	4
C				?			

(c) Use the graph to complete the table and observe the behavior of the function as t approaches 3.

t	2	2.5	2.9	3	3.1	3.5	4
C				?			

Does the limit of $C(t)$ as t approaches 3 exist? Explain.

30. Repeat Exercise 29 for

$$C(t) = 0.35 - 0.12[\![-(t - 1)]\!].$$

31. The graph of $f(x) = x + 1$ is shown in the figure. Find δ such that if $0 < |x - 2| < \delta$, then $|f(x) - 3| < 0.4$.

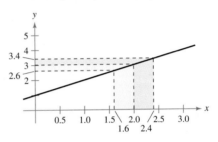

32. The graph of

$$f(x) = \frac{1}{x - 1}$$

is shown in the figure. Find δ such that if $0 < |x - 2| < \delta$, then $|f(x) - 1| < 0.01$.

33. The graph of

$$f(x) = 2 - \frac{1}{x}$$

is shown in the figure. Find δ such that if $0 < |x - 1| < \delta$, then $|f(x) - 1| < 0.1$.

34. The graph of

$$f(x) = x^2 - 1$$

is shown in the figure. Find δ such that if $0 < |x - 2| < \delta$, then $|f(x) - 3| < 0.2$.

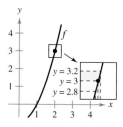

In Exercises 35–38, find the limit L. Then find $\delta > 0$ such that $|f(x) - L| < 0.01$ whenever $0 < |x - c| < \delta$.

35. $\displaystyle\lim_{x \to 2} (3x + 2)$

36. $\displaystyle\lim_{x \to 4} \left(4 - \frac{x}{2}\right)$

37. $\displaystyle\lim_{x \to 2} (x^2 - 3)$

38. $\displaystyle\lim_{x \to 5} (x^2 + 4)$

In Exercises 39–50, find the limit L. Then use the ε-δ definition to prove that the limit is L.

39. $\displaystyle\lim_{x \to 2} (x + 3)$

40. $\displaystyle\lim_{x \to -3} (2x + 5)$

41. $\displaystyle\lim_{x \to -4} \left(\tfrac{1}{2}x - 1\right)$

42. $\displaystyle\lim_{x \to 1} \left(\tfrac{2}{3}x + 9\right)$

43. $\displaystyle\lim_{x \to 6} 3$

44. $\displaystyle\lim_{x \to 2} (-1)$

45. $\displaystyle\lim_{x \to 0} \sqrt[3]{x}$

46. $\displaystyle\lim_{x \to 4} \sqrt{x}$

47. $\displaystyle\lim_{x \to -2} |x - 2|$

48. $\displaystyle\lim_{x \to 3} |x - 3|$

49. $\displaystyle\lim_{x \to 1} (x^2 + 1)$

50. $\displaystyle\lim_{x \to -3} (x^2 + 3x)$

Writing In Exercises 51–54, use a graphing utility to graph the function and estimate the limit (if it exists). What is the domain of the function? Can you detect a possible error in determining the domain of a function solely by analyzing the graph generated by a graphing utility? Write a short paragraph about the importance of examining a function analytically as well as graphically.

51. $f(x) = \dfrac{\sqrt{x + 5} - 3}{x - 4}$

$\displaystyle\lim_{x \to 4} f(x)$

52. $f(x) = \dfrac{x - 3}{x^2 - 4x + 3}$

$\displaystyle\lim_{x \to 3} f(x)$

53. $f(x) = \dfrac{x - 9}{\sqrt{x} - 3}$

$\displaystyle\lim_{x \to 9} f(x)$

54. $f(x) = \dfrac{e^{x/2} - 1}{x}$

$\displaystyle\lim_{x \to 0} f(x)$

Writing About Concepts

55. Write a brief description of the meaning of the notation $\displaystyle\lim_{x \to 8} f(x) = 25$.

56. If $f(2) = 4$, can you conclude anything about the limit of $f(x)$ as x approaches 2? Explain your reasoning.

57. If the limit of $f(x)$ as x approaches 2 is 4, can you conclude anything about $f(2)$? Explain your reasoning.

58. Identify three types of behavior associated with the nonexistence of a limit. Illustrate each type with a graph of a function.

59. *Jewelry*   A jeweler resizes a ring so that its inner circumference is 6 centimeters.

(a) What is the radius of the ring?

(b) If the ring's inner circumference can vary between 5.5 centimeters and 6.5 centimeters, how can the radius vary?

(c) Use the ε-δ definition of a limit to describe this situation. Identify ε and δ.

60. *Sports*   A sporting goods manufacturer designs a golf ball with a volume of 2.48 cubic inches.

(a) What is the radius of the golf ball?

(b) If the ball's volume can vary between 2.45 cubic inches and 2.51 cubic inches, how can the radius vary?

(c) Use the ε-δ definition of a limit to describe this situation. Identify ε and δ.

61. Consider the function $f(x) = (1 + x)^{1/x}$. Estimate the limit

$$\lim_{x \to 0} (1 + x)^{1/x}$$

by evaluating f at x-values near 0. Sketch the graph of f.

62. Consider the function

$$f(x) = \frac{|x + 1| - |x - 1|}{x}.$$

Estimate

$$\lim_{x \to 0} \frac{|x + 1| - |x - 1|}{x}$$

by evaluating f at x-values near 0. Sketch the graph of f.

The symbol *indicates an exercise in which you are instructed to use graphing technology or a symbolic computer algebra system. The solutions of other exercises may also be facilitated by use of appropriate technology.*

63. Graphical Analysis The statement

$$\lim_{x \to 2} \frac{x^2 - 4}{x - 2} = 4$$

means that for each $\varepsilon > 0$ there corresponds a $\delta > 0$ such that if $0 < |x - 2| < \delta$, then

$$\left| \frac{x^2 - 4}{x - 2} - 4 \right| < \varepsilon.$$

If $\varepsilon = 0.001$, then

$$\left| \frac{x^2 - 4}{x - 2} - 4 \right| < 0.001.$$

Use a graphing utility to graph each side of this inequality. Use the *zoom* feature to find an interval $(2 - \delta, 2 + \delta)$ such that the graph of the left side is below the graph of the right side of the inequality.

64. Graphical Analysis The statement

$$\lim_{x \to 3} \frac{x^2 - 3x}{x - 3}$$

means that for each $\varepsilon > 0$ there corresponds a $\delta > 0$ such that if $0 < |x - 3| < \delta$, then

$$\left| \frac{x^2 - 3x}{x - 3} - 3 \right| < \varepsilon.$$

If $\varepsilon = 0.001$, then

$$\left| \frac{x^2 - 3x}{x - 3} - 3 \right| < 0.001.$$

Use a graphing utility to graph each side of this inequality. Use the *zoom* feature to find an interval $(3 - \delta, 3 + \delta)$ such that the graph of the left side is below the graph of the right side of the inequality.

True or False? In Exercises 65–68, determine whether the statement is true or false. If it is false, explain why or give an example that shows it is false.

65. If f is undefined at $x = c$, then the limit of $f(x)$ as x approaches c does not exist.

66. If the limit of $f(x)$ as x approaches c is 0, then there must exist a number k such that $f(k) < 0.001$.

67. If $f(c) = L$, then $\lim_{x \to c} f(x) = L$.

68. If $\lim_{x \to c} f(x) = L$, then $f(c) = L$.

69. Consider the function $f(x) = \sqrt{x}$.

(a) Is $\lim_{x \to 0.25} \sqrt{x} = 0.5$ a true statement? Explain.

(b) Is $\lim_{x \to 0} \sqrt{x} = 0$ a true statement? Explain.

70. Writing The definition of limit on page 72 requires that f is a function defined on an open interval containing c, except possibly at c. Why is this requirement necessary?

71. Prove that if the limit of $f(x)$ as $x \to c$ exists, then the limit must be unique. [*Hint:* Let

$$\lim_{x \to c} f(x) = L_1 \quad \text{and} \quad \lim_{x \to c} f(x) = L_2$$

and prove that $L_1 = L_2$.]

72. Consider the line $f(x) = mx + b$, where $m \neq 0$. Use the ε-δ definition of a limit to prove that $\lim_{x \to c} f(x) = mc + b$.

73. Prove that $\lim_{x \to c} f(x) = L$ is equivalent to $\lim_{x \to c} [f(x) - L] = 0$.

74. (a) Given that

$$\lim_{x \to 0} (3x + 1)(3x - 1)x^2 + 0.01 = 0.01$$

prove that there exists an open interval (a, b) containing 0 such that $(3x + 1)(3x - 1)x^2 + 0.01 > 0$ for all $x \neq 0$ in (a, b).

(b) Given that $\lim_{x \to c} g(x) = L$, where $L > 0$, prove that there exists an open interval (a, b) containing c such that $g(x) > 0$ for all $x \neq c$ in (a, b).

75. Programming Use the programming capabilities of a graphing utility to write a program for approximating $\lim_{x \to c} f(x)$.

Assume the program will be applied only to functions whose limits exist as x approaches c. Let $y_1 = f(x)$ and generate two lists whose entries form the ordered pairs

$$(c \pm [0.1]^n, f(c \pm [0.1]^n))$$

for $n = 0, 1, 2, 3,$ and 4.

76. Programming Use the program you created in Exercise 75 to approximate the limit

$$\lim_{x \to 4} \frac{x^2 - x - 12}{x - 4}.$$

Putnam Exam Challenge

77. Inscribe a rectangle of base b and height h and an isosceles triangle of base b in a circle of radius one as shown. For what value of h do the rectangle and triangle have the same area?

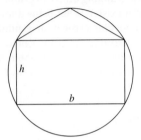

78. A right circular cone has base of radius 1 and height 3. A cube is inscribed in the cone so that one face of the cube is contained in the base of the cone. What is the side-length of the cube?

Evaluating Limits Analytically

- Evaluate a limit using properties of limits.
- Develop and use a strategy for finding limits.
- Evaluate a limit using dividing out and rationalizing techniques.
- Evaluate a limit using the Squeeze Theorem.

Properties of Limits

In Section 2.2, you learned that the limit of $f(x)$ as x approaches c does not depend on the value of f at $x = c$. It may happen, however, that the limit is precisely $f(c)$. In such cases, the limit can be evaluated by **direct substitution.** That is,

$$\lim_{x \to c} f(x) = f(c). \qquad \text{Substitute } c \text{ for } x.$$

Such *well-behaved* functions are **continuous at c.** You will examine this concept more closely in Section 2.4.

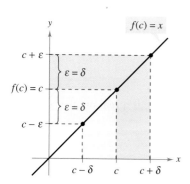

Figure 2.16

THEOREM 2.1   Some Basic Limits

Let b and c be real numbers and let n be a positive integer.

1. $\lim\limits_{x \to c} b = b$ **2.** $\lim\limits_{x \to c} x = c$ **3.** $\lim\limits_{x \to c} x^n = c^n$

Proof   To prove Property 2 of Theorem 2.1, you need to show that for each $\varepsilon > 0$ there exists a $\delta > 0$ such that $|x - c| < \varepsilon$ whenever $0 < |x - c| < \delta$. Because the second inequality is a stricter version of the first, you can simply choose $\delta = \varepsilon$, as shown in Figure 2.16. This completes the proof. (Proofs of the other properties of limits in this section are listed in Appendix A or are discussed in the exercises.)

NOTE   When you encounter new notations or symbols in mathematics, be sure you know how the notations are read. For instance, the limit in Example 1(c) is read as "the limit of x^2 as x approaches 2 is 4."

EXAMPLE 1   Evaluating Basic Limits

a. $\lim\limits_{x \to 2} 3 = 3$ **b.** $\lim\limits_{x \to -4} x = -4$ **c.** $\lim\limits_{x \to 2} x^2 = 2^2 = 4$

THEOREM 2.2   Properties of Limits

Let b and c be real numbers, let n be a positive integer, and let f and g be functions with the following limits.

$$\lim_{x \to c} f(x) = L \qquad \text{and} \qquad \lim_{x \to c} g(x) = K$$

1. Scalar multiple: $\lim\limits_{x \to c} [b f(x)] = bL$

2. Sum or difference: $\lim\limits_{x \to c} [f(x) \pm g(x)] = L \pm K$

3. Product: $\lim\limits_{x \to c} [f(x)g(x)] = LK$

4. Quotient: $\lim\limits_{x \to c} \dfrac{f(x)}{g(x)} = \dfrac{L}{K}, \qquad$ provided $K \neq 0$

5. Power: $\lim\limits_{x \to c} [f(x)]^n = L^n$

EXAMPLE 2 The Limit of a Polynomial

$$\lim_{x \to 2} (4x^2 + 3) = \lim_{x \to 2} 4x^2 + \lim_{x \to 2} 3 \qquad \text{Property 2}$$

$$= 4\left(\lim_{x \to 2} x^2\right) + \lim_{x \to 2} 3 \qquad \text{Property 1}$$

$$= 4(2^2) + 3 \qquad \text{Example 1}$$

$$= 19 \qquad \text{Simplify.}$$

In Example 2, note that the limit (as $x \to 2$) of the *polynomial function* $p(x) = 4x^2 + 3$ is simply the value of p at $x = 2$.

$$\lim_{x \to 2} p(x) = p(2) = 4(2^2) + 3 = 19$$

This *direct substitution* property is valid for all polynomial and rational functions with nonzero denominators.

THE SQUARE ROOT SYMBOL

The first use of a symbol to denote the square root can be traced to the sixteenth century. Mathematicians first used the symbol $\sqrt{}$, which had only two strokes. This symbol was chosen because it resembled a lowercase r, to stand for the Latin word *radix*, meaning root.

> **THEOREM 2.3 Limits of Polynomial and Rational Functions**
>
> If p is a polynomial function and c is a real number, then
>
> $$\lim_{x \to c} p(x) = p(c).$$
>
> If r is a rational function given by $r(x) = p(x)/q(x)$ and c is a real number such that $q(c) \neq 0$, then
>
> $$\lim_{x \to c} r(x) = r(c) = \frac{p(c)}{q(c)}.$$

NOTE Your goal in this section is to become familiar with limits that can be evaluated by direct substitution. In the following library of elementary functions, what are the values of c for which

$$\lim_{x \to c} f(x) = f(c)?$$

Polynomial function:

$$f(x) = a_n x^n + \cdots + a_1 x + a_0$$

Rational function: (p and q are polynomials):

$$f(x) = \frac{p(x)}{q(x)}$$

Trigonometric functions:

$$f(x) = \sin x, \quad f(x) = \cos x$$

$$f(x) = \tan x, \quad f(x) = \cot x$$

$$f(x) = \sec x, \quad f(x) = \csc x$$

Exponential functions:

$$f(x) = a^x, \quad f(x) = e^x$$

Natural logarithmic function:

$$f(x) = \ln x$$

EXAMPLE 3 The Limit of a Rational Function

Find the limit: $\displaystyle\lim_{x \to 1} \frac{x^2 + x + 2}{x + 1}$.

Solution Because the denominator is not 0 when $x = 1$, you can apply Theorem 2.3 to obtain

$$\lim_{x \to 1} \frac{x^2 + x + 2}{x + 1} = \frac{1^2 + 1 + 2}{1 + 1} = \frac{4}{2} = 2.$$

Polynomial functions and rational functions are two of the three basic types of algebraic functions. The following theorem deals with the limit of the third type of algebraic function—one that involves a radical. See Appendix A for a proof of this theorem.

> **THEOREM 2.4 The Limit of a Function Involving a Radical**
>
> Let n be a positive integer. The following limit is valid for all c if n is odd, and is valid for $c > 0$ if n is even.
>
> $$\lim_{x \to c} \sqrt[n]{x} = \sqrt[n]{c}$$

The following theorem greatly expands your ability to evaluate limits because it shows how to analyze the limit of a composite function. See Appendix A for a proof of this theorem.

THEOREM 2.5   The Limit of a Composite Function

If f and g are functions such that $\lim_{x \to c} g(x) = L$ and $\lim_{x \to L} f(x) = f(L)$, then

$$\lim_{x \to c} f(g(x)) = f\left(\lim_{x \to c} g(x)\right) = f(L).$$

 EXAMPLE 4   The Limit of a Composite Function

Because

$$\lim_{x \to 0} (x^2 + 4) = 0^2 + 4 = 4 \qquad \text{and} \qquad \lim_{x \to 4} \sqrt{x} = 2$$

it follows that

$$\lim_{x \to 0} \sqrt{x^2 + 4} = \sqrt{4} = 2.$$

You have seen that the limits of many algebraic functions can be evaluated by direct substitution. The basic transcendental functions (trigonometric, exponential, and logarithmic) also possess this desirable quality, as shown in the next theorem (presented without proof).

THEOREM 2.6   Limits of Transcendental Functions

Let c be a real number in the domain of the given trigonometric function.

1. $\lim_{x \to c} \sin x = \sin c$ **2.** $\lim_{x \to c} \cos x = \cos c$

3. $\lim_{x \to c} \tan x = \tan c$ **4.** $\lim_{x \to c} \cot x = \cot c$

5. $\lim_{x \to c} \sec x = \sec c$ **6.** $\lim_{x \to c} \csc x = \csc c$

7. $\lim_{x \to c} a^x = a^c, \; (a > 0)$ **8.** $\lim_{x \to c} \ln x = \ln c$

EXAMPLE 5   Limits of Transcendental Functions

a. $\lim_{x \to 0} \sin x = \sin(0) = 0$ **b.** $\lim_{x \to 2} (2 + \ln x) = 2 + \ln 2$

c. $\lim_{x \to \pi} (x \cos x) = \left(\lim_{x \to \pi} x\right)\left(\lim_{x \to \pi} \cos x\right) = \pi \cos(\pi) = -\pi$

d. $\lim_{x \to 0} \dfrac{\tan x}{x^2 + 1} = \dfrac{\lim_{x \to 0} \tan x}{\lim_{x \to 0} x^2 + 1} = \dfrac{0}{0^2 + 1} = 0$

e. $\lim_{x \to -1} xe^x = \left(\lim_{x \to -1} x\right)\left(\lim_{x \to -1} e^x\right) = (-1)(e^{-1}) = -e^{-1}$

f. $\lim_{x \to e} \ln x^3 = \lim_{x \to e} 3 \ln x = 3(1) = 3$

A Strategy for Finding Limits

On the previous three pages, you studied several types of functions whose limits can be evaluated by direct substitution. This knowledge, together with the following theorem, can be used to develop a strategy for finding limits. A proof of this theorem is given in Appendix A.

THEOREM 2.7   Functions That Agree at All But One Point

Let c be a real number and let $f(x) = g(x)$ for all $x \neq c$ in an open interval containing c. If the limit of $g(x)$ as x approaches c exists, then the limit of $f(x)$ also exists and

$$\lim_{x \to c} f(x) = \lim_{x \to c} g(x).$$

EXAMPLE 6   Finding the Limit of a Function

Find the limit: $\displaystyle\lim_{x \to 1} \frac{x^3 - 1}{x - 1}$.

Solution   Let $f(x) = (x^3 - 1)/(x - 1)$. By factoring and dividing out like factors, you can rewrite f as

$$f(x) = \frac{\cancel{(x - 1)}(x^2 + x + 1)}{\cancel{(x - 1)}} = x^2 + x + 1 = g(x), \qquad x \neq 1.$$

So, for all x-values other than $x = 1$, the functions f and g agree, as shown in Figure 2.17. Because $\lim\limits_{x \to 1} g(x)$ exists, you can apply Theorem 2.7 to conclude that f and g have the same limit at $x = 1$.

$$\lim_{x \to 1} \frac{x^3 - 1}{x - 1} = \lim_{x \to 1} \frac{(x - 1)(x^2 + x + 1)}{x - 1} \qquad \text{Factor.}$$

$$= \lim_{x \to 1} \frac{\cancel{(x - 1)}(x^2 + x + 1)}{\cancel{x - 1}} \qquad \text{Divide out like factors.}$$

$$= \lim_{x \to 1} (x^2 + x + 1) \qquad \text{Apply Theorem 2.7.}$$

$$= 1^2 + 1 + 1 \qquad \text{Use direct substitution.}$$

$$= 3 \qquad \text{Simplify.}$$

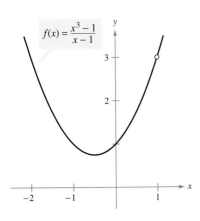

$f(x) = \dfrac{x^3 - 1}{x - 1}$

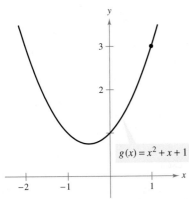

$g(x) = x^2 + x + 1$

f and g agree at all but one point.
Figure 2.17

STUDY TIP   When applying this strategy for finding a limit, remember that some functions do not have a limit (as x approaches c). For instance, the following limit does not exist.

$$\lim_{x \to 1} \frac{x^3 + 1}{x - 1}$$

A Strategy for Finding Limits

1. Learn to recognize which limits can be evaluated by direct substitution. (These limits are listed in Theorems 2.1 through 2.6.)

2. If the limit of $f(x)$ as x approaches c *cannot* be evaluated by direct substitution, try to find a function g that agrees with f for all x other than $x = c$. [Choose g such that the limit of $g(x)$ *can* be evaluated by direct substitution.]

3. Apply Theorem 2.7 to conclude *analytically* that

$$\lim_{x \to c} f(x) = \lim_{x \to c} g(x) = g(c).$$

4. Use a *graph* or *table* to reinforce your conclusion.

Dividing Out and Rationalizing Techniques

Two techniques for finding limits analytically are shown in Examples 7 and 8. The first technique involves dividing out common factors, and the second technique involves rationalizing the numerator of a fractional expression.

EXAMPLE 7   Dividing Out Technique

Find the limit: $\displaystyle\lim_{x \to -3} \frac{x^2 + x - 6}{x + 3}$.

Solution   Although you are taking the limit of a rational function, you *cannot* apply Theorem 2.3 because the limit of the denominator is 0.

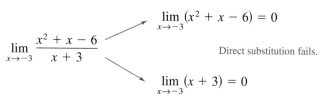

$$\lim_{x \to -3} \frac{x^2 + x - 6}{x + 3}$$

$$\lim_{x \to -3} (x^2 + x - 6) = 0$$

Direct substitution fails.

$$\lim_{x \to -3} (x + 3) = 0$$

Because the limit of the numerator is also 0, the numerator and denominator have a *common factor* of $(x + 3)$. So, for all $x \neq -3$, you can divide out this factor to obtain

$$f(x) = \frac{x^2 + x - 6}{x + 3} = \frac{(x + 3)(x - 2)}{x + 3} = x - 2 = g(x), \qquad x \neq -3.$$

Using Theorem 2.7, it follows that

$$\lim_{x \to -3} \frac{x^2 + x - 6}{x + 3} = \lim_{x \to -3} (x - 2) \qquad \text{Apply Theorem 2.7.}$$

$$= -5. \qquad \text{Use direct substitution.}$$

This result is shown graphically in Figure 2.18. Note that the graph of the function f coincides with the graph of the function $g(x) = x - 2$, except that the graph of f has a gap at the point $(-3, -5)$.

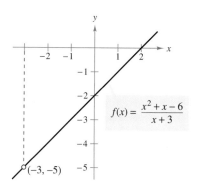

f is undefined when $x = -3$.
Figure 2.18

NOTE   In the solution of Example 7, be sure you see the usefulness of the Factor Theorem of Algebra. This theorem states that if c is a zero of a polynomial function, $(x - c)$ is a factor of the polynomial. So, if you apply direct substitution to a rational function and obtain

$$r(c) = \frac{p(c)}{q(c)} = \frac{0}{0}$$

you can conclude that $(x - c)$ must be a common factor to both $p(x)$ and $q(x)$.

In Example 7, direct substitution produced the meaningless fractional form $0/0$. An expression such as $0/0$ is called an **indeterminate form** because you cannot (from the form alone) determine the limit. When you try to evaluate a limit and encounter this form, remember that you must rewrite the fraction so that the new denominator does not have 0 as its limit. One way to do this is to *divide out common factors*, as shown in Example 7. A second way is to *rationalize the numerator*, as shown in Example 8.

TECHNOLOGY PITFALL   Because the graphs of

$$f(x) = \frac{x^2 + x - 6}{x + 3} \qquad \text{and} \qquad g(x) = x - 2$$

differ only at the point $(-3, -5)$, a standard graphing utility setting may not distinguish clearly between these graphs. However, because of the pixel configuration and rounding error of a graphing utility, it may be possible to find screen settings that distinguish between the graphs. Specifically, by repeatedly zooming in near the point $(-3, -5)$ on the graph of f, your graphing utility may show glitches or irregularities that do not exist on the actual graph. (See Figure 2.19.) By changing the screen settings on your graphing utility, you may obtain the correct graph of f.

Incorrect graph of f
Figure 2.19

EXAMPLE 8 **Rationalizing Technique**

Find the limit: $\lim\limits_{x \to 0} \dfrac{\sqrt{x+1}-1}{x}$.

Solution By direct substitution, you obtain the indeterminate form $0/0$.

$$\lim_{x \to 0} \left(\sqrt{x+1}-1\right) = 0$$

$$\lim_{x \to 0} \frac{\sqrt{x+1}-1}{x}$$

Direct substitution fails.

$$\lim_{x \to 0} x = 0$$

In this case, you can rewrite the fraction by rationalizing the numerator.

$$\frac{\sqrt{x+1}-1}{x} = \left(\frac{\sqrt{x+1}-1}{x}\right)\left(\frac{\sqrt{x+1}+1}{\sqrt{x+1}+1}\right)$$

$$= \frac{(x+1)-1}{x\left(\sqrt{x+1}+1\right)}$$

$$= \frac{\cancel{x}}{\cancel{x}\left(\sqrt{x+1}+1\right)}$$

$$= \frac{1}{\sqrt{x+1}+1}, \quad x \neq 0$$

Now, using Theorem 2.7, you can evaluate the limit as shown.

$$\lim_{x \to 0} \frac{\sqrt{x+1}-1}{x} = \lim_{x \to 0} \frac{1}{\sqrt{x+1}+1}$$

$$= \frac{1}{1+1}$$

$$= \frac{1}{2}$$

A table or a graph can reinforce your conclusion that the limit is $\frac{1}{2}$. (See Figure 2.20.)

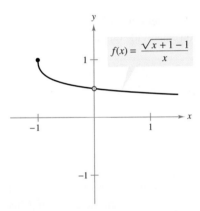

$f(x) = \dfrac{\sqrt{x+1}-1}{x}$

The limit of $f(x)$ as x approaches 0 is $\frac{1}{2}$.
Figure 2.20

		x approaches 0 from the left.				x approaches 0 from the right.			
x	−0.25	−0.1	−0.01	−0.001	0	0.001	0.01	0.1	0.25
f(x)	0.5359	0.5132	0.5013	0.5001	?	0.4999	0.4988	0.4881	0.4721

$f(x)$ approaches 0.5. $\qquad\qquad$ $f(x)$ approaches 0.5.

NOTE The rationalizing technique for evaluating limits is based on multiplication by a convenient form of 1. In Example 8, the convenient form is

$$1 = \frac{\sqrt{x+1}+1}{\sqrt{x+1}+1}.$$

The Squeeze Theorem

The next theorem concerns the limit of a function that is squeezed between two other functions, each of which has the same limit at a given x-value, as shown in Figure 2.21. (The proof of this theorem is given in Appendix A.)

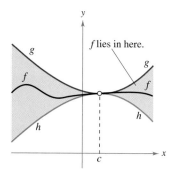

$h(x) \leq f(x) \leq g(x)$

f lies in here.

The Squeeze Theorem
Figure 2.21

THEOREM 2.8   The Squeeze Theorem

If $h(x) \leq f(x) \leq g(x)$ for all x in an open interval containing c, except possibly at c itself, and if

$$\lim_{x \to c} h(x) = L = \lim_{x \to c} g(x)$$

then $\lim_{x \to c} f(x)$ exists and is equal to L.

You can see the usefulness of the Squeeze Theorem in the proof of Theorem 2.9.

THEOREM 2.9   Three Special Limits

1. $\lim_{x \to 0} \dfrac{\sin x}{x} = 1$ **2.** $\lim_{x \to 0} \dfrac{1 - \cos x}{x} = 0$ **3.** $\lim_{x \to 0} (1 + x)^{1/x} = e$

FOR FURTHER INFORMATION
For more information on the function $f(x) = (\sin x)/x$, see the article "The Function $(\sin x)/x$" by William B. Gearhart and Harris S. Shultz in *The College Mathematics Journal*. To view this article, go to the website *www.matharticles.com*.

Proof   To avoid the confusion of two different uses of x, the proof of the first limit is presented using the variable θ, where θ is an acute positive angle *measured in radians*. Figure 2.22 shows a circular sector that is squeezed between two triangles.

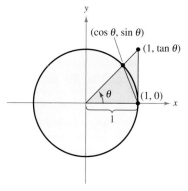

$(\cos \theta, \sin \theta)$

$(1, \tan \theta)$

$(1, 0)$

A circular sector is used to prove Theorem 2.9.
Figure 2.22

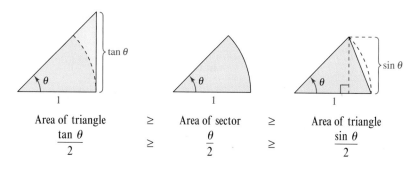

Area of triangle	$\geq$	Area of sector	$\geq$	Area of triangle
$\dfrac{\tan \theta}{2}$	$\geq$	$\dfrac{\theta}{2}$	$\geq$	$\dfrac{\sin \theta}{2}$

Multiplying each expression by $2/\sin \theta$ produces

$$\frac{1}{\cos \theta} \geq \frac{\theta}{\sin \theta} \geq 1$$

and taking reciprocals and reversing the inequalities yields

$$\cos \theta \leq \frac{\sin \theta}{\theta} \leq 1.$$

Because $\cos \theta = \cos(-\theta)$ and $(\sin \theta)/\theta = [\sin(-\theta)]/(-\theta)$, you can conclude that this inequality is valid for *all* nonzero θ in the open interval $(-\pi/2, \pi/2)$. Finally, because $\lim_{\theta \to 0} \cos \theta = 1$ and $\lim_{\theta \to 0} 1 = 1$, you can apply the Squeeze Theorem to conclude that $\lim_{\theta \to 0} (\sin \theta)/\theta = 1$. The proof of the second limit is left as an exercise (see Exercise 126). Recall from Section 1.6 that the third limit is actually the definition of the number e.

NOTE   The third limit of Theorem 2.9 will be used in Chapter 3 in the development of the formula for the derivative of the exponential function $f(x) = e^x$.

EXAMPLE 9   A Limit Involving a Trigonometric Function

Find the limit: $\lim\limits_{x \to 0} \dfrac{\tan x}{x}$.

Solution   Direct substitution yields the indeterminate form $0/0$. To solve this problem, you can write $\tan x$ as $(\sin x)/(\cos x)$ and obtain

$$\lim_{x \to 0} \frac{\tan x}{x} = \lim_{x \to 0} \left(\frac{\sin x}{x}\right)\left(\frac{1}{\cos x}\right).$$

Now, because

$$\lim_{x \to 0} \frac{\sin x}{x} = 1 \qquad \text{and} \qquad \lim_{x \to 0} \frac{1}{\cos x} = 1$$

you can obtain

$$\lim_{x \to 0} \frac{\tan x}{x} = \left(\lim_{x \to 0} \frac{\sin x}{x}\right)\left(\lim_{x \to 0} \frac{1}{\cos x}\right)$$

$$= (1)(1)$$

$$= 1.$$

(See Figure 2.23.)

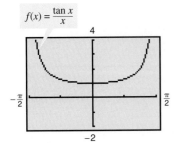

$f(x) = \dfrac{\tan x}{x}$

The limit of $f(x)$ as x approaches 0 is 1.
Figure 2.23

EXAMPLE 10   A Limit Involving a Trigonometric Function

Find the limit: $\lim\limits_{x \to 0} \dfrac{\sin 4x}{x}$.

Solution   Direct substitution yields the indeterminate form $0/0$. To solve this problem, you can rewrite the limit as

$$\lim_{x \to 0} \frac{\sin 4x}{x} = 4\left(\lim_{x \to 0} \frac{\sin 4x}{4x}\right).$$

Now, by letting $y = 4x$ and observing that $x \to 0$ if and only if $y \to 0$, you can write

$$\lim_{x \to 0} \frac{\sin 4x}{x} = 4\left(\lim_{x \to 0} \frac{\sin 4x}{4x}\right)$$

$$= 4\left(\lim_{y \to 0} \frac{\sin y}{y}\right)$$

$$= 4(1)$$

$$= 4.$$

(See Figure 2.24.)

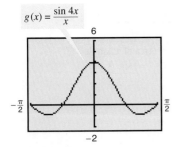

$g(x) = \dfrac{\sin 4x}{x}$

The limit of $g(x)$ as x approaches 0 is 4.
Figure 2.24

TECHNOLOGY   Try using a graphing utility to confirm the limits in the examples and exercise set. For instance, Figures 2.23 and 2.24 show the graphs of

$$f(x) = \frac{\tan x}{x} \qquad \text{and} \qquad g(x) = \frac{\sin 4x}{x}.$$

Note that the first graph appears to contain the point $(0, 1)$ and the second graph appears to contain the point $(0, 4)$, which lends support to the conclusions obtained in Examples 9 and 10.

Exercises for Section 2.3

See www.CalcChat.com for worked-out solutions to odd-numbered exercises.

In Exercises 1–4, use a graphing utility to graph the function and visually estimate the limits.

1. $h(x) = x^2 - 5x$

   (a) $\lim\limits_{x \to 5} h(x)$

   (b) $\lim\limits_{x \to -1} h(x)$

2. $g(x) = \dfrac{12(\sqrt{x} - 3)}{x - 9}$

   (a) $\lim\limits_{x \to 4} g(x)$

   (b) $\lim\limits_{x \to 0} g(x)$

3. $f(x) = x \cos x$

   (a) $\lim\limits_{x \to 0} f(x)$

   (b) $\lim\limits_{x \to \pi/3} f(x)$

4. $f(t) = t|t - 4|$

   (a) $\lim\limits_{t \to 4} f(t)$

   (b) $\lim\limits_{t \to -1} f(t)$

In Exercises 5–32, find the limit.

5. $\lim\limits_{x \to 2} x^4$

6. $\lim\limits_{x \to -2} x^5$

7. $\lim\limits_{x \to 0} (2x - 1)$

8. $\lim\limits_{x \to -3} (3x + 2)$

9. $\lim\limits_{x \to -3} (2x^2 + 4x + 1)$

10. $\lim\limits_{x \to 1} (3x^3 - 4x^2 + 3)$

11. $\lim\limits_{x \to 2} \dfrac{1}{x}$

12. $\lim\limits_{x \to -3} \dfrac{2}{x + 2}$

13. $\lim\limits_{x \to 1} \dfrac{x - 3}{x^2 + 4}$

14. $\lim\limits_{x \to 3} \dfrac{2x - 5}{x + 3}$

15. $\lim\limits_{x \to 7} \dfrac{5x}{\sqrt{x + 2}}$

16. $\lim\limits_{x \to 3} \dfrac{\sqrt{x + 1}}{x - 4}$

17. $\lim\limits_{x \to 3} \sqrt{x + 1}$

18. $\lim\limits_{x \to 4} \sqrt[3]{x + 23}$

19. $\lim\limits_{x \to \pi/2} \sin x$

20. $\lim\limits_{x \to \pi} \tan x$

21. $\lim\limits_{x \to 2} \cos\left(\dfrac{\pi x}{3}\right)$

22. $\lim\limits_{x \to 1} \sin\left(\dfrac{\pi x}{2}\right)$

23. $\lim\limits_{x \to 0} \sec 2x$

24. $\lim\limits_{x \to \pi} \cos 5x$

25. $\lim\limits_{x \to 5\pi/6} \sin x$

26. $\lim\limits_{x \to 5\pi/3} \cos x$

27. $\lim\limits_{x \to 3} \tan\left(\dfrac{\pi x}{4}\right)$

28. $\lim\limits_{x \to 7} \sec\left(\dfrac{\pi x}{6}\right)$

29. $\lim\limits_{x \to 0} e^x \cos 2x$

30. $\lim\limits_{x \to 0} e^{-x} \sin \pi x$

31. $\lim\limits_{x \to 1} (\ln 3x + e^x)$

32. $\lim\limits_{x \to 1} \ln\left(\dfrac{x}{e^x}\right)$

In Exercises 33–36, find the limits.

33. $f(x) = 5 - x, \ g(x) = x^3$

   (a) $\lim\limits_{x \to 1} f(x)$ (b) $\lim\limits_{x \to 4} g(x)$ (c) $\lim\limits_{x \to 1} g(f(x))$

34. $f(x) = x + 7, \ g(x) = x^2$

   (a) $\lim\limits_{x \to -3} f(x)$ (b) $\lim\limits_{x \to 4} g(x)$ (c) $\lim\limits_{x \to -3} g(f(x))$

35. $f(x) = 4 - x^2, \ g(x) = \sqrt{x + 1}$

   (a) $\lim\limits_{x \to 1} f(x)$ (b) $\lim\limits_{x \to 3} g(x)$ (c) $\lim\limits_{x \to 1} g(f(x))$

36. $f(x) = 2x^2 - 3x + 1, \ g(x) = \sqrt[3]{x + 6}$

   (a) $\lim\limits_{x \to 4} f(x)$ (b) $\lim\limits_{x \to 21} g(x)$ (c) $\lim\limits_{x \to 4} g(f(x))$

In Exercises 37–40, use the information to evaluate the limits.

37. $\lim\limits_{x \to c} f(x) = 2$

   $\lim\limits_{x \to c} g(x) = 3$

   (a) $\lim\limits_{x \to c} [5g(x)]$

   (b) $\lim\limits_{x \to c} [f(x) + g(x)]$

   (c) $\lim\limits_{x \to c} [f(x)g(x)]$

   (d) $\lim\limits_{x \to c} \dfrac{f(x)}{g(x)}$

38. $\lim\limits_{x \to c} f(x) = \frac{3}{2}$

   $\lim\limits_{x \to c} g(x) = \frac{1}{2}$

   (a) $\lim\limits_{x \to c} [4f(x)]$

   (b) $\lim\limits_{x \to c} [f(x) + g(x)]$

   (c) $\lim\limits_{x \to c} [f(x)g(x)]$

   (d) $\lim\limits_{x \to c} \dfrac{f(x)}{g(x)}$

39. $\lim\limits_{x \to c} f(x) = 4$

   (a) $\lim\limits_{x \to c} [f(x)]^3$

   (b) $\lim\limits_{x \to c} \sqrt{f(x)}$

   (c) $\lim\limits_{x \to c} [3f(x)]$

   (d) $\lim\limits_{x \to c} [f(x)]^{3/2}$

40. $\lim\limits_{x \to c} f(x) = 27$

   (a) $\lim\limits_{x \to c} \sqrt[3]{f(x)}$

   (b) $\lim\limits_{x \to c} \dfrac{f(x)}{18}$

   (c) $\lim\limits_{x \to c} [f(x)]^2$

   (d) $\lim\limits_{x \to c} [f(x)]^{2/3}$

In Exercises 41–44, use the graph to determine the limit visually (if it exists). Write a simpler function that agrees with the given function at all but one point.

41. $g(x) = \dfrac{-2x^2 + x}{x}$

   (a) $\lim\limits_{x \to 0} g(x)$

   (b) $\lim\limits_{x \to -1} g(x)$

42. $h(x) = \dfrac{x^2 - 3x}{x}$

   (a) $\lim\limits_{x \to -2} h(x)$

   (b) $\lim\limits_{x \to 0} h(x)$

43. $g(x) = \dfrac{x^3 - x}{x - 1}$

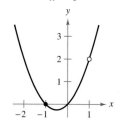

   (a) $\lim\limits_{x \to 1} g(x)$

   (b) $\lim\limits_{x \to -1} g(x)$

44. $f(x) = \dfrac{x}{x^2 - x}$

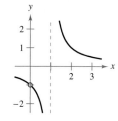

   (a) $\lim\limits_{x \to 1} f(x)$

   (b) $\lim\limits_{x \to 0} f(x)$

In Exercises 45–50, find the limit of the function (if it exists). Write a simpler function that agrees with the given function at all but one point. Use a graphing utility to confirm your result.

45. $\lim\limits_{x \to -1} \dfrac{x^2 - 1}{x + 1}$

46. $\lim\limits_{x \to -1} \dfrac{2x^2 - x - 3}{x + 1}$

47. $\lim\limits_{x \to 2} \dfrac{x^3 - 8}{x - 2}$

48. $\lim\limits_{x \to -1} \dfrac{x^3 + 1}{x + 1}$

49. $\lim\limits_{x \to -4} \dfrac{(x + 4) \ln(x + 6)}{x^2 - 16}$

50. $\lim\limits_{x \to 0} \dfrac{e^{2x} - 1}{e^x - 1}$

In Exercises 51–64, find the limit (if it exists).

51. $\lim\limits_{x \to 5} \dfrac{x - 5}{x^2 - 25}$

52. $\lim\limits_{x \to 3} \dfrac{3 - x}{x^2 - 9}$

53. $\lim\limits_{x \to -3} \dfrac{x^2 + x - 6}{x^2 - 9}$

54. $\lim\limits_{x \to 3} \dfrac{x^2 - x - 6}{x^2 - 5x + 6}$

55. $\lim\limits_{x \to 0} \dfrac{\sqrt{x + 5} - \sqrt{5}}{x}$

56. $\lim\limits_{x \to 0} \dfrac{\sqrt{3 + x} - \sqrt{3}}{x}$

57. $\lim\limits_{x \to 4} \dfrac{\sqrt{x + 5} - 3}{x - 4}$

58. $\lim\limits_{x \to 3} \dfrac{\sqrt{x + 1} - 2}{x - 3}$

59. $\lim\limits_{x \to 0} \dfrac{[1/(3 + x)] - (1/3)}{x}$

60. $\lim\limits_{x \to 0} \dfrac{[1/(x + 4)] - (1/4)}{x}$

61. $\lim\limits_{\Delta x \to 0} \dfrac{2(x + \Delta x) - 2x}{\Delta x}$

62. $\lim\limits_{\Delta x \to 0} \dfrac{(x + \Delta x)^2 - x^2}{\Delta x}$

63. $\lim\limits_{\Delta x \to 0} \dfrac{(x + \Delta x)^2 - 2(x + \Delta x) + 1 - (x^2 - 2x + 1)}{\Delta x}$

64. $\lim\limits_{\Delta x \to 0} \dfrac{(x + \Delta x)^3 - x^3}{\Delta x}$

⟳ Graphical, Numerical, and Analytic Analysis   In Exercises 65–68, use a graphing utility to graph the function and estimate the limit. Use a table to reinforce your conclusion. Then find the limit by analytic methods.

65. $\lim\limits_{x \to 0} \dfrac{\sqrt{x + 2} - \sqrt{2}}{x}$

66. $\lim\limits_{x \to 16} \dfrac{4 - \sqrt{x}}{x - 16}$

67. $\lim\limits_{x \to 0} \dfrac{[1/(2 + x)] - (1/2)}{x}$

68. $\lim\limits_{x \to 2} \dfrac{x^5 - 32}{x - 2}$

In Exercises 69–82, determine the limit of the transcendental function (if it exists).

69. $\lim\limits_{x \to 0} \dfrac{\sin x}{5x}$

70. $\lim\limits_{x \to 0} \dfrac{5(1 - \cos x)}{x}$

71. $\lim\limits_{x \to 0} \dfrac{\sin x(1 - \cos x)}{2x^2}$

72. $\lim\limits_{\theta \to 0} \dfrac{\cos \theta \tan \theta}{\theta}$

73. $\lim\limits_{x \to 0} \dfrac{\sin^2 x}{x}$

74. $\lim\limits_{x \to 0} \dfrac{2 \tan^2 x}{x}$

75. $\lim\limits_{h \to 0} \dfrac{(1 - \cos h)^2}{h}$

76. $\lim\limits_{\phi \to \pi} \phi \sec \phi$

77. $\lim\limits_{x \to \pi/2} \dfrac{\cos x}{\cot x}$

78. $\lim\limits_{x \to \pi/4} \dfrac{1 - \tan x}{\sin x - \cos x}$

79. $\lim\limits_{x \to 0} \dfrac{1 - e^{-x}}{e^x - 1}$

80. $\lim\limits_{x \to 0} \dfrac{4(e^{2x} - 1)}{e^x - 1}$

81. $\lim\limits_{t \to 0} \dfrac{\sin 3t}{2t}$

82. $\lim\limits_{x \to 0} \dfrac{\sin 2x}{\sin 3x}$ $\left[Hint\text{: Find } \lim\limits_{x \to 0} \left(\dfrac{2 \sin 2x}{2x} \right) \left(\dfrac{3x}{3 \sin 3x} \right). \right]$

⟳ Graphical, Numerical, and Analytic Analysis   In Exercises 83–88, use a graphing utility to graph the function and estimate the limit. Use a table to reinforce your conclusion. Then find the limit by analytic methods.

83. $\lim\limits_{t \to 0} \dfrac{\sin 3t}{t}$

84. $\lim\limits_{x \to 0} \dfrac{\cos x - 1}{2x^2}$

85. $\lim\limits_{x \to 0} \dfrac{\sin x^2}{x}$

86. $\lim\limits_{x \to 0} \dfrac{\sin x}{\sqrt[3]{x}}$

87. $\lim\limits_{x \to 1} \dfrac{\ln x}{x - 1}$

88. $\lim\limits_{x \to \ln 2} \dfrac{e^{3x} - 8}{e^{2x} - 4}$

In Exercises 89–92, find $\lim\limits_{\Delta x \to 0} \dfrac{f(x + \Delta x) - f(x)}{\Delta x}$.

89. $f(x) = 2x + 3$

90. $f(x) = \sqrt{x}$

91. $f(x) = \dfrac{4}{x}$

92. $f(x) = x^2 - 4x$

In Exercises 93 and 94, use the Squeeze Theorem to find $\lim\limits_{x \to c} f(x)$.

93. $c = 0$

$4 - x^2 \leq f(x) \leq 4 + x^2$

94. $c = a$

$b - |x - a| \leq f(x) \leq b + |x - a|$

⟳ In Exercises 95–100, use a graphing utility to graph the given function and the equations $y = |x|$ and $y = -|x|$ in the same viewing window. Using the graphs to visually observe the Squeeze Theorem, find $\lim\limits_{x \to 0} f(x)$.

95. $f(x) = x \cos x$

96. $f(x) = |x \sin x|$

97. $f(x) = |x| \sin x$

98. $f(x) = |x| \cos x$

99. $f(x) = x \sin \dfrac{1}{x}$

100. $h(x) = x \cos \dfrac{1}{x}$

Writing About Concepts

101. In the context of finding limits, discuss what is meant by two functions that agree at all but one point.

102. Give an example of two functions that agree at all but one point.

103. What is meant by an indeterminate form?

104. In your own words, explain the Squeeze Theorem.

105. *Writing*   Use a graphing utility to graph

$$f(x) = x, \quad g(x) = \sin x, \quad \text{and} \quad h(x) = \frac{\sin x}{x}$$

in the same viewing window. Compare the magnitudes of $f(x)$ and $g(x)$ when x is "close to" 0. Use the comparison to write a short paragraph explaining why $\lim_{x \to 0} h(x) = 1$.

106. *Writing*   Use a graphing utility to graph

$$f(x) = x, \quad g(x) = \sin^2 x, \quad \text{and } h(x) = \frac{\sin^2 x}{x}$$

in the same viewing window. Compare the magnitudes of $f(x)$ and $g(x)$ when x is "close to" 0. Use the comparison to write a short paragraph explaining why $\lim_{x \to 0} h(x) = 0$.

Free-Falling Object   In Exercises 107 and 108, use the position function $s(t) = -16t^2 + 1000$, which gives the height (in feet) of an object that has fallen for t seconds from a height of 1000 feet. The velocity at time $t = a$ seconds is given by

$$\lim_{t \to a} \frac{s(a) - s(t)}{a - t}.$$

107. If a construction worker drops a wrench from a height of 1000 feet, how fast will the wrench be falling after 5 seconds?

108. If a construction worker drops a wrench from a height of 1000 feet, when will the wrench hit the ground? At what velocity will the wrench impact the ground?

Free-Falling Object   In Exercises 109 and 110, use the position function $s(t) = -4.9t^2 + 150$, which gives the height (in meters) of an object that has fallen from a height of 150 meters. The velocity at time $t = a$ seconds is given by

$$\lim_{t \to a} \frac{s(a) - s(t)}{a - t}.$$

109. Find the velocity of the object when $t = 3$.

110. At what velocity will the object impact the ground?

111. Find two functions f and g such that $\lim_{x \to 0} f(x)$ and $\lim_{x \to 0} g(x)$ do not exist, but $\lim_{x \to 0} [f(x) + g(x)]$ does exist.

112. Prove that if $\lim_{x \to c} f(x)$ exists and $\lim_{x \to c} [f(x) + g(x)]$ does not exist, then $\lim_{x \to c} g(x)$ does not exist.

113. Prove Property 1 of Theorem 2.1.

114. Prove Property 3 of Theorem 2.1. (You may use Property 3 of Theorem 2.2.)

115. Prove Property 1 of Theorem 2.2.

116. Prove that if $\lim_{x \to c} f(x) = 0$, then $\lim_{x \to c} |f(x)| = 0$.

117. Prove that if $\lim_{x \to c} f(x) = 0$ and $|g(x)| \le M$ for a fixed number M and all $x \ne c$, then $\lim_{x \to c} f(x)g(x) = 0$.

118. (a) Prove that if $\lim_{x \to c} |f(x)| = 0$, then $\lim_{x \to c} f(x) = 0$.
(*Note:* This is the converse of Exercise 116.)

(b) Prove that if $\lim_{x \to c} f(x) = L$, then $\lim_{x \to c} |f(x)| = |L|$.
[*Hint:* Use the inequality $\big| |f(x)| - |L| \big| \le |f(x) - L|$.]

True or False?   **In Exercises 119–124, determine whether the statement is true or false. If it is false, explain why or give an example that shows it is false.**

119. $\lim_{x \to 0} \dfrac{|x|}{x} = 1$

120. $\lim_{x \to \pi} \dfrac{\sin x}{x} = 1$

121. If $f(x) = g(x)$ for all real numbers other than $x = 0$, and $\lim_{x \to 0} f(x) = L$, then $\lim_{x \to 0} g(x) = L$.

122. If $\lim_{x \to c} f(x) = L$, then $f(c) = L$.

123. $\lim_{x \to 2} f(x) = 3$, where $f(x) = \begin{cases} 3, & x \le 2 \\ 0, & x > 2 \end{cases}$

124. If $f(x) < g(x)$ for all $x \ne a$, then $\lim_{x \to a} f(x) < \lim_{x \to a} g(x)$.

125. *Think About It*   Find a function f to show that the converse of Exercise 118(b) is not true. [*Hint:* Find a function f such that $\lim_{x \to c} |f(x)| = |L|$ but $\lim_{x \to c} f(x)$ does not exist.]

126. Prove the second part of Theorem 2.9 by proving that

$$\lim_{x \to 0} \frac{1 - \cos x}{x} = 0.$$

127. Let $f(x) = \begin{cases} 0, & \text{if } x \text{ is rational} \\ 1, & \text{if } x \text{ is irrational} \end{cases}$
and
$g(x) = \begin{cases} 0, & \text{if } x \text{ is rational} \\ x, & \text{if } x \text{ is irrational.} \end{cases}$

Find (if possible) $\lim_{x \to 0} f(x)$ and $\lim_{x \to 0} g(x)$.

 128. *Graphical Reasoning*   Consider $f(x) = \dfrac{\sec x - 1}{x^2}$.

(a) Find the domain of f.

(b) Use a graphing utility to graph f. Is the domain of f obvious from the graph? If not, explain.

(c) Use the graph of f to approximate $\lim_{x \to 0} f(x)$.

(d) Confirm the answer in part (c) analytically.

129. *Approximation*

(a) Find $\lim_{x \to 0} \dfrac{1 - \cos x}{x^2}$.

(b) Use the result in part (a) to derive the approximation $\cos x \approx 1 - \frac{1}{2}x^2$ for x near 0.

(c) Use the result in part (b) to approximate $\cos(0.1)$.

(d) Use a calculator to approximate $\cos(0.1)$ to four decimal places. Compare the result with part (c).

130. *Think About It*   When using a graphing utility to generate a table to approximate $\lim_{x \to 0} [(\sin x)/x]$, a student concluded that the limit was 0.01745 rather than 1. Determine the probable cause of the error.

(a) Limit from right

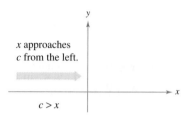

(b) Limit from left
Figure 2.28

One-Sided Limits and Continuity on a Closed Interval

To understand continuity on a closed interval, you first need to look at a different type of limit called a **one-sided limit.** For example, the **limit from the right** means that x approaches c from values greater than c [see Figure 2.28(a)]. This limit is denoted as

$$\lim_{x \to c^+} f(x) = L. \qquad \text{Limit from the right}$$

Similarly, the **limit from the left** means that x approaches c from values less than c [see Figure 2.28(b)]. This limit is denoted as

$$\lim_{x \to c^-} f(x) = L. \qquad \text{Limit from the left}$$

One-sided limits are useful in taking limits of functions involving radicals. For instance, if n is an even integer,

$$\lim_{x \to 0^+} \sqrt[n]{x} = 0.$$

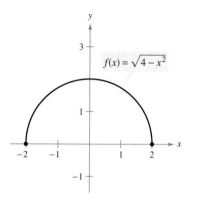

The limit of $f(x)$ as x approaches -2 from the right is 0.
Figure 2.29

EXAMPLE 2   A One-Sided Limit

Find the limit of $f(x) = \sqrt{4 - x^2}$ as x approaches -2 from the right.

Solution   As shown in Figure 2.29, the limit as x approaches -2 from the right is

$$\lim_{x \to -2^+} \sqrt{4 - x^2} = 0.$$

One-sided limits can be used to investigate the behavior of **step functions.** One common type of step function is the **greatest integer function** $[\![x]\!]$, defined by

$$[\![x]\!] = \text{greatest integer } n \text{ such that } n \leq x. \qquad \text{Greatest integer function}$$

For instance, $[\![2.5]\!] = 2$ and $[\![-2.5]\!] = -3$.

EXAMPLE 3   The Greatest Integer Function

Find the limit of the greatest integer function $f(x) = [\![x]\!]$ as x approaches 0 from the left and from the right.

Solution   As shown in Figure 2.30, the limit as x approaches 0 *from the left* is given by

$$\lim_{x \to 0^-} [\![x]\!] = -1$$

and the limit as x approaches 0 *from the right* is given by

$$\lim_{x \to 0^+} [\![x]\!] = 0.$$

The greatest integer function has a discontinuity at zero because the left and right limits at zero are different. By similar reasoning, you can see that the greatest integer function has a discontinuity at any integer n.

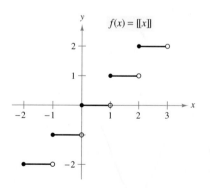

Greatest integer function
Figure 2.30

When the limit from the left is not equal to the limit from the right, the (two-sided) limit *does not exist*. The next theorem makes this more explicit. The proof of this theorem follows directly from the definition of a one-sided limit.

THEOREM 2.10   The Existence of a Limit

Let f be a function and let c and L be real numbers. The limit of $f(x)$ as x approaches c is L if and only if

$$\lim_{x \to c^-} f(x) = L \quad \text{and} \quad \lim_{x \to c^+} f(x) = L.$$

The concept of a one-sided limit allows you to extend the definition of continuity to closed intervals. Basically, a function is continuous on a closed interval if it is continuous in the interior of the interval and exhibits one-sided continuity at the endpoints. This is stated formally as follows.

Definition of Continuity on a Closed Interval

A function f is **continuous on the closed interval** $[a, b]$ if it is continuous on the open interval (a, b) and

$$\lim_{x \to a^+} f(x) = f(a) \quad \text{and} \quad \lim_{x \to b^-} f(x) = f(b).$$

The function f is **continuous from the right** at a and **continuous from the left** at b (see Figure 2.31).

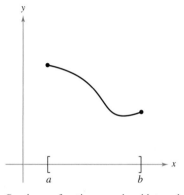

Continuous function on a closed interval
Figure 2.31

Similar definitions can be made to cover continuity on intervals of the form $(a, b]$ and $[a, b)$ that are neither open nor closed, or on infinite intervals. For example, the function

$$f(x) = \sqrt{x}$$

is continuous on the infinite interval $[0, \infty)$, and the function

$$g(x) = \sqrt{2 - x}$$

is continuous on the infinite interval $(-\infty, 2]$.

EXAMPLE 4   **Continuity on a Closed Interval**

Discuss the continuity of $f(x) = \sqrt{1 - x^2}$.

Solution   The domain of f is the closed interval $[-1, 1]$. At all points in the open interval $(-1, 1)$, the continuity of f follows from Theorems 2.4 and 2.5. Moreover, because

$$\lim_{x \to -1^+} \sqrt{1 - x^2} = 0 = f(-1) \qquad \text{Continuous from the right}$$

and

$$\lim_{x \to 1^-} \sqrt{1 - x^2} = 0 = f(1) \qquad \text{Continuous from the left}$$

you can conclude that f is continuous on the closed interval $[-1, 1]$, as shown in Figure 2.32.

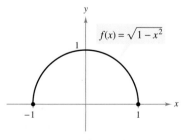

f is continuous on $[-1, 1]$.
Figure 2.32

The next example shows how a one-sided limit can be used to determine the value of absolute zero on the Kelvin scale.

EXAMPLE 5 Charles's Law and Absolute Zero

On the Kelvin scale, *absolute zero* is the temperature 0 K. Although temperatures of approximately 0.0001 K have been produced in laboratories, absolute zero has never been attained. In fact, evidence suggests that absolute zero *cannot* be attained. How did scientists determine that 0 K is the "lower limit" of the temperature of matter? What is absolute zero on the Celsius scale?

Solution The determination of absolute zero stems from the work of the French physicist Jacques Charles (1746–1823). Charles discovered that the volume of gas at a constant pressure increases linearly with the temperature of the gas. The table illustrates this relationship between volume and temperature. In the table, one mole of hydrogen is held at a constant pressure of one atmosphere. The volume V is measured in liters and the temperature T is measured in degrees Celsius.

T	-40	-20	0	20	40	60	80
V	19.1482	20.7908	22.4334	24.0760	25.7186	27.3612	29.0038

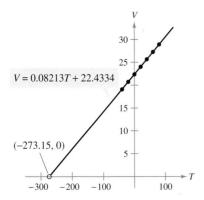

$V = 0.08213T + 22.4334$

$(-273.15, 0)$

The volume of hydrogen gas depends on its temperature.
Figure 2.33

The points represented by the table are shown in Figure 2.33. Moreover, by using the points in the table, you can determine that T and V are related by the linear equation

$$V = 0.08213T + 22.4334 \qquad \text{or} \qquad T = \frac{V - 22.4334}{0.08213}.$$

By reasoning that the volume of the gas can approach 0 (but never equal or go below 0), you can determine that the "least possible temperature" is given by

$$\lim_{V \to 0^+} T = \lim_{V \to 0^+} \frac{V - 22.4334}{0.08213}$$

$$= \frac{0 - 22.4334}{0.08213} \qquad \text{Use direct substitution.}$$

$$\approx -273.15.$$

So, absolute zero on the Kelvin scale (0 K) is approximately $-273.15°$ on the Celsius scale.

In 1995, physicists Carl Wieman and Eric Cornell of the University of Colorado at Boulder used lasers and evaporation to produce a supercold gas in which atoms overlap. This gas is called a Bose-Einstein condensate. "We get to within a billionth of a degree of absolute zero," reported Wieman. (*Source:* Time *magazine, April 10, 2000*)

The following table shows the temperatures in Example 5, converted to the Fahrenheit scale. Try repeating the solution shown in Example 5 using these temperatures and volumes. Use the result to find the value of absolute zero on the Fahrenheit scale.

T	-40	-4	32	68	104	140	176
V	19.1482	20.7908	22.4334	24.0760	25.7186	27.3612	29.0038

NOTE Charles's Law for gases (assuming constant pressure) can be stated as

$$V = RT \qquad \text{Charles's Law}$$

where V is volume, R is constant, and T is temperature. In the statement of this law, what property must the temperature scale have?

Properties of Continuity

In Section 2.3, you studied several properties of limits. Each of those properties yields a corresponding property pertaining to the continuity of a function. For instance, Theorem 2.11 follows directly from Theorem 2.2.

AUGUSTIN-LOUIS CAUCHY (1789–1857)

The concept of a continuous function was first introduced by Augustin-Louis Cauchy in 1821. The definition given in his text *Cours d'Analyse* stated that indefinite small changes in y were the result of indefinite small changes in x. "…$f(x)$ will be called a *continuous* function if … the numerical values of the difference $f(x + \alpha) - f(x)$ decrease indefinitely with those of α …."

THEOREM 2.11   Properties of Continuity

If b is a real number and f and g are continuous at $x = c$, then the following functions are also continuous at c.

1. Scalar multiple: bf
2. Sum and difference: $f \pm g$
3. Product: fg
4. Quotient: $\dfrac{f}{g}$,   if $g(c) \neq 0$

The following types of functions are continuous at every point in their domains.

1. Polynomial: $p(x) = a_n x^n + a_{n-1} x^{n-1} + \cdots + a_1 x + a_0$
2. Rational: $r(x) = \dfrac{p(x)}{q(x)}, \quad q(x) \neq 0$
3. Radical: $f(x) = \sqrt[n]{x}$
4. Trigonometric: $\sin x, \cos x, \tan x, \cot x, \sec x, \csc x$
5. Exponential and logarithmic: $f(x) = a^x, f(x) = e^x, f(x) = \ln x$

By combining Theorem 2.11 with this summary, you can conclude that a wide variety of elementary functions are continuous at every point in their domains.

 EXAMPLE 6   **Applying Properties of Continuity**

By Theorem 2.11, it follows that each of the following functions is continuous at every point in its domain.

$$f(x) = x + e^x, \quad f(x) = 3 \tan x, \quad f(x) = \frac{x^2 + 1}{\cos x}$$

For instance, the first function is continuous at every real number because the functions $y = x$ and $y = e^x$ are continuous at every real number and the sum of continuous functions is continuous.   ▬▬▬

The next theorem, which is a consequence of Theorem 2.5, allows you to determine the continuity of *composite* functions such as

$$f(x) = \sin 3x, \quad f(x) = \sqrt{x^2 + 1}, \quad f(x) = \tan \frac{1}{x}.$$

NOTE   One consequence of Theorem 2.12 is that if f and g satisfy the given conditions, you can determine the limit of $f(g(x))$ as x approaches c to be

$$\lim_{x \to c} f(g(x)) = f(g(c)).$$

THEOREM 2.12   Continuity of a Composite Function

If g is continuous at c and f is continuous at $g(c)$, then the composite function given by $(f \circ g)(x) = f(g(x))$ is continuous at c.

EXAMPLE 7   Testing for Continuity

Describe the interval(s) on which each function is continuous.

a. $f(x) = \tan x$ **b.** $g(x) = \begin{cases} \sin \dfrac{1}{x}, & x \neq 0 \\ 0, & x = 0 \end{cases}$ **c.** $h(x) = \begin{cases} x \sin \dfrac{1}{x}, & x \neq 0 \\ 0, & x = 0 \end{cases}$

Solution

a. The tangent function $f(x) = \tan x$ is undefined at

$$x = \frac{\pi}{2} + n\pi, \qquad n \text{ is an integer.}$$

At all other points it is continuous. So, $f(x) = \tan x$ is continuous on the open intervals

$$\cdots, \left(-\frac{3\pi}{2}, -\frac{\pi}{2} \right), \left(-\frac{\pi}{2}, \frac{\pi}{2} \right), \left(\frac{\pi}{2}, \frac{3\pi}{2} \right), \cdots$$

as shown in Figure 2.34(a).

b. Because $y = 1/x$ is continuous except at $x = 0$ and the sine function is continuous for all real values of x, it follows that $y = \sin(1/x)$ is continuous at all real values except $x = 0$. At $x = 0$, the limit of $g(x)$ does not exist (see Example 5, Section 2.2). So, g is continuous on the intervals $(-\infty, 0)$ and $(0, \infty)$, as shown in Figure 2.34(b).

c. This function is similar to that in part (b) except that the oscillations are damped by the factor x. Using the Squeeze Theorem, you obtain

$$-|x| \leq x \sin \frac{1}{x} \leq |x|, \qquad x \neq 0$$

and you can conclude that

$$\lim_{x \to 0} h(x) = 0.$$

So, h is continuous on the entire real number line, as shown in Figure 2.34(c).

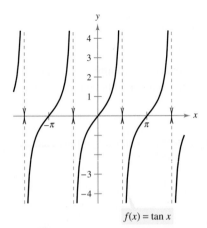

$f(x) = \tan x$

(a) f is continuous on each open interval in its domain.

Figure 2.34

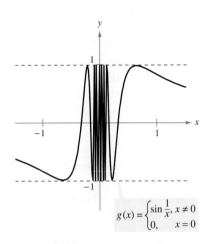

$g(x) = \begin{cases} \sin \frac{1}{x}, & x \neq 0 \\ 0, & x = 0 \end{cases}$

(b) g is continuous on $(-\infty, 0)$ and $(0, \infty)$.

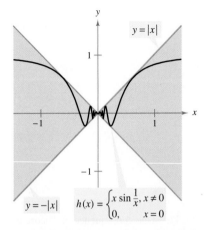

$y = -|x| \qquad h(x) = \begin{cases} x \sin \frac{1}{x}, & x \neq 0 \\ 0, & x = 0 \end{cases}$

(c) h is continuous on the entire real number line.

The Intermediate Value Theorem

Theorem 2.13 is an important theorem concerning the behavior of functions that are continuous on a closed interval.

THEOREM 2.13   Intermediate Value Theorem

If f is continuous on the closed interval $[a, b]$ and k is any number between $f(a)$ and $f(b)$, then there is at least one number c in $[a, b]$ such that

$$f(c) = k.$$

NOTE   The Intermediate Value Theorem tells you that at least one c exists, but it does not give a method for finding c. Such theorems are called **existence theorems.** By referring to a text on advanced calculus, you will find that a proof of this theorem is based on a property of real numbers called *completeness*. The Intermediate Value Theorem states that for a continuous function f, if x takes on all values between a and b, $f(x)$ must take on all values between $f(a)$ and $f(b)$.

As a simple example of this theorem, consider a person's height. Suppose that a girl is 5 feet tall on her thirteenth birthday and 5 feet 7 inches tall on her fourteenth birthday. Then, for any height h between 5 feet and 5 feet 7 inches, there must have been a time t when her height was exactly h. This seems reasonable because human growth is continuous and a person's height does not abruptly change from one value to another.

The Intermediate Value Theorem guarantees the existence of *at least one* number c in the closed interval $[a, b]$. There may, of course, be more than one number c such that $f(c) = k$, as shown in Figure 2.35. A function that is not continuous does not necessarily possess the intermediate value property. For example, the graph of the function shown in Figure 2.36 jumps over the horizontal line given by $y = k$, and for this function there is no value of c in $[a, b]$ such that $f(c) = k$.

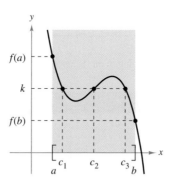

f is continuous on $[a, b]$.
[There exist three c's such that $f(c) = k$.]
Figure 2.35

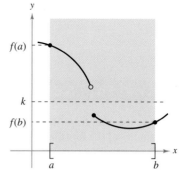

f is not continuous on $[a, b]$.
[There are no c's such that $f(c) = k$.]
Figure 2.36

The Intermediate Value Theorem often can be used to locate the zeros of a function that is continuous on a closed interval. Specifically, if f is continuous on $[a, b]$ and $f(a)$ and $f(b)$ differ in sign, the Intermediate Value Theorem guarantees the existence of at least one zero of f in the closed interval $[a, b]$.

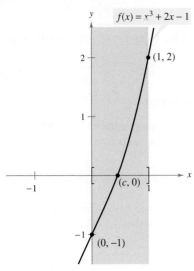

f is continuous on $[0, 1]$ with $f(0) < 0$ and $f(1) > 0$.

Figure 2.37

EXAMPLE 8   An Application of the Intermediate Value Theorem

Use the Intermediate Value Theorem to show that the polynomial function $f(x) = x^3 + 2x - 1$ has a zero in the interval $[0, 1]$.

Solution   Note that f is continuous on the closed interval $[0, 1]$. Because

$$f(0) = 0^3 + 2(0) - 1 = -1 \quad \text{and} \quad f(1) = 1^3 + 2(1) - 1 = 2$$

it follows that $f(0) < 0$ and $f(1) > 0$. You can therefore apply the Intermediate Value Theorem to conclude that there must be some c in $[0, 1]$ such that

$$f(c) = 0 \qquad \text{\small f has a zero in the closed interval $[0, 1]$.}$$

as shown in Figure 2.37.

The **bisection method** for approximating the real zeros of a continuous function is similar to the method used in Example 8. If you know that a zero exists in the closed interval $[a, b]$, the zero must lie in the interval $[a, (a + b)/2]$ or $[(a + b)/2, b]$. From the sign of $f([a + b]/2)$, you can determine which interval contains the zero. By repeatedly bisecting the interval, you can "close in" on the zero of the function.

TECHNOLOGY   You can also use the *zoom* feature of a graphing utility to approximate the real zeros of a continuous function. By repeatedly zooming in on the point where the graph crosses the x-axis, and adjusting the x-axis scale, you can approximate the zero of the function to any desired accuracy. The zero of $x^3 + 2x - 1$ is approximately 0.453, as shown in Figure 2.38.

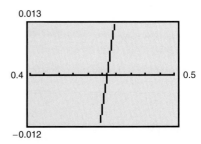

Figure 2.38   Zooming in on the zero of $f(x) = x^3 + 2x - 1$

Exercises for Section 2.4

In Exercises 1–6, use the graph to determine the limit, and discuss the continuity of the function.

(a) $\displaystyle\lim_{x \to c^+} f(x)$ **(b)** $\displaystyle\lim_{x \to c^-} f(x)$ **(c)** $\displaystyle\lim_{x \to c} f(x)$

1.

2.

3.

4.

5.

6.

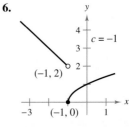

In Exercises 7–28, find the limit (if it exists). If it does not exist, explain why.

7. $\lim\limits_{x \to 5^+} \dfrac{x - 5}{x^2 - 25}$

8. $\lim\limits_{x \to 2^+} \dfrac{2 - x}{x^2 - 4}$

9. $\lim\limits_{x \to -3^-} \dfrac{x}{\sqrt{x^2 - 9}}$

10. $\lim\limits_{x \to 4^-} \dfrac{\sqrt{x} - 2}{x - 4}$

11. $\lim\limits_{x \to 0^-} \dfrac{|x|}{x}$

12. $\lim\limits_{x \to 3^+} \dfrac{|x - 3|}{x - 3}$

13. $\lim\limits_{\Delta x \to 0^-} \dfrac{\dfrac{1}{x + \Delta x} - \dfrac{1}{x}}{\Delta x}$

14. $\lim\limits_{\Delta x \to 0^+} \dfrac{(x + \Delta x)^2 + x + \Delta x - (x^2 + x)}{\Delta x}$

15. $\lim\limits_{x \to 3^-} f(x)$, where $f(x) = \begin{cases} \dfrac{x + 2}{2}, & x \le 3 \\ \dfrac{12 - 2x}{3}, & x > 3 \end{cases}$

16. $\lim\limits_{x \to 2^-} f(x)$, where $f(x) = \begin{cases} x^2 - 4x + 6, & x < 2 \\ -x^2 + 4x - 2, & x \ge 2 \end{cases}$

17. $\lim\limits_{x \to 1} f(x)$, where $f(x) = \begin{cases} x^3 + 1, & x < 1 \\ x + 1, & x \ge 1 \end{cases}$

18. $\lim\limits_{x \to 1^+} f(x)$, where $f(x) = \begin{cases} x, & x \le 1 \\ 1 - x, & x > 1 \end{cases}$

19. $\lim\limits_{x \to \pi} \cot x$

20. $\lim\limits_{x \to \pi/2} \sec x$

21. $\lim\limits_{x \to 4^-} (3[\![x]\!] - 5)$

22. $\lim\limits_{x \to 3^+} (3x - [\![x]\!])$

23. $\lim\limits_{x \to 3} (2 - [\![-x]\!])$

24. $\lim\limits_{x \to 1} \left(1 - \left[\!\left[-\dfrac{x}{2}\right]\!\right]\right)$

25. $\lim\limits_{x \to 3^+} \ln(x - 3)$

26. $\lim\limits_{x \to 6^-} \ln(6 - x)$

27. $\lim\limits_{x \to 2^-} \ln[x^2(3 - x)]$

28. $\lim\limits_{x \to 5^+} \ln \dfrac{x}{\sqrt{x - 4}}$

In Exercises 29–32, discuss the continuity of each function.

29. $f(x) = \dfrac{1}{x^2 - 4}$

30. $f(x) = \dfrac{x^2 - 1}{x + 1}$

31. $f(x) = \frac{1}{2}[\![x]\!] + x$

32. $f(x) = \begin{cases} x, & x < 1 \\ 2, & x = 1 \\ 2x - 1, & x > 1 \end{cases}$

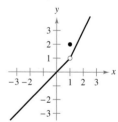

In Exercises 33–36, discuss the continuity of the function on the closed interval.

Function	Interval
33. $g(x) = \sqrt{25 - x^2}$	$[-5, 5]$
34. $f(t) = 2 - \sqrt{9 - t^2}$	$[-2, 2]$
35. $f(x) = \begin{cases} 3 - x, & x \le 0 \\ 3 + \frac{1}{2}x, & x > 0 \end{cases}$	$[-1, 4]$
36. $g(x) = \dfrac{1}{x^2 - 4}$	$[-1, 2]$

In Exercises 37–60, find the x-values (if any) at which f is not continuous. Which of the discontinuities are removable?

37. $f(x) = x^2 - 2x + 1$

38. $f(x) = \dfrac{1}{x^2 + 1}$

39. $f(x) = 3x - \cos x$

40. $f(x) = \cos \dfrac{\pi x}{4}$

41. $f(x) = \dfrac{x}{x^2 - x}$

42. $f(x) = \dfrac{x}{x^2 - 1}$

43. $f(x) = \dfrac{x}{x^2 + 1}$

44. $f(x) = \dfrac{x - 3}{x^2 - 9}$

45. $f(x) = \dfrac{x + 2}{x^2 - 3x - 10}$

46. $f(x) = \dfrac{x - 1}{x^2 + x - 2}$

47. $f(x) = \dfrac{|x + 2|}{x + 2}$

48. $f(x) = \dfrac{|x - 3|}{x - 3}$

49. $f(x) = \begin{cases} x, & x \le 1 \\ x^2, & x > 1 \end{cases}$

50. $f(x) = \begin{cases} -2x + 3, & x < 1 \\ x^2, & x \ge 1 \end{cases}$

51. $f(x) = \begin{cases} \frac{1}{2}x + 1, & x \le 2 \\ 3 - x, & x > 2 \end{cases}$

52. $f(x) = \begin{cases} -2x, & x \le 2 \\ x^2 - 4x + 1, & x > 2 \end{cases}$

53. $f(x) = \begin{cases} \tan \dfrac{\pi x}{4}, & |x| < 1 \\ x, & |x| \ge 1 \end{cases}$

54. $f(x) = \begin{cases} \csc \dfrac{\pi x}{6}, & |x - 3| \le 2 \\ 2, & |x - 3| > 2 \end{cases}$

55. $f(x) = \begin{cases} \ln(x + 1), & x \ge 0 \\ 1 - x^2, & x < 0 \end{cases}$

56. $f(x) = \begin{cases} 10 - 3e^{5-x}, & x > 5 \\ 10 - \frac{3}{5}x, & x \le 5 \end{cases}$

57. $f(x) = \csc 2x$

58. $f(x) = \tan \dfrac{\pi x}{4}$

59. $f(x) = [\![x - 1]\!]$

60. $f(x) = 3 - [\![x]\!]$

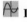 In Exercises 61 and 62, use a graphing utility to graph the function. From the graph, estimate

$$\lim_{x \to 0^+} f(x) \quad \text{and} \quad \lim_{x \to 0^-} f(x).$$

Is the function continuous on the entire real number line? Explain.

61. $f(x) = \dfrac{|x^2 - 4|x}{x + 2}$

62. $f(x) = \dfrac{|x^2 + 4x|(x + 2)}{x + 4}$

In Exercises 63–66, find the constants a and b such that the function is continuous on the entire real number line.

63. $f(x) = \begin{cases} x^3, & x \le 2 \\ ax^2, & x > 2 \end{cases}$

64. $g(x) = \begin{cases} \dfrac{4 \sin x}{x}, & x < 0 \\ a - 2x, & x \ge 0 \end{cases}$

65. $f(x) = \begin{cases} 2, & x \le -1 \\ ax + b, & -1 < x < 3 \\ -2, & x \ge 3 \end{cases}$

66. $g(x) = \begin{cases} \dfrac{x^2 - a^2}{x - a}, & x \ne a \\ 8, & x = a \end{cases}$

In Exercises 67–70, discuss the continuity of the composite function $h(x) = f(g(x))$.

67. $f(x) = x^2$

$g(x) = x - 1$

68. $f(x) = \dfrac{1}{\sqrt{x}}$

$g(x) = x - 1$

69. $f(x) = \dfrac{1}{x - 6}$

$g(x) = x^2 + 5$

70. $f(x) = \sin x$

$g(x) = x^2$

 In Exercises 71–74, use a graphing utility to graph the function. Use the graph to determine any x-values at which the function is not continuous.

71. $f(x) = [\![x]\!] - x$

72. $h(x) = \dfrac{1}{x^2 - x - 2}$

73. $g(x) = \begin{cases} 2x - 4, & x \le 3 \\ x^2 - 2x, & x > 3 \end{cases}$

74. $f(x) = \begin{cases} \dfrac{\cos x - 1}{x}, & x < 0 \\ 5x, & x \ge 0 \end{cases}$

In Exercises 75–78, describe the interval(s) on which the function is continuous.

75. $f(x) = \dfrac{x}{x^2 + 1}$

76. $f(x) = x\sqrt{x + 3}$

77. $f(x) = \sec \dfrac{\pi x}{4}$

78. $f(x) = \dfrac{x + 1}{\sqrt{x}}$

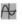 *Writing* In Exercises 79–82, use a graphing utility to graph the function on the interval $[-4, 4]$. Does the graph of the function appear continuous on this interval? Is the function continuous on $[-4, 4]$? Write a short paragraph about the importance of examining a function analytically as well as graphically.

79. $f(x) = \dfrac{\sin x}{x}$

80. $f(x) = \dfrac{x^3 - 8}{x - 2}$

81. $f(x) = \dfrac{\ln(x^2 + 1)}{x}$

82. $f(x) = \dfrac{e^{-x} + 1}{e^x - 1}$

Writing In Exercises 83–86, explain why the function has a zero in the given interval.

Function	Interval
83. $f(x) = x^2 - 4x + 3$	$[2, 4]$
84. $f(x) = x^3 + 3x - 2$	$[0, 1]$
85. $h(x) = -2e^{-x/2} \cos 2x$	$\left[0, \dfrac{\pi}{2}\right]$
86. $g(t) = (t^3 + 2t - 2) \ln(t^2 + 4)$	$[0, 1]$

 In Exercises 87–90, use the Intermediate Value Theorem and a graphing utility to approximate the zero of the function in the interval $[0, 1]$. Repeatedly "zoom in" on the graph of the function to approximate the zero accurate to two decimal places. Use the *zero* or *root* feature of the graphing utility to approximate the zero accurate to four decimal places.

87. $f(x) = x^3 + x - 1$

88. $f(x) = x^3 + 3x - 3$

89. $g(t) = 2 \cos t - 3t$

90. $h(\theta) = 1 + \theta - 3 \tan \theta$

In Exercises 91–94, verify that the Intermediate Value Theorem applies to the indicated interval and find the value of *c* guaranteed by the theorem.

91. $f(x) = x^2 + x - 1$, $[0, 5]$, $f(c) = 11$

92. $f(x) = x^2 - 6x + 8$, $[0, 3]$, $f(c) = 0$

93. $f(x) = x^3 - x^2 + x - 2$, $[0, 3]$, $f(c) = 4$

94. $f(x) = \dfrac{x^2 + x}{x - 1}$, $\left[\dfrac{5}{2}, 4\right]$, $f(c) = 6$

Writing About Concepts

95. State how continuity is destroyed at $x = c$ for each of the following.

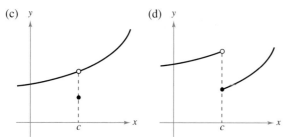

96. Describe the difference between a discontinuity that is removable and one that is nonremovable. In your explanation, give examples of the following.

(a) A function with a nonremovable discontinuity at $x = 2$

(b) A function with a removable discontinuity at $x = -2$

(c) A function that has both of the characteristics described in parts (a) and (b)

97. Sketch the graph of any function f such that

$$\lim_{x \to 3^+} f(x) = 1 \quad \text{and} \quad \lim_{x \to 3^-} f(x) = 0.$$

Is the function continuous at $x = 3$? Explain.

98. If the functions f and g are continuous for all real x, is $f + g$ always continuous for all real x? Is f/g always continuous for all real x? If either is not continuous, give an example to verify your conclusion.

True or False? **In Exercises 99–102, determine whether the statement is true or false. If it is false, explain why or give an example that shows it is false.**

99. If $\lim\limits_{x \to c} f(x) = L$ and $f(c) = L$, then f is continuous at c.

100. If $f(x) = g(x)$ for $x \ne c$ and $f(c) \ne g(c)$, then either f or g is not continuous at c.

101. A rational function can have infinitely many x-values at which it is not continuous.

102. The function $f(x) = |x - 1|/(x - 1)$ is continuous on $(-\infty, \infty)$.

103. ***Swimming Pool*** Every day you dissolve 28 ounces of chlorine in a swimming pool. The graph shows the amount of chlorine $f(t)$ in the pool after t days.

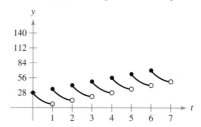

Estimate and interpret $\lim\limits_{t \to 4^-} f(t)$ and $\lim\limits_{t \to 4^+} f(t)$.

104. ***Think About It*** Describe how the functions

$$f(x) = 3 + [\![x]\!]$$

and

$$g(x) = 3 - [\![-x]\!]$$

differ.

105. ***Telephone Charges*** A dial-direct long distance call between two cities costs $1.04 for the first 2 minutes and $0.36 for each additional minute or fraction thereof. Use the greatest integer function to write the cost C of a call in terms of time t (in minutes). Sketch the graph of this function and discuss its continuity.

106. ***Inventory Management*** The number of units in inventory in a small company is given by

$$N(t) = 25\left(2\left[\!\left[\dfrac{t + 2}{2}\right]\!\right] - t\right)$$

where t is the time in months. Sketch the graph of this function and discuss its continuity. How often must this company replenish its inventory?

107. ***Déjà Vu*** At 8:00 A.M. on Saturday a man begins running up the side of a mountain to his weekend campsite (see figure on next page). On Sunday morning at 8:00 A.M. he runs back down the mountain. It takes him 20 minutes to run up, but only 10 minutes to run down. At some point on the way down, he realizes that he passed the same place at exactly the same time on Saturday. Prove that he is correct. [*Hint:* Let $s(t)$ and $r(t)$ be the position functions for the runs up and down, and apply the Intermediate Value Theorem to the function $f(t) = s(t) - r(t)$.]

Saturday 8:00 A.M.   Sunday 8:00 A.M.

Not drawn to scale

Figure for 107

108. Volume Use the Intermediate Value Theorem to show that for all spheres with radii in the interval $[1, 5]$, there is one with a volume of 275 cubic centimeters.

109. Prove that if f is continuous and has no zeros on $[a, b]$, then either

$$f(x) > 0 \text{ for all } x \text{ in } [a, b] \text{ or } f(x) < 0 \text{ for all } x \text{ in } [a, b].$$

110. Show that the Dirichlet function

$$f(x) = \begin{cases} 0, & \text{if } x \text{ is rational} \\ 1, & \text{if } x \text{ is irrational} \end{cases}$$

is not continuous at any real number.

111. Show that the function

$$f(x) = \begin{cases} 0, & \text{if } x \text{ is rational} \\ kx, & \text{if } x \text{ is irrational} \end{cases}$$

is continuous only at $x = 0$. (Assume that k is any nonzero real number.)

112. The **signum function** is defined by

$$\operatorname{sgn}(x) = \begin{cases} -1, & x < 0 \\ 0, & x = 0 \\ 1, & x > 0. \end{cases}$$

Sketch a graph of $\operatorname{sgn}(x)$ and find the following (if possible).

(a) $\lim\limits_{x \to 0^-} \operatorname{sgn}(x)$ (b) $\lim\limits_{x \to 0^+} \operatorname{sgn}(x)$ (c) $\lim\limits_{x \to 0} \operatorname{sgn}(x)$

113. Modeling Data After an object falls for t seconds, the speed S (in feet per second) of the object is recorded in the table.

t	0	5	10	15	20	25	30
S	0	48.2	53.5	55.2	55.9	56.2	56.3

(a) Create a line graph of the data.

(b) Does there appear to be a limiting speed of the object? If there is a limiting speed, identify a possible cause.

114. Creating Models A swimmer crosses a pool of width b by swimming in a straight line from $(0, 0)$ to $(2b, b)$. (See figure.)

(a) Let f be a function defined as the y-coordinate of the point on the long side of the pool that is nearest the swimmer at any given time during the swimmer's path across the pool. Determine the function f and sketch its graph. Is it continuous? Explain.

(b) Let g be the minimum distance between the swimmer and the long sides of the pool. Determine the function g and sketch its graph. Is it continuous? Explain.

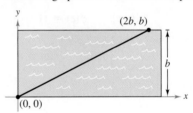

115. Find all values of c such that f is continuous on $(-\infty, \infty)$.

$$f(x) = \begin{cases} 1 - x^2, & x \le c \\ x, & x > c \end{cases}$$

116. Prove that for any real number y there exists x in $(-\pi/2, \pi/2)$ such that $\tan x = y$.

117. Let $f(x) = \left(\sqrt{x + c^2} - c\right)/x$, $c > 0$. What is the domain of f? How can you define f at $x = 0$ in order for f to be continuous there?

118. Prove that if $\lim\limits_{\Delta x \to 0} f(c + \Delta x) = f(c)$, then f is continuous at c.

119. Discuss the continuity of the function $h(x) = x[\![x]\!]$.

120. (a) Let $f_1(x)$ and $f_2(x)$ be continuous on the closed interval $[a, b]$. If $f_1(a) < f_2(a)$ and $f_1(b) > f_2(b)$, prove that there exists c between a and b such that $f_1(c) = f_2(c)$.

(b) Show that there exists c in $\left[0, \frac{\pi}{2}\right]$ such that $\cos x = x$. Use a graphing utility to approximate c to three decimal places.

121. Think About It Consider the function

$$f(x) = \frac{4}{1 + 2^{4/x}}.$$

(a) What is the domain of the function?

(b) Use a graphing utility to graph the function.

(c) Determine $\lim\limits_{x \to 0^-} f(x)$ and $\lim\limits_{x \to 0^+} f(x)$.

(d) Use your knowledge of the exponential function to explain the behavior of f near $x = 0$.

Putnam Exam Challenge

122. Prove or disprove: if x and y are real numbers with $y \ge 0$ and $y(y + 1) \le (x + 1)^2$, then $y(y - 1) \le x^2$.

123. Determine all polynomials $P(x)$ such that $P(x^2 + 1) = (P(x))^2 + 1$ and $P(0) = 0$.

- Determine infinite limits from the left and from the right.
- Find and sketch the vertical asymptotes of the graph of a function.

Infinite Limits

Let f be the function given by

$$f(x) = \frac{3}{x-2}.$$

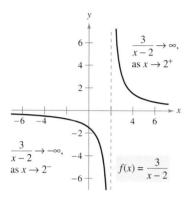

$f(x)$ increases and decreases without bound as x approaches 2.

Figure 2.39

From Figure 2.39 and the table, you can see that $f(x)$ *decreases without bound* as x approaches 2 from the left, and $f(x)$ *increases without bound* as x approaches 2 from the right. This behavior is denoted as

$$\lim_{x \to 2^-} \frac{3}{x-2} = -\infty \qquad f(x) \text{ decreases without bound as } x \text{ approaches 2 from the left.}$$

and

$$\lim_{x \to 2^+} \frac{3}{x-2} = \infty. \qquad f(x) \text{ increases without bound as } x \text{ approaches 2 from the right.}$$

x	1.5	1.9	1.99	1.999	2	2.001	2.01	2.1	2.5
$f(x)$	-6	-30	-300	-3000	?	3000	300	30	6

$f(x)$ decreases without bound. $\qquad$ $f(x)$ increases without bound.

A limit in which $f(x)$ increases or decreases without bound as x approaches c is called an **infinite limit.**

Infinite limits

Figure 2.40

Definition of Infinite Limits

Let f be a function that is defined at every real number in some open interval containing c (except possibly at c itself). The statement

$$\lim_{x \to c} f(x) = \infty$$

means that for each $M > 0$ there exists a $\delta > 0$ such that $f(x) > M$ whenever $0 < |x - c| < \delta$ (see Figure 2.40). Similarly, the statement

$$\lim_{x \to c} f(x) = -\infty$$

means that for each $N < 0$ there exists a $\delta > 0$ such that $f(x) < N$ whenever $0 < |x - c| < \delta$.

To define the **infinite limit from the left,** replace $0 < |x - c| < \delta$ by $c - \delta < x < c$. To define the **infinite limit from the right,** replace $0 < |x - c| < \delta$ by $c < x < c + \delta$.

Be sure you see that the equal sign in the statement $\lim f(x) = \infty$ does not mean that the limit exists. On the contrary, it tells you how the limit *fails to exist* by denoting the unbounded behavior of $f(x)$ as x approaches c.

Use a graphing utility to graph each function. For each function, analytically find the single real number c that is not in the domain. Then graphically find the limit of $f(x)$ as x approaches c from the left and from the right.

a. $f(x) = \dfrac{3}{x - 4}$ **b.** $f(x) = \dfrac{1}{2 - x}$

c. $f(x) = \dfrac{2}{(x - 3)^2}$ **d.** $f(x) = \dfrac{-3}{(x + 2)^2}$

EXAMPLE I   Determining Infinite Limits from a Graph

Use Figure 2.41 to determine the limit of each function as x approaches 1 from the left and from the right.

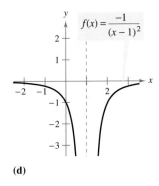

(a)   **(b)**   **(c)**   **(d)**

Figure 2.41 Each graph has an asymptote at $x = 1$.

Solution

a. $\displaystyle\lim_{x \to 1^-} \frac{1}{x - 1} = -\infty$ and $\displaystyle\lim_{x \to 1^+} \frac{1}{x - 1} = \infty$

b. $\displaystyle\lim_{x \to 1} \frac{1}{(x - 1)^2} = \infty$ Limit from each side is ∞.

c. $\displaystyle\lim_{x \to 1^-} \frac{-1}{x - 1} = \infty$ and $\displaystyle\lim_{x \to 1^+} \frac{-1}{x - 1} = -\infty$

d. $\displaystyle\lim_{x \to 1} \frac{-1}{(x - 1)^2} = -\infty$ Limit from each side is $-\infty$.

Vertical Asymptotes

If it were possible to extend the graphs in Figure 2.41 toward positive and negative infinity, you would see that each graph becomes arbitrarily close to the vertical line $x = 1$. This line is a **vertical asymptote** of the graph of f. (You will study other types of asymptotes in Sections 4.5 and 4.6.)

NOTE If a function f has a vertical asymptote at $x = c$, then f is *not continuous* at c.

Definition of a Vertical Asymptote

If $f(x)$ approaches infinity (or negative infinity) as x approaches c from the right or the left, then the line $x = c$ is a **vertical asymptote** of the graph of f.

In Example 1, note that each of the functions is a *quotient* and that the vertical asymptote occurs at a number where the denominator is 0 (and the numerator is not 0). The next theorem generalizes this observation. (A proof of this theorem is given in Appendix A.)

THEOREM 2.14   Vertical Asymptotes

Let f and g be continuous on an open interval containing c. If $f(c) \neq 0$, $g(c) = 0$, and there exists an open interval containing c such that $g(x) \neq 0$ for all $x \neq c$ in the interval, then the graph of the function given by

$$h(x) = \frac{f(x)}{g(x)}$$

has a vertical asymptote at $x = c$.

EXAMPLE 2   Finding Vertical Asymptotes

Determine all vertical asymptotes of the graph of each function.

a. $f(x) = \dfrac{1}{2(x + 1)}$ **b.** $f(x) = \dfrac{x^2 + 1}{x^2 - 1}$ **c.** $f(x) = \cot x$

Solution

a. When $x = -1$, the denominator of

$$f(x) = \frac{1}{2(x + 1)}$$

is 0 and the numerator is not 0. So, by Theorem 2.14, you can conclude that $x = -1$ is a vertical asymptote, as shown in Figure 2.42(a).

b. By factoring the denominator as

$$f(x) = \frac{x^2 + 1}{x^2 - 1} = \frac{x^2 + 1}{(x - 1)(x + 1)}$$

you can see that the denominator is 0 at $x = -1$ and $x = 1$. Moreover, because the numerator is not 0 at these two points, you can apply Theorem 2.14 to conclude that the graph of f has two vertical asymptotes, as shown in Figure 2.42(b).

c. By writing the cotangent function in the form

$$f(x) = \cot x = \frac{\cos x}{\sin x}$$

you can apply Theorem 2.14 to conclude that vertical asymptotes occur at all values of x such that $\sin x = 0$ and $\cos x \neq 0$, as shown in Figure 2.42(c). So, the graph of this function has infinitely many vertical asymptotes. These asymptotes occur when $x = n\pi$, where n is an integer.

Theorem 2.14 requires that the value of the numerator at $x = c$ be nonzero. If both the numerator and the denominator are 0 at $x = c$, you obtain the *indeterminate form* $0/0$, and you cannot determine the limit behavior at $x = c$ without further investigation, as illustrated in Example 3.

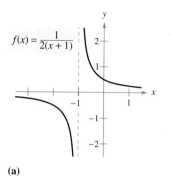

$f(x) = \dfrac{1}{2(x + 1)}$

(a)

$f(x) = \dfrac{x^2 + 1}{x^2 - 1}$

(b)

$f(x) = \cot x$

(c)

Functions with vertical asymptotes
Figure 2.42

7. Consider the function $f(x) = \dfrac{\sqrt{3 + x^{1/3}} - 2}{x - 1}$.

    (a) Find the domain of f.

    (b) Use a graphing utility to graph the function.

    (c) Calculate $\lim\limits_{x \to -27^{+}} f(x)$.

    (d) Calculate $\lim\limits_{x \to 1} f(x)$.

8. Determine all values of the constant a such that the following function is continuous for all real numbers.

    $$f(x) = \begin{cases} \dfrac{ax}{\tan x}, & x \ge 0 \\ a^2 - 2, & x < 0 \end{cases}$$

9. Consider the graphs of the four functions g_1, g_2, g_3, and g_4.

    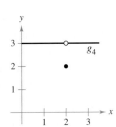

    For each given condition of the function f, which of the graphs could be the graph of f?

    (a) $\lim\limits_{x \to 2} f(x) = 3$

    (b) f is continuous at 2.

    (c) $\lim\limits_{x \to 2^{-}} f(x) = 3$

10. Sketch the graph of the function $f(x) = \left[\!\left[\dfrac{1}{x} \right]\!\right]$.

    (a) Evaluate $f\!\left(\tfrac{1}{4}\right)$, $f(3)$, and $f(1)$.

    (b) Evaluate the limits $\lim\limits_{x \to 1^{-}} f(x)$, $\lim\limits_{x \to 1^{+}} f(x)$, $\lim\limits_{x \to 0^{-}} f(x)$, and $\lim\limits_{x \to 0^{+}} f(x)$.

    (c) Discuss the continuity of the function.

11. Sketch the graph of the function $f(x) = [\![x]\!] + [\![-x]\!]$.

    (a) Evaluate $f(1)$, $f(0)$, $f\!\left(\tfrac{1}{2}\right)$, and $f(-2.7)$.

    (b) Evaluate the limits $\lim\limits_{x \to 1^{-}} f(x)$, $\lim\limits_{x \to 1^{+}} f(x)$, and $\lim\limits_{x \to \frac{1}{2}} f(x)$.

    (c) Discuss the continuity of the function.

12. To escape Earth's gravitational field, a rocket must be launched with an initial velocity called the **escape velocity.** A rocket launched from the surface of Earth has velocity v (in miles per second) given by

    $$v = \sqrt{\dfrac{2GM}{r} + v_0^2 - \dfrac{2GM}{R}} \approx \sqrt{\dfrac{192{,}000}{r} + v_0^2 - 48}$$

    where v_0 is the initial velocity, r is the distance from the rocket to the center of Earth, G is the gravitational constant, M is the mass of Earth, and R is the radius of Earth (approximately 4000 miles).

    (a) Find the value of v_0 for which you obtain an infinite limit for r as v tends to zero. This value of v_0 is the escape velocity for Earth.

    (b) A rocket launched from the surface of the moon has velocity v (in miles per second) given by

    $$v = \sqrt{\dfrac{1920}{r} + v_0^2 - 2.17}.$$

    Find the escape velocity for the moon.

    (c) A rocket launched from the surface of a planet has velocity v (in miles per second) given by

    $$v = \sqrt{\dfrac{10{,}600}{r} + v_0^2 - 6.99}.$$

    Find the escape velocity for this planet. Is the mass of this planet larger or smaller than that of Earth? (Assume that the mean density of this planet is the same as that of Earth.)

13. For positive numbers $a < b$, the **pulse function** is defined as

    $$P_{a,b}(x) = H(x - a) - H(x - b) = \begin{cases} 0, & x < a \\ 1, & a \le x < b \\ 0, & x \ge b \end{cases}$$

    where $H(x) = \begin{cases} 1, & x \ge 0 \\ 0, & x < 0 \end{cases}$ is the Heaviside function.

    (a) Sketch the graph of the pulse function.

    (b) Find the following limits:

    (i) $\lim\limits_{x \to a^{+}} P_{a,b}(x)$ (ii) $\lim\limits_{x \to a^{-}} P_{a,b}(x)$

    (iii) $\lim\limits_{x \to b^{+}} P_{a,b}(x)$ (iv) $\lim\limits_{x \to b^{-}} P_{a,b}(x)$

    (c) Discuss the continuity of the pulse function.

    (d) Why is

    $$U(x) = \dfrac{1}{b - a} P_{a,b}(x)$$

    called the **unit** pulse function?

14. Let a be a nonzero constant. Prove that if $\lim\limits_{x \to 0} f(x) = L$, then $\lim\limits_{x \to 0} f(ax) = L$. Show by means of an example that a must be nonzero.

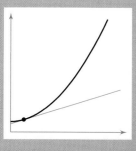

3

Differentiation

You pump air at a steady rate into a deflated balloon until the balloon bursts. Does the diameter of the balloon change faster when you first start pumping the air, or just before the balloon bursts? Why?

Dr. Gary Settles/SPL/Photo Researchers

To approximate the slope of a tangent line to a graph at a given point, find the slope of the secant line through the given point and a second point on the graph. As the second point approaches the given point, the approximation tends to become more accurate. In Section 3.1, you will use limits to find slopes of tangent lines to graphs. This process is called *differentiation*.

The Derivative and the Tangent Line Problem

- Find the slope of the tangent line to a curve at a point.
- Use the limit definition to find the derivative of a function.
- Understand the relationship between differentiability and continuity.

The Tangent Line Problem

Calculus grew out of four major problems that European mathematicians were working on during the seventeenth century.

1. The tangent line problem (Section 2.1 and this section)
2. The velocity and acceleration problem (Sections 3.2 and 3.3)
3. The minimum and maximum problem (Section 4.1)
4. The area problem (Sections 2.1 and 5.2)

Each problem involves the notion of a limit, and calculus can be introduced with any of the four problems.

A brief introduction to the tangent line problem is given in Section 2.1. Although partial solutions to this problem were given by Pierre de Fermat (1601–1665), René Descartes (1596–1650), Christian Huygens (1629–1695), and Isaac Barrow (1630–1677), credit for the first general solution is usually given to Isaac Newton (1642–1727) and Gottfried Leibniz (1646–1716). Newton's work on this problem stemmed from his interest in optics and light refraction.

What does it mean to say that a line is tangent to a curve at a point? For a circle, the tangent line at a point P is the line that is perpendicular to the radial line at point P, as shown in Figure 3.1.

For a general curve, however, the problem is more difficult. For example, how would you define the tangent lines shown in Figure 3.2? You might say that a line is tangent to a curve at a point P if it touches, but does not cross, the curve at point P. This definition would work for the first curve shown in Figure 3.2, but not for the second. *Or* you might say that a line is tangent to a curve if the line touches or intersects the curve at exactly one point. This definition would work for a circle but not for more general curves, as the third curve in Figure 3.2 shows.

ISAAC NEWTON (1642–1727)

In addition to his work in calculus, Newton made revolutionary contributions to physics, including the Law of Universal Gravitation and his three laws of motion.

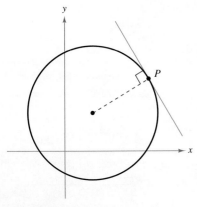

Tangent line to a circle
Figure 3.1

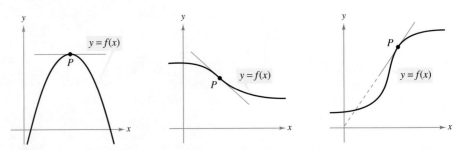

Tangent line to a curve at a point
Figure 3.2

FOR FURTHER INFORMATION For more information on the crediting of mathematical discoveries to the first "discoverer," see the article "Mathematical Firsts—Who Done It?" by Richard H. Williams and Roy D. Mazzagatti in *Mathematics Teacher*. To view this article, go to the website *www.matharticles.com*.

EXPLORATION

Identifying a Tangent Line Use a graphing utility to graph the function $f(x) = 2x^3 - 4x^2 + 3x - 5$. On the same screen, graph $y = x - 5$, $y = 2x - 5$, and $y = 3x - 5$. Which of these lines, if any, appears to be tangent to the graph of f at the point $(0, -5)$? Explain your reasoning.

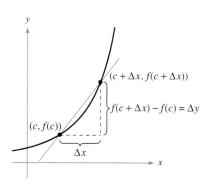

The secant line through $(c, f(c))$ and $(c + \Delta x, f(c + \Delta x))$

Figure 3.3

Essentially, the problem of finding the tangent line at a point P boils down to the problem of finding the *slope* of the tangent line at point P. You can approximate this slope using a **secant line*** through the point of tangency and a second point on the curve, as shown in Figure 3.3. If $(c, f(c))$ is the point of tangency and $(c + \Delta x, f(c + \Delta x))$ is a second point on the graph of f, the slope of the secant line through the two points is given by substitution into the slope formula

$$m = \frac{y_2 - y_1}{x_2 - x_1}$$

$$m_{\text{sec}} = \frac{f(c + \Delta x) - f(c)}{(c + \Delta x) - c} \qquad \frac{\text{Change in } y}{\text{Change in } x}$$

$$m_{\text{sec}} = \frac{f(c + \Delta x) - f(c)}{\Delta x}. \qquad \text{Slope of secant line}$$

The right-hand side of this equation is a **difference quotient.** The denominator Δx is the **change in x,** and the numerator $\Delta y = f(c + \Delta x) - f(c)$ is the **change in y.**

The beauty of this procedure is that you can obtain more and more accurate approximations of the slope of the tangent line by choosing points closer and closer to the point of tangency, as shown in Figure 3.4.

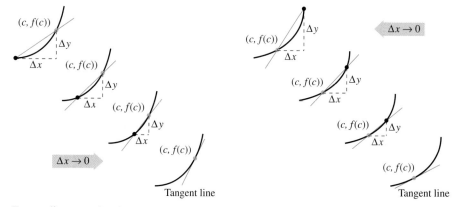

Tangent line approximations
Figure 3.4

Definition of Tangent Line with Slope _m_

If f is defined on an open interval containing c, and if the limit

$$\lim_{\Delta x \to 0} \frac{\Delta y}{\Delta x} = \lim_{\Delta x \to 0} \frac{f(c + \Delta x) - f(c)}{\Delta x} = m$$

exists, then the line passing through $(c, f(c))$ with slope m is the **tangent line** to the graph of f at the point $(c, f(c))$.

The slope of the tangent line to the graph of f at the point $(c, f(c))$ is also called the **slope of the graph of f at $x = c$.**

---

* *This use of the word* secant *comes from the Latin* secare, *meaning to cut, and is not a reference to the trigonometric function of the same name.*

www ▷ **EXAMPLE 3** **Finding the Derivative by the Limit Process**

Find the derivative of $f(x) = x^3 + 2x$.

Solution

$$f'(x) = \lim_{\Delta x \to 0} \frac{f(x + \Delta x) - f(x)}{\Delta x} \qquad \text{Definition of derivative}$$

$$= \lim_{\Delta x \to 0} \frac{(x + \Delta x)^3 + 2(x + \Delta x) - (x^3 + 2x)}{\Delta x}$$

$$= \lim_{\Delta x \to 0} \frac{x^3 + 3x^2\Delta x + 3x(\Delta x)^2 + (\Delta x)^3 + 2x + 2\Delta x - x^3 - 2x}{\Delta x}$$

$$= \lim_{\Delta x \to 0} \frac{3x^2\Delta x + 3x(\Delta x)^2 + (\Delta x)^3 + 2\Delta x}{\Delta x}$$

$$= \lim_{\Delta x \to 0} \frac{\Delta x[3x^2 + 3x\Delta x + (\Delta x)^2 + 2]}{\Delta x}$$

$$= \lim_{\Delta x \to 0} [3x^2 + 3x\Delta x + (\Delta x)^2 + 2]$$

$$= 3x^2 + 2$$

STUDY TIP When using the definition to find a derivative of a function, the key is to rewrite the difference quotient so that Δx does not occur as a factor of the denominator.

Remember that the derivative of a function f is itself a function, which can be used to find the slope of the tangent line at the point $(x, f(x))$ on the graph of f.

EXAMPLE 4 **Using the Derivative to Find the Slope at a Point**

Find $f'(x)$ for $f(x) = \sqrt{x}$. Then find the slope of the graph of f at the points $(1, 1)$ and $(4, 2)$. Discuss the behavior of f at $(0, 0)$.

Solution Use the procedure for rationalizing numerators, as discussed in Section 2.3.

$$f'(x) = \lim_{\Delta x \to 0} \frac{f(x + \Delta x) - f(x)}{\Delta x} \qquad \text{Definition of derivative}$$

$$= \lim_{\Delta x \to 0} \frac{\sqrt{x + \Delta x} - \sqrt{x}}{\Delta x}$$

$$= \lim_{\Delta x \to 0} \left(\frac{\sqrt{x + \Delta x} - \sqrt{x}}{\Delta x}\right)\left(\frac{\sqrt{x + \Delta x} + \sqrt{x}}{\sqrt{x + \Delta x} + \sqrt{x}}\right)$$

$$= \lim_{\Delta x \to 0} \frac{(x + \Delta x) - x}{\Delta x(\sqrt{x + \Delta x} + \sqrt{x})}$$

$$= \lim_{\Delta x \to 0} \frac{\Delta x}{\Delta x(\sqrt{x + \Delta x} + \sqrt{x})}$$

$$= \lim_{\Delta x \to 0} \frac{1}{\sqrt{x + \Delta x} + \sqrt{x}}$$

$$= \frac{1}{2\sqrt{x}}, \quad x > 0$$

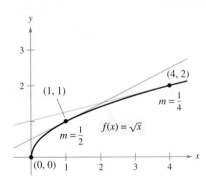

The slope of f at $(x, f(x))$, $x > 0$, is $m = 1/(2\sqrt{x})$.

Figure 3.8

At the point $(1, 1)$, the slope is $f'(1) = \frac{1}{2}$. At the point $(4, 2)$, the slope is $f'(4) = \frac{1}{4}$. See Figure 3.8. At the point $(0, 0)$, the slope is undefined. Moreover, the graph of f has a vertical tangent line at $(0, 0)$.

www ▷ *indicates that in the HM mathSpace® CD-ROM and the online* Eduspace® *system for this text, you will find an Open Exploration, which further explores this example using the computer algebra systems* Maple, Mathcad, Mathematica, *and* Derive.

In many applications, it is convenient to use a variable other than x as the independent variable, as shown in Example 5.

 EXAMPLE 5 **Finding the Derivative of a Function**

Find the derivative with respect to t for the function $y = 2/t$.

Solution   Considering $y = f(t)$, you obtain

$$\frac{dy}{dt} = \lim_{\Delta t \to 0} \frac{f(t + \Delta t) - f(t)}{\Delta t}$$ Definition of derivative

$$= \lim_{\Delta t \to 0} \frac{\dfrac{2}{t + \Delta t} - \dfrac{2}{t}}{\Delta t}$$ $f(t + \Delta t) = 2/(t + \Delta t)$ and $f(t) = 2/t$

$$= \lim_{\Delta t \to 0} \frac{\dfrac{2t - 2(t + \Delta t)}{t(t + \Delta t)}}{\Delta t}$$ Combine fractions in numerator.

$$= \lim_{\Delta t \to 0} \frac{-2\Delta t}{\Delta t(t)(t + \Delta t)}$$ Divide out common factor of Δt.

$$= \lim_{\Delta t \to 0} \frac{-2}{t(t + \Delta t)}$$ Simplify.

$$= -\frac{2}{t^2}.$$ Evaluate limit as $\Delta t \to 0$.

TECHNOLOGY   A graphing utility can be used to reinforce the result given in Example 5. For instance, using the formula $dy/dt = -2/t^2$, you know that the slope of the graph of $y = 2/t$ at the point $(1, 2)$ is $m = -2$. This implies that an equation of the tangent line to the graph at $(1, 2)$ is

$$y - 2 = -2(t - 1) \quad \text{or} \quad y = -2t + 4$$

as shown in Figure 3.9.

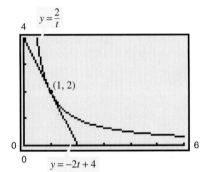

At the point $(1, 2)$, the line $y = -2t + 4$ is tangent to the graph of $y = 2/t$.
Figure 3.9

Differentiability and Continuity

The following alternative limit form of the derivative is useful in investigating the relationship between differentiability and continuity. The derivative of f at c is

$$f'(c) = \lim_{x \to c} \frac{f(x) - f(c)}{x - c}$$ Alternative form of derivative

provided this limit exists (see Figure 3.10). (A proof of the equivalence of this form is given in Appendix A.) Note that the existence of the limit in this alternative form requires that the one-sided limits

$$\lim_{x \to c^-} \frac{f(x) - f(c)}{x - c} \quad \text{and} \quad \lim_{x \to c^+} \frac{f(x) - f(c)}{x - c}$$

exist and are equal. These one-sided limits are called the **derivatives from the left and from the right,** respectively. It follows that f is **differentiable on the closed interval $[a, b]$** if it is differentiable on (a, b) and if the derivative from the right at a and the derivative from the left at b both exist.

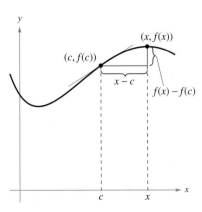

As x approaches c, the secant line approaches the tangent line.
Figure 3.10

The greatest integer function is not differentiable at $x = 0$, because it is not continuous at $x = 0$.
Figure 3.11

If a function is not continuous at $x = c$, it is also not differentiable at $x = c$. For instance, the greatest integer function

$$f(x) = [\![x]\!]$$

is not continuous at $x = 0$, and so it is not differentiable at $x = 0$ (see Figure 3.11). You can verify this by observing that

$$\lim_{x \to 0^-} \frac{f(x) - f(0)}{x - 0} = \lim_{x \to 0^-} \frac{[\![x]\!] - 0}{x} = \infty \qquad \text{Derivative from the left}$$

and

$$\lim_{x \to 0^+} \frac{f(x) - f(0)}{x - 0} = \lim_{x \to 0^+} \frac{[\![x]\!] - 0}{x} = 0. \qquad \text{Derivative from the right}$$

Although it is true that differentiability implies continuity (as we will show in Theorem 3.1), the converse is not true. That is, it is possible for a function to be continuous at $x = c$ and *not* differentiable at $x = c$. Examples 6 and 7 illustrate this possibility.

EXAMPLE 6 A Graph with a Sharp Turn

The function

$$f(x) = |x - 2|$$

shown in Figure 3.12 is continuous at $x = 2$. But, the one-sided limits

$$\lim_{x \to 2^-} \frac{f(x) - f(2)}{x - 2} = \lim_{x \to 2^-} \frac{|x - 2| - 0}{x - 2} = -1 \qquad \text{Derivative from the left}$$

and

$$\lim_{x \to 2^+} \frac{f(x) - f(2)}{x - 2} = \lim_{x \to 2^+} \frac{|x - 2| - 0}{x - 2} = 1 \qquad \text{Derivative from the right}$$

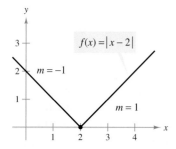

f is not differentiable at $x = 2$, because the derivatives from the left and from the right are not equal.
Figure 3.12

are not equal. So, f is not differentiable at $x = 2$ and the graph of f does not have a tangent line at the point $(2, 0)$.

EXAMPLE 7 A Graph with a Vertical Tangent Line

The function

$$f(x) = x^{1/3}$$

is continuous at $x = 0$, as shown in Figure 3.13. But, because the limit

$$\lim_{x \to 0} \frac{f(x) - f(0)}{x - 0} = \lim_{x \to 0} \frac{x^{1/3} - 0}{x}$$

$$= \lim_{x \to 0} \frac{1}{x^{2/3}}$$

$$= \infty$$

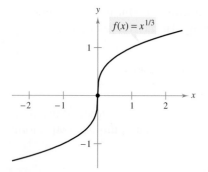

f is not differentiable at $x = 0$, because f has a vertical tangent line at $x = 0$.
Figure 3.13

is infinite, you can conclude that the tangent line is vertical at $x = 0$. So, f is not differentiable at $x = 0$.

From Examples 6 and 7, you can see that a function is not differentiable at a point at which its graph has a sharp turn *or* a vertical tangent line.

TECHNOLOGY   Some graphing utilities, such as *Derive*, *Maple*, *Mathcad*, *Mathematica*, and the *TI-89*, perform symbolic differentiation. Others perform *numerical differentiation* by finding values of derivatives using the formula

$$f'(x) \approx \frac{f(x + \Delta x) - f(x - \Delta x)}{2\Delta x}$$

where Δx is a small number such as 0.001. Can you see any problems with this definition? For instance, using this definition, what is the value of the derivative of $f(x) = |x|$ when $x = 0$?

THEOREM 3.1   Differentiability Implies Continuity

If f is differentiable at $x = c$, then f is continuous at $x = c$.

Proof   You can prove that f is continuous at $x = c$ by showing that $f(x)$ approaches $f(c)$ as $x \to c$. To do this, use the differentiability of f at $x = c$ and consider the following limit.

$$
\begin{aligned}
\lim_{x \to c} [f(x) - f(c)] &= \lim_{x \to c} \left[(x - c)\left(\frac{f(x) - f(c)}{x - c} \right) \right] \\
&= \left[\lim_{x \to c} (x - c) \right]\left[\lim_{x \to c} \frac{f(x) - f(c)}{x - c} \right] \\
&= (0)[f'(c)] \\
&= 0
\end{aligned}
$$

Because the difference $f(x) - f(c)$ approaches zero as $x \to c$, you can conclude that $\lim_{x \to c} f(x) = f(c)$. So, f is continuous at $x = c$.   ▬▬▬

You can summarize the relationship between continuity and differentiability as follows

1. If a function is differentiable at $x = c$, then it is continuous at $x = c$. So, differentiability implies continuity.

2. It is possible for a function to be continuous at $x = c$ and not be differentiable at $x = c$. So, continuity does not imply differentiability.

Exercises for Section 3.1

See www.CalcChat.com for worked-out solutions to odd-numbered exercises.

In Exercises 1 and 2, estimate the slope of the graph at the points (x_1, y_1) and (x_2, y_2).

1. (a)   (b)

2. (a)   (b)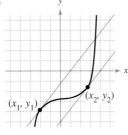

In Exercises 3 and 4, use the graph shown in the figure. To print an enlarged copy of the graph, go to the website www.mathgraphs.com.

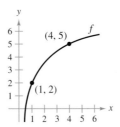

3. Identify or sketch each of the quantities on the figure.

   (a) $f(1)$ and $f(4)$ (b) $f(4) - f(1)$

   (c) $y = \dfrac{f(4) - f(1)}{4 - 1}(x - 1) + f(1)$

4. Insert the proper inequality symbol ($<$ or $>$) between the given quantities.

   (a) $\dfrac{f(4) - f(1)}{4 - 1}$ ▨ $\dfrac{f(4) - f(3)}{4 - 3}$

   (b) $\dfrac{f(4) - f(1)}{4 - 1}$ ▨ $f'(1)$

In Exercises 5–10, find the slope of the tangent line to the graph of the function at the given point.

5. $f(x) = 3 - 2x$, $(-1, 5)$ **6.** $g(x) = \frac{3}{2}x + 1$, $(-2, -2)$

7. $g(x) = x^2 - 4$, $(1, -3)$ **8.** $g(x) = 5 - x^2$, $(2, 1)$

9. $f(t) = 3t - t^2$, $(0, 0)$ **10.** $h(t) = t^2 + 3$, $(-2, 7)$

In Exercises 11–24, find the derivative by the limit process.

11. $f(x) = 3$ **12.** $g(x) = -5$

13. $f(x) = -5x$ **14.** $f(x) = 3x + 2$

15. $h(s) = 3 + \frac{2}{3}s$ **16.** $f(x) = 9 - \frac{1}{2}x$

17. $f(x) = 2x^2 + x - 1$ **18.** $f(x) = 1 - x^2$

19. $f(x) = x^3 - 12x$ **20.** $f(x) = x^3 + x^2$

21. $f(x) = \dfrac{1}{x - 1}$ **22.** $f(x) = \dfrac{1}{x^2}$

23. $f(x) = \sqrt{x + 1}$ **24.** $f(x) = \dfrac{4}{\sqrt{x}}$

In Exercises 25–32, (a) find an equation of the tangent line to the graph of f at the given point, (b) use a graphing utility to graph the function and its tangent line at the point, and (c) use the *derivative* feature of a graphing utility to confirm your results.

25. $f(x) = x^2 + 1$, $(2, 5)$

26. $f(x) = x^2 + 2x + 1$, $(-3, 4)$

27. $f(x) = x^3$, $(2, 8)$ **28.** $f(x) = x^3 + 1$, $(1, 2)$

29. $f(x) = \sqrt{x}$, $(1, 1)$ **30.** $f(x) = \sqrt{x - 1}$, $(5, 2)$

31. $f(x) = x + \dfrac{4}{x}$, $(4, 5)$ **32.** $f(x) = \dfrac{1}{x + 1}$, $(0, 1)$

In Exercises 33–36, find an equation of the line that is tangent to the graph of f *and* parallel to the given line.

Function	Line
33. $f(x) = x^3$	$3x - y + 1 = 0$
34. $f(x) = x^3 + 2$	$3x - y - 4 = 0$
35. $f(x) = \dfrac{1}{\sqrt{x}}$	$x + 2y - 6 = 0$
36. $f(x) = \dfrac{1}{\sqrt{x - 1}}$	$x + 2y + 7 = 0$

In Exercises 37–40, the graph of f is given. Select the graph of f'.

37.

38.

39.

40.

(a)

(b)

(c)

(d)

41. The tangent line to the graph of $y = g(x)$ at the point $(5, 2)$ passes through the point $(9, 0)$. Find $g(5)$ and $g'(5)$.

42. The tangent line to the graph of $y = h(x)$ at the point $(-1, 4)$ passes through the point $(3, 6)$. Find $h(-1)$ and $h'(-1)$.

Writing About Concepts

In Exercises 43–46, sketch the graph of f'. Explain how you found your answer.

43.

44.

45.

46.
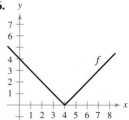

47. Sketch a graph of a function whose derivative is always negative.

48. Sketch a graph of a function whose derivative is always positive.

In Exercises 49–52, the limit represents $f'(c)$ for a function f and a number c. Find f and c.

49. $\displaystyle\lim_{\Delta x \to 0} \frac{[5 - 3(1 + \Delta x)] - 2}{\Delta x}$ **50.** $\displaystyle\lim_{\Delta x \to 0} \frac{(-2 + \Delta x)^3 + 8}{\Delta x}$

51. $\displaystyle\lim_{x \to 6} \frac{-x^2 + 36}{x - 6}$ **52.** $\displaystyle\lim_{x \to 9} \frac{2\sqrt{x} - 6}{x - 9}$

In Exercises 53–55, identify a function f that has the following characteristics. Then sketch the function.

53. $f(0) = 2$;
$f'(x) = -3, \ -\infty < x < \infty$

54. $f(0) = 4; f'(0) = 0$;
$f'(x) < 0$ for $x < 0$;
$f'(x) > 0$ for $x > 0$

55. $f(0) = 0; f'(0) = 0; f'(x) > 0$ if $x \neq 0$

56. Assume that $f'(c) = 3$. Find $f'(-c)$ if (a) f is an odd function and (b) f is an even function.

In Exercises 57 and 58, find equations of the two tangent lines to the graph of f that pass through the indicated point.

57. $f(x) = 4x - x^2$ **58.** $f(x) = x^2$

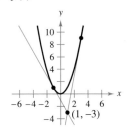

59. *Graphical Reasoning* The figure shows the graph of g'.

(a) $g'(0) =$ (b) $g'(3) =$

(c) What can you conclude about the graph of g knowing that $g'(1) = -\frac{8}{3}$?

(d) What can you conclude about the graph of g knowing that $g'(-4) = \frac{7}{3}$?

(e) Is $g(6) - g(4)$ positive or negative? Explain.

(f) Is it possible to find $g(2)$ from the graph? Explain.

 60. *Graphical Reasoning* Use a graphing utility to graph each function and its tangent lines at $x = -1$, $x = 0$, and $x = 1$. Based on the results, determine whether the slopes of tangent lines to the graph of a function at different values of x are always distinct.

(a) $f(x) = x^2$ (b) $g(x) = x^3$

 Graphical, Numerical, and Analytic Analysis **In Exercises 61 and 62, use a graphing utility to graph f on the interval $[-2, 2]$. Complete the table by graphically estimating the slope of the graph at the indicated points. Then evaluate the slopes analytically and compare your results with those obtained graphically.**

x	-2	-1.5	-1	-0.5	0	0.5	1	1.5	2
$f(x)$									
$f'(x)$									

61. $f(x) = \frac{1}{4}x^3$ **62.** $f(x) = \frac{1}{2}x^2$

 Graphical Reasoning **In Exercises 63 and 64, use a graphing utility to graph the functions f and g in the same viewing window where**

$$g(x) = \frac{f(x + 0.01) - f(x)}{0.01}.$$

Label the graphs and describe the relationship between them.

63. $f(x) = 2x - x^2$ **64.** $f(x) = 3\sqrt{x}$

In Exercises 65 and 66, evaluate $f(2)$ and $f(2.1)$ and use the results to approximate $f'(2)$.

65. $f(x) = x(4 - x)$ **66.** $f(x) = \frac{1}{4}x^3$

Graphical Reasoning **In Exercises 67 and 68, use a graphing utility to graph the function and its derivative in the same viewing window. Label the graphs and describe the relationship between them.**

67. $f(x) = \dfrac{1}{\sqrt{x}}$ **68.** $f(x) = \dfrac{x^3}{4} - 3x$

Writing **In Exercises 69 and 70, consider the functions f and $S_{\Delta x}$ where**

$$S_{\Delta x}(x) = \frac{f(2 + \Delta x) - f(2)}{\Delta x}(x - 2) + f(2).$$

(a) Use a graphing utility to graph f and $S_{\Delta x}$ in the same viewing window for $\Delta x = 1$, 0.5, and 0.1.

(b) Give a written description of the graphs of S for the different values of Δx in part (a).

69. $f(x) = 4 - (x - 3)^2$ **70.** $f(x) = x + \dfrac{1}{x}$

In Exercises 71–80, use the alternative form of the derivative to find the derivative at $x = c$ (if it exists).

71. $f(x) = x^2 - 1$, $c = 2$ **72.** $g(x) = x(x - 1)$, $c = 1$

73. $f(x) = x^3 + 2x^2 + 1$, $c = -2$

74. $f(x) = x^3 + 2x$, $c = 1$

75. $g(x) = \sqrt{|x|}$, $c = 0$

76. $f(x) = 1/x$, $c = 3$

77. $f(x) = (x - 6)^{2/3}$, $c = 6$

78. $g(x) = (x + 3)^{1/3}$, $c = -3$

79. $h(x) = |x + 5|$, $c = -5$ **80.** $f(x) = |x - 4|$, $c = 4$

In Exercises 81–86, describe the x-values at which f is differentiable.

81. $f(x) = \dfrac{1}{x + 1}$ **82.** $f(x) = |x^2 - 9|$

83. $f(x) = (x - 3)^{2/3}$ **84.** $f(x) = \dfrac{x^2}{x^2 - 4}$

85. $f(x) = \sqrt{x - 1}$ **86.** $f(x) = \begin{cases} x^2 - 4, & x \le 0 \\ 4 - x^2, & x > 0 \end{cases}$

 Graphical Analysis In Exercises 87–90, use a graphing utility to find the x-values at which f is differentiable.

87. $f(x) = |x + 3|$ **88.** $f(x) = \dfrac{2x}{x - 1}$

89. $f(x) = x^{2/5}$

90. $f(x) = \begin{cases} x^3 - 3x^2 + 3x, & x \le 1 \\ x^2 - 2x, & x > 1 \end{cases}$

In Exercises 91–94, find the derivatives from the left and from the right at $x = 1$ (if they exist). Is the function differentiable at $x = 1$?

91. $f(x) = |x - 1|$ **92.** $f(x) = \sqrt{1 - x^2}$

93. $f(x) = \begin{cases} (x - 1)^3, & x \le 1 \\ (x - 1)^2, & x > 1 \end{cases}$ **94.** $f(x) = \begin{cases} x, & x \le 1 \\ x^2, & x > 1 \end{cases}$

In Exercises 95 and 96, determine whether the function is differentiable at $x = 2$.

95. $f(x) = \begin{cases} x^2 + 1, & x \le 2 \\ 4x - 3, & x > 2 \end{cases}$ **96.** $f(x) = \begin{cases} \frac{1}{2}x + 1, & x < 2 \\ \sqrt{2x}, & x \ge 2 \end{cases}$

97. *Graphical Reasoning* A line with slope m passes through the point $(0, 4)$ and has the equation $y = mx + 4$.

(a) Write the distance d between the line and the point $(3, 1)$ as a function of m.

 (b) Use a graphing utility to graph the function d in part (a). Based on the graph, is the function differentiable at every value of m? If not, where is it not differentiable?

98. *Conjecture* Consider the functions $f(x) = x^2$ and $g(x) = x^3$.

(a) Graph f and f' on the same set of axes.

(b) Graph g and g' on the same set of axes.

(c) Identify a pattern between f and g and their respective derivatives. Use the pattern to make a conjecture about $h'(x)$ if $h(x) = x^n$, where n is an integer and $n \ge 2$.

(d) Find $f'(x)$ if $f(x) = x^4$. Compare the result with the conjecture in part (c). Is this a proof of your conjecture? Explain.

True or False? In Exercises 99–102, determine whether the statement is true or false. If it is false, explain why or give an example that shows it is false.

99. The slope of the tangent line to the differentiable function f at the point $(2, f(2))$ is $\dfrac{f(2 + \Delta x) - f(2)}{\Delta x}$.

100. If a function is continuous at a point, then it is differentiable at that point.

101. If a function has derivatives from both the right and the left at a point, then it is differentiable at that point.

102. If a function is differentiable at a point, then it is continuous at that point.

103. Let $f(x) = \begin{cases} x \sin \dfrac{1}{x}, & x \ne 0 \\ 0, & x = 0 \end{cases}$ and $g(x) = \begin{cases} x^2 \sin \dfrac{1}{x}, & x \ne 0 \\ 0, & x = 0 \end{cases}$.

Show that f is continuous, but not differentiable, at $x = 0$. Show that g is differentiable at 0, and find $g'(0)$.

 104. *Writing* Use a graphing utility to graph the two functions $f(x) = x^2 + 1$ and $g(x) = |x| + 1$ in the same viewing window. Use the *zoom* and *trace* features to analyze the graphs near the point $(0, 1)$. What do you observe? Which function is differentiable at this point? Write a short paragraph describing the geometric significance of differentiability at a point.

Basic Differentiation Rules and Rates of Change

- Find the derivative of a function using the Constant Rule.
- Find the derivative of a function using the Power Rule.
- Find the derivative of a function using the Constant Multiple Rule.
- Find the derivative of a function using the Sum and Difference Rules.
- Find the derivative of the sine, cosine, and exponential functions.
- Use derivatives to find rates of change.

The Constant Rule

In Section 3.1 you used the limit definition to find derivatives. In this and the next two sections, you will be introduced to several "differentiation rules" that allow you to find derivatives without the *direct* use of the limit definition.

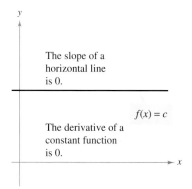

The slope of a horizontal line is 0.

$f(x) = c$

The derivative of a constant function is 0.

The Constant Rule
Figure 3.14

NOTE   In Figure 3.14, note that the Constant Rule is equivalent to saying that the slope of a horizontal line is 0. This demonstrates the relationship between slope and derivative.

> **THEOREM 3.2   The Constant Rule**
>
> The derivative of a constant function is 0. That is, if c is a real number, then
>
> $$\frac{d}{dx}[c] = 0.$$

Proof   Let $f(x) = c$. Then, by the limit definition of the derivative,

$$\frac{d}{dx}[c] = f'(x)$$

$$= \lim_{\Delta x \to 0} \frac{f(x + \Delta x) - f(x)}{\Delta x}$$

$$= \lim_{\Delta x \to 0} \frac{c - c}{\Delta x}$$

$$= \lim_{\Delta x \to 0} 0$$

$$= 0.$$

EXAMPLE 1   Using the Constant Rule

Function	*Derivative*
a. $y = 7$	$\dfrac{dy}{dx} = 0$
b. $f(x) = 0$	$f'(x) = 0$
c. $s(t) = -3$	$s'(t) = 0$
d. $y = k\pi^2$, k is constant	$y' = 0$

EXPLORATION

Writing a Conjecture   Use the definition of the derivative given in Section 3.1 to find the derivative of each of the following. What patterns do you see? Use your results to write a conjecture about the derivative of $f(x) = x^n$.

a. $f(x) = x^1$ **b.** $f(x) = x^2$ **c.** $f(x) = x^3$

d. $f(x) = x^4$ **e.** $f(x) = x^{1/2}$ **f.** $f(x) = x^{-1}$

The Constant Multiple Rule

> **THEOREM 3.4   The Constant Multiple Rule**
>
> If f is a differentiable function and c is a real number, then cf is also differentiable and $\dfrac{d}{dx}[cf(x)] = cf'(x)$.

Proof

$$\frac{d}{dx}[cf(x)] = \lim_{\Delta x \to 0} \frac{cf(x + \Delta x) - cf(x)}{\Delta x} \qquad \text{Definition of derivative}$$

$$= \lim_{\Delta x \to 0} c\left[\frac{f(x + \Delta x) - f(x)}{\Delta x}\right]$$

$$= c\left[\lim_{\Delta x \to 0} \frac{f(x + \Delta x) - f(x)}{\Delta x}\right]$$

$$= cf'(x)$$

Informally, the Constant Multiple Rule states that constants can be factored out of the differentiation process, even if the constants appear in the denominator.

$$\frac{d}{dx}[cf(x)] = c\frac{d}{dx}[f(x)] = cf'(x)$$

$$\frac{d}{dx}\left[\frac{f(x)}{c}\right] = \frac{d}{dx}\left[\left(\frac{1}{c}\right)f(x)\right]$$

$$= \left(\frac{1}{c}\right)\frac{d}{dx}[f(x)] = \left(\frac{1}{c}\right)f'(x)$$

EXAMPLE 5   Using the Constant Multiple Rule

Function	*Derivative*
a. $y = \dfrac{2}{x}$	$\dfrac{dy}{dx} = \dfrac{d}{dx}[2x^{-1}] = 2\dfrac{d}{dx}[x^{-1}] = 2(-1)x^{-2} = -\dfrac{2}{x^2}$
b. $f(t) = \dfrac{4t^2}{5}$	$f'(t) = \dfrac{d}{dt}\left[\dfrac{4}{5}t^2\right] = \dfrac{4}{5}\dfrac{d}{dt}[t^2] = \dfrac{4}{5}(2t) = \dfrac{8}{5}t$
c. $y = 2\sqrt{x}$	$\dfrac{dy}{dx} = \dfrac{d}{dx}[2x^{1/2}] = 2\left(\dfrac{1}{2}x^{-1/2}\right) = x^{-1/2} = \dfrac{1}{\sqrt{x}}$
d. $y = \dfrac{1}{2\sqrt[3]{x^2}}$	$\dfrac{dy}{dx} = \dfrac{d}{dx}\left[\dfrac{1}{2}x^{-2/3}\right] = \dfrac{1}{2}\left(-\dfrac{2}{3}\right)x^{-5/3} = -\dfrac{1}{3x^{5/3}}$
e. $y = -\dfrac{3x}{2}$	$y' = \dfrac{d}{dx}\left[-\dfrac{3}{2}x\right] = -\dfrac{3}{2}(1) = -\dfrac{3}{2}$

The Constant Multiple Rule and the Power Rule can be combined into one rule. The combination rule is

$$D_x[cx^n] = cnx^{n-1}.$$

EXAMPLE 6   **Using Parentheses When Differentiating**

Original Function	Rewrite	Differentiate	Simplify
a. $y = \dfrac{5}{2x^3}$	$y = \dfrac{5}{2}(x^{-3})$	$y' = \dfrac{5}{2}(-3x^{-4})$	$y' = -\dfrac{15}{2x^4}$
b. $y = \dfrac{5}{(2x)^3}$	$y = \dfrac{5}{8}(x^{-3})$	$y' = \dfrac{5}{8}(-3x^{-4})$	$y' = -\dfrac{15}{8x^4}$
c. $y = \dfrac{7}{3x^{-2}}$	$y = \dfrac{7}{3}(x^2)$	$y' = \dfrac{7}{3}(2x)$	$y' = \dfrac{14x}{3}$
d. $y = \dfrac{7}{(3x)^{-2}}$	$y = 63(x^2)$	$y' = 63(2x)$	$y' = 126x$

The Sum and Difference Rules

> **THEOREM 3.5   The Sum and Difference Rules**
>
> The sum (or difference) of two differentiable functions f and g is itself differentiable. Moreover, the derivative of $f + g$ (or $f - g$) is the sum (or difference) of the derivatives of f and g.
>
> $$\frac{d}{dx}[f(x) + g(x)] = f'(x) + g'(x) \qquad \text{Sum Rule}$$
>
> $$\frac{d}{dx}[f(x) - g(x)] = f'(x) - g'(x) \qquad \text{Difference Rule}$$

Proof   A proof of the Sum Rule follows from Theorem 2.2. (The Difference Rule can be proved in a similar way.)

$$\frac{d}{dx}[f(x) + g(x)] = \lim_{\Delta x \to 0} \frac{[f(x + \Delta x) + g(x + \Delta x)] - [f(x) + g(x)]}{\Delta x}$$

$$= \lim_{\Delta x \to 0} \frac{f(x + \Delta x) + g(x + \Delta x) - f(x) - g(x)}{\Delta x}$$

$$= \lim_{\Delta x \to 0} \left[\frac{f(x + \Delta x) - f(x)}{\Delta x} + \frac{g(x + \Delta x) - g(x)}{\Delta x} \right]$$

$$= \lim_{\Delta x \to 0} \frac{f(x + \Delta x) - f(x)}{\Delta x} + \lim_{\Delta x \to 0} \frac{g(x + \Delta x) - g(x)}{\Delta x}$$

$$= f'(x) + g'(x)$$

The Sum and Difference Rules can be extended to any finite number of functions. For instance, if $F(x) = f(x) + g(x) - h(x)$, then $F'(x) = f'(x) + g'(x) - h'(x)$.

EXAMPLE 7   **Using the Sum and Difference Rules**

Function	Derivative
a. $f(x) = x^3 - 4x + 5$	$f'(x) = 3x^2 - 4$
b. $g(x) = -\dfrac{x^4}{2} + 3x^3 - 2x$	$g'(x) = -2x^3 + 9x^2 - 2$

EXPLORATION

Use a graphing utility to graph the function

$$f(x) = \frac{\sin(x + \Delta x) - \sin x}{\Delta x}$$

for $\Delta x = 0.01$. What does this function represent? Compare this graph with that of the cosine function. What do you think the derivative of the sine function equals?

FOR FURTHER INFORMATION For the outline of a geometric proof of the derivatives of the sine and cosine functions, see the article "The Spider's Spacewalk Derivation of sin' and cos'" by Tim Hesterberg in *The College Mathematics Journal*. To view this article, go to the website *www.matharticles.com*.

Derivatives of Sine and Cosine Functions

In Section 2.3, you studied the following limits.

$$\lim_{\Delta x \to 0} \frac{\sin \Delta x}{\Delta x} = 1 \quad \text{and} \quad \lim_{\Delta x \to 0} \frac{1 - \cos \Delta x}{\Delta x} = 0$$

These two limits can be used to prove differentiation rules for the sine and cosine functions. (The derivatives of the other four trigonometric functions are discussed in Section 3.3.)

THEOREM 3.6   Derivatives of Sine and Cosine Functions

$$\frac{d}{dx}[\sin x] = \cos x \qquad\qquad \frac{d}{dx}[\cos x] = -\sin x$$

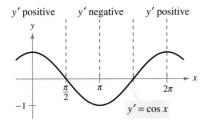

The derivative of the sine function is the cosine function.
Figure 3.18

Proof

$$\begin{aligned}
\frac{d}{dx}[\sin x] &= \lim_{\Delta x \to 0} \frac{\sin(x + \Delta x) - \sin x}{\Delta x} & \text{Definition of derivative}\\[2mm]
&= \lim_{\Delta x \to 0} \frac{\sin x \cos \Delta x + \cos x \sin \Delta x - \sin x}{\Delta x}\\[2mm]
&= \lim_{\Delta x \to 0} \frac{\cos x \sin \Delta x - (\sin x)(1 - \cos \Delta x)}{\Delta x}\\[2mm]
&= \lim_{\Delta x \to 0} \left[(\cos x)\left(\frac{\sin \Delta x}{\Delta x}\right) - (\sin x)\left(\frac{1 - \cos \Delta x}{\Delta x}\right)\right]\\[2mm]
&= \cos x\left(\lim_{\Delta x \to 0} \frac{\sin \Delta x}{\Delta x}\right) - \sin x\left(\lim_{\Delta x \to 0} \frac{1 - \cos \Delta x}{\Delta x}\right)\\[2mm]
&= (\cos x)(1) - (\sin x)(0)\\[2mm]
&= \cos x
\end{aligned}$$

This differentiation rule is shown graphically in Figure 3.18. Note that for each x, the *slope* of the sine curve is equal to the value of the cosine. The proof of the second rule is left as an exercise (see Exercise 116). ■

 EXAMPLE 8   Derivatives Involving Sines and Cosines

Function	*Derivative*
a. $y = 2 \sin x$	$y' = 2 \cos x$
b. $y = \dfrac{\sin x}{2} = \dfrac{1}{2} \sin x$	$y' = \dfrac{1}{2} \cos x = \dfrac{\cos x}{2}$
c. $y = x + \cos x$	$y' = 1 - \sin x$

TECHNOLOGY   A graphing utility can provide insight into the interpretation of a derivative. For instance, Figure 3.19 shows the graphs of

$$y = a \sin x$$

for $a = \frac{1}{2}, 1, \frac{3}{2}$, and 2. Estimate the slope of each graph at the point $(0, 0)$. Then verify your estimates analytically by evaluating the derivative of each function when $x = 0$.

$$\frac{d}{dx}[a \sin x] = a \cos x$$

Figure 3.19

Use a graphing utility to graph the function

$$f(x) = \frac{e^{x+\Delta x} - e^x}{\Delta x}$$

for $\Delta x = 0.01$. What does this function represent? Compare this graph with that of the exponential function. What do you think the derivative of the exponential function equals?

STUDY TIP The key to the formula for the derivative of $f(x) = e^x$ is the limit

$$\lim_{x \to 0} (1 + x)^{1/x} = e.$$

This important limit was introduced on page 51 and formalized later on page 85. It is used to conclude that for $\Delta x \approx 0$,

$$(1 + \Delta x)^{1/\Delta x} \approx e.$$

Derivatives of Exponential Functions

One of the most intriguing (and useful) characteristics of the natural exponential function is that *it is its own derivative*. Consider the following.

Let $f(x) = e^x$.

$$f'(x) = \lim_{\Delta x \to 0} \frac{f(x + \Delta x) - f(x)}{\Delta x}$$

$$= \lim_{\Delta x \to 0} \frac{e^{x+\Delta x} - e^x}{\Delta x}$$

$$= \lim_{\Delta x \to 0} \frac{e^x(e^{\Delta x} - 1)}{\Delta x}$$

The definition of e

$$\lim_{\Delta x \to 0} (1 + \Delta x)^{1/\Delta x} = e$$

tells you that for small values of Δx, you have $e \approx (1 + \Delta x)^{1/\Delta x}$, which implies that $e^{\Delta x} \approx 1 + \Delta x$. Replacing $e^{\Delta x}$ by this approximation produces the following.

$$f'(x) = \lim_{\Delta x \to 0} \frac{e^x[e^{\Delta x} - 1]}{\Delta x}$$

$$= \lim_{\Delta x \to 0} \frac{e^x[(1 + \Delta x) - 1]}{\Delta x}$$

$$= \lim_{\Delta x \to 0} \frac{e^x \Delta x}{\Delta x}$$

$$= e^x$$

This result is stated in the next theorem.

THEOREM 3.7   Derivative of the Natural Exponential Function

$$\frac{d}{dx}[e^x] = e^x$$

You can interpret Theorem 3.7 graphically by saying that the slope of the graph of $f(x) = e^x$ at any point (x, e^x) is equal to the y-coordinate of the point, as shown in Figure 3.20.

EXAMPLE 9   Derivatives of Exponential Functions

Find the derivative of each function.

a. $f(x) = 3e^x$ **b.** $f(x) = x^2 + e^x$ **c.** $f(x) = \sin x - e^x$

Solution

a. $f'(x) = 3\dfrac{d}{dx}[e^x] = 3e^x$

b. $f'(x) = \dfrac{d}{dx}[x^2] + \dfrac{d}{dx}[e^x] = 2x + e^x$

c. $f'(x) = \dfrac{d}{dx}[\sin x] - \dfrac{d}{dx}[e^x] = \cos x - e^x$

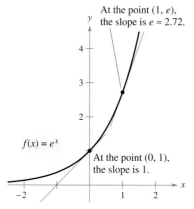

At the point $(1, e)$, the slope is $e \approx 2.72$.

At the point $(0, 1)$, the slope is 1.

$f(x) = e^x$

Figure 3.20

Exercises for Section 3.2

See www.CalcChat.com for worked-out solutions to odd-numbered exercises.

In Exercises 1 and 2, use the graph to estimate the slope of the tangent line to $y = x^n$ at the point $(1, 1)$. Verify your answer analytically. To print an enlarged copy of the graph, go to the website *www.mathgraphs.com*.

1. (a) $y = x^{1/2}$

(b) $y = x^3$

2. (a) $y = x^{-1/2}$

(b) $y = x^{-1}$

In Exercises 3–24, find the derivative of the function.

3. $y = 8$ **4.** $f(x) = -6$

5. $y = x^6$ **6.** $y = \dfrac{1}{x^8}$

7. $f(x) = \sqrt[5]{x}$ **8.** $g(x) = \sqrt[6]{x}$

9. $f(x) = x + 1$ **10.** $g(x) = 3x - 1$

11. $f(t) = -2t^2 + 3t - 6$ **12.** $y = t^2 + 2t - 3$

13. $g(x) = x^2 + 4x^3$ **14.** $y = 8 - x^3$

15. $s(t) = t^3 - 2t + 4$ **16.** $f(x) = 2x^3 - 4x^2 + 3x$

17. $f(x) = 6x - 5e^x$ **18.** $h(t) = t^3 + 2e^t$

19. $y = \dfrac{\pi}{2} \sin \theta - \cos \theta$ **20.** $g(t) = \pi \cos t$

21. $y = x^2 - \tfrac{1}{2} \cos x$ **22.** $y = 5 + \sin x$

23. $y = \tfrac{1}{2} e^x - 3 \sin x$ **24.** $y = \tfrac{3}{4} e^x + 2 \cos x$

In Exercises 25–30, complete the table using Example 6 as a model.

Original Function	Rewrite	Differentiate	Simplify
25. $y = \dfrac{5}{2x^2}$			
26. $y = \dfrac{4}{3x^2}$			
27. $y = \dfrac{3}{(2x)^3}$			
28. $y = \dfrac{\pi}{(5x)^2}$			
29. $y = \dfrac{\sqrt{x}}{x}$			
30. $y = \dfrac{4}{x^{-3}}$			

In Exercises 31–38, find the slope of the graph of the function at the indicated point. Use the *derivative* feature of a graphing utility to confirm your results.

Function	Point
31. $f(x) = \dfrac{3}{x^2}$	$(1, 3)$
32. $f(t) = 3 - \dfrac{3}{5t}$	$\left(\tfrac{3}{5}, 3\right)$
33. $f(x) = -\tfrac{1}{2} + \tfrac{7}{5}x^3$	$\left(0, -\tfrac{1}{2}\right)$
34. $f(x) = 3(5 - x)^2$	$(5, 0)$
35. $f(\theta) = 4 \sin \theta - \theta$	$(0, 0)$
36. $g(t) = 2 + 3 \cos t$	$(\pi, -1)$
37. $f(t) = \tfrac{3}{4} e^t$	$\left(0, \tfrac{3}{4}\right)$
38. $g(x) = -4e^x$	$(1, -4e)$

In Exercises 39–52, find the derivative of the function.

39. $g(t) = t^2 - \dfrac{4}{t^3}$ **40.** $f(x) = x + \dfrac{1}{x^2}$

41. $f(x) = \dfrac{x^3 - 3x^2 + 4}{x^2}$ **42.** $h(x) = \dfrac{2x^2 - 3x + 1}{x}$

43. $y = x(x^2 + 1)$ **44.** $y = 3x(6x - 5x^2)$

45. $f(x) = \sqrt{x} - 6\sqrt[3]{x}$ **46.** $f(x) = \sqrt[3]{x} + \sqrt[5]{x}$

47. $h(s) = s^{4/5} - s^{2/3}$ **48.** $f(t) = t^{2/3} - t^{1/3} + 4$

49. $f(x) = 6\sqrt{x} + 5 \cos x$ **50.** $f(x) = \dfrac{2}{\sqrt[3]{x}} + 5 \cos x$

51. $f(x) = x^{-2} - 2e^x$ **52.** $g(x) = \sqrt{x} - 3e^x$

In Exercises 53–56, (a) find an equation of the tangent line to the graph of f at the given point, (b) use a graphing utility to graph the function and its tangent line at the point, and (c) use the *derivative* feature of a graphing utility to confirm your results.

Function	Point
53. $y = x^4 - x$	$(-1, 2)$
54. $f(x) = \dfrac{2}{\sqrt[4]{x^3}}$	$(1, 2)$
55. $g(x) = x + e^x$	$(0, 1)$
56. $h(t) = \sin t + \tfrac{1}{2} e^t$	$\left(\pi, \tfrac{1}{2} e^\pi\right)$

In Exercises 57–62, determine the point(s) (if any) at which the graph of the function has a horizontal tangent line.

57. $y = x^4 - 8x^2 + 2$

58. $y = \dfrac{1}{x^2}$

59. $y = x + \sin x, \quad 0 \le x < 2\pi$

60. $y = \sqrt{3}x + 2\cos x, \quad 0 \le x < 2\pi$

61. $y = -4x + e^x$

62. $y = x + 4e^x$

In Exercises 63–66, find k such that the line is tangent to the graph of the function.

Function	*Line*
63. $f(x) = x^2 - kx$	$y = 4x - 9$
64. $f(x) = k - x^2$	$y = -4x + 7$
65. $f(x) = \dfrac{k}{x}$	$y = -\dfrac{3}{4}x + 3$
66. $f(x) = k\sqrt{x}$	$y = x + 4$

Writing About Concepts

67. Use the graph of f to answer each question. To print an enlarged copy of the graph, go to the website *www.mathgraphs.com*.

(a) Between which two consecutive points is the average rate of change of the function greatest?

(b) Is the average rate of change of the function between A and B greater than or less than the instantaneous rate of change at B?

(c) Sketch a tangent line to the graph between C and D such that the slope of the tangent line is the same as the average rate of change of the function between C and D.

68. Sketch the graph of a function f such that $f' > 0$ for all x and the rate of change of the function is decreasing.

In Exercises 69 and 70, the relationship between f and g is given. Explain the relationship between f' and g'.

69. $g(x) = f(x) + 6$

70. $g(x) = -5f(x)$

Writing About Concepts (continued)

In Exercises 71 and 72, the graphs of a function f and its derivative f' are shown on the same set of coordinate axes. Label the graphs as f or f' and write a short paragraph stating the criteria used in making the selection. To print an enlarged copy of the graph, go to the website *www.mathgraphs.com*.

71.   **72.**

73. Sketch the graphs of $y = x^2$ and $y = -x^2 + 6x - 5$, and sketch the two lines that are tangent to both graphs. Find equations of these lines.

74. Show that the graphs of the two equations $y = x$ and $y = 1/x$ have tangent lines that are perpendicular to each other at their point of intersection.

75. Show that the graph of the function

$$f(x) = 3x + \sin x + 2$$

does not have a horizontal tangent line.

76. Show that the graph of the function

$$f(x) = x^5 + 3x^3 + 5x$$

does not have a tangent line with a slope of 3.

In Exercises 77 and 78, find an equation of the tangent line to the graph of the function f through the point (x_0, y_0) not on the graph. To find the point of tangency (x, y) on the graph of f, solve the equation

$$f'(x) = \frac{y_0 - y}{x_0 - x}.$$

77. $f(x) = \sqrt{x}$

$(x_0, y_0) = (-4, 0)$

78. $f(x) = \dfrac{2}{x}$

$(x_0, y_0) = (5, 0)$

79. *Linear Approximation* Use a graphing utility (in square mode) to zoom in on the graph of $f(x) = 4 - \frac{1}{2}x^2$ to approximate $f'(1)$. Use the derivative to find $f'(1)$.

80. *Linear Approximation* Use a graphing utility (in square mode) to zoom in on the graph of $f(x) = 4\sqrt{x} + 1$ to approximate $f'(4)$. Use the derivative to find $f'(4)$.

81. *Linear Approximation* Consider the function $f(x) = x^{3/2}$ with the solution point $(4, 8)$.

(a) Use a graphing utility to obtain the graph of f. Use the *zoom* feature to obtain successive magnifications of the graph in the neighborhood of the point $(4, 8)$. After zooming in a few times, the graph should appear nearly linear. Use the *trace* feature to determine the coordinates of a point near $(4, 8)$. Find an equation of the secant line $S(x)$ through the two points.

(b) Find the equation of the line

$$T(x) = f'(4)(x - 4) + f(4)$$

tangent to the graph of f passing through the given point. Why are the linear functions S and T nearly the same?

(c) Use a graphing utility to graph f and T on the same set of coordinate axes. Note that T is a good approximation of f when x is close to 4. What happens to the accuracy of the approximation as you move farther away from the point of tangency?

(d) Demonstrate the conclusion in part (c) by completing the table.

Δx	-3	-2	-1	-0.5	-0.1	0
$f(4 + \Delta x)$						
$T(4 + \Delta x)$						

Δx	0.1	0.5	1	2	3
$f(4 + \Delta x)$					
$T(4 + \Delta x)$					

82. *Linear Approximation* Repeat Exercise 81 for the function $f(x) = x^3$, where $T(x)$ is the line tangent to the graph at the point $(1, 1)$. Explain why the accuracy of the linear approximation decreases more rapidly than in Exercise 81.

True or False? **In Exercises 83–88, determine whether the statement is true or false. If it is false, explain why or give an example that shows it is false.**

83. If $f'(x) = g'(x)$, then $f(x) = g(x)$.

84. If $f(x) = g(x) + c$, then $f'(x) = g'(x)$.

85. If $y = \pi^2$, then $dy/dx = 2\pi$.

86. If $y = x/\pi$, then $dy/dx = 1/\pi$.

87. If $g(x) = 3f(x)$, then $g'(x) = 3f'(x)$.

88. If $f(x) = 1/x^n$, then $f'(x) = 1/(nx^{n-1})$.

In Exercises 89–92, find the average rate of change of the function over the given interval. Compare this average rate of change with the instantaneous rates of change at the endpoints of the interval.

89. $f(x) = \dfrac{-1}{x}, \quad [1, 2]$

90. $f(x) = \cos x, \quad \left[0, \dfrac{\pi}{3}\right]$

91. $g(x) = x^2 + e^x, \quad [0, 1]$

92. $h(x) = x^3 - \frac{1}{2}e^x, \quad [0, 2]$

Vertical Motion **In Exercises 93 and 94, use the position function $s(t) = -16t^2 + v_0 t + s_0$ for free-falling objects.**

93. A silver dollar is dropped from the top of a building that is 1362 feet tall.

(a) Determine the position and velocity functions for the coin.

(b) Determine the average velocity on the interval $[1, 2]$.

(c) Find the instantaneous velocities when $t = 1$ and $t = 2$.

(d) Find the time required for the coin to reach ground level.

(e) Find the velocity of the coin at impact.

94. A ball is thrown straight down from the top of a 220-foot building with an initial velocity of -22 feet per second. What is its velocity after 3 seconds? What is its velocity after falling 108 feet?

Vertical Motion **In Exercises 95 and 96, use the position function $s(t) = -4.9t^2 + v_0 t + s_0$ for free-falling objects.**

95. A projectile is shot upward from the surface of Earth with an initial velocity of 120 meters per second. What is its velocity after 5 seconds? After 10 seconds?

96. To estimate the height of a building, a stone is dropped from the top of the building into a pool of water at ground level. How high is the building if the splash is seen 6.8 seconds after the stone is dropped?

Think About It **In Exercises 97 and 98, the graph of a position function is shown. It represents the distance in miles that a person drives during a 10-minute trip to work. Make a sketch of the corresponding velocity function.**

97.

98.

Think About It **In Exercises 99 and 100, the graph of a velocity function is shown. It represents the velocity in miles per hour during a 10-minute drive to work. Make a sketch of the corresponding position function.**

99.

100.

101. *Modeling Data*   The stopping distance of an automobile, on dry, level pavement, traveling at a speed v (kilometers per hour) is the distance R (meters) the car travels during the reaction time of the driver plus the distance B (meters) the car travels after the brakes are applied (see figure). The table shows the results of an experiment.

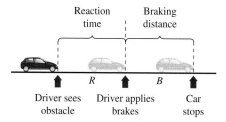

Reaction time   Braking distance

R B

Driver sees obstacle   Driver applies brakes   Car stops

Speed, v	20	40	60	80	100
Reaction Time Distance, R	8.3	16.7	25.0	33.3	41.7
Braking Time Distance, B	2.3	9.0	20.2	35.8	55.9

(a) Use the regression capabilities of a graphing utility to find a linear model for reaction time distance.

(b) Use the regression capabilities of a graphing utility to find a quadratic model for braking distance.

(c) Determine the polynomial giving the total stopping distance T.

(d) Use a graphing utility to graph the functions R, B, and T in the same viewing window.

(e) Find the derivative of T and the rates of change of the total stopping distance for $v = 40$, $v = 80$, and $v = 100$.

(f) Use the results of this exercise to draw conclusions about the total stopping distance as speed increases.

102. *Fuel Cost*   A car is driven 15,000 miles a year and gets x miles per gallon. Assume that the average fuel cost is \$1.55 per gallon. Find the annual cost of fuel C as a function of x and use this function to complete the table.

x	10	15	20	25	30	35	40
C							
dC/dx							

Who would benefit more from a one-mile-per-gallon increase in fuel efficiency—the driver of a car that gets 15 miles per gallon or the driver of a car that gets 35 miles per gallon? Explain.

103. *Volume*   The volume of a cube with sides of length s is given by $V = s^3$. Find the rate of change of the volume with respect to s when $s = 4$ centimeters.

104. *Area*   The area of a square with sides of length s is given by $A = s^2$. Find the rate of change of the area with respect to s when $s = 4$ meters.

105. *Velocity*   Verify that the average velocity over the time interval $[t_0 - \Delta t, t_0 + \Delta t]$ is the same as the instantaneous velocity at $t = t_0$ for the position function

$$s(t) = -\tfrac{1}{2}at^2 + c.$$

106. *Inventory Management*   The annual inventory cost C for a manufacturer is

$$C = \frac{1{,}008{,}000}{Q} + 6.3Q$$

where Q is the order size when the inventory is replenished. Find the change in annual cost when Q is increased from 350 to 351, and compare this with the instantaneous rate of change when $Q = 350$.

107. *Writing*   The number of gallons N of regular unleaded gasoline sold by a gasoline station at a price of p dollars per gallon is given by $N = f(p)$.

(a) Describe the meaning of $f'(1.479)$.

(b) Is $f'(1.479)$ usually positive or negative? Explain.

108. *Newton's Law of Cooling*   This law states that the rate of change of the temperature of an object is proportional to the difference between the object's temperature T and the temperature T_a of the surrounding medium. Write an equation for this law.

109. Find an equation of the parabola $y = ax^2 + bx + c$ that passes through $(0, 1)$ and is tangent to the line $y = x - 1$ at $(1, 0)$.

110. Let (a, b) be an arbitrary point on the graph of $y = 1/x$, $x > 0$. Prove that the area of the triangle formed by the tangent line through (a, b) and the coordinate axes is 2.

111. Find the tangent line(s) to the curve $y = x^3 - 9x$ through the point $(1, -9)$.

112. Find the equation(s) of the tangent line(s) to the parabola $y = x^2$ through the given point.

(a) $(0, a)$ (b) $(a, 0)$

Are there any restrictions on the constant a?

In Exercises 113 and 114, find a and b such that f is differentiable everywhere.

113. $f(x) = \begin{cases} ax^3, & x \le 2 \\ x^2 + b, & x > 2 \end{cases}$

114. $f(x) = \begin{cases} \cos x, & x < 0 \\ ax + b, & x \ge 0 \end{cases}$

115. Where are the functions $f_1(x) = |\sin x|$ and $f_2(x) = \sin |x|$ differentiable?

116. Prove that $\dfrac{d}{dx}[\cos x] = -\sin x$.

FOR FURTHER INFORMATION   For a geometric interpretation of the derivatives of trigonometric functions, see the article "Sines and Cosines of the Times" by Victor J. Katz in *Math Horizons*. To view this article, go to the website *www.matharticles.com*.

Product and Quotient Rules and Higher-Order Derivatives

- Find the derivative of a function using the Product Rule.
- Find the derivative of a function using the Quotient Rule.
- Find the derivative of a trigonometric function.
- Find a higher-order derivative of a function.

The Product Rule

In Section 3.2 you learned that the derivative of the sum of two functions is simply the sum of their derivatives. The rules for the derivatives of the product and quotient of two functions are not as simple.

NOTE A version of the Product Rule that some people prefer is

$$\frac{d}{dx}[f(x)g(x)] = f'(x)g(x) + f(x)g'(x).$$

The advantage of this form is that it generalizes easily to products involving three or more factors.

THEOREM 3.8 The Product Rule

The product of two differentiable functions f and g is itself differentiable. Moreover, the derivative of fg is the first function times the derivative of the second, plus the second function times the derivative of the first.

$$\frac{d}{dx}[f(x)g(x)] = f(x)g'(x) + g(x)f'(x)$$

Proof Some mathematical proofs, such as the proof of the Sum Rule, are straightforward. Others involve clever steps that may appear unmotivated to a reader. This proof involves such a step—subtracting and adding the same quantity—which is shown in color.

$$\frac{d}{dx}[f(x)g(x)] = \lim_{\Delta x \to 0} \frac{f(x + \Delta x)g(x + \Delta x) - f(x)g(x)}{\Delta x}$$

$$= \lim_{\Delta x \to 0} \frac{f(x + \Delta x)g(x + \Delta x) - f(x + \Delta x)g(x) + f(x + \Delta x)g(x) - f(x)g(x)}{\Delta x}$$

$$= \lim_{\Delta x \to 0} \left[f(x + \Delta x)\frac{g(x + \Delta x) - g(x)}{\Delta x} + g(x)\frac{f(x + \Delta x) - f(x)}{\Delta x} \right]$$

$$= \lim_{\Delta x \to 0} \left[f(x + \Delta x)\frac{g(x + \Delta x) - g(x)}{\Delta x} \right] + \lim_{\Delta x \to 0} \left[g(x)\frac{f(x + \Delta x) - f(x)}{\Delta x} \right]$$

$$= \lim_{\Delta x \to 0} f(x + \Delta x) \cdot \lim_{\Delta x \to 0} \frac{g(x + \Delta x) - g(x)}{\Delta x} + \lim_{\Delta x \to 0} g(x) \cdot \lim_{\Delta x \to 0} \frac{f(x + \Delta x) - f(x)}{\Delta x}$$

$$= f(x)g'(x) + g(x)f'(x)$$

THE PRODUCT RULE

When Leibniz originally wrote a formula for the Product Rule, he was motivated by the expression

$$(x + dx)(y + dy) - xy$$

from which he subtracted $dx\,dy$ (as being negligible) and obtained the differential form $x\,dy + y\,dx$. This derivation resulted in the traditional form of the Product Rule. (*Source: The History of Mathematics by David M. Burton*)

Note that $\lim_{\Delta x \to 0} f(x + \Delta x) = f(x)$ because f is given to be differentiable and therefore is continuous.

The Product Rule can be extended to cover products involving more than two factors. For example, if f, g, and h are differentiable functions of x, then

$$\frac{d}{dx}[f(x)g(x)h(x)] = f'(x)g(x)h(x) + f(x)g'(x)h(x) + f(x)g(x)h'(x).$$

For instance, the derivative of $y = x^2 \sin x \cos x$ is

$$\frac{dy}{dx} = 2x \sin x \cos x + x^2 \cos x \cos x + x^2 \sin x(-\sin x)$$

$$= 2x \sin x \cos x + x^2(\cos^2 x - \sin^2 x).$$

The derivative of a product of two functions is not (in general) given by the product of the derivatives of the two functions. To see this, try comparing the product of the derivatives of $f(x) = 3x - 2x^2$ and $g(x) = 5 + 4x$ with the derivative in Example 1.

EXAMPLE 1   Using the Product Rule

Find the derivative of $h(x) = (3x - 2x^2)(5 + 4x)$.

Solution

$$h'(x) = \overbrace{(3x - 2x^2)}^{\text{First}} \overbrace{\frac{d}{dx}[5 + 4x]}^{\substack{\text{Derivative} \\ \text{of second}}} + \overbrace{(5 + 4x)}^{\text{Second}} \overbrace{\frac{d}{dx}[3x - 2x^2]}^{\substack{\text{Derivative} \\ \text{of first}}} \qquad \text{Apply Product Rule.}$$

$$= (3x - 2x^2)(4) + (5 + 4x)(3 - 4x)$$
$$= (12x - 8x^2) + (15 - 8x - 16x^2)$$
$$= -24x^2 + 4x + 15$$

In Example 1, you have the option of finding the derivative with or without the Product Rule. To find the derivative without the Product Rule, you can write

$$D_x[(3x - 2x^2)(5 + 4x)] = D_x[-8x^3 + 2x^2 + 15x]$$
$$= -24x^2 + 4x + 15.$$

In the next example, you must use the Product Rule.

EXAMPLE 2   Using the Product Rule

Find the derivative of $y = xe^x$.

Solution

$$\frac{d}{dx}[xe^x] = x\frac{d}{dx}[e^x] + e^x\frac{d}{dx}[x] \qquad \text{Apply Product Rule.}$$

$$= xe^x + e^x(1)$$
$$= e^x(x + 1)$$

EXAMPLE 3   Using the Product Rule

Find the derivative of $y = 2x \cos x - 2 \sin x$.

Solution

NOTE   In Example 3, notice that you use the Product Rule when both factors of the product are variable, and you use the Constant Multiple Rule when one of the factors is a constant.

$$\frac{dy}{dx} = \overbrace{(2x)\left(\frac{d}{dx}[\cos x]\right) + (\cos x)\left(\frac{d}{dx}[2x]\right)}^{\text{Product Rule}} - \overbrace{2\frac{d}{dx}[\sin x]}^{\text{Constant Multiple Rule}}$$

$$= (2x)(-\sin x) + (\cos x)(2) - 2(\cos x)$$
$$= -2x \sin x$$

The Quotient Rule

THEOREM 3.9 The Quotient Rule

The quotient f/g of two differentiable functions f and g is itself differentiable at all values of x for which $g(x) \neq 0$. Moreover, the derivative of f/g is given by the denominator times the derivative of the numerator minus the numerator times the derivative of the denominator, all divided by the square of the denominator.

$$\frac{d}{dx}\left[\frac{f(x)}{g(x)}\right] = \frac{g(x)f'(x) - f(x)g'(x)}{[g(x)]^2}, \qquad g(x) \neq 0$$

Proof As with the proof of Theorem 3.8, the key to this proof is subtracting and adding the same quantity.

$$\frac{d}{dx}\left[\frac{f(x)}{g(x)}\right] = \lim_{\Delta x \to 0} \frac{\dfrac{f(x + \Delta x)}{g(x + \Delta x)} - \dfrac{f(x)}{g(x)}}{\Delta x} \qquad \text{Definition of derivative}$$

$$= \lim_{\Delta x \to 0} \frac{g(x)f(x + \Delta x) - f(x)g(x + \Delta x)}{\Delta x\, g(x)g(x + \Delta x)}$$

$$= \lim_{\Delta x \to 0} \frac{g(x)f(x + \Delta x) - f(x)g(x) + f(x)g(x) - f(x)g(x + \Delta x)}{\Delta x\, g(x)g(x + \Delta x)}$$

$$= \frac{\displaystyle\lim_{\Delta x \to 0} \frac{g(x)[f(x + \Delta x) - f(x)]}{\Delta x} - \lim_{\Delta x \to 0}\frac{f(x)[g(x + \Delta x) - g(x)]}{\Delta x}}{\displaystyle\lim_{\Delta x \to 0}[g(x)g(x + \Delta x)]}$$

$$= \frac{g(x)\left[\displaystyle\lim_{\Delta x \to 0}\frac{f(x + \Delta x) - f(x)}{\Delta x}\right] - f(x)\left[\displaystyle\lim_{\Delta x \to 0}\frac{g(x + \Delta x) - g(x)}{\Delta x}\right]}{\displaystyle\lim_{\Delta x \to 0}[g(x)g(x + \Delta x)]}$$

$$= \frac{g(x)f'(x) - f(x)g'(x)}{[g(x)]^2}$$

Note that $\displaystyle\lim_{\Delta x \to 0} g(x + \Delta x) = g(x)$ because g is given to be differentiable and therefore is continuous.

TECHNOLOGY Graphing utilities can be used to compare the graph of a function with the graph of its derivative. For instance, in Figure 3.23, the graph of the function in Example 4 appears to have two points that have horizontal tangent lines. What are the values of y' at these two points?

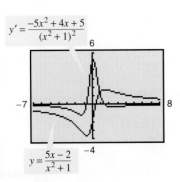

$$y' = \frac{-5x^2 + 4x + 5}{(x^2 + 1)^2}$$

$$y = \frac{5x - 2}{x^2 + 1}$$

Graphical comparison of a function and its derivative
Figure 3.23

EXAMPLE 4 Using the Quotient Rule

Find the derivative of $y = \dfrac{5x - 2}{x^2 + 1}$.

Solution

$$\frac{d}{dx}\left[\frac{5x - 2}{x^2 + 1}\right] = \frac{(x^2 + 1)\dfrac{d}{dx}[5x - 2] - (5x - 2)\dfrac{d}{dx}[x^2 + 1]}{(x^2 + 1)^2} \qquad \text{Apply Quotient Rule.}$$

$$= \frac{(x^2 + 1)(5) - (5x - 2)(2x)}{(x^2 + 1)^2}$$

$$= \frac{(5x^2 + 5) - (10x^2 - 4x)}{(x^2 + 1)^2}$$

$$= \frac{-5x^2 + 4x + 5}{(x^2 + 1)^2}$$

Note the use of parentheses in Example 4. A liberal use of parentheses is recommended for *all* types of differentiation problems. For instance, with the Quotient Rule, it is a good idea to enclose all factors and derivatives in parentheses, and to pay special attention to the subtraction required in the numerator.

When differentiation rules were introduced in the preceding section, the need for rewriting *before* differentiating was emphasized. The next example illustrates this point with the Quotient Rule.

EXAMPLE 5   Rewriting Before Differentiating

Find an equation of the tangent line to the graph of $f(x) = \dfrac{3 - (1/x)}{x + 5}$ at $(-1, 1)$.

Solution   Begin by rewriting the function.

$$f(x) = \frac{3 - (1/x)}{x + 5} \qquad \text{Write original function.}$$

$$= \frac{x\left(3 - \dfrac{1}{x}\right)}{x(x + 5)} \qquad \text{Multiply numerator and denominator by } x.$$

$$= \frac{3x - 1}{x^2 + 5x} \qquad \text{Rewrite.}$$

$$f'(x) = \frac{(x^2 + 5x)(3) - (3x - 1)(2x + 5)}{(x^2 + 5x)^2} \qquad \text{Apply Quotient Rule.}$$

$$= \frac{(3x^2 + 15x) - (6x^2 + 13x - 5)}{(x^2 + 5x)^2}$$

$$= \frac{-3x^2 + 2x + 5}{(x^2 + 5x)^2} \qquad \text{Simplify.}$$

To find the slope at $(-1, 1)$, evaluate $f'(-1)$.

$$f'(-1) = 0 \qquad \text{Slope of graph at } (-1, 1)$$

Then, using the point-slope form of the equation of a line, you can determine that the equation of the tangent line at $(-1, 1)$ is $y = 1$. See Figure 3.24.

The line $y = 1$ is tangent to the graph of $f(x)$ at the point $(-1, 1)$.
Figure 3.24

Not every quotient needs to be differentiated by the Quotient Rule. For example, each quotient in the next example can be considered as the product of a constant times a function of x. In such cases it is more convenient to use the Constant Multiple Rule.

EXAMPLE 6   Using the Constant Multiple Rule

	Original Function	*Rewrite*	*Differentiate*	*Simplify*
a.	$y = \dfrac{x^2 + 3x}{6}$	$y = \dfrac{1}{6}(x^2 + 3x)$	$y' = \dfrac{1}{6}(2x + 3)$	$y' = \dfrac{2x + 3}{6}$
b.	$y = \dfrac{5x^4}{8}$	$y = \dfrac{5}{8}x^4$	$y' = \dfrac{5}{8}(4x^3)$	$y' = \dfrac{5}{2}x^3$
c.	$y = \dfrac{-3(3x - 2x^2)}{7x}$	$y = -\dfrac{3}{7}(3 - 2x)$	$y' = -\dfrac{3}{7}(-2)$	$y' = \dfrac{6}{7}$
d.	$y = \dfrac{9}{5x^2}$	$y = \dfrac{9}{5}(x^{-2})$	$y' = \dfrac{9}{5}(-2x^{-3})$	$y' = -\dfrac{18}{5x^3}$

NOTE   To see the benefit of using the Constant Multiple Rule for some quotients, try using the Quotient Rule to differentiate the functions in Example 6—you should obtain the same results but with more work.

In Section 3.2, the Power Rule was proved only for the case where the exponent n is a positive integer greater than 1. The next example extends the proof to include negative integer exponents.

EXAMPLE 7 Proof of the Power Rule (Negative Integer Exponents)

If n is a negative integer, there exists a positive integer k such that $n = -k$. So, by the Quotient Rule, you can write

$$\frac{d}{dx}[x^n] = \frac{d}{dx}\left[\frac{1}{x^k}\right]$$

$$= \frac{x^k(0) - (1)(kx^{k-1})}{(x^k)^2} \qquad \text{Quotient Rule and Power Rule}$$

$$= \frac{0 - kx^{k-1}}{x^{2k}}$$

$$= -kx^{-k-1}$$

$$= nx^{n-1}. \qquad n = -k$$

So, the Power Rule

$$D_x[x^n] = nx^{n-1} \qquad \text{Power Rule}$$

is valid for any integer. In Exercise 91 in Section 3.5, you are asked to prove the case for which n is any rational number.

Derivatives of Trigonometric Functions

Knowing the derivatives of the sine and cosine functions, you can use the Quotient Rule to find the derivatives of the four remaining trigonometric functions.

THEOREM 3.10 Derivatives of Trigonometric Functions

$$\frac{d}{dx}[\tan x] = \sec^2 x \qquad\qquad \frac{d}{dx}[\cot x] = -\csc^2 x$$

$$\frac{d}{dx}[\sec x] = \sec x \tan x \qquad\qquad \frac{d}{dx}[\csc x] = -\csc x \cot x$$

Proof Considering $\tan x = (\sin x)/(\cos x)$ and applying the Quotient Rule, you obtain

$$\frac{d}{dx}[\tan x] = \frac{(\cos x)(\cos x) - (\sin x)(-\sin x)}{\cos^2 x} \qquad \text{Apply Quotient Rule.}$$

$$= \frac{\cos^2 x + \sin^2 x}{\cos^2 x}$$

$$= \frac{1}{\cos^2 x}$$

$$= \sec^2 x.$$

The proofs of the other three parts of the theorem are left as an exercise (see Exercise 93).

EXAMPLE 8 **Differentiating Trigonometric Functions**

NOTE Because of trigonometric identities, the derivative of a trigonometric function can take many forms. This presents a challenge when you are trying to match your answers to those given in the back of the text.

Function	_Derivative_
a. $y = x - \tan x$	$\dfrac{dy}{dx} = 1 - \sec^2 x$
b. $y = x \sec x$	$y' = x(\sec x \tan x) + (\sec x)(1)$
	$= (\sec x)(1 + x \tan x)$

EXAMPLE 9 **Different Forms of a Derivative**

Differentiate both forms of $y = \dfrac{1 - \cos x}{\sin x} = \csc x - \cot x.$

Solution

First form: $y = \dfrac{1 - \cos x}{\sin x}$

$$y' = \frac{(\sin x)(\sin x) - (1 - \cos x)(\cos x)}{\sin^2 x}$$

$$= \frac{\sin^2 x + \cos^2 x - \cos x}{\sin^2 x}$$

$$= \frac{1 - \cos x}{\sin^2 x}$$

Second form: $y = \csc x - \cot x$

$$y' = -\csc x \cot x + \csc^2 x$$

To verify that the two derivatives are equal, you can write

$$\frac{1 - \cos x}{\sin^2 x} = \frac{1}{\sin^2 x} - \left(\frac{1}{\sin x}\right)\left(\frac{\cos x}{\sin x}\right)$$

$$= \csc^2 x - \csc x \cot x.$$

The summary below shows that much of the work in obtaining a simplified form of a derivative occurs _after_ differentiating. Note that two characteristics of a simplified form are the absence of negative exponents and the combining of like terms.

	f′(x) After Differentiating	_f′(x) After Simplifying_
Example 1	$(3x - 2x^2)(4) + (5 + 4x)(3 - 4x)$	$-24x^2 + 4x + 15$
Example 3	$(2x)(-\sin x) + (\cos x)(2) - 2(\cos x)$	$-2x \sin x$
Example 4	$\dfrac{(x^2 + 1)(5) - (5x - 2)(2x)}{(x^2 + 1)^2}$	$\dfrac{-5x^2 + 4x + 5}{(x^2 + 1)^2}$
Example 5	$\dfrac{(x^2 + 5x)(3) - (3x - 1)(2x + 5)}{(x^2 + 5x)^2}$	$\dfrac{-3x^2 + 2x + 5}{(x^2 + 5x)^2}$
Example 9	$\dfrac{(\sin x)(\sin x) - (1 - \cos x)(\cos x)}{\sin^2 x}$	$\dfrac{1 - \cos x}{\sin^2 x}$

EXPLORATION

For which of the functions

$$y = e^x, \qquad y = \frac{1}{e^x}$$

$$y = \sin x, \qquad y = \cos x$$

are the following equations true?

a. $y = y'$ b. $y = y''$

c. $y = y'''$ d. $y = y^{(4)}$

Without determining the actual derivative, is $y = y^{(8)}$ for $y = \sin x$ true? What conclusion can you draw from this?

Higher-Order Derivatives

Just as you can obtain a velocity function by differentiating a position function, you can obtain an **acceleration** function by differentiating a velocity function. Another way of looking at this is that you can obtain an acceleration function by differentiating a position function *twice*.

$$s(t) \qquad\qquad \text{Position function}$$
$$v(t) = s'(t) \qquad\qquad \text{Velocity function}$$
$$a(t) = v'(t) = s''(t) \qquad\qquad \text{Acceleration function}$$

The function given by $a(t)$ is the **second derivative** of $s(t)$ and is denoted by $s''(t)$.

The second derivative is an example of a **higher-order derivative.** You can define derivatives of any positive integer order. For instance, the **third derivative** is the derivative of the second derivative. Higher-order derivatives are denoted as follows.

First derivative: y', $f'(x)$, $\dfrac{dy}{dx}$, $\dfrac{d}{dx}[f(x)]$, $D_x[y]$

Second derivative: y'', $f''(x)$, $\dfrac{d^2y}{dx^2}$, $\dfrac{d^2}{dx^2}[f(x)]$, $D_x^2[y]$

Third derivative: y''', $f'''(x)$, $\dfrac{d^3y}{dx^3}$, $\dfrac{d^3}{dx^3}[f(x)]$, $D_x^3[y]$

Fourth derivative: $y^{(4)}$, $f^{(4)}(x)$, $\dfrac{d^4y}{dx^4}$, $\dfrac{d^4}{dx^4}[f(x)]$, $D_x^4[y]$

$$\vdots$$

nth derivative: $y^{(n)}$, $f^{(n)}(x)$, $\dfrac{d^ny}{dx^n}$, $\dfrac{d^n}{dx^n}[f(x)]$, $D_x^n[y]$

EXAMPLE 10   Finding the Acceleration Due to Gravity

Because the moon has no atmosphere, a falling object on the moon encounters no air resistance. In 1971, astronaut David Scott demonstrated that a feather and a hammer fall at the same rate on the moon. The position function for each of these falling objects is given by

$$s(t) = -0.81t^2 + 2$$

where $s(t)$ is the height in meters and t is the time in seconds. What is the ratio of Earth's gravitational force to the moon's?

Solution   To find the acceleration, differentiate the position function twice.

$$s(t) = -0.81t^2 + 2 \qquad\qquad \text{Position function}$$
$$s'(t) = -1.62t \qquad\qquad \text{Velocity function}$$
$$s''(t) = -1.62 \qquad\qquad \text{Acceleration function}$$

So, the acceleration due to gravity on the moon is -1.62 meters per second per second. Because the acceleration due to gravity on Earth is -9.8 meters per second per second, the ratio of Earth's gravitational force to the moon's is

$$\frac{\text{Earth's gravitational force}}{\text{Moon's gravitational force}} = \frac{-9.8}{-1.62}$$

$$\approx 6.05.$$

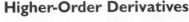

THE MOON

The moon's mass is 7.349×10^{22} kilograms, and Earth's mass is 5.976×10^{24} kilograms. The moon's radius is 1737 kilometers, and Earth's radius is 6378 kilometers. Because the gravitational force on the surface of a planet is directly proportional to its mass and inversely proportional to the square of its radius, the ratio of the gravitational force on Earth to the gravitational force on the moon is

$$\frac{(5.976 \times 10^{24})/6378^2}{(7.349 \times 10^{22})/1737^2} \approx 6.03.$$

NASA

See www.CalcChat.com for worked-out solutions to odd-numbered exercises.

In Exercises 1–6, use the Product Rule to differentiate the function.

1. $g(x) = (x^2 + 1)(x^2 - 2x)$

2. $f(x) = (6x + 5)(x^3 - 2)$

3. $h(t) = \sqrt[3]{t}(t^2 + 4)$

4. $g(s) = \sqrt{s}(4 - s^2)$

5. $f(x) = x^3 \cos x$

6. $g(x) = \sqrt{x} \sin x$

In Exercises 7–12, use the Quotient Rule to differentiate the function.

7. $f(x) = \dfrac{x}{x^2 + 1}$

8. $g(t) = \dfrac{t^2 + 2}{2t - 7}$

9. $h(x) = \dfrac{\sqrt[3]{x}}{x^3 + 1}$

10. $h(s) = \dfrac{s}{\sqrt{s} - 1}$

11. $g(x) = \dfrac{\sin x}{x^2}$

12. $f(t) = \dfrac{\cos t}{t^3}$

In Exercises 13–20, find $f'(x)$ and $f'(c)$.

Function	Value of c
13. $f(x) = (x^3 - 3x)(2x^2 + 3x + 5)$	$c = 0$
14. $f(x) = (x^2 - 2x + 1)(x^3 - 1)$	$c = 1$
15. $f(x) = \dfrac{x^2 - 4}{x - 3}$	$c = 1$
16. $f(x) = \dfrac{x + 1}{x - 1}$	$c = 2$
17. $f(x) = x \cos x$	$c = \dfrac{\pi}{4}$
18. $f(x) = \dfrac{\sin x}{x}$	$c = \dfrac{\pi}{6}$
19. $f(x) = e^x \sin x$	$c = 0$
20. $f(x) = \dfrac{\cos x}{e^x}$	$c = 0$

In Exercises 21–26, complete the table without using the Quotient Rule (see Example 6).

Function	Rewrite	Differentiate	Simplify
21. $y = \dfrac{x^2 + 2x}{3}$			
22. $y = \dfrac{5x^2 - 3}{4}$			
23. $y = \dfrac{7}{3x^3}$			
24. $y = \dfrac{5}{4x^2}$			
25. $y = \dfrac{4x^{3/2}}{x}$			
26. $y = \dfrac{3x^2 - 5}{7}$			

In Exercises 27–40, find the derivative of the algebraic function.

27. $f(x) = \dfrac{3 - 2x - x^2}{x^2 - 1}$

28. $f(x) = \dfrac{x^3 + 3x + 2}{x^2 + 1}$

29. $f(x) = x\left(1 - \dfrac{4}{x + 3}\right)$

30. $f(x) = x^4\left(1 - \dfrac{2}{x + 1}\right)$

31. $f(x) = \dfrac{2x + 5}{\sqrt{x}}$

32. $f(x) = \sqrt[3]{x}(\sqrt{x} + 3)$

33. $h(s) = (s^3 - 2)^2$

34. $h(x) = (x^2 + 1)^2$

35. $f(x) = \dfrac{2 - \dfrac{1}{x}}{x - 3}$

36. $g(x) = x^2\left(\dfrac{2}{x} - \dfrac{1}{x + 1}\right)$

37. $f(x) = (3x^3 + 4x)(x - 5)(x + 1)$

38. $f(x) = (x^2 - x)(x^2 + 1)(x^2 + x + 1)$

39. $f(x) = \dfrac{x^2 + c^2}{x^2 - c^2}$, c is a constant

40. $f(x) = \dfrac{c^2 - x^2}{c^2 + x^2}$, c is a constant

In Exercises 41–58, find the derivative of the transcendental function.

41. $f(t) = t^2 \sin t$

42. $f(\theta) = (\theta + 1) \cos \theta$

43. $f(t) = \dfrac{\cos t}{t}$

44. $f(x) = \dfrac{\sin x}{x}$

45. $f(x) = -e^x + \tan x$

46. $y = e^x - \cot x$

47. $g(t) = \sqrt[4]{t} + 8 \sec t$

48. $h(s) = \dfrac{1}{s} - 10 \csc s$

49. $y = \dfrac{3(1 - \sin x)}{2 \cos x}$

50. $y = \dfrac{\sec x}{x}$

51. $y = -\csc x - \sin x$

52. $y = x \cos x + \sin x$

53. $f(x) = x^2 \tan x$

54. $f(x) = 2 \sin x \cos x$

55. $y = 2x \sin x + x^2 e^x$

56. $h(x) = 2e^x \cos x$

57. $y = \dfrac{e^x}{4\sqrt{x}}$

58. $y = \dfrac{2e^x}{x^2 + 1}$

In Exercises 59–62, use a computer algebra system to differentiate the function.

59. $g(x) = \left(\dfrac{x + 1}{x + 2}\right)(2x - 5)$

60. $f(x) = \left(\dfrac{x^2 - x - 3}{x^2 + 1}\right)(x^2 + x + 1)$

61. $g(\theta) = \dfrac{\theta}{1 - \sin \theta}$

62. $f(\theta) = \dfrac{\sin \theta}{1 - \cos \theta}$

In Exercises 63–66, evaluate the derivative of the function at the indicated point. Use a graphing utility to verify your result.

Function	Point
63. $y = \dfrac{1 + \csc x}{1 - \csc x}$	$\left(\dfrac{\pi}{6}, -3\right)$

Function	Point
64. $f(x) = \tan x \cot x$	$(1, 1)$
65. $h(t) = \dfrac{\sec t}{t}$	$\left(\pi, -\dfrac{1}{\pi}\right)$
66. $f(x) = \sin x(\sin x + \cos x)$	$\left(\dfrac{\pi}{4}, 1\right)$

In Exercises 67–72, **(a) find an equation of the tangent line to the graph of f at the given point, (b) use a graphing utility to graph the function and its tangent line at the point, and (c) use the *derivative* feature of a graphing utility to confirm your results.**

Function	Point
67. $f(x) = (x^3 - 3x + 1)(x + 2)$	$(1, -3)$
68. $f(x) = \dfrac{(x - 1)}{(x + 1)}$	$\left(2, \dfrac{1}{3}\right)$
69. $f(x) = \tan x$	$\left(\dfrac{\pi}{4}, 1\right)$
70. $f(x) = \sec x$	$\left(\dfrac{\pi}{3}, 2\right)$
71. $f(x) = (x - 1)e^x$	$(1, 0)$
72. $f(x) = \dfrac{e^x}{x + 4}$	$\left(0, \dfrac{1}{4}\right)$

Famous Curves In Exercises 73–76, find an equation of the tangent line to the graph at the given point. (The graphs in Exercises 73 and 74 are called *witches of Agnesi*. The graphs in Exercises 75 and 76 are called *serpentines*.)

73.

$f(x) = \dfrac{8}{x^2 + 4}$

$(2, 1)$

74.

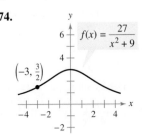

$f(x) = \dfrac{27}{x^2 + 9}$

$\left(-3, \dfrac{3}{2}\right)$

75.

$f(x) = \dfrac{16x}{x^2 + 16}$

$\left(-2, -\dfrac{8}{5}\right)$

76.

$\left(2, \dfrac{4}{5}\right)$

$f(x) = \dfrac{4x}{x^2 + 6}$

In Exercises 77–80, determine the point(s) at which the graph of the function has a horizontal tangent.

77. $f(x) = \dfrac{x^2}{x - 1}$

78. $f(x) = \dfrac{x^2}{x^2 + 1}$

79. $g(x) = \dfrac{8(x - 2)}{e^x}$

80. $f(x) = e^x \sin x, \quad [0, \pi]$

81. *Tangent Lines* Find equations of the tangent lines to the graph of $f(x) = \dfrac{x + 1}{x - 1}$ that are parallel to the line $2y + x = 6$. Then graph the function and the tangent lines.

82. *Tangent Lines* Find equations of the tangent lines to the graph of $f(x) = \dfrac{x}{x - 1}$ that pass through the point $(-1, 5)$. Then graph the function and the tangent lines.

In Exercises 83 and 84, verify that $f'(x) = g'(x)$, and explain the relationship between f and g.

83. $f(x) = \dfrac{3x}{x + 2}, \quad g(x) = \dfrac{5x + 4}{x + 2}$

84. $f(x) = \dfrac{\sin x - 3x}{x}, \quad g(x) = \dfrac{\sin x + 2x}{x}$

In Exercises 85 and 86, use the graphs of f and g. Let $p(x) = f(x)g(x)$ and $q(x) = \dfrac{f(x)}{g(x)}$.

85. (a) Find $p'(1)$.
    (b) Find $q'(4)$.

86. (a) Find $p'(4)$.
    (b) Find $q'(7)$.

87. *Area* The length of a rectangle is given by $2t + 1$ and its height is $\sqrt{t}$, where t is time in seconds and the dimensions are in centimeters. Find the rate of change of the area with respect to time.

88. *Volume* The radius of a right circular cylinder is given by $\sqrt{t + 2}$ and its height is $\frac{1}{2}\sqrt{t}$, where t is time in seconds and the dimensions are in inches. Find the rate of change of the volume with respect to time.

89. *Inventory Replenishment* The ordering and transportation cost C for the components used in manufacturing a product is

$$C = \dfrac{375{,}000 + 6x^2}{x}, \quad x \geq 1$$

where C is measured in dollars and x is the order size. Find the rate of change of C with respect to x when (a) $x = 200$, (b) $x = 250$, and (c) $x = 300$. Interpret the meaning of these values.

90. *Boyle's Law* This law states that if the temperature of a gas remains constant, its pressure is inversely proportional to its volume. Use the derivative to show that the rate of change of the pressure is inversely proportional to the square of the volume.

91. Population Growth   A population of 500 bacteria is introduced into a culture and grows in number according to the equation

$$P(t) = 500\left(1 + \frac{4t}{50 + t^2}\right)$$

where t is measured in hours. Find the rate at which the population is growing when $t = 2$.

92. Gravitational Force   Newton's Law of Universal Gravitation states that the force F between two masses, m_1 and m_2, is

$$F = \frac{Gm_1m_2}{d^2}$$

where G is a constant and d is the distance between the masses. Find an equation that gives the instantaneous rate of change of F with respect to d. (Assume m_1 and m_2 represent moving points.)

93. Prove the following differentiation rules.

(a) $\dfrac{d}{dx}[\sec x] = \sec x \tan x$ (b) $\dfrac{d}{dx}[\csc x] = -\csc x \cot x$

(c) $\dfrac{d}{dx}[\cot x] = -\csc^2 x$

94. Rate of Change   Determine whether there exist any values of x in the interval $[0, 2\pi)$ such that the rate of change of $f(x) = \sec x$ and the rate of change of $g(x) = \csc x$ are equal.

95. Modeling Data   The table shows the numbers n (in thousands) of motor homes sold in the United States and the retail values v (in billions of dollars) of these motor homes for the years 1996 through 2001. The year is represented by t, with $t = 6$ corresponding to 1996. *(Source: Recreation Vehicle Industry Association)*

Year, t	6	7	8	9	10	11
n	247.5	254.5	292.7	321.2	300.1	256.8
v	6.3	6.9	8.4	10.4	9.5	8.6

(a) Use a graphing utility to find cubic models for the number of motor homes sold $n(t)$ and the total retail value $v(t)$ of the motor homes.

(b) Graph each model found in part (a).

(c) Find $A = v(t)/n(t)$, then graph A. What does this function represent?

(d) Interpret $A'(t)$ in the context of these data.

96. Satellites   When satellites observe Earth, they can scan only part of Earth's surface. Some satellites have sensors that can measure the angle θ shown in the figure. Let h represent the satellite's distance from Earth's surface and let r represent Earth's radius.

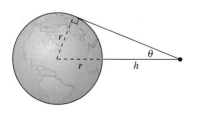

(a) Show that $h = r(\csc \theta - 1)$.

(b) Find the rate at which h is changing with respect to θ when $\theta = 30°$. (Assume $r = 3960$ miles.)

In Exercises 97–104, find the second derivative of the function.

97. $f(x) = 4x^{3/2}$

98. $f(x) = x + 32x^{-2}$

99. $f(x) = \dfrac{x}{x - 1}$

100. $f(x) = \dfrac{x^2 + 2x - 1}{x}$

101. $f(x) = 3 \sin x$

102. $f(x) = \sec x$

103. $g(x) = \dfrac{e^x}{x}$

104. $h(t) = e^t \sin t$

In Exercises 105–108, find the given higher-order derivative.

Given	*Find*
105. $f'(x) = x^2$	$f''(x)$
106. $f''(x) = 2 - \dfrac{2}{x}$	$f'''(x)$
107. $f'''(x) = 2\sqrt{x}$	$f^{(4)}(x)$
108. $f^{(4)}(x) = 2x + 1$	$f^{(6)}(x)$

Writing About Concepts

109. Sketch the graph of a differentiable function f such that $f(2) = 0$, $f' < 0$ for $-\infty < x < 2$, and $f' > 0$ for $2 < x < \infty$.

110. Sketch the graph of a differentiable function f such that $f > 0$ and $f' < 0$ for all real numbers x.

In Exercises 111–114, use the given information to find $f'(2)$.

$g(2) = 3$ and $g'(2) = -2$

$h(2) = -1$ and $h'(2) = 4$

111. $f(x) = 2g(x) + h(x)$ **112.** $f(x) = 4 - h(x)$

113. $f(x) = \dfrac{g(x)}{h(x)}$ **114.** $f(x) = g(x)h(x)$

In Exercises 115 and 116, the graphs of f, f', and f'' are shown on the same set of coordinate axes. Which is which? Explain your reasoning. To print an enlarged copy of the graph, go to the website www.mathgraphs.com.

115.

116.

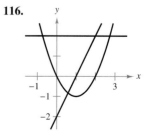

In Exercises 117–120, the graph of f is shown. Sketch the graphs of f' and f''. To print an enlarged copy of the graph, go to the website *www.mathgraphs.com*.

117.

118.

119.

120.

121. *Acceleration* The velocity of an object in meters per second is $v(t) = 36 - t^2$, $0 \le t \le 6$. Find the velocity and acceleration of the object when $t = 3$. What can be said about the speed of the object when the velocity and acceleration have opposite signs?

122. *Particle Motion* The figure shows the graphs of the position, velocity, and acceleration functions of a particle.

(a) Copy the graphs of the functions shown. Identify each graph. Explain your reasoning. To print an enlarged copy of the graph, go to the website *www.mathgraphs.com*.

(b) On your sketch, identify when the particle speeds up and when it slows down. Explain your reasoning.

Finding a Pattern In Exercises 123 and 124, develop a general rule for $f^{(n)}(x)$ given $f(x)$.

123. $f(x) = x^n$

124. $f(x) = \dfrac{1}{x}$

125. *Finding a Pattern* Consider the function $f(x) = g(x)h(x)$.

(a) Use the Product Rule to generate rules for finding $f''(x)$, $f'''(x)$, and $f^{(4)}(x)$.

(b) Use the results in part (a) to write a general rule for $f^{(n)}(x)$.

126. *Finding a Pattern* Develop a general rule for $[xf(x)]^{(n)}$, where f is a differentiable function of x.

Differential Equations In Exercises 127–130, verify that the function satisfies the differential equation.

Function	Differential Equation
127. $y = \dfrac{1}{x}$, $x > 0$	$x^3 y'' + 2x^2 y' = 0$
128. $y = 2x^3 - 6x + 10$	$-y''' - xy'' - 2y' = -24x^2$
129. $y = 2\sin x + 3$	$y'' + y = 3$
130. $y = 3\cos x + \sin x$	$y'' + y = 0$

 Linear and Quadratic Approximations The linear and quadratic approximations of a function f at $x = a$ are

$$P_1(x) = f'(a)(x - a) + f(a) \text{ and}$$

$$P_2(x) = \tfrac{1}{2}f''(a)(x - a)^2 + f'(a)(x - a) + f(a).$$

In Exercises 131 and 132, (a) find the specified linear and quadratic approximations of f, (b) use a graphing utility to graph f and the approximations, (c) determine whether P_1 or P_2 is the better approximation, and (d) state how the accuracy changes as you move farther from $x = a$.

131. $f(x) = \ln x$

$a = 1$

132. $f(x) = e^x$

$a = 0$

True or False? In Exercises 133–138, determine whether the statement is true or false. If it is false, explain why or give an example that shows it is false.

133. If $y = f(x)g(x)$, then $dy/dx = f'(x)g'(x)$.

134. If $y = (x + 1)(x + 2)(x + 3)(x + 4)$, then $d^5y/dx^5 = 0$.

135. If $f'(c)$ and $g'(c)$ are zero and $h(x) = f(x)g(x)$, then $h'(c) = 0$.

136. If $f(x)$ is an nth-degree polynomial, then $f^{(n+1)}(x) = 0$.

137. The second derivative represents the rate of change of the first derivative.

138. If the velocity of an object is constant, then its acceleration is zero.

139. Find a second-degree polynomial $f(x) = ax^2 + bx + c$ such that its graph has a tangent line with slope 10 at the point $(2, 7)$ and an x-intercept at $(1, 0)$.

140. Consider the third-degree polynomial

$$f(x) = ax^3 + bx^2 + cx + d, \quad a \ne 0.$$

Determine conditions for a, b, c, and d if the graph of f has (a) no horizontal tangents, (b) exactly one horizontal tangent, and (c) exactly two horizontal tangents. Give an example for each case.

141. Find the derivative of $f(x) = x|x|$. Does $f''(0)$ exist?

142. *Think About It* Let f and g be functions whose first and second derivatives exist on an interval I. Which of the following formulas is (are) true?

(a) $fg'' - f''g = (fg' - f'g)'$

(b) $fg'' + f''g = (fg)''$

The Chain Rule

- Find the derivative of a composite function using the Chain Rule.
- Find the derivative of a function using the General Power Rule.
- Simplify the derivative of a function using algebra.
- Find the derivative of a transcendental function using the Chain Rule.
- Find the derivative of a function involving the natural logarithmic function.
- Define and differentiate exponential functions that have bases other than e.

The Chain Rule

This text has yet to discuss one of the most powerful differentiation rules—the **Chain Rule.** This rule deals with composite functions and adds a surprising versatility to the rules discussed in the two previous sections. For example, compare the following functions. Those on the left can be differentiated without the Chain Rule, and those on the right are best done with the Chain Rule.

Without the Chain Rule	*With the Chain Rule*
$y = x^2 + 1$	$y = \sqrt{x^2 + 1}$
$y = \sin x$	$y = \sin 6x$
$y = 3x + 2$	$y = (3x + 2)^5$
$y = e^x + \tan x$	$y = e^{5x} + \tan x^2$

Basically, the Chain Rule states that if y changes dy/du times as fast as u, and u changes du/dx times as fast as x, then y changes $(dy/du)(du/dx)$ times as fast as x.

EXAMPLE 1   The Derivative of a Composite Function

A set of gears is constructed, as shown in Figure 3.25, such that the second and third gears are on the same axle. As the first axle revolves, it drives the second axle, which in turn drives the third axle. Let y, u, and x represent the numbers of revolutions per minute of the first, second, and third axles, respectively. Find dy/du, du/dx, and dy/dx, and show that

$$\frac{dy}{dx} = \frac{dy}{du} \cdot \frac{du}{dx}.$$

Solution   Because the circumference of the second gear is three times that of the first, the first axle must make three revolutions to turn the second axle once. Similarly, the second axle must make two revolutions to turn the third axle once, and you can write

$$\frac{dy}{du} = 3 \quad \text{and} \quad \frac{du}{dx} = 2.$$

Combining these two results, you know that the first axle must make six revolutions to turn the third axle once. So, you can write

$$\frac{dy}{dx} = \boxed{\begin{array}{c}\text{Rate of change of first axle}\\\text{with respect to second axle}\end{array}} \cdot \boxed{\begin{array}{c}\text{Rate of change of second axle}\\\text{with respect to third axle}\end{array}}$$

$$= \frac{dy}{du} \cdot \frac{du}{dx} = 3 \cdot 2 = 6 = \boxed{\begin{array}{c}\text{Rate of change of first axle}\\\text{with respect to third axle}\end{array}}$$

In other words, the rate of change of y with respect to x is the product of the rate of change of y with respect to u and the rate of change of u with respect to x.

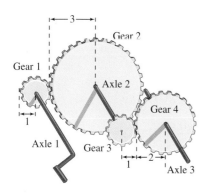

Axle 1: y revolutions per minute
Axle 2: u revolutions per minute
Axle 3: x revolutions per minute
Figure 3.25

Example 1 illustrates a simple case of the Chain Rule. The general rule is stated below.

EXPLORATION

Using the Chain Rule Each of the following functions can be differentiated using rules that you studied in Sections 3.2 and 3.3. For each function, find the derivative using those rules. Then find the derivative using the Chain Rule. Compare your results. Which method is simpler?

a. $\dfrac{2}{3x + 1}$

b. $(x + 2)^3$

c. $\sin 2x$

THEOREM 3.11 The Chain Rule

If $y = f(u)$ is a differentiable function of u and $u = g(x)$ is a differentiable function of x, then $y = f(g(x))$ is a differentiable function of x and

$$\frac{dy}{dx} = \frac{dy}{du} \cdot \frac{du}{dx}$$

or, equivalently,

$$\frac{d}{dx}[f(g(x))] = f'(g(x))g'(x).$$

Proof Let $h(x) = f(g(x))$. Then, using the alternative form of the derivative, you need to show that, for $x = c$,

$$h'(c) = f'(g(c))g'(c).$$

An important consideration in this proof is the behavior of g as x approaches c. A problem occurs if there are values of x, other than c, such that $g(x) = g(c)$. Appendix A shows how to use the differentiability of f and g to overcome this problem. For now, assume that $g(x) \neq g(c)$ for values of x other than c. In the proofs of the Product Rule and the Quotient Rule, the same quantity was added and subtracted to obtain the desired form. This proof uses a similar technique—multiplying and dividing by the same (nonzero) quantity. Note that because g is differentiable, it is also continuous, and it follows that $g(x) \to g(c)$ as $x \to c$.

$$h'(c) = \lim_{x \to c} \frac{f(g(x)) - f(g(c))}{x - c}$$

$$= \lim_{x \to c} \left[\frac{f(g(x)) - f(g(c))}{g(x) - g(c)} \cdot \frac{g(x) - g(c)}{x - c} \right], \quad g(x) \neq g(c)$$

$$= \left[\lim_{x \to c} \frac{f(g(x)) - f(g(c))}{g(x) - g(c)} \right] \left[\lim_{x \to c} \frac{g(x) - g(c)}{x - c} \right]$$

$$= f'(g(c))g'(c)$$

When applying the Chain Rule, it is helpful to think of the composite function $f \circ g$ as having two parts—an inner part and an outer part.

Outer function

$$y = f(g(x)) = f(u)$$

Inner function

The derivative of $y = f(u)$ is the derivative of the outer function (at the inner function u) *times* the derivative of the inner function.

$$y' = f'(u) \cdot u'$$

Derivative of outer function   Derivative of inner function

EXAMPLE 2   Decomposition of a Composite Function

$y = f(g(x))$	$u = g(x)$	$y = f(u)$
a. $y = \dfrac{1}{x+1}$	$u = x + 1$	$y = \dfrac{1}{u}$
b. $y = \sin 2x$	$u = 2x$	$y = \sin u$
c. $y = \sqrt{3x^2 - x + 1}$	$u = 3x^2 - x + 1$	$y = \sqrt{u}$
d. $y = \tan^2 x$	$u = \tan x$	$y = u^2$

EXAMPLE 3   Using the Chain Rule

Find dy/dx for $y = (x^2 + 1)^3$.

Solution   For this function, you can consider the inside function to be $u = x^2 + 1$. By the Chain Rule, you obtain

$$\frac{dy}{dx} = \underbrace{3(x^2+1)^2}_{\frac{dy}{du}}\underbrace{(2x)}_{\frac{du}{dx}} = 6x(x^2+1)^2.$$

STUDY TIP   You could also solve the problem in Example 3 without using the Chain Rule by observing that

$$y = x^6 + 3x^4 + 3x^2 + 1$$

and

$$y' = 6x^5 + 12x^3 + 6x.$$

Verify that this is the same result as the derivative in Example 3. Which method would you use to find

$$\frac{d}{dx}(x^2 + 1)^{50}?$$

The General Power Rule

The function in Example 3 is an example of one of the most common types of composite functions, $y = [u(x)]^n$. The rule for differentiating such functions is called the **General Power Rule**, and it is a special case of the Chain Rule.

THEOREM 3.12   The General Power Rule

If $y = [u(x)]^n$, where u is a differentiable function of x and n is a rational number, then

$$\frac{dy}{dx} = n[u(x)]^{n-1}\frac{du}{dx}$$

or, equivalently,

$$\frac{d}{dx}[u^n] = nu^{n-1}\,u'.$$

Proof   Because $y = u^n$, you apply the Chain Rule to obtain

$$\frac{dy}{dx} = \left(\frac{dy}{du}\right)\left(\frac{du}{dx}\right)$$

$$= \frac{d}{du}[u^n]\frac{du}{dx}.$$

By the (Simple) Power Rule in Section 3.2, you have $D_u[u^n] = nu^{n-1}$, and it follows that

$$\frac{dy}{dx} = n[u(x)]^{n-1}\frac{du}{dx}.$$

EXAMPLE 4 Applying the General Power Rule

Find the derivative of $f(x) = (3x - 2x^2)^3$.

Solution Let $u = 3x - 2x^2$. Then

$$f(x) = (3x - 2x^2)^3 = u^3$$

and, by the General Power Rule, the derivative is

$$
\overset{\displaystyle n}{\underset{\displaystyle |}{}}\;\overbrace{(3x - 2x^2)^2}^{u^{n-1}}\;\overbrace{\frac{d}{dx}[3x - 2x^2]}^{u'}
$$

$$f'(x) = 3(3x - 2x^2)^2 \frac{d}{dx}[3x - 2x^2] \qquad \text{Apply General Power Rule.}$$

$$= 3(3x - 2x^2)^2(3 - 4x). \qquad \text{Differentiate } 3x - 2x^2.$$

EXAMPLE 5 Differentiating Functions Involving Radicals

Find all points on the graph of $f(x) = \sqrt[3]{(x^2 - 1)^2}$ for which $f'(x) = 0$ and those for which $f'(x)$ does not exist.

Solution Begin by rewriting the function as

$$f(x) = (x^2 - 1)^{2/3}.$$

Then, applying the General Power Rule (with $u = x^2 - 1$) produces

$$
\overset{\displaystyle n}{\underset{\displaystyle |}{}}\;\overbrace{(x^2 - 1)^{-1/3}}^{u^{n-1}}\;\overbrace{(2x)}^{u'}
$$

$$f'(x) = \frac{2}{3}(x^2 - 1)^{-1/3}(2x) \qquad \text{Apply General Power Rule.}$$

$$= \frac{4x}{3\sqrt[3]{x^2 - 1}}. \qquad \text{Write in radical form.}$$

So, $f'(x) = 0$ when $x = 0$ and $f'(x)$ does not exist when $x = \pm 1$, as shown in Figure 3.26.

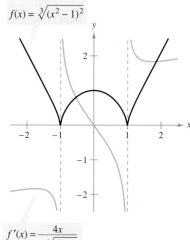

$f(x) = \sqrt[3]{(x^2 - 1)^2}$

$f'(x) = \dfrac{4x}{3\sqrt[3]{x^2 - 1}}$

The derivative of f is 0 at $x = 0$ and is undefined at $x = \pm 1$.
Figure 3.26

EXAMPLE 6 Differentiating Quotients with Constant Numerators

Differentiate $g(t) = \dfrac{-7}{(2t - 3)^2}$.

Solution Begin by rewriting the function as

$$g(t) = -7(2t - 3)^{-2}.$$

Then, applying the General Power Rule produces

$$
\overbrace{(-7)}^{\;}\;\overset{\displaystyle n}{\overbrace{(-2)}}\;\overbrace{(2t - 3)^{-3}}^{u^{n-1}}\;\overbrace{(2)}^{u'}
$$

$$g'(t) = (-7)(-2)(2t - 3)^{-3}(2) \qquad \text{Apply General Power Rule.}$$

$$\underbrace{}_{\substack{\text{Constant} \\ \text{Multiple Rule}}}$$

$$= 28(2t - 3)^{-3} \qquad \text{Simplify.}$$

$$= \frac{28}{(2t - 3)^3}. \qquad \text{Write with positive exponent.}$$

NOTE Try differentiating the function in Example 6 using the Quotient Rule. You should obtain the same result, but using the Quotient Rule is less efficient than using the General Power Rule.

Simplifying Derivatives

The next three examples illustrate some techniques for simplifying the "raw derivatives" of functions involving products, quotients, and composites.

EXAMPLE 7   **Simplifying by Factoring Out the Least Powers**

$$f(x) = x^2\sqrt{1 - x^2} \qquad \text{Original function}$$

$$= x^2(1 - x^2)^{1/2} \qquad \text{Rewrite.}$$

$$f'(x) = x^2\frac{d}{dx}\left[(1 - x^2)^{1/2}\right] + (1 - x^2)^{1/2}\frac{d}{dx}\left[x^2\right] \qquad \text{Product Rule}$$

$$= x^2\left[\frac{1}{2}(1 - x^2)^{-1/2}(-2x)\right] + (1 - x^2)^{1/2}(2x) \qquad \text{General Power Rule}$$

$$= -x^3(1 - x^2)^{-1/2} + 2x(1 - x^2)^{1/2} \qquad \text{Simplify.}$$

$$= x(1 - x^2)^{-1/2}\left[-x^2(1) + 2(1 - x^2)\right] \qquad \text{Factor.}$$

$$= \frac{x(2 - 3x^2)}{\sqrt{1 - x^2}} \qquad \text{Simplify.}$$

EXAMPLE 8   **Simplifying the Derivative of a Quotient**

$$f(x) = \frac{x}{\sqrt[3]{x^2 + 4}} \qquad \text{Original function}$$

$$= \frac{x}{(x^2 + 4)^{1/3}} \qquad \text{Rewrite.}$$

$$f'(x) = \frac{(x^2 + 4)^{1/3}(1) - x(1/3)(x^2 + 4)^{-2/3}(2x)}{(x^2 + 4)^{2/3}} \qquad \text{Quotient Rule}$$

$$= \frac{1}{3}(x^2 + 4)^{-2/3}\left[\frac{3(x^2 + 4) - (2x^2)(1)}{(x^2 + 4)^{2/3}}\right] \qquad \text{Factor.}$$

$$= \frac{x^2 + 12}{3(x^2 + 4)^{4/3}} \qquad \text{Simplify.}$$

EXAMPLE 9   **Simplifying the Derivative of a Power**

$$y = \left(\frac{3x - 1}{x^2 + 3}\right)^2 \qquad \text{Original function}$$

$$y' = 2\overbrace{\left(\frac{3x - 1}{x^2 + 3}\right)}^{u^{n-1}}\overbrace{\frac{d}{dx}\left[\frac{3x - 1}{x^2 + 3}\right]}^{u'} \qquad \text{General Power Rule}$$

$$= \left[\frac{2(3x - 1)}{x^2 + 3}\right]\left[\frac{(x^2 + 3)(3) - (3x - 1)(2x)}{(x^2 + 3)^2}\right] \qquad \text{Quotient Rule}$$

$$= \frac{2(3x - 1)(3x^2 + 9 - 6x^2 + 2x)}{(x^2 + 3)^3} \qquad \text{Multiply.}$$

$$= \frac{2(3x - 1)(-3x^2 + 2x + 9)}{(x^2 + 3)^3} \qquad \text{Simplify.}$$

Trigonometric Functions and the Chain Rule

The "Chain Rule versions" of the derivatives of the six trigonometric functions and the natural exponential function are as follows.

$$\frac{d}{dx}[\sin u] = (\cos u)\,u'$$ $$\frac{d}{dx}[\cos u] = -(\sin u)\,u'$$

$$\frac{d}{dx}[\tan u] = (\sec^2 u)\,u'$$ $$\frac{d}{dx}[\cot u] = -(\csc^2 u)\,u'$$

$$\frac{d}{dx}[\sec u] = (\sec u \tan u)\,u'$$ $$\frac{d}{dx}[\csc u] = -(\csc u \cot u)\,u'$$

$$\frac{d}{dx}[e^u] = e^u u'$$

EXAMPLE 10   Applying the Chain Rule to Transcendental Functions

NOTE   Be sure that you understand the mathematical conventions regarding parentheses and trigonometric functions. For instance, in Example 10(a), $\sin 2x$ is written to mean $\sin(2x)$.

a. $y = \sin 2x$ $\overset{\cos u}{\overbrace{}} \overset{u'}{\overbrace{}}$

$y' = \cos 2x \dfrac{d}{dx}[2x] = (\cos 2x)(2) = 2\cos 2x$

b. $y = \cos(x - 1)$ $\overset{-\sin u}{\overbrace{}} \overset{u'}{\overbrace{}}$

$y' = -\sin(x - 1)\dfrac{d}{dx}[x - 1] = -\sin(x - 1)$

c. $y = e^{3x}$ $\overset{e^u}{\overbrace{}} \overset{u'}{\overbrace{}}$

$y' = e^{3x}\dfrac{d}{dx}[3x] = 3e^{3x}$

EXAMPLE 11   Parentheses and Trigonometric Functions

a. $y = \cos 3x^2 = \cos(3x^2)$ $y' = (-\sin 3x^2)(6x) = -6x \sin 3x^2$

b. $y = (\cos 3)x^2$ $y' = (\cos 3)(2x) = 2x \cos 3$

c. $y = \cos(3x)^2 = \cos(9x^2)$ $y' = (-\sin 9x^2)(18x) = -18x \sin 9x^2$

d. $y = \cos^2 x = (\cos x)^2$ $y' = 2(\cos x)(-\sin x) = -2 \cos x \sin x$

To find the derivative of a function of the form $k(x) = f(g(h(x)))$, you need to apply the Chain Rule twice, as shown in Example 12.

EXAMPLE 12   Repeated Application of the Chain Rule

$f(t) = \sin^3 4t$ Original function

$\quad = (\sin 4t)^3$ Rewrite.

$f'(t) = 3(\sin 4t)^2 \dfrac{d}{dt}[\sin 4t]$ Apply Chain Rule once.

$\quad = 3(\sin 4t)^2(\cos 4t) \dfrac{d}{dt}[4t]$ Apply Chain Rule a second time.

$\quad = 3(\sin 4t)^2(\cos 4t)(4)$

$\quad = 12 \sin^2 4t \cos 4t$ Simplify.

Bases Other than e

The **base** of the natural exponential function is e. This "natural" base can be used to assign a meaning to a general base a.

Definition of Exponential Function to Base *a*

If a is a positive real number ($a \neq 1$) and x is any real number, then the **exponential function to the base a** is denoted by a^x and is defined by

$$a^x = e^{(\ln a)x}.$$

If $a = 1$, then $y = 1^x = 1$ is a constant function.

Logarithmic functions to bases other than e can be defined in much the same way as exponential functions to other bases are defined.

Definition of Logarithmic Function to Base *a*

If a is a positive real number ($a \neq 1$) and x is any positive real number, then the **logarithmic function to the base a** is denoted by $\log_a x$ and is defined as

$$\log_a x = \frac{1}{\ln a} \ln x.$$

To differentiate exponential and logarithmic functions to other bases, you have two options: (1) use the definitions of a^x and $\log_a x$ and differentiate using the rules for the natural exponential and logarithmic functions, or (2) use the following differentiation rules for bases other than e.

NOTE   These differentiation rules are similar to those for the natural exponential function and the natural logarithmic function. In fact, they differ only by the constant factors $\ln a$ and $1/\ln a$. This points out one reason why, for calculus, e is the most convenient base.

THEOREM 3.15   Derivatives for Bases Other than e

Let a be a positive real number ($a \neq 1$) and let u be a differentiable function of x.

1. $\dfrac{d}{dx}[a^x] = (\ln a)a^x$ **2.** $\dfrac{d}{dx}[a^u] = (\ln a)a^u \dfrac{du}{dx}$

3. $\dfrac{d}{dx}[\log_a x] = \dfrac{1}{(\ln a)x}$ **4.** $\dfrac{d}{dx}[\log_a u] = \dfrac{1}{(\ln a)u} \dfrac{du}{dx}$

Proof   By definition, $a^x = e^{(\ln a)x}$. Therefore, you can prove the first rule by letting $u = (\ln a)x$ and differentiating with base e to obtain

$$\frac{d}{dx}[a^x] = \frac{d}{dx}[e^{(\ln a)x}] = e^u \frac{du}{dx} = e^{(\ln a)x}(\ln a) = (\ln a)a^x.$$

To prove the third rule, you can write

$$\frac{d}{dx}[\log_a x] = \frac{d}{dx}\left[\frac{1}{\ln a} \ln x\right] = \frac{1}{\ln a}\left(\frac{1}{x}\right) = \frac{1}{(\ln a)x}.$$

The second and fourth rules are simply the Chain Rule versions of the first and third rules.

EXAMPLE 16 Differentiating Functions to Other Bases

Find the derivative of each of the following.

a. $y = 2^x$ **b.** $y = 2^{3x}$ **c.** $y = \log_{10} \cos x$

Solution

a. $y' = \dfrac{d}{dx}[2^x] = (\ln 2)2^x$

b. $y' = \dfrac{d}{dx}[2^{3x}] = (\ln 2)2^{3x}(3) = (3 \ln 2)2^{3x}$

STUDY TIP To become skilled at differentiation, you should memorize each rule. As an aid to memorization, note that the cofunctions (cosine, cotangent, and cosecant) require a negative sign as part of their derivatives.

Try writing 2^{3x} as 8^x and differentiating to see that you obtain the same result.

c. $y' = \dfrac{d}{dx}[\log_{10} \cos x] = \dfrac{-\sin x}{(\ln 10) \cos x} = -\dfrac{1}{\ln 10} \tan x$

This section conludes with a summary of the differentiation rules studied so far.

Summary of Differentiation Rules

General Differentiation Rules

Let u and v be differentiable functions of x.

Constant Rule:

$\dfrac{d}{dx}[c] = 0$

(Simple) Power Rule:

$\dfrac{d}{dx}[x^n] = nx^{n-1} \quad \dfrac{d}{dx}[x] = 1$

Constant Multiple Rule:

$\dfrac{d}{dx}[cu] = cu'$

Sum or Difference Rule:

$\dfrac{d}{dx}[u \pm v] = u' \pm v'$

Product Rule:

$\dfrac{d}{dx}[uv] = uv' + vu'$

Quotient Rule:

$\dfrac{d}{dx}\left[\dfrac{u}{v}\right] = \dfrac{vu' - uv'}{v^2}$

Chain Rule:

$\dfrac{d}{dx}[f(u)] = f'(u)\, u'$

General Power Rule:

$\dfrac{d}{dx}[u^n] = nu^{n-1}\, u'$

Derivatives of Trigonometric Functions

$\dfrac{d}{dx}[\sin x] = \cos x$ $\dfrac{d}{dx}[\tan x] = \sec^2 x$ $\dfrac{d}{dx}[\sec x] = \sec x \tan x$

$\dfrac{d}{dx}[\cos x] = -\sin x$ $\dfrac{d}{dx}[\cot x] = -\csc^2 x$ $\dfrac{d}{dx}[\csc x] = -\csc x \cot x$

Derivatives of Exponential and Logarithmic Functions

$\dfrac{d}{dx}[e^x] = e^x$ $\dfrac{d}{dx}[\ln x] = \dfrac{1}{x}$

$\dfrac{d}{dx}[a^x] = (\ln a)a^x$ $\dfrac{d}{dx}[\log_a x] = \dfrac{1}{(\ln a)x}$

Exercises for Section 3.4

See www.CalcChat.com for worked-out solutions to odd-numbered exercises.

In Exercises 1–8, complete the table using Example 2 as a model.

$y = f(g(x))$	$u = g(x)$	$y = f(u)$
1. $y = (6x - 5)^4$		
2. $y = \dfrac{1}{\sqrt{x + 2}}$		
3. $y = \sqrt{x^2 - 1}$		
4. $y = 3\tan(\pi x^2)$		
5. $y = \csc^3 x$		
6. $y = \cos\dfrac{3x}{2}$		
7. $y = e^{-2x}$		
8. $y = (\ln x)^3$		

In Exercises 9–36, find the derivative of the function.

9. $y = (2x - 7)^3$

10. $y = (2x^3 + 1)^2$

11. $g(x) = 3(4 - 9x)^4$

12. $y = 3(5 - x^2)^5$

13. $f(x) = (9 - x^2)^{2/3}$

14. $f(t) = (9t + 7)^{2/3}$

15. $f(t) = \sqrt{1 - t}$

16. $g(x) = \sqrt{5 - 3x}$

17. $y = \sqrt[3]{9x^2 + 4}$

18. $g(x) = \sqrt{x^2 - 2x + 1}$

19. $y = 2\sqrt[4]{4 - x^2}$

20. $f(x) = -3\sqrt[4]{2 - 9x}$

21. $y = \dfrac{1}{x - 2}$

22. $s(t) = \dfrac{1}{t^2 + 3t - 1}$

23. $f(t) = \left(\dfrac{1}{t - 3}\right)^2$

24. $y = -\dfrac{8}{(t + 3)^3}$

25. $y = \dfrac{1}{\sqrt{x + 2}}$

26. $g(t) = \sqrt{\dfrac{1}{t^2 - 2}}$

27. $f(x) = x^2(x - 2)^4$

28. $f(x) = x(3x - 7)^3$

29. $y = x\sqrt{1 - x^2}$

30. $y = \frac{1}{2}x^2\sqrt{16 - x^2}$

31. $y = \dfrac{x}{\sqrt{x^2 + 1}}$

32. $y = \dfrac{x}{\sqrt{x^4 + 2}}$

33. $g(x) = \left(\dfrac{x + 5}{x^2 + 2}\right)^2$

34. $h(t) = \left(\dfrac{t^2}{t^3 + 2}\right)^2$

35. $f(v) = \left(\dfrac{1 - 2v}{1 + v}\right)^3$

36. $g(x) = \left(\dfrac{3x^2 - 1}{2x + 5}\right)^3$

In Exercises 37–46, use a computer algebra system to find the derivative of the function. Then use the utility to graph the function and its derivative on the same set of coordinate axes. Describe the behavior of the function that corresponds to any zeros of the graph of the derivative.

37. $y = \dfrac{\sqrt{x + 1}}{x^2 + 1}$

38. $y = \sqrt{\dfrac{2x}{x + 1}}$

39. $g(t) = \dfrac{3t^2}{\sqrt{t^2 + 2t - 1}}$

40. $f(x) = \sqrt{x}(2 - x)^2$

41. $y = \sqrt{\dfrac{x + 1}{x}}$

42. $y = (t^2 - 9)\sqrt{t + 2}$

43. $s(t) = \dfrac{-2(2 - t)\sqrt{1 + t}}{3}$

44. $g(x) = \sqrt{x - 1} + \sqrt{x + 1}$

45. $y = \dfrac{\cos\pi x + 1}{x}$

46. $y = x^2\tan\dfrac{1}{x}$

In Exercises 47 and 48, find the slope of the tangent line to the sine function at the origin. Compare this value with the number of complete cycles in the interval $[0, 2\pi]$. What can you conclude about the slope of the sine function $\sin ax$ at the origin?

47. (a)

$y = \sin x$

(b)

$y = \sin 2x$

48. (a)

$y = \sin 3x$

(b)

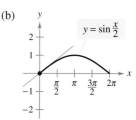

$y = \sin\dfrac{x}{2}$

In Exercises 49 and 50, find the slope of the tangent line to the graph of the function at the point $(0, 1)$.

49. (a) $y = e^{3x}$

(b) $y = e^{-3x}$

$(0, 1)$

50. (a) $y = e^{2x}$

(b) $y = e^{-2x}$

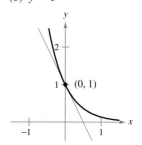

$(0, 1)$

In Exercises 51–54, find the slope of the tangent line to the graph of the logarithmic function at the point (1, 0).

51. $y = \ln x^3$

52. $y = \ln x^{3/2}$

53. $y = \ln x^2$

54. $y = \ln x^{1/2}$

In Exercises 55–100, find the derivative of the function.

55. $y = \cos 3x$

56. $y = \sin \pi x$

57. $g(x) = 3 \tan 4x$

58. $h(x) = \sec x^3$

59. $f(\theta) = \frac{1}{4} \sin^2 2\theta$

60. $g(t) = 5 \cos^3 \pi t$

61. $y = \sqrt{x} + \frac{1}{4} \sin(2x)^2$

62. $y = 3x - 5 \cos(2x)^2$

63. $y = \sin(\cos x)$

64. $y = \sin \sqrt[3]{x} + \sqrt[3]{\sin x}$

65. $f(x) = e^{2x}$

66. $y = e^{-x^2}$

67. $y = e^{\sqrt{x}}$

68. $y = x^2 e^{-x}$

69. $g(t) = (e^{-t} + e^t)^3$

70. $g(t) = e^{-3/t^2}$

71. $y = \ln(e^{x^2})$

72. $y = \ln\left(\dfrac{1 + e^x}{1 - e^x}\right)$

73. $y = \dfrac{2}{e^x + e^{-x}}$

74. $y = \dfrac{e^x - e^{-x}}{2}$

75. $y = x^2 e^x - 2xe^x + 2e^x$

76. $y = xe^x - e^x$

77. $f(x) = e^{-x} \ln x$

78. $f(x) = e^3 \ln x$

79. $y = e^x(\sin x + \cos x)$

80. $y = \ln e^x$

81. $g(x) = \ln x^2$

82. $h(x) = \ln(2x^2 + 3)$

83. $y = (\ln x)^4$

84. $y = x \ln x$

85. $y = \ln\left(x \sqrt{x^2 - 1}\right)$

86. $y = \ln \sqrt{x^2 - 9}$

87. $f(x) = \ln\left(\dfrac{x}{x^2 + 1}\right)$

88. $f(x) = \ln\left(\dfrac{2x}{x + 3}\right)$

89. $g(t) = \dfrac{\ln t}{t^2}$

90. $h(t) = \dfrac{\ln t}{t}$

91. $y = \ln \sqrt{\dfrac{x + 1}{x - 1}}$

92. $y = \ln \sqrt[3]{\dfrac{x - 2}{x + 2}}$

93. $y = \dfrac{-\sqrt{x^2 + 1}}{x} + \ln\left(x + \sqrt{x^2 + 1}\right)$

94. $y = \dfrac{-\sqrt{x^2 + 4}}{2x^2} - \dfrac{1}{4} \ln\left(\dfrac{2 + \sqrt{x^2 + 4}}{x}\right)$

95. $y = \ln|\sin x|$

96. $y = \ln|\csc x|$

97. $y = \ln\left|\dfrac{\cos x}{\cos x - 1}\right|$

98. $y = \ln|\sec x + \tan x|$

99. $y = \ln\left|\dfrac{-1 + \sin x}{2 + \sin x}\right|$

100. $y = \ln \sqrt{1 + \sin^2 x}$

In Exercises 101–106, find the second derivative of the function.

101. $f(x) = 2(x^2 - 1)^3$

102. $f(x) = \dfrac{1}{x - 2}$

103. $f(x) = \sin x^2$

104. $f(x) = \sec^2 \pi x$

105. $f(x) = (3 + 2x)e^{-3x}$

106. $g(x) = \sqrt{x} + e^x \ln x$

In Exercises 107–114, evaluate the derivative of the function at the indicated point. Use a graphing utility to verify your result.

Function	Point
107. $s(t) = \sqrt{t^2 + 2t + 8}$	$(2, 4)$
108. $y = \sqrt[5]{3x^3 + 4x}$	$(2, 2)$
109. $f(x) = \dfrac{3}{x^3 - 4}$	$\left(-1, -\dfrac{3}{5}\right)$
110. $f(x) = \dfrac{1}{(x^2 - 3x)^2}$	$\left(4, \dfrac{1}{16}\right)$
111. $f(t) = \dfrac{3t + 2}{t - 1}$	$(0, -2)$
112. $f(x) = \dfrac{x + 1}{2x - 3}$	$(2, 3)$
113. $y = 37 - \sec^3(2x)$	$(0, 36)$
114. $y = \dfrac{1}{x} + \sqrt{\cos x}$	$\left(\dfrac{\pi}{2}, \dfrac{2}{\pi}\right)$

In Exercises 115–122, (a) find an equation of the tangent line to the graph of f at the indicated point, (b) use a graphing utility to graph the function and its tangent line at the point, and (c) use the derivative feature of a graphing utility to confirm your results.

Function	Point
115. $f(x) = \sqrt{3x^2 - 2}$	$(3, 5)$
116. $f(x) = \frac{1}{3}x \sqrt{x^2 + 5}$	$(2, 2)$
117. $f(x) = \sin 2x$	$(\pi, 0)$
118. $y = \cos 3x$	$\left(\dfrac{\pi}{4}, -\dfrac{\sqrt{2}}{2}\right)$
119. $y = 2 \tan^3 x$	$\left(\dfrac{\pi}{4}, 2\right)$
120. $f(x) = \tan^2 x$	$\left(\dfrac{\pi}{4}, 1\right)$
121. $y = 4 - x^2 - \ln(\frac{1}{2}x + 1)$	$(0, 4)$
122. $y = 2e^{1 - x^2}$	$(1, 2)$

In Exercises 123–138, find the derivative of the function.

123. $f(x) = 4^x$

124. $g(x) = 5^{-x}$

125. $y = 5^{x-2}$

126. $y = x(6^{-2x})$

127. $g(t) = t^2 2^t$

128. $f(t) = \dfrac{3^{2t}}{t}$

129. $h(\theta) = 2^{-\theta} \cos \pi\theta$

130. $g(\alpha) = 5^{-\alpha/2} \sin 2\alpha$

131. $y = \log_3 x$

132. $y = \log_{10} 2x$

133. $f(x) = \log_2 \dfrac{x^2}{x-1}$

134. $h(x) = \log_3 \dfrac{x\sqrt{x-1}}{2}$

135. $y = \log_5 \sqrt{x^2 - 1}$

136. $y = \log_{10} \dfrac{x^2 - 1}{x}$

137. $g(t) = \dfrac{10 \log_4 t}{t}$

138. $f(t) = t^{3/2} \log_2 \sqrt{t+1}$

Writing About Concepts

In Exercises 139–142, the graphs of a function f and its derivative f' are shown. Label the graphs as f or f' and write a short paragraph stating the criteria used in making the selection. To print an enlarged copy of the graph, go to the website *www.mathgraphs.com*.

139.

140.

141.

142.

In Exercises 143 and 144, the relationship between f and g is given. Explain the relationship between f' and g'.

143. $g(x) = f(3x)$

144. $g(x) = f(x^2)$

145. Given that $g(5) = -3$, $g'(5) = 6$, $h(5) = 3$, and $h'(5) = -2$, find $f'(5)$ (if possible) for each of the following. If it is not possible, state what additional information is required.

(a) $f(x) = g(x)h(x)$

(b) $f(x) = g(h(x))$

(c) $f(x) = \dfrac{g(x)}{h(x)}$

(d) $f(x) = [g(x)]^3$

146. (a) Find the derivative of the function $g(x) = \sin^2 x + \cos^2 x$ in two ways.

(b) For $f(x) = \sec^2 x$ and $g(x) = \tan^2 x$, show that
$$f'(x) = g'(x).$$

 In Exercises 147–150, (a) use a graphing utility to find the derivative of the function at the given point, (b) find an equation of the tangent line to the graph of the function at the given point, and (c) use the utility to graph the function and its tangent line in the same viewing window.

147. $g(t) = \dfrac{3t^2}{\sqrt{t^2 + 2t - 1}}$, $\left(\dfrac{1}{2}, \dfrac{3}{2}\right)$

148. $f(x) = \sqrt{x}(2-x)^2$, $(4, 8)$

149. $s(t) = \dfrac{(4-2t)\sqrt{1+t}}{3}$, $\left(0, \dfrac{4}{3}\right)$

150. $y = (t^2 - 9)\sqrt{t+2}$, $(2, -10)$

 Famous Curves **In Exercises 151 and 152, find an equation of the tangent line to the graph at the given point. Then use a graphing utility to graph the function and its tangent line in the same viewing window.**

151. Top half of circle

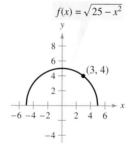

$f(x) = \sqrt{25 - x^2}$

$(3, 4)$

152. Bullet-nose curve

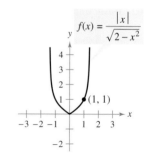

$f(x) = \dfrac{|x|}{\sqrt{2 - x^2}}$

$(1, 1)$

153. *Horizontal Tangent Line* Determine the point(s) in the interval $(0, 2\pi)$ at which the graph of $f(x) = 2 \cos x + \sin 2x$ has a horizontal tangent line.

154. *Horizontal Tangent Line* Determine the point(s) at which the graph of $f(x) = \dfrac{x}{\sqrt{2x - 1}}$ has a horizontal tangent line.

In Exercises 155–158, evaluate the second derivative of the function at the given point. Use a computer algebra system to verify your result.

155. $h(x) = \frac{1}{9}(3x + 1)^3$, $\left(1, \dfrac{64}{9}\right)$

156. $f(x) = \dfrac{1}{\sqrt{x + 4}}$, $\left(0, \dfrac{1}{2}\right)$

157. $f(x) = \cos(x^2)$, $(0, 1)$

158. $g(t) = \tan 2t$, $\left(\dfrac{\pi}{6}, \sqrt{3}\right)$

159. Doppler Effect The frequency F of a fire truck siren heard by a stationary observer is

$$F = \frac{132{,}400}{331 \pm v}$$

where $\pm v$ represents the velocity of the accelerating fire truck in meters per second. Find the rate of change of F with respect to v when

(a) the fire truck is approaching at a velocity of 30 meters per second (use $-v$).

(b) the fire truck is moving away at a velocity of 30 meters per second (use $+v$).

160. Harmonic Motion The displacement from equilibrium of an object in harmonic motion on the end of a spring is

$$y = \tfrac{1}{3}\cos 12t - \tfrac{1}{4}\sin 12t$$

where y is measured in feet and t is the time in seconds. Determine the position and velocity of the object when $t = \pi/8$.

161. Pendulum A 15-centimeter pendulum moves according to the equation $\theta = 0.2 \cos 8t$, where θ is the angular displacement from the vertical in radians and t is the time in seconds. Determine the maximum angular displacement and the rate of change of θ when $t = 3$ seconds.

162. Wave Motion A buoy oscillates in simple harmonic motion $y = A \cos \omega t$ as waves move past it. The buoy moves a total of 3.5 feet (vertically) from its low point to its high point. It returns to its high point every 10 seconds.

(a) Write an equation describing the motion of the buoy if it is at its high point at $t = 0$.

(b) Determine the velocity of the buoy as a function of t.

163. Circulatory System The speed S of blood that is r centimeters from the center of an artery is

$$S = C(R^2 - r^2)$$

where C is a constant, R is the radius of the artery, and S is measured in centimeters per second. Suppose a drug is administered and the artery begins to dilate at a rate of dR/dt. At a constant distance r, find the rate at which S changes with respect to t for $C = 1.76 \times 10^5$, $R = 1.2 \times 10^{-2}$, and $dR/dt = 10^{-5}$.

164. Modeling Data The normal daily maximum temperatures T (in degrees Fahrenheit) for Denver, Colorado, are shown in the table. (*Source: National Oceanic and Atmospheric Administration*)

Month	Jan	Feb	Mar	Apr	May	Jun
Temperature	43.2	47.2	53.7	60.9	70.5	82.1

Month	Jul	Aug	Sep	Oct	Nov	Dec
Temperature	88.0	86.0	77.4	66.0	51.5	44.1

(a) Use a graphing utility to plot the data and find a model for the data of the form

$$T(t) = a + b \sin(\pi t/6 - c)$$

where T is the temperature and t is the time in months, with $t = 1$ corresponding to January.

(b) Use a graphing utility to graph the model. How well does the model fit the data?

(c) Find T' and use a graphing utility to graph the derivative.

(d) Based on the graph of the derivative, during what times does the temperature change most rapidly? Most slowly? Do your answers agree with your observations of the temperature changes? Explain.

165. Volume Air is being pumped into a spherical balloon so that the radius is increasing at the rate of $dr/dt = 3$ inches per second. What is the rate of change of the volume of the balloon, in cubic inches per second, when $r = 8$ inches? $\left[Hint:\ V = \tfrac{4}{3}\pi r^3 \right]$

166. Think About It The table shows some values of the derivative of an unknown function f. Complete the table by finding (if possible) the derivative of each transformation of f.

(a) $g(x) = f(x) - 2$ (b) $h(x) = 2f(x)$

(c) $r(x) = f(-3x)$ (d) $s(x) = f(x + 2)$

x	-2	-1	0	1	2	3
$f'(x)$	4	$\tfrac{2}{3}$	$-\tfrac{1}{3}$	-1	-2	-4
$g'(x)$						
$h'(x)$						
$r'(x)$						
$s'(x)$						

167. Modeling Data The table shows the temperature T (°F) at which water boils at selected pressures p (pounds per square inch). (*Source: Standard Handbook of Mechanical Engineers*)

p	5	10	14.696 (1 atm)	20
T	162.24°	193.21°	212.00°	227.96°

p	30	40	60	80	100
T	250.33°	267.25°	292.71°	312.03°	327.81°

A model that approximates the data is

$$T = 87.97 + 34.96 \ln p + 7.91 \sqrt{p}.$$

(a) Use a graphing utility to plot the data and graph the model.

(b) Find the rate of change of T with respect to p when $p = 10$ and $p = 70$.

168. Depreciation   After t years, the value of a car purchased for $20,000 is

$$V(t) = 20,000\left(\tfrac{3}{4}\right)^t.$$

 (a) Use a graphing utility to graph the function and determine the value of the car 2 years after it was purchased.

(b) Find the rate of change of V with respect to t when $t = 1$ and $t = 4$.

169. Inflation   If the annual rate of inflation averages 5% over the next 10 years, the approximate cost C of goods or services during any year in that decade is $C(t) = P(1.05)^t$, where t is the time in years and P is the present cost.

(a) If the price of an oil change for your car is presently $24.95, estimate the price 10 years from now.

(b) Find the rate of change of C with respect to t when $t = 1$ and $t = 8$.

(c) Verify that the rate of change of C is proportional to C. What is the constant of proportionality?

170. Finding a Pattern   Consider the function $f(x) = \sin \beta x$, where β is a constant.

(a) Find the first-, second-, third-, and fourth-order derivatives of the function.

(b) Verify that the function and its second derivative satisfy the equation $f''(x) + \beta^2 f(x) = 0$.

(c) Use the results in part (a) to write general rules for the even- and odd-order derivatives

$$f^{(2k)}(x) \text{ and } f^{(2k-1)}(x).$$

[*Hint:* $(-1)^k$ is positive if k is even and negative if k is odd.]

171. Conjecture   Let f be a differentiable function of period p.

(a) Is the function f' periodic? Verify your answer.

(b) Consider the function $g(x) = f(2x)$. Is the function $g'(x)$ periodic? Verify your answer.

172. Think About It   Let $r(x) = f(g(x))$ and $s(x) = g(f(x))$, where f and g are shown in the figure. Find (a) $r'(1)$ and (b) $s'(4)$.

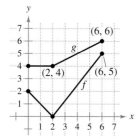

173. (a) Show that the derivative of an odd function is even. That is, if $f(-x) = -f(x)$, then $f'(-x) = f'(x)$.

(b) Show that the derivative of an even function is odd. That is, if $f(-x) = f(x)$, then $f'(-x) = -f'(x)$.

174. Let u be a differentiable function of x. Use the fact that $|u| = \sqrt{u^2}$ to prove that

$$\frac{d}{dx}[|u|] = u' \frac{u}{|u|}, \quad u \neq 0.$$

In Exercises 175–178, use the result of Exercise 174 to find the derivative of the function.

175. $g(x) = |2x - 3|$

176. $f(x) = |x^2 - 4|$

177. $h(x) = |x| \cos x$

178. $f(x) = |\sin x|$

 Linear and Quadratic Approximations   The linear and quadratic approximations of a function f at $x = a$ are

$$P_1(x) = f'(a)(x - a) + f(a) \text{ and }$$

$$P_2(x) = \tfrac{1}{2}f''(a)(x - a)^2 + f'(a)(x - a) + f(a).$$

In Exercises 179–182, (a) find the specified linear and quadratic approximations of f, (b) use a graphing utility to graph f and the approximations, (c) determine whether P_1 or P_2 is the better approximation, and (d) state how the accuracy changes as you move farther from $x = a$.

179. $f(x) = \tan \dfrac{\pi x}{4}$

$a = 1$

180. $f(x) = \sec 2x$

$a = \dfrac{\pi}{6}$

181. $f(x) = e^{-x^2/2}$

$a = 0$

182. $f(x) = x \ln x$

$a = 1$

True or False?   **In Exercises 183–185, determine whether the statement is true or false. If it is false, explain why or give an example that shows it is false.**

183. If $y = (1 - x)^{1/2}$, then $y' = \tfrac{1}{2}(1 - x)^{-1/2}$.

184. If $f(x) = \sin^2(2x)$, then $f'(x) = 2(\sin 2x)(\cos 2x)$.

185. If y is a differentiable function of u, u is a differentiable function of v, and v is a differentiable function of x, then

$$\frac{dy}{dx} = \frac{dy}{du}\frac{du}{dv}\frac{dv}{dx}.$$

Putnam Exam Challenge

186. Let $f(x) = a_1 \sin x + a_2 \sin 2x + \cdots + a_n \sin nx$, where $a_1, a_2, \ldots, a_n$ are real numbers and where n is a positive integer. Given that $|f(x)| \leq |\sin x|$ for all real x, prove that $|a_1 + 2a_2 + \cdots + na_n| \leq 1$.

187. Let k be a fixed positive integer. The nth derivative of $\dfrac{1}{x^k - 1}$ has the form

$$\frac{P_n(x)}{(x^k - 1)^{n+1}}$$

where $P_n(x)$ is a polynomial. Find $P_n(1)$.

These problems were composed by the Committee on the Putnam Prize Competition. © The Mathematical Association of America. All rights reserved.

Implicit Differentiation

- Distinguish between functions written in implicit form and explicit form.
- Use implicit differentiation to find the derivative of a function.
- Find derivatives of functions using logarithmic differentiation.

Implicit and Explicit Functions

Up to this point in the text, most functions have been expressed in **explicit form**. For example, in the equation

$$y = 3x^2 - 5 \qquad \text{Explicit form}$$

the variable y is explicitly written as a function of x. Some functions, however, are only *implied* by an equation. For instance, the function $y = 1/x$ is defined **implicitly** by the equation $xy = 1$. Suppose you were asked to find dy/dx for this equation. You could begin by writing y explicitly as a function of x and then differentiating.

Implicit Form	Explicit Form	Derivative
$xy = 1$	$y = \dfrac{1}{x} = x^{-1}$	$\dfrac{dy}{dx} = -x^{-2} = -\dfrac{1}{x^2}$

This strategy works whenever you can solve for the function explicitly. You cannot, however, use this procedure when you are unable to solve for y as a function of x. For instance, how would you find dy/dx for the equation $x^2 - 2y^3 + 4y = 2$, where it is very difficult to express y as a function of x explicitly? To do this, you can use **implicit differentiation**.

To understand how to find dy/dx implicitly, you must realize that the differentiation is taking place *with respect to x*. This means that when you differentiate terms involving x alone, you can differentiate as usual. However, when you differentiate terms involving y, you must apply the Chain Rule, because you are assuming that y is defined implicitly as a differentiable function of x.

EXAMPLE 1 Differentiating with Respect to x

a. $\dfrac{d}{dx}[x^3] = 3x^2$ Variables agree: Use Simple Power Rule.

    Variables agree

b. $\dfrac{d}{dx}[y^3] = 3y^2\dfrac{dy}{dx}$ Variables disagree: Use Chain Rule.

    Variables disagree

c. $\dfrac{d}{dx}[x + 3y] = 1 + 3\dfrac{dy}{dx}$ Chain Rule: $\dfrac{d}{dx}[3y] = 3y'$

d. $\dfrac{d}{dx}[xy^2] = x\dfrac{d}{dx}[y^2] + y^2\dfrac{d}{dx}[x]$ Product Rule

$\qquad\qquad = x\left(2y\dfrac{dy}{dx}\right) + y^2(1)$ Chain Rule

$\qquad\qquad = 2xy\dfrac{dy}{dx} + y^2$ Simplify.

Implicit Differentiation

> **Guidelines for Implicit Differentiation**
>
> 1. Differentiate both sides of the equation *with respect to x*.
> 2. Collect all terms involving dy/dx on the left side of the equation and move all other terms to the right side of the equation.
> 3. Factor dy/dx out of the left side of the equation.
> 4. Solve for dy/dx by dividing both sides of the equation by the left-hand factor that does not contain dy/dx.

EXAMPLE 2   Implicit Differentiation

Find dy/dx given that $y^3 + y^2 - 5y - x^2 = -4$.

Solution

1. Differentiate both sides of the equation with respect to x.

$$\frac{d}{dx}[y^3 + y^2 - 5y - x^2] = \frac{d}{dx}[-4]$$

$$\frac{d}{dx}[y^3] + \frac{d}{dx}[y^2] - \frac{d}{dx}[5y] - \frac{d}{dx}[x^2] = \frac{d}{dx}[-4]$$

$$3y^2\frac{dy}{dx} + 2y\frac{dy}{dx} - 5\frac{dy}{dx} - 2x = 0$$

NOTE   In Example 2, note that implicit differentiation can produce an expression for dy/dx that contains both x and y.

2. Collect the dy/dx terms on the left side of the equation.

$$3y^2\frac{dy}{dx} + 2y\frac{dy}{dx} - 5\frac{dy}{dx} = 2x$$

3. Factor dy/dx out of the left side of the equation.

$$\frac{dy}{dx}(3y^2 + 2y - 5) = 2x$$

4. Solve for dy/dx by dividing by $(3y^2 + 2y - 5)$.

$$\frac{dy}{dx} = \frac{2x}{3y^2 + 2y - 5}$$

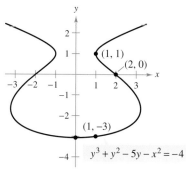

Point on Graph	Slope of Graph
$(2, 0)$	$-\frac{4}{5}$
$(1, -3)$	$\frac{1}{8}$
$x = 0$	0
$(1, 1)$	Undefined

The implicit equation

$$y^3 + y^2 - 5y - x^2 = -4$$

has the derivative

$$\frac{dy}{dx} = \frac{2x}{3y^2 + 2y - 5}.$$

Figure 3.27

To see how you can use an *implicit derivative*, consider the graph shown in Figure 3.27. From the graph, you can see that y is not a function of x. Even so, the derivative found in Example 2 gives a formula for the slope of the tangent line at a point on this graph. The slopes at several points on the graph are shown below the graph.

TECHNOLOGY   With most graphing utilities, it is easy to graph an equation that explicitly represents y as a function of x. Graphing other equations, however, can require some ingenuity. For instance, to graph the equation given in Example 2, use a graphing utility, set in *parametric* mode, to graph the parametric representations $x = \sqrt{t^3 + t^2 - 5t + 4}$, $y = t$, and $x = -\sqrt{t^3 + t^2 - 5t + 4}$, $y = t$, for $-5 \le t \le 5$. How does the result compare with the graph shown in Figure 3.27?

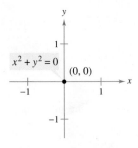

$x^2 + y^2 = 0$

(0, 0)

(a)

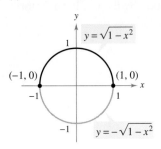

$y = \sqrt{1 - x^2}$

$(-1, 0)$ $(1, 0)$

$y = -\sqrt{1 - x^2}$

(b)

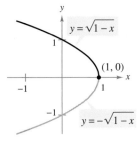

$y = \sqrt{1 - x}$

(1, 0)

$y = -\sqrt{1 - x}$

(c)

Some graph segments can be represented by differentiable functions.

Figure 3.28

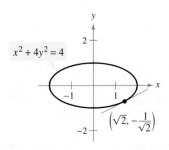

$x^2 + 4y^2 = 4$

$\left(\sqrt{2}, -\dfrac{1}{\sqrt{2}}\right)$

Figure 3.29

It is meaningless to solve for dy/dx in an equation that has no solution points. (For example, $x^2 + y^2 = -4$ has no solution points.) If, however, a segment of a graph can be represented by a differentiable function, dy/dx will have meaning as the slope at each point on the segment. Recall that a function is not differentiable at (1) points with vertical tangents and (2) points at which the function is not continuous.

EXAMPLE 3 Representing a Graph by Differentiable Functions

If possible, represent y as a differentiable function of x (see Figure 3.28).

a. $x^2 + y^2 = 0$ **b.** $x^2 + y^2 = 1$ **c.** $x + y^2 = 1$

Solution

a. The graph of this equation is a single point. So, the equation does not define y as a differentiable function of x.

b. The graph of this equation is the unit circle, centered at $(0, 0)$. The upper semicircle is given by the differentiable function

$$y = \sqrt{1 - x^2}, \quad -1 < x < 1$$

and the lower semicircle is given by the differentiable function

$$y = -\sqrt{1 - x^2}, \quad -1 < x < 1.$$

At the points $(-1, 0)$ and $(1, 0)$, the slope of the graph is undefined.

c. The upper half of this parabola is given by the differentiable function

$$y = \sqrt{1 - x}, \quad x < 1$$

and the lower half of this parabola is given by the differentiable function

$$y = -\sqrt{1 - x}, \quad x < 1.$$

At the point $(1, 0)$, the slope of the graph is undefined.

EXAMPLE 4 Finding the Slope of a Graph Implicitly

Determine the slope of the tangent line to the graph of

$$x^2 + 4y^2 = 4$$

at the point $\left(\sqrt{2}, -1/\sqrt{2}\right)$. See Figure 3.29.

Solution

$x^2 + 4y^2 = 4$	Write original equation.
$2x + 8y\dfrac{dy}{dx} = 0$	Differentiate with respect to x.
$\dfrac{dy}{dx} = \dfrac{-2x}{8y} = \dfrac{-x}{4y}$	Solve for $\dfrac{dy}{dx}$.

So, at $\left(\sqrt{2}, -1/\sqrt{2}\right)$, the slope is

$$\frac{dy}{dx} = \frac{-\sqrt{2}}{-4/\sqrt{2}} = \frac{1}{2}. \qquad \text{Evaluate } \frac{dy}{dx} \text{ when } x = \sqrt{2} \text{ and } y = -\frac{1}{\sqrt{2}}.$$

NOTE To see the benefit of implicit differentiation, try doing Example 4 using the explicit function $y = -\frac{1}{2}\sqrt{4 - x^2}$.

EXAMPLE 5   Finding the Slope of a Graph Implicitly

Determine the slope of the graph of $3(x^2 + y^2)^2 = 100xy$ at the point $(3, 1)$.

Solution

$$\frac{d}{dx}[3(x^2 + y^2)^2] = \frac{d}{dx}[100xy]$$

$$3(2)(x^2 + y^2)\left(2x + 2y\frac{dy}{dx}\right) = 100\left[x\frac{dy}{dx} + y(1)\right]$$

$$12y(x^2 + y^2)\frac{dy}{dx} - 100x\frac{dy}{dx} = 100y - 12x(x^2 + y^2)$$

$$[12y(x^2 + y^2) - 100x]\frac{dy}{dx} = 100y - 12x(x^2 + y^2)$$

$$\frac{dy}{dx} = \frac{100y - 12x(x^2 + y^2)}{-100x + 12y(x^2 + y^2)}$$

$$= \frac{25y - 3x(x^2 + y^2)}{-25x + 3y(x^2 + y^2)}$$

At the point $(3, 1)$, the slope of the graph is

$$\frac{dy}{dx} = \frac{25(1) - 3(3)(3^2 + 1^2)}{-25(3) + 3(1)(3^2 + 1^2)} = \frac{25 - 90}{-75 + 30} = \frac{-65}{-45} = \frac{13}{9}$$

as shown in Figure 3.30. This graph is called a **lemniscate.**

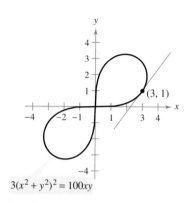

$3(x^2 + y^2)^2 = 100xy$

Lemniscate
Figure 3.30

EXAMPLE 6   Determining a Differentiable Function

Find dy/dx implicitly for the equation $\sin y = x$. Then find the largest interval of the form $-a < y < a$ on which y is a differentiable function of x (see Figure 3.31).

Solution

$$\frac{d}{dx}[\sin y] = \frac{d}{dx}[x]$$

$$\cos y\frac{dy}{dx} = 1$$

$$\frac{dy}{dx} = \frac{1}{\cos y}$$

The largest interval about the origin for which y is a differentiable function of x is $-\pi/2 < y < \pi/2$. To see this, note that $\cos y$ is positive for all y in this interval and is 0 at the endpoints. If you restrict y to the interval $-\pi/2 < y < \pi/2$, you should be able to write dy/dx explicitly as a function of x. To do this, you can use

$$\cos y = \sqrt{1 - \sin^2 y}$$

$$= \sqrt{1 - x^2}, \quad -\frac{\pi}{2} < y < \frac{\pi}{2}$$

and conclude that

$$\frac{dy}{dx} = \frac{1}{\sqrt{1 - x^2}}.$$

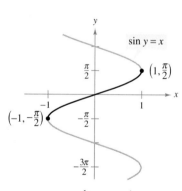

The derivative is $\dfrac{dy}{dx} = \dfrac{1}{\sqrt{1 - x^2}}$.

Figure 3.31

With implicit differentiation, the form of the derivative often can be simplified (as in Example 6) by an appropriate use of the *original* equation. A similar technique can be used to find and simplify higher-order derivatives obtained implicitly.

EXAMPLE 7 Finding the Second Derivative Implicitly

Given $x^2 + y^2 = 25$, find $\dfrac{d^2y}{dx^2}$.

Solution Differentiating each term with respect to x produces

$$2x + 2y\frac{dy}{dx} = 0$$

$$2y\frac{dy}{dx} = -2x$$

$$\frac{dy}{dx} = \frac{-2x}{2y} = -\frac{x}{y}.$$

Differentiating a second time with respect to x yields

$$\frac{d^2y}{dx^2} = -\frac{(y)(1) - (x)(dy/dx)}{y^2} \qquad \text{Quotient Rule}$$

$$= -\frac{y - (x)(-x/y)}{y^2} \qquad \text{Substitute } -x/y \text{ for } \frac{dy}{dx}.$$

$$= -\frac{y^2 + x^2}{y^3} \qquad \text{Simplify.}$$

$$= -\frac{25}{y^3}. \qquad \text{Substitute 25 for } x^2 + y^2.$$

EXAMPLE 8 Finding a Tangent Line to a Graph

Find the tangent line to the graph given by $x^2(x^2 + y^2) = y^2$ at the point $\left(\sqrt{2}/2, \sqrt{2}/2\right)$, as shown in Figure 3.32.

Solution By rewriting and differentiating implicitly, you obtain

$$x^4 + x^2y^2 - y^2 = 0$$

$$4x^3 + x^2\left(2y\frac{dy}{dx}\right) + 2xy^2 - 2y\frac{dy}{dx} = 0$$

$$2y(x^2 - 1)\frac{dy}{dx} = -2x(2x^2 + y^2)$$

$$\frac{dy}{dx} = \frac{x(2x^2 + y^2)}{y(1 - x^2)}.$$

At the point $\left(\sqrt{2}/2, \sqrt{2}/2\right)$, the slope is

$$\frac{dy}{dx} = \frac{\left(\sqrt{2}/2\right)[2(1/2) + (1/2)]}{\left(\sqrt{2}/2\right)[1 - (1/2)]} = \frac{3/2}{1/2} = 3$$

and the equation of the tangent line at this point is

$$y - \frac{\sqrt{2}}{2} = 3\left(x - \frac{\sqrt{2}}{2}\right)$$

$$y = 3x - \sqrt{2}.$$

The Granger Collection

ISAAC BARROW (1630–1677)

The graph in Example 8 is called the **kappa curve** because it resembles the Greek letter kappa, κ. The general solution for the tangent line to this curve was discovered by the English mathematician Isaac Barrow. Newton was Barrow's student, and they corresponded frequently regarding their work in the early development of calculus.

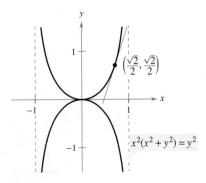

$x^2(x^2 + y^2) = y^2$

Kappa curve
Figure 3.32

Logarithmic Differentiation

On occasion, it is convenient to use logarithms as aids in differentiating nonlogarithmic functions. This procedure is called **logarithmic differentiation.**

EXAMPLE 9   Logarithmic Differentiation

Find the derivative of $y = \dfrac{(x-2)^2}{\sqrt{x^2+1}}$, $x \neq 2$.

Solution   Note that $y > 0$ and so $\ln y$ is defined. Begin by taking the natural logarithms of both sides of the equation. Then apply logarithmic properties and differentiate implicitly. Finally, solve for y'.

$$\ln y = \ln \frac{(x-2)^2}{\sqrt{x^2+1}} \qquad \text{Take ln of both sides.}$$

$$\ln y = 2\ln(x-2) - \frac{1}{2}\ln(x^2+1) \qquad \text{Logarithmic properties}$$

$$\frac{y'}{y} = 2\left(\frac{1}{x-2}\right) - \frac{1}{2}\left(\frac{2x}{x^2+1}\right) \qquad \text{Differentiate.}$$

$$= \frac{2}{x-2} - \frac{x}{x^2+1} \qquad \text{Simplify.}$$

$$y' = y\left(\frac{2}{x-2} - \frac{x}{x^2+1}\right) \qquad \text{Solve for } y'.$$

$$= \frac{(x-2)^2}{\sqrt{x^2+1}}\left[\frac{x^2+2x+2}{(x-2)(x^2+1)}\right] \qquad \text{Substitute for } y.$$

$$= \frac{(x-2)(x^2+2x+2)}{(x^2+1)^{3/2}} \qquad \text{Simplify.}$$

Exercises for Section 3.5

See www.CalcChat.com for worked-out solutions to odd-numbered exercises.

In Exercises 1–20, find dy/dx by implicit differentiation.

1. $x^2 + y^2 = 36$

2. $x^2 - y^2 = 81$

3. $x^{1/2} + y^{1/2} = 9$

4. $x^3 + y^3 = 8$

5. $x^3 - xy + y^2 = 4$

6. $x^2 y + y^2 x = -3$

7. $xe^y - 10x + 3y = 0$

8. $e^{xy} + x^2 - y^2 = 10$

9. $x^3 y^3 - y = x$

10. $\sqrt{xy} = x - 2y$

11. $x^3 - 2x^2 y + 3xy^2 = 38$

12. $2\sin x \cos y = 1$

13. $\sin x + 2\cos 2y = 1$

14. $(\sin \pi x + \cos \pi y)^2 = 2$

15. $\sin x = x(1 + \tan y)$

16. $\cot y = x - y$

17. $y = \sin(xy)$

18. $x = \sec \dfrac{1}{y}$

19. $x^2 - 3\ln y + y^2 = 10$

20. $\ln xy + 5x = 30$

In Exercises 21–24, (a) find two explicit functions by solving the equation for y in terms of x, (b) sketch the graph of the equation and label the parts given by the corresponding explicit functions, (c) differentiate the explicit functions, and (d) find dy/dx implicitly and show that the result is equivalent to that of part (c).

21. $x^2 + y^2 = 16$

22. $x^2 + y^2 - 4x + 6y + 9 = 0$

23. $9x^2 + 16y^2 = 144$

24. $4y^2 - x^2 = 4$

In Exercises 25–34, find dy/dx by implicit differentiation and evaluate the derivative at the indicated point.

Equation	Point
25. $xy = 4$	$(-4, -1)$
26. $x^3 - y^2 = 0$	$(1, 1)$
27. $y^2 = \dfrac{x^2 - 9}{x^2 + 9}$	$(3, 0)$
28. $(x + y)^3 = x^3 + y^3$	$(-1, 1)$
29. $x^{2/3} + y^{2/3} = 5$	$(8, 1)$
30. $x^3 + y^3 = 2xy$	$(1, 1)$
31. $\tan(x + y) = x$	$(0, 0)$
32. $x \cos y = 1$	$\left(2, \dfrac{\pi}{3}\right)$
33. $3e^{xy} - x = 0$	$(3, 0)$
34. $y^2 = \ln x$	$(e, 1)$

Famous Curves **In Exercises 35–38, find the slope of the tangent line to the graph at the indicated point.**

35. Witch of Agnesi:

$(x^2 + 4)y = 8$

Point: $(2, 1)$

36. Cissoid:

$(4 - x)y^2 = x^3$

Point: $(2, 2)$

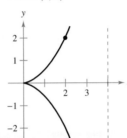

37. Bifolium:

$(x^2 + y^2)^2 = 4x^2y$

Point: $(1, 1)$

38. Folium of Descartes:

$x^3 + y^3 - 6xy = 0$

Point: $\left(\frac{4}{3}, \frac{8}{3}\right)$

Famous Curves **In Exercises 39–46, find an equation of the tangent line to the graph at the given point. To print an enlarged copy of the graph, go to the website *www.mathgraphs.com*.**

39. Parabola

$(y - 2)^2 = 4(x - 3)$

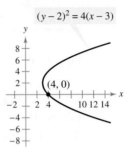

40. Circle

$(x + 1)^2 + (y - 2)^2 = 20$

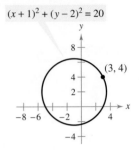

41. Rotated hyperbola

$xy = 1$

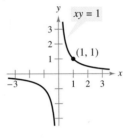

42. Rotated ellipse

$7x^2 - 6\sqrt{3}xy + 13y^2 - 16 = 0$

43. Cruciform

$x^2y^2 - 9x^2 - 4y^2 = 0$

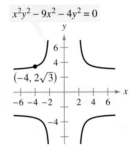

44. Astroid

$x^{2/3} + y^{2/3} = 5$

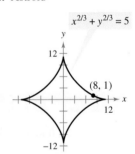

45. Lemniscate

$3(x^2 + y^2)^2 = 100(x^2 - y^2)$

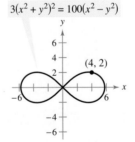

46. Kappa curve

$y^2(x^2 + y^2) = 2x^2$

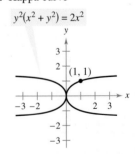

47. (a) Use implicit differentiation to find an equation of the tangent line to the ellipse $\dfrac{x^2}{2} + \dfrac{y^2}{8} = 1$ at $(1, 2)$.

(b) Show that the equation of the tangent line to the ellipse $\dfrac{x^2}{a^2} + \dfrac{y^2}{b^2} = 1$ at (x_0, y_0) is $\dfrac{x_0 x}{a^2} + \dfrac{y_0 y}{b^2} = 1$.

48. (a) Use implicit differentiation to find an equation of the tangent line to the hyperbola $\dfrac{x^2}{6} - \dfrac{y^2}{8} = 1$ at $(3, -2)$.

(b) Show that the equation of the tangent line to the hyperbola $\dfrac{x^2}{a^2} - \dfrac{y^2}{b^2} = 1$ at (x_0, y_0) is $\dfrac{x_0 x}{a^2} - \dfrac{y_0 y}{b^2} = 1$.

In Exercises 49 and 50, find dy/dx implicitly and find the largest interval of the form $-a < y < a$ or $0 < y < a$ such that y is a differentiable function of x. Write dy/dx as a function of x.

49. $\tan y = x$ **50.** $\cos y = x$

In Exercises 51–56, find d^2y/dx^2 in terms of x and y.

51. $x^2 + y^2 = 36$ **52.** $x^2 y^2 - 2x = 3$

53. $x^2 - y^2 = 16$ **54.** $1 - xy = x - y$

55. $y^2 = x^3$ **56.** $y^2 = 4x$

 In Exercises 57 and 58, use a graphing utility to graph the equation. Find an equation of the tangent line to the graph at the given point and graph the tangent line in the same viewing window.

57. $\sqrt{x} + \sqrt{y} = 4$, $(9, 1)$ **58.** $y^2 = \dfrac{x - 1}{x^2 + 1}$, $\left(2, \dfrac{\sqrt{5}}{5}\right)$

 In Exercises 59 and 60, find equations for the tangent line and normal line to the circle at the given points. (The _normal line_ at a point is perpendicular to the tangent line at the point.) Use a graphing utility to graph the equation, tangent line, and normal line.

59. $x^2 + y^2 = 25$ **60.** $x^2 + y^2 = 9$

$(4, 3), (-3, 4)$ $(0, 3), \left(2, \sqrt{5}\right)$

61. Show that the normal line at any point on the circle $x^2 + y^2 = r^2$ passes through the origin.

62. Two circles of radius 4 are tangent to the graph of $y^2 = 4x$ at the point $(1, 2)$. Find equations of these two circles.

In Exercises 63 and 64, find the points at which the graph of the equation has a vertical or horizontal tangent line.

63. $25x^2 + 16y^2 + 200x - 160y + 400 = 0$

64. $4x^2 + y^2 - 8x + 4y + 4 = 0$

In Exercises 65–74, find dy/dx using logarithmic differentiation.

65. $y = x\sqrt{x^2 - 1}$

66. $y = \sqrt{(x - 1)(x - 2)(x - 3)}$

67. $y = \dfrac{x^2\sqrt{3x - 2}}{(x - 1)^2}$ **68.** $y = \sqrt{\dfrac{x^2 - 2}{x^2 + 2}}$

69. $y = \dfrac{x(x - 1)^{3/2}}{\sqrt{x + 1}}$ **70.** $y = \dfrac{(x + 1)(x + 2)}{(x - 1)(x - 2)}$

71. $y = x^{2/x}$ **72.** $y = x^{x - 1}$

73. $y = (x - 2)^{x + 1}$ **74.** $y = (1 + x)^{1/x}$

 Orthogonal Trajectories **In Exercises 75–78, use a graphing utility to sketch the intersecting graphs of the equations and show that they are orthogonal. [Two graphs are _orthogonal_ if at their point(s) of intersection their tangent lines are perpendicular to each other.]**

75. $2x^2 + y^2 = 6$ **76.** $y^2 = x^3$

$y^2 = 4x$ $2x^2 + 3y^2 = 5$

77. $x + y = 0$ **78.** $x^3 = 3(y - 1)$

$x = \sin y$ $x(3y - 29) = 3$

Orthogonal Trajectories **In Exercises 79 and 80, verify that the two families of curves are orthogonal, where C and K are real numbers. Use a graphing utility to graph the two families for two values of C and two values of K.**

79. $xy = C$, $x^2 - y^2 = K$ **80.** $x^2 + y^2 = C^2$, $y = Kx$

In Exercises 81–84, differentiate (a) with respect to x (y is a function of x) and (b) with respect to t (x and y are functions of t).

81. $2y^2 - 3x^4 = 0$ **82.** $x^2 - 3xy^2 + y^3 = 10$

83. $\cos \pi y - 3 \sin \pi x = 1$ **84.** $4 \sin x \cos y = 1$

Writing About Concepts

85. Describe the difference between the explicit form of a function and an implicit equation. Give an example of each.

86. In your own words, state the guidelines for implicit differentiation.

---

87. *Orthogonal Trajectories* The figure below shows the topographic map carried by a group of hikers. The hikers are in a wooded area on top of the hill shown on the map and they decide to follow a path of steepest descent (orthogonal trajectories to the contours on the map). Draw their routes if they start from point A and if they start from point B. If their goal is to reach the road along the top of the map, which starting point should they use? To print an enlarged copy of the map, go to the website *www.mathgraphs.com*.

88. *Weather Map* The weather map shows several *isobars*—curves that represent areas of constant air pressure. Three high pressures *H* and one low pressure *L* are shown on the map. Given that wind speed is greatest along the orthogonal trajectories of the isobars, use the map to determine the areas having high wind speed.

89. Consider the equation $x^4 = 4(4x^2 - y^2)$.

(a) Use a graphing utility to graph the equation.

(b) Find and graph the four tangent lines to the curve for $y = 3$.

(c) Find the exact coordinates of the point of intersection of the two tangent lines in the first quadrant.

90. Let *L* be any tangent line to the curve $\sqrt{x} + \sqrt{y} = \sqrt{c}$. Show that the sum of the *x*- and *y*-intercepts of *L* is *c*.

91. Prove (Theorem 3.3) that

$$\frac{d}{dx}[x^n] = nx^{n-1}$$

for the case in which *n* is a rational number. (*Hint:* Write $y = x^{p/q}$ in the form $y^q = x^p$ and differentiate implicitly. Assume that *p* and *q* are integers, where $q > 0$.)

92. *Slope* Find all points on the circle $x^2 + y^2 = 25$ where the slope is $\frac{3}{4}$.

93. *Horizontal Tangent* Determine the point(s) at which the graph of $y^4 = y^2 - x^2$ has a horizontal tangent.

94. *Tangent Lines* Find equations of both tangent lines to the ellipse $\frac{x^2}{4} + \frac{y^2}{9} = 1$ that passes through the point $(4, 0)$.

95. *Normals to a Parabola* The graph shows the normal lines from the point $(2, 0)$ to the graph of the parabola $x = y^2$. How many normal lines are there from the point $(x_0, 0)$ to the graph of the parabola if (a) $x_0 = \frac{1}{4}$, (b) $x_0 = \frac{1}{2}$, and (c) $x_0 = 1$? For what value of x_0 are two of the normal lines perpendicular to each other?

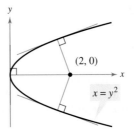

96. *Normal Lines* (a) Find an equation of the normal line to the ellipse

$$\frac{x^2}{32} + \frac{y^2}{8} = 1$$

at the point $(4, 2)$. (b) Use a graphing utility to graph the ellipse and the normal line. (c) At what other point does the normal line intersect the ellipse?

Section Project: Optical Illusions

In each graph below, an optical illusion is created by having lines intersect a family of curves. In each case, the lines appear to be curved. Find the value of dy/dx for the given values of x and y.

(a) Circles: $x^2 + y^2 = C^2$
$x = 3, y = 4, C = 5$

(b) Hyperbolas: $xy = C$
$x = 1, y = 4, C = 4$

(c) Lines: $ax = by$
$x = \sqrt{3}, y = 3,$
$a = \sqrt{3}, b = 1$

(d) Cosine curves: $y = C \cos x$
$x = \frac{\pi}{3}, y = \frac{1}{3}, C = \frac{2}{3}$

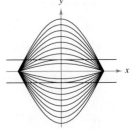

FOR FURTHER INFORMATION For more information on the mathematics of optical illusions, see the article "Descriptive Models for Perception of Optical Illusions" by David A. Smith in *The UMAP Journal.*

Derivatives of Inverse Functions

- Find the derivative of an inverse function.
- Differentiate an inverse trigonometric function.
- Review the basic differentiation formulas for elementary functions.

Derivative of an Inverse Function

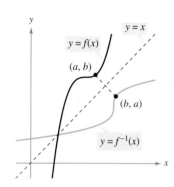

The next two theorems discuss the derivative of an inverse function. The reasonableness of Theorem 3.16 follows from the reflective property of inverse functions, as shown in Figure 3.33. Proofs of the two theorems are given in Appendix A.

The graph of f^{-1} is a reflection of the graph of f in the line $y = x$.
Figure 3.33

THEOREM 3.16   Continuity and Differentiability of Inverse Functions

Let f be a function whose domain is an interval I. If f has an inverse function, then the following statements are true.

1. If f is continuous on its domain, then f^{-1} is continuous on its domain.
2. If f is differentiable on an interval containing c and $f'(c) \neq 0$, then f^{-1} is differentiable at $f(c)$.

THEOREM 3.17   The Derivative of an Inverse Function

Let f be a function that is differentiable on an interval I. If f has an inverse function g, then g is differentiable at any x for which $f'(g(x)) \neq 0$. Moreover,

$$g'(x) = \frac{1}{f'(g(x))}, \quad f'(g(x)) \neq 0.$$

EXAMPLE 1   **Evaluating the Derivative of an Inverse Function**

Let $f(x) = \frac{1}{4}x^3 + x - 1$.

a. What is the value of $f^{-1}(x)$ when $x = 3$?

b. What is the value of $(f^{-1})'(x)$ when $x = 3$?

Solution   Notice that f is one-to-one and therefore has an inverse function.

a. Because $f(2) = 3$, you know that $f^{-1}(3) = 2$.

b. Because the function f is differentiable and has an inverse function, you can apply Theorem 3.17 (with $g = f^{-1}$) to write

$$(f^{-1})'(3) = \frac{1}{f'(f^{-1}(3))} = \frac{1}{f'(2)}.$$

Moreover, using $f'(x) = \frac{3}{4}x^2 + 1$, you can conclude that

$$(f^{-1})'(3) = \frac{1}{f'(2)}$$

$$= \frac{1}{\frac{3}{4}(2^2) + 1}$$

$$= \frac{1}{4}. \qquad \text{(See Figure 3.34.)}$$

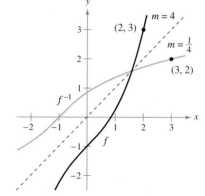

The graphs of the inverse functions f and f^{-1} have reciprocal slopes at points (a, b) and (b, a).
Figure 3.34

In Example 1, note that at the point $(2, 3)$ the slope of the graph of f is 4 and at the point $(3, 2)$ the slope of the graph of f^{-1} is $\frac{1}{4}$ (see Figure 3.34). This reciprocal relationship (which follows from Theorem 3.17) is sometimes written as

$$\frac{dy}{dx} = \frac{1}{dx/dy}.$$

EXAMPLE 2   Graphs of Inverse Functions Have Reciprocal Slopes

Let $f(x) = x^2$ (for $x \geq 0$) and let $f^{-1}(x) = \sqrt{x}$. Show that the slopes of the graphs of f and f^{-1} are reciprocals at each of the following points.

a. $(2, 4)$ and $(4, 2)$ **b.** $(3, 9)$ and $(9, 3)$

Solution   The derivatives of f and f^{-1} are $f'(x) = 2x$ and $(f^{-1})'(x) = \dfrac{1}{2\sqrt{x}}$.

a. At $(2, 4)$, the slope of the graph of f is $f'(2) = 2(2) = 4$. At $(4, 2)$, the slope of the graph of f^{-1} is

$$(f^{-1})'(4) = \frac{1}{2\sqrt{4}} = \frac{1}{2(2)} = \frac{1}{4}.$$

b. At $(3, 9)$, the slope of the graph of f is $f'(3) = 2(3) = 6$. At $(9, 3)$, the slope of the graph of f^{-1} is

$$(f^{-1})'(9) = \frac{1}{2\sqrt{9}} = \frac{1}{2(3)} = \frac{1}{6}.$$

So, in both cases, the slopes are reciprocals, as shown in Figure 3.35.

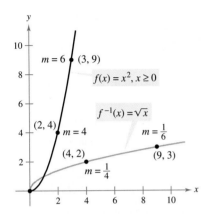

At $(0, 0)$, the derivative of f is 0 and the derivative of f^{-1} does not exist.
Figure 3.35

When determining the derivative of an inverse function, you have two options: (1) you can apply Theorem 3.17, or (2) you can use implicit differentiation. The first approach is illustrated in Example 3, and the second in the proof of Theorem 3.18.

EXAMPLE 3   Finding the Derivative of an Inverse Function

Find the derivative of the inverse tangent function.

Solution   Let $f(x) = \tan x$, $-\pi/2 < x < \pi/2$. Then let $g(x) = \arctan x$ be the inverse tangent function. To find the derivative of $g(x)$, use the fact that $f'(x) = \sec^2 x = \tan^2 x + 1$, and apply Theorem 3.17 as follows.

$$g'(x) = \frac{1}{f'(g(x))} = \frac{1}{f'(\arctan x)} = \frac{1}{[\tan(\arctan x)]^2 + 1} = \frac{1}{x^2 + 1}$$

Derivatives of Inverse Trigonometric Functions

In Section 3.4, you saw that the derivative of the *transcendental* function $f(x) = \ln x$ is the *algebraic* function $f'(x) = 1/x$. You will now see that the derivatives of the inverse trigonometric functions also are algebraic (even though the inverse trigonometric functions are themselves transcendental).

The following theorem lists the derivatives of the six inverse trigonometric functions. Note that the derivatives of arccos u, arccot u, and arccsc u are the *negatives* of the derivatives of arcsin u, arctan u, and arcsec u, respectively.

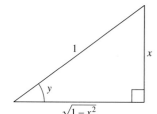

$y = \arcsin x$
Figure 3.36

> **THEOREM 3.18   Derivatives of Inverse Trigonometric Functions**
>
> Let u be a differentiable function of x.
>
> $$\frac{d}{dx}[\arcsin u] = \frac{u'}{\sqrt{1 - u^2}} \qquad \frac{d}{dx}[\arccos u] = \frac{-u'}{\sqrt{1 - u^2}}$$
>
> $$\frac{d}{dx}[\arctan u] = \frac{u'}{1 + u^2} \qquad \frac{d}{dx}[\operatorname{arccot} u] = \frac{-u'}{1 + u^2}$$
>
> $$\frac{d}{dx}[\operatorname{arcsec} u] = \frac{u'}{|u|\sqrt{u^2 - 1}} \qquad \frac{d}{dx}[\operatorname{arccsc} u] = \frac{-u'}{|u|\sqrt{u^2 - 1}}$$

Proof   Let $y = \arcsin x$, $-\pi/2 \le y \le \pi/2$ (see Figure 3.36). So, $\sin y = x$, and you can use implicit differentiation as follows.

$$\sin y = x$$
$$(\cos y)\left(\frac{dy}{dx}\right) = 1$$
$$\frac{dy}{dx} = \frac{1}{\cos y} = \frac{1}{\sqrt{1 - \sin^2 y}} = \frac{1}{\sqrt{1 - x^2}}$$

Because u is a differentiable function of x, you can use the Chain Rule to write

$$\frac{d}{dx}[\arcsin u] = \frac{u'}{\sqrt{1 - u^2}}, \quad \text{where } u' = \frac{du}{dx}.$$

Proofs of the other differentiation rules are left as exercises (see Exercise 71).

TECHNOLOGY   If your graphing utility does not have the arcsecant function, you can obtain its graph using

$$f(x) = \operatorname{arcsec} x = \arccos \frac{1}{x}.$$

There is no common agreement on the definition of $\operatorname{arcsec} x$ (or $\operatorname{arccsc} x$) for negative values of x. When we defined the range of the arcsecant, we chose to preserve the reciprocal identity $\operatorname{arcsec} x = \arccos(1/x)$. For example, to evaluate $\operatorname{arcsec}(-2)$, you can write

$$\operatorname{arcsec}(-2) = \arccos(-0.5)$$
$$\approx 2.09.$$

One of the consequences of the definition of the inverse secant function given in this text is that its graph has a positive slope at every x-value in its domain. This accounts for the absolute value sign in the formula for the derivative of $\operatorname{arcsec} x$.

EXPLORATION

Suppose that you want to find a linear approximation to the graph of the function in Example 4. You decide to use the tangent line at the origin, as shown below. Use a graphing utility to describe an interval about the origin where the tangent line is within 0.01 unit of the graph of the function. What might a person mean by saying that the original function is "locally linear"?

EXAMPLE 4   **A Derivative That Can Be Simplified**

Differentiate $y = \arcsin x + x\sqrt{1 - x^2}$.

Solution

$$y' = \frac{1}{\sqrt{1 - x^2}} + x\left(\frac{1}{2}\right)(-2x)(1 - x^2)^{-1/2} + \sqrt{1 - x^2}$$
$$= \frac{1}{\sqrt{1 - x^2}} - \frac{x^2}{\sqrt{1 - x^2}} + \sqrt{1 - x^2}$$
$$= \sqrt{1 - x^2} + \sqrt{1 - x^2}$$
$$= 2\sqrt{1 - x^2}$$

EXAMPLE 5 **Differentiating Inverse Trigonometric Functions**

a. $\dfrac{d}{dx}[\arcsin(2x)] = \dfrac{2}{\sqrt{1-(2x)^2}}$ $u = 2x$

$\qquad\qquad\qquad\;\; = \dfrac{2}{\sqrt{1-4x^2}}$

b. $\dfrac{d}{dx}[\arctan(3x)] = \dfrac{3}{1+(3x)^2}$ $u = 3x$

$\qquad\qquad\qquad\;\; = \dfrac{3}{1+9x^2}$

c. $\dfrac{d}{dx}[\arcsin\sqrt{x}] = \dfrac{(1/2)x^{-1/2}}{\sqrt{1-x}}$ $u = \sqrt{x}$

$\qquad\qquad\qquad\;\; = \dfrac{1}{2\sqrt{x}\sqrt{1-x}}$

$\qquad\qquad\qquad\;\; = \dfrac{1}{2\sqrt{x-x^2}}$

d. $\dfrac{d}{dx}[\operatorname{arcsec} e^{2x}] = \dfrac{2e^{2x}}{e^{2x}\sqrt{(e^{2x})^2 - 1}}$ $u = e^{2x}$

$\qquad\qquad\qquad\;\; = \dfrac{2e^{2x}}{e^{2x}\sqrt{e^{4x}-1}}$

$\qquad\qquad\qquad\;\; = \dfrac{2}{\sqrt{e^{4x}-1}}$

In part (d), the absolute value sign is not necessary because $e^{2x} > 0$.

Review of Basic Differentiation Rules

In the 1600s, Europe was ushered into the scientific age by such great thinkers as Descartes, Galileo, Huygens, Newton, and Kepler. These men believed that nature is governed by basic laws—laws that can, for the most part, be written in terms of mathematical equations. One of the most influential publications of this period—*Dialogue on the Great World Systems*, by Galileo Galilei—has become a classic description of modern scientific thought.

As mathematics has developed during the past few hundred years, a small number of elementary functions has proven sufficient for modeling most* phenomena in physics, chemistry, biology, engineering, economics, and a variety of other fields. An **elementary function** is a function from the following list or one that can be formed as the sum, product, quotient, or composition of functions in the list.

Algebraic Functions	Transcendental Functions
Polynomial functions	Logarithmic functions
Rational functions	Exponential functions
Functions involving radicals	Trigonometric functions
	Inverse trigonometric functions

With the differentiation rules introduced so far in the text, you can differentiate any elementary function. For convenience, these differentiation rules are summarized on the next page.

The Granger Collection

GALILEO GALILEI (1564–1642)

Galileo's approach to science departed from the accepted Aristotelian view that nature had describable *qualities*, such as "fluidity" and "potentiality." He chose to describe the physical world in terms of measurable *quantities*, such as time, distance, force, and mass.

---

* *Some important functions used in engineering and science (such as Bessel functions and gamma functions) are not elementary functions.*

Basic Differentiation Rules for Elementary Functions

1. $\dfrac{d}{dx}[cu] = cu'$

2. $\dfrac{d}{dx}[u \pm v] = u' \pm v'$

3. $\dfrac{d}{dx}[uv] = uv' + vu'$

4. $\dfrac{d}{dx}\left[\dfrac{u}{v}\right] = \dfrac{vu' - uv'}{v^2}$

5. $\dfrac{d}{dx}[c] = 0$

6. $\dfrac{d}{dx}[u^n] = nu^{n-1}u'$

7. $\dfrac{d}{dx}[x] = 1$

8. $\dfrac{d}{dx}[|u|] = \dfrac{u}{|u|}(u'), \quad u \neq 0$

9. $\dfrac{d}{dx}[\ln u] = \dfrac{u'}{u}$

10. $\dfrac{d}{dx}[e^u] = e^u u'$

11. $\dfrac{d}{dx}[\log_a u] = \dfrac{u'}{(\ln a)u}$

12. $\dfrac{d}{dx}[a^u] = (\ln a)a^u u'$

13. $\dfrac{d}{dx}[\sin u] = (\cos u)u'$

14. $\dfrac{d}{dx}[\cos u] = -(\sin u)u'$

15. $\dfrac{d}{dx}[\tan u] = (\sec^2 u)u'$

16. $\dfrac{d}{dx}[\cot u] = -(\csc^2 u)u'$

17. $\dfrac{d}{dx}[\sec u] = (\sec u \tan u)u'$

18. $\dfrac{d}{dx}[\csc u] = -(\csc u \cot u)u'$

19. $\dfrac{d}{dx}[\arcsin u] = \dfrac{u'}{\sqrt{1 - u^2}}$

20. $\dfrac{d}{dx}[\arccos u] = \dfrac{-u'}{\sqrt{1 - u^2}}$

21. $\dfrac{d}{dx}[\arctan u] = \dfrac{u'}{1 + u^2}$

22. $\dfrac{d}{dx}[\operatorname{arccot} u] = \dfrac{-u'}{1 + u^2}$

23. $\dfrac{d}{dx}[\operatorname{arcsec} u] = \dfrac{u'}{|u|\sqrt{u^2 - 1}}$

24. $\dfrac{d}{dx}[\operatorname{arccsc} u] = \dfrac{-u'}{|u|\sqrt{u^2 - 1}}$

Exercises for Section 3.6

See www.CalcChat.com for worked-out solutions to odd-numbered exercises.

In Exercises 1–6, find $(f^{-1})'(a)$ for the function f and real number a.

Function	Real Number
1. $f(x) = x^3 + 2x - 1$	$a = 2$
2. $f(x) = \frac{1}{27}(x^5 + 2x^3)$	$a = -11$
3. $f(x) = \sin x, \quad -\frac{\pi}{2} \leq x \leq \frac{\pi}{2}$	$a = \frac{1}{2}$
4. $f(x) = \cos 2x, \quad 0 \leq x \leq \frac{\pi}{2}$	$a = 1$
5. $f(x) = x^3 - \frac{4}{x}$	$a = 6$
6. $f(x) = \sqrt{x - 4}$	$a = 2$

In Exercises 7–10, show that the slopes of the graphs of f and f^{-1} are reciprocals at the indicated points.

Function	Point
7. $f(x) = x^3$	$\left(\frac{1}{2}, \frac{1}{8}\right)$
$f^{-1}(x) = \sqrt[3]{x}$	$\left(\frac{1}{8}, \frac{1}{2}\right)$
8. $f(x) = 3 - 4x$	$(1, -1)$
$f^{-1}(x) = \dfrac{3 - x}{4}$	$(-1, 1)$
9. $f(x) = \sqrt{x - 4}$	$(5, 1)$
$f^{-1}(x) = x^2 + 4, \quad x \geq 0$	$(1, 5)$

Function	Point
10. $f(x) = \dfrac{4}{1 + x^2}, \quad x \geq 0$	$(1, 2)$
$f^{-1}(x) = \sqrt{\dfrac{4 - x}{x}}$	$(2, 1)$

In Exercises 11–14, (a) find an equation of the tangent line to the graph of f at the indicated point and (b) use a graphing utility to graph the function and its tangent line at the point.

Function	Point
11. $f(x) = \arcsin 2x$	$\left(\dfrac{\sqrt{2}}{4}, \dfrac{\pi}{4}\right)$
12. $f(x) = \arctan x$	$\left(-1, -\dfrac{\pi}{4}\right)$
13. $f(x) = \arccos x^2$	$\left(0, \dfrac{\pi}{2}\right)$
14. $f(x) = \operatorname{arcsec} x$	$\left(\sqrt{2}, \dfrac{\pi}{4}\right)$

In Exercises 15–18, find dy/dx at the indicated point for the equation.

15. $x = y^3 - 7y^2 + 2, \quad (-4, 1)$

16. $x = 2\ln(y^2 - 3), \quad (0, 4)$

17. $x \arctan x = e^y, \left(1, \ln\dfrac{\pi}{4}\right)$

18. $\arcsin xy = \frac{2}{3}\arctan 2x, \left(\frac{1}{2}, 1\right)$

In Exercises 19–44, find the derivative of the function.

19. $f(x) = 2 \arcsin(x - 1)$ **20.** $f(t) = \arcsin t^2$

21. $g(x) = 3 \arccos(x/2)$ **22.** $f(x) = \operatorname{arcsec} 3x$

23. $f(x) = \arctan(x/a)$ **24.** $f(x) = \arctan \sqrt{x}$

25. $g(x) = \dfrac{\arcsin 3x}{x}$ **26.** $h(x) = x^2 \arctan x$

27. $g(x) = \dfrac{\arccos x}{x + 1}$ **28.** $g(x) = e^x \arcsin x$

29. $h(x) = \operatorname{arccot} 6x$ **30.** $f(x) = \operatorname{arcsec} 2x$

31. $h(t) = \sin(\arccos t)$ **32.** $f(x) = \arcsin x + \arccos x$

33. $y = x \arccos x - \sqrt{1 - x^2}$

34. $y = 2 \ln(t^2 + 4) - \arctan \dfrac{t}{2}$

35. $y = \dfrac{1}{2}\left(\dfrac{1}{2} \ln \dfrac{x + 1}{x - 1} + \arctan x\right)$

36. $y = \dfrac{1}{2}\left[x\sqrt{4 - x^2} + 4 \arcsin\left(\dfrac{x}{2}\right)\right]$

37. $g(t) = \tan(\arcsin t)$ **38.** $f(x) = \operatorname{arcsec} x + \operatorname{arccsc} x$

39. $y = x \arcsin x + \sqrt{1 - x^2}$

40. $y = x \arctan 2x - \dfrac{1}{4} \ln(1 + 4x^2)$

41. $y = 8 \arcsin \dfrac{x}{4} - \dfrac{x\sqrt{16 - x^2}}{2}$

42. $y = 25 \arcsin \dfrac{x}{5} - x\sqrt{25 - x^2}$

43. $y = \arctan x + \dfrac{x}{1 + x^2}$ **44.** $y = \arctan \dfrac{x}{2} - \dfrac{1}{2(x^2 + 4)}$

In Exercises 45–50, find an equation of the tangent line to the graph of the function at the given point.

45. $y = 2 \arcsin x$ **46.** $y = \dfrac{1}{2} \arccos x$

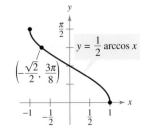

47. $y = \arctan \dfrac{x}{2}$ **48.** $y = \operatorname{arcsec} 4x$

49. $y = 4x \arccos(x - 1)$ **50.** $y = 3x \arcsin x$

51. Find equations of all tangent lines to the graph of $f(x) = \arccos x$ that have slope -2.

52. Find an equation of the tangent line to the graph of $g(x) = \arctan x$ when $x = 1$.

Linear and Quadratic Approximations   **In Exercises 53–56, use a computer algebra system to find the linear approximation**

$$P_1(x) = f(a) + f'(a)(x - a)$$

and the quadratic approximation

$$P_2(x) = f(a) + f'(a)(x - a) + \dfrac{1}{2}f''(a)(x - a)^2$$

to the function f at $x = a$. Sketch the graph of the function and its linear and quadratic approximations.

53. $f(x) = \arcsin x$, $a = \dfrac{1}{2}$ **54.** $f(x) = \arctan x$, $a = 1$

55. $f(x) = \arctan x$, $a = 0$ **56.** $f(x) = \arccos x$, $a = 0$

Implicit Differentiation   **In Exercises 57–60, find an equation of the tangent line to the graph of the equation at the given point.**

57. $x^2 + x \arctan y = y - 1$, $\left(-\dfrac{\pi}{4}, 1\right)$

58. $\arctan(xy) = \arcsin(x + y)$, $(0, 0)$

59. $\arcsin x + \arcsin y = \dfrac{\pi}{2}$, $\left(\dfrac{\sqrt{2}}{2}, \dfrac{\sqrt{2}}{2}\right)$

60. $\arctan(x + y) = y^2 + \dfrac{\pi}{4}$, $(1, 0)$

Writing About Concepts

In Exercises 61 and 62, the derivative of the function has the same sign for all x in its domain, but the function is not one-to-one. Explain.

61. $f(x) = \tan x$ **62.** $f(x) = \dfrac{x}{x^2 - 4}$

63. State the theorem that gives the method for finding the derivative of an inverse function.

64. Are the derivatives of the inverse trigonometric functions algebraic or transcendental functions? List the derivatives of the inverse trigonometric functions.

65. *Angular Rate of Change* An airplane flies at an altitude of 5 miles toward a point directly over an observer. Consider θ and x as shown in the figure.

(a) Write θ as a function of x.

(b) The speed of the plane is 400 miles per hour. Find $d\theta/dt$ when $x = 10$ miles and $x = 3$ miles.

5 mi

θ

x

Not drawn to scale

66. *Writing* Repeat Exercise 65 if the altitude of the plane is 3 miles and describe how the altitude affects the rate of change of θ.

67. *Angular Rate of Change* In a free-fall experiment, an object is dropped from a height of 256 feet. A camera on the ground 500 feet from the point of impact records the fall of the object (see figure).

(a) Find the position function giving the height of the object at time t, assuming the object is released at time $t = 0$. At what time will the object reach ground level?

(b) Find the rates of change of the angle of elevation of the camera when $t = 1$ and $t = 2$.

256 ft

θ

500 ft

Figure for 67

s

θ

750 m

h

Figure for 68

68. *Angular Rate of Change* A television camera at ground level is filming the lift-off of a space shuttle at a point 750 meters from the launch pad. Let θ be the angle of elevation of the shuttle and let s be the distance between the camera and the shuttle (see figure). Write θ as a function of s for the period of time when the shuttle is moving vertically. Differentiate the result to find $d\theta/dt$ in terms of s and ds/dt.

69. *Angular Rate of Change* An observer is standing 300 feet from the point at which a balloon is released. The balloon rises at a rate of 5 feet per second. How fast is the angle of elevation of the observer's line of sight increasing when the balloon is 100 feet high?

70. *Farm Workers* The table gives the number of farm workers (in millions) in the United States for selected years from 1910 to 2000. *(Source: U.S. Department of Agriculture)*

Year	1910	1920	1930	1940	1950
Workers	13.56	13.43	12.50	10.98	9.93

Year	1960	1970	1980	1990	2000
Workers	7.06	4.52	3.73	2.91	2.95

A model for the data is

$$y = 24.760 - 0.361t + 0.001t^2 - 79.564 \text{ arccot } t$$

where t is time in years, with $t = 0$ corresponding to the year 1900, and y is the number of workers in millions.

(a) Use a graphing utility to plot the data and graph the model.

(b) Find the rate of change of the number of workers when $t = 20$ and $t = 60$.

71. Verify each differentiation formula.

(a) $\dfrac{d}{dx}[\arctan u] = \dfrac{u'}{1 + u^2}$

(b) $\dfrac{d}{dx}[\text{arcsec } u] = \dfrac{u'}{|u|\sqrt{u^2 - 1}}$

(c) $\dfrac{d}{dx}[\arccos u] = \dfrac{-u'}{\sqrt{1 - u^2}}$

(d) $\dfrac{d}{dx}[\text{arccot } u] = \dfrac{-u'}{1 + u^2}$

(e) $\dfrac{d}{dx}[\text{arccsc } u] = \dfrac{-u'}{|u|\sqrt{u^2 - 1}}$

72. *Existence of an Inverse* Determine the values of k such that the function $f(x) = kx + \sin x$ has an inverse function.

True or False? **In Exercises 73 and 74, determine whether the statement is true or false. If it is false, explain why or give an example that shows it is false.**

73. The slope of the graph of the inverse tangent function is positive for all x.

74. $\dfrac{d}{dx}[\arctan(\tan x)] = 1$ for all x in the domain.

75. Prove that $\arcsin x = \arctan\left(\dfrac{x}{\sqrt{1 - x^2}}\right), \; |x| < 1.$

76. Prove that $\arccos x = \dfrac{\pi}{2} - \arctan\left(\dfrac{x}{\sqrt{1 - x^2}}\right), \; |x| < 1.$

77. Some calculus textbooks define the inverse secant function using the range $[0, \pi/2) \cup [\pi, 3\pi/2)$.

(a) Sketch the graph of $y = \text{arcsec } x$ using this range.

(b) Show that $y' = \dfrac{1}{x\sqrt{x^2 - 1}}.$

78. Compare the graphs of $y_1 = \sin(\arcsin x)$ and $y_2 = \arcsin(\sin x)$. What are the domains and ranges of y_1 and y_2?

79. Show that the function

$$f(x) = \arcsin\dfrac{x - 2}{2} - 2\arcsin\left(\dfrac{\sqrt{x}}{2}\right)$$

is constant for $0 \le x \le 4$.

80. Use a graphing utility to graph the functions $y_1 = \arctan x$ and $y_2 = \arctan e^x$. Compare the functions and explain why the ranges are not the same.

Related Rates

- Find a related rate.
- Use related rates to solve real-life problems.

Finding Related Rates

You have seen how the Chain Rule can be used to find dy/dx implicitly. Another important use of the Chain Rule is to find the rates of change of two or more related variables that are changing with respect to *time*.

For example, when water is drained out of a conical tank (see Figure 3.37), the volume V, the radius r, and the height h of the water level are all functions of time t. Knowing that these variables are related by the equation

$$V = \frac{\pi}{3} r^2 h \qquad \text{Original equation}$$

you can differentiate implicitly with respect to t to obtain the **related-rate** equation

$$\frac{d}{dt}(V) = \frac{d}{dt}\left(\frac{\pi}{3}r^2 h\right)$$

$$\frac{dV}{dt} = \frac{\pi}{3}\left[r^2 \frac{dh}{dt} + h\left(2r\frac{dr}{dt}\right)\right] \qquad \text{Differentiate with respect to } t.$$

$$= \frac{\pi}{3}\left(r^2 \frac{dh}{dt} + 2rh\frac{dr}{dt}\right).$$

From this equation, you can see that the rate of change of V is related to the rates of change of both h and r.

Volume is related to radius and height.
Figure 3.37

EXPLORATION

Finding a Related Rate In the conical tank shown in Figure 3.37, suppose that the height is changing at a rate of -0.2 foot per minute and the radius is changing at a rate of -0.1 foot per minute. What is the rate of change of the volume when the radius is $r = 1$ foot and the height is $h = 2$ feet? Does the rate of change of the volume depend on the values of r and h? Explain.

EXAMPLE 1 Two Rates That Are Related

Suppose x and y are both differentiable functions of t and are related by the equation $y = x^2 + 3$. Find dy/dt when $x = 1$, given that $dx/dt = 2$ when $x = 1$.

Solution Using the Chain Rule, you can differentiate both sides of the equation *with respect to t*.

$$y = x^2 + 3 \qquad \text{Write original equation.}$$

$$\frac{d}{dt}[y] = \frac{d}{dt}[x^2 + 3] \qquad \text{Differentiate with respect to } t.$$

$$\frac{dy}{dt} = 2x\frac{dx}{dt} \qquad \text{Chain Rule}$$

When $x = 1$ and $dx/dt = 2$, you have

$$\frac{dy}{dt} = 2(1)(2) = 4.$$

FOR FURTHER INFORMATION To learn more about the history of related-rate problems, see the article "The Lengthening Shadow: The Story of Related Rates" by Bill Austin, Don Barry, and David Berman in *Mathematics Magazine*. To view this article, go to the website *www.matharticles.com*.

Problem Solving with Related Rates

In Example 1, you were *given* an equation that related the variables x and y and were asked to find the rate of change of y when $x = 1$.

Equation: $y = x^2 + 3$

Given rate: $\dfrac{dx}{dt} = 2$ when $x = 1$

Find: $\dfrac{dy}{dt}$ when $x = 1$

In each of the remaining examples in this section, you must *create* a mathematical model from a verbal description.

EXAMPLE 2   Ripples in a Pond

A pebble is dropped into a calm pond, causing ripples in the form of concentric circles, as shown in Figure 3.38. The radius r of the outer ripple is increasing at a constant rate of 1 foot per second. When the radius is 4 feet, at what rate is the total area A of the disturbed water changing?

Solution   The variables r and A are related by $A = \pi r^2$. The rate of change of the radius r is $dr/dt = 1$.

Equation: $A = \pi r^2$

Given rate: $\dfrac{dr}{dt} = 1$

Find: $\dfrac{dA}{dt}$ when $r = 4$

With this information, you can proceed as in Example 1.

$$\frac{d}{dt}[A] = \frac{d}{dt}[\pi r^2]$$ Differentiate with respect to t.

$$\frac{dA}{dt} = 2\pi r \frac{dr}{dt}$$ Chain Rule

$$\frac{dA}{dt} = 2\pi(4)(1) = 8\pi$$ Substitute 4 for r and 1 for dr/dt.

When the radius is 4 feet, the area is changing at a rate of 8π square feet per second.

Total area increases as the outer radius increases.
Figure 3.38

Guidelines For Solving Related-Rate Problems

1. Identify all *given* quantities and quantities *to be determined*. Make a sketch and label the quantities.
2. Write an equation involving the variables whose rates of change either are given or are to be determined.
3. Using the Chain Rule, implicitly differentiate both sides of the equation *with respect to time t*.
4. *After* completing Step 3, substitute into the resulting equation all known values for the variables and their rates of change. Then solve for the required rate of change.

NOTE   When using these guidelines, be sure you perform Step 3 before Step 4. Substituting the known values of the variables before differentiating will produce an inappropriate derivative.

The table below lists examples of mathematical models involving rates of change. For instance, the rate of change in the first example is the velocity of a car.

Verbal Statement	Mathematical Model
The velocity of a car after traveling for 1 hour is 50 miles per hour.	x = distance traveled $\dfrac{dx}{dt} = 50$ when $t = 1$
Water is being pumped into a swimming pool at a rate of 10 cubic meters per hour.	V = volume of water in pool $\dfrac{dV}{dt} = 10$ m³/hr
A gear is revolving at a rate of 25 revolutions per minute (1 revolution $= 2\pi$ radians).	θ = angle of revolution $\dfrac{d\theta}{dt} = 25(2\pi)$ rad/min

EXAMPLE 3 An Inflating Balloon

Air is being pumped into a spherical balloon (see Figure 3.39) at a rate of 4.5 cubic feet per minute. Find the rate of change of the radius when the radius is 2 feet.

Solution Let V be the volume of the balloon and let r be its radius. Because the volume is increasing at a rate of 4.5 cubic feet per minute, you know that at time t the rate of change of the volume is $dV/dt = \frac{9}{2}$. So, the problem can be stated as shown.

Given rate: $\quad \dfrac{dV}{dt} = \dfrac{9}{2}$ (constant rate)

Find: $\qquad \dfrac{dr}{dt}$ when $r = 2$

To find the rate of change of the radius, you must find an equation that relates the radius r to the volume V.

Equation: $\quad V = \dfrac{4}{3}\pi r^3 \qquad$ Volume of a sphere

Differentiating both sides of the equation with respect to t produces

$$\dfrac{dV}{dt} = 4\pi r^2 \dfrac{dr}{dt} \qquad \text{Differentiate with respect to } t.$$

$$\dfrac{dr}{dt} = \dfrac{1}{4\pi r^2}\left(\dfrac{dV}{dt}\right). \qquad \text{Solve for } dr/dt.$$

Finally, when $r = 2$, the rate of change of the radius is

$$\dfrac{dr}{dt} = \dfrac{1}{16\pi}\left(\dfrac{9}{2}\right) \approx 0.09 \text{ foot per minute.}$$

In Example 3, note that the volume is increasing at a *constant* rate but the radius is increasing at a *variable* rate. Just because two rates are related does not mean that they are proportional. In this particular case, the radius is growing more and more slowly as t increases. Do you see why?

Inflating a balloon
Figure 3.39

EXAMPLE 4   **The Speed of an Airplane Tracked by Radar**

An airplane is flying on a flight path that will take it directly over a radar tracking station, as shown in Figure 3.40. If s is decreasing at a rate of 400 miles per hour when $s = 10$ miles, what is the speed of the plane?

Solution   Let x be the horizontal distance from the station, as shown in Figure 3.40. Notice that when $s = 10$, $x = \sqrt{10^2 - 36} = 8$.

Given rate: $ds/dt = -400$ when $s = 10$

Find: dx/dt when $s = 10$ and $x = 8$

You can find the velocity of the plane as shown.

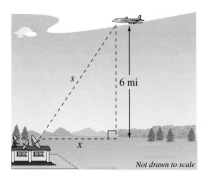

An airplane is flying at an altitude of 6 miles, s miles from the station.
Figure 3.40

Equation: $x^2 + 6^2 = s^2$ Pythagorean Theorem

$$2x\frac{dx}{dt} = 2s\frac{ds}{dt}$$ Differentiate with respect to t.

$$\frac{dx}{dt} = \frac{s}{x}\left(\frac{ds}{dt}\right)$$ Solve for dx/dt.

$$\frac{dx}{dt} = \frac{10}{8}(-400)$$ Substitute for s, x, and ds/dt.

$$= -500 \text{ miles per hour}$$ Simplify.

Because the velocity is -500 miles per hour, the *speed* is 500 miles per hour.

EXAMPLE 5   **A Changing Angle of Elevation**

Find the rate of change of the angle of elevation of the camera shown in Figure 3.41 at 10 seconds after lift-off.

Solution   Let θ be the angle of elevation, as shown in Figure 3.41. When $t = 10$, the height s of the rocket is $s = 50t^2 = 50(10)^2 = 5000$ feet.

Given rate: $ds/dt = 100t =$ velocity of rocket

Find: $d\theta/dt$ when $t = 10$ and $s = 5000$

Using Figure 3.41, you can relate s and θ by the equation $\tan\theta = s/2000$.

A television camera at ground level is filming the lift-off of a space shuttle that is rising vertically according to the position equation $s = 50t^2$, where s is measured in feet and t is measured in seconds. The camera is 2000 feet from the launch pad.
Figure 3.41

Equation: $\tan\theta = \dfrac{s}{2000}$ See Figure 3.41.

$$(\sec^2\theta)\frac{d\theta}{dt} = \frac{1}{2000}\left(\frac{ds}{dt}\right)$$ Differentiate with respect to t.

$$\frac{d\theta}{dt} = \cos^2\theta\,\frac{100t}{2000}$$ Substitute $100t$ for ds/dt.

$$= \left(\frac{2000}{\sqrt{s^2 + 2000^2}}\right)^2 \frac{100t}{2000}$$ $\cos\theta = 2000/\sqrt{s^2 + 2000^2}$

When $t = 10$ and $s = 5000$, you have

$$\frac{d\theta}{dt} = \frac{2000(100)(10)}{5000^2 + 2000^2} = \frac{2}{29} \text{ radian per second.}$$

So, when $t = 10$, θ is changing at a rate of $\frac{2}{29}$ radian per second.

EXAMPLE 6   **The Velocity of a Piston**

In the engine shown in Figure 3.42, a 7-inch connecting rod is fastened to a crank of radius 3 inches. The crankshaft rotates counterclockwise at a constant rate of 200 revolutions per minute. Find the velocity of the piston when $\theta = \pi/3$.

The velocity of a piston is related to the angle of the crankshaft.
Figure 3.42

Solution   Label the distances as shown in Figure 3.42. Because a complete revolution corresponds to 2π radians, it follows that $d\theta/dt = 200(2\pi) = 400\pi$ radians per minute.

Given rate: $\dfrac{d\theta}{dt} = 400\pi$ (constant rate)

Find: $\dfrac{dx}{dt}$ when $\theta = \dfrac{\pi}{3}$

You can use the Law of Cosines (Figure 3.43) to find an equation that relates x and θ.

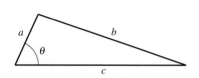

Law of Cosines:
$b^2 = a^2 + c^2 - 2ac \cos \theta$
Figure 3.43

Equation:

$$7^2 = 3^2 + x^2 - 2(3)(x) \cos \theta$$

$$0 = 2x \frac{dx}{dt} - 6\left(-x \sin \theta \frac{d\theta}{dt} + \cos \theta \frac{dx}{dt}\right)$$

$$(6 \cos \theta - 2x)\frac{dx}{dt} = 6x \sin \theta \frac{d\theta}{dt}$$

$$\frac{dx}{dt} = \frac{6x \sin \theta}{6 \cos \theta - 2x}\left(\frac{d\theta}{dt}\right)$$

When $\theta = \pi/3$, you can solve for x as shown.

$$7^2 = 3^2 + x^2 - 2(3)(x) \cos \frac{\pi}{3}$$

$$49 = 9 + x^2 - 6x\left(\frac{1}{2}\right)$$

$$0 = x^2 - 3x - 40$$

$$0 = (x - 8)(x + 5)$$

$$x = 8 \qquad\qquad\qquad \text{Choose positive solution.}$$

So, when $x = 8$ and $\theta = \pi/3$, the velocity of the piston is

$$\frac{dx}{dt} = \frac{6(8)\left(\sqrt{3}/2\right)}{6(1/2) - 16}(400\pi)$$

$$= \frac{9600\pi\sqrt{3}}{-13}$$

$$\approx -4018 \text{ inches per minute.}$$

NOTE   The velocity in Example 6 is negative because x represents a distance that is decreasing.

In Exercises 1–4, assume that x and y are both differentiable functions of t and find the required values of dy/dt and dx/dt.

Equation	Find	Given
1. $y = \sqrt{x}$	(a) $\dfrac{dy}{dt}$ when $x = 4$	$\dfrac{dx}{dt} = 3$
	(b) $\dfrac{dx}{dt}$ when $x = 25$	$\dfrac{dy}{dt} = 2$
2. $y = 2(x^2 - 3x)$	(a) $\dfrac{dy}{dt}$ when $x = 3$	$\dfrac{dx}{dt} = 2$
	(b) $\dfrac{dx}{dt}$ when $x = 1$	$\dfrac{dy}{dt} = 5$
3. $xy = 4$	(a) $\dfrac{dy}{dt}$ when $x = 8$	$\dfrac{dx}{dt} = 10$
	(b) $\dfrac{dx}{dt}$ when $x = 1$	$\dfrac{dy}{dt} = -6$
4. $x^2 + y^2 = 25$	(a) $\dfrac{dy}{dt}$ when $x = 3, y = 4$	$\dfrac{dx}{dt} = 8$
	(b) $\dfrac{dx}{dt}$ when $x = 4, y = 3$	$\dfrac{dy}{dt} = -2$

In Exercises 5–8, a point is moving along the graph of the given function such that dx/dt is 2 centimeters per second. Find dy/dt for the given values of x.

Function	Values of x
5. $y = x^2 + 1$	(a) $x = -1$ (b) $x = 0$ (c) $x = 1$
6. $y = \dfrac{1}{1 + x^2}$	(a) $x = -2$ (b) $x = 0$ (c) $x = 2$
7. $y = \tan x$	(a) $x = -\dfrac{\pi}{3}$ (b) $x = -\dfrac{\pi}{4}$ (c) $x = 0$
8. $y = \sin x$	(a) $x = \dfrac{\pi}{6}$ (b) $x = \dfrac{\pi}{4}$ (c) $x = \dfrac{\pi}{3}$

Writing About Concepts

In Exercises 9 and 10, use the graph of f to (a) determine whether dy/dt is positive or negative given that dx/dt is negative, and (b) determine whether dx/dt is positive or negative given that dy/dt is positive.

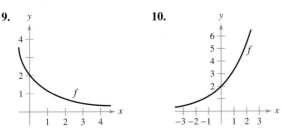

11. Consider the linear function $y = ax + b$. If x changes at a constant rate, does y change at a constant rate? If so, does it change at the same rate as x? Explain.

Writing About Concepts *(continued)*

12. In your own words, state the guidelines for solving related-rate problems.

13. Find the rate of change of the distance between the origin and a moving point on the graph of $y = x^2 + 1$ if $dx/dt = 2$ centimeters per second.

14. Find the rate of change of the distance between the origin and a moving point on the graph of $y = \sin x$ if $dx/dt = 2$ centimeters per second.

15. *Area* The radius r of a circle is increasing at a rate of 3 centimeters per minute. Find the rate of change of the area when (a) $r = 6$ centimeters and (b) $r = 24$ centimeters.

16. *Area* Let A be the area of a circle of radius r that is changing with respect to time. If dr/dt is constant, is dA/dt constant? Explain.

17. *Area* The included angle of the two sides of constant equal length s of an isosceles triangle is θ.

(a) Show that the area of the triangle is given by $A = \frac{1}{2}s^2 \sin \theta$.

(b) If θ is increasing at the rate of $\frac{1}{2}$ radian per minute, find the rates of change of the area when $\theta = \pi/6$ and $\theta = \pi/3$.

(c) Explain why the rate of change of the area of the triangle is not constant even though $d\theta/dt$ is constant.

18. *Volume* The radius r of a sphere is increasing at a rate of 2 inches per minute.

(a) Find the rate of change of the volume when $r = 6$ inches and $r = 24$ inches.

(b) Explain why the rate of change of the volume of the sphere is not constant even though dr/dt is constant.

19. *Volume* A hemispherical water tank with radius 6 meters is filled to a depth of h meters. The volume of water in the tank is given by $V = \frac{1}{3}\pi h(108 - h^2)$, $0 < h < 6$. If water is being pumped into the tank at the rate of 3 cubic meters per minute, find the rate of change of the depth of the water when $h = 2$ meters.

20. *Volume* All edges of a cube are expanding at a rate of 3 centimeters per second. How fast is the volume changing when each edge is (a) 1 centimeter and (b) 10 centimeters?

21. *Surface Area* The conditions are the same as in Exercise 20. Determine how fast the *surface area* is changing when each edge is (a) 1 centimeter and (b) 10 centimeters.

22. *Volume* The formula for the volume of a cone is $V = \frac{1}{3}\pi r^2 h$. Find the rate of change of the volume if dr/dt is 2 inches per minute and $h = 3r$ when (a) $r = 6$ inches and (b) $r = 24$ inches.

23. *Volume* At a sand and gravel plant, sand is falling off a conveyor and onto a conical pile at a rate of 10 cubic feet per minute. The diameter of the base of the cone is approximately three times the altitude. At what rate is the height of the pile changing when the pile is 15 feet high?

24. Depth A conical tank (with vertex down) is 10 feet across the top and 12 feet deep. If water is flowing into the tank at a rate of 10 cubic feet per minute, find the rate of change of the depth of the water when the water is 8 feet deep.

25. Depth A swimming pool is 12 meters long, 6 meters wide, 1 meter deep at the shallow end, and 3 meters deep at the deep end (see figure). Water is being pumped into the pool at $\frac{1}{4}$ cubic meter per minute, and there is 1 meter of water at the deep end.

(a) What percent of the pool is filled?

(b) At what rate is the water level rising?

Figure for 25   **Figure for 26**

26. Depth A trough is 12 feet long and 3 feet across the top (see figure). Its ends are isosceles triangles with altitudes of 3 feet.

(a) If water is being pumped into the trough at 2 cubic feet per minute, how fast is the water level rising when h is 1 foot deep?

(b) If the water is rising at a rate of $\frac{3}{8}$ inch per minute when $h = 2$, determine the rate at which water is being pumped into the trough.

27. Moving Ladder A ladder 25 feet long is leaning against the wall of a house (see figure). The base of the ladder is pulled away from the wall at a rate of 2 feet per second.

(a) How fast is the top of the ladder moving down the wall when its base is 7 feet, 15 feet, and 24 feet from the wall?

(b) Consider the triangle formed by the side of the house, the ladder, and the ground. Find the rate at which the area of the triangle is changing when the base of the ladder is 7 feet from the wall.

(c) Find the rate at which the angle between the ladder and the wall of the house is changing when the base of the ladder is 7 feet from the wall.

Figure for 27

Figure for 28

FOR FURTHER INFORMATION For more information on the mathematics of moving ladders, see the article "The Falling Ladder Paradox" by Paul Scholten and Andrew Simoson in *The College Mathematics Journal.* To view this article, go to the website *www.matharticles.com.*

28. Construction A construction worker pulls a five-meter plank up the side of a building under construction by means of a rope tied to one end of the plank (see figure). Assume the opposite end of the plank follows a path perpendicular to the wall of the building and the worker pulls the rope at a rate of 0.15 meter per second. How fast is the end of the plank sliding along the ground when it is 2.5 meters from the wall of the building?

29. Construction A winch at the top of a 12-meter building pulls a pipe of the same length to a vertical position, as shown in the figure. The winch pulls in rope at a rate of -0.2 meter per second. Find the rate of vertical change and the rate of horizontal change at the end of the pipe when $y = 6$.

Figure for 29   **Figure for 30**

30. Boating A boat is pulled into a dock by means of a winch 12 feet above the deck of the boat (see figure).

(a) The winch pulls in rope at a rate of 4 feet per second. Determine the speed of the boat when there is 13 feet of rope out. What happens to the speed of the boat as it gets closer to the dock?

(b) Suppose the boat is moving at a constant rate of 4 feet per second. Determine the speed at which the winch pulls in rope when there is a total of 13 feet of rope out. What happens to the speed at which the winch pulls in rope as the boat gets closer to the dock?

31. Air Traffic Control An air traffic controller spots two planes at the same altitude converging on a point as they fly at right angles to each other (see figure). One plane is 150 miles from the point moving at 450 miles per hour. The other plane is 200 miles from the point moving at 600 miles per hour.

(a) At what rate is the distance between the planes decreasing?

(b) How much time does the air traffic controller have to get one of the planes on a different flight path?

Figure for 31   **Figure for 32**

32. *Air Traffic Control*   An airplane is flying at an altitude of 5 miles and passes directly over a radar antenna (see figure on previous page). When the plane is 10 miles away ($s = 10$), the radar detects that the distance s is changing at a rate of 240 miles per hour. What is the speed of the plane?

33. *Baseball*   A baseball diamond has the shape of a square with sides 90 feet long (see figure). A player running from second base to third base at a speed of 28 feet per second is 30 feet from third base. At what rate is the player's distance s from home plate changing?

Figure for 33 and 34   **Figure for 35**

34. *Baseball*   For the baseball diamond in Exercise 33, suppose the player is running from first to second at a speed of 28 feet per second. Find the rate at which the distance from home plate is changing when the player is 30 feet from second base.

35. *Shadow Length*   A man 6 feet tall walks at a rate of 5 feet per second away from a light that is 15 feet above the ground (see figure). When he is 10 feet from the base of the light,

(a) at what rate is the tip of his shadow moving?

(b) at what rate is the length of his shadow changing?

36. *Shadow Length*   Repeat Exercise 35 for a man 6 feet tall walking at a rate of 5 feet per second *toward* a light that is 20 feet above the ground (see figure).

Figure for 36   **Figure for 37**

37. *Machine Design*   The endpoints of a movable rod of length 1 meter have coordinates $(x, 0)$ and $(0, y)$ (see figure). The position of the end of the rod on the x-axis is

$$x(t) = \frac{1}{2} \sin \frac{\pi t}{6}$$

where t is the time in seconds.

(a) Find the time of one complete cycle of the rod.

(b) What is the lowest point reached by the end of the rod on the y-axis?

(c) Find the speed of the y-axis endpoint when the x-axis endpoint is $\left(\frac{1}{4}, 0\right)$.

38. *Machine Design*   Repeat Exercise 37 for a position function of $x(t) = \frac{3}{5} \sin \pi t$. Use the point $\left(\frac{3}{10}, 0\right)$ for part (c).

39. *Evaporation*   As a spherical raindrop falls, it reaches a layer of dry air and begins to evaporate at a rate that is proportional to its surface area ($S = 4\pi r^2$). Show that the radius of the raindrop decreases at a constant rate.

40. *Electricity*   The combined electrical resistance R of R_1 and R_2, connected in parallel, is given by

$$\frac{1}{R} = \frac{1}{R_1} + \frac{1}{R_2}$$

where R, R_1, and R_2 are measured in ohms. R_1 and R_2 are increasing at rates of 1 and 1.5 ohms per second, respectively. At what rate is R changing when $R_1 = 50$ ohms and $R_2 = 75$ ohms?

41. *Adiabatic Expansion*   When a certain polyatomic gas undergoes adiabatic expansion, its pressure p and volume V satisfy the equation $pV^{1.3} = k$, where k is a constant. Find the relationship between the related rates dp/dt and dV/dt.

42. *Roadway Design*   Cars on a certain roadway travel on a circular arc of radius r. In order not to rely on friction alone to overcome the centrifugal force, the road is banked at an angle of magnitude θ from the horizontal (see figure). The banking angle must satisfy the equation $rg \tan \theta = v^2$, where v is the velocity of the cars and $g = 32$ feet per second per second is the acceleration due to gravity. Find the relationship between the related rates dv/dt and $d\theta/dt$.

43. *Angle of Elevation*   A balloon rises at a rate of 3 meters per second from a point on the ground 30 meters from an observer. Find the rate of change of the angle of elevation of the balloon from the observer when the balloon is 30 meters above the ground.

44. *Angle of Elevation*   A fish is reeled in at a rate of 1 foot per second from a point 10 feet above the water (see figure). At what rate is the angle between the line and the water changing when there is a total of 25 feet of line out?

45. *Relative Humidity* When the dewpoint is 65° Fahrenheit, the relative humidity H is

$$H = \frac{4347}{400,000,000} e^{369,444/(50t + 19,793)}$$

where t is the temperature in degrees Fahrenheit.

(a) Determine the relative humidity when $t = 65°$ and $t = 80°$.

(b) At 10 A.M., the temperature is 75° and increasing at the rate of 2° per hour. Find the rate at which the relative humidity is changing.

46. *Linear vs. Angular Speed* A patrol car is parked 50 feet from a long warehouse (see figure). The revolving light on top of the car turns at a rate of 30 revolutions per minute. How fast is the light beam moving along the wall when the beam makes angles of (a) $\theta = 30°$, (b) $\theta = 60°$, and (c) $\theta = 70°$ with the line perpendicular from the light to the wall?

Figure for 46

Figure for 47

 47. *Linear vs. Angular Speed* A wheel of radius 20 centimeters revolves at a rate of 10 revolutions per second. A dot is painted at a point P on the rim of the wheel (see figure).

(a) Find dx/dt as a function of θ.

(b) Use a graphing utility to graph the function in part (a).

(c) When is the absolute value of the rate of change of x greatest? When is it least?

(d) Find dx/dt when $\theta = 30°$ and $\theta = 60°$.

48. *Flight Control* An airplane is flying in still air with an airspeed of 240 miles per hour. If it is climbing at an angle of 22°, find the rate at which it is gaining altitude.

49. *Security Camera* A security camera is centered 50 feet above a 100-foot hallway (see figure). It is easiest to design the camera with a constant angular rate of rotation, but this results in a variable rate at which the images of the surveillance area are recorded. So, it is desirable to design a system with a variable rate of rotation and a constant rate of movement of the scanning beam along the hallway. Find a model for the variable rate of rotation if $|dx/dt| = 2$ feet per second.

50. *Think About It* Describe the relationship between the rate of change of y and the rate of change of x in each expression. Assume all variables and derivatives are positive.

(a) $\dfrac{dy}{dt} = 3\dfrac{dx}{dt}$ (b) $\dfrac{dy}{dt} = x(L - x)\dfrac{dx}{dt}$, $0 \le x \le L$

51. *Angle of Elevation* An airplane flies at an altitude of 5 miles toward a point directly over an observer (see figure). The speed of the plane is 600 miles per hour. Find the rates at which the angle of elevation θ is changing when the angle is (a) $\theta = 30°$, (b) $\theta = 60°$, and (c) $\theta = 75°$.

Figure for 51   **Figure for 52**

52. *Moving Shadow* A ball is dropped from a height of 20 meters, 12 meters away from the top of a 20-meter lamppost (see figure). The ball's shadow, caused by the light at the top of the lamppost, is moving along the level ground. How fast is the shadow moving 1 second after the ball is released? *(Submitted by Dennis Gittinger, St. Philips College, San Antonio, TX)*

Acceleration **In Exercises 53 and 54, find the acceleration of the specified object. (*Hint:* Recall that if a variable is changing at a constant rate, its acceleration is zero.)**

53. Find the acceleration of the top of the ladder described in Exercise 27 when the base of the ladder is 7 feet from the wall.

54. Find the acceleration of the boat in Exercise 30(a) when there is a total of 13 feet of rope out.

55. *Modeling Data* The table shows the numbers (in millions) of single women (never married) s and married women m in the civilian work force in the United States for the years 1993 through 2001. *(Source: U.S. Bureau of Labor Statistics)*

Year	1993	1994	1995	1996	1997	1998	1999	2000	2001
s	15.0	15.3	15.5	15.8	16.5	17.1	17.6	17.8	18.0
m	32.0	32.9	33.4	33.6	33.8	33.9	34.4	34.6	34.7

 (a) Use the regression capabilities of a graphing utility to find a model of the form $m(s) = as^3 + bs^2 + cs + d$ for the data, where t is the time in years, with $t = 3$ corresponding to 1993.

(b) Find dm/dt. Then use the model to estimate dm/dt for $t = 10$ if it is predicted that the number of single women in the work force will increase at the rate of 0.75 million per year.

Newton's Method

- **Approximate a zero of a function using Newton's Method.**

Newton's Method

In this section you will study a technique for approximating the real zeros of a function. The technique is called **Newton's Method,** and it uses tangent lines to approximate the graph of the function near its x-intercepts.

To see how Newton's Method works, consider a function f that is continuous on the interval $[a, b]$ and differentiable on the interval (a, b). If $f(a)$ and $f(b)$ differ in sign, then, by the Intermediate Value Theorem, f must have at least one zero in the interval (a, b). Suppose you estimate this zero to occur at

$$x = x_1 \qquad \text{First estimate}$$

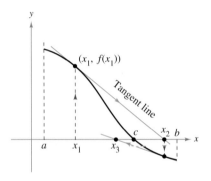

(a)

as shown in Figure 3.44(a). Newton's Method is based on the assumption that the graph of f and the tangent line at $(x_1, f(x_1))$ both cross the x-axis at *about* the same point. Because you can easily calculate the x-intercept for this tangent line, you can use it as a second (and, usually, better) estimate for the zero of f. The tangent line passes through the point $(x_1, f(x_1))$ with a slope of $f'(x_1)$. In point-slope form, the equation of the tangent line is therefore

$$y - f(x_1) = f'(x_1)(x - x_1)$$
$$y = f'(x_1)(x - x_1) + f(x_1).$$

Letting $y = 0$ and solving for x produces

$$x = x_1 - \frac{f(x_1)}{f'(x_1)}.$$

So, from the initial estimate x_1 you obtain a new estimate

$$x_2 = x_1 - \frac{f(x_1)}{f'(x_1)}. \qquad \text{Second estimate [see Figure 3.44(b)]}$$

(b)

The x-intercept of the tangent line approximates the zero of f.

Figure 3.44

You can improve on x_2 and calculate yet a third estimate

$$x_3 = x_2 - \frac{f(x_2)}{f'(x_2)}. \qquad \text{Third estimate}$$

Repeated application of this process is called Newton's Method.

NEWTON'S METHOD

Isaac Newton first described the method for approximating the real zeros of a function in his text *Method of Fluxions*. Although the book was written in 1671, it was not published until 1736. Meanwhile, in 1690, Joseph Raphson (1648–1715) published a paper describing a method for approximating the real zeros of a function that was very similar to Newton's. For this reason, the method is often referred to as the Newton-Raphson method.

Newton's Method for Approximating the Zeros of a Function

Let $f(c) = 0$, where f is differentiable on an open interval containing c. Then, to approximate c, use the following steps.

1. Make an initial estimate x_1 that is close to c. (A graph is helpful.)

2. Determine a new approximation

$$x_{n+1} = x_n - \frac{f(x_n)}{f'(x_n)}.$$

3. If $|x_n - x_{n+1}|$ is within the desired accuracy, let x_{n+1} serve as the final approximation. Otherwise, return to Step 2 and calculate a new approximation.

Each successive application of this procedure is called an **iteration.**

NOTE For many functions, just a few iterations of Newton's Method will produce approximations having very small errors, as shown in Example 1.

EXAMPLE 1 Using Newton's Method

Calculate three iterations of Newton's Method to approximate a zero of $f(x) = x^2 - 2$. Use $x_1 = 1$ as the initial guess.

Solution Because $f(x) = x^2 - 2$, you have $f'(x) = 2x$, and the iterative process is given by the formula

$$x_{n+1} = x_n - \frac{f(x_n)}{f'(x_n)} = x_n - \frac{x_n^2 - 2}{2x_n}.$$

The calculations for three iterations are shown in the table.

n	x_n	$f(x_n)$	$f'(x_n)$	$\dfrac{f(x_n)}{f'(x_n)}$	$x_n - \dfrac{f(x_n)}{f'(x_n)}$
1	1.000000	-1.000000	2.000000	-0.500000	1.500000
2	1.500000	0.250000	3.000000	0.083333	1.416667
3	1.416667	0.006945	2.833334	0.002451	1.414216
4	1.414216				

Of course, in this case you know that the two zeros of the function are $\pm\sqrt{2}$. To six decimal places, $\sqrt{2} = 1.414214$. So, after only three iterations of Newton's Method, you have obtained an approximation that is within 0.000002 of an actual root. The first iteration of this process is shown in Figure 3.45.

The first iteration of Newton's Method
Figure 3.45

EXAMPLE 2 Using Newton's Method

Use Newton's Method to approximate the zeros of

$$f(x) = e^x + x.$$

Continue the iterations until two successive approximations differ by less than 0.0001.

Solution Begin by sketching a graph of f, as shown in Figure 3.46. From the graph, you can observe that the function has only one zero, which occurs near $x = -0.6$. Next, differentiate f and form the iterative formula

$$x_{n+1} = x_n - \frac{f(x_n)}{f'(x_n)} = x_n - \frac{e^{x_n} + x_n}{e^{x_n} + 1}.$$

The calculations are shown in the table.

n	x_n	$f(x_n)$	$f'(x_n)$	$\dfrac{f(x_n)}{f'(x_n)}$	$x_n - \dfrac{f(x_n)}{f'(x_n)}$
1	-0.60000	-0.05119	1.54881	-0.03305	-0.56695
2	-0.56695	0.00030	1.56725	0.00019	-0.56714
3	-0.56714	0.00000	1.56714	0.00000	-0.56714
4	-0.56714				

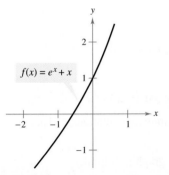

After three iterations of Newton's Method, the zero of f is approximated to the desired accuracy.
Figure 3.46

Because two successive approximations differ by less than the required 0.0001, you can estimate the zero of f to be -0.56714.

When, as in Examples 1 and 2, the approximations approach a limit, the sequence $x_1, x_2, x_3, \ldots, x_n, \ldots$ is said to **converge.** Moreover, if the limit is c, it can be shown that c must be a zero of f.

Newton's Method does not always yield a convergent sequence. One way it can fail to do so is shown in Figure 3.47. Because Newton's Method involves division by $f'(x_n)$, it is clear that the method will fail if the derivative is zero for any x_n in the sequence. When you encounter this problem, you can usually overcome it by choosing a different value for x_1. Another way Newton's Method can fail is shown in the next example.

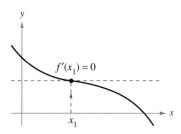

$f'(x_1) = 0$

x_1

Newton's Method fails to converge if $f'(x_n) = 0$.
Figure 3.47

EXAMPLE 3   An Example in Which Newton's Method Fails

The function $f(x) = x^{1/3}$ is not differentiable at $x = 0$. Show that Newton's Method fails to converge using $x_1 = 0.1$.

Solution   Because $f'(x) = \frac{1}{3}x^{-2/3}$, the iterative formula is

$$
\begin{aligned}
x_{n+1} &= x_n - \frac{f(x_n)}{f'(x_n)} \\
&= x_n - \frac{x_n^{1/3}}{\frac{1}{3}x_n^{-2/3}} \\
&= x_n - 3x_n \\
&= -2x_n.
\end{aligned}
$$

The calculations are shown in the table. This table and Figure 3.48 indicate that x_n continues to increase in magnitude as $n \to \infty$, and so the limit of the sequence does not exist.

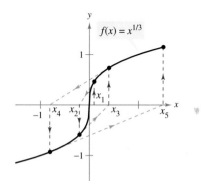

$f(x) = x^{1/3}$

Newton's Method fails to converge for every x-value other than the actual zero of f.
Figure 3.48

n	x_n	$f(x_n)$	$f'(x_n)$	$\dfrac{f(x_n)}{f'(x_n)}$	$x_n - \dfrac{f(x_n)}{f'(x_n)}$
1	0.10000	0.46416	1.54720	0.30000	-0.20000
2	-0.20000	-0.58480	0.97467	-0.60000	0.40000
3	0.40000	0.73681	0.61401	1.20000	-0.80000
4	-0.80000	-0.92832	0.38680	-2.40000	1.60000

NOTE   In Example 3, the initial estimate $x_1 = 0.1$ fails to produce a convergent sequence. Try showing that Newton's Method also fails for every other choice of x_1 (other than the actual zero).

It can be shown that a condition sufficient to produce convergence of Newton's Method to a zero of f is that

$$\left| \frac{f(x)\,f''(x)}{[f'(x)]^2} \right| < 1$$

Condition for convergence

on an open interval containing the zero. For instance, in Example 1 this test would yield $f(x) = x^2 - 2, f'(x) = 2x, f''(x) = 2$, and

$$\left| \frac{f(x)\,f''(x)}{[f'(x)]^2} \right| = \left| \frac{(x^2 - 2)(2)}{4x^2} \right| = \left| \frac{1}{2} - \frac{1}{x^2} \right|.$$

Example 1

On the interval $(1, 3)$, this quantity is less than 1 and therefore the convergence of Newton's Method is guaranteed. On the other hand, in Example 3, you have $f(x) = x^{1/3}, f'(x) = \frac{1}{3}x^{-2/3}, f''(x) = -\frac{2}{9}x^{-5/3}$, and

$$\left| \frac{f(x)\,f''(x)}{[f'(x)]^2} \right| = \left| \frac{x^{1/3}(-2/9)(x^{-5/3})}{(1/9)(x^{-4/3})} \right| = 2$$

Example 3

which is not less than 1 for any value of x, so you cannot conclude that Newton's Method will converge.

Algebraic Solutions of Polynomial Equations

The zeros of some functions, such as

$$f(x) = x^3 - 2x^2 - x + 2$$

can be found by simple algebraic techniques, such as factoring. The zeros of other functions, such as

$$f(x) = x^3 - x + 1$$

cannot be found by *elementary* algebraic methods. This particular function has only one real zero, and by using more advanced algebraic techniques you can determine the zero to be

$$x = -\sqrt[3]{\frac{3 - \sqrt{23/3}}{6}} - \sqrt[3]{\frac{3 + \sqrt{23/3}}{6}}.$$

Because the *exact* solution is written in terms of square roots and cube roots, it is called a **solution by radicals.**

NOTE   Try approximating the real zero of $f(x) = x^3 - x + 1$ and compare your result with the exact solution shown above.

NIELS HENRIK ABEL (1802–1829)

EVARISTE GALOIS (1811–1832)

Although the lives of both Abel and Galois were brief, their work in the fields of analysis and abstract algebra was far-reaching.

The determination of radical solutions of a polynomial equation is one of the fundamental problems of algebra. The earliest such result is the Quadratic Formula, which dates back at least to Babylonian times. The general formula for the zeros of a cubic function was developed much later. In the sixteenth century an Italian mathematician, Jerome Cardan, published a method for finding radical solutions to cubic and quartic equations. Then, for 300 years, the problem of finding a general quintic formula remained open. Finally, in the nineteenth century, the problem was answered independently by two young mathematicians. Niels Henrik Abel, a Norwegian mathematician, and Evariste Galois, a French mathematician, proved that it is not possible to solve a *general* fifth- (or higher-) degree polynomial equation by radicals. Of course, you can solve particular fifth-degree equations such as $x^5 - 1 = 0$, but Abel and Galois were able to show that no general *radical* solution exists.

Exercises for Section 3.8

In Exercises 1–4, complete two iterations of Newton's Method for the function using the given initial guess.

1. $f(x) = x^2 - 3$, $x_1 = 1.7$ **2.** $f(x) = 2x^2 - 3$, $x_1 = 1$

3. $f(x) = \sin x$, $x_1 = 3$ **4.** $f(x) = \tan x$, $x_1 = 0.1$

In Exercises 5–16, approximate the zero(s) of the function. Use Newton's Method and continue the process until two successive approximations differ by less than 0.001. Then find the zero(s) using a graphing utility and compare the results.

5. $f(x) = x^3 + x - 1$ **6.** $f(x) = x^5 + x - 1$

7. $f(x) = 3\sqrt{x - 1} - x$ **8.** $f(x) = x - 2\sqrt{x + 1}$

9. $f(x) = x - e^{-x}$ **10.** $f(x) = x - 3 + \ln x$

11. $f(x) = x^3 + 3$ **12.** $f(x) = 3 - 2x^3$

13. $f(x) = x^3 - 3.9x^2 + 4.79x - 1.881$

14. $f(x) = \frac{1}{2}x^4 - 3x - 3$

15. $f(x) = x + \sin(x + 1)$ **16.** $f(x) = x^3 - \cos x$

In Exercises 17–24, apply Newton's Method to approximate the x-value(s) of the indicated point(s) of intersection of the two graphs. Continue the process until two successive approximations differ by less than 0.001. [*Hint:* Let $h(x) = f(x) - g(x)$.]

17. $f(x) = 2x + 1$
    $g(x) = \sqrt{x + 4}$

18. $f(x) = 3 - x$
    $g(x) = 1/(x^2 + 1)$

19. $f(x) = x$
    $g(x) = \tan x$

20. $f(x) = x^2$
    $g(x) = \cos x$

21. $f(x) = -x$
    $g(x) = \ln x$

22. $f(x) = 2 - x^2$
    $g(x) = e^{x/2}$

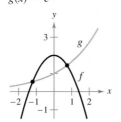

23. $f(x) = \arccos x$
    $g(x) = \arctan x$

24. $f(x) = 1 - x$
    $g(x) = \arcsin x$

25. *Mechanic's Rule* The Mechanic's Rule for approximating $\sqrt{a}$, $a > 0$, is

$$x_{n+1} = \frac{1}{2}\left(x_n + \frac{a}{x_n}\right), \quad n = 1, 2, 3 \ldots$$

where x_1 is an approximation of $\sqrt{a}$.

(a) Use Newton's Method and the function $f(x) = x^2 - a$ to derive the Mechanic's Rule.

(b) Use the Mechanic's Rule to approximate $\sqrt{5}$ and $\sqrt{7}$ to three decimal places.

26. (a) Use Newton's Method and the function $f(x) = x^n - a$ to obtain a general rule for approximating $x = \sqrt[n]{a}$.

(b) Use the general rule found in part (a) to approximate $\sqrt[4]{6}$ and $\sqrt[3]{15}$ to three decimal places.

In Exercises 27–30, apply Newton's Method using the given initial guess, and explain why the method fails.

27. $y = 2x^3 - 6x^2 + 6x - 1$, $x_1 = 1$

28. $y = 4x^3 - 12x^2 + 12x - 3$, $x_1 = \frac{3}{2}$

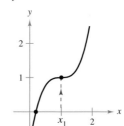

Figure for 27   **Figure for 28**

29. $f(x) = -x^3 + 6x^2 - 10x + 6$, $x_1 = 2$

30. $f(x) = 2\sin x + \cos 2x$, $x_1 = \dfrac{3\pi}{2}$

Figure for 29   **Figure for 30**

Writing About Concepts

31. In your own words and using a sketch, describe Newton's Method for approximating the zeros of a function.

32. Under what conditions will Newton's Method fail?

Fixed Point In Exercises 33–36, approximate the fixed point of the function to two decimal places. [A *fixed point* x_0 of a function f is a value of x such that $f(x_0) = x_0$.]

33. $f(x) = \cos x$

34. $f(x) = \cot x, \quad 0 < x < \pi$

35. $f(x) = e^{x/10}$

36. $f(x) = -\ln x$

 37. *Writing* Consider the function $f(x) = x^3 - 3x^2 + 3$.

(a) Use a graphing utility to graph f.

(b) Use Newton's Method with $x_1 = 1$ as an initial guess.

(c) Repeat part (b) using $x_1 = \frac{1}{4}$ as an initial guess and observe that the result is different.

(d) To understand why the results in parts (b) and (c) are different, sketch the tangent lines to the graph of f at the points $(1, f(1))$ and $\left(\frac{1}{4}, f\left(\frac{1}{4}\right)\right)$. Find the x-intercept of each tangent line and compare the intercepts with the first iteration of Newton's Method using the respective initial guesses.

(e) Write a short paragraph summarizing how Newton's Method works. Use the results of this exercise to describe why it is important to select the initial guess carefully.

 38. *Writing* Repeat the steps in Exercise 37 for the function $f(x) = \sin x$ with initial guesses of $x_1 = 1.8$ and $x_1 = 3$.

39. Use Newton's Method to show that the equation $x_{n+1} = x_n(2 - ax_n)$ can be used to approximate $1/a$ if x_1 is an initial guess for the reciprocal of a. Note that this method of approximating reciprocals uses only the operations of multiplication and subtraction. [*Hint:* Consider $f(x) = (1/x) - a$.]

40. Use the result of Exercise 39 to approximate the indicated reciprocal to three decimal places.

(a) $\frac{1}{3}$ (b) $\frac{1}{11}$

41. *Advertising Costs* A company that produces portable CD players estimates that the profit for selling a particular model is

$$P = -76x^3 + 4830x^2 - 320,000, \quad 0 \le x \le 60$$

where P is the profit in dollars and x is the advertising expense in 10,000s of dollars (see figure). According to this model, find the smaller of two advertising amounts that yield a profit P of $2,500,000.

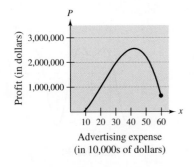

Advertising expense
(in 10,000s of dollars)

42. *Engine Power* The torque produced by a compact automobile engine is approximated by the model

$$T = 0.808x^3 - 17.974x^2 + 71.248x + 110.843, \quad 1 \le x \le 5$$

where T is the torque in foot-pounds and x is the engine speed in thousands of revolutions per minute (see figure). Approximate the two engine speeds that yield a torque T of 170 foot-pounds.

Engine speed
(in thousands of rpm)

True or False? In Exercises 43–46, determine whether the statement is true or false. If it is false, explain why or give an example that shows it is false.

43. The zeros of $f(x) = p(x)/q(x)$ coincide with the zeros of $p(x)$.

44. If the coefficients of a polynomial function are all positive, then the polynomial has no positive zeros.

45. If $f(x)$ is a cubic polynomial such that $f'(x)$ is never zero, then any initial guess will force Newton's Method to converge to the zero of f.

46. The roots of $\sqrt{f(x)} = 0$ coincide with the roots of $f(x) = 0$.

 In Exercises 47 and 48, *write a computer program or use a spreadsheet to find the zeros of a function using Newton's Method. Approximate the zeros of the function accurate to three decimal places. The output should be a table with the following headings.*

$$n, \quad x_n, \quad f(x_n), \quad f'(x_n), \quad \frac{f(x_n)}{f'(x_n)}, \quad x_n - \frac{f(x_n)}{f'(x_n)}$$

47. $f(x) = \frac{1}{4}x^3 - 3x^2 + \frac{3}{4}x - 2$

48. $f(x) = \sqrt{4 - x^2}\sin(x - 2)$

49. *Tangent Lines* The graph of $f(x) = -\sin x$ has infinitely many tangent lines that pass through the origin. Use Newton's Method to approximate the slope of the tangent line having the greatest slope to three decimal places.

50. Consider the function $f(x) = 2x^3 - 20x^2 - 12x - 24$.

(a) Use a graphing utility to determine the number of zeros of f.

(b) Use Newton's Method with an initial estimate of $x_1 = 2$ to approximate the zero of f to four decimal places.

(c) Repeat part (b) using initial estimates of $x_1 = 10$ and $x_1 = 100$.

(d) Discuss the results of parts (b) and (c). What can you conclude?

In Exercises 1–4, find the derivative of the function by using the definition of the derivative.

1. $f(x) = x^2 - 2x + 3$

2. $f(x) = \dfrac{x+1}{x-1}$

3. $f(x) = \sqrt{x} + 1$

4. $f(x) = 2/x$

In Exercises 5 and 6, describe the x-values at which f is differentiable.

5. $f(x) = (x+1)^{2/3}$

6. $f(x) = 4x/(x+3)$

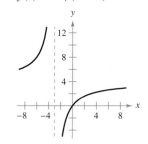

7. Sketch the graph of $f(x) = 4 - |x - 2|$.

   (a) Is f continuous at $x = 2$?

   (b) Is f differentiable at $x = 2$? Explain.

8. Sketch the graph of $f(x) = \begin{cases} x^2 + 4x + 2, & x < -2 \\ 1 - 4x - x^2, & x \geq -2. \end{cases}$

   (a) Is f continuous at $x = -2$?

   (b) Is f differentiable at $x = -2$? Explain.

In Exercises 9 and 10, find the slope of the tangent line to the graph of the function at the specified point.

9. $g(x) = \dfrac{2}{3}x^2 - \dfrac{x}{6}$, $\left(-1, \dfrac{5}{6}\right)$

10. $h(x) = \dfrac{3x}{8} - 2x^2$, $\left(-2, -\dfrac{35}{4}\right)$

In Exercises 11 and 12, (a) find an equation of the tangent line to the graph of f at the indicated point, (b) use a graphing utility to graph the function and its tangent line at the point, and (c) use the *derivative* feature of the graphing utility to confirm your results.

11. $f(x) = x^3 - 1$, $(-1, -2)$

12. $f(x) = \dfrac{2}{x+1}$, $(0, 2)$

In Exercises 13 and 14, use the alternative form of the derivative to find the derivative at x = c (if it exists).

13. $g(x) = x^2(x - 1)$, $c = 2$

14. $f(x) = \dfrac{1}{x+1}$, $c = 2$

Writing **In Exercises 15 and 16, the figure shows the graphs of a function and its derivative. Label the graphs as f or f′ and write a short paragraph stating the criteria used in making the selection. To print an enlarged copy of the graph, go to the website *www.mathgraphs.com*.**

15.

16.

In Exercises 17–32, find the derivative of the function.

17. $y = 25$

18. $y = -12$

19. $f(x) = x^8$

20. $g(x) = x^{12}$

21. $h(t) = 3t^4$

22. $f(t) = -8t^5$

23. $f(x) = x^3 - 3x^2$

24. $g(s) = 4s^4 - 5s^2$

25. $h(x) = 6\sqrt{x} + 3\sqrt[3]{x}$

26. $f(x) = x^{1/2} - x^{-1/2}$

27. $g(t) = \dfrac{2}{3t^2}$

28. $h(x) = \dfrac{2}{(3x)^2}$

29. $f(\theta) = 2\theta - 3\sin\theta$

30. $g(\alpha) = 4\cos\alpha + 6$

31. $f(t) = 3\cos t - 4e^t$

32. $g(s) = \dfrac{5}{3}\sin s - 2e^s$

33. ***Vibrating String*** When a guitar string is plucked, it vibrates with a frequency of $F = 200\sqrt{T}$, where F is measured in vibrations per second and the tension T is measured in pounds. Find the rate of change of F when (a) $T = 4$ and (b) $T = 9$.

34. ***Vertical Motion*** A ball is dropped from a height of 100 feet. One second later, another ball is dropped from a height of 75 feet. Which ball hits the ground first?

35. ***Vertical Motion*** To estimate the height of a building, a weight is dropped from the top of the building into a pool at ground level. How high is the building if the splash is seen 9.2 seconds after the weight is dropped?

36. ***Vertical Motion*** A bomb is dropped from an airplane at an altitude of 14,400 feet. How long will it take for the bomb to reach the ground? (Because of the motion of the plane, the fall will not be vertical, but the time will be the same as that for a vertical fall.) The plane is moving at 600 miles per hour. How far will the bomb move horizontally after it is released from the plane?

37. ***Projectile Motion*** A thrown ball follows a path described by $y = x - 0.02x^2$.

   (a) Sketch a graph of the path.

   (b) Find the total horizontal distance the ball was thrown.

   (c) At what x-value does the ball reach its maximum height? (Use the symmetry of the path.)

   (d) Find an equation that gives the instantaneous rate of change of the height of the ball with respect to the horizontal change. Evaluate the equation at $x = 0$, 10, 25, 30, and 50.

   (e) What is the instantaneous rate of change of the height when the ball reaches its maximum height?

38. Projectile Motion The path of a projectile thrown at an angle of 45° with level ground is

$$y = x - \frac{32}{v_0^2}(x^2)$$

where the initial velocity is v_0 feet per second.

(a) Find the x-coordinate of the point where the projectile strikes the ground. Use the symmetry of the path of the projectile to locate the x-coordinate of the point where the projectile reaches its maximum height.

(b) What is the instantaneous rate of change of the height when the projectile is at its maximum height?

(c) Show that doubling the initial velocity of the projectile multiplies both the maximum height and the range by a factor of 4.

(d) Find the maximum height and range of a projectile thrown with an initial velocity of 70 feet per second. Use a graphing utility to sketch the path of the projectile.

39. Horizontal Motion The position function of a particle moving along the x-axis is

$$x(t) = t^2 - 3t + 2$$

for $-\infty < t < \infty$.

(a) Find the velocity of the particle.

(b) Find the open t-interval(s) in which the particle is moving to the left.

(c) Find the position of the particle when the velocity is 0.

(d) Find the speed of the particle when the position is 0.

40. Modeling Data The speed of a car in miles per hour and its stopping distance in feet are recorded in the table.

Speed (x)	20	30	40	50	60
Stopping Distance (y)	25	55	105	188	300

(a) Use the regression capabilities of a graphing utility to find a quadratic model for the data.

(b) Use a graphing utility to plot the data and graph the model.

(c) Use a graphing utility to graph dy/dx.

(d) Use the model to approximate the stopping distance at a speed of 65 miles per hour.

(e) Use the graphs in parts (b) and (c) to explain the change in stopping distance as the speed increases.

In Exercises 41–57, find the derivative of the function.

41. $f(x) = (3x^2 + 7)(x^2 - 2x + 3)$

42. $g(x) = (x^3 - 3x)(x + 2)$

43. $h(x) = \sqrt{x} \sin x$

44. $f(t) = t^3 \cos t$

45. $f(x) = \dfrac{2x^3 - 1}{x^2}$

46. $f(x) = \dfrac{x + 1}{x - 1}$

47. $f(x) = \dfrac{x^2 + x - 1}{x^2 - 1}$

48. $f(x) = \dfrac{6x - 5}{x^2 + 1}$

49. $f(x) = \dfrac{1}{4 - 3x^2}$

50. $f(x) = \dfrac{9}{3x^2 - 2x}$

51. $y = \dfrac{x^2}{\cos x}$

52. $y = \dfrac{\sin x}{x^2}$

53. $y = 3x^2 \sec x$

54. $y = 2x - x^2 \tan x$

55. $y = -x \tan x$

56. $y = \dfrac{1 + \sin x}{1 - \sin x}$

57. $y = 4xe^x$

58. Acceleration The velocity of an object in meters per second is $v(t) = 36 - t^2$, $0 \le t \le 6$. Find the velocity and acceleration of the object when $t = 4$.

In Exercises 59–62, find the second derivative of the function.

59. $g(t) = t^3 - 3t + 2$

60. $f(x) = 12\sqrt[4]{x}$

61. $f(\theta) = 3 \tan \theta$

62. $h(t) = 4 \sin t - 5 \cos t$

In Exercises 63 and 64, show that the function satisfies the equation.

Function	Equation
63. $y = 2 \sin x + 3 \cos x$	$y'' + y = 0$
64. $y = \dfrac{10 - \cos x}{x}$	$xy' + y = \sin x$

65. Rate of Change Determine whether there exist any values of x in the interval $[0, 2\pi)$ such that the rate of change of $f(x) = \sec x$ and the rate of change of $g(x) = \csc x$ are equal.

66. Volume The radius of a right circular cylinder is given by $\sqrt{t + 2}$ and its height is $\frac{1}{2}\sqrt{t}$, where t is time in seconds and the dimensions are in inches. Find the rate of change of the volume with respect to time.

In Exercises 67–96, find the derivative of the function.

67. $f(x) = \sqrt{1 - x^3}$

68. $f(x) = \sqrt[3]{x^2 - 1}$

69. $h(x) = \left(\dfrac{x - 3}{x^2 + 1}\right)^2$

70. $f(x) = \left(x^2 + \dfrac{1}{x}\right)^5$

71. $f(s) = (s^2 - 1)^{5/2}(s^3 + 5)$

72. $h(\theta) = \dfrac{\theta}{(1 - \theta)^3}$

73. $y = 3 \cos(3x + 1)$

74. $y = 1 - \cos 2x + 2 \cos^2 x$

75. $y = \frac{1}{2} \csc 2x$

76. $y = \csc 3x + \cot 3x$

77. $y = \dfrac{x}{2} - \dfrac{\sin 2x}{4}$

78. $y = \dfrac{\sec^7 x}{7} - \dfrac{\sec^5 x}{5}$

79. $y = \dfrac{2}{3} \sin^{3/2} x - \dfrac{2}{7} \sin^{7/2} x$

80. $f(x) = \dfrac{3x}{\sqrt{x^2 + 1}}$

81. $y = \dfrac{\sin \pi x}{x + 2}$

82. $y = \dfrac{\cos(x - 1)}{x - 1}$

83. $g(t) = t^2 e^{t/4}$

84. $h(z) = e^{-z^2/2}$

85. $y = \sqrt{e^{2x} + e^{-2x}}$

86. $y = 3e^{-3/t}$

87. $g(x) = \dfrac{x^2}{e^x}$

88. $f(\theta) = \dfrac{1}{2} e^{\sin 2\theta}$

89. $g(x) = \ln \sqrt{x}$

90. $h(x) = \ln \dfrac{x(x - 1)}{x - 2}$

91. $f(x) = x\sqrt{\ln x}$

92. $f(x) = \ln[x(x^2 - 2)^{2/3}]$

93. $y = \dfrac{1}{b^2}\left[\ln(a + bx) + \dfrac{a}{a + bx}\right]$

94. $y = \dfrac{1}{b^2}[a + bx - a\ln(a + bx)]$

95. $y = -\dfrac{1}{a}\ln\dfrac{a + bx}{x}$

96. $y = -\dfrac{1}{ax} + \dfrac{b}{a^2}\ln\dfrac{a + bx}{x}$

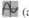 **In Exercises 97–104, use a computer algebra system to find the derivative of the function. Use the utility to graph the function and its derivative on the same set of coordinate axes. Describe the behavior of the function that corresponds to any zeros of the graph of the derivative.**

97. $f(t) = t^2(t - 1)^5$

98. $f(x) = [(x - 2)(x + 4)]^2$

99. $g(x) = \dfrac{2x}{\sqrt{x + 1}}$

100. $g(x) = x\sqrt{x^2 + 1}$

101. $f(t) = \sqrt{t + 1}\sqrt[3]{t + 1}$

102. $y = \sqrt{3x}(x + 2)^3$

103. $y = \tan\sqrt{1 - x}$

104. $y = 2\csc^3\left(\sqrt{x}\right)$

In Exercises 105–108, find the second derivative of the function.

105. $y = 2x^2 + \sin 2x$

106. $y = \dfrac{1}{x} + \tan x$

107. $f(x) = \cot x$

108. $y = \sin^2 x$

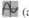 **In Exercises 109–114, use a computer algebra system to find the second derivative of the function.**

109. $f(t) = \dfrac{t}{(1 - t)^2}$

110. $g(x) = \dfrac{6x - 5}{x^2 + 1}$

111. $g(\theta) = \tan 3\theta - \sin(\theta - 1)$

112. $h(x) = x\sqrt{x^2 - 1}$

113. $g(x) = x^3\ln x$

114. $f(x) = 6x^2\,e^{-x/3}$

115. *Refrigeration* The temperature T of food put in a freezer is

$$T = \dfrac{700}{t^2 + 4t + 10}$$

where t is the time in hours. Find the rate of change of T with respect to t at each of the following times.

(a) $t = 1$ (b) $t = 3$ (c) $t = 5$ (d) $t = 10$

116. *Fluid Flow* The emergent velocity v of a liquid flowing from a hole in the bottom of a tank is given by $v = \sqrt{2gh}$, where g is the acceleration due to gravity (32 feet per second per second) and h is the depth of the liquid in the tank. Find the rate of change of v with respect to h when (a) $h = 9$ and (b) $h = 4$. (Note that $g = +32$ feet per second per second. The sign of g depends on how a problem is modeled. In this case, letting g be negative would produce an imaginary value for v.)

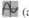 **117.** *Modeling Data* The atmospheric pressure decreases with increasing altitude. At sea level, the average air pressure is one atmosphere (1.033227 kilograms per square centimeter). The table gives the pressure p (in atmospheres) at a given altitude h (in kilometers).

h	0	5	10	15	20	25
p	1	0.55	0.25	0.12	0.06	0.02

(a) Use a graphing utility to find a model of the form $p = a + b\ln h$ for the data. Explain why the result is an error message.

(b) Use a graphing utility to find the logarithmic model $h = a + b\ln p$ for the data.

(c) Use a graphing utility to plot the data and graph the logarithmic model.

(d) Use the model to estimate the altitude at which the pressure is 0.75 atmosphere.

(e) Use the model to estimate the pressure at an altitude of 13 kilometers.

(f) Find the rate of change of pressure when $h = 5$ and $h = 20$. Interpret the results in the context of the problem.

118. *Tractrix* A person walking along a dock drags a boat by a 10-meter rope. The boat travels along a path known as a *tractrix* (see figure). The equation of this path is

$$y = 10\ln\left(\dfrac{10 + \sqrt{100 - x^2}}{x}\right) - \sqrt{100 - x^2}.$$

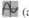 (a) Use a graphing utility to graph the function.

(b) What is the slope of the path when $x = 5$ and $x = 9$?

(c) What does the slope of the path approach as $x \to 10$?

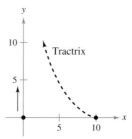

Tractrix

In Exercises 119–126, use implicit differentiation to find dy/dx.

119. $x^2 + 3xy + y^3 = 10$

120. $x^2 + 9y^2 - 4x + 3y = 0$

121. $\cos x^2 = xe^y$

122. $ye^x + xe^y = xy$

123. $y\sqrt{x} - x\sqrt{y} = 16$

124. $y^2 = (x - y)(x^2 + y)$

125. $x\sin y = y\cos x$

126. $\cos(x + y) = x$

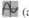 **In Exercises 127–130, find the equations of the tangent line and the normal line to the graph of the equation at the indicated point. Use a graphing utility to graph the equation, the tangent line, and the normal line.**

127. $x^2 + y^2 = 20$, $(2, 4)$

128. $x^2 - y^2 = 16$, $(5, 3)$

129. $y\ln x + y^2 = 0$, $(e, -1)$

130. $\ln(x + y) = x$, $(0, 1)$

In Exercises 131 and 132, use logarithmic differentiation to find dy/dx.

131. $y = \dfrac{x\sqrt{x^2 + 1}}{x + 4}$ **132.** $y = \dfrac{(2x + 1)^3(x^2 - 1)^2}{x + 3}$

In Exercise 133–136, find $(f^{-1})'(a)$ for the function f and real number a.

Function	Real number
133. $f(x) = x^3 + 2$	$a = -1$
134. $f(x) = x\sqrt{x - 3}$	$a = 4$
135. $f(x) = \tan x, \ -\dfrac{\pi}{4} \le x \le \dfrac{\pi}{4}$	$a = \dfrac{\sqrt{3}}{3}$
136. $f(x) = \ln x$	$a = 0$

In Exercises 137–142, find the derivative of the function.

137. $y = \tan(\arcsin x)$ **138.** $y = \arctan(x^2 - 1)$

139. $y = x \operatorname{arcsec} x$ **140.** $y = \frac{1}{2} \arctan e^{2x}$

141. $y = x(\arcsin x)^2 - 2x + 2\sqrt{1 - x^2} \arcsin x$

142. $y = \sqrt{x^2 - 4} - 2 \operatorname{arcsec} \dfrac{x}{2}, \quad 2 < x < 4$

143. A point moves along the curve $y = \sqrt{x}$ in such a way that the y-value is increasing at a rate of 2 units per second. At what rate is x changing for each of the following values?

(a) $x = \frac{1}{2}$ (b) $x = 1$ (c) $x = 4$

144. *Surface Area* The edges of a cube are expanding at a rate of 5 centimeters per second. How fast is the surface area changing when each edge is 4.5 centimeters?

145. *Changing Depth* The cross section of a 5-meter trough is an isosceles trapezoid with a 2-meter lower base, a 3-meter upper base, and an altitude of 2 meters. Water is running into the trough at a rate of 1 cubic meter per minute. How fast is the water level rising when the water is 1 meter deep?

146. *Linear and Angular Velocity* A rotating beacon is located 1 kilometer off a straight shoreline (see figure). If the beacon rotates at a rate of 3 revolutions per minute, how fast (in kilometers per hour) does the beam of light appear to be moving to a viewer who is $\frac{1}{2}$ kilometer down the shoreline?

147. *Moving Shadow* A sandbag is dropped from a balloon at a height of 60 meters when the angle of elevation to the sun is $30°$ (see figure). Find the rate at which the shadow of the sandbag is traveling along the ground when the sandbag is at a height of 35 meters. [*Hint:* The position of the sandbag is given by $s(t) = 60 - 4.9t^2$.]

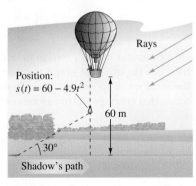

148. *Geometry* Consider the rectangle shown in the figure.

(a) Find the area of the rectangle as a function of x.

(b) Find the rate of change of the area when $x = 4$ centimeters if $dx/dt = 4$ centimeters per minute.

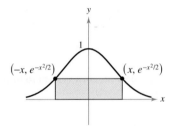

In Exercises 149–152, use Newton's Method to approximate any real zeros of the function accurate to three decimal places. Use the root-finding capabilities of a graphing utility to verify your results.

149. $f(x) = x^3 - 3x - 1$

150. $f(x) = x^3 + 2x + 1$

151. $g(x) = xe^x - 4$

152. $f(x) = 3 - x \ln x$

In Exercises 153 and 154, use Newton's Method to approximate, to three decimal places, the x-values of any points of intersection of the graphs of the equations. Use a graphing utility to verify your results.

153. $y = x^4$

$y = x + 3$

154. $y = \sin \pi x$

$y = 1 - x$

1. Consider the graph of the parabola $y = x^2$.

(a) Find the radius r of the largest possible circle centered on the y-axis that is tangent to the parabola at the origin, as indicated in the figure. This circle is called the **circle of curvature** (see Section 12.5). Use a graphing utility to graph the circle and parabola in the same viewing window.

(b) Find the center $(0, b)$ of the circle of radius 1 centered on the y-axis that is tangent to the parabola at two points, as indicated in the figure. Use a graphing utility to graph the circle and parabola in the same viewing window.

        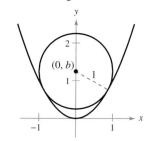

Figure for 1(a)   **Figure for 1(b)**

2. Graph the two parabolas $y = x^2$ and $y = -x^2 + 2x - 5$ in the same coordinate plane. Find equations of the two lines simultaneously tangent to both parabolas.

3. (a) Find the polynomial $P_1(x) = a_0 + a_1x$ whose value and slope agree with the value and slope of $f(x) = \cos x$ at the point $x = 0$.

(b) Find the polynomial $P_2(x) = a_0 + a_1x + a_2x^2$ whose value and first two derivatives agree with the value and first two derivatives of $f(x) = \cos x$ at the point $x = 0$. This polynomial is called the second-degree **Taylor polynomial** of $f(x) = \cos x$ at $x = 0$.

(c) Complete the table comparing the values of f and P_2. What do you observe?

x	-1.0	-0.1	-0.001	0	0.001	0.1	1.0
$\cos x$							
$P_2(x)$							

(d) Find the third-degree Taylor polynomial of $f(x) = \sin x$ at $x = 0$.

4. (a) Find an equation of the tangent line to the parabola $y = x^2$ at the point $(2, 4)$.

(b) Find an equation of the normal line to $y = x^2$ at the point $(2, 4)$. (The normal line is perpendicular to the tangent line.) Where does this line intersect the parabola a second time?

(c) Find equations of the tangent line and normal line to $y = x^2$ at the point $(0, 0)$.

(d) Prove that for any point $(a, b) \neq (0, 0)$ on the parabola $y = x^2$, the normal line intersects the graph a second time.

5. Find a third-degree polynomial $p(x)$ that is tangent to the line $y = 14x - 13$ at the point $(1, 1)$, and tangent to the line $y = -2x - 5$ at the point $(-1, -3)$.

6. Find a function of the form $f(x) = a + b \cos cx$ that is tangent to the line $y = 1$ at the point $(0, 1)$, and tangent to the line

$$y = x + \frac{3}{2} - \frac{\pi}{4}$$

at the point $\left(\frac{\pi}{4}, \frac{3}{2}\right)$.

 7. The graph of the **eight curve,**

$$x^4 = a^2(x^2 - y^2), a \neq 0,$$

is shown below.

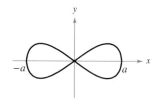

(a) Explain how you could use a graphing utility to obtain the graph of this curve.

(b) Use a graphing utility to graph the curve for various values of the constant a. Describe how a affects the shape of the curve.

(c) Determine the points on the curve where the tangent line is horizontal.

 8. The graph of the **pear-shaped quartic,**

$$b^2y^2 = x^3(a - x), a, b > 0,$$

is shown below.

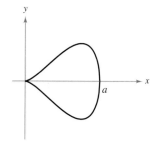

(a) Explain how you could use a graphing utility to obtain the graph of this curve.

(b) Use a graphing utility to graph the curve for various values of the constants a and b. Describe how a and b affect the shape of the curve.

(c) Determine the points on the curve where the tangent line is horizontal.

9. A man 6 feet tall walks at a rate of 5 feet per second toward a streetlight that is 30 feet high (see figure). The man's 3-foot-tall child follows at the same speed, but 10 feet behind the man. At times, the shadow behind the child is caused by the man, and at other times, by the child.

(a) Suppose the man is 90 feet from the streetlight. Show that the man's shadow extends beyond the child's shadow.

(b) Suppose the man is 60 feet from the streetlight. Show that the child's shadow extends beyond the man's shadow.

(c) Determine the distance d from the man to the streetlight at which the tips of the two shadows are exactly the same distance from the streetlight.

(d) Determine how fast the tip of the shadow is moving as a function of x, the distance between the man and the streetlight. Discuss the continuity of this shadow speed function.

Figure for 9

Figure for 10

10. A particle is moving along the graph of $y = \sqrt[3]{x}$ (see figure). When $x = 8$, the y-component of its position is increasing at the rate of 1 centimeter per second.

(a) How fast is the x-component changing at this moment?

(b) How fast is the distance from the origin changing at this moment?

(c) How fast is the angle of inclination θ changing at this moment?

11. Let L be the tangent line to the graph of the function $y = \ln x$ at the point (a, b). Show that the distance between b and c is always equal to 1.

12. Let L be the tangent line to the graph of the function $y = e^x$ at the point (a, b). Show that the distance between a and c is always equal to 1.

Figure for 11   **Figure for 12**

13. The fundamental limit

$$\lim_{x \to 0} \frac{\sin x}{x} = 1$$

assumes that x is measured in radians. What happens if we assume that x is measured in degrees instead of radians?

(a) Set your calculator to degree mode and complete the table.

z **(in degrees)**	0.1	0.01	0.0001
$\dfrac{\sin z}{z}$			

(b) Use the table to estimate $\lim\limits_{z \to 0} \dfrac{\sin z}{z}$ for z in degrees. What is the exact value of this limit? (*Hint:* $180° = \pi$ radians)

(c) Use the limit definition of the derivative to find $\dfrac{d}{dz} \sin z$ for z in degrees.

(d) Define the new functions $S(z) = \sin(cz)$ and $C(z) = \cos(cz)$, where $c = \pi/180$. Find $S(90)$ and $C(180)$. Use the Chain Rule to calculate $\dfrac{d}{dz} S(z)$.

(e) Explain why calculus is made easier by using radians instead of degrees.

14. An astronaut standing on the moon throws a rock into the air. The height of the rock is $s = -\frac{27}{10}t^2 + 27t + 6$, where s is measured in feet and t is measured in seconds.

(a) Find expressions for the velocity and acceleration of the rock.

(b) Find the time when the rock is at its highest point by finding the time when the velocity is zero. What is the rock's height at this time?

(c) How does the acceleration of the rock compare with the acceleration due to gravity on Earth?

15. If a is the acceleration of an object, the *jerk* j is defined by $j = a'(t)$.

(a) Use this definition to give a physical interpretation of j.

(b) The figure shows the graphs of the position, velocity, acceleration, and jerk functions of a vehicle. Identify each graph and explain your reasoning.

Applications of Differentiation

In Chapter 4, you will use calculus to analyze graphs of functions. For example, you can use the derivative of a function to determine the function's maximum and minimum values. You can use limits to identify any asymptotes of the function's graph. In Section 4.6, you will combine these techniques to sketch the graph of a function.

When a glassblower removes a glowing "blob" of molten glass from a kiln, its temperature is about 1700°F. At first, the molten glass cools rapidly. Then, as the temperature of the glass approaches room temperature, it cools more and more slowly. Will the temperature of the glass ever actually reach room temperature? Why?

www.shawnolson.net/a/507

Extrema on an Interval

- Understand the definition of extrema of a function on an interval.
- Understand the definition of relative extrema of a function on an open interval.
- Find extrema on a closed interval.

Extrema of a Function

In calculus, much effort is devoted to determining the behavior of a function f on an interval I. Does f have a maximum value on I? Does it have a minimum value? Where is the function increasing? Where is it decreasing? In this chapter you will learn how derivatives can be used to answer these questions. You will also see why these questions are important in real-life applications.

(a) f is continuous, $[-1, 2]$ is closed.

Definition of Extrema

Let f be defined on an interval I containing c.

1. $f(c)$ is the **minimum of f on I** if $f(c) \leq f(x)$ for all x in I.
2. $f(c)$ is the **maximum of f on I** if $f(c) \geq f(x)$ for all x in I.

The minimum and maximum of a function on an interval are the **extreme values,** or **extrema** (the singular form of extrema is extremum), of the function on the interval. The minimum and maximum of a function on an interval are also called the **absolute minimum** and **absolute maximum** on the interval.

A function need not have a minimum or a maximum on an interval. For instance, in Figure 4.1(a) and (b), you can see that the function $f(x) = x^2 + 1$ has both a minimum and a maximum on the closed interval $[-1, 2]$, but does not have a maximum on the open interval $(-1, 2)$. Moreover, in Figure 4.1(c), you can see that continuity (or the lack of it) can affect the existence of an extremum on the interval. This suggests the following theorem. (Although the Extreme Value Theorem is intuitively plausible, a proof of this theorem is not within the scope of this text.)

(b) f is continuous, $(-1, 2)$ is open.

THEOREM 4.1   The Extreme Value Theorem

If f is continuous on a closed interval $[a, b]$, then f has both a minimum and a maximum on the interval.

EXPLORATION

Finding Minimum and Maximum Values   The Extreme Value Theorem (like the Intermediate Value Theorem) is an *existence theorem* because it tells of the existence of minimum and maximum values but does not show how to find these values. Use the extreme-value capability of a graphing utility to find the minimum and maximum values of each of the following functions. In each case, do you think the x-values are exact or approximate? Explain your reasoning.

a. $f(x) = x^2 - 4x + 5$ on the closed interval $[-1, 3]$

b. $f(x) = x^3 - 2x^2 - 3x - 2$ on the closed interval $[-1, 3]$

(c) g is not continuous, $[-1, 2]$ is closed. Extrema can occur at interior points or endpoints of an interval. Extrema that occur at the endpoints are called **endpoint extrema.**

Figure 4.1

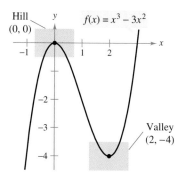

Hill (0, 0)

$f(x) = x^3 - 3x^2$

Valley (2, -4)

f has a relative maximum at $(0, 0)$ and a relative minimum at $(2, -4)$.

Figure 4.2

Relative Extrema and Critical Numbers

In Figure 4.2, the graph of $f(x) = x^3 - 3x^2$ has a **relative maximum** at the point $(0, 0)$ and a **relative minimum** at the point $(2, -4)$. Informally, you can think of a relative maximum as occurring on a "hill" on the graph, and a relative minimum as occurring in a "valley" on the graph. Such a hill and valley can occur in two ways. If the hill (or valley) is smooth and rounded, the graph has a horizontal tangent line at the high point (or low point). If the hill (or valley) is sharp and peaked, the graph represents a function that is not differentiable at the high point (or low point).

Definition of Relative Extrema

1. If there is an open interval containing c on which $f(c)$ is a maximum, then $f(c)$ is called a **relative maximum** of f, or you can say that f has a **relative maximum at $(c, f(c))$**.

2. If there is an open interval containing c on which $f(c)$ is a minimum, then $f(c)$ is called a **relative minimum** of f, or you can say that f has a **relative minimum at $(c, f(c))$**.

The plural of relative maximum is relative maxima, and the plural of relative minimum is relative minima.

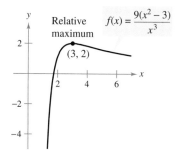

Relative maximum

$f(x) = \dfrac{9(x^2 - 3)}{x^3}$

(3, 2)

(a) $f'(3) = 0$

Example 1 examines the derivatives of functions at *given* relative extrema. (Much more is said about *finding* the relative extrema of a function in Section 4.3.)

EXAMPLE I   The Value of the Derivative at Relative Extrema

Find the value of the derivative at each of the relative extrema shown in Figure 4.3.

Solution

a. The derivative of $f(x) = \dfrac{9(x^2 - 3)}{x^3}$ is

$$f'(x) = \frac{x^3(18x) - (9)(x^2 - 3)(3x^2)}{(x^3)^2} \qquad \text{Differentiate using Quotient Rule.}$$

$$= \frac{9(9 - x^2)}{x^4}. \qquad \text{Simplify.}$$

At the point $(3, 2)$, the value of the derivative is $f'(3) = 0$ [see Figure 4.3(a)].

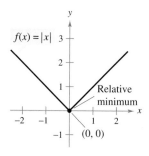

$f(x) = |x|$

Relative minimum

(0, 0)

(b) $f'(0)$ does not exist.

b. At $x = 0$, the derivative of $f(x) = |x|$ *does not exist* because the following one-sided limits differ [see Figure 4.3(b)].

$$\lim_{x \to 0^-} \frac{f(x) - f(0)}{x - 0} = \lim_{x \to 0^-} \frac{|x|}{x} = -1 \qquad \text{Limit from the left}$$

$$\lim_{x \to 0^+} \frac{f(x) - f(0)}{x - 0} = \lim_{x \to 0^+} \frac{|x|}{x} = 1 \qquad \text{Limit from the right}$$

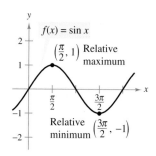

$f(x) = \sin x$

$\left(\dfrac{\pi}{2}, 1\right)$ Relative maximum

Relative minimum $\left(\dfrac{3\pi}{2}, -1\right)$

(c) $f'\left(\dfrac{\pi}{2}\right) = 0$; $f'\left(\dfrac{3\pi}{2}\right) = 0$

Figure 4.3

c. The derivative of $f(x) = \sin x$ is

$$f'(x) = \cos x.$$

At the point $(\pi/2, 1)$, the value of the derivative is $f'(\pi/2) = \cos(\pi/2) = 0$. At the point $(3\pi/2, -1)$, the value of the derivative is $f'(3\pi/2) = \cos(3\pi/2) = 0$ [see Figure 4.3(c)].

Note in Example 1 that at each relative extremum, the derivative is either zero or does not exist. The x-values at these special points are called **critical numbers.** Figure 4.4 illustrates the two types of critical numbers.

Definition of a Critical Number

Let f be defined at c. If $f'(c) = 0$ or if f is not differentiable at c, then c is a **critical number** of f.

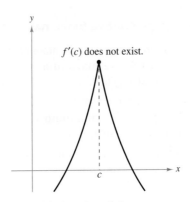

c is a critical number of f.
Figure 4.4

THEOREM 4.2 Relative Extrema Occur Only at Critical Numbers

If f has a relative minimum or relative maximum at $x = c$, then c is a critical number of f.

Proof

Case 1: If f is *not* differentiable at $x = c$, then, by definition, c is a critical number of f and the theorem is valid.

Case 2: If f is differentiable at $x = c$, then $f'(c)$ must be positive, negative, or 0. Suppose $f'(c)$ is positive. Then

$$f'(c) = \lim_{x \to c} \frac{f(x) - f(c)}{x - c} > 0$$

which implies that there exists an interval (a, b) containing c such that

$$\frac{f(x) - f(c)}{x - c} > 0, \text{ for all } x \neq c \text{ in } (a, b). \qquad \text{[See Exercise 74(b), Section 2.2.]}$$

Because this quotient is positive, the signs of the denominator and numerator must agree. This produces the following inequalities for x-values in the interval (a, b).

> *Left of c:* $x < c$ and $f(x) < f(c)$ ⟹ $f(c)$ is not a relative minimum
> *Right of c:* $x > c$ and $f(x) > f(c)$ ⟹ $f(c)$ is not a relative maximum

So, the assumption that $f'(c) > 0$ contradicts the hypothesis that $f(c)$ is a relative extremum. Assuming that $f'(c) < 0$ produces a similar contradiction, you are left with only one possibility—namely, $f'(c) = 0$. So, by definition, c is a critical number of f and the theorem is valid. ■

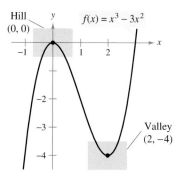

f has a relative maximum at $(0, 0)$ and a relative minimum at $(2, -4)$.
Figure 4.2

(a) $f'(3) = 0$

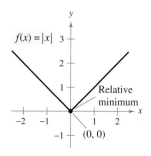

(b) $f'(0)$ does not exist.

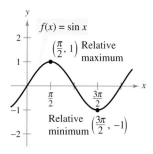

(c) $f'\left(\dfrac{\pi}{2}\right) = 0$; $f'\left(\dfrac{3\pi}{2}\right) = 0$

Figure 4.3

Relative Extrema and Critical Numbers

In Figure 4.2, the graph of $f(x) = x^3 - 3x^2$ has a **relative maximum** at the point $(0, 0)$ and a **relative minimum** at the point $(2, -4)$. Informally, you can think of a relative maximum as occurring on a "hill" on the graph, and a relative minimum as occurring in a "valley" on the graph. Such a hill and valley can occur in two ways. If the hill (or valley) is smooth and rounded, the graph has a horizontal tangent line at the high point (or low point). If the hill (or valley) is sharp and peaked, the graph represents a function that is not differentiable at the high point (or low point).

Definition of Relative Extrema

1. If there is an open interval containing c on which $f(c)$ is a maximum, then $f(c)$ is called a **relative maximum** of f, or you can say that f has a **relative maximum at** $(c, f(c))$.

2. If there is an open interval containing c on which $f(c)$ is a minimum, then $f(c)$ is called a **relative minimum** of f, or you can say that f has a **relative minimum at** $(c, f(c))$.

The plural of relative maximum is relative maxima, and the plural of relative minimum is relative minima.

Example 1 examines the derivatives of functions at *given* relative extrema. (Much more is said about *finding* the relative extrema of a function in Section 4.3.)

EXAMPLE 1   The Value of the Derivative at Relative Extrema

Find the value of the derivative at each of the relative extrema shown in Figure 4.3.

Solution

a. The derivative of $f(x) = \dfrac{9(x^2 - 3)}{x^3}$ is

$$f'(x) = \frac{x^3(18x) - (9)(x^2 - 3)(3x^2)}{(x^3)^2} \qquad \text{Differentiate using Quotient Rule.}$$

$$= \frac{9(9 - x^2)}{x^4}. \qquad \text{Simplify.}$$

At the point $(3, 2)$, the value of the derivative is $f'(3) = 0$ [see Figure 4.3(a)].

b. At $x = 0$, the derivative of $f(x) = |x|$ *does not exist* because the following one-sided limits differ [see Figure 4.3(b)].

$$\lim_{x \to 0^-} \frac{f(x) - f(0)}{x - 0} = \lim_{x \to 0^-} \frac{|x|}{x} = -1 \qquad \text{Limit from the left}$$

$$\lim_{x \to 0^+} \frac{f(x) - f(0)}{x - 0} = \lim_{x \to 0^+} \frac{|x|}{x} = 1 \qquad \text{Limit from the right}$$

c. The derivative of $f(x) = \sin x$ is

$$f'(x) = \cos x.$$

At the point $(\pi/2, 1)$, the value of the derivative is $f'(\pi/2) = \cos(\pi/2) = 0$. At the point $(3\pi/2, -1)$, the value of the derivative is $f'(3\pi/2) = \cos(3\pi/2) = 0$ [see Figure 4.3(c)].

Note in Example 1 that at each relative extremum, the derivative is either zero or does not exist. The x-values at these special points are called **critical numbers.** Figure 4.4 illustrates the two types of critical numbers.

Definition of a Critical Number

Let f be defined at c. If $f'(c) = 0$ or if f is not differentiable at c, then c is a **critical number** of f.

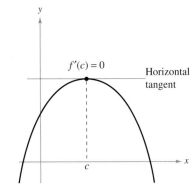

c is a critical number of f.
Figure 4.4

THEOREM 4.2 Relative Extrema Occur Only at Critical Numbers

If f has a relative minimum or relative maximum at $x = c$, then c is a critical number of f.

Proof

Case 1: If f is *not* differentiable at $x = c$, then, by definition, c is a critical number of f and the theorem is valid.

Case 2: If f is differentiable at $x = c$, then $f'(c)$ must be positive, negative, or 0. Suppose $f'(c)$ is positive. Then

$$f'(c) = \lim_{x \to c} \frac{f(x) - f(c)}{x - c} > 0$$

which implies that there exists an interval (a, b) containing c such that

$$\frac{f(x) - f(c)}{x - c} > 0, \text{ for all } x \neq c \text{ in } (a, b). \qquad \text{[See Exercise 74(b), Section 2.2.]}$$

Because this quotient is positive, the signs of the denominator and numerator must agree. This produces the following inequalities for x-values in the interval (a, b).

Left of c: $x < c$ and $f(x) < f(c) \implies f(c)$ is not a relative minimum
Right of c: $x > c$ and $f(x) > f(c) \implies f(c)$ is not a relative maximum

So, the assumption that $f'(c) > 0$ contradicts the hypothesis that $f(c)$ is a relative extremum. Assuming that $f'(c) < 0$ produces a similar contradiction, you are left with only one possibility—namely, $f'(c) = 0$. So, by definition, c is a critical number of f and the theorem is valid.

PIERRE DE FERMAT (1601–1665)

For Fermat, who was trained as a lawyer, mathematics was more of a hobby than a profession. Nevertheless, Fermat made many contributions to analytic geometry, number theory, calculus, and probability. In letters to friends, he wrote of many of the fundamental ideas of calculus, long before Newton or Leibniz. For instance, the theorem at the right is sometimes attributed to Fermat.

Finding Extrema on a Closed Interval

Theorem 4.2 states that the relative extrema of a function can occur *only* at the critical numbers of the function. Knowing this, you can use the following guidelines to find extrema on a closed interval.

Guidelines for Finding Extrema on a Closed Interval

To find the extrema of a continuous function f on a closed interval $[a, b]$, use the following steps.

1. Find the critical numbers of f in (a, b).
2. Evaluate f at each critical number in (a, b).
3. Evaluate f at each endpoint of $[a, b]$.
4. The least of these values is the minimum. The greatest is the maximum.

The next three examples show how to apply these guidelines. Be sure you see that finding the critical numbers of the function is only part of the procedure. Evaluating the function at the critical numbers *and* the endpoints is the other part.

EXAMPLE 2   Finding Extrema on a Closed Interval

Find the extrema of $f(x) = 3x^4 - 4x^3$ on the interval $[-1, 2]$.

Solution   Begin by differentiating the function.

$$f(x) = 3x^4 - 4x^3 \qquad \text{Write original function.}$$
$$f'(x) = 12x^3 - 12x^2 \qquad \text{Differentiate.}$$

To find the critical numbers of f, you must find all x-values for which $f'(x) = 0$ and all x-values for which $f'(x)$ does not exist.

$$f'(x) = 12x^3 - 12x^2 = 0 \qquad \text{Set } f'(x) \text{ equal to 0.}$$
$$12x^2(x - 1) = 0 \qquad \text{Factor.}$$
$$x = 0, 1 \qquad \text{Critical numbers}$$

Because f' is defined for all x, you can conclude that these are the only critical numbers of f. By evaluating f at these two critical numbers and at the endpoints of $[-1, 2]$, you can determine that the maximum is $f(2) = 16$ and the minimum is $f(1) = -1$, as shown in the table. The graph of f is shown in Figure 4.5.

Left Endpoint	Critical Number	Critical Number	Right Endpoint
$f(-1) = 7$	$f(0) = 0$	$f(1) = -1$ Minimum	$f(2) = 16$ Maximum

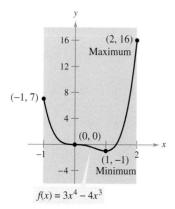

$f(x) = 3x^4 - 4x^3$

On the closed interval $[-1, 2]$, f has a minimum at $(1, -1)$ and a maximum at $(2, 16)$.
Figure 4.5

In Figure 4.5, note that the critical number $x = 0$ does not yield a relative minimum or a relative maximum. This tells you that the converse of Theorem 4.2 is not true. In other words, *the critical numbers of a function need not produce relative extrema.*

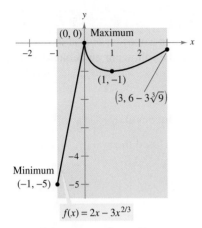

On the closed interval $[-1, 3]$, f has a minimum at $(-1, -5)$ and a maximum at $(0, 0)$.

Figure 4.6

EXAMPLE 3   **Finding Extrema on a Closed Interval**

Find the extrema of $f(x) = 2x - 3x^{2/3}$ on the interval $[-1, 3]$.

Solution   Begin by differentiating the function.

$$f(x) = 2x - 3x^{2/3} \qquad \text{Write original function.}$$

$$f'(x) = 2 - \frac{2}{x^{1/3}} = 2\left(\frac{x^{1/3} - 1}{x^{1/3}}\right) \qquad \text{Differentiate.}$$

From this derivative, you can see that the function has two critical numbers in the interval $[-1, 3]$. The number 1 is a critical number because $f'(1) = 0$, and the number 0 is a critical number because $f'(0)$ does not exist. By evaluating f at these two numbers and at the endpoints of the interval, you can conclude that the minimum is $f(-1) = -5$ and the maximum is $f(0) = 0$, as shown in the table. The graph of f is shown in Figure 4.6.

Left Endpoint	Critical Number	Critical Number	Right Endpoint
$f(-1) = -5$ Minimum	$f(0) = 0$ Maximum	$f(1) = -1$	$f(3) = 6 - 3\sqrt[3]{9} \approx -0.24$

 EXAMPLE 4   **Finding Extrema on a Closed Interval**

Find the extrema of $f(x) = 2 \sin x - \cos 2x$ on the interval $[0, 2\pi]$.

Solution   This function is differentiable for all real x, so you can find all critical numbers by differentiating the function and setting $f'(x)$ equal to zero, as shown.

$$f(x) = 2 \sin x - \cos 2x \qquad \text{Write original function.}$$

$$f'(x) = 2 \cos x + 2 \sin 2x = 0 \qquad \text{Set } f'(x) \text{ equal to 0.}$$

$$2 \cos x + 4 \cos x \sin x = 0 \qquad \sin 2x = 2 \cos x \sin x$$

$$2(\cos x)(1 + 2 \sin x) = 0 \qquad \text{Factor.}$$

In the interval $[0, 2\pi]$, the factor $\cos x$ is zero when $x = \pi/2$ and when $x = 3\pi/2$. The factor $(1 + 2 \sin x)$ is zero when $x = 7\pi/6$ and when $x = 11\pi/6$. By evaluating f at these four critical numbers and at the endpoints of the interval, you can conclude that the maximum is $f(\pi/2) = 3$ and the minimum occurs at *two* points, $f(7\pi/6) = -3/2$ and $f(11\pi/6) = -3/2$, as shown in the table. The graph is shown in Figure 4.7.

On the closed interval $[0, 2\pi]$, f has minima at $(7\pi/6, -3/2)$ and $(11\pi/6, -3/2)$ and a maximum at $(\pi/2, 3)$.

Figure 4.7

Left Endpoint	Critical Number	Critical Number	Critical Number	Critical Number	Right Endpoint
$f(0) = -1$	$f\left(\dfrac{\pi}{2}\right) = 3$ Maximum	$f\left(\dfrac{7\pi}{6}\right) = -\dfrac{3}{2}$ Minimum	$f\left(\dfrac{3\pi}{2}\right) = -1$	$f\left(\dfrac{11\pi}{6}\right) = -\dfrac{3}{2}$ Minimum	$f(2\pi) = -1$

indicates that in the HM mathSpace® *CD-ROM and the online* Eduspace® *system for this text, you will find an Open Exploration, which further explores this example using the computer algebra systems* Maple, Mathcad, Mathematica, *and* Derive.

Exercises for Section 4.1

In Exercises 1 and 2, decide whether each labeled point is an absolute maximum, an absolute minimum, a relative maximum, a relative minimum, or none of these.

1.

2.
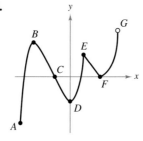

In Exercises 3–8, find the value of the derivative (if it exists) at each indicated extremum.

3. $f(x) = \dfrac{x^2}{x^2 + 4}$

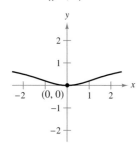

4. $f(x) = \cos \dfrac{\pi x}{2}$

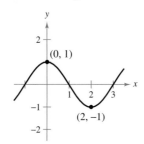

5. $f(x) = x + \dfrac{27}{2x^2}$

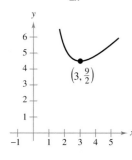

6. $f(x) = -3x\sqrt{x + 1}$

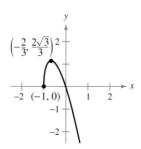

7. $f(x) = (x + 2)^{2/3}$

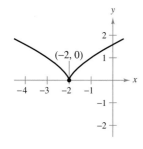

8. $f(x) = 4 - |x|$

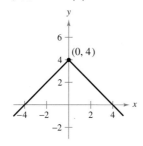

In Exercises 9–12, approximate the critical numbers of the function shown in the graph. Determine whether the function has a relative maximum, a relative minimum, an absolute maximum, an absolute minimum, or none of these at each critical number on the interval shown.

9.

10.

11.

12.
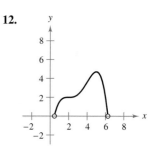

In Exercises 13–20, find any critical numbers of the function.

13. $f(x) = x^2(x - 3)$

14. $g(x) = x^2(x^2 - 4)$

15. $g(t) = t\sqrt{4 - t}, \ t < 3$

16. $f(x) = \dfrac{4x}{x^2 + 1}$

17. $h(x) = \sin^2 x + \cos x$
$0 < x < 2\pi$

18. $f(\theta) = 2 \sec \theta + \tan \theta$
$0 < \theta < 2\pi$

19. $f(x) = x^2 \log_2(x^2 + 1)$

20. $g(x) = 4x^2(3^x)$

In Exercises 21–38, locate the absolute extrema of the function on the closed interval.

21. $f(x) = 2(3 - x), \ [-1, 2]$

22. $f(x) = \dfrac{2x + 5}{3}, \ [0, 5]$

23. $f(x) = -x^2 + 3x, \ [0, 3]$

24. $f(x) = x^2 + 2x - 4, \ [-1, 1]$

25. $f(x) = x^3 - \dfrac{3}{2}x^2, \ [-1, 2]$

26. $f(x) = x^3 - 12x, \ [0, 4]$

27. $y = 3x^{2/3} - 2x, \ [-1, 1]$

28. $g(x) = \sqrt[3]{x}, \ [-1, 1]$

29. $g(t) = \dfrac{t^2}{t^2 + 3}, \ [-1, 1]$

30. $y = 3 - |t - 3|, \ [-1, 5]$

31. $h(s) = \dfrac{1}{s - 2}, \ [0, 1]$

32. $h(t) = \dfrac{t}{t - 2}, \ [3, 5]$

33. $y = e^x \sin x, \ [0, \pi]$

34. $y = x \ln(x + 3), \ [0, 3]$

35. $f(x) = \cos \pi x, \ \left[0, \dfrac{1}{6}\right]$

36. $g(x) = \sec x, \ \left[-\dfrac{\pi}{6}, \dfrac{\pi}{3}\right]$

37. $y = \dfrac{4}{x} + \tan\left(\dfrac{\pi x}{8}\right), \ [1, 2]$

38. $y = x^2 - 2 - \cos x, \ [-1, 3]$

In Exercises 39 and 40, locate the absolute extrema of the function (if any exist) over each interval.

39. $f(x) = 2x - 3$

(a) $[0, 2]$ (b) $[0, 2)$

(c) $(0, 2]$ (d) $(0, 2)$

40. $f(x) = \sqrt{4 - x^2}$

(a) $[-2, 2]$ (b) $[-2, 0)$

(c) $(-2, 2)$ (d) $[1, 2)$

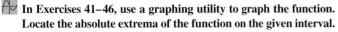 **In Exercises 41–46, use a graphing utility to graph the function. Locate the absolute extrema of the function on the given interval.**

Function	Interval
41. $f(x) = \begin{cases} 2x + 2, & 0 \le x \le 1 \\ 4x^2, & 1 < x \le 3 \end{cases}$	$[0, 3]$
42. $f(x) = \begin{cases} 2 - x^2, & 1 \le x < 3 \\ 2 - 3x, & 3 \le x \le 5 \end{cases}$	$[1, 5]$
43. $f(x) = \dfrac{3}{x - 1}$	$(1, 4]$
44. $f(x) = \dfrac{2}{2 - x}$	$[0, 2)$
45. $f(x) = x^4 - 2x^3 + x + 1$	$[-1, 3]$
46. $f(x) = \sqrt{x} + \cos\dfrac{x}{2}$	$[0, 2\pi]$

 In Exercises 47–52, (a) use a computer algebra system to graph the function and approximate any absolute extrema on the indicated interval. (b) Use the utility to find any critical numbers, and use them to find any absolute extrema not located at the endpoints. Compare the results with those in part (a).

Function	Interval
47. $f(x) = 3.2x^5 + 5x^3 - 3.5x$	$[0, 1]$
48. $f(x) = \dfrac{4}{3}x\sqrt{3 - x}$	$[0, 3]$
49. $f(x) = (x^2 - 2x)\ln(x + 3)$	$[0, 3]$
50. $f(x) = \sqrt{x + 4}\, e^{x^2/10}$	$[-2, 2]$
51. $f(x) = 2x \arctan(x - 1)$	$[0, 2]$
52. $f(x) = (x - 4) \arcsin\dfrac{x}{4}$	$[-2, 4]$

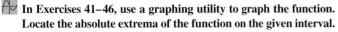 **In Exercises 53–56, use a computer algebra system to find the maximum value of $|f''(x)|$ on the closed interval. (This value is used in the error estimate for the Trapezoidal Rule, as discussed in Section 5.6.)**

Function	Interval	Function	Interval
53. $f(x) = \sqrt{1 + x^3}$	$[0, 2]$	**54.** $f(x) = \dfrac{1}{x^2 + 1}$	$\left[\dfrac{1}{2}, 3\right]$
55. $f(x) = e^{-x^2/2}$	$[0, 1]$	**56.** $f(x) = x\ln(x + 1)$	$[0, 2]$

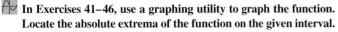 **In Exercises 57 and 58, use a computer algebra system to find the maximum value of $|f^4(x)|$ on the closed interval. (This value is used in the error estimate for Simpson's Rule, as discussed in Section 5.6.)**

Function	Interval	Function	Interval
57. $f(x) = (x + 1)^{2/3}$	$[0, 2]$	**58.** $f(x) = \dfrac{1}{x^2 + 1}$	$[-1, 1]$

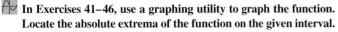 **59.** Explain why the function $f(x) = \tan x$ has a maximum on $[0, \pi/4]$ but not on $[0, \pi]$.

60. *Writing* Write a short paragraph explaining why a continuous function on an open interval may not have a maximum or minimum. Illustrate your explanation with a sketch of the graph of such a function.

Writing About Concepts

In Exercises 61 and 62, graph a function on the interval $[-2, 5]$ having the given characteristics.

61. Absolute maximum at $x = -2$
Absolute minimum at $x = 1$
Relative maximum at $x = 3$

62. Relative minimum at $x = -1$
Critical number at $x = 0$, but no extrema
Absolute maximum at $x = 2$
Absolute minimum at $x = 5$

In Exercises 63–66, determine from the graph whether f has a minimum in the open interval (a, b).

63. (a)   (b)

64. (a)   (b)

65. (a)   (b)

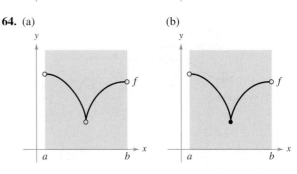

Writing About Concepts (continued)

66. (a)   (b)

67. *Lawn Sprinkler*   A lawn sprinkler is constructed in such a way that $d\theta/dt$ is constant, where θ ranges between 45° and 135° (see figure). The distance the water travels horizontally is

$$x = \frac{v^2 \sin 2\theta}{32}, \quad 45° \le \theta \le 135°$$

where v is the speed of the water. Find dx/dt and explain why this lawn sprinkler does not water evenly. What part of the lawn receives the most water?

Water sprinkler: $45° \le \theta \le 135°$

FOR FURTHER INFORMATION   For more information on the "calculus of lawn sprinklers," see the article "Design of an Oscillating Sprinkler" by Bart Braden in *Mathematics Magazine*. To view this article, go to the website *www.matharticles.com*.

68. *Honeycomb*   The surface area of a cell in a honeycomb is

$$S = 6hs + \frac{3s^2}{2}\left(\frac{\sqrt{3} - \cos \theta}{\sin \theta}\right)$$

where h and s are positive constants and θ is the angle at which the upper faces meet the altitude of the cell (see figure). Find the angle θ ($\pi/6 \le \theta \le \pi/2$) that minimizes the surface area S.

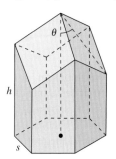

FOR FURTHER INFORMATION   For more information on the geometric structure of a honeycomb cell, see the article "The Design of Honeycombs" by Anthony L. Peressini in UMAP Module 502, published by COMAP, Inc., Suite 210, 57 Bedford Street, Lexington, MA.

True or False?   **In Exercises 69–72, determine whether the statement is true or false. If it is false, explain why or give an example that shows it is false.**

69. The maximum of a function that is continuous on a closed interval can occur at two different values in the interval.

70. If a function is continuous on a closed interval, then it must have a minimum on the interval.

71. If $x = c$ is a critical number of the function f, then it is also a critical number of the function $g(x) = f(x) + k$, where k is a constant.

72. If $x = c$ is a critical number of the function f, then it is also a critical number of the function $g(x) = f(x - k)$, where k is a constant.

73. Let the function f be differentiable on an interval I containing c. If f has a maximum value at $x = c$, show that $-f$ has a minimum value at $x = c$.

74. Consider the cubic function $f(x) = ax^3 + bx^2 + cx + d$ where $a \ne 0$. Show that f can have zero, one, or two critical numbers and give an example of each case.

75. *Highway Design*   In order to build a highway, it is necessary to fill a section of a valley where the grades (slopes) of the sides are 9% and 6% (see figure). The top of the filled region will have the shape of a parabolic arc that is tangent to the two slopes at the points A and B. The horizontal distance between the points A and B is 1000 feet.

(a) Find a quadratic function $y = ax^2 + bx + c$, $-500 \le x \le 500$, that describes the top of the filled region.

(b) Construct a table giving the depths d of the fill for $x = -500, -400, -300, -200, -100, 0, 100, 200, 300, 400,$ and 500.

(c) What will be the lowest point on the completed highway? Will it be directly over the point where the two hillsides come together?

Rolle's Theorem and the Mean Value Theorem

• Understand and use Rolle's Theorem.
• Understand and use the Mean Value Theorem.

Rolle's Theorem

The Extreme Value Theorem (Section 4.1) states that a continuous function on a closed interval $[a, b]$ must have both a minimum and a maximum on the interval. Both of these values, however, can occur at the endpoints. **Rolle's Theorem,** named after the French mathematician Michel Rolle (1652–1719), gives conditions that guarantee the existence of an extreme value in the *interior* of a closed interval.

> **ROLLE'S THEOREM**
>
> French mathematician Michel Rolle first published the theorem that bears his name in 1691. Before this time, however, Rolle was one of the most vocal critics of calculus, stating that it gave erroneous results and was based on unsound reasoning. Later in life, Rolle came to see the usefulness of calculus.

EXPLORATION

Extreme Values in a Closed Interval Sketch a rectangular coordinate plane on a piece of paper. Label the points $(1, 3)$ and $(5, 3)$. Using a pencil or pen, draw the graph of a differentiable function f that starts at $(1, 3)$ and ends at $(5, 3)$. Is there at least one point on the graph for which the derivative is zero? Would it be possible to draw the graph so that there *isn't* a point for which the derivative is zero? Explain your reasoning.

> **THEOREM 4.3 Rolle's Theorem**
>
> Let f be continuous on the closed interval $[a, b]$ and differentiable on the open interval (a, b). If
>
> $$f(a) = f(b)$$
>
> then there is at least one number c in (a, b) such that $f'(c) = 0$.

Proof Let $f(a) = d = f(b)$.

Case 1: If $f(x) = d$ for all x in $[a, b]$, then f is constant on the interval and, by Theorem 3.2, $f'(x) = 0$ for all x in (a, b).

Case 2: Suppose $f(x) > d$ for some x in (a, b). By the Extreme Value Theorem, you know that f has a maximum at some c in the interval. Moreover, because $f(c) > d$, this maximum does not occur at either endpoint. So, f has a maximum in the *open* interval (a, b). This implies that $f(c)$ is a *relative* maximum and, by Theorem 4.2, c is a critical number of f. Finally, because f is differentiable at c, you can conclude that $f'(c) = 0$.

Case 3: If $f(x) < d$ for some x in (a, b), you can use an argument similar to that in Case 2, but involving the minimum instead of the maximum. ∎

From Rolle's Theorem, you can see that if a function f is continuous on $[a, b]$ and differentiable on (a, b), and if $f(a) = f(b)$, then there must be at least one x-value between a and b at which the graph of f has a horizontal tangent, as shown in Figure 4.8(a). If the differentiability requirement is dropped from Rolle's Theorem, f will still have a critical number in (a, b), but it may not yield a horizontal tangent. Such a case is shown in Figure 4.8(b).

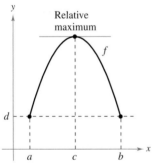

(a) f is continuous on $[a, b]$ and differentiable on (a, b).

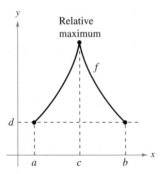

(b) f is continuous on $[a, b]$.
Figure 4.8

EXAMPLE 1   Illustrating Rolle's Theorem

Find the two x-intercepts of

$$f(x) = x^2 - 3x + 2$$

and show that $f'(x) = 0$ at some point between the two x-intercepts.

Solution   Note that f is differentiable on the entire real number line. Setting $f(x)$ equal to 0 produces

$x^2 - 3x + 2 = 0$	Set $f(x)$ equal to 0.
$(x - 1)(x - 2) = 0.$	Factor.

So, $f(1) = f(2) = 0$, and from Rolle's Theorem you know that there *exists* at least one c in the interval $(1, 2)$ such that $f'(c) = 0$. To *find* such a c, you can solve the equation

$f'(x) = 2x - 3 = 0$	Set $f'(x)$ equal to 0.

and determine that $f'(x) = 0$ when $x = \frac{3}{2}$. Note that the x-value lies in the open interval $(1, 2)$, as shown in Figure 4.9.

Rolle's Theorem states that if f satisfies the conditions of the theorem, there must be *at least* one point between a and b at which the derivative is 0. There may of course be more than one such point, as shown in the next example.

EXAMPLE 2   Illustrating Rolle's Theorem

Let $f(x) = x^4 - 2x^2$. Find all values of c in the interval $(-2, 2)$ such that $f'(c) = 0$.

Solution   To begin, note that the function satisfies the conditions of Rolle's Theorem. That is, f is continuous on the interval $[-2, 2]$ and differentiable on the interval $(-2, 2)$. Moreover, because $f(-2) = f(2) = 8$, you can conclude that there exists at least one c in $(-2, 2)$ such that $f'(c) = 0$. Setting the derivative equal to 0 produces

$f'(x) = 4x^3 - 4x = 0$	Set $f'(x)$ equal to 0.
$4x(x - 1)(x + 1) = 0$	Factor.
$x = 0, 1, -1.$	x-values for which $f'(x) = 0$

So, in the interval $(-2, 2)$, the derivative is zero at three different values of x, as shown in Figure 4.10.

TECHNOLOGY PITFALL   A graphing utility can be used to indicate whether the points on the graphs in Examples 1 and 2 are relative minima or relative maxima of the functions. When using a graphing utility, however, you should keep in mind that it can give misleading pictures of graphs. For example, use a graphing utility to graph

$$f(x) = 1 - (x - 1)^2 - \frac{1}{1000(x - 1)^{1/7} + 1}.$$

With most viewing windows, it appears that the function has a maximum of 1 when $x = 1$ (see Figure 4.11). By evaluating the function at $x = 1$, however, you can see that $f(1) = 0$. To determine the behavior of this function near $x = 1$, you need to examine the graph analytically to get the complete picture.

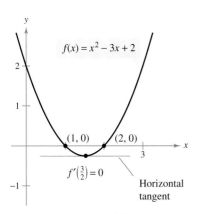

The x-value for which $f'(x) = 0$ is between the two x-intercepts.
Figure 4.9

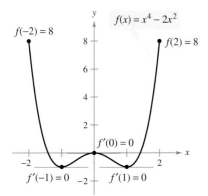

$f'(x) = 0$ for more than one x-value in the interval $(-2, 2)$.
Figure 4.10

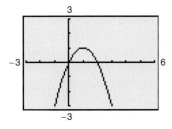

Figure 4.11

The Mean Value Theorem

Rolle's Theorem can be used to prove another theorem—the **Mean Value Theorem.**

> ### THEOREM 4.4 The Mean Value Theorem
>
> If f is continuous on the closed interval $[a, b]$ and differentiable on the open interval (a, b), then there exists a number c in (a, b) such that
>
> $$f'(c) = \frac{f(b) - f(a)}{b - a}.$$

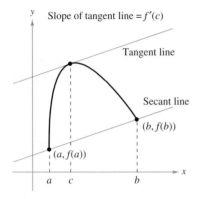

Slope of tangent line $= f'(c)$

Tangent line

Secant line

$(b, f(b))$

$(a, f(a))$

Figure 4.12

Proof   Refer to Figure 4.12. The equation of the secant line containing the points $(a, f(a))$ and $(b, f(b))$ is

$$y = \left[\frac{f(b) - f(a)}{b - a} \right] (x - a) + f(a).$$

Let $g(x)$ be the difference between $f(x)$ and y. Then

$$g(x) = f(x) - y$$
$$= f(x) - \left[\frac{f(b) - f(a)}{b - a} \right](x - a) - f(a).$$

By evaluating g at a and b, you can see that $g(a) = 0 = g(b)$. Because f is continuous on $[a, b]$, it follows that g is also continuous on $[a, b]$. Furthermore, because f is differentiable, g is also differentiable, and you can apply Rolle's Theorem to the function g. So, there exists a number c in (a, b) such that $g'(c) = 0$, which implies that

$$0 = g'(c)$$
$$= f'(c) - \frac{f(b) - f(a)}{b - a}.$$

So, there exists a number c in (a, b) such that

$$f'(c) = \frac{f(b) - f(a)}{b - a}.$$

NOTE   The "mean" in the Mean Value Theorem refers to the mean (or average) rate of change of f in the interval $[a, b]$.

Although the Mean Value Theorem can be used directly in problem solving, it is used more often to prove other theorems. In fact, some people consider this to be the most important theorem in calculus—it is closely related to the Fundamental Theorem of Calculus discussed in Chapter 5. For now, you can get an idea of the versatility of this theorem by looking at the results stated in Exercises 83–91 in this section.

The Mean Value Theorem has implications for both basic interpretations of the derivative. Geometrically, the theorem guarantees the existence of a tangent line that is parallel to the secant line through the points $(a, f(a))$ and $(b, f(b))$, as shown in Figure 4.12. Example 3 illustrates this geometric interpretation of the Mean Value Theorem. In terms of rates of change, the Mean Value Theorem implies that there must be a point in the open interval (a, b) at which the instantaneous rate of change is equal to the average rate of change over the interval $[a, b]$. This is illustrated in Example 4.

JOSEPH-LOUIS LAGRANGE (1736–1813)

The Mean Value Theorem was first proved by the famous mathematician Joseph-Louis Lagrange. Born in Italy, Lagrange held a position in the court of Frederick the Great in Berlin for 20 years. Afterward, he moved to France, where he met emperor Napoleon Bonaparte, who is quoted as saying, "Lagrange is the lofty pyramid of the mathematical sciences."

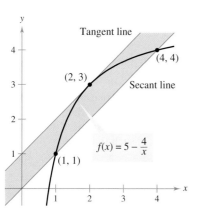

Tangent line

(2, 3)

(4, 4)

Secant line

$f(x) = 5 - \dfrac{4}{x}$

(1, 1)

The tangent line at $(2, 3)$ is parallel to the secant line through $(1, 1)$ and $(4, 4)$.
Figure 4.13

At some time t, the instantaneous velocity is equal to the average velocity over 4 minutes.
Figure 4.14

EXAMPLE 3 **Finding a Tangent Line**

Given $f(x) = 5 - (4/x)$, find all values of c in the open interval $(1, 4)$ such that

$$f'(c) = \frac{f(4) - f(1)}{4 - 1}.$$

Solution   The slope of the secant line through $(1, f(1))$ and $(4, f(4))$ is

$$\frac{f(4) - f(1)}{4 - 1} = \frac{4 - 1}{4 - 1} = 1.$$

Because f satisfies the conditions of the Mean Value Theorem, there exists at least one number c in $(1, 4)$ such that $f'(c) = 1$. Solving the equation $f'(x) = 1$ yields

$$f'(x) = \frac{4}{x^2} = 1$$

which implies that $x = \pm 2$. So, in the interval $(1, 4)$, you can conclude that $c = 2$, as shown in Figure 4.13.

EXAMPLE 4 **Finding an Instantaneous Rate of Change**

Two stationary patrol cars equipped with radar are 5 miles apart on a highway, as shown in Figure 4.14. As a truck passes the first patrol car, its speed is clocked at 55 miles per hour. Four minutes later, when the truck passes the second patrol car, its speed is clocked at 50 miles per hour. Prove that the truck must have exceeded the speed limit (of 55 miles per hour) at some time during the 4 minutes.

Solution   Let $t = 0$ be the time (in hours) when the truck passes the first patrol car. The time when the truck passes the second patrol car is

$$t = \frac{4}{60} = \frac{1}{15} \text{ hour.}$$

By letting $s(t)$ represent the distance (in miles) traveled by the truck, you have $s(0) = 0$ and $s\left(\frac{1}{15}\right) = 5$. So, the average velocity of the truck over the five-mile stretch of highway is

$$\text{Average velocity} = \frac{s(1/15) - s(0)}{(1/15) - 0}$$

$$= \frac{5}{1/15} = 75 \text{ miles per hour.}$$

Assuming that the position function is differentiable, you can apply the Mean Value Theorem to conclude that the truck must have been traveling at a rate of 75 miles per hour sometime during the 4 minutes.

A useful alternative form of the Mean Value Theorem is as follows: If f is continuous on $[a, b]$ and differentiable on (a, b), then there exists a number c in (a, b) such that

$$f(b) = f(a) + (b - a)f'(c). \qquad \text{Alternative form of Mean Value Theorem}$$

NOTE   When doing the exercises for this section, keep in mind that polynomial functions, rational functions, and transcendental functions are differentiable at all points in their domains.

Exercises for Section 4.2 See www.CalcChat.com for worked-out solutions to odd-numbered exercises.

In Exercises 1–4, explain why Rolle's Theorem does not apply to the function even though there exist a and b such that $f(a) = f(b)$.

1. $f(x) = 1 - |x - 1|$

2. $f(x) = \cot \dfrac{x}{2}$

3. $f(x) = \left| \dfrac{1}{x} \right|$,

$[-1, 1]$

4. $f(x) = \sqrt{(2 - x^{2/3})^3}$,

$[-1, 1]$

In Exercises 5–8, find the two x-intercepts of the function f and show that $f'(x) = 0$ at some point between the two x-intercepts.

5. $f(x) = x^2 - x - 2$

6. $f(x) = x(x - 3)$

7. $f(x) = x\sqrt{x + 4}$

8. $f(x) = -3x\sqrt{x + 1}$

Rolle's Theorem In Exercises 9 and 10, the graph of f is shown. Apply Rolle's Theorem and find all values of c such that $f'(c) = 0$ at some point between the labeled intercepts.

9. $f(x) = x^2 + 3x - 4$

10.

In Exercises 11–26, determine whether Rolle's Theorem can be applied to f on the closed interval $[a, b]$. If Rolle's Theorem can be applied, find all values of c in the open interval (a, b) such that $f'(c) = 0$.

11. $f(x) = x^2 - 2x, [0, 2]$

12. $f(x) = x^2 - 5x + 4, [1, 4]$

13. $f(x) = (x - 1)(x - 2)(x - 3), [1, 3]$

14. $f(x) = (x - 3)(x + 1)^2, [-1, 3]$

15. $f(x) = x^{2/3} - 1, [-8, 8]$

16. $f(x) = 3 - |x - 3|, [0, 6]$

17. $f(x) = \dfrac{x^2 - 2x - 3}{x + 2}, [-1, 3]$

18. $f(x) = \dfrac{x^2 - 1}{x}, [-1, 1]$

19. $f(x) = (x^2 - 2x)e^x, [0, 2]$

20. $f(x) = x - 2 \ln x, [1, 3]$

21. $f(x) = \sin x, [0, 2\pi]$

22. $f(x) = \cos x, [0, 2\pi]$

23. $f(x) = \dfrac{6x}{\pi} - 4 \sin^2 x, \left[0, \dfrac{\pi}{6}\right]$

24. $f(x) = \cos 2x, \left[-\dfrac{\pi}{12}, \dfrac{\pi}{6}\right]$

25. $f(x) = \tan x, [0, \pi]$

26. $f(x) = \sec x, \left[-\dfrac{\pi}{4}, \dfrac{\pi}{4}\right]$

In Exercises 27–32, use a graphing utility to graph the function on the closed interval $[a, b]$. Determine whether Rolle's Theorem can be applied to f on the interval and, if so, find all values of c in the open interval (a, b) such that $f'(c) = 0$.

27. $f(x) = |x| - 1, \quad [-1, 1]$

28. $f(x) = x - x^{1/3}, \quad [0, 1]$

29. $f(x) = 4x - \tan \pi x, \quad \left[-\dfrac{1}{4}, \dfrac{1}{4}\right]$

30. $f(x) = \dfrac{x}{2} - \sin \dfrac{\pi x}{6}, \quad [-1, 0]$

31. $f(x) = 2 + \arcsin(x^2 - 1), \quad [-1, 1]$

32. $f(x) = 2 + (x^2 - 4x)(2^{-x/4}), \quad [0, 4]$

33. *Vertical Motion* The height of a ball t seconds after it is thrown upward from a height of 32 feet and with an initial velocity of 48 feet per second is $f(t) = -16t^2 + 48t + 32$.

(a) Verify that $f(1) = f(2)$.

(b) According to Rolle's Theorem, what must be the velocity at some time in the interval $(1, 2)$? Find that time.

34. *Reorder Costs* The ordering and transportation cost C of components used in a manufacturing process is approximated by

$$C(x) = 10\left(\dfrac{1}{x} + \dfrac{x}{x + 3}\right)$$

where C is measured in thousands of dollars and x is the order size in hundreds.

(a) Verify that $C(3) = C(6)$.

(b) According to Rolle's Theorem, the rate of change of cost must be 0 for some order size in the interval $(3, 6)$. Find that order size.

In Exercises 35 and 36, copy the graph and sketch the secant line to the graph through the points $(a, f(a))$ and $(b, f(b))$. Then sketch any tangent lines to the graph for each value of c guaranteed by the Mean Value Theorem. To print an enlarged copy of the graph, go to the website *www.mathgraphs.com*.

35.

36.

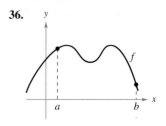

Writing In Exercises 37–40, explain why the Mean Value Theorem does not apply to the function *f* on the interval [0, 6].

37.

38.

39. $f(x) = \dfrac{1}{x - 3}$

40. $f(x) = |x - 3|$

41. *Mean Value Theorem* Consider the graph of the function $f(x) = x^2 + 1$. (a) Find the equation of the secant line joining the points $(-1, 2)$ and $(2, 5)$. (b) Use the Mean Value Theorem to determine a point *c* in the interval $(-1, 2)$ such that the tangent line at *c* is parallel to the secant line. (c) Find the equation of the tangent line through *c*. (d) Use a graphing utility to graph *f*, the secant line, and the tangent line.

$f(x) = x^2 + 1$

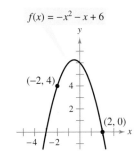
$f(x) = -x^2 - x + 6$

Figure for 41   **Figure for 42**

42. *Mean Value Theorem* Consider the graph of the function $f(x) = -x^2 - x + 6$. (a) Find the equation of the secant line joining the points $(-2, 4)$ and $(2, 0)$. (b) Use the Mean Value Theorem to determine a point *c* in the interval $(-2, 2)$ such that the tangent line at *c* is parallel to the secant line. (c) Find the equation of the tangent line through *c*. (d) Use a graphing utility to graph *f*, the secant line, and the tangent line.

In Exercises 43–52, determine whether the Mean Value Theorem can be applied to *f* on the closed interval [*a*, *b*]. If the Mean Value Theorem can be applied, find all values of *c* in the open interval (*a*, *b*) such that $f'(c) = \dfrac{f(b) - f(a)}{b - a}$.

43. $f(x) = x^2$, $[-2, 1]$

44. $f(x) = x(x^2 - x - 2)$, $[-1, 1]$

45. $f(x) = x^{2/3}$, $[0, 1]$ **46.** $f(x) = \dfrac{x + 1}{x}$, $\left[\frac{1}{2}, 2\right]$

47. $f(x) = \sqrt{2 - x}$, $[-7, 2]$ **48.** $f(x) = x^3$, $[0, 1]$

49. $f(x) = \sin x$, $[0, \pi]$

50. $f(x) = 2\sin x + \sin 2x$, $[0, \pi]$

51. $f(x) = x \log_2 x$, $[1, 2]$

52. $f(x) = \arctan(1 - x)$, $[0, 1]$

In Exercises 53–58, use a graphing utility to (a) graph the function *f* on the given interval, (b) find and graph the secant line through points on the graph of *f* at the endpoints of the given interval, and (c) find and graph any tangent lines to the graph of *f* that are parallel to the secant line.

53. $f(x) = \dfrac{x}{x + 1}$, $\left[-\frac{1}{2}, 2\right]$ **54.** $f(x) = x - 2\sin x$, $[-\pi, \pi]$

55. $f(x) = \sqrt{x}$, $[1, 9]$

56. $f(x) = -x^4 + 4x^3 + 8x^2 + 5$, $[0, 5]$

57. $f(x) = 2e^{x/4} \cos \dfrac{\pi x}{4}$, $[0, 2]$ **58.** $f(x) = \ln|\sec \pi x|$, $\left[0, \frac{1}{4}\right]$

Writing About Concepts

59. Let *f* be continuous on [*a*, *b*] and differentiable on (*a*, *b*). If there exists *c* in (*a*, *b*) such that $f'(c) = 0$, does it follow that $f(a) = f(b)$? Explain.

60. Let *f* be continuous on the closed interval [*a*, *b*] and differentiable on the open interval (*a*, *b*). Also, suppose that $f(a) = f(b)$ and that *c* is a real number in the interval such that $f'(c) = 0$. Find an interval for the function *g* over which Rolle's Theorem can be applied, and find the corresponding critical number of *g* (*k* is a constant).

(a) $g(x) = f(x) + k$ (b) $g(x) = f(x - k)$

(c) $g(x) = f(kx)$

61. The function

$$f(x) = \begin{cases} 0, & x = 0 \\ 1 - x, & 0 < x \le 1 \end{cases}$$

is differentiable on (0, 1) and satisfies $f(0) = f(1)$. However, its derivative is never zero on (0, 1). Does this contradict Rolle's Theorem? Explain.

62. Can you find a function *f* such that $f(-2) = -2$, $f(2) = 6$, and $f'(x) < 1$ for all *x*? Why or why not?

63. *Speed* A plane begins its takeoff at 2:00 P.M. on a 2500-mile flight. The plane arrives at its destination at 7:30 P.M. Explain why there are at least two times during the flight when the speed of the plane is 400 miles per hour.

64. *Temperature* When an object is removed from a furnace and placed in an environment with a constant temperature of 90°F, its core temperature is 1500°F. Five hours later the core temperature is 390°F. Explain why there must exist a time in the interval when the temperature is decreasing at a rate of 222°F per hour.

65. *Velocity* Two bicyclists begin a race at 8:00 A.M. They both finish the race 2 hours and 15 minutes later. Prove that at some time during the race, the bicyclists are traveling at the same velocity.

66. *Acceleration* At 9:13 A.M., a sports car is traveling 35 miles per hour. Two minutes later, the car is traveling 85 miles per hour. Prove that at some time during this two-minute interval, the car's acceleration is exactly 1500 miles per hour squared.

67. *Graphical Reasoning*   The figure shows two parts of the graph of a continuous differentiable function f on $[-10, 4]$. The derivative f' is also continuous. To print an enlarged copy of the graph, go to the website *www.mathgraphs.com*.

(a) Explain why f must have at least one zero in $[-10, 4]$.

(b) Explain why f' must also have at least one zero in the interval $[-10, 4]$. What are these zeros called?

(c) Make a possible sketch of the function with one zero of f' on the interval $[-10, 4]$.

(d) Make a possible sketch of the function with two zeros of f' on the interval $[-10, 4]$.

(e) Were the conditions of continuity of f and f' necessary to do parts (a) through (d)? Explain.

68. Consider the function $f(x) = 3\cos^2\left(\dfrac{\pi x}{2}\right)$.

(a) Use a graphing utility to graph f and f'.

(b) Is f a continuous function? Is f' a continuous function?

(c) Does Rolle's Theorem apply on the interval $[-1, 1]$? Does it apply on the interval $[1, 2]$? Explain.

(d) Evaluate, if possible, $\displaystyle\lim_{x\to 3^-} f'(x)$ and $\displaystyle\lim_{x\to 3^+} f'(x)$.

Think About It   In Exercises 69 and 70, sketch the graph of an arbitrary function f that satisfies the given condition but does not satisfy the conditions of the Mean Value Theorem on the interval $[-5, 5]$.

69. f is continuous on $[-5, 5]$.

70. f is not continuous on $[-5, 5]$.

In Exercises 71 and 72, use the Intermediate Value Theorem and Rolle's Theorem to prove that the equation has exactly one real solution.

71. $x^5 + x^3 + x + 1 = 0$ **72.** $2x - 2 - \cos x = 0$

73. Determine the values of a, b, and c such that the function f satisfies the hypotheses of the Mean Value Theorem on the interval $[0, 3]$.

$$f(x) = \begin{cases} 1, & x = 0 \\ ax + b, & 0 < x \le 1 \\ x^2 + 4x + c, & 1 < x \le 3 \end{cases}$$

74. Determine the values of a, b, c, and d such that the function f satisfies the hypotheses of the Mean Value Theorem on the interval $[-1, 2]$.

$$f(x) = \begin{cases} a, & x = -1 \\ 2, & -1 < x \le 0 \\ bx^2 + c, & 0 < x \le 1 \\ dx + 4, & 1 < x \le 2 \end{cases}$$

Differential Equations   **In Exercises 75–78, find a function f that has the derivative $f'(x)$ and whose graph passes through the given point. Explain your reasoning.**

75. $f'(x) = 0, \quad (2, 5)$ **76.** $f'(x) = 4, \quad (0, 1)$

77. $f'(x) = 2x, \quad (1, 0)$ **78.** $f'(x) = 2x + 3, \quad (1, 0)$

True or False?   **In Exercises 79–82, determine whether the statement is true or false. If it is false, explain why or give an example that shows it is false.**

79. The Mean Value Theorem can be applied to $f(x) = 1/x$ on the interval $[-1, 1]$.

80. If the graph of a function has three x-intercepts, then it must have at least two points at which its tangent line is horizontal.

81. If the graph of a polynomial function has three x-intercepts, then it must have at least two points at which its tangent line is horizontal.

82. If $f'(x) = 0$ for all x in the domain of f, then f is a constant function.

83. Prove that if $a > 0$ and n is any positive integer, then the polynomial function $p(x) = x^{2n+1} + ax + b$ cannot have two real roots.

84. Prove that if $f'(x) = 0$ for all x in an interval (a, b), then f is constant on (a, b).

85. Let $p(x) = Ax^2 + Bx + C$. Prove that for any interval $[a, b]$, the value c guaranteed by the Mean Value Theorem is the midpoint of the interval.

86. (a) Let $f(x) = x^2$ and $g(x) = -x^3 + x^2 + 3x + 2$. Then $f(-1) = g(-1)$ and $f(2) = g(2)$. Show that there is at least one value c in the interval $(-1, 2)$ where the tangent line to f at $(c, f(c))$ is parallel to the tangent line to g at $(c, g(c))$. Identify c.

(b) Let f and g be differentiable functions on $[a, b]$ where $f(a) = g(a)$ and $f(b) = g(b)$. Show that there is at least one value c in the interval (a, b) where the tangent line to f at $(c, f(c))$ is parallel to the tangent line to g at $(c, g(c))$.

87. Prove that if f is differentiable on $(-\infty, \infty)$ and $f'(x) < 1$ for all real numbers, then f has at most one fixed point. A fixed point of a function f is a real number c such that $f(c) = c$.

88. Use the result of Exercise 87 to show that $f(x) = \frac{1}{2}\cos x$ has at most one fixed point.

89. Prove that $|\cos a - \cos b| \le |a - b|$ for all a and b.

90. Prove that $|\sin a - \sin b| \le |a - b|$ for all a and b.

91. Let $0 < a < b$. Use the Mean Value Theorem to show that

$$\sqrt{b} - \sqrt{a} < \frac{b - a}{2\sqrt{a}}.$$

Increasing and Decreasing Functions and the First Derivative Test

- Determine intervals on which a function is increasing or decreasing.
- Apply the First Derivative Test to find relative extrema of a function.

Increasing and Decreasing Functions

In this section you will learn how derivatives can be used to *classify* relative extrema as either relative minima or relative maxima. First, it is important to define increasing and decreasing functions.

The derivative is related to the slope of a function.

Figure 4.15

Definitions of Increasing and Decreasing Functions

A function f is **increasing** on an interval if for any two numbers x_1 and x_2 in the interval, $x_1 < x_2$ implies $f(x_1) < f(x_2)$.

A function f is **decreasing** on an interval if for any two numbers x_1 and x_2 in the interval, $x_1 < x_2$ implies $f(x_1) > f(x_2)$.

A function is increasing if, *as x moves to the right*, its graph moves up, and is decreasing if its graph moves down. For example, the function in Figure 4.15 is decreasing on the interval $(-\infty, a)$, is constant on the interval (a, b), and is increasing on the interval (b, ∞). As shown in Theorem 4.5 below, a positive derivative implies that the function is increasing; a negative derivative implies that the function is decreasing; and a zero derivative on an entire interval implies that the function is constant on that interval.

THEOREM 4.5   Test for Increasing and Decreasing Functions

Let f be a function that is continuous on the closed interval $[a, b]$ and differentiable on the open interval (a, b).

1. If $f'(x) > 0$ for all x in (a, b), then f is increasing on $[a, b]$.

2. If $f'(x) < 0$ for all x in (a, b), then f is decreasing on $[a, b]$.

3. If $f'(x) = 0$ for all x in (a, b), then f is constant on $[a, b]$.

Proof   To prove the first case, assume that $f'(x) > 0$ for all x in the interval (a, b) and let $x_1 < x_2$ be any two points in the interval. By the Mean Value Theorem, you know that there exists a number c such that $x_1 < c < x_2$, and

$$f'(c) = \frac{f(x_2) - f(x_1)}{x_2 - x_1}.$$

Because $f'(c) > 0$ and $x_2 - x_1 > 0$, you know that

$$f(x_2) - f(x_1) > 0$$

which implies that $f(x_1) < f(x_2)$. So, f is increasing on the interval. The second case has a similar proof (see Exercise 107), and the third case was given as Exercise 84 in Section 4.2.

NOTE   The conclusions in the first two cases of Theorem 4.5 are valid even if $f'(x) = 0$ at a finite number of x-values in (a, b).

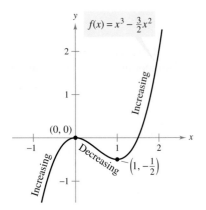

Figure 4.16

EXAMPLE 1   Intervals on Which f Is Increasing or Decreasing

Find the open intervals on which $f(x) = x^3 - \frac{3}{2}x^2$ is increasing or decreasing.

Solution   Note that f is differentiable on the entire real number line. To determine the critical numbers of f, set $f'(x)$ equal to zero.

$$f(x) = x^3 - \frac{3}{2}x^2 \qquad \text{Write original function.}$$

$$f'(x) = 3x^2 - 3x = 0 \qquad \text{Differentiate and set } f'(x) \text{ equal to 0.}$$

$$3(x)(x - 1) = 0 \qquad \text{Factor.}$$

$$x = 0, 1 \qquad \text{Critical numbers}$$

Because there are no points for which f' does not exist, you can conclude that $x = 0$ and $x = 1$ are the only critical numbers. The table summarizes the testing of the three intervals determined by these two critical numbers.

Interval	$-\infty < x < 0$	$0 < x < 1$	$1 < x < \infty$
Test Value	$x = -1$	$x = \frac{1}{2}$	$x = 2$
Sign of $f'(x)$	$f'(-1) = 6 > 0$	$f'(\frac{1}{2}) = -\frac{3}{4} < 0$	$f'(2) = 6 > 0$
Conclusion	Increasing	Decreasing	Increasing

So, f is increasing on the intervals $(-\infty, 0)$ and $(1, \infty)$ and decreasing on the interval $(0, 1)$, as shown in Figure 4.16.

Example 1 gives you one example of how to find intervals on which a function is increasing or decreasing. The guidelines below summarize the steps followed in the example.

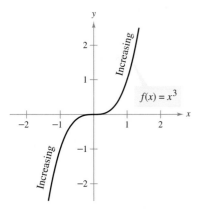

(a) Strictly monotonic function

Guidelines for Finding Intervals on Which a Function Is Increasing or Decreasing

Let f be continuous on the interval (a, b). To find the open intervals on which f is increasing or decreasing, use the following steps.

1. Locate the critical numbers of f in (a, b), and use these numbers to determine test intervals.
2. Determine the sign of $f'(x)$ at one test value in each of the intervals.
3. Use Theorem 4.5 to determine whether f is increasing or decreasing on each interval.

These guidelines are also valid if the interval (a, b) is replaced by an interval of the form $(-\infty, b)$, (a, ∞), or $(-\infty, \infty)$.

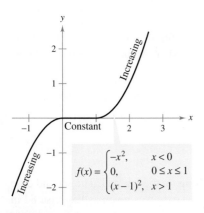

(b) Not strictly monotonic

Figure 4.17

A function is **strictly monotonic** on an interval if it is either increasing on the entire interval or decreasing on the entire interval. For instance, the function $f(x) = x^3$ is strictly monotonic on the entire real number line because it is increasing on the entire real number line, as shown in Figure 4.17(a). The function shown in Figure 4.17(b) is not strictly monotonic on the entire real number line because it is constant on the interval $[0, 1]$.

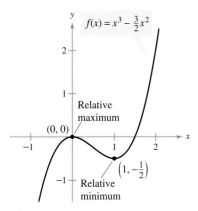

$f(x) = x^3 - \frac{3}{2}x^2$

Relative maximum

$(0, 0)$

$\left(1, -\frac{1}{2}\right)$

Relative minimum

Relative extrema of f
Figure 4.18

The First Derivative Test

After you have determined the intervals on which a function is increasing or decreasing, it is not difficult to locate the relative extrema of the function. For instance, in Figure 4.18 (from Example 1), the function

$$f(x) = x^3 - \frac{3}{2}x^2$$

has a relative maximum at the point $(0, 0)$ because f is increasing immediately to the left of $x = 0$ and decreasing immediately to the right of $x = 0$. Similarly, f has a relative minimum at the point $\left(1, -\frac{1}{2}\right)$ because f is decreasing immediately to the left of $x = 1$ and increasing immediately to the right of $x = 1$. The following theorem, called the First Derivative Test, makes this more explicit.

THEOREM 4.6   The First Derivative Test

Let c be a critical number of a function f that is continuous on an open interval I containing c. If f is differentiable on the interval, except possibly at c, then $f(c)$ can be classified as follows.

1. If $f'(x)$ changes from negative to positive at c, then f has a *relative minimum* at $(c, f(c))$.

2. If $f'(x)$ changes from positive to negative at c, then f has a *relative maximum* at $(c, f(c))$.

3. If $f'(x)$ is positive on both sides of c or negative on both sides of c, then $f(c)$ is neither a relative minimum nor a relative maximum.

Proof   Assume that $f'(x)$ changes from negative to positive at c. Then there exist a and b in I such that

$$f'(x) < 0 \text{ for all } x \text{ in } (a, c)$$

and

$$f'(x) > 0 \text{ for all } x \text{ in } (c, b).$$

By Theorem 4.5, f is decreasing on $[a, c]$ and increasing on $[c, b]$. So, $f(c)$ is a minimum of f on the open interval (a, b) and, consequently, a relative minimum of f. This proves the first case of the theorem. The second case can be proved in a similar way (see Exercise 108).

<u>EXAMPLE 2</u>   **Applying the First Derivative Test**

Find the relative extrema of the function $f(x) = \frac{1}{2}x - \sin x$ in the interval $(0, 2\pi)$.

Solution   Note that f is continuous on the interval $(0, 2\pi)$. To determine the critical numbers of f in this interval, set $f'(x)$ equal to 0.

$$f'(x) = \frac{1}{2} - \cos x = 0 \qquad \text{Set } f'(x) \text{ equal to 0.}$$

$$\cos x = \frac{1}{2}$$

$$x = \frac{\pi}{3}, \frac{5\pi}{3} \qquad \text{Critical numbers}$$

Because there are no points for which f' does not exist, you can conclude that $x = \pi/3$ and $x = 5\pi/3$ are the only critical numbers. The table summarizes the testing of the three intervals determined by these two critical numbers.

Interval	$0 < x < \dfrac{\pi}{3}$	$\dfrac{\pi}{3} < x < \dfrac{5\pi}{3}$	$\dfrac{5\pi}{3} < x < 2\pi$
Test Value	$x = \dfrac{\pi}{4}$	$x = \pi$	$x = \dfrac{7\pi}{4}$
Sign of $f'(x)$	$f'\left(\dfrac{\pi}{4}\right) < 0$	$f'(\pi) > 0$	$f'\left(\dfrac{7\pi}{4}\right) < 0$
Conclusion	Decreasing	Increasing	Decreasing

By applying the First Derivative Test, you can conclude that f has a relative minimum at the point where

$$x = \frac{\pi}{3} \qquad \textit{x-value where relative minimum occurs}$$

and a relative maximum at the point where

$$x = \frac{5\pi}{3} \qquad \textit{x-value where relative maximum occurs}$$

as shown in Figure 4.19.

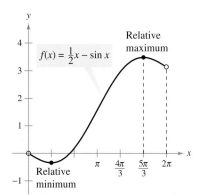

$f(x) = \frac{1}{2}x - \sin x$

A relative minimum occurs where f changes from decreasing to increasing, and a relative maximum occurs where f changes from increasing to decreasing.
Figure 4.19

EXPLORATION

Comparing Graphical and Analytic Approaches   From Section 4.2, you know that, *by itself*, a graphing utility can give misleading information about the relative extrema of a graph. *Used in conjunction with an analytic approach*, however, a graphing utility can provide a good way to reinforce your conclusions. Try using a graphing utility to graph the function in Example 2. Then use the *zoom* and *trace* features to estimate the relative extrema. How close are your graphical approximations?

Note that in Examples 1 and 2 the given functions are differentiable on the entire real number line. For such functions, the only critical numbers are those for which $f'(x) = 0$. Example 3 concerns a function that has two types of critical numbers—those for which $f'(x) = 0$ and those for which f is not differentiable.

EXAMPLE 3   Applying the First Derivative Test

Find the relative extrema of

$$f(x) = (x^2 - 4)^{2/3}.$$

Solution   Begin by noting that f is continuous on the entire real number line. The derivative of f

$$f'(x) = \frac{2}{3}(x^2 - 4)^{-1/3}(2x) \qquad \text{General Power Rule}$$

$$= \frac{4x}{3(x^2 - 4)^{1/3}} \qquad \text{Simplify.}$$

is 0 when $x = 0$ and does not exist when $x = \pm 2$. So, the critical numbers are $x = -2$, $x = 0$, and $x = 2$. The table summarizes the testing of the four intervals determined by these three critical numbers.

Interval	$-\infty < x < -2$	$-2 < x < 0$	$0 < x < 2$	$2 < x < \infty$
Test Value	$x = -3$	$x = -1$	$x = 1$	$x = 3$
Sign of $f'(x)$	$f'(-3) < 0$	$f'(-1) > 0$	$f'(1) < 0$	$f'(3) > 0$
Conclusion	Decreasing	Increasing	Decreasing	Increasing

By applying the First Derivative Test, you can conclude that f has a relative minimum at the point $(-2, 0)$, a relative maximum at the point $\left(0, \sqrt[3]{16}\right)$, and another relative minimum at the point $(2, 0)$, as shown in Figure 4.20.

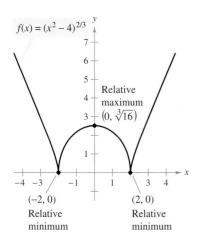

$f(x) = (x^2 - 4)^{2/3}$

Relative maximum $\left(0, \sqrt[3]{16}\right)$

$(-2, 0)$ Relative minimum

$(2, 0)$ Relative minimum

You can apply the First Derivative Test to find relative extrema.
Figure 4.20

TECHNOLOGY PITFALL   When using a graphing utility to graph a function involving radicals or rational exponents, be sure you understand the way the utility evaluates radical expressions. For instance, even though

$$f(x) = (x^2 - 4)^{2/3}$$

and

$$g(x) = [(x^2 - 4)^2]^{1/3}$$

are the same algebraically, some graphing utilities distinguish between these two functions. Which of the graphs shown in Figure 4.21 is incorrect? Why did the graphing utility produce an incorrect graph?

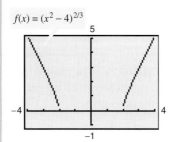

$f(x) = (x^2 - 4)^{2/3}$

$g(x) = [(x^2 - 4)^2]^{1/3}$

Which graph is incorrect?
Figure 4.21

When using the First Derivative Test, be sure to consider the domain of the function. For instance, in the next example, the function

$$f(x) = \frac{x^4 + 1}{x^2}$$

is not defined when $x = 0$. This x-value must be used with the critical numbers to determine the test intervals.

 EXAMPLE 4   Applying the First Derivative Test

Find the relative extrema of $f(x) = \dfrac{x^4 + 1}{x^2}$.

Solution

$f(x) = x^2 + x^{-2}$	Rewrite original function.
$f'(x) = 2x - 2x^{-3}$	Differentiate.
$\qquad = 2x - \dfrac{2}{x^3}$	Rewrite with positive exponent.
$\qquad = \dfrac{2(x^4 - 1)}{x^3}$	Simplify.
$\qquad = \dfrac{2(x^2 + 1)(x - 1)(x + 1)}{x^3}$	Factor.

So, $f'(x)$ is zero at $x = \pm 1$. Moreover, because $x = 0$ is not in the domain of f, you should use this x-value along with the critical numbers to determine the test intervals.

$x = \pm 1$	Critical numbers, $f'(\pm 1) = 0$
$x = 0$	0 is not in the domain of f.

The table summarizes the testing of the four intervals determined by these three x-values.

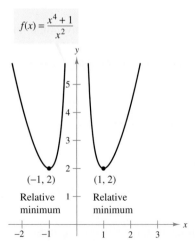

$f(x) = \dfrac{x^4 + 1}{x^2}$

$(-1, 2)$ Relative minimum

$(1, 2)$ Relative minimum

x-values that are not in the domain of f, as well as critical numbers, determine test intervals for f'.

Figure 4.22

Interval	$-\infty < x < -1$	$-1 < x < 0$	$0 < x < 1$	$1 < x < \infty$
Test Value	$x = -2$	$x = -\frac{1}{2}$	$x = \frac{1}{2}$	$x = 2$
Sign of $f'(x)$	$f'(-2) < 0$	$f'(-\frac{1}{2}) > 0$	$f'(\frac{1}{2}) < 0$	$f'(2) > 0$
Conclusion	Decreasing	Increasing	Decreasing	Increasing

By applying the First Derivative Test, you can conclude that f has one relative minimum at the point $(-1, 2)$ and another at the point $(1, 2)$, as shown in Figure 4.22.

TECHNOLOGY   The most difficult step in applying the First Derivative Test is finding the values for which the derivative is equal to 0. For instance, the values of x for which the derivative of

$$f(x) = \frac{x^4 + 1}{x^2 + 1}$$

is equal to zero are $x = 0$ and $x = \pm\sqrt{\sqrt{2} - 1}$. If you have access to technology that can perform symbolic differentiation and solve equations, use it to apply the First Derivative Test to this function.

If a projectile is propelled from ground level and air resistance is neglected, the object will travel farthest with an initial angle of 45°. If, however, the projectile is propelled from a point above ground level, the angle that yields a maximum horizontal distance is not 45° (see Example 5).

EXAMPLE 5   **The Path of a Projectile**

Neglecting air resistance, the path of a projectile that is propelled at an angle θ is

$$y = \frac{g \sec^2 \theta}{2v_0^2}x^2 + (\tan \theta)x + h, \quad 0 \le \theta \le \frac{\pi}{2}$$

where y is the height, x is the horizontal distance, g is the acceleration due to gravity, v_0 is the initial velocity, and h is the initial height. (This equation is derived in Section 12.3.) Let $g = -32$ feet per second per second, $v_0 = 24$ feet per second, and $h = 9$ feet. What value of θ will produce a maximum horizontal distance?

Solution   To find the distance the projectile travels, let $y = 0$ and use the Quadratic Formula to solve for x.

$$\frac{g \sec^2 \theta}{2v_0^2}x^2 + (\tan \theta)x + h = 0$$

$$\frac{-32 \sec^2 \theta}{2(24^2)}x^2 + (\tan \theta)x + 9 = 0$$

$$-\frac{\sec^2 \theta}{36}x^2 + (\tan \theta)x + 9 = 0$$

$$x = \frac{-\tan \theta \pm \sqrt{\tan^2 \theta + \sec^2 \theta}}{-\sec^2 \theta/18}$$

$$x = 18 \cos \theta \left(\sin \theta + \sqrt{\sin^2 \theta + 1}\right), \quad x \ge 0$$

At this point, you need to find the value of θ that produces a maximum value of x. Applying the First Derivative Test by hand would be very tedious. Using technology to solve the equation $dx/d\theta = 0$, however, eliminates most of the messy computations. The result is that the maximum value of x occurs when

$$\theta \approx 0.61548 \text{ radian, or } 35.3°.$$

This conclusion is reinforced by sketching the path of the projectile for different values of θ, as shown in Figure 4.23. Of the three paths shown, note that the distance traveled is greatest for $\theta = 35°$.

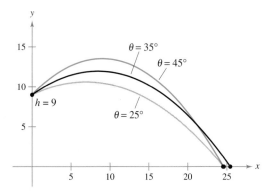

The path of a projectile with initial angle θ
Figure 4.23

NOTE   A computer simulation of this example is given in the *HM mathSpace®* CD-ROM and the online *Eduspace®* system for this text. Using that simulation, you can experimentally discover that the maximum value of x occurs when $\theta \approx 35.3°$.

Exercises for Section 4.3

See www.CalcChat.com for worked-out solutions to odd-numbered exercises.

In Exercises 1 and 2, use the graph of f to find (a) the largest open interval on which f is increasing, and (b) the largest open interval on which f is decreasing.

1.

2.

In Exercises 3–16, identify the open intervals on which the function is increasing or decreasing.

3. $f(x) = x^2 - 6x + 8$

4. $y = -(x + 1)^2$

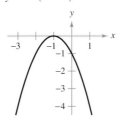

5. $y = \dfrac{x^3}{4} - 3x$

6. $f(x) = x^4 - 2x^2$

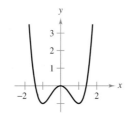

7. $f(x) = \sin x + 2, \ 0 < x < 2\pi$

8. $h(x) = \cos \dfrac{x}{2}, \ 0 < x < 2\pi$

9. $f(x) = \dfrac{1}{x^2}$

10. $y = \dfrac{x^2}{x + 1}$

11. $g(x) = x^2 - 2x - 8$

12. $h(x) = 27x - x^3$

13. $y = x\sqrt{16 - x^2}$

14. $y = x + \dfrac{4}{x}$

15. $y = x - 2\cos x, \ 0 < x < 2\pi$

16. $f(x) = \cos^2 x - \cos x, \ 0 < x < 2\pi$

In Exercises 17–46, find the critical numbers of f (if any). Find the open intervals on which the function is increasing or decreasing and locate all relative extrema. Use a graphing utility to confirm your results.

17. $f(x) = x^2 - 6x$

18. $f(x) = x^2 + 8x + 10$

19. $f(x) = -2x^2 + 4x + 3$

20. $f(x) = -(x^2 + 8x + 12)$

21. $f(x) = 2x^3 + 3x^2 - 12x$

22. $f(x) = x^3 - 6x^2 + 15$

23. $f(x) = x^2(3 - x)$

24. $f(x) = (x + 2)^2(x - 1)$

25. $f(x) = \dfrac{x^5 - 5x}{5}$

26. $f(x) = x^4 - 32x + 4$

27. $f(x) = x^{1/3} + 1$

28. $f(x) = x^{2/3} - 4$

29. $f(x) = (x - 1)^{2/3}$

30. $f(x) = (x - 1)^{1/3}$

31. $f(x) = 5 - |x - 5|$

32. $f(x) = |x + 3| - 1$

33. $f(x) = x + \dfrac{1}{x}$

34. $f(x) = \dfrac{x}{x + 1}$

35. $f(x) = \dfrac{x^2}{x^2 - 9}$

36. $f(x) = \dfrac{x + 3}{x^2}$

37. $f(x) = \dfrac{x^2 - 2x + 1}{x + 1}$

38. $f(x) = \dfrac{x^2 - 3x - 4}{x - 2}$

39. $f(x) = (3 - x)e^{x-3}$

40. $f(x) = (x - 1)e^x$

41. $f(x) = 4(x - \arcsin x)$

42. $f(x) = x \arctan x$

43. $g(x) = (x)3^{-x}$

44. $f(x) = 2^{x^2 - 3}$

45. $f(x) = x - \log_4 x$

46. $f(x) = \dfrac{x^3}{3} - \ln x$

In Exercises 47–54, consider the function on the interval $(0, 2\pi)$. For each function, (a) find the open interval(s) on which the function is increasing or decreasing, (b) apply the First Derivative Test to identify all relative extrema, and (c) use a graphing utility to confirm your results.

47. $f(x) = \dfrac{x}{2} + \cos x$

48. $f(x) = \sin x \cos x$

49. $f(x) = \sin x + \cos x$

50. $f(x) = x + 2\sin x$

51. $f(x) = \cos^2(2x)$

52. $f(x) = \sqrt{3}\sin x + \cos x$

53. $f(x) = \sin^2 x + \sin x$

54. $f(x) = \dfrac{\sin x}{1 + \cos^2 x}$

In Exercises 55–60, (a) use a computer algebra system to differentiate the function, (b) sketch the graphs of f and f' on the same set of coordinate axes over the indicated interval, (c) find the critical numbers of f in the open interval, and (d) find the interval(s) on which f' is positive and the interval(s) on which it is negative. Compare the behavior of f and the sign of f'.

55. $f(x) = 2x\sqrt{9 - x^2}, \ [-3, 3]$

56. $f(x) = 10\big(5 - \sqrt{x^2 - 3x + 16}\big), \ [0, 5]$

57. $f(t) = t^2 \sin t, [0, 2\pi]$ **58.** $f(x) = \dfrac{x}{2} + \cos \dfrac{x}{2}, [0, 4\pi]$

59. $f(x) = \frac{1}{2}(x^2 - \ln x), (0, 3]$ **60.** $f(x) = (4 - x^2)e^x, [0, 2]$

In Exercises 61 and 62, use symmetry, extrema, and zeros to sketch the graph of *f*. How do the functions *f* and *g* differ? Explain.

61. $f(x) = \dfrac{x^5 - 4x^3 + 3x}{x^2 - 1}, \quad g(x) = x(x^2 - 3)$

62. $f(t) = \cos^2 t - \sin^2 t, \quad g(t) = 1 - 2\sin^2 t, \quad (-2, 2)$

Think About It   In Exercises 63–68, the graph of *f* is shown in the figure. Sketch a graph of the derivative of *f*. To print an enlarged copy of the graph, go to the website *www.mathgraphs.com*.

63.

64.

65.

66.

67.

68.

In Exercises 69–72, use the graph of *f′* to (a) identify the interval(s) on which *f* is increasing or decreasing, and (b) estimate the values of *x* at which *f* has a relative maximum or minimum.

69.

70.

71.

72.

In Exercises 73–78, assume that *f* is differentiable for all *x*. The signs of *f′* are as follows.

$f'(x) > 0$ on $(-\infty, -4)$

$f'(x) < 0$ on $(-4, 6)$

$f'(x) > 0$ on $(6, \infty)$

Supply the appropriate inequality for the indicated value of *c*.

Function	Sign of $g'(c)$
73. $g(x) = f(x) + 5$	$g'(0)$ ▨ 0
74. $g(x) = 3f(x) - 3$	$g'(-5)$ ▨ 0
75. $g(x) = -f(x)$	$g'(-6)$ ▨ 0
76. $g(x) = -f(x)$	$g'(0)$ ▨ 0
77. $g(x) = f(x - 10)$	$g'(0)$ ▨ 0
78. $g(x) = f(x - 10)$	$g'(8)$ ▨ 0

79. Sketch the graph of the arbitrary function *f* such that

$$f'(x) \begin{cases} > 0, & x < 4 \\ \text{undefined}, & x = 4. \\ < 0, & x > 4 \end{cases}$$

80. A differentiable function *f* has one critical number at $x = 5$. Identify the relative extrema of *f* at the critical number if $f'(4) = -2.5$ and $f'(6) = 3$.

81. *Think About It*   The function *f* is differentiable on the interval $[-1, 1]$. The table shows the values of *f′* for selected values of *x*. Sketch the graph of *f*, approximate the critical numbers, and identify the relative extrema.

x	-1	-0.75	-0.50	-0.25
$f'(x)$	-10	-3.2	-0.5	0.8

x	0	0.25	0.50	0.75	1
$f'(x)$	5.6	3.6	-0.2	-6.7	-20.1

82. Think About It The function f is differentiable on the interval $[0, \pi]$. The table shows the values of f' for selected values of x. Sketch the graph of f, approximate the critical numbers, and identify the relative extrema.

x	0	$\pi/6$	$\pi/4$	$\pi/3$	$\pi/2$
$f'(x)$	3.14	-0.23	-2.45	-3.11	0.69

x	$2\pi/3$	$3\pi/4$	$5\pi/6$	π
$f'(x)$	3.00	1.37	-1.14	-2.84

83. Rolling a Ball Bearing A ball bearing is placed on an inclined plane and begins to roll. The angle of elevation of the plane is θ. The distance (in meters) the ball bearing rolls in t seconds is $s(t) = 4.9(\sin \theta)t^2$.

(a) Determine the speed of the ball bearing after t seconds.

(b) Complete the table and use it to determine the value of θ that produces the maximum speed at a particular time.

θ	0	$\pi/4$	$\pi/3$	$\pi/2$	$2\pi/3$	$3\pi/4$	π
$s'(t)$							

84. Numerical, Graphical, and Analytic Analysis The concentration C of a chemical in the bloodstream t hours after injection into muscle tissue is

$$C(t) = \frac{3t}{27 + t^3}, \quad t \geq 0.$$

(a) Complete the table and use it to approximate the time when the concentration is greatest.

t	0	0.5	1	1.5	2	2.5	3
$C(t)$							

(b) Use a graphing utility to graph the concentration function and use the graph to approximate the time when the concentration is greatest.

(c) Use calculus to determine analytically the time when the concentration is greatest.

85. Numerical, Graphical, and Analytic Analysis Consider the functions $f(x) = x$ and $g(x) = \sin x$ on the interval $(0, \pi)$.

(a) Complete the table and make a conjecture about which is the greater function on the interval $(0, \pi)$.

x	0.5	1	1.5	2	2.5	3
$f(x)$						
$g(x)$						

(b) Use a graphing utility to graph the functions and use the graphs to make a conjecture about which is the greater function on the interval $(0, \pi)$.

(c) Prove that $f(x) > g(x)$ on the interval $(0, \pi)$. [Hint: Show that $h'(x) > 0$ where $h = f - g$.]

86. Numerical, Graphical, and Analytic Analysis Consider the functions $f(x) = x$ and $g(x) = \tan x$ on the interval $(0, \pi/2)$.

(a) Complete the table and make a conjecture about which is the greater function on the interval $(0, \pi/2)$.

x	0.25	0.5	0.75	1	1.25	1.5
$f(x)$						
$g(x)$						

(b) Use a graphing utility to graph the functions and use the graphs to make a conjecture about which is the greater function on the interval $(0, \pi/2)$.

(c) Prove that $f(x) < g(x)$ on the interval $(0, \pi/2)$. [Hint: Show that $h'(x) > 0$, where $h = g - f$.]

87. Trachea Contraction Coughing forces the trachea (windpipe) to contract, which affects the velocity v of the air passing through the trachea. The velocity of the air during coughing is

$$v = k(R - r)r^2, \quad 0 \leq r < R$$

where k is constant, R is the normal radius of the trachea, and r is the radius during coughing. What radius will produce the maximum air velocity?

88. Profit The profit P (in dollars) made by a fast-food restaurant selling x hamburgers is

$$P = 40{,}000(e^{-x} - 1) - 3x + 850\sqrt{x}, \quad 0 \leq x \leq 35{,}000.$$

Find the open intervals on which P is increasing or decreasing.

89. Modeling Data The end-of-year assets for the Medicare Hospital Insurance Trust Fund (in billions of dollars) for the years 1995 through 2001 are shown.

1995: 130.3; 1996: 124.9; 1997: 115.6; 1998: 120.4;

1999: 141.4; 2000: 177.5; 2001: 208.7

(Source: U.S. Centers for Medicare and Medicaid Services)

(a) Use the regression capabilities of a graphing utility to find a model of the form $M = at^2 + bt + c$ for the data. (Let $t = 5$ represent 1995.)

(b) Use a graphing utility to plot the data and graph the model.

(c) Analytically find the minimum of the model and compare the result with the actual data.

90. Modeling Data The number of bankruptcies (in thousands) for the years 1988 through 2001 are shown.

1988: 594.6; 1989: 643.0; 1990: 725.5; 1991: 880.4;

1992: 972.5; 1993: 918.7; 1994: 845.3; 1995: 858.1;

1996: 1042.1; 1997: 1317.0; 1998: 1429.5;

1999: 1392.0; 2000: 1277.0; 2001: 1386.6

(Source: Administrative Office of the U.S. Courts)

(a) Use the regression capabilities of a graphing utility to find a model of the form $B = at^4 + bt^3 + ct^2 + dt + e$ for the data. (Let $t = 8$ represent 1988.)

(b) Use a graphing utility to plot the data and graph the model.

(c) Find the maximum of the model and compare the result with the actual data.

Motion Along a Line **In Exercises 91–94, the function $s(t)$ describes the motion of a particle moving along a line. For each function, (a) find the velocity function of the particle at any time $t \geq 0$, (b) identify the time interval(s) when the particle is moving in a positive direction, (c) identify the time interval(s) when the particle is moving in a negative direction, and (d) identify the time(s) when the particle changes its direction.**

91. $s(t) = 6t - t^2$

92. $s(t) = t^2 - 7t + 10$

93. $s(t) = t^3 - 5t^2 + 4t$

94. $s(t) = t^3 - 20t^2 + 128t - 280$

Motion Along a Line **In Exercises 95 and 96, the graph shows the position of a particle moving along a line. Describe how the particle's position changes with respect to time.**

95.

96.

 Creating Polynomial Functions **In Exercises 97–100, find a polynomial function**

$$f(x) = a_n x^n + a_{n-1} x^{n-1} + \cdots + a_2 x^2 + a_1 x + a_0$$

that has only the specified extrema. (a) Determine the minimum degree of the function and give the criteria you used in determining the degree. (b) Using the fact that the coordinates of the extrema are solution points of the function, and that the x-coordinates are critical numbers, determine a system of linear equations whose solution yields the coefficients of the required function. (c) Use a graphing utility to solve the system of equations and determine the function. (d) Use a graphing utility to confirm your result graphically.

97. Relative minimum: $(0, 0)$; Relative maximum: $(2, 2)$

98. Relative minimum: $(0, 0)$; Relative maximum: $(4, 1000)$

99. Relative minima: $(0, 0)$, $(4, 0)$

Relative maximum: $(2, 4)$

100. Relative minimum: $(1, 2)$

Relative maxima: $(-1, 4)$, $(3, 4)$

True or False? **In Exercises 101–106, determine whether the statement is true or false. If it is false, explain why or give an example that shows it is false.**

101. The sum of two increasing functions is increasing.

102. The product of two increasing functions is increasing.

103. Every nth-degree polynomial has $(n - 1)$ critical numbers.

104. An nth-degree polynomial has at most $(n - 1)$ critical numbers.

105. There is a relative maximum or minimum at each critical number.

106. The relative maxima of the function f are $f(1) = 4$ and $f(3) = 10$. So, f has at least one minimum for some x in the interval $(1, 3)$.

107. Prove the second case of Theorem 4.5.

108. Prove the second case of Theorem 4.6.

109. Let $x > 0$ and $n > 1$ be real numbers. Prove that $(1 + x)^n > 1 + nx$.

110. Use the definitions of increasing and decreasing functions to prove that $f(x) = x^3$ is increasing on $(-\infty, \infty)$.

111. Use the definitions of increasing and decreasing functions to prove that $f(x) = 1/x$ is decreasing on $(0, \infty)$.

Section Project: Rainbows

Rainbows are formed when light strikes raindrops and is reflected and refracted, as shown in the figure. (This figure shows a cross section of a spherical raindrop.) The Law of Refraction states that $(\sin \alpha)/(\sin \beta) = k$, where $k \approx 1.33$ (for water). The angle of deflection is given by $D = \pi + 2\alpha - 4\beta$.

(a) Use a graphing utility to graph

$D = \pi + 2\alpha - 4 \sin^{-1}(1/k \sin \alpha)$,

$0 \leq \alpha \leq \pi/2$.

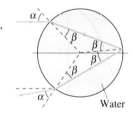

(b) Prove that the minimum angle of deflection occurs when

$$\cos \alpha = \sqrt{\frac{k^2 - 1}{3}}.$$

For water, what is the minimum angle of deflection, D_{min}? (The angle $\pi - D_{min}$ is called the *rainbow angle*.) What value of α produces this minimum angle? (A ray of sunlight that strikes a raindrop at this angle, α, is called a *rainbow ray*.)

FOR FURTHER INFORMATION For more information about the mathematics of rainbows, see the article "Somewhere Within the Rainbow" by Steven Janke in *The UMAP Journal*.

5. *Graphical Reasoning* The graph of f is shown in the figure.

(a) For which values of x is $f'(x)$ zero? Positive? Negative?

(b) For which values of x is $f''(x)$ zero? Positive? Negative?

(c) On what interval is f' an increasing function?

(d) For which value of x is $f'(x)$ minimum? For this value of x, how does the rate of change of f compare with the rate of change of f for other values of x? Explain.

 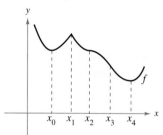

Figure for 5   **Figure for 6**

6. *Graphical Reasoning* Identify the real numbers $x_0, x_1, x_2, x_3,$ and x_4 in the figure such that each of the following is true.

(a) $f'(x) = 0$ (b) $f''(x) = 0$

(c) $f'(x)$ does not exist.   (d) f has a relative maximum.

(e) f has a point of inflection.

In Exercises 7–50, analyze and sketch a graph of the function. Label any intercepts, relative extrema, points of inflection, and asymptotes. Use a graphing utility to verify your results.

7. $y = \dfrac{x^2}{x^2 + 3}$ **8.** $y = \dfrac{x}{x^2 + 1}$

9. $y = \dfrac{1}{x - 2} - 3$ **10.** $y = \dfrac{x^2 + 1}{x^2 - 9}$

11. $y = \dfrac{2x}{x^2 - 1}$ **12.** $f(x) = \dfrac{x + 2}{x}$

13. $g(x) = x + \dfrac{4}{x^2 + 1}$ **14.** $f(x) = x + \dfrac{32}{x^2}$

15. $f(x) = \dfrac{x^2 + 1}{x}$ **16.** $f(x) = \dfrac{x^3}{x^2 - 4}$

17. $y = \dfrac{x^2 - 6x + 12}{x - 4}$ **18.** $y = \dfrac{2x^2 - 5x + 5}{x - 2}$

19. $y = x\sqrt{4 - x}$ **20.** $g(x) = x\sqrt{9 - x}$

21. $h(x) = x\sqrt{9 - x^2}$ **22.** $y = x\sqrt{16 - x^2}$

23. $y = 3x^{2/3} - 2x$

24. $y = 3(x - 1)^{2/3} - (x - 1)^2$

25. $y = x^3 - 3x^2 + 3$ **26.** $y = -\frac{1}{3}(x^3 - 3x + 2)$

27. $y = 2 - x - x^3$ **28.** $f(x) = \frac{1}{3}(x - 1)^3 + 2$

29. $f(x) = 3x^3 - 9x + 1$

30. $f(x) = (x + 1)(x - 2)(x - 5)$

31. $y = 3x^4 + 4x^3$ **32.** $y = 3x^4 - 6x^2 + \frac{5}{3}$

33. $f(x) = x^4 - 4x^3 + 16x$

34. $f(x) = x^4 - 8x^3 + 18x^2 - 16x + 5$

35. $y = x^5 - 5x$ **36.** $y = (x - 1)^5$

37. $y = |2x - 3|$ **38.** $y = |x^2 - 6x + 5|$

39. $f(x) = e^{3x}(2 - x)$ **40.** $f(x) = -2 + e^{3x}(4 - 2x)$

41. $g(t) = \dfrac{10}{1 + 4e^{-t}}$ **42.** $h(x) = \dfrac{10}{2 + 3e^{-x/2}}$

43. $y = (x - 1)\ln(x - 1)$ **44.** $y = \frac{1}{24}x^3 - \ln x$

45. $g(x) = 6 \arcsin\left(\dfrac{x - 2}{2}\right)^2$

46. $h(x) = 7 \arctan(x + 1) - \ln(x^2 + 2x + 2)$

47. $f(x) = \dfrac{x}{2^{x-2}}$ **48.** $h(t) = (3 - t)3^t$

49. $g(x) = \log_4(x - x^2)$ **50.** $f(x) = \log_2|x^2 - 4x|$

In Exercises 51–58, sketch a graph of the function over the indicated interval. Use a graphing utility to verify your graph.

Function	Interval
51. $y = \sin x - \frac{1}{18}\sin 3x$	$0 \le x \le 2\pi$
52. $y = \cos x - \frac{1}{2}\cos 2x$	$0 \le x \le 2\pi$
53. $y = 2x - \tan x$	$-\dfrac{\pi}{2} < x < \dfrac{\pi}{2}$
54. $y = 2(x - 2) + \cot x$	$0 < x < \pi$
55. $y = 2(\csc x + \sec x)$	$0 < x < \dfrac{\pi}{2}$
56. $y = \sec^2\left(\dfrac{\pi x}{8}\right) - 2\tan\left(\dfrac{\pi x}{8}\right) - 1$	$-3 < x < 3$
57. $g(x) = x \tan x$	$-\dfrac{3\pi}{2} < x < \dfrac{3\pi}{2}$
58. $g(x) = x \cot x$	$-2\pi < x < 2\pi$

In Exercises 59–64, use a computer algebra system to analyze and graph the function. Identify any relative extrema, points of inflection, and asymptotes.

59. $f(x) = \dfrac{20x}{x^2 + 1} - \dfrac{1}{x}$ **60.** $f(x) = 5\left(\dfrac{1}{x - 4} - \dfrac{1}{x + 2}\right)$

61. $f(x) = \dfrac{x}{\sqrt{x^2 + 7}}$ **62.** $f(x) = \dfrac{4x}{\sqrt{x^2 + 15}}$

63. $y = \dfrac{x}{2} + \ln\left(\dfrac{x}{x + 3}\right)$ **64.** $y = \dfrac{3x}{2}(1 + 4e^{-x/3})$

Think About It **In Exercises 65–68, create a function whose graph has the indicated characteristics. (There is more than one correct answer.)**

65. Vertical asymptote: $x = 5$

    Horizontal asymptote: $y = 0$

66. Vertical asymptote: $x = -3$

    Horizontal asymptote: None

67. Vertical asymptote: $x = 5$

    Slant asymptote: $y = 3x + 2$

68. Vertical asymptote: $x = 0$

    Slant asymptote: $y = -x$

Writing About Concepts

In Exercises 69 and 70, the graphs of f, f', and f'' are shown on the same set of coordinate axes. Which is which? Explain your reasoning. To print an enlarged copy of the graph, go to the website *www.mathgraphs.com*.

69.

70.

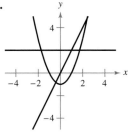

In Exercises 71–74, use the graph of f' to sketch a graph of f and the graph of f''. To print an enlarged copy of the graph, go to the website *www.mathgraphs.com*.

71.

72.

73.

74.

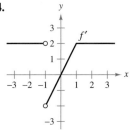

(Submitted by Bill Fox, Moberly Area Community College, Moberly, MO)

75. Suppose $f'(t) < 0$ for all t in the interval $(2, 8)$. Explain why $f(3) > f(5)$.

76. Suppose $f(0) = 3$ and $2 \le f'(x) \le 4$ for all x in the interval $[-5, 5]$. Determine the greatest and least possible values of $f(2)$.

 In Exercises 77–80, use a graphing utility to graph the function. Use the graph to determine whether it is possible for the graph of a function to cross its horizontal asymptote. Do you think it is possible for the graph of a function to cross its vertical asymptote? Why or why not?

77. $f(x) = \dfrac{4(x-1)^2}{x^2 - 4x + 5}$

78. $g(x) = \dfrac{3x^4 - 5x + 3}{x^4 + 1}$

79. $h(x) = \dfrac{\sin 2x}{x}$

80. $f(x) = \dfrac{\cos 3x}{4x}$

 Writing   In Exercises 81 and 82, use a graphing utility to graph the function. Explain why there is no vertical asymptote when a superficial examination of the function may indicate that there should be one.

81. $h(x) = \dfrac{6 - 2x}{3 - x}$

82. $g(x) = \dfrac{x^2 + x - 2}{x - 1}$

 Writing   In Exercises 83–86, use a graphing utility to graph the function and determine the slant asymptote of the graph. Zoom out repeatedly and describe how the graph on the display appears to change. Why does this occur?

83. $f(x) = -\dfrac{x^2 - 3x - 1}{x - 2}$

84. $g(x) = \dfrac{2x^2 - 8x - 15}{x - 5}$

85. $f(x) = \dfrac{x^3}{x^2 + 1}$

86. $h(x) = \dfrac{-x^3 + x^2 + 4}{x^2}$

87. *Graphical Reasoning*   Consider the function

$$f(x) = \frac{\cos^2 \pi x}{\sqrt{x^2 + 1}}, \quad 0 < x < 4.$$

(a) Use a computer algebra system to graph the function and use the graph to approximate the critical numbers visually.

(b) Use a computer algebra system to find f' and approximate the critical numbers. Are the results the same as the visual approximation in part (a)? Explain.

88. *Graphical Reasoning*   Consider the function

$$f(x) = \tan(\sin \pi x).$$

(a) Use a graphing utility to graph the function.

(b) Identify any symmetry of the graph.

(c) Is the function periodic? If so, what is the period?

(d) Identify any extrema on $(-1, 1)$.

(e) Use a graphing utility to determine the concavity of the graph on $(0, 1)$.

89. *Graphical Reasoning*   Consider the function

$$f(x) = \frac{ax}{(x - b)^2}.$$

(a) Determine the effect on the graph of f if $b \ne 0$ and a is varied. Consider cases where a is positive and a is negative.

(b) Determine the effect on the graph of f if $a \ne 0$ and b is varied.

90. Consider the function $f(x) = \frac{1}{2}(ax)^2 - (ax)$, $a \ne 0$.

(a) Determine the changes (if any) in the intercepts, extrema, and concavity of the graph of f when a is varied.

(b) In the same viewing window, use a graphing utility to graph the function for four different values of a.

91. *Investigation* Consider the function $f(x) = \dfrac{3x^n}{x^4 + 1}$ for nonnegative integer values of n.

(a) Discuss the relationship between the value of n and the symmetry of the graph.

(b) For which values of n will the x-axis be the horizontal asymptote?

(c) For which value of n will $y = 3$ be the horizontal asymptote?

(d) What is the asymptote of the graph when $n = 5$?

(e) Use a graphing utility to graph f for the indicated values of n in the table. Use the graph to determine the number of extrema M and the number of inflection points N of the graph.

n	0	1	2	3	4	5
M						
N						

92. *Investigation* Let $P(x_0, y_0)$ be an arbitrary point on the graph of f such that $f'(x_0) \neq 0$, as shown in the figure. Verify each statement.

(a) The x-intercept of the tangent line is $\left(x_0 - \dfrac{f(x_0)}{f'(x_0)}, 0\right)$.

(b) The y-intercept of the tangent line is $(0, f(x_0) - x_0 f'(x_0))$.

(c) The x-intercept of the normal line is $(x_0 + f(x_0)f'(x_0), 0)$.

(d) The y-intercept of the normal line is $\left(0, y_0 + \dfrac{x_0}{f'(x_0)}\right)$.

(e) $|BC| = \left|\dfrac{f(x_0)}{f'(x_0)}\right|$ (f) $|PC| = \left|\dfrac{f(x_0)\sqrt{1 + [f'(x_0)]^2}}{f'(x_0)}\right|$

(g) $|AB| = |f(x_0)f'(x_0)|$

(h) $|AP| = |f(x_0)|\sqrt{1 + [f'(x_0)]^2}$

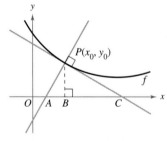

93. *Modeling Data* The data in the table show the number N of bacteria in a culture at time t, where t is measured in days.

t	1	2	3	4	5	6	7	8
N	25	200	804	1756	2296	2434	2467	2473

A model for these data is given by

$$N = \frac{24{,}670 - 35{,}153t + 13{,}250t^2}{100 - 39t + 7t^2}, \quad 1 \leq t \leq 8.$$

(a) Use a graphing utility to plot the data and graph the model.

(b) Use the model to estimate the number of bacteria when $t = 10$.

(c) Approximate the day when the number of bacteria is greatest.

(d) Use a computer algebra system to determine the time when the rate of increase in the number of bacteria is greatest.

(e) Find $\lim\limits_{t \to \infty} N(t)$.

94. *Modeling Data* A meteorologist measures the atmospheric pressure P (in kilograms per square meter) at altitude h (in kilometers). The data are shown below.

h	0	5	10	15	20
P	10,332	5583	2376	1240	517

(a) Use a graphing utility to plot the points $(h, \ln P)$. Use the regression capabilities of the graphing utility to find a linear model for the revised data points.

(b) The line in part (a) has the form $\ln P = ah + b$. Write the equation in exponential form.

(c) Use a graphing utility to plot the original data and graph the exponential model in part (b).

(d) Find the rate of change of the pressure when $h = 5$ and $h = 18$.

95. Let f be a function that is positive and differentiable on the entire real number line. Let $g(x) = \ln f(x)$.

(a) If g is increasing, must f be increasing? Explain.

(b) If the graph of f is concave upward, must the graph of g be concave upward? Explain.

96. *Conjecture* Use a graphing utility to graph f and g in the same viewing window and determine which is increasing at the faster rate for "large" values of x. What can you conclude about the rate of growth of the natural logarithmic function?

(a) $f(x) = \ln x$, $g(x) = \sqrt{x}$ (b) $f(x) = \ln x$, $g(x) = \sqrt[4]{x}$

Slant Asymptotes **In Exercises 97 and 98, the graph of the function has two slant asymptotes. Identify each slant asymptote. Then graph the function and its asymptotes.**

97. $y = \sqrt{4 + 16x^2}$ **98.** $y = \sqrt{x^2 + 6x}$

Putnam Exam Challenge

99. Let $f(x)$ be defined for $a \leq x \leq b$. Assuming appropriate properties of continuity and derivability, prove for $a < x < b$ that

$$\frac{\dfrac{f(x) - f(a)}{x - a} - \dfrac{f(b) - f(a)}{b - a}}{x - b} = \frac{1}{2} f''(\beta)$$

where β is some number between a and b.

Optimization Problems

• Solve applied minimum and maximum problems.

Applied Minimum and Maximum Problems

One of the most common applications of calculus involves the determination of minimum and maximum values. Consider how frequently you hear or read terms such as greatest profit, least cost, least time, greatest voltage, optimum size, least size, greatest strength, and greatest distance. Before outlining a general problem-solving strategy for such problems, let's look at an example.

EXAMPLE 1   Finding Maximum Volume

A manufacturer wants to design an open box having a square base and a surface area of 108 square inches, as shown in Figure 4.54. What dimensions will produce a box with maximum volume?

Solution   Because the box has a square base, its volume is

$$V = x^2h.$$ Primary equation

This equation is called the **primary equation** because it gives a formula for the quantity to be optimized. The surface area of the box is

$$S = (\text{area of base}) + (\text{area of four sides})$$
$$S = x^2 + 4xh = 108.$$ Secondary equation

Because V is to be maximized, you want to write V as a function of just one variable. To do this, you can solve the equation $x^2 + 4xh = 108$ for h in terms of x to obtain $h = (108 - x^2)/(4x)$. Substituting into the primary equation produces

$$V = x^2h$$ Function of two variables

$$= x^2\left(\frac{108 - x^2}{4x}\right)$$ Substitute for h.

$$= 27x - \frac{x^3}{4}.$$ Function of one variable

Before finding which x-value will yield a maximum value of V, you should determine the *feasible domain*. That is, what values of x make sense in this problem? You know that $V \geq 0$. You also know that x must be nonnegative and that the area of the base ($A = x^2$) is at most 108. So, the feasible domain is

$$0 \leq x \leq \sqrt{108}.$$ Feasible domain

To maximize V, find the critical numbers of the volume function on the interval $\left[0, \sqrt{108}\right]$.

$$\frac{dV}{dx} = 27 - \frac{3x^2}{4} = 0$$ Set derivative equal to 0.

$$3x^2 = 108$$ Simplify.

$$x = \pm 6$$ Critical numbers

So, the critical numbers are $x = \pm 6$. You do not need to consider $x = -6$ because it is outside the domain. Evaluating V at the critical number $x = 6$ and at the endpoints of the domain produces $V(0) = 0$, $V(6) = 108$, and $V\left(\sqrt{108}\right) = 0$. So, V is maximum when $x = 6$ and the dimensions of the box are $6 \times 6 \times 3$ inches.

Open box with square base:
$S = x^2 + 4xh = 108$
Figure 4.54

TECHNOLOGY   You can verify your answer by using a graphing utility to graph the volume function

$$V = 27x - \frac{x^3}{4}.$$

Use a viewing window in which $0 \leq x \leq \sqrt{108} \approx 10.4$ and $0 \leq y \leq 120$ and the *trace* feature to determine the maximum value of V.

In Example 1, you should realize that there are infinitely many open boxes having 108 square inches of surface area. To begin solving the problem, you might ask yourself which basic shape would seem to yield a maximum volume. Should the box be tall, squat, or nearly cubical?

You might even try calculating a few volumes, as shown in Figure 4.55, to see if you can get a better feeling for what the optimum dimensions should be. Remember that you are not ready to begin solving a problem until you have clearly identified what the problem is.

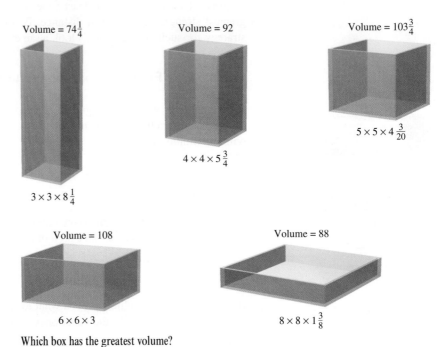

Volume = $74\frac{1}{4}$

$3 \times 3 \times 8\frac{1}{4}$

Volume = 92

$4 \times 4 \times 5\frac{3}{4}$

Volume = $103\frac{3}{4}$

$5 \times 5 \times 4\frac{3}{20}$

Volume = 108

$6 \times 6 \times 3$

Volume = 88

$8 \times 8 \times 1\frac{3}{8}$

Which box has the greatest volume?
Figure 4.55

Example 1 illustrates the following guidelines for solving applied minimum and maximum problems.

Guidelines for Solving Applied Minimum and Maximum Problems

1. Identify all *given* quantities and quantities *to be determined*. If possible, make a sketch.

2. Write a **primary equation** for the quantity that is to be maximized or minimized. (A review of several useful formulas from geometry is presented inside the back cover.)

3. Reduce the primary equation to one having a *single independent variable*. This may involve the use of **secondary equations** relating the independent variables of the primary equation.

4. Determine the feasible domain of the primary equation. That is, determine the values for which the stated problem makes sense.

5. Determine the desired maximum or minimum value by the calculus techniques discussed in Sections 4.1 through 4.4.

NOTE When performing Step 5, recall that to determine the maximum or minimum value of a continuous function f on a closed interval, you should compare the values of f at its critical numbers with the values of f at the endpoints of the interval.

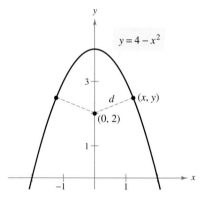

The quantity to be minimized is distance:
$d = \sqrt{(x - 0)^2 + (y - 2)^2}$.
Figure 4.56

The quantity to be minimized is area:
$A = (x + 3)(y + 2)$.
Figure 4.57

EXAMPLE 2   Finding Minimum Distance

Which points on the graph of $y = 4 - x^2$ are closest to the point $(0, 2)$?

Solution   Figure 4.56 shows that there are two points at a minimum distance from the point $(0, 2)$. The distance between the point $(0, 2)$ and a point (x, y) on the graph of $y = 4 - x^2$ is given by

$$d = \sqrt{(x - 0)^2 + (y - 2)^2}. \qquad \text{Primary equation}$$

Using the secondary equation $y = 4 - x^2$, you can rewrite the primary equation as

$$d = \sqrt{x^2 + (4 - x^2 - 2)^2} = \sqrt{x^4 - 3x^2 + 4}.$$

Because d is smallest when the expression inside the radical is smallest, you need only find the critical numbers of $f(x) = x^4 - 3x^2 + 4$. Note that the domain of f is the entire real number line. So, there are no endpoints of the domain to consider. Moreover, setting $f'(x)$ equal to 0 yields

$$f'(x) = 4x^3 - 6x = 2x(2x^2 - 3) = 0$$

$$x = 0, \ \sqrt{\frac{3}{2}}, \ -\sqrt{\frac{3}{2}}.$$

The First Derivative Test verifies that $x = 0$ yields a relative maximum, whereas both $x = \sqrt{3/2}$ and $x = -\sqrt{3/2}$ yield a minimum distance. So, the closest points are $\left(\sqrt{3/2}, 5/2\right)$ and $\left(-\sqrt{3/2}, 5/2\right)$.

EXAMPLE 3   Finding Minimum Area

A rectangular page is to contain 24 square inches of print. The margins at the top and bottom of the page are to be $1\frac{1}{2}$ inches, and the margins on the left and right are to be 1 inch (see Figure 4.57). What should the dimensions of the page be so that the least amount of paper is used?

Solution   Let A be the area to be minimized.

$$A = (x + 3)(y + 2) \qquad \text{Primary equation}$$

The printed area inside the margins is given by

$$24 = xy. \qquad \text{Secondary equation}$$

Solving this equation for y produces $y = 24/x$. Substitution into the primary equation produces

$$A = (x + 3)\left(\frac{24}{x} + 2\right) = 30 + 2x + \frac{72}{x}. \qquad \text{Function of one variable}$$

Because x must be positive, you are interested only in values of A for $x > 0$. To find the critical numbers, differentiate with respect to x.

$$\frac{dA}{dx} = 2 - \frac{72}{x^2} = 0 \quad \Longrightarrow \quad x^2 = 36$$

So, the critical numbers are $x = \pm 6$. You do not have to consider $x = -6$ because it is outside the domain. The First Derivative Test confirms that A is a minimum when $x = 6$. So, $y = \frac{24}{6} = 4$ and the dimensions of the page should be $x + 3 = 9$ inches by $y + 2 = 6$ inches.

EXAMPLE 4   Finding Minimum Length

Two posts, one 12 feet high and the other 28 feet high, stand 30 feet apart. They are to be stayed by two wires, attached to a single stake, running from ground level to the top of each post. Where should the stake be placed to use the least amount of wire?

The quantity to be minimized is length. From the diagram, you can see that x varies between 0 and 30.
Figure 4.58

Solution   Let W be the wire length to be minimized. Using Figure 4.58, you can write

$$W = y + z. \qquad \text{Primary equation}$$

In this problem, rather than solving for y in terms of z (or vice versa), you can solve for both y and z in terms of a third variable x, as shown in Figure 4.58. From the Pythagorean Theorem, you obtain

$$x^2 + 12^2 = y^2$$
$$(30 - x)^2 + 28^2 = z^2$$

which implies that

$$y = \sqrt{x^2 + 144}$$
$$z = \sqrt{x^2 - 60x + 1684}.$$

So, W is given by

$$\begin{aligned} W &= y + z \\ &= \sqrt{x^2 + 144} + \sqrt{x^2 - 60x + 1684}, \qquad 0 \le x \le 30. \end{aligned}$$

Differentiating W with respect to x yields

$$\frac{dW}{dx} = \frac{x}{\sqrt{x^2 + 144}} + \frac{x - 30}{\sqrt{x^2 - 60x + 1684}}.$$

By letting $dW/dx = 0$, you obtain

$$\frac{x}{\sqrt{x^2 + 144}} + \frac{x - 30}{\sqrt{x^2 - 60x + 1684}} = 0$$
$$x\sqrt{x^2 - 60x + 1684} = (30 - x)\sqrt{x^2 + 144}$$
$$x^2(x^2 - 60x + 1684) = (30 - x)^2(x^2 + 144)$$
$$x^4 - 60x^3 + 1684x^2 = x^4 - 60x^3 + 1044x^2 - 8640x + 129{,}600$$
$$640x^2 + 8640x - 129{,}600 = 0$$
$$320(x - 9)(2x + 45) = 0$$
$$x = 9, \, -22.5.$$

Because $x = -22.5$ is not in the domain and

$$W(0) \approx 53.04, \qquad W(9) = 50, \qquad \text{and} \qquad W(30) \approx 60.31$$

you can conclude that the wire should be staked at 9 feet from the 12-foot pole.

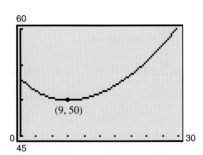

You can confirm the minimum value of W with a graphing utility.
Figure 4.59

TECHNOLOGY   From Example 4, you can see that applied optimization problems can involve a lot of algebra. If you have access to a graphing utility, you can confirm that $x = 9$ yields a minimum value of W by graphing

$$W = \sqrt{x^2 + 144} + \sqrt{x^2 - 60x + 1684}$$

as shown in Figure 4.59.

In each of the first four examples, the extreme value occurred at a critical number. Although this happens often, remember that an extreme value can also occur at an endpoint of an interval, as shown in Example 5.

EXAMPLE 5   An Endpoint Maximum

Four feet of wire is to be used to form a square and a circle. How much of the wire should be used for the square and how much should be used for the circle to enclose the maximum total area?

Solution   The total area (see Figure 4.60) is given by

$$A = \text{(area of square)} + \text{(area of circle)}$$
$$A = x^2 + \pi r^2. \qquad \text{Primary equation}$$

Because the total length of wire is 4 feet, you obtain

$$4 = \text{(perimeter of square)} + \text{(circumference of circle)}$$
$$4 = 4x + 2\pi r.$$

So, $r = 2(1 - x)/\pi$, and by substituting into the primary equation you have

$$A = x^2 + \pi \left[\frac{2(1 - x)}{\pi} \right]^2$$
$$= x^2 + \frac{4(1 - x)^2}{\pi}$$
$$= \frac{1}{\pi} \left[(\pi + 4)x^2 - 8x + 4 \right].$$

The feasible domain is $0 \le x \le 1$ restricted by the square's perimeter. Because

$$\frac{dA}{dx} = \frac{2(\pi + 4)x - 8}{\pi}$$

the only critical number in $(0, 1)$ is $x = 4/(\pi + 4) \approx 0.56$. So, using

$$A(0) \approx 1.273, \qquad A(0.56) \approx 0.56, \qquad \text{and} \qquad A(1) = 1$$

you can conclude that the maximum area occurs when $x = 0$. That is, *all* the wire is used for the circle.

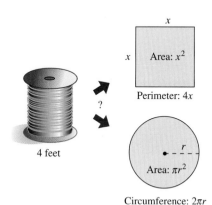

The quantity to be maximized is area:
$A = x^2 + \pi r^2.$
Figure 4.60

EXPLORATION

What would the answer be if Example 5 asked for the dimensions needed to enclose the *minimum* total area?

Let's review the primary equations developed in the first five examples. As applications go, these five examples are fairly simple, and yet the resulting primary equations are quite complicated.

$$V = 27x - \frac{x^3}{4} \qquad\qquad W = \sqrt{x^2 + 144} + \sqrt{x^2 - 60x + 1684}$$

$$d = \sqrt{x^4 - 3x^2 + 4} \qquad\qquad A = \frac{1}{\pi} \left[(\pi + 4)x^2 - 8x + 4 \right]$$

$$A = 30 + 2x + \frac{72}{x}$$

You must expect that real-life applications often involve equations that are *at least as complicated* as these five. Remember that one of the main goals of this course is to learn to use calculus to analyze equations that initially seem formidable.

The camera should be 2.236 feet from the painting to maximize the angle β.
Figure 4.61

EXAMPLE 6 Maximizing an Angle

A photographer is taking a picture of a 4-foot painting hung in an art gallery. The camera lens is 1 foot below the lower edge of the painting, as shown in Figure 4.61. How far should the camera be from the painting to maximize the angle subtended by the camera lens?

Solution In Figure 4.61, let β be the angle to be maximized.

$$\beta = \theta - \alpha \qquad \text{Primary equation}$$

From Figure 4.61, you can see that $\cot \theta = \dfrac{x}{5}$ and $\cot \alpha = \dfrac{x}{1}$. Therefore, $\theta = \operatorname{arccot}\dfrac{x}{5}$ and $\alpha = \operatorname{arccot} x$. So,

$$\beta = \operatorname{arccot}\frac{x}{5} - \operatorname{arccot} x.$$

Differentiating β with respect to x produces

$$\frac{d\beta}{dx} = \frac{-1/5}{1 + (x^2/25)} - \frac{-1}{1 + x^2}$$

$$= \frac{-5}{25 + x^2} + \frac{1}{1 + x^2}$$

$$= \frac{4(5 - x^2)}{(25 + x^2)(1 + x^2)}.$$

Because $d\beta/dx = 0$ when $x = \sqrt{5}$, you can conclude from the First Derivative Test that this distance yields a maximum value of β. So, the distance is $x \approx 2.236$ feet and the angle is $\beta \approx 0.7297$ radian $\approx 41.81°$.

EXAMPLE 7 Finding a Maximum Revenue

The demand function for a product is modeled by

$$p = 56e^{-0.000012x} \qquad \text{Demand function}$$

where p is the price per unit (in dollars) and x is the number of units. What price will yield a maximum revenue?

Solution Substituting for p (from the demand function) produces the revenue function

$$R = xp = 56xe^{-0.000012x}. \qquad \text{Primary equation}$$

The rate of change of revenue R with respect to the number of units sold x is called the *marginal revenue*. Setting the marginal revenue equal to zero,

$$\frac{dR}{dx} = 56x(e^{-0.000012x})(-0.000012) + e^{-0.000012x}(56) = 0$$

yields $x \approx 83,333$ units. From this, you can conclude that the maximum revenue occurs when the price is

$$p = 56e^{-0.000012(83,333)} \approx \$20.60.$$

So, a price of about $20.60 will yield a maximum revenue (see Figure 4.62).

Figure 4.62

Exercises for Section 4.7

See www.CalcChat.com for worked-out solutions to odd-numbered exercises.

1. *Numerical, Graphical, and Analytic Analysis* Find two positive numbers whose sum is 110 and whose product is a maximum.

(a) Analytically complete six rows of a table such as the one below. (The first two rows are shown.)

First Number x	Second Number	Product P
10	$110 - 10$	$10(110 - 10) = 1000$
20	$110 - 20$	$20(110 - 20) = 1800$

(b) Use a graphing utility to generate additional rows of the table. Use the table to estimate the solution. (*Hint:* Use the *table* feature of the graphing utility.)

(c) Write the product P as a function of x.

(d) Use a graphing utility to graph the function in part (c) and estimate the solution from the graph.

(e) Use calculus to find the critical number of the function in part (c). Then find the two numbers.

2. *Numerical, Graphical, and Analytic Analysis* An open box of maximum volume is to be made from a square piece of material, 24 inches on a side, by cutting equal squares from the corners and turning up the sides (see figure).

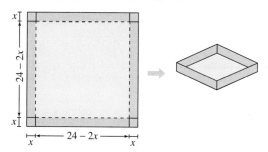

(a) Analytically complete six rows of a table such as the one below. (The first two rows are shown.) Use the table to guess the maximum volume.

Height	Length and Width	Volume
1	$24 - 2(1)$	$1[24 - 2(1)]^2 = 484$
2	$24 - 2(2)$	$2[24 - 2(2)]^2 = 800$

(b) Write the volume V as a function of x.

(c) Use calculus to find the critical number of the function in part (b) and find the maximum value.

(d) Use a graphing utility to graph the function in part (b) and verify the maximum volume from the graph.

In Exercises 3–8, find two positive numbers that satisfy the given requirements.

3. The sum is S and the product is a maximum.

4. The product is 192 and the sum is a minimum.

5. The product is 192 and the sum of the first plus three times the second is a minimum.

6. The second number is the reciprocal of the first and the sum is a minimum.

7. The sum of the first and twice the second is 100 and the product is a maximum.

8. The sum of the first number squared and the second is 27 and the product is a maximum.

In Exercises 9 and 10, find the length and width of a rectangle that has the given perimeter and a maximum area.

9. Perimeter: 100 meters   **10.** Perimeter: P units

In Exercises 11 and 12, find the length and width of a rectangle that has the given area and a minimum perimeter.

11. Area: 64 square feet   **12.** Area: A square centimeters

In Exercises 13–16, find the point on the graph of the function that is closest to the given point.

	Function	*Point*		*Function*	*Point*
13.	$f(x) = \sqrt{x}$	$(4, 0)$	**14.**	$f(x) = \sqrt{x - 8}$	$(2, 0)$
15.	$f(x) = x^2$	$(2, \frac{1}{2})$	**16.**	$f(x) = (x + 1)^2$	$(5, 3)$

17. *Chemical Reaction* In an autocatalytic chemical reaction, the product formed is a catalyst for the reaction. If Q_0 is the amount of the original substance and x is the amount of catalyst formed, the rate of chemical reaction is

$$\frac{dQ}{dx} = kx(Q_0 - x).$$

For what value of x will the rate of chemical reaction be greatest?

18. *Traffic Control* On a given day, the flow rate F (in cars per hour) on a congested roadway is

$$F = \frac{v}{22 + 0.02v^2}$$

where v is the speed of the traffic in miles per hour. What speed will maximize the flow rate on the road?

19. *Area* A farmer plans to fence a rectangular pasture adjacent to a river. The pasture must contain 180,000 square meters in order to provide enough grass for the herd. What dimensions would require the least amount of fencing if no fencing is needed along the river?

20. *Maximum Area* A rancher has 200 feet of fencing with which to enclose two adjacent rectangular corrals (see figure). What dimensions should be used so that the enclosed area will be a maximum?

21. *Maximum Volume*

(a) Verify that each of the rectangular solids shown in the figure has a surface area of 150 square inches.

(b) Find the volume of each solid.

(c) Determine the dimensions of a rectangular solid (with a square base) of maximum volume if its surface area is 150 square inches.

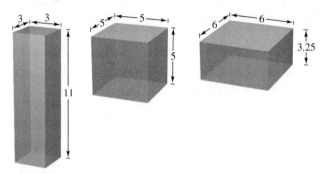

22. *Maximum Volume* Determine the dimensions of a rectangular solid (with a square base) with maximum volume if its surface area is 337.5 square centimeters.

23. *Maximum Area* A Norman window is constructed by adjoining a semicircle to the top of an ordinary rectangular window (see figure). Find the dimensions of a Norman window of maximum area if the total perimeter is 16 feet.

24. *Maximum Area* A rectangle is bounded by the x- and y-axes and the graph of $y = (6 - x)/2$ (see figure). What length and width should the rectangle have so that its area is a maximum?

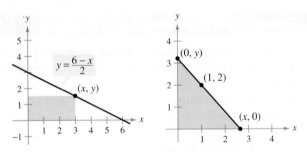

Figure for 24   **Figure for 25**

25. *Minimum Length* A right triangle is formed in the first quadrant by the x- and y-axes and a line through the point $(1, 2)$ (see figure).

(a) Write the length L of the hypotenuse as a function of x.

(b) Use a graphing utility to approximate x graphically such that the length of the hypotenuse is a minimum.

(c) Find the vertices of the triangle such that its area is a minimum.

26. *Maximum Area* Find the area of the largest isosceles triangle that can be inscribed in a circle of radius 4 (see figure).

(a) Solve by writing the area as a function of h.

(b) Solve by writing the area as a function of α.

(c) Identify the type of triangle of maximum area.

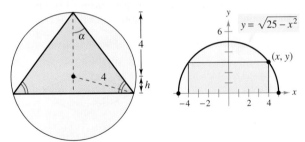

Figure for 26   **Figure for 27**

27. *Maximum Area* A rectangle is bounded by the x-axis and the semicircle $y = \sqrt{25 - x^2}$ (see figure). What length and width should the rectangle have so that its area is a maximum?

28. *Area* Find the dimensions of the largest rectangle that can be inscribed in a semicircle of radius r (see Exercise 27).

29. *Area* A rectangular page is to contain 30 square inches of print. The margins on each side are 1 inch. Find the dimensions of the page such that the least amount of paper is used.

30. *Area* A rectangular page is to contain 36 square inches of print. The margins on each side are to be $1\frac{1}{2}$ inches. Find the dimensions of the page such that the least amount of paper is used.

31. *Numerical, Graphical, and Analytic Analysis*   An exercise room consists of a rectangle with a semicircle on each end. A 200-meter running track runs around the outside of the room.

(a) Draw a figure to represent the problem. Let x and y represent the length and width of the rectangle.

(b) Analytically complete six rows of a table such as the one below. (The first two rows are shown.) Use the table to guess the maximum area of the rectangular region.

Length x	Width y	Area
10	$\dfrac{2}{\pi}(100 - 10)$	$(10)\dfrac{2}{\pi}(100 - 10) \approx 573$
20	$\dfrac{2}{\pi}(100 - 20)$	$(20)\dfrac{2}{\pi}(100 - 20) \approx 1019$

(c) Write the area A as a function of x.

(d) Use calculus to find the critical number of the function in part (c) and find the maximum value.

(e) Use a graphing utility to graph the function in part (c) and verify the maximum area from the graph.

32. *Numerical, Graphical, and Analytic Analysis*   A right circular cylinder is to be designed to hold 22 cubic inches of a soft drink (approximately 12 fluid ounces).

(a) Analytically complete six rows of a table such as the one below. (The first two rows are shown.)

Radius r	Height	Surface Area
0.2	$\dfrac{22}{\pi(0.2)^2}$	$2\pi(0.2)\left[0.2 + \dfrac{22}{\pi(0.2)^2}\right] \approx 220.3$
0.4	$\dfrac{22}{\pi(0.4)^2}$	$2\pi(0.4)\left[0.4 + \dfrac{22}{\pi(0.4)^2}\right] \approx 111.0$

(b) Use a graphing utility to generate additional rows of the table. Use the table to estimate the minimum surface area. (*Hint:* Use the *table* feature of the graphing utility.)

(c) Write the surface area S as a function of r.

(d) Use a graphing utility to graph the function in part (c) and estimate the minimum surface area from the graph.

(e) Use calculus to find the critical number of the function in part (c) and find dimensions that will yield the minimum surface area.

33. *Maximum Volume*   A rectangular package to be sent by a postal service can have a maximum combined length and girth (perimeter of a cross section) of 108 inches (see figure). Find the dimensions of the package of maximum volume that can be sent. (Assume the cross section is square.)

34. *Maximum Volume*   Rework Exercise 33 for a cylindrical package. (The cross section is circular.)

35. *Maximum Volume*   Find the volume of the largest right circular cone that can be inscribed in a sphere of radius r.

36. *Maximum Volume*   Find the volume of the largest right circular cylinder that can be inscribed in a sphere of radius r.

Writing About Concepts

37. The perimeter of a rectangle is 20 feet. Of all possible dimensions, the maximum area is 25 square feet when its length and width are both 5 feet. Are there dimensions that yield a minimum area? Explain.

38. A shampoo bottle is a right circular cylinder. Because the surface area of the bottle does not change when it is squeezed, is it true that the volume remains the same? Explain.

39. *Minimum Surface Area*   A solid is formed by adjoining two hemispheres to the ends of a right circular cylinder. The total volume of the solid is 12 cubic centimeters. Find the radius of the cylinder that produces the minimum surface area.

40. *Minimum Cost*   An industrial tank of the shape described in Exercise 39 must have a volume of 3000 cubic feet. The hemispherical ends cost twice as much per square foot of surface area as the sides. Find the dimensions that will minimize cost.

41. *Minimum Area*   The sum of the perimeters of an equilateral triangle and a square is 10. Find the dimensions of the triangle and the square that produce a minimum total area.

42. *Maximum Area*   Twenty feet of wire is to be used to form two figures. In each of the following cases, how much wire should be used for each figure so that the total enclosed area is a maximum?

(a) Equilateral triangle and square

(b) Square and regular pentagon

(c) Regular pentagon and regular hexagon

(d) Regular hexagon and circle

What can you conclude from this pattern? {*Hint:* The area of a regular polygon with n sides of length x is $A = (n/4)[\cot(\pi/n)]x^2$.}

43. *Beam Strength*   A wooden beam has a rectangular cross section of height h and width w (see figure on the next page). The strength S of the beam is directly proportional to the width and the square of the height. What are the dimensions of the strongest beam that can be cut from a round log of diameter 24 inches? (*Hint:* $S = kh^2w$, where k is the proportionality constant.)

Minimum Distance In Exercises 63–65, consider a fuel distribution center located at the origin of the rectangular coordinate system (units in miles; see figures). The center supplies three factories with coordinates (4, 1), (5, 6), and (10, 3). A trunk line will run from the distribution center along the line $y = mx$, and feeder lines will run to the three factories. The objective is to find m such that the lengths of the feeder lines are minimized.

63. Minimize the sum of the squares of the lengths of vertical feeder lines given by

$$S_1 = (4m - 1)^2 + (5m - 6)^2 + (10m - 3)^2.$$

Find the equation for the trunk line by this method and then determine the sum of the lengths of the feeder lines.

64. Minimize the sum of the absolute values of the lengths of vertical feeder lines given by

$$S_2 = |4m - 1| + |5m - 6| + |10m - 3|.$$

Find the equation for the trunk line by this method and then determine the sum of the lengths of the feeder lines. (*Hint:* Use a graphing utility to graph the function S_2 and approximate the required critical number.)

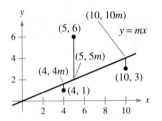

Figure for 63 and 64   **Figure for 65**

65. Minimize the sum of the perpendicular distances (see Exercises 85–90 in Section 1.2) from the trunk line to the factories given by $S_3 = \dfrac{|4m - 1|}{\sqrt{m^2 + 1}} + \dfrac{|5m - 6|}{\sqrt{m^2 + 1}} + \dfrac{|10m - 3|}{\sqrt{m^2 + 1}}$. Find the equation for the trunk line by this method and then determine the sum of the lengths of the feeder lines. (*Hint:* Use a graphing utility to graph the function S_3 and approximate the required critical number.)

66. **Maximum Area** Consider a symmetric cross inscribed in a circle of radius r (see figure).

(a) Write the area A of the cross as a function of x and find the value of x that maximizes the area.

(b) Write the area A of the cross as a function of θ and find the value of θ that maximizes the area.

(c) Show that the critical numbers of parts (a) and (b) yield the same maximum area. What is that area?

Section Project: Connecticut River

Whenever the Connecticut River reaches a level of 105 feet above sea level, two Northampton, Massachusetts flood control station operators begin a round-the-clock river watch. Every 2 hours, they check the height of the river, using a scale marked off in tenths of a foot, and record the data in a log book. In the spring of 1996, the flood watch lasted from April 4, when the river reached 105 feet and was rising at 0.2 foot per hour, until April 25, when the level subsided again to 105 feet. Between those dates, their log shows that the river rose and fell several times, at one point coming close to the 115-foot mark. If the river had reached 115 feet, the city would have closed down Mount Tom Road (Route 5, south of Northampton).

The graph below shows the rate of change of the level of the river during one portion of the flood watch. Use the graph to answer each question.

Day (0 ↔ 12:01 A.M. April 14)

(a) On what date was the river rising most rapidly? How do you know?

(b) On what date was the river falling most rapidly? How do you know?

(c) There were two dates in a row on which the river rose, then fell, then rose again during the course of the day. On which days did this occur, and how do you know?

(d) At 1 minute past midnight, April 14, the river level was 111.0 feet. Estimate the river's height 24 hours later and 48 hours later. Explain how you made your estimates.

(e) The river crested at 114.4 feet. On what date do you think this occurred?

(Submitted by Mary Murphy, Smith College, Northampton, MA)

Differentials

- Understand the concept of a tangent line approximation.
- Compare the value of the differential, *dy*, with the actual change in *y*, **Δy**.
- Estimate a propagated error using a differential.
- Find the differential of a function using differentiation formulas.

Tangent Line Approximations

EXPLORATION

Tangent Line Approximation   Use a graphing utility to graph

$$f(x) = x^2.$$

In the same viewing window, graph the tangent line to the graph of *f* at the point $(1, 1)$. Zoom in twice on the point of tangency. Does your graphing utility distinguish between the two graphs? Use the *trace* feature to compare the two graphs. As the *x*-values get closer to 1, what can you say about the *y*-values?

Newton's Method (Section 3.8) is an example of the use of a tangent line to a graph to approximate the graph. In this section, you will study other situations in which the graph of a function can be approximated by a straight line.

To begin, consider a function *f* that is differentiable at *c*. The equation for the tangent line at the point $(c, f(c))$ is given by

$$y - f(c) = f'(c)(x - c)$$
$$y = f(c) + f'(c)(x - c)$$

and is called the **tangent line approximation** (or **linear approximation**) **of *f* at *c*.** Because *c* is a constant, *y* is a linear function of *x*. Moreover, by restricting the values of *x* to be sufficiently close to *c*, the values of *y* can be used as approximations (to any desired accuracy) of the values of the function *f*. In other words, as $x \to c$, the limit of *y* is $f(c)$.

EXAMPLE 1   **Using a Tangent Line Approximation**

Find the tangent line approximation of

$$f(x) = 1 + \sin x$$

at the point $(0, 1)$. Then use a table to compare the *y*-values of the linear function with those of $f(x)$ on an open interval containing $x = 0$.

Solution   The derivative of *f* is

$$f'(x) = \cos x. \qquad \text{First derivative}$$

So, the equation of the tangent line to the graph of *f* at the point $(0, 1)$ is

$$y - f(0) = f'(0)(x - 0)$$
$$y - 1 = (1)(x - 0)$$
$$y = 1 + x. \qquad \text{Tangent line approximation}$$

The table compares the values of *y* given by this linear approximation with the values of $f(x)$ near $x = 0$. Notice that the closer *x* is to 0, the better the approximation is. This conclusion is reinforced by the graph shown in Figure 4.63.

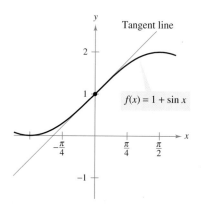

The tangent line approximation of *f* at the point $(0, 1)$

Figure 4.63

x	-0.5	-0.1	-0.01	0	0.01	0.1	0.5
$f(x) = 1 + \sin x$	0.521	0.9002	0.9900002	1	1.0099998	1.0998	1.479
$y = 1 + x$	0.5	0.9	0.99	1	1.01	1.1	1.5

NOTE   Be sure you see that this linear approximation of $f(x) = 1 + \sin x$ depends on the point of tangency. At a different point on the graph of *f*, you would obtain a different tangent line approximation.

When Δx is small, $\Delta y = f(c + \Delta x) - f(c)$ is approximated by $f'(c)\Delta x$.

Figure 4.64

Differentials

When the tangent line to the graph of f at the point $(c, f(c))$

$$y = f(c) + f'(c)(x - c)$$ Tangent line at $(c, f(c))$

is used as an approximation of the graph of f, the quantity $x - c$ is called the change in x, and is denoted by Δx, as shown in Figure 4.64. When Δx is small, the change in y (denoted by Δy) can be approximated as shown.

$$\Delta y = f(c + \Delta x) - f(c)$$ Actual change in y
$$\approx f'(c)\Delta x$$ Approximate change in y

For such an approximation, the quantity Δx is traditionally denoted by dx, and is called the **differential of x.** The expression $f'(x)\,dx$ is denoted by dy, and is called the **differential of y.**

Definition of Differentials

Let $y = f(x)$ represent a function that is differentiable on an open interval containing x. The **differential of x** (denoted by dx) is any nonzero real number. The **differential of y** (denoted by dy) is

$$dy = f'(x)\,dx.$$

In many types of applications, the differential of y can be used as an approximation of the change in y. That is,

$$\Delta y \approx dy \qquad \text{or} \qquad \Delta y \approx f'(x)dx.$$

EXAMPLE 2   Comparing Δy and dy

Let $y = x^2$. Find dy when $x = 1$ and $dx = 0.01$. Compare this value with Δy for $x = 1$ and $\Delta x = 0.01$.

Solution   Because $y = f(x) = x^2$, you have $f'(x) = 2x$, and the differential dy is given by

$$dy = f'(x)\,dx = f'(1)(0.01) = 2(0.01) = 0.02.$$ Differential of y

Now, using $\Delta x = 0.01$, the change in y is

$$\Delta y = f(x + \Delta x) - f(x) = f(1.01) - f(1) = (1.01)^2 - 1^2 = 0.0201.$$

Figure 4.65 shows the geometric comparison of dy and Δy. Try comparing other values of dy and Δy. You will see that the values become closer to each other as dx (or Δx) approaches 0. ▬▬▬

In Example 2, the tangent line to the graph of $f(x) = x^2$ at $x = 1$ is

$$y = 2x - 1 \qquad \text{or} \qquad g(x) = 2x - 1.$$ Tangent line to the graph of f at $x = 1$

For x-values near 1, this line is close to the graph of f, as shown in Figure 4.65. For instance,

$$f(1.01) = 1.01^2 = 1.0201 \qquad \text{and} \qquad g(1.01) = 2(1.01) - 1 = 1.02.$$

The change in y, Δy, is approximated by the differential of y, dy.

Figure 4.65

Error Propagation

Physicists and engineers tend to make liberal use of the approximation of Δy by dy. One way this occurs in practice is in the estimation of errors propagated by physical measuring devices. For example, if you let x represent the measured value of a variable and let $x + \Delta x$ represent the exact value, then Δx is the *error in measurement*. Finally, if the measured value x is used to compute another value $f(x)$, the difference between $f(x + \Delta x)$ and $f(x)$ is the **propagated error.**

$$\underbrace{f(x + \overbrace{\Delta x}^{\text{Measurement error}})}_{\substack{\text{Exact} \\ \text{value}}} - \underbrace{f(x)}_{\substack{\text{Measured} \\ \text{value}}} = \overbrace{\Delta y}^{\substack{\text{Propagated} \\ \text{error}}}$$

EXAMPLE 3   Estimation of Error

The radius of a ball bearing is measured to be 0.7 inch, as shown in Figure 4.66. If the measurement is correct to within 0.01 inch, estimate the propagated error in the volume V of the ball bearing.

Solution   The formula for the volume of a sphere is $V = \frac{4}{3}\pi r^3$, where r is the radius of the sphere. So, you can write

$$r = 0.7 \qquad\qquad \text{Measured radius}$$

and

$$-0.01 \le \Delta r \le 0.01. \qquad \text{Possible error}$$

To approximate the propagated error in the volume, differentiate V to obtain $dV/dr = 4\pi r^2$ and write

$$
\begin{aligned}
\Delta V &\approx dV && \text{Approximate } \Delta V \text{ by } dV. \\
&= 4\pi r^2\, dr \\
&= 4\pi(0.7)^2(\pm 0.01) && \text{Substitute for } r \text{ and } dr. \\
&\approx \pm 0.06158 \text{ in}^3.
\end{aligned}
$$

So the volume has a propagated error of about 0.06 cubic inch.   ▬▬▬

Would you say that the propagated error in Example 3 is large or small? The answer is best given in *relative* terms by comparing dV with V. The ratio

$$
\begin{aligned}
\frac{dV}{V} &= \frac{4\pi r^2\, dr}{\frac{4}{3}\pi r^3} && \text{Ratio of } dV \text{ to } V \\
&= \frac{3\, dr}{r} && \text{Simplify.} \\
&\approx \frac{3}{0.7}(\pm 0.01) && \text{Substitute for } dr \text{ and } r. \\
&\approx \pm 0.0429
\end{aligned}
$$

is called the **relative error.** The corresponding **percent error** is approximately 4.29%.

Ball bearing with measured radius that is correct to within 0.01 inch

Figure 4.66

Calculating Differentials

Each of the differentiation rules that you studied in Chapter 3 can be written in **differential form.** For example, suppose u and v are differentiable functions of x. By the definition of differentials, you have

$$du = u'\, dx \quad \text{and} \quad dv = v'\, dx.$$

So, you can write the differential form of the Product Rule as follows.

$$d[uv] = \frac{d}{dx}[uv]\, dx \qquad \text{Differential of } uv$$

$$= [uv' + vu']\, dx \qquad \text{Product Rule}$$

$$= uv'\, dx + vu'\, dx$$

$$= u\, dv + v\, du$$

Differential Formulas

Let u and v be differentiable functions of x.

Constant multiple: $\quad d[cu] = c\, du$

Sum or difference: $\quad d[u \pm v] = du \pm dv$

Product: $\quad d[uv] = u\, dv + v\, du$

Quotient: $\quad d\left[\dfrac{u}{v}\right] = \dfrac{v\, du - u\, dv}{v^2}$

EXAMPLE 4 Finding Differentials

Function	Derivative	Differential
a. $y = x^2$	$\dfrac{dy}{dx} = 2x$	$dy = 2x\, dx$
b. $y = 2\sin x$	$\dfrac{dy}{dx} = 2\cos x$	$dy = 2\cos x\, dx$
c. $y = xe^x$	$\dfrac{dy}{dx} = e^x(x+1)$	$dy = e^x(x+1)\, dx$
d. $y = \dfrac{1}{x}$	$\dfrac{dy}{dx} = -\dfrac{1}{x^2}$	$dy = -\dfrac{dx}{x^2}$

The notation in Example 4 is called the **Leibniz notation** for derivatives and differentials, named after the German mathematician Gottfried Wilhelm Leibniz. The beauty of this notation is that it provides an easy way to remember several important calculus formulas by making it seem as though the formulas were derived from algebraic manipulations of differentials. For instance, in Leibniz notation, the *Chain Rule*

$$\frac{dy}{dx} = \frac{dy}{du}\frac{du}{dx}$$

would appear to be true because the du's divide out. Even though this reasoning is *incorrect*, the notation does help one remember the Chain Rule.

Mary Evans Picture Library

GOTTFRIED WILHELM LEIBNIZ (1646–1716)

Both Leibniz and Newton are credited with creating calculus. It was Leibniz, however, who tried to broaden calculus by developing rules and formal notation. He often spent days choosing an appropriate notation for a new concept.

EXAMPLE 5   Finding the Differential of a Composite Function

$$y = f(x) = \sin 3x \qquad \text{Original function}$$

$$f'(x) = 3 \cos 3x \qquad \text{Apply Chain Rule.}$$

$$dy = f'(x) \, dx = 3 \cos 3x \, dx \qquad \text{Differential form}$$

EXAMPLE 6   Finding the Differential of a Composite Function

$$y = f(x) = (x^2 + 1)^{1/2} \qquad \text{Original function}$$

$$f'(x) = \frac{1}{2}(x^2 + 1)^{-1/2}(2x) = \frac{x}{\sqrt{x^2 + 1}} \qquad \text{Apply Chain Rule.}$$

$$dy = f'(x) \, dx = \frac{x}{\sqrt{x^2 + 1}} \, dx \qquad \text{Differential form}$$

Differentials can be used to approximate function values. To do this for the function given by $y = f(x)$, you use the formula

$$f(x + \Delta x) \approx f(x) + dy = f(x) + f'(x) \, dx$$

which is derived from the approximation $\Delta y = f(x + \Delta x) - f(x) \approx dy$. The key to using this formula is to choose a value for x that makes the calculations easier, as shown in Example 7.

EXAMPLE 7   Approximating Function Values

Use differentials to approximate $\sqrt{16.5}$.

Solution   Using $f(x) = \sqrt{x}$, you can write

$$f(x + \Delta x) \approx f(x) + f'(x) \, dx = \sqrt{x} + \frac{1}{2\sqrt{x}} \, dx.$$

Now, choosing $x = 16$ and $dx = 0.5$, you obtain the following approximation.

$$f(x + \Delta x) = \sqrt{16.5} \approx \sqrt{16} + \frac{1}{2\sqrt{16}}(0.5) = 4 + \left(\frac{1}{8}\right)\left(\frac{1}{2}\right) = 4.0625$$

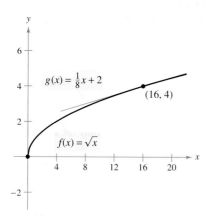

$g(x) = \frac{1}{8}x + 2$

$(16, 4)$

$f(x) = \sqrt{x}$

Figure 4.67

The tangent line approximation to $f(x) = \sqrt{x}$ at $x = 16$ is the line $g(x) = \frac{1}{8}x + 2$. For x-values near 16, the graphs of f and g are close together, as shown in Figure 4.67. For instance,

$$f(16.5) = \sqrt{16.5} \approx 4.0620 \quad \text{and} \quad g(16.5) = \frac{1}{8}(16.5) + 2 = 4.0625.$$

In fact, if you use a graphing utility to zoom in near the point of tangency $(16, 4)$, you will see that the two graphs appear to coincide. Notice also that as you move farther away from the point of tangency, the linear approximation is less accurate.

Review Exercises for Chapter 4

See www.CalcChat.com for worked-out solutions to odd-numbered exercises.

1. Give the definition of a critical number, and graph a function f showing the different types of critical numbers.

2. Consider the odd function f that is continuous, differentiable, and has the functional values shown in the table.

x	-5	-4	-1	0	2	3	6
$f(x)$	1	3	2	0	-1	-4	0

(a) Determine $f(4)$.

(b) Determine $f(-3)$.

(c) Plot the points and make a possible sketch of the graph of f on the interval $[-6, 6]$. What is the smallest number of critical points in the interval? Explain.

(d) Does there exist at least one real number c in the interval $(-6, 6)$ where $f'(c) = -1$? Explain.

(e) Is it possible that $\lim_{x \to 0} f(x)$ does not exist? Explain.

(f) Is it necessary that $f'(x)$ exists at $x = 2$? Explain.

In Exercises 3 and 4, find the absolute extrema of the function on the closed interval. Use a graphing utility to graph the function over the indicated interval to confirm your results.

3. $g(x) = 2x + 5 \cos x$, $[0, 2\pi]$

4. $f(x) = \dfrac{x}{\sqrt{x^2 + 1}}$, $[0, 2]$

In Exercises 5 and 6, determine whether Rolle's Theorem can be applied to f on the closed interval $[a, b]$. If Rolle's Theorem can be applied, find all values of c in the open interval (a, b) such that $f'(c) = 0$.

5. $f(x) = (x - 2)(x + 3)^2$, $[-3, 2]$

6. $f(x) = |x - 2| - 2$, $[0, 4]$

7. Consider the function $f(x) = 3 - |x - 4|$.

(a) Graph the function and verify that $f(1) = f(7)$.

(b) Note that $f'(x)$ is not equal to zero for any x in $[1, 7]$. Explain why this does not contradict Rolle's Theorem.

8. Can the Mean Value Theorem be applied to the function $f(x) = 1/x^2$ on the interval $[-2, 1]$? Explain.

In Exercises 9–12, find the point(s) guaranteed by the Mean Value Theorem for the closed interval $[a, b]$.

9. $f(x) = x^{2/3}$, $[1, 8]$

10. $f(x) = \dfrac{1}{x}$, $[1, 4]$

11. $f(x) = x - \cos x$, $\left[-\dfrac{\pi}{2}, \dfrac{\pi}{2}\right]$

12. $f(x) = x \log_2 x$, $[1, 2]$

13. For the function $f(x) = Ax^2 + Bx + C$, determine the value of c guaranteed by the Mean Value Theorem in the interval $[x_1, x_2]$.

14. Demonstrate the result of Exercise 13 for $f(x) = 2x^2 - 3x + 1$ on the interval $[0, 4]$.

In Exercises 15–20, find the critical numbers (if any) and the open intervals on which the function is increasing or decreasing.

15. $f(x) = (x - 1)^2(x - 3)$

16. $g(x) = (x + 1)^3$

17. $h(x) = \sqrt{x}(x - 3)$, $x > 0$

18. $f(x) = \sin x + \cos x$, $[0, 2\pi]$

19. $f(t) = (2 - t)2^t$

20. $g(x) = 2x \ln x$

In Exercises 21 and 22, use the First Derivative Test to find any relative extrema of the function. Use a graphing utility to verify your results.

21. $h(t) = \dfrac{1}{4}t^4 - 8t$

22. $g(x) = \dfrac{3}{2} \sin\left(\dfrac{\pi x}{2} - 1\right)$, $[0, 4]$

23. *Harmonic Motion* The height of an object attached to a spring is given by the harmonic equation

$$y = \tfrac{1}{3} \cos 12t - \tfrac{1}{4} \sin 12t$$

where y is measured in inches and t is measured in seconds.

(a) Calculate the height and velocity of the object when $t = \pi/8$ second.

(b) Show that the maximum displacement of the object is $\dfrac{5}{12}$ inch.

(c) Find the period P of y. Also, find the frequency f (number of oscillations per second) if $f = 1/P$.

24. *Writing* The general equation giving the height of an oscillating object attached to a spring is

$$y = A \sin \sqrt{\dfrac{k}{m}}\, t + B \cos \sqrt{\dfrac{k}{m}}\, t$$

where k is the spring constant and m is the mass of the object.

(a) Show that the maximum displacement of the object is $\sqrt{A^2 + B^2}$.

(b) Show that the object oscillates with a frequency of

$$f = \dfrac{1}{2\pi} \sqrt{\dfrac{k}{m}}.$$

In Exercises 25 and 26, determine the points of inflection of the function.

25. $f(x) = x + \cos x$, $[0, 2\pi]$

26. $f(x) = (x + 2)^2(x - 4)$

In Exercises 27 and 28, use the Second Derivative Test to find all relative extrema.

27. $g(x) = 2x^2(1 - x^2)$

28. $h(t) = t - 4\sqrt{t + 1}$

Think About It In Exercises 29 and 30, sketch the graph of a function f having the indicated characteristics.

29. $f(0) = f(6) = 0$

$f'(3) = f'(5) = 0$

$f'(x) > 0$ if $x < 3$

$f'(x) > 0$ if $3 < x < 5$

$f'(x) < 0$ if $x > 5$

$f''(x) < 0$ if $x < 3$ and $x > 4$

$f''(x) > 0$, $3 < x < 4$

30. $f(0) = 4$, $f(6) = 0$

$f'(x) < 0$ if $x < 2$ and $x > 4$

$f'(2)$ does not exist.

$f'(4) = 0$

$f'(x) > 0$ if $2 < x < 4$

$f''(x) < 0$, $x \neq 2$

31. *Writing* A newspaper headline states that "The rate of growth of the national deficit is decreasing." What does this mean? What does it imply about the graph of the deficit as a function of time?

32. *Inventory Cost* The cost of inventory depends on the ordering and storage costs according to the inventory model

$$C = \left(\frac{Q}{x}\right)s + \left(\frac{x}{2}\right)r.$$

Determine the order size that will minimize the cost, assuming that sales occur at a constant rate, Q is the number of units sold per year, r is the cost of storing one unit for 1 year, s is the cost of placing an order, and x is the number of units per order.

33. *Modeling Data* Outlays for national defense D (in billions of dollars) for selected years from 1970 through 1999 are shown in the table, where t is time in years, with $t = 0$ corresponding to 1970. (*Source: U.S. Office of Management and Budget*)

t	0	5	10	15	20
D	90.4	103.1	155.1	279.0	328.3

t	25	26	27	28	29
D	309.9	302.7	309.8	310.3	320.2

(a) Use the regression capabilities of a graphing utility to fit a model of the form $D = at^4 + bt^3 + ct^2 + dt + e$ to the data.

(b) Use a graphing utility to plot the data and graph the model.

(c) For the years shown in the table, when does the model indicate that the outlay for national defense is at a maximum? When is it at a minimum?

(d) For the years shown in the table, when does the model indicate that the outlay for national defense is increasing at the greatest rate?

34. *Climb Rate* The time t (in minutes) for a small plane to climb to an altitude of h feet is

$$t = 50 \log_{10} \frac{18,000}{18,000 - h}$$

where 18,000 feet is the plane's absolute ceiling.

(a) Determine the domain of the function appropriate for the context of the problem.

(b) Use a graphing utility to graph the time function and identify any asymptotes.

(c) Find the time when the altitude is increasing at the greatest rate.

In Exercises 35–42, find the limit.

35. $\lim\limits_{x \to \infty} \dfrac{2x^2}{3x^2 + 5}$

36. $\lim\limits_{x \to \infty} \dfrac{2x}{3x^2 + 5}$

37. $\lim\limits_{x \to -\infty} \dfrac{3x^2}{x + 5}$

38. $\lim\limits_{x \to -\infty} \dfrac{\sqrt{x^2 + x}}{-2x}$

39. $\lim\limits_{x \to \infty} \dfrac{5 \cos x}{x}$

40. $\lim\limits_{x \to \infty} \dfrac{3x}{\sqrt{x^2 + 4}}$

41. $\lim\limits_{x \to -\infty} \dfrac{6x}{x + \cos x}$

42. $\lim\limits_{x \to -\infty} \dfrac{x}{2 \sin x}$

In Exercises 43–50, find any vertical and horizontal asymptotes of the graph of the function. Use a graphing utility to verify your results.

43. $h(x) = \dfrac{2x + 3}{x - 4}$

44. $g(x) = \dfrac{5x^2}{x^2 + 2}$

45. $f(x) = \dfrac{3}{x} - 2$

46. $f(x) = \dfrac{3x}{\sqrt{x^2 + 2}}$

47. $f(x) = \dfrac{5}{3 + 2e^{-x}}$

48. $g(x) = 30xe^{-2x}$

49. $g(x) = 3 \ln(1 + e^{-x/4})$

50. $h(x) = 10 \ln\left(\dfrac{x}{x + 1}\right)$

In Exercises 51–54, use a graphing utility to graph the function. Use the graph to approximate any relative extrema or asymptotes.

51. $f(x) = x^3 + \dfrac{243}{x}$

52. $f(x) = |x^3 - 3x^2 + 2x|$

53. $f(x) = \dfrac{x - 1}{1 + 3x^2}$

54. $g(x) = \dfrac{\pi^2}{3} - 4 \cos x + \cos 2x$

In Exercises 55–80, analyze and sketch the graph of the function.

55. $f(x) = 4x - x^2$

56. $f(x) = 4x^3 - x^4$

57. $f(x) = x\sqrt{16 - x^2}$

58. $f(x) = (x^2 - 4)^2$

59. $f(x) = (x - 1)^3(x - 3)^2$

60. $f(x) = (x - 3)(x + 2)^3$

61. $f(x) = x^{1/3}(x + 3)^{2/3}$

62. $f(x) = (x - 2)^{1/3}(x + 1)^{2/3}$

63. $f(x) = \dfrac{x+1}{x-1}$

64. $f(x) = \dfrac{2x}{1+x^2}$

65. $f(x) = \dfrac{4}{1+x^2}$

66. $f(x) = \dfrac{x^2}{1+x^4}$

67. $f(x) = x^3 + x + \dfrac{4}{x}$

68. $f(x) = x^2 + \dfrac{1}{x}$

69. $f(x) = |x^2 - 9|$

70. $f(x) = |x-1| + |x-3|$

71. $h(x) = (1-x)e^x$

72. $g(x) = 5xe^{-x^2}$

73. $g(x) = (x+3)\ln(x+3)$

74. $h(t) = \dfrac{\ln t}{t^2}$

75. $f(x) = \dfrac{10\log_4 x}{x}$

76. $g(x) = 100x(3^{-x})$

77. $f(x) = x + \cos x, \quad 0 \le x \le 2\pi$

78. $f(x) = \dfrac{1}{\pi}(2\sin \pi x - \sin 2\pi x), \quad -1 \le x \le 1$

79. $y = 4x - 6\arctan x$

80. $y = \dfrac{1}{2}x^2 - \arcsin \dfrac{x}{2}$

81. Find the maximum and minimum points on the graph of

$$x^2 + 4y^2 - 2x - 16y + 13 = 0$$

(a) without using calculus.

(b) using calculus.

82. Consider the function $f(x) = x^n$ for positive integer values of n.

(a) For what values of n does the function have a relative minimum at the origin?

(b) For what values of n does the function have a point of inflection at the origin?

83. *Minimum Distance* At noon, ship A is 100 kilometers due east of ship B. Ship A is sailing west at 12 kilometers per hour, and ship B is sailing south at 10 kilometers per hour. At what time will the ships be nearest to each other, and what will this distance be?

84. *Maximum Area* Find the dimensions of the rectangle of maximum area, with sides parallel to the coordinate axes, that can be inscribed in the ellipse given by

$$\frac{x^2}{144} + \frac{y^2}{16} = 1.$$

85. *Minimum Length* A right triangle in the first quadrant has the coordinate axes as sides, and the hypotenuse passes through the point $(1, 8)$. Find the vertices of the triangle such that the length of the hypotenuse is minimum.

86. *Minimum Length* The wall of a building is to be braced by a beam that must pass over a parallel fence 5 feet high and 4 feet from the building. Find the length of the shortest beam that can be used.

87. *Maximum Area* Three sides of a trapezoid have the same length s. Of all such possible trapezoids, show that the one of maximum area has a fourth side of length $2s$.

88. *Maximum Area* Show that the greatest area of any rectangle inscribed in a triangle is one-half that of the triangle.

89. *Maximum Length* Find the length of the longest pipe that can be carried level around a right-angle corner at the intersection of two corridors of widths 4 feet and 6 feet. (Do not use trigonometry.)

90. *Maximum Length* Rework Exercise 89, given corridors of widths a meters and b meters.

91. *Maximum Length* A hallway of width 6 feet meets a hallway of width 9 feet at right angles. Find the length of the longest pipe that can be carried level around this corner. [*Hint:* If L is the length of the pipe, show that

$$L = 6\csc\theta + 9\csc\left(\frac{\pi}{2} - \theta\right)$$

where θ is the angle between the pipe and the wall of the narrower hallway.]

92. *Maximum Length* Rework Exercise 91, given that one hallway is of width a meters and the other is of width b meters. Show that the result is the same as in Exercise 90.

Minimum Cost **In Exercises 93 and 94, find the speed v (in miles per hour) that will minimize costs on a 110-mile delivery trip. The cost per hour for fuel is C dollars, and the driver is paid W dollars per hour. (Assume there are no costs other than wages and fuel.)**

93. Fuel cost: $C = \dfrac{v^2}{600}$

Driver: $W = \$5$

94. Fuel cost: $C = \dfrac{v^2}{500}$

Driver: $W = \$7.50$

In Exercises 95 and 96, find the differential dy.

95. $y = x(1 - \cos x)$

96. $y = \sqrt{36 - x^2}$

97. *Surface Area and Volume* The diameter of a sphere is measured to be 18 centimeters, with a maximum possible error of 0.05 centimeter. Use differentials to approximate the possible propagated error and percent error in calculating the surface area and the volume of the sphere.

98. *Demand Function* A company finds that the demand for its commodity is $p = 75 - \frac{1}{4}x$. If x changes from 7 to 8, find and compare the values of Δp and dp.

1. Graph the fourth-degree polynomial $p(x) = x^4 + ax^2 + 1$ for various values of the constant a.

   (a) Determine the values of a for which p has exactly one relative minimum.

   (b) Determine the values of a for which p has exactly one relative maximum.

   (c) Determine the values of a for which p has exactly two relative minima.

   (d) Show that the graph of p cannot have exactly two relative extrema.

2. (a) Graph the fourth-degree polynomial $p(x) = ax^4 - 6x^2$ for $a = -3, -2, -1, 0, 1, 2,$ and 3. For what values of the constant a does p have a relative minimum or relative maximum?

   (b) Show that p has a relative maximum for all values of the constant a.

   (c) Determine analytically the values of a for which p has a relative minimum.

   (d) Let $(x, y) = (x, p(x))$ be a relative extremum of p. Show that (x, y) lies on the graph of $y = -3x^2$. Verify this result graphically by graphing $y = -3x^2$ together with the seven curves from part (a).

3. Let $f(x) = \dfrac{c}{x} + x^2$. Determine all values of the constant c such that f has a relative minimum, but no relative maximum.

4. (a) Let $f(x) = ax^2 + bx + c$, $a \neq 0$, be a quadratic polynomial. How many points of inflection does the graph of f have?

   (b) Let $f(x) = ax^3 + bx^2 + cx + d$, $a \neq 0$, be a cubic polynomial. How many points of inflection does the graph of f have?

   (c) Suppose the function $y = f(x)$ satisfies the equation $\dfrac{dy}{dx} = ky\left(1 - \dfrac{y}{L}\right)$ where k and L are positive constants. Show that the graph of f has a point of inflection at the point where $y = \dfrac{L}{2}$. (This equation is called the **logistic differential equation.**)

5. Prove Darboux's Theorem: Let f be differentiable on the closed interval $[a, b]$ such that $f'(a) = y_1$ and $f'(b) = y_2$. If d lies between y_1 and y_2, then there exists c in (a, b) such that $f'(c) = d$.

6. Let f and g be functions that are continuous on $[a, b]$ and differentiable on (a, b). Prove that if $f(a) = g(a)$ and $g'(x) > f'(x)$ for all x in (a, b), then $g(b) > f(b)$.

7. Prove the following **Extended Mean Value Theorem:** If f and f' are continuous on the closed interval $[a, b]$, and if f'' exists on the open interval (a, b), then there exists a number c in (a, b) such that

$$f(b) = f(a) + f'(a)(b - a) + \frac{1}{2}f''(c)(b - a)^2.$$

8. (a) Let $V = x^3$. Find dV and ΔV. Show that for small values of x, the difference $\Delta V - dV$ is very small in the sense that there exists ε such that $\Delta V - dV = \varepsilon \Delta x$, where $\varepsilon \to 0$ as $\Delta x \to 0$.

   (b) Generalize this result by showing that if $y = f(x)$ is a differentiable function, then $\Delta y - dy = \varepsilon \Delta x$, where $\varepsilon \to 0$ as $\Delta x \to 0$.

9. The amount of illumination of a surface is proportional to the intensity of the light source, inversely proportional to the square of the distance from the light source, and proportional to $\sin \theta$, where θ is the angle at which the light strikes the surface. A rectangular room measures 10 feet by 24 feet, with a 10-foot ceiling. Determine the height at which the light should be placed to allow the corners of the floor to receive as much light as possible.

10. Consider a room in the shape of a cube, 4 meters on each side. A bug at point P wants to walk to point Q at the opposite corner, as shown in the figure. Use calculus to determine the shortest path. Can you solve the problem without calculus?

11. The line joining P and Q crosses two parallel lines, as shown in the figure. The point R is d units from P. How far from Q should the point S be so that the sum of the areas of the two shaded triangles is a minimum? So that the sum is a maximum?

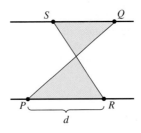

12. The figures show a rectangle, a circle, and a semicircle inscribed in a triangle bounded by the coordinate axes and the first-quadrant portion of the line with intercepts $(3, 0)$ and $(0, 4)$. Find the dimensions of each inscribed figure such that its area is maximum. State whether calculus was helpful in finding the required dimensions. Explain your reasoning.

13. (a) Prove that $\lim_{x \to \infty} x^2 = \infty$.

    (b) Prove that $\lim_{x \to \infty} \left(\frac{1}{x^2}\right) = 0$.

    (c) Let L be a real number. Prove that if $\lim_{x \to \infty} f(x) = L$, then
    $$\lim_{y \to 0^+} f\left(\frac{1}{y}\right) = L.$$

14. Find the point on the graph of $y = \dfrac{1}{1 + x^2}$ (see figure) where the tangent line has the greatest slope, and the point where the tangent line has the least slope.

15. (a) Let x be a positive number. Use the *table* feature of a graphing utility to verify that $\sqrt{1 + x} < \frac{1}{2}x + 1$.

    (b) Use the Mean Value Theorem to prove that $\sqrt{1 + x} < \frac{1}{2}x + 1$ for all positive real numbers x.

16. (a) Let x be a positive number. Use the *table* feature of a graphing utility to verify that $\sin x < x$.

    (b) Use the Mean Value Theorem to prove that $\sin x < x$ for all positive real numbers x.

17. The police department must determine the speed limit on a bridge such that the flow rate of cars is maximized per unit time. The greater the speed limit, the farther apart the cars must be in order to keep a safe stopping distance. Experimental data on the stopping distance d (in meters) for various speeds v (in kilometers per hour) are shown in the table.

v	20	40	60	80	100
d	5.1	13.7	27.2	44.2	66.4

(a) Convert the speeds v in the table to the speeds s in meters per second. Use the regression capabilities of a graphing utility to find a model of the form $d(s) = as^2 + bs + c$ for the data.

(b) Consider two consecutive vehicles of average length 5.5 meters, traveling at a safe speed on the bridge. Let T be the difference between the times (in seconds) when the front bumpers of the vehicles pass a given point on the bridge. Verify that this difference in times is given by
$$T = \frac{d(s)}{s} + \frac{5.5}{s}.$$

(c) Use a graphing utility to graph the function T and estimate the speed s that minimizes the time between vehicles.

(d) Use calculus to determine the speed that minimizes T. What is the minimum value of T? Convert the required speed to kilometers per hour.

(e) Find the optimal distance between vehicles for the posted speed limit determined in part (d).

18. A legal-sized sheet of paper (8.5 inches by 14 inches) is folded so that corner P touches the opposite 14-inch edge at R. $\left(Note: PQ = \sqrt{C^2 - x^2}.\right)$

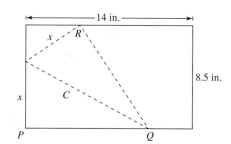

8.5 in.

(a) Show that $C^2 = \dfrac{2x^3}{2x - 8.5}$.

(b) What is the domain of C?

(c) Determine the x-value that minimizes C.

(d) Determine the minimum length of C.

19. Let $f(x) = \sin(\ln x)$.

(a) Determine the domain of the function f.

(b) Find two values of x satisfying $f(x) = 1$.

(c) Find two values of x satisfying $f(x) = -1$.

(d) What is the range of the function f?

(e) Calculate $f'(x)$ and use calculus to find the maximum value of f on the interval $[1, 10]$.

(f) Use a graphing utility to graph f in the viewing window $[0, 5] \times [-2, 2]$ and estimate $\lim_{x \to 0^+} f(x)$, if it exists.

(g) Determine $\lim_{x \to 0^+} f(x)$ analytically, if it exists.

5 Integration

This photo of a jet breaking the sound barrier was taken by Ensign John Gay. At the time the photo was taken, was the jet's velocity constant or changing? Why?

©Corbis Sygma

The area of a parabolic region can be approximated by the sum of the areas of rectangles. As you increase the number of rectangles, the approximation tends to become more and more accurate. In Section 5.2, you will learn how the limit process can be used to find the areas of a wide variety of regions. This process is called *integration* and is closely related to differentiation.

Antiderivatives and Indefinite Integration

- Write the general solution of a differential equation.
- Use indefinite integral notation for antiderivatives.
- Use basic integration rules to find antiderivatives.
- Find a particular solution of a differential equation.

Antiderivatives

Suppose you were asked to find a function F whose derivative is $f(x) = 3x^2$. From your knowledge of derivatives, you would probably say that

$$F(x) = x^3 \text{ because } \frac{d}{dx}[x^3] = 3x^2.$$

The function F is an *antiderivative* of f.

Definition of an Antiderivative

A function F is an **antiderivative** of f on an interval I if $F'(x) = f(x)$ for all x in I.

Note that F is called *an* antiderivative of f, rather than *the* antiderivative of f. To see why, observe that

$$F_1(x) = x^3, \quad F_2(x) = x^3 - 5, \quad \text{and} \quad F_3(x) = x^3 + 97$$

are all antiderivatives of $f(x) = 3x^2$. In fact, for any constant C, the function given by $F(x) = x^3 + C$ is an antiderivative of f.

THEOREM 5.1   Representation of Antiderivatives

If F is an antiderivative of f on an interval I, then G is an antiderivative of f on the interval I if and only if G is of the form $G(x) = F(x) + C$, for all x in I, where C is a constant.

Proof   The proof of Theorem 5.1 in one direction is straightforward. That is, if $G(x) = F(x) + C$, $F'(x) = f(x)$, and C is a constant, then

$$G'(x) = \frac{d}{dx}[F(x) + C] = F'(x) + 0 = f(x).$$

To prove this theorem in the other direction, assume that G is an antiderivative of f. Define a function H such that

$$H(x) = G(x) - F(x).$$

If H is not constant on the interval I, then there must exist a and b $(a < b)$ in the interval such that $H(a) \neq H(b)$. Moreover, because H is differentiable on (a, b), you can apply the Mean Value Theorem to conclude that there exists some c in (a, b) such that

$$H'(c) = \frac{H(b) - H(a)}{b - a}.$$

Because $H(b) \neq H(a)$, it follows that $H'(c) \neq 0$. However, because $G'(c) = F'(c)$, you know that $H'(c) = G'(c) - F'(c) = 0$, which contradicts the fact that $H'(c) \neq 0$. Consequently, you can conclude that $H(x)$ is a constant, C. So, $G(x) - F(x) = C$ and it follows that $G(x) = F(x) + C$.

Using Theorem 5.1, you can represent the entire family of antiderivatives of a function by adding a constant to a *known* antiderivative. For example, knowing that $D_x[x^2] = 2x$, you can represent the family of *all* antiderivatives of $f(x) = 2x$ by

$$G(x) = x^2 + C \qquad \text{Family of all antiderivatives of } f(x) = 2x$$

where C is a constant. The constant C is called the **constant of integration.** The family of functions represented by G is the **general antiderivative** of f, and $G(x) = x^2 + C$ is the **general solution** of the *differential equation*

$$G'(x) = 2x. \qquad \text{Differential equation}$$

A **differential equation** in x and y is an equation that involves x, y, and derivatives of y. For instance, $y' = 3x$ and $y' = x^2 + 1$ are examples of differential equations.

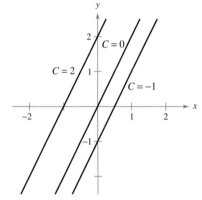

Functions of the form $y = 2x + C$
Figure 5.1

EXAMPLE 1   Solving a Differential Equation

Find the general solution of the differential equation $y' = 2$.

Solution   To begin, you need to find a function whose derivative is 2. One such function is

$$y = 2x. \qquad \text{2x is } an \text{ antiderivative of 2.}$$

Now, you can use Theorem 5.1 to conclude that the general solution of the differential equation is

$$y = 2x + C. \qquad \text{General solution}$$

The graphs of several functions of the form $y = 2x + C$ are shown in Figure 5.1.

Notation for Antiderivatives

When solving a differential equation of the form

$$\frac{dy}{dx} = f(x)$$

it is convenient to write it in the equivalent differential form

$$dy = f(x)\, dx.$$

The operation of finding all solutions of this equation is called **antidifferentiation** (or **indefinite integration**) and is denoted by an integral sign $\int$. The general solution is denoted by

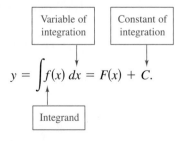

$$y = \int f(x)\, dx = F(x) + C.$$

The expression $\int f(x)\, dx$ is read as the *antiderivative of f with respect to x.* So, the differential dx serves to identify x as the variable of integration. The term **indefinite integral** is a synonym for antiderivative.

NOTE   In this text, the notation $\int f(x)\, dx = F(x) + C$ means that F is an antiderivative of f *on an interval.*

Basic Integration Rules

The inverse nature of integration and differentiation can be verified by substituting $F'(x)$ for $f(x)$ in the indefinite integration definition to obtain

$$\int F'(x)\,dx = F(x) + C.$$

Integration is the "inverse" of differentiation.

Moreover, if $\int f(x)\,dx = F(x) + C$, then

$$\frac{d}{dx}\left[\int f(x)\,dx\right] = f(x).$$

Differentiation is the "inverse" of integration.

NOTE   The Power Rule for integration has the restriction that $n \neq -1$. To evaluate $\int x^{-1}\,dx$, you must use the natural log rule. (See Exercise 106)

These two equations allow you to obtain integration formulas directly from differentiation formulas, as shown in the following summary.

Basic Integration Rules

Differentiation Formula	*Integration Formula*
$\dfrac{d}{dx}[C] = 0$	$\displaystyle\int 0\,dx = C$
$\dfrac{d}{dx}[kx] = k$	$\displaystyle\int k\,dx = kx + C$
$\dfrac{d}{dx}[kf(x)] = kf'(x)$	$\displaystyle\int kf(x)\,dx = k\int f(x)\,dx$
$\dfrac{d}{dx}[f(x) \pm g(x)] = f'(x) \pm g'(x)$	$\displaystyle\int [f(x) \pm g(x)]\,dx = \int f(x)\,dx \pm \int g(x)\,dx$
$\dfrac{d}{dx}[x^n] = nx^{n-1}$	$\displaystyle\int x^n\,dx = \dfrac{x^{n+1}}{n+1} + C, \quad n \neq -1$ Power Rule
$\dfrac{d}{dx}[\sin x] = \cos x$	$\displaystyle\int \cos x\,dx = \sin x + C$
$\dfrac{d}{dx}[\cos x] = -\sin x$	$\displaystyle\int \sin x\,dx = -\cos x + C$
$\dfrac{d}{dx}[\tan x] = \sec^2 x$	$\displaystyle\int \sec^2 x\,dx = \tan x + C$
$\dfrac{d}{dx}[\sec x] = \sec x \tan x$	$\displaystyle\int \sec x \tan x\,dx = \sec x + C$
$\dfrac{d}{dx}[\cot x] = -\csc^2 x$	$\displaystyle\int \csc^2 x\,dx = -\cot x + C$
$\dfrac{d}{dx}[\csc x] = -\csc x \cot x$	$\displaystyle\int \csc x \cot x\,dx = -\csc x + C$
$\dfrac{d}{dx}[e^x] = e^x$	$\displaystyle\int e^x\,dx = e^x + C$
$\dfrac{d}{dx}[a^x] = (\ln a)a^x$	$\displaystyle\int a^x\,dx = \left(\dfrac{1}{\ln a}\right)a^x + C$
$\dfrac{d}{dx}[\ln x] = \dfrac{1}{x}, \quad x > 0$	$\displaystyle\int \dfrac{1}{x}\,dx = \ln|x| + C$

EXAMPLE 2   **Applying the Basic Integration Rules**

Describe the antiderivatives of $3x$.

Solution	$\displaystyle\int 3x \, dx = 3\int x \, dx$	Constant Multiple Rule
	$\displaystyle = 3\int x^1 \, dx$	Rewrite x as x^1.
	$\displaystyle = 3\left(\frac{x^2}{2}\right) + C$	Power Rule ($n = 1$)
	$\displaystyle = \frac{3}{2}x^2 + C$	Simplify.

When indefinite integrals are evaluated, a strict application of the basic integration rules tends to produce complicated constants of integration. For instance, in Example 2, you could have written

$$\int 3x \, dx = 3\int x \, dx = 3\left(\frac{x^2}{2} + C\right) = \frac{3}{2}x^2 + 3C.$$

However, because C represents *any* constant, it is both cumbersome and unnecessary to write $3C$ as the constant of integration. So, $\frac{3}{2}x^2 + 3C$ is written in the simpler form $\frac{3}{2}x^2 + C$.

In Example 2, note that the general pattern of integration is similar to that of differentiation.

Original integral	⇒	Rewrite	⇒	Integrate	⇒	Simplify

EXAMPLE 3   **Rewriting Before Integrating**

	Original Integral	*Rewrite*	*Integrate*	*Simplify*
a.	$\displaystyle\int \frac{1}{x^3} \, dx$	$\displaystyle\int x^{-3} \, dx$	$\dfrac{x^{-2}}{-2} + C$	$-\dfrac{1}{2x^2} + C$
b.	$\displaystyle\int \sqrt{x} \, dx$	$\displaystyle\int x^{1/2} \, dx$	$\dfrac{x^{3/2}}{3/2} + C$	$\dfrac{2}{3}x^{3/2} + C$
c.	$\displaystyle\int 2\sin x \, dx$	$\displaystyle 2\int \sin x \, dx$	$2(-\cos x) + C$	$-2\cos x + C$
d.	$\displaystyle\int \frac{3}{x} \, dx$	$\displaystyle 3\int \frac{1}{x} \, dx$	$3(\ln|x|) + C$	$3\ln|x| + C$

TECHNOLOGY   Some software programs, such as *Derive*, *Maple*, *Mathcad*, *Mathematica*, and the *TI-89*, are capable of performing integration symbolically. If you have access to such a symbolic integration utility, try using it to evaluate the indefinite integrals in Example 3.

NOTE   The properties of logarithms presented on page 53 can be used to rewrite anitderivatives in different forms. For instant, the antiderivative in Example 3(d) can be rewritten as

$$3\ln|x| + C = \ln|x|^3 + C.$$

Remember that you can check your answer to an antidifferentiation problem by differentiating. For instance, in Example 3(b), you can check that $\frac{2}{3}x^{3/2} + C$ is the correct antiderivative by differentiating the answer to obtain

$$D_x\left[\frac{2}{3}x^{3/2} + C\right] = \left(\frac{2}{3}\right)\left(\frac{3}{2}\right)x^{1/2} = \sqrt{x}.$$ Use differentiation to check antiderivative.

indicates that in the HM mathSpace® *CD-ROM and the online* Eduspace® *system for this text, you will find an Open Exploration, which further explores this example using the computer algebra systems* Maple, Mathcad, Mathematica, *and* Derive.

The basic integration rules listed earlier in this section allow you to integrate any polynomial function, as shown in Example 4.

EXAMPLE 4 **Integrating Polynomial Functions**

a. $\displaystyle\int dx = \int 1\, dx$ Integrand is understood to be 1.

$\qquad\quad = x + C$ Integrate.

b. $\displaystyle\int (x + 2)\, dx = \int x\, dx + \int 2\, dx$

$\qquad\qquad\qquad\quad = \dfrac{x^2}{2} + C_1 + 2x + C_2$ Integrate.

$\qquad\qquad\qquad\quad = \dfrac{x^2}{2} + 2x + C$ $C = C_1 + C_2$

The second line in the solution is usually omitted.

c. $\displaystyle\int (3x^4 - 5x^2 + x)\, dx = 3\left(\dfrac{x^5}{5}\right) - 5\left(\dfrac{x^3}{3}\right) + \dfrac{x^2}{2} + C$ Integrate.

$\qquad\qquad\qquad\qquad\qquad = \dfrac{3}{5}x^5 - \dfrac{5}{3}x^3 + \dfrac{1}{2}x^2 + C$ Simplify.

EXAMPLE 5 **Rewriting Before Integrating**

$\displaystyle\int \dfrac{x + 1}{\sqrt{x}}\, dx = \int \left(\dfrac{x}{\sqrt{x}} + \dfrac{1}{\sqrt{x}}\right) dx$ Rewrite as two fractions.

$\qquad\qquad\quad = \int (x^{1/2} + x^{-1/2})\, dx$ Rewrite with fractional exponents.

$\qquad\qquad\quad = \dfrac{x^{3/2}}{3/2} + \dfrac{x^{1/2}}{1/2} + C$ Integrate.

$\qquad\qquad\quad = \dfrac{2}{3}x^{3/2} + 2x^{1/2} + C$ Simplify.

$\qquad\qquad\quad = \dfrac{2}{3}\sqrt{x}(x + 3) + C$ Factor.

NOTE When integrating quotients, do not integrate the numerator and denominator separately. This is no more valid in integration than it is in differentiation. For instance, in Example 5, be sure you understand that

$\displaystyle\int \dfrac{x + 1}{\sqrt{x}}\, dx = \dfrac{2}{3}\sqrt{x}(x + 3) + C$ is not the same as $\dfrac{\int (x + 1)\, dx}{\int \sqrt{x}\, dx} = \dfrac{\frac{1}{2}x^2 + x + C_1}{\frac{2}{3}x\sqrt{x} + C_2}$.

EXAMPLE 6 **Rewriting Before Integrating**

$\displaystyle\int \dfrac{\sin x}{\cos^2 x}\, dx = \int \left(\dfrac{1}{\cos x}\right)\left(\dfrac{\sin x}{\cos x}\right) dx$ Rewrite as a product.

$\qquad\qquad\quad = \int \sec x \tan x\, dx$ Rewrite using trigonometric identities.

$\qquad\qquad\quad = \sec x + C$ Integrate.

Initial Conditions and Particular Solutions

You have already seen that the equation $y = \int f(x)\,dx$ has many solutions (each differing from the others by a constant). This means that the graphs of any two antiderivatives of f are vertical translations of each other. For example, Figure 5.2 shows the graphs of several antiderivatives of the form

$$y = \int (3x^2 - 1)\,dx$$

$$= x^3 - x + C \qquad \text{General solution}$$

for various integer values of C. Each of these antiderivatives is a solution of the differential equation

$$\frac{dy}{dx} = 3x^2 - 1.$$

In many applications of integration, you are given enough information to determine a **particular solution.** To do this, you need only know the value of $y = F(x)$ for one value of x. This information is called an **initial condition.** For example, in Figure 5.2, only one curve passes through the point $(2, 4)$. To find this curve, you can use the following information.

$$F(x) = x^3 - x + C \qquad \text{General solution}$$
$$F(2) = 4 \qquad \text{Initial condition}$$

By using the initial condition in the general solution, you can determine that $F(2) = 8 - 2 + C = 4$, which implies that $C = -2$. So, you obtain

$$F(x) = x^3 - x - 2. \qquad \text{Particular solution}$$

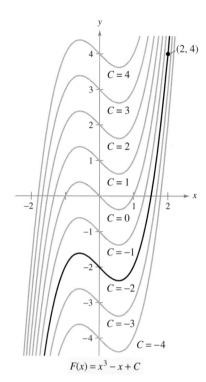

The particular solution that satisfies the initial condition $F(2) = 4$ is $F(x) = x^3 - x - 2$.

Figure 5.2

EXAMPLE 7   Finding a Particular Solution

Find the general solution of

$$F'(x) = e^x$$

and find the particular solution that satisfies the initial condition $F(0) = 3$.

Solution   To find the general solution, integrate to obtain

$$F(x) = \int e^x\,dx$$

$$= e^x + C. \qquad \text{General solution}$$

Using the initial condition $F(0) = 3$, you can solve for C as follows.

$$F(0) = e^0 + C$$
$$3 = 1 + C$$
$$2 = C$$

So, the particular solution, as shown in Figure 5.3, is

$$F(x) = e^x + 2. \qquad \text{Particular solution}$$

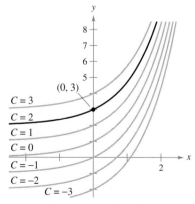

The particular solution that satisfies the initial condition $F(0) = 3$ is $F(x) = e^x + 2$.

Figure 5.3

So far in this section you have been using x as the variable of integration. In applications, it is often convenient to use a different variable. For instance, in the following example involving *time*, the variable of integration is t.

EXAMPLE 8 Solving a Vertical Motion Problem

A ball is thrown upward with an initial velocity of 64 feet per second from an initial height of 80 feet.

a. Find the position function giving the height s as a function of the time t.

b. When does the ball hit the ground?

Solution

a. Let $t = 0$ represent the initial time. The two given initial conditions can be written as follows.

$$s(0) = 80 \qquad \text{Initial height is 80 feet.}$$

$$s'(0) = 64 \qquad \text{Initial velocity is 64 feet per second.}$$

Using -32 feet per second per second as the acceleration due to gravity, you can write

$$s''(t) = -32$$

$$s'(t) = \int s''(t)\, dt$$

$$= \int -32\, dt = -32t + C_1.$$

Using the initial velocity, you obtain $s'(0) = 64 = -32(0) + C_1$, which implies that $C_1 = 64$. Next, by integrating $s'(t)$, you obtain

$$s(t) = \int s'(t)\, dt$$

$$= \int (-32t + 64)\, dt$$

$$= -16t^2 + 64t + C_2.$$

Using the initial height, you obtain

$$s(0) = 80 = -16(0^2) + 64(0) + C_2$$

which implies that $C_2 = 80$. So, the position function is

$$s(t) = -16t^2 + 64t + 80. \qquad \text{See Figure 5.4.}$$

b. Using the position function found in part (a), you can find the time that the ball hits the ground by solving the equation $s(t) = 0$.

$$s(t) = -16t^2 + 64t + 80 = 0$$

$$-16(t + 1)(t - 5) = 0$$

$$t = -1, 5$$

Because t must be positive, you can conclude that the ball hit the ground 5 seconds after it was thrown.

$s(t) = -16t^2 + 64t + 80$

Height of a ball at time t
Figure 5.4

NOTE In Example 8, note that the position function has the form

$$s(t) = \tfrac{1}{2}gt^2 + v_0 t + s_0$$

where $g = -32$, v_0 is the initial velocity, and s_0 is the initial height, as presented in Section 3.2.

Example 8 shows how to use calculus to analyze vertical motion problems in which the acceleration is determined by a gravitational force. You can use a similar strategy to analyze other linear motion problems (vertical or horizontal) in which the acceleration (or deceleration) is the result of some other force, as you will see in Exercises 87–94.

Before you begin the exercise set, be sure you realize that one of the most important steps in integration is *rewriting the integrand* in a form that fits the basic integration rules. To further illustrate this point, here are some additional examples.

Original Integral	Rewrite	Integrate	Simplify
$\displaystyle\int \frac{2}{\sqrt{x}}\,dx$	$\displaystyle 2\int x^{-1/2}\,dx$	$2\left(\dfrac{x^{1/2}}{1/2}\right) + C$	$4x^{1/2} + C$
$\displaystyle\int (t^2 + 1)^2\,dt$	$\displaystyle\int (t^4 + 2t^2 + 1)\,dt$	$\dfrac{t^5}{5} + 2\left(\dfrac{t^3}{3}\right) + t + C$	$\dfrac{1}{5}t^5 + \dfrac{2}{3}t^3 + t + C$
$\displaystyle\int \frac{x^3 + 3}{x^2}\,dx$	$\displaystyle\int (x + 3x^{-2})\,dx$	$\dfrac{x^2}{2} + 3\left(\dfrac{x^{-1}}{-1}\right) + C$	$\dfrac{1}{2}x^2 - \dfrac{3}{x} + C$
$\displaystyle\int \sqrt[3]{x}(x - 4)\,dx$	$\displaystyle\int (x^{4/3} - 4x^{1/3})\,dx$	$\dfrac{x^{7/3}}{7/3} - 4\left(\dfrac{x^{4/3}}{4/3}\right) + C$	$\dfrac{3}{7}x^{4/3}(x - 7) + C$

Exercises for Section 5.1

See www.CalcChat.com for worked-out solutions to odd-numbered exercises.

In Exercises 1–4, verify the statement by showing that the derivative of the right side equals the integrand of the left side.

1. $\displaystyle\int \left(-\frac{9}{x^4}\right) dx = \frac{3}{x^3} + C$

2. $\displaystyle\int \left(4x^3 - \frac{1}{x^2}\right) dx = x^4 + \frac{1}{x} + C$

3. $\displaystyle\int (x - 2)(x + 2)\,dx = \frac{1}{3}x^3 - 4x + C$

4. $\displaystyle\int \frac{x^2 - 1}{x^{3/2}}\,dx = \frac{2(x^2 + 3)}{3\sqrt{x}} + C$

In Exercises 5–8, find the general solution of the differential equation and check the result by differentiation.

5. $\dfrac{dy}{dt} = 3t^2$

6. $\dfrac{dr}{d\theta} = \pi$

7. $\dfrac{dy}{dx} = x^{3/2}$

8. $\dfrac{dy}{dx} = 2x^{-3}$

In Exercises 9–14, complete the table using Example 3 and the examples at the top of this page as a model.

Original Integral	Rewrite	Integrate	Simplify
9. $\displaystyle\int \sqrt[3]{x}\,dx$			
10. $\displaystyle\int \frac{1}{x^2}\,dx$			
11. $\displaystyle\int \frac{1}{x\sqrt{x}}\,dx$			
12. $\displaystyle\int x(x^2 + 3)\,dx$			
13. $\displaystyle\int \frac{1}{2x^3}\,dx$			
14. $\displaystyle\int \frac{1}{(3x)^2}\,dx$			

In Exercises 15–44, find the indefinite integral and check the result by differentiation.

15. $\displaystyle\int (x + 3)\,dx$

16. $\displaystyle\int (5 - x)\,dx$

17. $\displaystyle\int (x^3 + 5)\,dx$

18. $\displaystyle\int (4x^3 + 6x^2 - 1)\,dx$

19. $\displaystyle\int (x^{3/2} + 2x + 1)\,dx$

20. $\displaystyle\int (\sqrt[4]{x^3} + 1)\,dx$

21. $\displaystyle\int \frac{1}{x^3}\,dx$

22. $\displaystyle\int \frac{1}{x^4}\,dx$

23. $\displaystyle\int \frac{x^2 + x + 1}{\sqrt{x}}\,dx$

24. $\displaystyle\int \frac{x^2 + 2x - 3}{x^4}\,dx$

25. $\displaystyle\int (x + 1)(3x - 2)\,dx$

26. $\displaystyle\int (2t^2 - 1)^2\,dt$

27. $\displaystyle\int y^2\sqrt{y}\,dy$

28. $\displaystyle\int (1 + 3t)t^2\,dt$

29. $\displaystyle\int dx$

30. $\displaystyle\int 3\,dt$

31. $\displaystyle\int (2\sin x + 3\cos x)\,dx$

32. $\displaystyle\int (t^2 - \sin t)\,dt$

33. $\displaystyle\int (1 - \csc t \cot t)\,dt$

34. $\displaystyle\int (\theta^2 + \sec^2 \theta)\,d\theta$

35. $\displaystyle\int (2\sin x - 5e^x)\,dx$

36. $\displaystyle\int (3x^2 + 2e^x)\,dx$

37. $\displaystyle\int (\sec^2 \theta - \sin \theta)\,d\theta$

38. $\displaystyle\int \sec y(\tan y - \sec y)\,dy$

39. $\displaystyle\int (\tan^2 y + 1)\,dy$

40. $\displaystyle\int \frac{\cos x}{1 - \cos^2 x}\,dx$

41. $\displaystyle\int (2x - 4^x)\,dx$

42. $\displaystyle\int (\cos x + 3^x)\,dx$

43. $\displaystyle\int \left(x - \frac{5}{x}\right) dx$

44. $\displaystyle\int \left(\frac{4}{x} + \sec^2 x\right) dx$

In Exercises 45–48, sketch the graphs of the function $g(x) = f(x) + C$ for $C = -2$, $C = 0$, and $C = 3$ on the same set of coordinate axes.

45. $f(x) = \cos x$ **46.** $f(x) = \sqrt{x}$

47. $f(x) = \ln x$ **48.** $f(x) = \frac{1}{2} e^x$

In Exercises 49–52, the graph of the derivative of a function is given. Sketch the graphs of *two* functions that have the given derivative. (There is more than one correct answer.) To print an enlarged copy of the graph, go to the website *www.mathgraphs.com*.

49.

50.

51.

52.
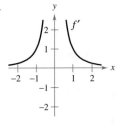

In Exercises 53–56, find the equation for y, given the derivative and the indicated point on the curve.

53. $\dfrac{dy}{dx} = 2x - 1$

54. $\dfrac{dy}{dx} = 2(x - 1)$

55. $\dfrac{dy}{dx} = \cos x$

56. $\dfrac{dy}{dx} = \dfrac{3}{x}$, $x > 0$

 Slope Fields In Exercises 57–60, a differential equation, a point, and a slope field are given. A *slope field* (or *direction field*) consists of line segments with slopes given by the differential equation. These line segments give a visual perspective of the slopes of the solutions of the differential equation. (a) Sketch two approximate solutions of the differential equation on the slope field, one of which passes through the indicated point. (To print an enlarged copy of the graph, go to the website *www.mathgraphs.com*.) (b) Use integration to find the particular solution of the differential equation and use a graphing utility to graph the solution. Compare the result with the sketches in part (a).

57. $\dfrac{dy}{dx} = \dfrac{1}{2}x - 1$, $(4, 2)$ **58.** $\dfrac{dy}{dx} = x^2 - 1$, $(-1, 3)$

59. $\dfrac{dy}{dx} = \cos x$, $(0, 4)$ **60.** $\dfrac{dy}{dx} = -\dfrac{1}{x^2}$, $x > 0$, $(1, 3)$

 *Slope Fields* In Exercises 61 and 62, (a) use a graphing utility to graph a slope field for the differential equation, (b) use integration and the given point to find the particular solution of the differential equation, and (c) graph the solution and the slope field in the same viewing window.

61. $\dfrac{dy}{dx} = 2x$, $(-2, -2)$ **62.** $\dfrac{dy}{dx} = 2\sqrt{x}$, $(4, 12)$

In Exercises 63–72, solve the differential equation.

63. $f'(x) = 4x$, $f(0) = 6$ **64.** $g'(x) = 6x^2$, $g(0) = -1$

65. $h'(t) = 8t^3 + 5$, $h(1) = -4$ **66.** $f'(s) = 6s - 8s^3$, $f(2) = 3$

67. $f''(x) = 2$, $f'(2) = 5$, $f(2) = 10$

68. $f''(x) = x^2$, $f'(0) = 6$, $f(0) = 3$

69. $f''(x) = x^{-3/2}$, $f'(4) = 2$, $f(0) = 0$

70. $f''(x) = \sin x$, $f'(0) = 1$, $f(0) = 6$

71. $f''(x) = e^x$, $f'(0) = 2$, $f(0) = 5$

72. $f''(x) = \dfrac{2}{x^2}$, $f'(1) = 4$, $f(1) = 3$

73. *Tree Growth*   An evergreen nursery usually sells a certain shrub after 6 years of growth and shaping. The growth rate during those 6 years is approximated by $dh/dt = 1.5t + 5$, where t is the time in years and h is the height in centimeters. The seedlings are 12 centimeters tall when planted ($t = 0$).

   (a) Find the height after t years.

   (b) How tall are the shrubs when they are sold?

74. *Population Growth*   The rate of growth dP/dt of a population of bacteria is proportional to the square root of t, where P is the population size and t is the time in days ($0 \le t \le 10$). That is,

$$\frac{dP}{dt} = k\sqrt{t}.$$

The initial size of the population is 500. After 1 day the population has grown to 600. Estimate the population after 7 days.

Writing About Concepts

75. Use the graph of f' shown in the figure to answer the following, given that $f(0) = -4$.

   (a) Approximate the slope of f at $x = 4$. Explain.

   (b) Is it possible that $f(2) = -1$? Explain.

   (c) Is $f(5) - f(4) > 0$? Explain.

   (d) Approximate the value of x where f is maximum. Explain.

   (e) Approximate any intervals in which the graph of f is concave upward and any intervals in which it is concave downward. Approximate the x-coordinates of any points of inflection.

   (f) Approximate the x-coordinate of the minimum of $f''(x)$.

   (g) Sketch an approximate graph of f. To print an enlarged copy of the graph, go to the website *www.mathgraphs.com*.

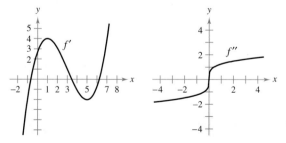

Figure for 75   **Figure for 76**

76. The graphs of f and f' each pass through the origin. Use the graph of f'' shown in the figure to sketch the graphs of f and f'. To print an enlarged copy of the graph, go to the website *www.mathgraphs.com*.

Vertical Motion   In Exercises 77–80, use $a(t) = -32$ feet per second per second as the acceleration due to gravity. (Neglect air resistance.)

77. A ball is thrown vertically upward from a height of 6 feet with an initial velocity of 60 feet per second. How high will the ball go?

78. Show that the height above the ground of an object thrown upward from a point s_0 feet above the ground with an initial velocity of v_0 feet per second is given by the function

$$f(t) = -16t^2 + v_0 t + s_0.$$

79. With what initial velocity must an object be thrown upward (from ground level) to reach the top of the Washington Monument (approximately 550 feet)?

80. A balloon, rising vertically with a velocity of 8 feet per second, releases a sandbag at the instant it is 64 feet above the ground.

   (a) How many seconds after its release will the bag strike the ground?

   (b) At what velocity will it hit the ground?

Vertical Motion   In Exercises 81 and 82, use $a(t) = -9.8$ meters per second per second as the acceleration due to gravity. (Neglect air resistance.)

81. Show that the height above the ground of an object thrown upward from a point s_0 meters above the ground with an initial velocity of v_0 meters per second is given by the function

$$f(t) = -4.9t^2 + v_0 t + s_0.$$

82. The Grand Canyon is 1600 meters deep at its deepest point. A rock is dropped from the rim above this point. Express the height of the rock as a function of the time t in seconds. How long will it take the rock to hit the canyon floor?

83. A baseball is thrown upward from a height of 2 meters with an initial velocity of 10 meters per second. Determine its maximum height.

84. With what initial velocity must an object be thrown upward (from a height of 2 meters) to reach a maximum height of 200 meters?

85. *Lunar Gravity*   On the moon, the acceleration due to gravity is -1.6 meters per second per second. A stone is dropped from a cliff on the moon and hits the surface of the moon 20 seconds later. How far did it fall? What was its velocity at impact?

86. *Escape Velocity*   The minimum velocity required for an object to escape Earth's gravitational pull is obtained from the solution of the equation

$$\int v \, dv = -GM \int \frac{1}{y^2} \, dy$$

where v is the velocity of the object projected from Earth, y is the distance from the center of Earth, G is the gravitational constant, and M is the mass of Earth. Show that v and y are related by the equation

$$v^2 = v_0^2 + 2GM\left(\frac{1}{y} - \frac{1}{R}\right)$$

where v_0 is the initial velocity of the object and R is the radius of Earth.

Rectilinear Motion In Exercises 87–90, consider a particle moving along the x-axis where x(t) is the position of the particle at time t, x'(t) is its velocity, and x''(t) is its acceleration.

87. $x(t) = t^3 - 6t^2 + 9t - 2, \quad 0 \le t \le 5$

(a) Find the velocity and acceleration of the particle.

(b) Find the open t-intervals on which the particle is moving to the right.

(c) Find the velocity of the particle when the acceleration is 0.

88. Repeat Exercise 87 for the position function

$x(t) = (t - 1)(t - 3)^2, \quad 0 \le t \le 5.$

89. A particle moves along the x-axis at a velocity of $v(t) = 1/\sqrt{t}$, $t > 0$. At time $t = 1$, its position is $x = 4$. Find the acceleration and position functions for the particle.

90. A particle, initially at rest, moves along the x-axis such that its acceleration at time $t > 0$ is given by $a(t) = \cos t$. At the time $t = 0$, its position is $x = 3$.

(a) Find the velocity and position functions for the particle.

(b) Find the values of t for which the particle is at rest.

91. *Acceleration* The maker of an automobile advertises that it takes 13 seconds to accelerate from 25 kilometers per hour to 80 kilometers per hour. Assuming constant acceleration, compute the following.

(a) The acceleration in meters per second per second

(b) The distance the car travels during the 13 seconds

92. *Deceleration* A car traveling at 45 miles per hour is brought to a stop, at constant deceleration, 132 feet from where the brakes are applied.

(a) How far has the car moved when its speed has been reduced to 30 miles per hour?

(b) How far has the car moved when its speed has been reduced to 15 miles per hour?

(c) Draw the real number line from 0 to 132, and plot the points found in parts (a) and (b). What can you conclude?

93. *Acceleration* At the instant the traffic light turns green, a car that has been waiting at an intersection starts with a constant acceleration of 6 feet per second per second. At the same instant, a truck traveling with a constant velocity of 30 feet per second passes the car.

(a) How far beyond its starting point will the car pass the truck?

(b) How fast will the car be traveling when it passes the truck?

94. *Modeling Data* The table shows the velocities (in miles per hour) of two cars on an entrance ramp to an interstate highway. The time t is in seconds.

t	0	5	10	15	20	25	30
v_1	0	2.5	7	16	29	45	65
v_2	0	21	38	51	60	64	65

(a) Rewrite the table, converting miles per hour to feet per second.

(b) Use the regression capabilities of a graphing utility to find quadratic models for the data in part (a).

(c) Approximate the distance traveled by each car during the 30 seconds. Explain the difference in the distances.

True or False? In Exercises 95–100, determine whether the statement is true or false. If it is false, explain why or give an example that shows it is false.

95. Each antiderivative of an nth-degree polynomial function is an $(n + 1)$th-degree polynomial function.

96. If $p(x)$ is a polynomial function, then p has exactly one antiderivative whose graph contains the origin.

97. If $F(x)$ and $G(x)$ are antiderivatives of $f(x)$, then $F(x) = G(x) + C$.

98. If $f'(x) = g(x)$, then $\int g(x)\, dx = f(x) + C$.

99. $\int f(x)g(x)\, dx = \int f(x)\, dx \int g(x)\, dx$

100. The antiderivative of $f(x)$ is unique.

101. Find a function f such that the graph of f has a horizontal tangent at $(2, 0)$ and $f''(x) = 2x$.

102. The graph of f' is shown. Sketch the graph of f given that f is continuous and $f(0) = 1$.

103. If $f'(x) = \begin{cases} 1, & 0 \le x < 2 \\ 3x, & 2 \le x \le 5 \end{cases}$, f is continuous, and $f(1) = 3$, find f. Is f differentiable at $x = 2$?

104. Let $s(x)$ and $c(x)$ be two functions satisfying $s'(x) = c(x)$ and $c'(x) = -s(x)$ for all x. If $s(0) = 0$ and $c(0) = 1$, prove that $[s(x)]^2 + [c(x)]^2 = 1$.

105. *Verification* Verify the natural log rule $\int \dfrac{1}{x}\, dx = \ln|Cx|$, $C \ne 0$, by showing that the derivative of $\ln|Cx|$ is $1/x$.

106. *Verification* Verify the natural log rule $\int \dfrac{1}{x}\, dx = \ln|x| + C$ by showing that the derivative of $\ln|x| + C$ is $1/x$.

Putnam Exam Challenge

107. Suppose f and g are nonconstant, differentiable, real-valued functions on R. Furthermore, suppose that for each pair of real numbers x and y, $f(x + y) = f(x)f(y) - g(x)g(y)$ and $g(x + y) = f(x)g(y) + g(x)f(y)$. If $f'(0) = 0$, prove that $(f(x))^2 + (g(x))^2 = 1$ for all x.

This problem was composed by the Committee on the Putnam Prize Competition.
© The Mathematical Association of America. All rights reserved.

Area

- Use sigma notation to write and evaluate a sum.
- Understand the concept of area.
- Approximate the area of a plane region.
- Find the area of a plane region using limits.

Sigma Notation

In the preceding section, you studied antidifferentiation. In this section, you will look further into a problem introduced in Section 2.1—that of finding the area of a region in the plane. At first glance, these two ideas may seem unrelated, but you will discover in Section 5.4 that they are closely related by an extremely important theorem called the Fundamental Theorem of Calculus.

This section begins by introducing a concise notation for sums. This notation is called **sigma notation** because it uses the uppercase Greek letter sigma, written as Σ.

Sigma Notation

The sum of n terms $a_1, a_2, a_3, \ldots, a_n$ is written as

$$\sum_{i=1}^{n} a_i = a_1 + a_2 + a_3 + \cdots + a_n$$

where i is the **index of summation**, a_i is the ***i*th term** of the sum, and the **upper and lower bounds of summation** are n and 1.

NOTE   The upper and lower bounds must be constant with respect to the index of summation. However, the lower bound doesn't have to be 1. Any integer less than or equal to the upper bound is legitimate.

EXAMPLE 1   **Examples of Sigma Notation**

a. $\displaystyle\sum_{i=1}^{6} i = 1 + 2 + 3 + 4 + 5 + 6$

b. $\displaystyle\sum_{i=0}^{5} (i + 1) = 1 + 2 + 3 + 4 + 5 + 6$

c. $\displaystyle\sum_{j=3}^{7} j^2 = 3^2 + 4^2 + 5^2 + 6^2 + 7^2$

d. $\displaystyle\sum_{k=1}^{n} \frac{1}{n}(k^2 + 1) = \frac{1}{n}(1^2 + 1) + \frac{1}{n}(2^2 + 1) + \cdots + \frac{1}{n}(n^2 + 1)$

e. $\displaystyle\sum_{i=1}^{n} f(x_i)\,\Delta x = f(x_1)\,\Delta x + f(x_2)\,\Delta x + \cdots + f(x_n)\,\Delta x$

From parts (a) and (b), notice that the same sum can be represented in different ways using sigma notation.

Although any variable can be used as the index of summation, i, j, and k are often used. Notice in Example 1 that the index of summation does not appear in the terms of the expanded sum.

FOR FURTHER INFORMATION For a geometric interpretation of summation formulas, see the article, "Looking at $\displaystyle\sum_{k=1}^{n} k$ and $\displaystyle\sum_{k=1}^{n} k^2$ Geometrically" by Eric Hegblom in *Mathematics Teacher*. To view this article, go to the website *www.matharticles.com*.

The following properties of summation can be derived using the associative and commutative properties of addition and the distributive property of multiplication over addition. (In the first property, k is a constant.)

1. $\displaystyle\sum_{i=1}^{n} ka_i = k \sum_{i=1}^{n} a_i$

2. $\displaystyle\sum_{i=1}^{n} (a_i \pm b_i) = \sum_{i=1}^{n} a_i \pm \sum_{i=1}^{n} b_i$

The next theorem lists some useful formulas for sums of powers. A proof of this theorem is given in Appendix A.

THEOREM 5.2   Summation Formulas

1. $\displaystyle\sum_{i=1}^{n} c = cn$ **2.** $\displaystyle\sum_{i=1}^{n} i = \frac{n(n+1)}{2}$

3. $\displaystyle\sum_{i=1}^{n} i^2 = \frac{n(n+1)(2n+1)}{6}$ **4.** $\displaystyle\sum_{i=1}^{n} i^3 = \frac{n^2(n+1)^2}{4}$

EXAMPLE 2   **Evaluating a Sum**

Evaluate $\displaystyle\sum_{i=1}^{n} \frac{i+1}{n^2}$ for $n = 10$, 100, 1000, and $10,000$.

Solution   Applying Theorem 5.2, you can write

$$\sum_{i=1}^{n} \frac{i+1}{n^2} = \frac{1}{n^2} \sum_{i=1}^{n} (i+1) \qquad \text{Factor constant } 1/n^2 \text{ out of sum.}$$

$$= \frac{1}{n^2} \left(\sum_{i=1}^{n} i + \sum_{i=1}^{n} 1 \right) \qquad \text{Write as two sums.}$$

$$= \frac{1}{n^2} \left[\frac{n(n+1)}{2} + n \right] \qquad \text{Apply Theorem 5.2.}$$

$$= \frac{1}{n^2} \left[\frac{n^2 + 3n}{2} \right] \qquad \text{Simplify.}$$

$$= \frac{n+3}{2n}. \qquad \text{Simplify.}$$

Now you can evaluate the sum by substituting the appropriate values of n, as shown in the table at the left.

n	$\displaystyle\sum_{i=1}^{n} \frac{i+1}{n^2} = \frac{n+3}{2n}$
10	0.65000
100	0.51500
1000	0.50150
10,000	0.50015

In the table, note that the sum appears to approach a limit as n increases. Although the discussion of limits at infinity in Section 4.5 applies to a variable x, where x can be any real number, many of the same results hold true for limits involving the variable n, where n is restricted to positive integer values. So, to find the limit of $(n+3)/2n$ as n approaches infinity, you can write

$$\lim_{n \to \infty} \frac{n+3}{2n} = \frac{1}{2}.$$

Rectangle: $A = bh$
Figure 5.5

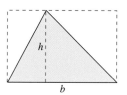

Triangle: $A = \frac{1}{2}bh$
Figure 5.6

ARCHIMEDES (287–212 B.C.)

Archimedes used the method of exhaustion to derive formulas for the areas of ellipses, parabolic segments, and sectors of a spiral. He is considered to have been the greatest applied mathematician of antiquity.

Mary Evans Picture Library

FOR FURTHER INFORMATION For an alternative development of the formula for the area of a circle, see the article "Proof Without Words: Area of a Disk is πR^2" by Russell Jay Hendel in *Mathematics Magazine*. To view this article, go to the website *www.matharticles.com*.

Area

In Euclidean geometry, the simplest type of plane region is a rectangle. Although people often say that the *formula* for the area of a rectangle is $A = bh$, as shown in Figure 5.5, it is actually more proper to say that this is the *definition* of the **area of a rectangle.**

From this definition, you can develop formulas for the areas of many other plane regions. For example, to determine the area of a triangle, you can form a rectangle whose area is twice that of the triangle, as shown in Figure 5.6. Once you know how to find the area of a triangle, you can determine the area of any polygon by subdividing the polygon into triangular regions, as shown in Figure 5.7.

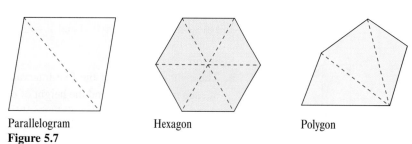

Parallelogram   Hexagon   Polygon
Figure 5.7

Finding the areas of regions other than polygons is more difficult. The ancient Greeks were able to determine formulas for the areas of some general regions (principally those bounded by conics) by the *exhaustion* method. The clearest description of this method was given by Archimedes. Essentially, the method is a limiting process in which the area is squeezed between two polygons—one inscribed in the region and one circumscribed about the region.

For instance, in Figure 5.8, the area of a circular region is approximated by an n-sided inscribed polygon and an n-sided circumscribed polygon. For each value of n, the area of the inscribed polygon is less than the area of the circle, and the area of the circumscribed polygon is greater than the area of the circle. Moreover, as n increases, the areas of both polygons become better and better approximations of the area of the circle.

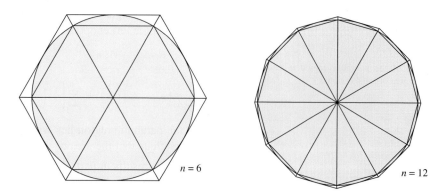

The exhaustion method for finding the area of a circular region
Figure 5.8

A process that is similar to that used by Archimedes to determine the area of a plane region is used in the remaining examples in this section.

The Area of a Plane Region

Recall from Section 2.1 that the origins of calculus are connected to two classic problems: the tangent line problem and the area problem. Example 3 begins the investigation of the area problem.

EXAMPLE 3 Approximating the Area of a Plane Region

Use the five rectangles in Figure 5.9(a) and (b) to find *two* approximations of the area of the region lying between the graph of

$$f(x) = -x^2 + 5$$

and the x-axis between $x = 0$ and $x = 2$.

Solution

a. The right endpoints of the five intervals are $\frac{2}{5}i$, where $i = 1, 2, 3, 4, 5$. The width of each rectangle is $\frac{2}{5}$, and the height of each rectangle can be obtained by evaluating f at the right endpoint of each interval.

$$\left[0, \frac{2}{5}\right], \left[\frac{2}{5}, \frac{4}{5}\right], \left[\frac{4}{5}, \frac{6}{5}\right], \left[\frac{6}{5}, \frac{8}{5}\right], \left[\frac{8}{5}, \frac{10}{5}\right]$$

Evaluate f at the right endpoints of these intervals.

The sum of the areas of the five rectangles is

$$\sum_{i=1}^{5} f\left(\frac{2i}{5}\right)\left(\frac{2}{5}\right) = \sum_{i=1}^{5}\left[-\left(\frac{2i}{5}\right)^2 + 5\right]\left(\frac{2}{5}\right) = \frac{162}{25} = 6.48.$$

Because each of the five rectangles lies inside the parabolic region, you can conclude that the area of the parabolic region is greater than 6.48.

b. The left endpoints of the five intervals are $\frac{2}{5}(i-1)$, where $i = 1, 2, 3, 4, 5$. The width of each rectangle is $\frac{2}{5}$, and the height of each rectangle can be obtained by evaluating f at the left endpoint of each interval.

$$\sum_{i=1}^{5} f\left(\frac{2i-2}{5}\right)\left(\frac{2}{5}\right) = \sum_{i=1}^{5}\left[-\left(\frac{2i-2}{5}\right)^2 + 5\right]\left(\frac{2}{5}\right) = \frac{202}{25} = 8.08.$$

Because the parabolic region lies within the union of the five rectangular regions, you can conclude that the area of the parabolic region is less than 8.08.

By combining the results in parts (a) and (b), you can conclude that

$$6.48 < (\text{Area of region}) < 8.08.$$

NOTE By increasing the number of rectangles used in Example 3, you can obtain closer and closer approximations of the area of the region. For instance, using 25 rectangles of width $\frac{2}{25}$ each, you can conclude that

$$7.17 < (\text{Area of region}) < 7.49.$$

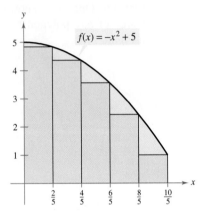

(a) The area of the parabolic region is greater than the area of the rectangles.

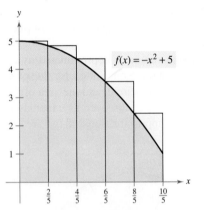

(b) The area of the parabolic region is less than the area of the rectangles.

Figure 5.9

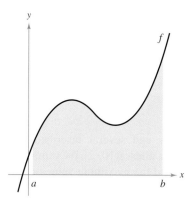

The region under a curve
Figure 5.10

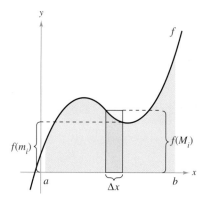

The interval $[a, b]$ is divided into n
subintervals of width $\Delta x = \dfrac{b - a}{n}$.

Figure 5.11

Upper and Lower Sums

The procedure used in Example 3 can be generalized as follows. Consider a plane region bounded above by the graph of a nonnegative, continuous function $y = f(x)$, as shown in Figure 5.10. The region is bounded below by the x-axis, and the left and right boundaries of the region are the vertical lines $x = a$ and $x = b$.

To approximate the area of the region, begin by subdividing the interval $[a, b]$ into n subintervals, each of width

$$\Delta x = (b - a)/n$$

as shown in Figure 5.11. The endpoints of the intervals are as follows.

$$\overbrace{a = x_0}\qquad \overbrace{x_1}\qquad \overbrace{x_2}\qquad\qquad \overbrace{x_n = b}$$
$$a + 0(\Delta x) < a + 1(\Delta x) < a + 2(\Delta x) < \cdots < a + n(\Delta x)$$

Because f is continuous, the Extreme Value Theorem guarantees the existence of a minimum and a maximum value of $f(x)$ in *each* subinterval.

$f(m_i) = $ Minimum value of $f(x)$ in ith subinterval

$f(M_i) = $ Maximum value of $f(x)$ in ith subinterval

Next, define an **inscribed rectangle** lying *inside* the ith subregion and a **circumscribed rectangle** extending *outside* the ith subregion. The height of the ith inscribed rectangle is $f(m_i)$ and the height of the ith circumscribed rectangle is $f(M_i)$. For *each* i, the area of the inscribed rectangle is less than or equal to the area of the circumscribed rectangle.

$$\left(\begin{array}{c}\text{Area of inscribed}\\ \text{rectangle}\end{array}\right) = f(m_i)\,\Delta x \le f(M_i)\,\Delta x = \left(\begin{array}{c}\text{Area of circumscribed}\\ \text{rectangle}\end{array}\right)$$

The sum of the areas of the inscribed rectangles is called a **lower sum,** and the sum of the areas of the circumscribed rectangles is called an **upper sum.**

Lower sum $= s(n) = \displaystyle\sum_{i=1}^{n} f(m_i)\,\Delta x$ Area of inscribed rectangles

Upper sum $= S(n) = \displaystyle\sum_{i=1}^{n} f(M_i)\,\Delta x$ Area of circumscribed rectangles

From Figure 5.12, you can see that the lower sum $s(n)$ is less than or equal to the upper sum $S(n)$. Moreover, the actual area of the region lies between these two sums.

$$s(n) \le (\text{Area of region}) \le S(n)$$

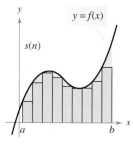

Area of inscribed rectangles is less than area of region.

Area of region

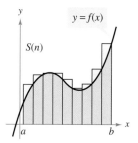

Area of circumscribed rectangles is greater than area of region.

Figure 5.12

Inscribed rectangles

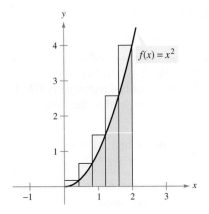

Circumscribed rectangles
Figure 5.13

EXAMPLE 4 **Finding Upper and Lower Sums for a Region**

Find the upper and lower sums for the region bounded by the graph of $f(x) = x^2$ and the x-axis between $x = 0$ and $x = 2$.

Solution To begin, partition the interval $[0, 2]$ into n subintervals, each of width

$$\Delta x = \frac{b - a}{n} = \frac{2 - 0}{n} = \frac{2}{n}.$$

Figure 5.13 shows the endpoints of the subintervals and several inscribed and circumscribed rectangles. Because f is increasing on the interval $[0, 2]$, the minimum value on each subinterval occurs at the left endpoint, and the maximum value occurs at the right endpoint.

Left Endpoints

$$m_i = 0 + (i - 1)\left(\frac{2}{n}\right) = \frac{2(i - 1)}{n}$$

Right Endpoints

$$M_i = 0 + i\left(\frac{2}{n}\right) = \frac{2i}{n}$$

Using the left endpoints, the lower sum is

$$
\begin{aligned}
s(n) &= \sum_{i=1}^{n} f(m_i) \, \Delta x = \sum_{i=1}^{n} f\left[\frac{2(i - 1)}{n}\right]\left(\frac{2}{n}\right) \\
&= \sum_{i=1}^{n} \left[\frac{2(i - 1)}{n}\right]^2 \left(\frac{2}{n}\right) \\
&= \sum_{i=1}^{n} \left(\frac{8}{n^3}\right)(i^2 - 2i + 1) \\
&= \frac{8}{n^3}\left(\sum_{i=1}^{n} i^2 - 2\sum_{i=1}^{n} i + \sum_{i=1}^{n} 1\right) \\
&= \frac{8}{n^3}\left\{\frac{n(n + 1)(2n + 1)}{6} - 2\left[\frac{n(n + 1)}{2}\right] + n\right\} \\
&= \frac{4}{3n^3}(2n^3 - 3n^2 + n) \\
&= \frac{8}{3} - \frac{4}{n} + \frac{4}{3n^2}. \qquad \text{Lower sum}
\end{aligned}
$$

Using the right endpoints, the upper sum is

$$
\begin{aligned}
S(n) &= \sum_{i=1}^{n} f(M_i) \, \Delta x = \sum_{i=1}^{n} f\left(\frac{2i}{n}\right)\left(\frac{2}{n}\right) \\
&= \sum_{i=1}^{n} \left(\frac{2i}{n}\right)^2 \left(\frac{2}{n}\right) \\
&= \sum_{i=1}^{n} \left(\frac{8}{n^3}\right) i^2 \\
&= \frac{8}{n^3}\left[\frac{n(n + 1)(2n + 1)}{6}\right] \\
&= \frac{4}{3n^3}(2n^3 + 3n^2 + n) \\
&= \frac{8}{3} + \frac{4}{n} + \frac{4}{3n^2}. \qquad \text{Upper sum}
\end{aligned}
$$

For the region given in Example 4, evaluate the lower sum

$$s(n) = \frac{8}{3} - \frac{4}{n} + \frac{4}{3n^2}$$

and the upper sum

$$S(n) = \frac{8}{3} + \frac{4}{n} + \frac{4}{3n^2}$$

for $n = 10$, 100, and 1000. Use your results to determine the area of the region.

Example 4 illustrates some important things about lower and upper sums. First, notice that for any value of n, the lower sum is less than (or equal to) the upper sum.

$$s(n) = \frac{8}{3} - \frac{4}{n} + \frac{4}{3n^2} < \frac{8}{3} + \frac{4}{n} + \frac{4}{3n^2} = S(n)$$

Second, the difference between these two sums lessens as n increases. In fact, if you take the limits as $n \to \infty$, both the upper sum and the lower sum approach $\frac{8}{3}$.

$$\lim_{n \to \infty} s(n) = \lim_{n \to \infty} \left(\frac{8}{3} - \frac{4}{n} + \frac{4}{3n^2} \right) = \frac{8}{3} \qquad \text{Lower sum limit}$$

$$\lim_{n \to \infty} S(n) = \lim_{n \to \infty} \left(\frac{8}{3} + \frac{4}{n} + \frac{4}{3n^2} \right) = \frac{8}{3} \qquad \text{Upper sum limit}$$

The next theorem shows that the equivalence of the limits (as $n \to \infty$) of the upper and lower sums is not mere coincidence. It is true for all functions that are continuous and nonnegative on the closed interval $[a, b]$. The proof of this theorem is best left to a course in advanced calculus.

THEOREM 5.3   Limits of the Lower and Upper Sums

Let f be continuous and nonnegative on the interval $[a, b]$. The limits as $n \to \infty$ of both the lower and upper sums exist and are equal to each other. That is,

$$\lim_{n \to \infty} s(n) = \lim_{n \to \infty} \sum_{i=1}^{n} f(m_i) \, \Delta x$$

$$= \lim_{n \to \infty} \sum_{i=1}^{n} f(M_i) \, \Delta x$$

$$= \lim_{n \to \infty} S(n)$$

where $\Delta x = (b - a)/n$ and $f(m_i)$ and $f(M_i)$ are the minimum and maximum values of f on the subinterval.

Because the same limit is attained for both the minimum value $f(m_i)$ and the maximum value $f(M_i)$, it follows from the Squeeze Theorem (Theorem 2.8) that the choice of x in the ith subinterval does not affect the limit. This means that you are free to choose an *arbitrary* x-value in the ith subinterval, as in the following *definition of the area of a region in the plane.*

Definition of the Area of a Region in the Plane

Let f be continuous and nonnegative on the interval $[a, b]$. The area of the region bounded by the graph of f, the x-axis, and the vertical lines $x = a$ and $x = b$ is

$$\text{Area} = \lim_{n \to \infty} \sum_{i=1}^{n} f(c_i) \, \Delta x, \qquad x_{i-1} \le c_i \le x_i$$

where $\Delta x = (b - a)/n$ (see Figure 5.14).

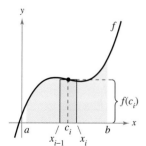

The width of the ith subinterval is $\Delta x = x_i - x_{i-1}$.

Figure 5.14

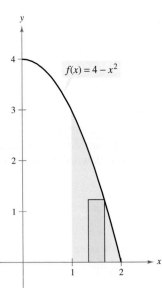

The area of the region bounded by the graph of f, the x-axis, $x = 0$, and $x = 1$ is $\frac{1}{4}$.

Figure 5.15

EXAMPLE 5 Finding Area by the Limit Definition

Find the area of the region bounded by the graph $f(x) = x^3$, the x-axis, and the vertical lines $x = 0$ and $x = 1$, as shown in Figure 5.15.

Solution Begin by noting that f is continuous and nonnegative on the interval $[0, 1]$. Next, partition the interval $[0, 1]$ into n subintervals, each of width $\Delta x = 1/n$. According to the definition of area, you can choose any x-value in the ith subinterval. For this example, the right endpoints $c_i = i/n$ are convenient.

$$\text{Area} = \lim_{n \to \infty} \sum_{i=1}^{n} f(c_i)\, \Delta x = \lim_{n \to \infty} \sum_{i=1}^{n} \left(\frac{i}{n}\right)^3 \left(\frac{1}{n}\right) \qquad \text{Right endpoints: } c_i = \frac{i}{n}$$

$$= \lim_{n \to \infty} \frac{1}{n^4} \sum_{i=1}^{n} i^3$$

$$= \lim_{n \to \infty} \frac{1}{n^4} \left[\frac{n^2(n+1)^2}{4}\right]$$

$$= \lim_{n \to \infty} \left(\frac{1}{4} + \frac{1}{2n} + \frac{1}{4n^2}\right)$$

$$= \frac{1}{4}$$

The area of the region is $\frac{1}{4}$.

EXAMPLE 6 Finding Area by the Limit Definition

Find the area of the region bounded by the graph of $f(x) = 4 - x^2$, the x-axis, and the vertical lines $x = 1$ and $x = 2$, as shown in Figure 5.16.

Solution The function f is continuous and nonnegative on the interval $[1, 2]$, so begin by partitioning the interval into n subintervals, each of width $\Delta x = 1/n$. Choosing the right endpoint

$$c_i = a + i\Delta x = 1 + \frac{i}{n} \qquad \text{Right endpoints}$$

of each subinterval, you obtain

$$\text{Area} = \lim_{n \to \infty} \sum_{i=1}^{n} f(c_i)\, \Delta x = \lim_{n \to \infty} \sum_{i=1}^{n} \left[4 - \left(1 + \frac{i}{n}\right)^2\right]\left(\frac{1}{n}\right)$$

$$= \lim_{n \to \infty} \sum_{i=1}^{n} \left(3 - \frac{2i}{n} - \frac{i^2}{n^2}\right)\left(\frac{1}{n}\right)$$

$$= \lim_{n \to \infty} \left(\frac{1}{n}\sum_{i=1}^{n} 3 - \frac{2}{n^2}\sum_{i=1}^{n} i - \frac{1}{n^3}\sum_{i=1}^{n} i^2\right)$$

$$= \lim_{n \to \infty} \left[3 - \left(1 + \frac{1}{n}\right) - \left(\frac{1}{3} + \frac{1}{2n} + \frac{1}{6n^2}\right)\right]$$

$$= 3 - 1 - \frac{1}{3}$$

$$= \frac{5}{3}.$$

The area of the region bounded by the graph of f, the x-axis, $x = 1$, and $x = 2$ is $\frac{5}{3}$.

Figure 5.16

The area of the region is $\frac{5}{3}$.

The last example in this section looks at a region that is bounded by the y-axis (rather than by the x-axis).

EXAMPLE 7   A Region Bounded by the y-axis

Find the area of the region bounded by the graph of $f(y) = y^2$ and the y-axis for $0 \le y \le 1$, as shown in Figure 5.17.

Solution   When f is a continuous, nonnegative function of y, you still can use the same basic procedure shown in Examples 5 and 6. Begin by partitioning the interval $[0, 1]$ into n subintervals, each of width $\Delta y = 1/n$. Then, using the upper endpoints $c_i = i/n$, you obtain

$$\text{Area} = \lim_{n \to \infty} \sum_{i=1}^{n} f(c_i)\,\Delta y = \lim_{n \to \infty} \sum_{i=1}^{n} \left(\frac{i}{n}\right)^2 \left(\frac{1}{n}\right) \qquad \text{Upper endpoints: } c_i = \frac{i}{n}$$

$$= \lim_{n \to \infty} \frac{1}{n^3} \sum_{i=1}^{n} i^2$$

$$= \lim_{n \to \infty} \frac{1}{n^3} \left[\frac{n(n+1)(2n+1)}{6}\right]$$

$$= \lim_{n \to \infty} \left(\frac{1}{3} + \frac{1}{2n} + \frac{1}{6n^2}\right)$$

$$= \frac{1}{3}.$$

The area of the region is $\frac{1}{3}$.

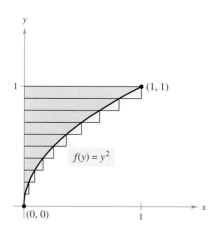

The area of the region bounded by the graph of f and the y-axis for $0 \le y \le 1$ is $\frac{1}{3}$.
Figure 5.17

Exercises for Section 5.2

In Exercises 1–6, find the sum. Use the summation capabilities of a graphing utility to verify your result.

1. $\displaystyle\sum_{i=1}^{5} (2i + 1)$

2. $\displaystyle\sum_{k=3}^{6} k(k - 2)$

3. $\displaystyle\sum_{k=0}^{4} \frac{1}{k^2 + 1}$

4. $\displaystyle\sum_{j=3}^{5} \frac{1}{j}$

5. $\displaystyle\sum_{k=1}^{4} c$

6. $\displaystyle\sum_{i=1}^{4} [(i - 1)^2 + (i + 1)^3]$

In Exercises 7–14, use sigma notation to write the sum.

7. $\dfrac{1}{3(1)} + \dfrac{1}{3(2)} + \dfrac{1}{3(3)} + \cdots + \dfrac{1}{3(9)}$

8. $\dfrac{5}{1 + 1} + \dfrac{5}{1 + 2} + \dfrac{5}{1 + 3} + \cdots + \dfrac{5}{1 + 15}$

9. $\left[5\left(\dfrac{1}{8}\right) + 3\right] + \left[5\left(\dfrac{2}{8}\right) + 3\right] + \cdots + \left[5\left(\dfrac{8}{8}\right) + 3\right]$

10. $\left[1 - \left(\dfrac{1}{4}\right)^2\right] + \left[1 - \left(\dfrac{2}{4}\right)^2\right] + \cdots + \left[1 - \left(\dfrac{4}{4}\right)^2\right]$

11. $\left[\left(\dfrac{2}{n}\right)^3 - \dfrac{2}{n}\right]\left(\dfrac{2}{n}\right) + \cdots + \left[\left(\dfrac{2n}{n}\right)^3 - \dfrac{2n}{n}\right]\left(\dfrac{2}{n}\right)$

12. $\left[1 - \left(\dfrac{2}{n} - 1\right)^2\right]\left(\dfrac{2}{n}\right) + \cdots + \left[1 - \left(\dfrac{2n}{n} - 1\right)^2\right]\left(\dfrac{2}{n}\right)$

13. $\left[2\left(1 + \dfrac{3}{n}\right)^2\right]\left(\dfrac{3}{n}\right) + \cdots + \left[2\left(1 + \dfrac{3n}{n}\right)^2\right]\left(\dfrac{3}{n}\right)$

14. $\left(\dfrac{1}{n}\right)\sqrt{1 - \left(\dfrac{0}{n}\right)^2} + \cdots + \left(\dfrac{1}{n}\right)\sqrt{1 - \left(\dfrac{n-1}{n}\right)^2}$

In Exercises 15–20, use the properties of summation and Theorem 5.2 to evaluate the sum. Use the summation capabilities of a graphing utility to verify your result.

15. $\displaystyle\sum_{i=1}^{20} 2i$

16. $\displaystyle\sum_{i=1}^{15} (2i - 3)$

17. $\displaystyle\sum_{i=1}^{20} (i - 1)^2$

18. $\displaystyle\sum_{i=1}^{10} (i^2 - 1)$

19. $\displaystyle\sum_{i=1}^{15} i(i - 1)^2$

20. $\displaystyle\sum_{i=1}^{10} i(i^2 + 1)$

 In Exercises 21 and 22, use the summation capabilities of a graphing utility to evaluate the sum. Then use the properties of summation and Theorem 5.2 to verify the sum.

21. $\displaystyle\sum_{i=1}^{20} (i^2 + 3)$

22. $\displaystyle\sum_{i=1}^{15} (i^3 - 2i)$

In Exercises 23–26, bound the area of the shaded region by approximating the upper and lower sums. Use rectangles of width 1.

23.

24.

25.

26.

In Exercises 27–30, use upper and lower sums to approximate the area of the region using the given number of subintervals (of equal width).

27. $y = \sqrt{x}$

28. $y = 4e^{-x}$

29. $y = \dfrac{1}{x}$

30. $y = \sqrt{1 - x^2}$

In Exercises 31–34, find the limit of $s(n)$ as $n \to \infty$.

31. $s(n) = \dfrac{81}{n^4}\left[\dfrac{n^2(n+1)^2}{4}\right]$

32. $s(n) = \dfrac{64}{n^3}\left[\dfrac{n(n+1)(2n+1)}{6}\right]$

33. $s(n) = \dfrac{18}{n^2}\left[\dfrac{n(n+1)}{2}\right]$

34. $s(n) = \dfrac{1}{n^2}\left[\dfrac{n(n+1)}{2}\right]$

In Exercises 35–38, use the summation formulas to rewrite the expression without the summation notation. Use the result to find the sum for $n = 10, 100, 1000,$ and $10,000$.

35. $\displaystyle\sum_{i=1}^{n} \dfrac{2i+1}{n^2}$

36. $\displaystyle\sum_{j=1}^{n} \dfrac{4j+1}{n^2}$

37. $\displaystyle\sum_{k=1}^{n} \dfrac{6k(k-1)}{n^3}$

38. $\displaystyle\sum_{i=1}^{n} \dfrac{4i^2(i-1)}{n^4}$

In Exercises 39–44, find a formula for the sum of n terms. Use the formula to find the limit as $n \to \infty$.

39. $\displaystyle\lim_{n\to\infty} \sum_{i=1}^{n} \dfrac{16i}{n^2}$

40. $\displaystyle\lim_{n\to\infty} \sum_{i=1}^{n} \left(\dfrac{2i}{n}\right)\left(\dfrac{2}{n}\right)$

41. $\displaystyle\lim_{n\to\infty} \sum_{i=1}^{n} \dfrac{1}{n^3}(i-1)^2$

42. $\displaystyle\lim_{n\to\infty} \sum_{i=1}^{n} \left(1 + \dfrac{2i}{n}\right)^2\left(\dfrac{2}{n}\right)$

43. $\displaystyle\lim_{n\to\infty} \sum_{i=1}^{n} \left(1 + \dfrac{i}{n}\right)\left(\dfrac{2}{n}\right)$

44. $\displaystyle\lim_{n\to\infty} \sum_{i=1}^{n} \left(1 + \dfrac{2i}{n}\right)^3\left(\dfrac{2}{n}\right)$

45. *Numerical Reasoning* Consider a triangle of area 2 bounded by the graphs of $y = x$, $y = 0$, and $x = 2$.

(a) Sketch the region.

(b) Divide the interval $[0, 2]$ into n subintervals of equal width and show that the endpoints are

$$0 < 1\left(\dfrac{2}{n}\right) < \cdots < (n-1)\left(\dfrac{2}{n}\right) < n\left(\dfrac{2}{n}\right).$$

(c) Show that $s(n) = \displaystyle\sum_{i=1}^{n}\left[(i-1)\left(\dfrac{2}{n}\right)\right]\left(\dfrac{2}{n}\right)$.

(d) Show that $S(n) = \displaystyle\sum_{i=1}^{n}\left[i\left(\dfrac{2}{n}\right)\right]\left(\dfrac{2}{n}\right)$.

(e) Complete the table.

n	5	10	50	100
$s(n)$				
$S(n)$				

(f) Show that $\displaystyle\lim_{n\to\infty} s(n) = \lim_{n\to\infty} S(n) = 2$.

46. *Numerical Reasoning* Consider a trapezoid of area 4 bounded by the graphs of $y = x$, $y = 0$, $x = 1$, and $x = 3$.

(a) Sketch the region.

(b) Divide the interval $[1, 3]$ into n subintervals of equal width and show that the endpoints are

$$1 < 1 + 1\left(\dfrac{2}{n}\right) < \cdots < 1 + (n-1)\left(\dfrac{2}{n}\right) < 1 + n\left(\dfrac{2}{n}\right).$$

(c) Show that $s(n) = \displaystyle\sum_{i=1}^{n}\left[1 + (i-1)\left(\dfrac{2}{n}\right)\right]\left(\dfrac{2}{n}\right)$.

(d) Show that $S(n) = \displaystyle\sum_{i=1}^{n}\left[1 + i\left(\dfrac{2}{n}\right)\right]\left(\dfrac{2}{n}\right)$.

(e) Complete the table.

n	5	10	50	100
$s(n)$				
$S(n)$				

(f) Show that $\displaystyle\lim_{n\to\infty} s(n) = \lim_{n\to\infty} S(n) = 4$.

In Exercises 47–56, use the limit process to find the area of the region between the graph of the function and the *x*-axis over the given interval. Sketch the region.

47. $y = -2x + 3$, $[0, 1]$

48. $y = 3x - 4$, $[2, 5]$

49. $y = x^2 + 2$, $[0, 1]$

50. $y = x^2 + 1$, $[0, 3]$

51. $y = 16 - x^2$, $[1, 3]$

52. $y = 1 - x^2$, $[-1, 1]$

53. $y = 64 - x^3$, $[1, 4]$

54. $y = 2x - x^3$, $[0, 1]$

55. $y = x^2 - x^3$, $[-1, 1]$

56. $y = x^2 - x^3$, $[-1, 0]$

In Exercises 57–62, use the limit process to find the area of the region between the graph of the function and the *y*-axis over the given *y*-interval. Sketch the region.

57. $f(y) = 3y, 0 \le y \le 2$

58. $g(y) = \frac{1}{2}y, 2 \le y \le 4$

59. $f(y) = y^2, 0 \le y \le 3$

60. $f(y) = 4y - y^2, 1 \le y \le 2$

61. $g(y) = 4y^2 - y^3, 1 \le y \le 3$

62. $h(y) = y^3 + 1, 1 \le y \le 2$

In Exercises 63–66, use the *Midpoint Rule*

$$\text{Area} \approx \sum_{i=1}^{n} f\left(\frac{x_i + x_{i-1}}{2}\right)\Delta x$$

with $n = 4$ to approximate the area of the region bounded by the graph of the function and the *x*-axis over the given interval.

63. $f(x) = x^2 + 3$, $[0, 2]$

64. $f(x) = x^2 + 4x$, $[0, 4]$

65. $f(x) = \tan x$, $\left[0, \frac{\pi}{4}\right]$

66. $f(x) = \sin x$, $\left[0, \frac{\pi}{2}\right]$

Programming Write a program for a graphing utility to approximate areas by using the Midpoint Rule. Assume that the function is positive over the given interval and the subintervals are of equal width. In Exercises 67–72, use the program to approximate the area of the region between the graph of the function and the *x*-axis over the given interval, and complete the table.

n	4	8	12	16	20
Approximate Area					

67. $f(x) = \sqrt{x}$, $[0, 4]$

68. $f(x) = \dfrac{8}{x^2 + 1}$, $[2, 6]$

69. $f(x) = \tan\left(\dfrac{\pi x}{8}\right)$, $[1, 3]$

70. $f(x) = \cos \sqrt{x}$, $[0, 2]$

71. $f(x) = \ln x$, $[1, 5]$

72. $f(x) = xe^x$, $[0, 2]$

Writing About Concepts

Approximation In Exercises 73 and 74, determine which value best approximates the area of the region between the *x*-axis and the graph of the function over the given interval. (Make your selection on the basis of a sketch of the region and not by performing calculations.)

73. $f(x) = 4 - x^2$, $[0, 2]$

(a) -2 (b) 6   (c) 10   (d) 3   (e) 8

Writing About Concepts (continued)

74. $f(x) = \sin\dfrac{\pi x}{4}$, $[0, 4]$

(a) 3   (b) 1   (c) -2 (d) 8   (e) 6

75. In your own words and using appropriate figures, describe the methods of upper sums and lower sums in approximating the area of a region.

76. Give the definition of the area of a region in the plane.

77. *Graphical Reasoning* Consider the region bounded by the graphs of

$$f(x) = \frac{8x}{x + 1},$$

$x = 0$, $x = 4$,   and $y = 0$,   as shown in the figure. To print an enlarged copy of the graph, go to the website *www.mathgraphs.com*.

(a) Redraw the figure, and complete and shade the rectangles representing the lower sum when $n = 4$. Find this lower sum.

(b) Redraw the figure, and complete and shade the rectangles representing the upper sum when $n = 4$. Find this upper sum.

(c) Redraw the figure, and complete and shade the rectangles whose heights are determined by the functional values at the midpoint of each subinterval when $n = 4$. Find this sum using the Midpoint Rule.

(d) Verify the following formulas for approximating the area of the region using n subintervals of equal width.

Lower sum: $s(n) = \displaystyle\sum_{i=1}^{n} f\left[(i - 1)\frac{4}{n}\right]\left(\frac{4}{n}\right)$

Upper sum: $S(n) = \displaystyle\sum_{i=1}^{n} f\left[(i)\frac{4}{n}\right]\left(\frac{4}{n}\right)$

Midpoint Rule: $M(n) = \displaystyle\sum_{i=1}^{n} f\left[\left(i - \frac{1}{2}\right)\frac{4}{n}\right]\left(\frac{4}{n}\right)$

(e) Use a graphing utility and the formulas in part (d) to complete the table.

n	4	8	20	100	200
$s(n)$					
$S(n)$					
$M(n)$					

(f) Explain why $s(n)$ increases and $S(n)$ decreases for increasing values of n, as shown in the table in part (e).

78. *Monte Carlo Method* The following computer program approximates the area of the region under the graph of a monotonic function and above the x-axis between $x = a$ and $x = b$. Run the program for $a = 0$ and $b = \pi/2$ for several values of N2. Explain why the Monte Carlo Method works. [*Adaptation of Monte Carlo Method program from James M. Sconyers, "Approximation of Area Under a Curve," MATHEMATICS TEACHER 77, no. 2 (February 1984). Copyright © 1984 by the National Council of Teachers of Mathematics. Reprinted with permission.*]

```
10   DEF FNF(X)=SIN(X)
20   A=0
30   B=π/2
40   PRINT "Input Number of Random Points"
50   INPUT N2
60   N1=0
70   IF  FNF(A)>FNF(B)  THEN  YMAX=FNF(A)  ELSE
     YMAX=FNF(B)
80   FOR I=1 TO N2
90   X=A+(B-A)*RND(1)
100  Y=YMAX*RND(1)
110  IF Y>=FNF(X) THEN GOTO 130
120  N1=N1+1
130  NEXT I
140  AREA=(N1/N2)*(B-A)*YMAX
150  PRINT "Approximate Area:"; AREA
160  END
```

True or False? **In Exercises 79 and 80, determine whether the statement is true or false. If it is false, explain why or give an example that shows it is false.**

79. The sum of the first n positive integers is $n(n + 1)/2$.

80. If f is continuous and nonnegative on $[a, b]$, then the limits as $n \to \infty$ of its lower sum $s(n)$ and upper sum $S(n)$ both exist and are equal.

81. *Writing* Use the figure to write a short paragraph explaining why the formula $1 + 2 + \cdots + n = \frac{1}{2}n(n + 1)$ is valid for all positive integers n.

Figure for 81

Figure for 82

82. *Graphical Reasoning* Consider an n-sided regular polygon inscribed in a circle of radius r. Join the vertices of the polygon to the center of the circle, forming n congruent triangles (see figure).

(a) Determine the central angle θ in terms of n.

(b) Show that the area of each triangle is $\frac{1}{2}r^2 \sin \theta$.

(c) Let A_n be the sum of the areas of the n triangles. Find $\lim\limits_{n \to \infty} A_n$.

83. *Modeling Data* The table lists the measurements of a lot bounded by a stream and two straight roads that meet at right angles, where x and y are measured in feet (see figure).

x	0	50	100	150	200	250	300
y	450	362	305	268	245	156	0

(a) Use the regression capabilities of a graphing utility to find a model of the form $y = ax^3 + bx^2 + cx + d$.

(b) Use a graphing utility to plot the data and graph the model.

(c) Use the model in part (a) to estimate the area of the lot.

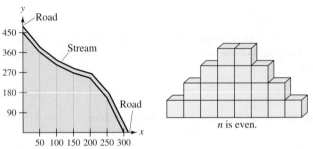

Figure for 83   **Figure for 84**

84. *Building Blocks* A child places n cubic building blocks in a row to form the base of a triangular design (see figure). Each successive row contains two fewer blocks than the preceding row. Find a formula for the number of blocks used in the design. (*Hint:* The number of building blocks in the design depends on whether n is odd or even.)

85. Prove each formula by mathematical induction. (You may need to review the method of proof by induction from a precalculus text.)

(a) $\displaystyle\sum_{i=1}^{n} 2i = n(n + 1)$

(b) $\displaystyle\sum_{i=1}^{n} i^3 = \frac{n^2(n + 1)^2}{4}$

Putnam Exam Challenge

86. A dart, thrown at random, hits a square target. Assuming that any two parts of the target of equal area are equally likely to be hit, find the probability that the point hit is nearer to the center than to any edge. Write your answer in the form $\left(a\sqrt{b} + c\right)/d$, where a, b, c, and d are positive integers.

This problem was composed by the Committee on the Putnam Prize Competition.
© The Mathematical Association of America. All rights reserved.

Riemann Sums and Definite Integrals

- Understand the definition of a Riemann sum.
- Evaluate a definite integral using limits.
- Evaluate a definite integral using properties of definite integrals.

Riemann Sums

In the definition of area given in Section 5.2, the partitions have subintervals of *equal width*. This was done only for computational convenience. The following example shows that it is not necessary to have subintervals of equal width.

EXAMPLE 1   A Partition with Subintervals of Unequal Widths

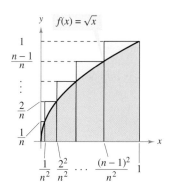

The subintervals do not have equal widths.
Figure 5.18

Consider the region bounded by the graph of $f(x) = \sqrt{x}$ and the x-axis for $0 \le x \le 1$, as shown in Figure 5.18. Evaluate the limit

$$\lim_{n \to \infty} \sum_{i=1}^{n} f(c_i) \, \Delta x_i$$

where c_i is the right endpoint of the partition given by $c_i = i^2/n^2$ and Δx_i is the width of the ith interval.

Solution   The width of the ith interval is given by

$$\Delta x_i = \frac{i^2}{n^2} - \frac{(i-1)^2}{n^2}$$

$$= \frac{i^2 - i^2 + 2i - 1}{n^2}$$

$$= \frac{2i - 1}{n^2}.$$

So, the limit is

$$\lim_{n \to \infty} \sum_{i=1}^{n} f(c_i) \, \Delta x_i = \lim_{n \to \infty} \sum_{i=1}^{n} \sqrt{\frac{i^2}{n^2}} \left(\frac{2i - 1}{n^2} \right)$$

$$= \lim_{n \to \infty} \frac{1}{n^3} \sum_{i=1}^{n} (2i^2 - i)$$

$$= \lim_{n \to \infty} \frac{1}{n^3} \left[2 \left(\frac{n(n+1)(2n+1)}{6} \right) - \frac{n(n+1)}{2} \right]$$

$$= \lim_{n \to \infty} \frac{4n^3 + 3n^2 - n}{6n^3}$$

$$= \frac{2}{3}.$$

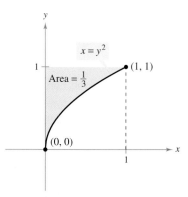

The area of the region bounded by the graph of $x = y^2$ and the y-axis for $0 \le y \le 1$ is $\frac{1}{3}$.
Figure 5.19

From Example 7 in Section 5.2, you know that the region shown in Figure 5.19 has an area of $\frac{1}{3}$. Because the square bounded by $0 \le x \le 1$ and $0 \le y \le 1$ has an area of 1, you can conclude that the area of the region shown in Figure 5.18 has an area of $\frac{2}{3}$. This agrees with the limit found in Example 1, even though that example used a partition having subintervals of unequal widths. The reason this particular partition gave the proper area is that as n increases, the *width of the largest subinterval approaches zero*. This is a key feature of the development of definite integrals.

GEORG FRIEDRICH BERNHARD RIEMANN (1826–1866)

German mathematician Riemann did his most famous work in the areas of non-Euclidean geometry, differential equations, and number theory. It was Riemann's results in physics and mathematics that formed the structure on which Einstein's theory of general relativity is based.

In the preceding section, the limit of a sum was used to define the area of a region in the plane. Finding area by this means is only one of *many* applications involving the limit of a sum. A similar approach can be used to determine quantities as diverse as arc lengths, average values, centroids, volumes, work, and surface areas. The following definition is named after Georg Friedrich Bernhard Riemann. Although the definite integral had been defined and used long before the time of Riemann, he generalized the concept to cover a broader category of functions.

In the following definition of a Riemann sum, note that the function f has no restrictions other than being defined on the interval $[a, b]$. (In the preceding section, the function f was assumed to be continuous and nonnegative because we were dealing with the area under a curve.)

Definition of a Riemann Sum

Let f be defined on the closed interval $[a, b]$, and let Δ be a partition of $[a, b]$ given by

$$a = x_0 < x_1 < x_2 < \cdots < x_{n-1} < x_n = b$$

where Δx_i is the width of the ith subinterval $[x_{i-1}, x_i]$. If c_i is *any* point in the ith subinterval, then the sum

$$\sum_{i=1}^{n} f(c_i)\, \Delta x_i, \qquad x_{i-1} \le c_i \le x_i$$

is called a **Riemann sum** of f for the partition Δ.

NOTE   The sums in Section 5.2 are examples of Riemann sums, but there are more general Riemann sums than those covered there.

The width of the largest subinterval of a partition Δ is the **norm** of the partition and is denoted by $\|\Delta\|$. If every subinterval is of equal width, the partition is **regular** and the norm is denoted by

$$\|\Delta\| = \Delta x = \frac{b - a}{n}. \qquad \text{Regular partition}$$

For a general partition, the norm is related to the number of subintervals of $[a, b]$ in the following way.

$$\frac{b - a}{\|\Delta\|} \le n \qquad \text{General partition}$$

So, the number of subintervals in a partition approaches infinity as the norm of the partition approaches 0. That is, $\|\Delta\| \to 0$ implies that $n \to \infty$.

The converse of this statement is not true. For example, let Δ_n be the partition of the interval $[0, 1]$ given by

$$0 < \frac{1}{2^n} < \frac{1}{2^{n-1}} < \cdots < \frac{1}{8} < \frac{1}{4} < \frac{1}{2} < 1.$$

As shown in Figure 5.20, for any positive value of n, the norm of the partition Δ_n is $\frac{1}{2}$. So, letting n approach infinity does not force $\|\Delta\|$ to approach 0. In a regular partition, however, the statements $\|\Delta\| \to 0$ and $n \to \infty$ are equivalent.

$\|\Delta\| = \frac{1}{2}$

$n \to \infty$ does not imply that $\|\Delta\| \to 0$.

Figure 5.20

Definite Integrals

To define the definite integral, consider the following limit.

$$\lim_{\|\Delta\| \to 0} \sum_{i=1}^{n} f(c_i) \, \Delta x_i = L$$

To say that this limit exists means that for $\varepsilon > 0$ there exists a $\delta > 0$ such that for every partition with $\|\Delta\| < \delta$ it follows that

$$\left| L - \sum_{i=1}^{n} f(c_i) \, \Delta x_i \right| < \varepsilon.$$

(This must be true for any choice of c_i in the ith subinterval of Δ.)

FOR FURTHER INFORMATION For insight into the history of the definite integral, see the article "The Evolution of Integration" by A. Shenitzer and J. Steprāns in The *American Mathematical Monthly*. To view this article, go to the website *www.matharticles.com*.

Definition of a Definite Integral

If f is defined on the closed interval $[a, b]$ and the limit

$$\lim_{\|\Delta\| \to 0} \sum_{i=1}^{n} f(c_i) \, \Delta x_i$$

exists (as described above), then f is **integrable** on $[a, b]$ and the limit is denoted by

$$\lim_{\|\Delta\| \to 0} \sum_{i=1}^{n} f(c_i) \, \Delta x_i = \int_{a}^{b} f(x) \, dx.$$

The limit is called the **definite integral** of f from a to b. The number a is the **lower limit** of integration, and the number b is the **upper limit** of integration.

It is not a coincidence that the notation used for definite integrals is similar to that used for indefinite integrals. You will see why in the next section when the Fundamental Theorem of Calculus is introduced. For now it is important to see that definite integrals and indefinite integrals are different identities. A definite integral is a *number*, whereas an indefinite integral is a *family of functions*.

A sufficient condition for a function f to be integrable on $[a, b]$ is that it is continuous on $[a, b]$. A proof of this theorem is beyond the scope of this text.

THEOREM 5.4   Continuity Implies Integrability

If a function f is continuous on the closed interval $[a, b]$, then f is integrable on $[a, b]$.

EXPLORATION

The Converse of Theorem 5.4   Is the converse of Theorem 5.4 true? That is, if a function is integrable, does it have to be continuous? Explain your reasoning and give examples.

Describe the relationships among continuity, differentiability, and integrability. Which is the strongest condition? Which is the weakest? Which conditions imply other conditions?

EXAMPLE 2   Evaluating a Definite Integral as a Limit

Evaluate the definite integral $\int_{-2}^{1} 2x\, dx$.

Solution   The function $f(x) = 2x$ is integrable on the interval $[-2, 1]$ because it is continuous on $[-2, 1]$. Moreover, the definition of integrability implies that any partition whose norm approaches 0 can be used to determine the limit. For computational convenience, define Δ by subdividing $[-2, 1]$ into n subintervals of equal width

$$\Delta x_i = \Delta x = \frac{b - a}{n} = \frac{3}{n}.$$

Choosing c_i as the right endpoint of each subinterval produces

$$c_i = a + i(\Delta x) = -2 + \frac{3i}{n}.$$

So, the definite integral is given by

$$\int_{-2}^{1} 2x\, dx = \lim_{\|\Delta\| \to 0} \sum_{i=1}^{n} f(c_i)\, \Delta x_i$$

$$= \lim_{n \to \infty} \sum_{i=1}^{n} f(c_i)\, \Delta x$$

$$= \lim_{n \to \infty} \sum_{i=1}^{n} 2\left(-2 + \frac{3i}{n}\right)\left(\frac{3}{n}\right)$$

$$= \lim_{n \to \infty} \frac{6}{n} \sum_{i=1}^{n} \left(-2 + \frac{3i}{n}\right)$$

$$= \lim_{n \to \infty} \frac{6}{n} \left\{-2n + \frac{3}{n}\left[\frac{n(n + 1)}{2}\right]\right\}$$

$$= \lim_{n \to \infty} \left(-12 + 9 + \frac{9}{n}\right)$$

$$= -3.$$

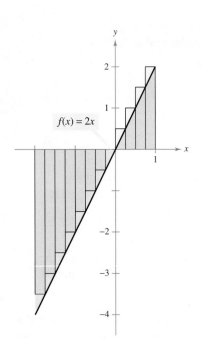

Because the definite integral is negative, it does not represent the area of the region.
Figure 5.21

Because the definite integral in Example 2 is negative, it *cannot* represent the area of the region shown in Figure 5.21. Definite integrals can be positive, negative, or zero. For a definite integral to be interpreted as an area (as defined in Section 5.2), the function f must be continuous and nonnegative on $[a, b]$, as stated in the following theorem. (The proof of this theorem is straightforward—you simply use the definition of area given in Section 5.2.)

THEOREM 5.5   The Definite Integral as the Area of a Region

If f is continuous and nonnegative on the closed interval $[a, b]$, then the area of the region bounded by the graph of f, the x-axis, and the vertical lines $x = a$ and $x = b$ is given by

$$\text{Area} = \int_{a}^{b} f(x)\, dx.$$

(See Figure 5.22.)

You can use a definite integral to find the area of the region bounded by the graph of f, the x-axis, $x = a$, and $x = b$.
Figure 5.22

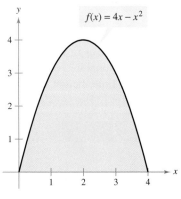

$$\text{Area} = \int_0^4 (4x - x^2)\, dx$$

Figure 5.23

As an example of Theorem 5.5, consider the region bounded by the graph of

$$f(x) = 4x - x^2$$

and the x-axis, as shown in Figure 5.23. Because f is continuous and nonnegative on the closed interval $[0, 4]$, the area of the region is

$$\text{Area} = \int_0^4 (4x - x^2)\, dx.$$

A straightforward technique for evaluating a definite integral such as this will be discussed in Section 5.4. For now, however, you can evaluate a definite integral in two ways—you can use the limit definition *or* you can check to see whether the definite integral represents the area of a common geometric region such as a rectangle, triangle, or semicircle.

EXAMPLE 3   Areas of Common Geometric Figures

Sketch the region corresponding to each definite integral. Then evaluate each integral using a geometric formula.

a. $\displaystyle\int_1^3 4\, dx$ **b.** $\displaystyle\int_0^3 (x + 2)\, dx$ **c.** $\displaystyle\int_{-2}^2 \sqrt{4 - x^2}\, dx$

Solution   A sketch of each region is shown in Figure 5.24.

a. This region is a rectangle of height 4 and width 2.

$$\int_1^3 4\, dx = (\text{Area of rectangle}) = 4(2) = 8$$

b. This region is a trapezoid with an altitude of 3 and parallel bases of lengths 2 and 5. The formula for the area of a trapezoid is $\frac{1}{2}h(b_1 + b_2)$.

$$\int_0^3 (x + 2)\, dx = (\text{Area of trapezoid}) = \frac{1}{2}(3)(2 + 5) = \frac{21}{2}$$

c. This region is a semicircle of radius 2. The formula for the area of a semicircle is $\frac{1}{2}\pi r^2$.

$$\int_{-2}^2 \sqrt{4 - x^2}\, dx = (\text{Area of semicircle}) = \frac{1}{2}\pi(2^2) = 2\pi$$

NOTE   The variable of integration in a definite integral is sometimes called a *dummy variable* because it can be replaced by any other variable without changing the value of the integral. For instance, the definite integrals

$$\int_0^3 (x + 2)\, dx$$

and

$$\int_0^3 (t + 2)\, dt$$

have the same value.

(a)

(b)

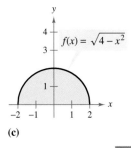

(c)

Figure 5.24

Properties of Definite Integrals

The definition of the definite integral of f on the interval $[a, b]$ specifies that $a < b$. Now, however, it is convenient to extend the definition to cover cases in which $a = b$ or $a > b$. Geometrically, the following two definitions seem reasonable. For instance, it makes sense to define the area of a region of zero width and finite height to be 0.

Definitions of Two Special Definite Integrals

1. If f is defined at $x = a$, then we define $\displaystyle\int_a^a f(x)\, dx = 0$.

2. If f is integrable on $[a, b]$, then we define $\displaystyle\int_b^a f(x)\, dx = -\int_a^b f(x)\, dx$.

EXAMPLE 4   **Evaluating Definite Integrals**

a. Because the sine function is defined at $x = \pi$, and the upper and lower limits of integration are equal, you can write

$$\int_\pi^\pi \sin x \, dx = 0.$$

b. The integral $\int_3^0 (x + 2)\, dx$ is the same as that given in Example 3(b) except that the upper and lower limits are interchanged. Because the integral in Example 3(b) has a value of $\frac{21}{2}$, you can write

$$\int_3^0 (x + 2)\, dx = -\int_0^3 (x + 2)\, dx = -\frac{21}{2}.$$

In Figure 5.25, the larger region can be divided at $x = c$ into two subregions whose intersection is a line segment. Because the line segment has zero area, it follows that the area of the larger region is equal to the sum of the areas of the two smaller regions.

THEOREM 5.6   Additive Interval Property

If f is integrable on the three closed intervals determined by a, b, and c, then

$$\int_a^b f(x)\, dx = \int_a^c f(x)\, dx + \int_c^b f(x)\, dx.$$

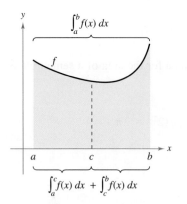

Figure 5.25

EXAMPLE 5   **Using the Additive Interval Property**

$$\int_{-1}^1 |x|\, dx = \int_{-1}^0 -x \, dx + \int_0^1 x \, dx \qquad \text{Theorem 5.6}$$

$$= \frac{1}{2} + \frac{1}{2} \qquad\qquad \text{Area of a triangle}$$

$$= 1$$

Because the definite integral is defined as the limit of a sum, it inherits the properties of summation given at the top of page 296.

THEOREM 5.7   Properties of Definite Integrals

If f and g are integrable on $[a, b]$ and k is a constant, then the functions of kf and $f \pm g$ are integrable on $[a, b]$, and

1. $\displaystyle\int_a^b kf(x)\, dx = k\int_a^b f(x)\, dx$

2. $\displaystyle\int_a^b [f(x) \pm g(x)]\, dx = \int_a^b f(x)\, dx \pm \int_a^b g(x)\, dx.$

Note that Property 2 of Theorem 5.7 can be extended to cover any finite number of functions. For example,

$$\int_a^b [f(x) + g(x) + h(x)]\, dx = \int_a^b f(x)\, dx + \int_a^b g(x)\, dx + \int_a^b h(x)\, dx.$$

EXAMPLE 6   Evaluation of a Definite Integral

Evaluate $\displaystyle\int_1^3 (-x^2 + 4x - 3)\, dx$ using each of the following values.

$$\int_1^3 x^2\, dx = \frac{26}{3}, \qquad \int_1^3 x\, dx = 4, \qquad \int_1^3 dx = 2$$

Solution

$$\int_1^3 (-x^2 + 4x - 3)\, dx = \int_1^3 (-x^2)\, dx + \int_1^3 4x\, dx + \int_1^3 (-3)\, dx$$

$$= -\int_1^3 x^2\, dx + 4\int_1^3 x\, dx - 3\int_1^3 dx$$

$$= -\left(\frac{26}{3}\right) + 4(4) - 3(2)$$

$$= \frac{4}{3}$$

If f and g are continuous on the closed interval $[a, b]$ and

$$0 \le f(x) \le g(x)$$

for $a \le x \le b$, the following properties are true. First, the area of the region bounded by the graph of f and the x-axis (between a and b) must be nonnegative. Second, this area must be less than or equal to the area of the region bounded by the graph of g and the x-axis (between a and b), as shown in Figure 5.26. These two results are generalized in Theorem 5.8. (A proof of this theorem is given in Appendix A.)

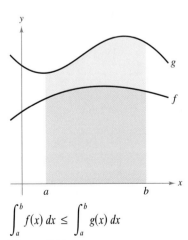

$$\int_a^b f(x)\, dx \le \int_a^b g(x)\, dx$$

Figure 5.26

> **THEOREM 5.8** **Preservation of Inequality**
>
> 1. If f is integrable and nonnegative on the closed interval $[a, b]$, then
>
> $$0 \le \int_a^b f(x)\, dx.$$
>
> 2. If f and g are integrable on the closed interval $[a, b]$ and $f(x) \le g(x)$ for every x in $[a, b]$, then
>
> $$\int_a^b f(x)\, dx \le \int_a^b g(x)\, dx.$$

Exercises for Section 5.3

See www.CalcChat.com for worked-out solutions to odd-numbered exercises.

In Exercises 1 and 2, use Example 1 as a model to evaluate the limit

$$\lim_{n \to \infty} \sum_{i=1}^{n} f(c_i)\, \Delta x_i$$

over the region bounded by the graphs of the equations.

1. $f(x) = \sqrt{x}, \quad y = 0, \quad x = 0, \quad x = 3$

(*Hint:* Let $c_i = 3i^2/n^2$.)

2. $f(x) = 2\sqrt[3]{x}, \quad y = 0, \quad x = 0, \quad x = 1$

(*Hint:* Let $c_i = i^3/n^3$.)

In Exercises 3–8, evaluate the definite integral by the limit definition.

3. $\displaystyle\int_4^{10} 6\, dx$

4. $\displaystyle\int_{-2}^{3} x\, dx$

5. $\displaystyle\int_{-1}^{1} x^3\, dx$

6. $\displaystyle\int_1^3 3x^2\, dx$

7. $\displaystyle\int_1^2 (x^2 + 1)\, dx$

8. $\displaystyle\int_{-1}^{2} (3x^2 + 2)\, dx$

In Exercises 9–14, write the limit as a definite integral on the interval $[a, b]$, where c_i is any point in the ith subinterval.

Limit	Interval
9. $\displaystyle\lim_{\|\Delta\| \to 0} \sum_{i=1}^{n} (3c_i + 10)\, \Delta x_i$	$[-1, 5]$
10. $\displaystyle\lim_{\|\Delta\| \to 0} \sum_{i=1}^{n} 6c_i(4 - c_i)^2\, \Delta x_i$	$[0, 4]$
11. $\displaystyle\lim_{\|\Delta\| \to 0} \sum_{i=1}^{n} \sqrt{c_i^2 + 4}\, \Delta x_i$	$[0, 3]$
12. $\displaystyle\lim_{\|\Delta\| \to 0} \sum_{i=1}^{n} \left(\frac{3}{c_i^2}\right) \Delta x_i$	$[1, 3]$
13. $\displaystyle\lim_{\|\Delta\| \to 0} \sum_{i=1}^{n} \left(1 + \frac{3}{c_i}\right) \Delta x_i$	$[1, 5]$
14. $\displaystyle\lim_{\|\Delta\| \to 0} \sum_{i=1}^{n} (2^{-c_i} \sin c_i)\, \Delta x_i$	$[0, \pi]$

In Exercises 15–22, set up a definite integral that yields the area of the region. (Do not evaluate the integral.)

15. $f(x) = 3$

16. $f(x) = x^2$

17. $f(x) = \dfrac{2}{x}$

18. $f(x) = 2e^{-x}$

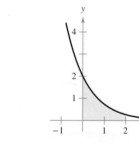

19. $f(x) = \sin x$

20. $f(x) = \tan x$

21. $g(y) = y^3$

22. $f(y) = (y - 2)^2$

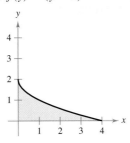

In Exercises 23–32, sketch the region whose area is given by the definite integral. Then use a geometric formula to evaluate the integral ($a > 0, r > 0$).

23. $\displaystyle\int_0^3 4\, dx$

24. $\displaystyle\int_{-a}^a 4\, dx$

25. $\displaystyle\int_0^4 x\, dx$

26. $\displaystyle\int_0^4 \frac{x}{2}\, dx$

27. $\displaystyle\int_0^2 (2x + 5)\, dx$

28. $\displaystyle\int_0^8 (8 - x)\, dx$

29. $\displaystyle\int_{-1}^1 (1 - |x|)\, dx$

30. $\displaystyle\int_{-a}^a (a - |x|)\, dx$

31. $\displaystyle\int_{-3}^3 \sqrt{9 - x^2}\, dx$

32. $\displaystyle\int_{-r}^r \sqrt{r^2 - x^2}\, dx$

In Exercises 33–40, evaluate the integral using the following values.

$$\int_2^4 x^3\, dx = 60, \qquad \int_2^4 x\, dx = 6, \qquad \int_2^4 dx = 2$$

33. $\displaystyle\int_4^2 x\, dx$

34. $\displaystyle\int_2^2 x^3\, dx$

35. $\displaystyle\int_2^4 4x\, dx$

36. $\displaystyle\int_2^4 15\, dx$

37. $\displaystyle\int_2^4 (x - 8)\, dx$

38. $\displaystyle\int_2^4 (x^3 + 4)\, dx$

39. $\displaystyle\int_2^4 \left(\tfrac{1}{2}x^3 - 3x + 2\right) dx$

40. $\displaystyle\int_2^4 (6 + 2x - x^3)\, dx$

41. Given $\displaystyle\int_0^5 f(x)\, dx = 10$ and $\displaystyle\int_5^7 f(x)\, dx = 3$, evaluate

(a) $\displaystyle\int_0^7 f(x)\, dx.$

(b) $\displaystyle\int_5^0 f(x)\, dx.$

(c) $\displaystyle\int_5^5 f(x)\, dx.$

(d) $\displaystyle\int_0^5 3f(x)\, dx.$

42. Given $\displaystyle\int_0^3 f(x)\, dx = 4$ and $\displaystyle\int_3^6 f(x)\, dx = -1$, evaluate

(a) $\displaystyle\int_0^6 f(x)\, dx.$

(b) $\displaystyle\int_6^3 f(x)\, dx.$

(c) $\displaystyle\int_3^3 f(x)\, dx.$

(d) $\displaystyle\int_3^6 -5f(x)\, dx.$

43. Given $\displaystyle\int_2^6 f(x)\, dx = 10$ and $\displaystyle\int_2^6 g(x)\, dx = -2$, evaluate

(a) $\displaystyle\int_2^6 [f(x) + g(x)]\, dx.$

(b) $\displaystyle\int_2^6 [g(x) - f(x)]\, dx.$

(c) $\displaystyle\int_2^6 2g(x)\, dx.$

(d) $\displaystyle\int_2^6 3f(x)\, dx.$

44. Given $\displaystyle\int_{-1}^1 f(x)\, dx = 0$ and $\displaystyle\int_0^1 f(x)\, dx = 5$, evaluate

(a) $\displaystyle\int_{-1}^0 f(x)\, dx.$

(b) $\displaystyle\int_0^1 f(x)\, dx - \int_{-1}^0 f(x)\, dx.$

(c) $\displaystyle\int_{-1}^1 3f(x)\, dx.$

(d) $\displaystyle\int_0^1 3f(x)\, dx.$

45. Use the table of values to find lower and upper estimates of

$$\int_0^{10} f(x)\, dx.$$

Assume that f is a decreasing function.

x	0	2	4	6	8	10
$f(x)$	32	24	12	-4	-20	-36

46. Use the table of values to estimate

$$\int_0^6 f(x)\, dx.$$

Use three equal subintervals and the (a) left endpoints, (b) right endpoints, and (c) midpoints. If f is an increasing function, how does each estimate compare with the actual value? Explain your reasoning.

x	0	1	2	3	4	5	6
$f(x)$	-6	0	8	18	30	50	80

47. *Think About It*   The graph of f consists of line segments and a semicircle, as shown in the figure. Evaluate each definite integral by using geometric formulas.

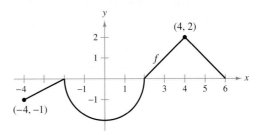

(a) $\displaystyle\int_0^2 f(x)\, dx$

(b) $\displaystyle\int_2^6 f(x)\, dx$

(c) $\displaystyle\int_{-4}^2 f(x)\, dx$

(d) $\displaystyle\int_{-4}^6 f(x)\, dx$

(e) $\displaystyle\int_{-4}^6 |f(x)|\, dx$

(f) $\displaystyle\int_{-4}^6 [f(x) + 2]\, dx$

48. ***Think About It***   The graph of f consists of line segments, as shown in the figure. Evaluate each definite integral by using geometric formulas.

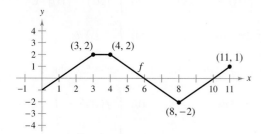

(a) $\displaystyle\int_0^1 -f(x)\,dx$

(b) $\displaystyle\int_3^4 3f(x)\,dx$

(c) $\displaystyle\int_0^7 f(x)\,dx$

(d) $\displaystyle\int_5^{11} f(x)\,dx$

(e) $\displaystyle\int_0^{11} f(x)\,dx$

(f) $\displaystyle\int_4^{10} f(x)\,dx$

49. ***Think About It***   Consider the function f that is continuous on the interval $[-5, 5]$ and for which

$$\int_0^5 f(x)\,dx = 4.$$

Evaluate each integral.

(a) $\displaystyle\int_0^5 [f(x) + 2]\,dx$

(b) $\displaystyle\int_{-2}^3 f(x + 2)\,dx$

(c) $\displaystyle\int_{-5}^5 f(x)\,dx$ (f is even.)

(d) $\displaystyle\int_{-5}^5 f(x)\,dx$ (f is odd.)

50. ***Think About It***   A function f is defined below. Use geometric formulas to find $\int_0^8 f(x)\,dx$.

$$f(x) = \begin{cases} 4, & x < 4 \\ x, & x \geq 4 \end{cases}$$

Writing About Concepts

In Exercises 51 and 52, use the figure to fill in the blank with the symbol <, >, or =.

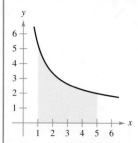

51. The interval $[1, 5]$ is partitioned into n subintervals of equal width Δx, and x_i is the left endpoint of the ith subinterval.

$$\sum_{i=1}^n f(x_i)\,\Delta x \quad\rule{1.5em}{0.8em}\quad \int_1^5 f(x)\,dx$$

Writing About Concepts   *(continued)*

52. The interval $[1, 5]$ is partitioned into n subintervals of equal width Δx, and x_i is the right endpoint of the ith subinterval.

$$\sum_{i=1}^n f(x_i)\,\Delta x \quad\rule{1.5em}{0.8em}\quad \int_1^5 f(x)\,dx$$

53. Determine whether the function $f(x) = \dfrac{1}{x - 4}$ is integrable on the interval $[3, 5]$. Explain.

54. Give an example of a function that is integrable on the interval $[-1, 1]$, but not continuous on $[-1, 1]$.

In Exercises 55–58, determine which value best approximates the definite integral. Make your selection on the basis of a sketch.

55. $\displaystyle\int_0^4 \sqrt{x}\,dx$

(a) 5 (b) -3 (c) 10 (d) 2 (e) 8

56. $\displaystyle\int_0^{1/2} 4\cos \pi x\,dx$

(a) 4 (b) $\frac{4}{3}$ (c) 16 (d) 2π (e) -6

57. $\displaystyle\int_0^2 2e^{-x^2}\,dx$

(a) $\frac{1}{3}$ (b) 6 (c) 2 (d) 4

58. $\displaystyle\int_1^2 \ln x\,dx$

(a) $\frac{1}{3}$ (b) 1 (c) 4 (d) 3

Programming   Write a program for your graphing utility to approximate a definite integral using the Riemann sum

$$\sum_{i=1}^n f(c_i)\,\Delta x_i$$

where the subintervals are of equal width. The output should give three approximations of the integral where c_i is the left-hand endpoint $L(n)$, midpoint $M(n)$, and right-hand endpoint $R(n)$ of each subinterval. In Exercises 59–64, use the program to approximate the definite integral and complete the table.

n	4	8	12	16	20
$L(n)$					
$M(n)$					
$R(n)$					

59. $\displaystyle\int_0^3 x\sqrt{3 - x}\,dx$

60. $\displaystyle\int_0^3 \dfrac{5}{x^2 + 1}\,dx$

61. $\displaystyle\int_1^3 \dfrac{1}{x}\,dx$

62. $\displaystyle\int_0^4 e^x\,dx$

63. $\int_0^{\pi/2} \sin^2 x \, dx$

64. $\int_0^3 x \sin x \, dx$

True or False? In Exercises 65–70, determine whether the statement is true or false. If it is false, explain why or give an example that shows it is false.

65. $\int_a^b [f(x) + g(x)] \, dx = \int_a^b f(x) \, dx + \int_a^b g(x) \, dx$

66. $\int_a^b f(x)g(x) \, dx = \left[\int_a^b f(x) \, dx \right]\left[\int_a^b g(x) \, dx \right]$

67. If the norm of a partition approaches zero, then the number of subintervals approaches infinity.

68. If f is increasing on $[a, b]$, then the minimum value of $f(x)$ on $[a, b]$ is $f(a)$.

69. The value of

$$\int_a^b f(x) \, dx$$

must be positive.

70. The value of

$$\int_2^2 \sin(x^2) \, dx$$

is 0.

71. Find the Riemann sum for $f(x) = x^2 + 3x$ over the interval $[0, 8]$, where $x_0 = 0$, $x_1 = 1$, $x_2 = 3$, $x_3 = 7$, and $x_4 = 8$, and where $c_1 = 1$, $c_2 = 2$, $c_3 = 5$, and $c_4 = 8$.

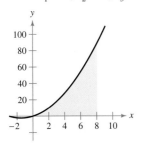

72. Find the Riemann sum for $f(x) = \sin x$ over the interval $[0, 2\pi]$, where $x_0 = 0$, $x_1 = \pi/4$, $x_2 = \pi/3$, $x_3 = \pi$, and $x_4 = 2\pi$, and where $c_1 = \pi/6$, $c_2 = \pi/3$, $c_3 = 2\pi/3$, and $c_4 = 3\pi/2$.

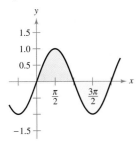

73. Prove that $\int_a^b x \, dx = \dfrac{b^2 - a^2}{2}$.

74. Prove that $\int_a^b x^2 \, dx = \dfrac{b^3 - a^3}{3}$.

75. ***Think About It*** Determine whether the Dirichlet function

$$f(x) = \begin{cases} 1, & x \text{ is rational} \\ 0, & x \text{ is irrational} \end{cases}$$

is integrable on the interval $[0, 1]$. Explain.

76. Suppose the function f is defined on $[0, 1]$, as shown in the figure.

$$f(x) = \begin{cases} 0, & x = 0 \\ \dfrac{1}{x}, & 0 < x \le 1 \end{cases}$$

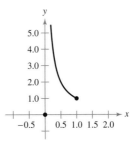

Show that $\int_0^1 f(x) \, dx$ does not exist. Why doesn't this contradict Theorem 5.4?

77. Find the constants a and b that maximize the value of

$$\int_a^b (1 - x^2) \, dx.$$

Explain your reasoning.

78. Evaluate, if possible, the integral $\int_0^2 [\![x]\!] \, dx$.

79. Determine

$$\lim_{n \to \infty} \frac{1}{n^3}[1^2 + 2^2 + 3^2 + \cdots + n^2]$$

by using an appropriate Riemann sum.

- Evaluate a definite integral using the Fundamental Theorem of Calculus.
- Understand and use the Mean Value Theorem for Integrals.
- Find the average value of a function over a closed interval.
- Understand and use the Second Fundamental Theorem of Calculus.

The Fundamental Theorem of Calculus

You have now been introduced to the two major branches of calculus: differential calculus (introduced with the tangent line problem) and integral calculus (introduced with the area problem). At this point, these two problems might seem unrelated—but there is a very close connection. The connection was discovered independently by Isaac Newton and Gottfried Leibniz and is stated in a theorem that is appropriately called the **Fundamental Theorem of Calculus.**

Informally, the theorem states that differentiation and (definite) integration are inverse operations, in the same sense that division and multiplication are inverse operations. To see how Newton and Leibniz might have anticipated this relationship, consider the approximations shown in Figure 5.27. The slope of the tangent line was defined using the *quotient* $\Delta y/\Delta x$ (the slope of the secant line). Similarly, the area of a region under a curve was defined using the *product* $\Delta y\Delta x$ (the area of a rectangle). So, at least in the primitive approximation stage, the operations of differentiation and definite integration appear to have an inverse relationship in the same sense that division and multiplication are inverse operations. The Fundamental Theorem of Calculus states that the limit processes (used to define the derivative and definite integral) preserve this inverse relationship.

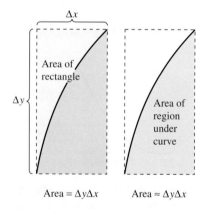

(a) Differentiation   **(b)** Definite integration

Differentiation and definite integration have an "inverse" relationship.
Figure 5.27

THEOREM 5.9   The Fundamental Theorem of Calculus

If a function f is continuous on the closed interval $[a, b]$ and F is an antiderivative of f on the interval $[a, b]$, then

$$\int_a^b f(x)\, dx = F(b) - F(a).$$

Proof   The key to the proof is in writing the difference $F(b) - F(a)$ in a convenient form. Let Δ be the following partition of $[a, b]$.

$$a = x_0 < x_1 < x_2 < \cdots < x_{n-1} < x_n = b$$

By pairwise subtraction and addition of like terms, you can write

$$F(b) - F(a) = F(x_n) - F(x_{n-1}) + F(x_{n-1}) - \cdots - F(x_1) + F(x_1) - F(x_0)$$
$$= \sum_{i=1}^{n} [F(x_i) - F(x_{i-1})].$$

By the Mean Value Theorem, you know that there exists a number c_i in the ith subinterval such that

$$F'(c_i) = \frac{F(x_i) - F(x_{i-1})}{x_i - x_{i-1}}.$$

Because $F'(c_i) = f(c_i)$, you can let $\Delta x_i = x_i - x_{i-1}$ and obtain

$$F(b) - F(a) = \sum_{i=1}^{n} f(c_i)\,\Delta x_i.$$

This important equation tells you that by applying the Mean Value Theorem, you can always find a collection of c_i's such that the *constant* $F(b) - F(a)$ is a Riemann sum of f on $[a, b]$. Taking the limit (as $\|\Delta\| \to 0$) produces

$$F(b) - F(a) = \int_a^b f(x)\,dx.$$

The following guidelines can help you understand the use of the Fundamental Theorem of Calculus.

Guidelines for Using the Fundamental Theorem of Calculus

1. *Provided you can find* an antiderivative of f, you now have a way to evaluate a definite integral without having to use the limit of a sum.

2. When applying the Fundamental Theorem of Calculus, the following notation is convenient.

$$\int_a^b f(x)\,dx = F(x) \Big]_a^b$$
$$= F(b) - F(a)$$

For instance, to evaluate $\int_1^3 x^3\,dx$, you can write

$$\int_1^3 x^3\,dx = \frac{x^4}{4} \Big]_1^3 = \frac{3^4}{4} - \frac{1^4}{4} = \frac{81}{4} - \frac{1}{4} = 20.$$

3. It is not necessary to include a constant of integration C in the antiderivative because

$$\int_a^b f(x)\,dx = \left[F(x) + C \right]_a^b$$
$$= [F(b) + C] - [F(a) + C]$$
$$= F(b) - F(a).$$

www   *EXAMPLE 1*   **Evaluating a Definite Integral**

Evaluate each definite integral.

a. $\displaystyle\int_1^2 (x^2 - 3)\, dx$ **b.** $\displaystyle\int_1^4 3\sqrt{x}\, dx$ **c.** $\displaystyle\int_0^{\pi/4} \sec^2 x\, dx$

Solution

a. $\displaystyle\int_1^2 (x^2 - 3)\, dx = \left[\frac{x^3}{3} - 3x\right]_1^2 = \left(\frac{8}{3} - 6\right) - \left(\frac{1}{3} - 3\right) = -\frac{2}{3}$

b. $\displaystyle\int_1^4 3\sqrt{x}\, dx = 3\int_1^4 x^{1/2}\, dx = 3\left[\frac{x^{3/2}}{3/2}\right]_1^4 = 2(4)^{3/2} - 2(1)^{3/2} = 14$

c. $\displaystyle\int_0^{\pi/4} \sec^2 x\, dx = \tan x \Big]_0^{\pi/4} = 1 - 0 = 1$

EXAMPLE 2   **A Definite Integral Involving Absolute Value**

Evaluate $\displaystyle\int_0^2 |2x - 1|\, dx$.

Solution   Using Figure 5.28 and the definition of absolute value, you can rewrite the integrand as shown.

$$|2x - 1| = \begin{cases} -(2x - 1), & x < \frac{1}{2} \\ 2x - 1, & x \geq \frac{1}{2} \end{cases}$$

From this, you can rewrite the integral in two parts.

$$\int_0^2 |2x - 1|\, dx = \int_0^{1/2} -(2x - 1)\, dx + \int_{1/2}^2 (2x - 1)\, dx$$

$$= \left[-x^2 + x\right]_0^{1/2} + \left[x^2 - x\right]_{1/2}^2$$

$$= \left(-\frac{1}{4} + \frac{1}{2}\right) - (0 + 0) + (4 - 2) - \left(\frac{1}{4} - \frac{1}{2}\right)$$

$$= \frac{5}{2}$$

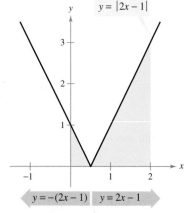

The definite integral of y on $[0, 2]$ is $\frac{5}{2}$.
Figure 5.28

EXAMPLE 3   **Using the Fundamental Theorem to Find Area**

Find the area of the region bounded by the graph of $y = 1/x$, the x-axis, and the vertical lines $x = 1$ and $x = e$, as shown in Figure 5.29.

Solution   Note that $y > 0$ on the interval $[1, e]$.

$$\text{Area} = \int_1^e \frac{1}{x}\, dx \qquad \text{Integrate between } x = 1 \text{ and } x = e.$$

$$= \left[\ln x\right]_1^e \qquad \text{Find antiderivative.}$$

$$= (\ln e) - (\ln 1) \qquad \text{Apply Fundamental Theorem of Calculus.}$$

$$= 1 \qquad \text{Simplify.}$$

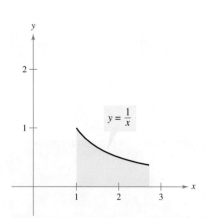

The area of the region bounded by the graph of $y = 1/x$, the x-axis, $x = 1$, and $x = e$ is 1.
Figure 5.29

The Mean Value Theorem for Integrals

In Section 5.2, you saw that the area of a region under a curve is greater than the area of an inscribed rectangle and less than the area of a circumscribed rectangle. The Mean Value Theorem for Integrals states that somewhere "between" the inscribed and circumscribed rectangles there is a rectangle whose area is precisely equal to the area of the region under the curve, as shown in Figure 5.30.

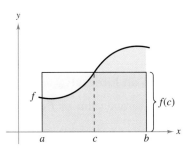

Mean value rectangle:

$$f(c)(b - a) = \int_a^b f(x)\, dx$$

Figure 5.30

THEOREM 5.10   Mean Value Theorem for Integrals

If f is continuous on the closed interval $[a, b]$, then there exists a number c in the closed interval $[a, b]$ such that

$$\int_a^b f(x)\, dx = f(c)(b - a).$$

Proof

Case 1: If f is constant on the interval $[a, b]$, the theorem is clearly valid because c can be any point in $[a, b]$.

Case 2: If f is not constant on $[a, b]$, then, by the Extreme Value Theorem, you can choose $f(m)$ and $f(M)$ to be the minimum and maximum values of f on $[a, b]$. Because $f(m) \le f(x) \le f(M)$ for all x in $[a, b]$, you can apply Theorem 5.8 to write the following.

$$\int_a^b f(m)\, dx \le \int_a^b f(x)\, dx \le \int_a^b f(M)\, dx \qquad \text{See Figure 5.31.}$$

$$f(m)(b - a) \le \int_a^b f(x)\, dx \le f(M)(b - a)$$

$$f(m) \le \frac{1}{b - a}\int_a^b f(x)\, dx \le f(M)$$

From the third inequality, you can apply the Intermediate Value Theorem to conclude that there exists some c in $[a, b]$ such that

$$f(c) = \frac{1}{b - a}\int_a^b f(x)\, dx \qquad \text{or} \qquad f(c)(b - a) = \int_a^b f(x)\, dx.$$

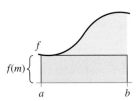

Inscribed rectangle
(less than actual area)

$$\int_a^b f(m)\, dx = f(m)(b - a)$$

Mean value rectangle
(equal to actual area)

$$\int_a^b f(x)\, dx$$

Circumscribed rectangle
(greater than actual area)

$$\int_a^b f(M)\, dx = f(M)(b - a)$$

Figure 5.31

NOTE   Notice that Theorem 5.10 does not specify how to determine c. It merely guarantees the existence of at least one number c in the interval.

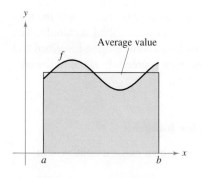

Average value $= \dfrac{1}{b-a}\displaystyle\int_a^b f(x)\,dx$

Figure 5.32

NOTE   Notice in Figure 5.32 that the area of the region under the graph of f is equal to the area of the rectangle whose height is the average value.

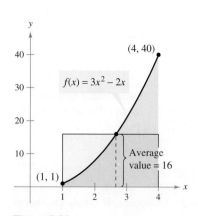

Figure 5.33

Average Value of a Function

The value of $f(c)$ given in the Mean Value Theorem for Integrals is called the **average value** of f on the interval $[a, b]$.

Definition of the Average Value of a Function on an Interval

If f is integrable on the closed interval $[a, b]$, then the **average value** of f on the interval is

$$\frac{1}{b-a}\int_a^b f(x)\,dx.$$

To see why the average value of f is defined in this way, suppose that you partition $[a, b]$ into n subintervals of equal width $\Delta x = (b-a)/n$. If c_i is any point in the ith subinterval, the arithmetic average (or mean) of the function values at the c_i's is given by

$$a_n = \frac{1}{n}\left[f(c_1) + f(c_2) + \cdots + f(c_n)\right]. \qquad \text{Average of } f(c_1),\ldots,f(c_n)$$

By multiplying and dividing by $(b-a)$, you can write the average as

$$a_n = \frac{1}{n}\sum_{i=1}^n f(c_i)\left(\frac{b-a}{b-a}\right) = \frac{1}{b-a}\sum_{i=1}^n f(c_i)\left(\frac{b-a}{n}\right)$$

$$= \frac{1}{b-a}\sum_{i=1}^n f(c_i)\,\Delta x.$$

Finally, taking the limit as $n\to\infty$ produces the average value of f on the interval $[a, b]$, as given in the definition above.

This development of the average value of a function on an interval is only one of many practical uses of definite integrals to represent summation processes. In Chapter 7, you will study other applications, such as volume, arc length, centers of mass, and work.

EXAMPLE 4   Finding the Average Value of a Function

Find the average value of $f(x) = 3x^2 - 2x$ on the interval $[1, 4]$.

Solution   The average value is given by

$$\frac{1}{b-a}\int_a^b f(x)\,dx = \frac{1}{3}\int_1^4 (3x^2 - 2x)\,dx$$

$$= \frac{1}{3}\left[x^3 - x^2\right]_1^4$$

$$= \frac{1}{3}[64 - 16 - (1 - 1)]$$

$$= \frac{48}{3}$$

$$= 16.$$

(See Figure 5.33.)

EXAMPLE 5   The Speed of Sound

The first person to fly at a speed greater than the speed of sound was Charles Yeager. On October 14, 1947, Yeager was clocked at 295.9 meters per second at an altitude of 12.2 kilometers. If Yeager had been flying at an altitude below 11.275 kilometers, this speed would not have "broken the sound barrier." The photo above shows an F-14 *Tomcat*, a supersonic, twin-engine strike fighter. Currently, the *Tomcat* can reach heights of 15.24 kilometers and speeds up to 2 mach (707.78 meters per second).

At different altitudes in Earth's atmosphere, sound travels at different speeds. The speed of sound $s(x)$ (in meters per second) can be modeled by

$$s(x) = \begin{cases} -4x + 341, & 0 \leq x < 11.5 \\ 295, & 11.5 \leq x < 22 \\ \frac{3}{4}x + 278.5, & 22 \leq x < 32 \\ \frac{3}{2}x + 254.5, & 32 \leq x < 50 \\ -\frac{3}{2}x + 404.5, & 50 \leq x \leq 80 \end{cases}$$

where x is the altitude in kilometers (see Figure 5.34). What is the average speed of sound over the interval $[0, 80]$?

Solution   Begin by integrating $s(x)$ over the interval $[0, 80]$. To do this, you can break the integral into five parts.

$$\int_0^{11.5} s(x)\, dx = \int_0^{11.5} (-4x + 341)\, dx = \left[-2x^2 + 341x \right]_0^{11.5} = 3657$$

$$\int_{11.5}^{22} s(x)\, dx = \int_{11.5}^{22} (295)\, dx = \left[295x \right]_{11.5}^{22} = 3097.5$$

$$\int_{22}^{32} s(x)\, dx = \int_{22}^{32} \left(\tfrac{3}{4}x + 278.5 \right) dx = \left[\tfrac{3}{8}x^2 + 278.5x \right]_{22}^{32} = 2987.5$$

$$\int_{32}^{50} s(x)\, dx = \int_{32}^{50} \left(\tfrac{3}{2}x + 254.5 \right) dx = \left[\tfrac{3}{4}x^2 + 254.5x \right]_{32}^{50} = 5688$$

$$\int_{50}^{80} s(x)\, dx = \int_{50}^{80} \left(-\tfrac{3}{2}x + 404.5 \right) dx = \left[-\tfrac{3}{4}x^2 + 404.5x \right]_{50}^{80} = 9210$$

By adding the values of the five integrals, you have

$$\int_0^{80} s(x)\, dx = 24{,}640.$$

So, the average speed of sound from an altitude of 0 kilometers to an altitude of 80 kilometers is

$$\text{Average speed} = \frac{1}{80} \int_0^{80} s(x)\, dx = \frac{24{,}640}{80} = 308 \text{ meters per second.}$$

The speed of sound depends on altitude.
Figure 5.34

The Second Fundamental Theorem of Calculus

Earlier you saw that the definite integral of f on the interval $[a, b]$ was defined using the constant b as the upper limit of integration and x as the variable of integration. However, a slightly different situation may arise in which the variable x is used as the upper limit of integration. To avoid the confusion of using x in two different ways, t is temporarily used as the variable of integration. (Remember that the definite integral is *not* a function of its variable of integration.)

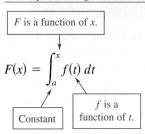

The Definite Integral as a Number

Constant

$$\int_a^b f(x)\,dx$$

Constant f is a function of x.

The Definite Integral as a Function of x

F is a function of x.

$$F(x) = \int_a^x f(t)\,dt$$

Constant f is a function of t.

EXAMPLE 6   The Definite Integral as a Function

Evaluate the function

$$F(x) = \int_0^x \cos t\,dt$$

at $x = 0$, $\pi/6$, $\pi/4$, $\pi/3$, and $\pi/2$.

Solution   You could evaluate five different definite integrals, one for each of the given upper limits. However, it is much simpler to fix x (as a constant) temporarily and apply the Fundamental Theorem once, to obtain

$$\int_0^x \cos t\,dt = \sin t \Big]_0^x = \sin x - \sin 0 = \sin x.$$

Now, using $F(x) = \sin x$, you can obtain the results shown in Figure 5.35.

    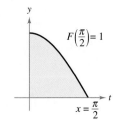

$F(x) = \int_0^x \cos t\,dt$ is the area under the curve $f(t) = \cos t$ from 0 to x.

Figure 5.35

You can think of the function $F(x)$ as *accumulating* the area under the curve $f(t) = \cos t$ from $t = 0$ to $t = x$. For $x = 0$, the area is 0 and $F(0) = 0$. For $x = \pi/2$, $F(\pi/2) = 1$ gives the accumulated area under the cosine curve on the entire interval $[0, \pi/2]$. This interpretation of an integral as an **accumulation function** is used often in applications of integration.

In Example 6, note that the derivative of F is the original integrand (with only the variable changed). That is,

$$\frac{d}{dx}[F(x)] = \frac{d}{dx}[\sin x] = \frac{d}{dx}\left[\int_0^x \cos t \, dt\right] = \cos x.$$

This result is generalized in the following theorem, called the **Second Fundamental Theorem of Calculus.**

THEOREM 5.11   The Second Fundamental Theorem of Calculus

If f is continuous on an open interval I containing a, then, for every x in the interval,

$$\frac{d}{dx}\left[\int_a^x f(t) \, dt\right] = f(x).$$

Proof   Begin by defining F as

$$F(x) = \int_a^x f(t) \, dt.$$

Then, by the definition of the derivative, you can write

$$F'(x) = \lim_{\Delta x \to 0} \frac{F(x + \Delta x) - F(x)}{\Delta x}$$

$$= \lim_{\Delta x \to 0} \frac{1}{\Delta x}\left[\int_a^{x+\Delta x} f(t) \, dt - \int_a^x f(t) \, dt\right]$$

$$= \lim_{\Delta x \to 0} \frac{1}{\Delta x}\left[\int_a^{x+\Delta x} f(t) \, dt + \int_x^a f(t) \, dt\right]$$

$$= \lim_{\Delta x \to 0} \frac{1}{\Delta x}\left[\int_x^{x+\Delta x} f(t) \, dt\right].$$

From the Mean Value Theorem for Integrals (assuming $\Delta x > 0$), you know there exists a number c in the interval $[x, x + \Delta x]$ such that the integral in the expression above is equal to $f(c) \Delta x$. Moreover, because $x \le c \le x + \Delta x$, it follows that $c \to x$ as $\Delta x \to 0$. So, you obtain

$$F'(x) = \lim_{\Delta x \to 0}\left[\frac{1}{\Delta x} f(c) \Delta x\right]$$

$$= \lim_{\Delta x \to 0} f(c)$$

$$= f(x).$$

A similar argument can be made for $\Delta x < 0$.   ▬▬▬▬

NOTE   Using the area model for definite integrals, you can view the approximation

$$f(x) \Delta x \approx \int_x^{x+\Delta x} f(t) \, dt$$

as saying that the area of the rectangle of height $f(x)$ and width Δx is approximately equal to the area of the region lying between the graph of f and the x-axis on the interval $[x, x + \Delta x]$, as shown in Figure 5.36.

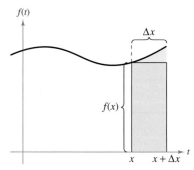

$$f(x) \Delta x \approx \int_x^{x+\Delta x} f(t) \, dt$$

Figure 5.36

Note that the Second Fundamental Theorem of Calculus tells you that if a function is continuous, you can be sure that it has an antiderivative. This antiderivative need not, however, be an elementary function. (Recall the discussion of elementary functions in Section 1.3.)

EXAMPLE 7 Using the Second Fundamental Theorem of Calculus

Evaluate $\dfrac{d}{dx}\left[\displaystyle\int_0^x \sqrt{t^2+1}\,dt\right]$.

Solution Note that $f(t) = \sqrt{t^2+1}$ is continuous on the entire real number line. So, using the Second Fundamental Theorem of Calculus, you can write

$$\frac{d}{dx}\left[\int_0^x \sqrt{t^2+1}\,dt\right] = \sqrt{x^2+1}.$$

The differentiation shown in Example 7 is a straightforward application of the Second Fundamental Theorem of Calculus. The next example shows how this theorem can be combined with the Chain Rule to find the derivative of a function.

EXAMPLE 8 Using the Second Fundamental Theorem of Calculus

Find the derivative of $F(x) = \displaystyle\int_{\pi/2}^{x^3} \cos t\,dt$.

Solution Using $u = x^3$, you can apply the Second Fundamental Theorem of Calculus with the Chain Rule as shown.

$$
\begin{aligned}
F'(x) &= \frac{dF}{du}\frac{du}{dx} &&\text{Chain Rule}\\[4pt]
&= \frac{d}{du}[F(x)]\frac{du}{dx} &&\text{Definition of } \frac{dF}{du}\\[4pt]
&= \frac{d}{du}\left[\int_{\pi/2}^{x^3} \cos t\,dt\right]\frac{du}{dx} &&\text{Substitute } \int_{\pi/2}^{x^3} \cos t\,dt \text{ for } F(x).\\[4pt]
&= \frac{d}{du}\left[\int_{\pi/2}^{u} \cos t\,dt\right]\frac{du}{dx} &&\text{Substitute } u \text{ for } x^3.\\[4pt]
&= (\cos u)(3x^2) &&\text{Apply Second Fundamental Theorem of Calculus.}\\[4pt]
&= (\cos x^3)(3x^2) &&\text{Rewrite as function of } x.
\end{aligned}
$$

Because the integrand in Example 8 is easily integrated, you can verify the derivative as follows.

$$F(x) = \int_{\pi/2}^{x^3} \cos t\,dt = \sin t\,\Big]_{\pi/2}^{x^3} = \sin x^3 - \sin\frac{\pi}{2} = (\sin x^3) - 1$$

In this form, you can apply the Power Rule to verify that the derivative is the same as that obtained in Example 8.

$$F'(x) = (\cos x^3)(3x^2)$$

Graphical Reasoning   In Exercises 1–4, use a graphing utility to graph the integrand. Use the graph to determine whether the definite integral is positive, negative, or zero.

1. $\displaystyle\int_0^\pi \frac{4}{x^2+1}\,dx$

2. $\displaystyle\int_0^\pi \cos x\,dx$

3. $\displaystyle\int_{-2}^2 x\sqrt{x^2+1}\,dx$

4. $\displaystyle\int_{-2}^2 x\sqrt{2-x}\,dx$

In Exercises 5–26, evaluate the definite integral of the algebraic function. Use a graphing utility to verify your result.

5. $\displaystyle\int_0^1 2x\,dx$

6. $\displaystyle\int_2^7 3\,dv$

7. $\displaystyle\int_{-1}^0 (x-2)\,dx$

8. $\displaystyle\int_2^5 (-3v+4)\,dv$

9. $\displaystyle\int_{-1}^1 (t^2-2)\,dt$

10. $\displaystyle\int_1^3 (3x^2+5x-4)\,dx$

11. $\displaystyle\int_0^1 (2t-1)^2\,dt$

12. $\displaystyle\int_{-1}^1 (t^3-9t)\,dt$

13. $\displaystyle\int_1^2 \left(\frac{3}{x^2}-1\right)dx$

14. $\displaystyle\int_{-2}^{-1}\left(u-\frac{1}{u^2}\right)du$

15. $\displaystyle\int_1^4 \frac{u-2}{\sqrt{u}}\,du$

16. $\displaystyle\int_{-3}^3 v^{1/3}\,dv$

17. $\displaystyle\int_{-1}^1 \left(\sqrt[3]{t}-2\right)dt$

18. $\displaystyle\int_1^8 \sqrt{\frac{2}{x}}\,dx$

19. $\displaystyle\int_0^1 \frac{x-\sqrt{x}}{3}\,dx$

20. $\displaystyle\int_0^2 (2-t)\sqrt{t}\,dt$

21. $\displaystyle\int_{-1}^0 \left(t^{1/3}-t^{2/3}\right)dt$

22. $\displaystyle\int_{-8}^{-1}\frac{x-x^2}{2\sqrt[3]{x}}\,dx$

23. $\displaystyle\int_0^3 |2x-3|\,dx$

24. $\displaystyle\int_2^5 (3-|x-4|)\,dx$

25. $\displaystyle\int_0^3 |x^2-4|\,dx$

26. $\displaystyle\int_0^4 |x^2-4x+3|\,dx$

In Exercises 27–38, evaluate the definite integral of the transcendental function. Use a graphing utility to verify your result.

27. $\displaystyle\int_0^\pi (1+\sin x)\,dx$

28. $\displaystyle\int_0^{\pi/4}\frac{1-\sin^2\theta}{\cos^2\theta}\,d\theta$

29. $\displaystyle\int_{-\pi/6}^{\pi/6}\sec^2 x\,dx$

30. $\displaystyle\int_{\pi/4}^{\pi/2}(2-\csc^2 x)\,dx$

31. $\displaystyle\int_1^e \left(2x-\frac{1}{x}\right)dx$

32. $\displaystyle\int_1^5 \frac{x+1}{x}\,dx$

33. $\displaystyle\int_{-\pi/3}^{\pi/3} 4\sec\theta\tan\theta\,d\theta$

34. $\displaystyle\int_{-\pi/2}^{\pi/2}(2t+\cos t)\,dt$

35. $\displaystyle\int_0^2 (2^x+6)\,dx$

36. $\displaystyle\int_0^3 (t-5^t)\,dt$

37. $\displaystyle\int_{-1}^1 (e^\theta+\sin\theta)\,d\theta$

38. $\displaystyle\int_e^{2e}\left(\cos x-\frac{1}{x}\right)dx$

In Exercises 39–44, determine the area of the given region.

39. $y = x - x^2$

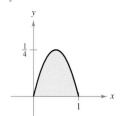

40. $y = 1 - x^4$

41. $y = (3-x)\sqrt{x}$

42. $y = \dfrac{1}{x^2}$

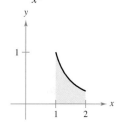

43. $y = \cos x$

44. $y = x + \sin x$

In Exercises 45–50, find the area of the region bounded by the graphs of the equations.

45. $y = 3x^2+1,$ $x = 0,$ $x = 2,$ $y = 0$

46. $y = 1+\sqrt[3]{x},$ $x = 0,$ $x = 8,$ $y = 0$

47. $y = x^3+x,$ $x = 2,$ $y = 0$

48. $y = -x^2+3x,$ $y = 0$

49. $y = \dfrac{4}{x},$ $x = 1,$ $x = e,$ $y = 0$

50. $y = e^x,$ $x = 0,$ $x = 2,$ $y = 0$

In Exercises 51–56, find the value(s) of c guaranteed by the Mean Value Theorem for Integrals for the function over the indicated interval.

Function	Interval
51. $f(x) = x - 2\sqrt{x}$	$[0, 2]$
52. $f(x) = 9/x^3$	$[1, 3]$
53. $f(x) = 2\sec^2 x$	$[-\pi/4, \pi/4]$
54. $f(x) = \cos x$	$[-\pi/3, \pi/3]$
55. $f(x) = 5 - \dfrac{1}{x}$	$[1, 4]$
56. $f(x) = 10 - 2^x$	$[0, 3]$

In Exercises 57–62, find the average value of the function over the given interval and all values of x in the interval for which the function equals its average value.

Function	Interval
57. $f(x) = 4 - x^2$	$[-2, 2]$
58. $f(x) = \dfrac{4(x^2 + 1)}{x^2}$	$[1, 3]$
59. $f(x) = 2e^x$	$[-1, 1]$
60. $f(x) = \dfrac{1}{2x}$	$[1, 4]$
61. $f(x) = \sin x$	$[0, \pi]$
62. $f(x) = \cos x$	$[0, \pi/2]$

63. Velocity The graph shows the velocity, in feet per second, of a car accelerating from rest. Use the graph to estimate the distance the car travels in 8 seconds.

Figure for 63

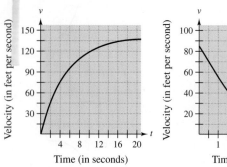

Figure for 64

64. Velocity The graph shows the velocity of a car as soon as the driver applies the brakes. Use the graph to estimate how far the car travels before it comes to a stop.

Writing About Concepts

65. State the Fundamental Theorem of Calculus.

66. The graph of f is shown in the figure.

(a) Evaluate $\int_1^7 f(x)\, dx$.

(b) Determine the average value of f on the interval $[1, 7]$.

(c) Determine the answers to parts (a) and (b) if the graph is translated two units upward.

Figure for 66

Figure for 67–72

Writing About Concepts *(continued)*

In Exercises 67–72, use the graph of f shown in the figure. The shaded region A has an area of 1.5, and $\int_0^6 f(x)\, dx = 3.5$. Use this information to fill in the blanks.

67. $\displaystyle\int_0^2 f(x)\, dx = $ ▢

68. $\displaystyle\int_2^6 f(x)\, dx = $ ▢

69. $\displaystyle\int_0^6 |f(x)|\, dx = $ ▢

70. $\displaystyle\int_0^2 -2f(x)\, dx = $ ▢

71. $\displaystyle\int_0^6 [2 + f(x)]\, dx = $ ▢

72. The average value of f over the interval $[0, 6]$ is ▢.

73. Force The force F (in newtons) of a hydraulic cylinder in a press is proportional to the square of $\sec x$, where x is the distance (in meters) that the cylinder is extended in its cycle. The domain of F is $[0, \pi/3]$, and $F(0) = 500$.

(a) Find F as a function of x.

(b) Find the average force exerted by the press over the interval $[0, \pi/3]$.

74. Blood Flow The velocity v of the flow of blood at a distance r from the central axis of an artery of radius R is

$$v = k(R^2 - r^2)$$

where k is the constant of proportionality. Find the average rate of flow of blood along a radius of the artery. (Use 0 and R as the limits of integration.)

75. Respiratory Cycle The volume V (in liters) of air in the lungs during a five-second respiratory cycle is approximated by the model

$$V = 0.1729t + 0.1522t^2 - 0.0374t^3$$

where t is the time in seconds. Approximate the average volume of air in the lungs during one cycle.

76. Average Sales A company fits a model to the monthly sales data of a seasonal product. The model is

$$S(t) = \frac{t}{4} + 1.8 + 0.5 \sin\left(\frac{\pi t}{6}\right), \quad 0 \le t \le 24$$

where S is sales (in thousands) and t is time in months.

(a) Use a graphing utility to graph $f(t) = 0.5 \sin(\pi t/6)$ for $0 \le t \le 24$. Use the graph to explain why the average value of $f(t)$ is 0 over the interval.

(b) Use a graphing utility to graph $S(t)$ and the line $g(t) = t/4 + 1.8$ in the same viewing window. Use the graph and the result of part (a) to explain why g is called the *trend line*.

77. *Modeling Data* An experimental vehicle is tested on a straight track. It starts from rest, and its velocity v (in meters per second) is recorded in the table every 10 seconds for 1 minute.

t	0	10	20	30	40	50	60
v	0	5	21	40	62	78	83

(a) Use a graphing utility to find a model of the form $v = at^3 + bt^2 + ct + d$ for the data.

(b) Use a graphing utility to plot the data and graph the model.

(c) Use the Fundamental Theorem of Calculus to approximate the distance traveled by the vehicle during the test.

78. *Modeling Data* A department store manager wants to estimate the number of customers that enter the store from noon until closing at 9 P.M. The table shows the number of customers N entering the store during a randomly selected minute each hour from $t - 1$ to t, with $t = 0$ corresponding to noon.

t	1	2	3	4	5	6	7	8	9
N	6	7	9	12	15	14	11	7	2

(a) Draw a histogram of the data.

(b) Estimate the total number of customers entering the store between noon and 9 P.M.

(c) Use the regression capabilities of a graphing utility to find a model of the form $N(t) = at^3 + bt^2 + ct + d$ for the data.

(d) Use a graphing utility to plot the data and graph the model.

(e) Use a graphing utility to evaluate $\int_0^9 N(t) \, dt$, and use the result to estimate the number of customers entering the store between noon and 9 P.M. Compare this with your answer in part (b).

(f) Estimate the average number of customers entering the store per minute between 3 P.M. and 7 P.M.

In Exercises 79–84, find F as a function of x and evaluate it at $x = 2$, $x = 5$, and $x = 8$.

79. $F(x) = \displaystyle\int_0^x (t - 5) \, dt$

80. $F(x) = \displaystyle\int_2^x (t^3 + 2t - 2) \, dt$

81. $F(x) = \displaystyle\int_1^x \frac{10}{v^2} \, dv$

82. $F(x) = \displaystyle\int_2^x -\frac{2}{t^3} \, dt$

83. $F(x) = \displaystyle\int_1^x \cos \theta \, d\theta$

84. $F(x) = \displaystyle\int_0^x \sin \theta \, d\theta$

85. Let $g(x) = \int_0^x f(t) \, dt$, where f is the function whose graph is shown.

(a) Estimate $g(0)$, $g(2)$, $g(4)$, $g(6)$, and $g(8)$.

(b) Find the largest open interval on which g is increasing. Find the largest open interval on which g is decreasing.

(c) Identify any extrema of g.

(d) Sketch a rough graph of g.

Figure for 85

Figure for 86

86. Let $g(x) = \int_0^x f(t) \, dt$, where f is the function whose graph is shown.

(a) Estimate $g(0)$, $g(2)$, $g(4)$, $g(6)$, and $g(8)$.

(b) Find the largest open interval on which g is increasing. Find the largest open interval on which g is decreasing.

(c) Identify any extrema of g.

(d) Sketch a rough graph of g.

In Exercises 87–94, (a) integrate to find F as a function of x and (b) demonstrate the Second Fundamental Theorem of Calculus by differentiating the result in part (a).

87. $F(x) = \displaystyle\int_0^x (t + 2) \, dt$

88. $F(x) = \displaystyle\int_0^x t(t^2 + 1) \, dt$

89. $F(x) = \displaystyle\int_8^x \sqrt[3]{t} \, dt$

90. $F(x) = \displaystyle\int_4^x \sqrt{t} \, dt$

91. $F(x) = \displaystyle\int_{\pi/4}^x \sec^2 t \, dt$

92. $F(x) = \displaystyle\int_{\pi/3}^x \sec t \tan t \, dt$

93. $F(x) = \displaystyle\int_{-1}^x e^t \, dt$

94. $F(x) = \displaystyle\int_1^x \frac{1}{t} \, dt$

In Exercises 95–100, use the Second Fundamental Theorem of Calculus to find $F'(x)$.

95. $F(x) = \displaystyle\int_{-2}^x (t^2 - 2t) \, dt$

96. $F(x) = \displaystyle\int_1^x \frac{t^2}{t^2 + 1} \, dt$

97. $F(x) = \displaystyle\int_{-1}^x \sqrt{t^4 + 1} \, dt$

98. $F(x) = \displaystyle\int_1^x \sqrt[4]{t} \, dt$

99. $F(x) = \displaystyle\int_0^x t \cos t \, dt$

100. $F(x) = \displaystyle\int_0^x \sec^3 t \, dt$

In Exercises 101–106, find $F'(x)$.

101. $F(x) = \displaystyle\int_x^{x+2} (4t + 1) \, dt$

102. $F(x) = \displaystyle\int_{-x}^x t^3 \, dt$

103. $F(x) = \displaystyle\int_0^{\sin x} \sqrt{t} \, dt$

104. $F(x) = \displaystyle\int_2^{x^2} \frac{1}{t^3} \, dt$

105. $F(x) = \displaystyle\int_0^{x^3} \sin t^2 \, dt$

106. $F(x) = \displaystyle\int_0^{x^2} \sin \theta^2 \, d\theta$

107. *Graphical Analysis* Approximate the graph of g on the interval $0 \le x \le 4$, where $g(x) = \int_0^x f(t)\, dt$. Identify the x-coordinate of an extremum of g. To print an enlarged copy of the graph, go to the website *www.mathgraphs.com*.

108. *Area* The area A between the graph of the function $g(t) = 4 - 4/t^2$ and the t-axis over the interval $[1, x]$ is

$$A(x) = \int_1^x \left(4 - \frac{4}{t^2}\right) dt.$$

(a) Find the horizontal asymptote of the graph of g.

(b) Integrate to find A as a function of x. Does the graph of A have a horizontal asymptote? Explain.

Rectilinear Motion In Exercises 109–111, consider a particle moving along the x-axis where $x(t)$ is the position of the particle at time t, $x'(t)$ is its velocity, and $\int_a^b |x'(t)|\, dt$ is the distance the particle travels in the interval of time.

109. The position function is given by $x(t) = t^3 - 6t^2 + 9t - 2$, $0 \le t \le 5$. Find the total distance the particle travels in 5 units of time.

110. Repeat Exercise 109 for the position function given by $x(t) = (t - 1)(t - 3)^2$, $0 \le t \le 5$.

111. A particle moves along the x-axis with velocity $v(t) = 1/\sqrt{t}$, $t > 0$. At time $t = 1$, its position is $x = 4$. Find the total distance traveled by the particle on the interval $1 \le t \le 4$.

112. *Buffon's Needle Experiment* A horizontal plane is ruled with parallel lines 2 inches apart. A two-inch needle is tossed randomly onto the plane. The probability that the needle will touch a line is

$$P = \frac{2}{\pi} \int_0^{\pi/2} \sin\theta\, d\theta$$

where θ is the acute angle between the needle and any one of the parallel lines. Find this probability.

True or False? **In Exercises 113 and 114, determine whether the statement is true or false. If it is false, explain why or give an example that shows it is false.**

113. If $F'(x) = G'(x)$ on the interval $[a, b]$, then $F(b) - F(a) = G(b) - G(a)$.

114. If f is continuous on $[a, b]$, then f is integrable on $[a, b]$.

115. *Find the Error* Describe why the statement is incorrect.

$$\int_{-1}^{1} x^{-2}\, dx = \left[-x^{-1}\right]_{-1}^{1} = (-1) - 1 = -2$$

116. Prove that $\dfrac{d}{dx}\left[\displaystyle\int_{u(x)}^{v(x)} f(t)\, dt\right] = f(v(x))v'(x) - f(u(x))u'(x)$.

117. Show that the function

$$f(x) = \int_0^{1/x} \frac{1}{t^2 + 1}\, dt + \int_0^{x} \frac{1}{t^2 + 1}\, dt$$

is constant for $x > 0$.

118. Let $G(x) = \displaystyle\int_0^x \left[s \int_0^s f(t)\,dt\right] ds$, where f is continuous for all real t. Find (a) $G(0)$, (b) $G'(0)$, (c) $G''(x)$, and (d) $G''(0)$.

Section Project: Demonstrating the Fundamental Theorem

Use a graphing utility to graph the function $y_1 = \sin^2 t$ on the interval $0 \le t \le \pi$. Let $F(x)$ be the following function of x.

$$F(x) = \int_0^x \sin^2 t\, dt$$

(a) Complete the table. Explain why the values of F are increasing.

x	0	$\pi/6$	$\pi/3$	$\pi/2$	$2\pi/3$	$5\pi/6$	π
$F(x)$							

(b) Use the integration capabilities of a graphing utility to graph F.

(c) Use the differentiation capabilities of a graphing utility to graph $F'(x)$. How is this graph related to the graph in part (b)?

(d) Verify that the derivative of $y = (1/2)t - (\sin 2t)/4$ is $\sin^2 t$. Graph y and write a short paragraph about how this graph is related to those in parts (b) and (c).

Integration by Substitution

- Use pattern recognition to find an indefinite integral.
- Use a change of variables to find an indefinite integral.
- Use the General Power Rule for Integration to find an indefinite integral.
- Use a change of variables to evaluate a definite integral.
- Evaluate a definite integral involving an even or odd function.

Pattern Recognition

In this section you will study techniques for integrating composite functions. The discussion is split into two parts—*pattern recognition* and *change of variables*. Both techniques involve a **u-substitution.** With pattern recognition you perform the substitution mentally, and with change of variables you write the substitution steps.

The role of substitution in integration is comparable to the role of the Chain Rule in differentiation. Recall that for differentiable functions given by $y = F(u)$ and $u = g(x)$, the Chain Rule states that

$$\frac{d}{dx}[F(g(x))] = F'(g(x))g'(x).$$

From the definition of an antiderivative, it follows that

$$\int F'(g(x))g'(x)\, dx = F(g(x)) + C$$

$$= F(u) + C.$$

These results are summarized in the following theorem.

NOTE The statement of Theorem 5.12 doesn't tell how to distinguish between $f(g(x))$ and $g'(x)$ in the integrand. As you become more experienced at integration, your skill in doing this will increase. Of course, part of the key is familiarity with derivatives.

THEOREM 5.12   Antidifferentiation of a Composite Function

Let g be a function whose range is an interval I, and let f be a function that is continuous on I. If g is differentiable on its domain and F is an antiderivative of f on I, then

$$\int f(g(x))g'(x)\, dx = F(g(x)) + C.$$

If $u = g(x)$, then $du = g'(x)\, dx$ and

$$\int f(u)\, du = F(u) + C.$$

EXPLORATION

STUDY TIP There are several techniques for applying substitution, each differing slightly from the others. However, you should remember that the goal is the same with every technique— *you are trying to find an antiderivative of the integrand.*

Recognizing Patterns   The integrand in each of the following integrals fits the pattern $f(g(x))g'(x)$. Identify the pattern and use the result to evaluate the integral.

a. $\displaystyle\int 2x(x^2 + 1)^4\, dx$ **b.** $\displaystyle\int 3x^2\sqrt{x^3 + 1}\, dx$ **c.** $\displaystyle\int \sec^2 x(\tan x + 3)\, dx$

The next three integrals are similar to the first three. Show how you can multiply and divide by a constant to evaluate these integrals.

d. $\displaystyle\int x(x^2 + 1)^4\, dx$ **e.** $\displaystyle\int x^2\sqrt{x^3 + 1}\, dx$ **f.** $\displaystyle\int 2\sec^2 x(\tan x + 3)\, dx$

Examples 1 and 2 show how to apply Theorem 5.12 *directly*, by recognizing the presence of $f(g(x))$ and $g'(x)$. Note that the composite function in the integrand has an *outside function* f and an *inside function* g. Moreover, the derivative $g'(x)$ is present as a factor of the integrand.

$$\int \underbrace{f(g(x))}_{\text{Inside function}} \underbrace{g'(x)}_{\substack{\text{Derivative of}\\\text{inside function}}} dx = F(g(x)) + C$$

Outside function

EXAMPLE 1 Recognizing the $f(g(x))g'(x)$ Pattern

Find $\displaystyle\int (x^2 + 1)^2(2x)\,dx.$

Solution Letting $g(x) = x^2 + 1$, you obtain

$$g'(x) = 2x$$

and

$$f(g(x)) = f(x^2 + 1) = (x^2 + 1)^2.$$

From this, you can recognize that the integrand follows the $f(g(x))g'(x)$ pattern. Using the Power Rule for Integration and Theorem 5.12, you can write

$$\int \overbrace{(x^2 + 1)^2}^{f(g(x))}\,\overbrace{(2x)}^{g'(x)}\,dx = \frac{1}{3}(x^2 + 1)^3 + C.$$

Try using the Chain Rule to check that the derivative of $\frac{1}{3}(x^2 + 1)^3 + C$ is the integrand of the original integral.

TECHNOLOGY Try using a computer algebra system, such as *Maple, Derive, Mathematica, Mathcad,* or the *TI-89,* to solve the integrals given in Examples 1 and 2. Do you obtain the same antiderivatives that are listed in the examples?

EXAMPLE 2 Recognizing the $f(g(x))g'(x)$ Pattern

Find $\displaystyle\int 5e^{5x}\,dx.$

Solution Letting $g(x) = 5x$, you obtain

$$g'(x) = 5 \qquad \text{and} \qquad f(g(x)) = f(5x) = e^{5x}.$$

From this, you can recognize that the integrand follows the $f(g(x))g'(x)$ pattern. Using the Exponential Rule for Integration and Theorem 5.12, you can write

$$\int \overset{f(g(x))}{e^{5x}}\overset{g'(x)}{(5)}\,dx = e^{5x} + C.$$

You can check this by differentiating $e^{5x} + C$ to obtain the original integrand.

The integrands in Examples 1 and 2 fit the $f(g(x))g'(x)$ pattern exactly—you only had to recognize the pattern. You can extend this technique considerably with the Constant Multiple Rule

$$\int kf(x)\,dx = k\int f(x)\,dx.$$

Many integrands contain the essential part (the variable part) of $g'(x)$ but are missing a constant multiple. In such cases, you can multiply and divide by the necessary constant multiple, as shown in Example 3.

EXAMPLE 3   Multiplying and Dividing by a Constant

Find $\int x(x^2 + 1)^2\,dx.$

Solution   This is similar to the integral given in Example 1, except that the integrand is missing a factor of 2. Recognizing that $2x$ is the derivative of $x^2 + 1$, you can let $g(x) = x^2 + 1$ and supply the $2x$ as follows.

$$\int x(x^2 + 1)^2\,dx = \int (x^2 + 1)^2 \left(\frac{1}{2}\right)(2x)\,dx \qquad \text{Multiply and divide by 2.}$$

$$= \frac{1}{2}\int \overbrace{(x^2 + 1)^2}^{f(g(x))}\ \overbrace{(2x)}^{g'(x)}\,dx \qquad \text{Constant Multiple Rule}$$

$$= \frac{1}{2}\left[\frac{(x^2 + 1)^3}{3}\right] + C \qquad \text{Integrate.}$$

$$= \frac{1}{6}(x^2 + 1)^3 + C \qquad \text{Simplify.}$$

In practice, most people would not write as many steps as are shown in Example 3. For instance, you could evaluate the integral by simply writing

$$\int x(x^2 + 1)^2\,dx = \frac{1}{2}\int (x^2 + 1)^2\ 2x\,dx$$

$$= \frac{1}{2}\left[\frac{(x^2 + 1)^3}{3}\right] + C$$

$$= \frac{1}{6}(x^2 + 1)^3 + C.$$

NOTE   Be sure you see that the *Constant* Multiple Rule applies only to *constants*. You cannot multiply and divide by a variable and then move the variable outside the integral sign. For instance,

$$\int (x^2 + 1)^2\,dx \neq \frac{1}{2x}\int (x^2 + 1)^2\,(2x)\,dx.$$

After all, if it were legitimate to move variable quantities outside the integral sign, you could move the entire integrand out and simplify the whole process. But the result would be incorrect.

Change of Variables

With a formal **change of variables,** you completely rewrite the integral in terms of u and du (or any other convenient variable). Although this procedure can involve more written steps than the pattern recognition illustrated in Examples 1 to 3, it is useful for complicated integrands. The change of variable technique uses the Leibniz notation for the differential. That is, if $u = g(x)$, then $du = g'(x)\,dx$, and the integral in Theorem 5.12 takes the form

$$\int f(g(x))g'(x)\,dx = \int f(u)\,du = F(u) + C.$$

EXAMPLE 4 Change of Variables

Find $\displaystyle\int \sqrt{2x - 1}\,dx$.

Solution First, let u be the inner function, $u = 2x - 1$. Then calculate the differential du to be $du = 2\,dx$. Now, using $\sqrt{2x - 1} = \sqrt{u}$ and $dx = du/2$, substitute to obtain

$$\int \sqrt{2x - 1}\,dx = \int \sqrt{u}\left(\frac{du}{2}\right) \qquad \text{Integral in terms of } u$$

$$= \frac{1}{2}\int u^{1/2}\,du \qquad \text{Constant Multiple Rule}$$

$$= \frac{1}{2}\left(\frac{u^{3/2}}{3/2}\right) + C \qquad \text{Antiderivative in terms of } u$$

$$= \frac{1}{3}u^{3/2} + C \qquad \text{Simplify.}$$

$$= \frac{1}{3}(2x - 1)^{3/2} + C. \qquad \text{Antiderivative in terms of } x$$

STUDY TIP Because integration is usually more difficult than differentiation, you should always check your answer to an integration problem by differentiating. For instance, in Example 4 you should differentiate $\frac{1}{3}(2x - 1)^{3/2} + C$ to verify that you obtain the original integrand.

EXAMPLE 5 Change of Variables

Find $\displaystyle\int x\sqrt{2x - 1}\,dx$.

Solution As in the previous example, let $u = 2x - 1$ and obtain $dx = du/2$. Because the integrand contains a factor of x, you must also solve for x in terms of u, as shown.

$$u = 2x - 1 \quad \Longrightarrow \quad x = (u + 1)/2 \qquad \text{Solve for } x \text{ in terms of } u.$$

Now, using substitution, you obtain

$$\int x\sqrt{2x - 1}\,dx = \int \left(\frac{u + 1}{2}\right)u^{1/2}\left(\frac{du}{2}\right)$$

$$= \frac{1}{4}\int (u^{3/2} + u^{1/2})\,du$$

$$= \frac{1}{4}\left(\frac{u^{5/2}}{5/2} + \frac{u^{3/2}}{3/2}\right) + C$$

$$= \frac{1}{10}(2x - 1)^{5/2} + \frac{1}{6}(2x - 1)^{3/2} + C.$$

To complete the change of variables in Example 5, you solved for x in terms of u. Sometimes this is very difficult. Fortunately it is not always necessary, as shown in the next example.

EXAMPLE 6   Change of Variables

Find $\int \sin^2 3x \cos 3x \, dx$.

Solution   Because $\sin^2 3x = (\sin 3x)^2$, you can let $u = \sin 3x$. Then

$$du = (\cos 3x)(3) \, dx.$$

Now, because $\cos 3x \, dx$ is part of the original integral, you can write

$$\frac{du}{3} = \cos 3x \, dx.$$

STUDY TIP   When making a change of variables, be sure that your answer is written using the same variables used in the original integrand. For instance, in Example 6, you should not leave your answer as

$$\tfrac{1}{9} u^3 + C$$

but rather, replace u by $\sin 3x$.

Substituting u and $du/3$ in the original integral yields

$$\int \sin^2 3x \cos 3x \, dx = \int u^2 \frac{du}{3}$$

$$= \frac{1}{3} \int u^2 \, du$$

$$= \frac{1}{3} \left(\frac{u^3}{3} \right) + C$$

$$= \frac{1}{9} \sin^3 3x + C.$$

You can check this by differentiating.

$$\frac{d}{dx} \left[\frac{1}{9} \sin^3 3x \right] = \left(\frac{1}{9} \right)(3)(\sin 3x)^2(\cos 3x)(3)$$

$$= \sin^2 3x \cos 3x$$

Because differentiation produces the original integrand, you know that you have obtained the correct antiderivative.

The steps used for integration by substitution are summarized in the following guidelines.

Guidelines for Making a Change of Variables

1. Choose a substitution $u = g(x)$. Usually, it is best to choose the *inner* part of a composite function, such as a quantity raised to a power.
2. Compute $du = g'(x) \, dx$.
3. Rewrite the integral in terms of the variable u.
4. Find the resulting integral in terms of u.
5. Replace u by $g(x)$ to obtain an antiderivative in terms of x.
6. Check your answer by differentiating.

The General Power Rule for Integration

One of the most common u-substitutions involves quantities in the integrand that are raised to a power. Because of the importance of this type of substitution, it is given a special name—the **General Power Rule for Integration.** A proof of this rule follows directly from the (simple) Power Rule for Integration, together with Theorem 5.12.

THEOREM 5.13   The General Power Rule for Integration

If g is a differentiable function of x, then

$$\int [g(x)]^n g'(x)\, dx = \frac{[g(x)]^{n+1}}{n+1} + C, \qquad n \neq -1.$$

Equivalently, if $u = g(x)$, then

$$\int u^n\, du = \frac{u^{n+1}}{n+1} + C, \qquad n \neq -1.$$

EXAMPLE 7   Substitution and the General Power Rule

a. $\displaystyle \int 3(3x-1)^4\, dx = \int \overbrace{(3x-1)^4}^{u^4}\,\overbrace{(3)\, dx}^{du} = \overbrace{\frac{(3x-1)^5}{5}}^{u^5/5} + C$

b. $\displaystyle \int (e^x+1)(e^x+x)\, dx = \int \overbrace{(e^x+x)}^{u^1}\,\overbrace{(e^x+1)\, dx}^{du} = \overbrace{\frac{(e^x+x)^2}{2}}^{u^2/2} + C$

c. $\displaystyle \int 3x^2\sqrt{x^3-2}\, dx = \int \overbrace{(x^3-2)^{1/2}}^{u^{1/2}}\,\overbrace{(3x^2)\, dx}^{du} = \overbrace{\frac{(x^3-2)^{3/2}}{3/2}}^{u^{3/2}/(3/2)} + C = \frac{2}{3}(x^3-2)^{3/2} + C$

d. $\displaystyle \int \frac{-4x}{(1-2x^2)^2}\, dx = \int \overbrace{(1-2x^2)^{-2}}^{u^{-2}}\,\overbrace{(-4x)\, dx}^{du} = \overbrace{\frac{(1-2x^2)^{-1}}{-1}}^{u^{-1}/(-1)} + C = -\frac{1}{1-2x^2} + C$

e. $\displaystyle \int \cos^2 x \sin x\, dx = -\int \overbrace{(\cos x)^2}^{u^2}\,\overbrace{(-\sin x)\, dx}^{du} = -\overbrace{\frac{(\cos x)^3}{3}}^{u^3/3} + C$

Some integrals whose integrands involve quantities raised to powers cannot be found by the General Power Rule. Consider the two integrals

$$\int x(x^2+1)^2\, dx \quad \text{and} \quad \int (x^2+1)^2\, dx.$$

The substitution $u = x^2 + 1$ works in the first integral but not in the second. In the second, the substitution fails because the integrand lacks the factor x needed for du. Fortunately, *for this particular integral*, you can expand the integrand as $(x^2+1)^2 = x^4 + 2x^2 + 1$ and use the (simple) Power Rule to integrate each term.

EXPLORATION

Suppose you were asked to find one of the following integrals. Which one would you choose? Explain your reasoning.

a. $\displaystyle \int \sqrt{x^3+1}\, dx$ or

$\displaystyle \int x^2\sqrt{x^3+1}\, dx$

b. $\displaystyle \int \tan(3x)\sec^2(3x)\, dx$ or

$\displaystyle \int \tan(3x)\, dx$

Change of Variables for Definite Integrals

When using u-substitution with a definite integral, it is often convenient to determine the limits of integration for the variable u rather than to convert the antiderivative back to the variable x and evaluate at the original limits. This change of variables is stated explicitly in the next theorem. The proof follows from Theorem 5.12 combined with the Fundamental Theorem of Calculus.

THEOREM 5.14   Change of Variables for Definite Integrals

If the function $u = g(x)$ has a continuous derivative on the closed interval $[a, b]$ and f is continuous on the range of g, then

$$\int_a^b f(g(x))g'(x)\,dx = \int_{g(a)}^{g(b)} f(u)\,du.$$

EXAMPLE 8   Change of Variables

Evaluate $\displaystyle\int_0^1 x(x^2 + 1)^3\,dx$.

Solution   To evaluate this integral, let $u = x^2 + 1$. Then, you obtain

$$u = x^2 + 1 \implies du = 2x\,dx.$$

Before substituting, determine the new upper and lower limits of integration.

Lower Limit	Upper Limit
When $x = 0$, $u = 0^2 + 1 = 1$.	When $x = 1$, $u = 1^2 + 1 = 2$.

Now, you can substitute to obtain

$$\int_0^1 x(x^2 + 1)^3\,dx = \frac{1}{2}\int_0^1 (x^2 + 1)^3(2x)\,dx \qquad \text{Integration limits for } x$$

$$= \frac{1}{2}\int_1^2 u^3\,du \qquad \text{Integration limits for } u$$

$$= \frac{1}{2}\left[\frac{u^4}{4}\right]_1^2$$

$$= \frac{1}{2}\left(4 - \frac{1}{4}\right)$$

$$= \frac{15}{8}.$$

Try rewriting the antiderivative $\frac{1}{2}(u^4/4)$ in terms of the variable x and then evaluate the definite integral at the original limits of integration, as shown.

$$\frac{1}{2}\left[\frac{u^4}{4}\right]_1^2 = \frac{1}{2}\left[\frac{(x^2 + 1)^4}{4}\right]_0^1$$

$$= \frac{1}{2}\left(4 - \frac{1}{4}\right) = \frac{15}{8}$$

Notice that you obtain the same result.

EXAMPLE 9 Change of Variables

Evaluate $A = \int_1^5 \dfrac{x}{\sqrt{2x-1}}\, dx$.

Solution To evaluate this integral, let $u = \sqrt{2x-1}$. Then, you obtain

$$u^2 = 2x - 1$$

$$u^2 + 1 = 2x$$

$$\frac{u^2+1}{2} = x$$

$$u\, du = dx. \qquad \text{Differentiate each side.}$$

Before substituting, determine the new upper and lower limits of integration.

Lower Limit	Upper Limit
When $x = 1$, $u = \sqrt{2-1} = 1$.	When $x = 5$, $u = \sqrt{10-1} = 3$.

Now, substitute to obtain

$$\int_1^5 \frac{x}{\sqrt{2x-1}}\, dx = \int_1^3 \frac{1}{u}\left(\frac{u^2+1}{2}\right) u\, du$$

$$= \frac{1}{2}\int_1^3 (u^2 + 1)\, du$$

$$= \frac{1}{2}\left[\frac{u^3}{3} + u\right]_1^3$$

$$= \frac{1}{2}\left(9 + 3 - \frac{1}{3} - 1\right)$$

$$= \frac{16}{3}.$$

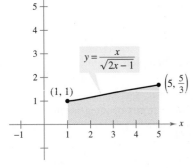

The region before substitution has an area of $\frac{16}{3}$.

Figure 5.37

The region after substitution has an area of $\frac{16}{3}$.

Figure 5.38

Geometrically, you can interpret the equation

$$\int_1^5 \frac{x}{\sqrt{2x-1}}\, dx = \int_1^3 \frac{u^2+1}{2}\, du$$

to mean that the two *different* regions shown in Figures 5.37 and 5.38 have the *same* area.

When evaluating definite integrals by substitution, it is possible for the upper limit of integration of the u-variable form to be smaller than the lower limit. If this happens, don't rearrange the limits. Simply evaluate as usual. For example, after substituting $u = \sqrt{1-x}$ in the integral

$$\int_0^1 x^2(1-x)^{1/2}\, dx$$

you obtain $u = \sqrt{1-1} = 0$ when $x = 1$, and $u = \sqrt{1-0} = 1$ when $x = 0$. So, the correct u-variable form of this integral is

$$-2\int_1^0 (1-u^2)^2 u^2\, du.$$

Integration of Even and Odd Functions

Even with a change of variables, integration can be difficult. Occasionally, you can simplify the evaluation of a definite integral (over an interval that is symmetric about the y-axis or about the origin) by recognizing the integrand to be an even or odd function (see Figure 5.39).

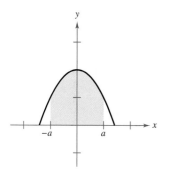

Even function

THEOREM 5.15   Integration of Even and Odd Functions

Let f be integrable on the closed interval $[-a, a]$.

1. If f is an *even* function, then $\displaystyle\int_{-a}^{a} f(x)\,dx = 2\int_{0}^{a} f(x)\,dx.$

2. If f is an *odd* function, then $\displaystyle\int_{-a}^{a} f(x)\,dx = 0.$

Proof   Because f is even, you know that $f(x) = f(-x)$. Using Theorem 5.12 with the substitution $u = -x$ produces

$$\int_{-a}^{0} f(x)\,dx = \int_{a}^{0} f(-u)(-du) = -\int_{a}^{0} f(u)\,du = \int_{0}^{a} f(u)\,du = \int_{0}^{a} f(x)\,dx.$$

Finally, using Theorem 5.6, you obtain

$$\int_{-a}^{a} f(x)\,dx = \int_{-a}^{0} f(x)\,dx + \int_{0}^{a} f(x)\,dx$$

$$= \int_{0}^{a} f(x)\,dx + \int_{0}^{a} f(x)\,dx = 2\int_{0}^{a} f(x)\,dx.$$

This proves the first property. The proof of the second property is left to you (see Exercise 169).

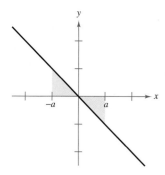

Odd function
Figure 5.39

EXAMPLE 10   Integration of an Odd Function

Evaluate $\displaystyle\int_{-\pi/2}^{\pi/2} (\sin^3 x \cos x + \sin x \cos x)\,dx.$

Solution   Letting $f(x) = \sin^3 x \cos x + \sin x \cos x$ produces

$$f(-x) = \sin^3(-x)\cos(-x) + \sin(-x)\cos(-x)$$

$$= -\sin^3 x \cos x - \sin x \cos x = -f(x).$$

So, f is an odd function, and because f is symmetric about the origin over $[-\pi/2, \pi/2]$, you can apply Theorem 5.15 to conclude that

$$\int_{-\pi/2}^{\pi/2} (\sin^3 x \cos x + \sin x \cos x)\,dx = 0.$$

$f(x) = \sin^3 x \cos x + \sin x \cos x$

Because f is an odd function,

$$\int_{-\pi/2}^{\pi/2} f(x)\,dx = 0.$$

Figure 5.40

NOTE   From Figure 5.40, you can see that the two regions on either side of the y-axis have the same area. However, because one lies below the x-axis and one lies above it, integration produces a cancellation effect. (More will be said about this in Section 7.1.)

Exercises for Section 5.5

See www.CalcChat.com for worked-out solutions to odd-numbered exercises.

In Exercises 1–6, complete the table by identifying u and du for the integral.

$\int f(g(x))g'(x)\,dx$	$u = g(x)$	$du = g'(x)\,dx$
1. $\int (5x^2 + 1)^2 (10x)\,dx$		
2. $\int x^2 \sqrt{x^3 + 1}\,dx$		
3. $\int \dfrac{x}{\sqrt{x^2 + 1}}\,dx$		
4. $\int \sec 2x \tan 2x\,dx$		
5. $\int \tan^2 x \sec^2 x\,dx$		
6. $\int \dfrac{\cos x}{\sin^3 x}\,dx$		

In Exercises 7–34, find the indefinite integral and check the result by differentiation.

7. $\int (1 + 2x)^4 (2)\,dx$

8. $\int (x^2 - 9)^3 (2x)\,dx$

9. $\int \sqrt{9 - x^2}\,(-2x)\,dx$

10. $\int \sqrt[3]{(1 - 2x^2)}(-4x)\,dx$

11. $\int x^3 (x^4 + 3)^2\,dx$

12. $\int x^2 (x^3 + 5)^4\,dx$

13. $\int x^2 (x^3 - 1)^4\,dx$

14. $\int x(4x^2 + 3)^2\,dx$

15. $\int t\sqrt{t^2 + 2}\,dt$

16. $\int t^3 \sqrt{t^4 + 5}\,dt$

17. $\int 5x \sqrt[3]{1 - x^2}\,dx$

18. $\int u^2 \sqrt{u^3 + 5}\,du$

19. $\int \dfrac{x}{(1 - x^2)^3}\,dx$

20. $\int \dfrac{x^3}{(1 + x^4)^2}\,dx$

21. $\int \dfrac{x^2}{(1 + x^3)^2}\,dx$

22. $\int \dfrac{x^2}{(9 - x^3)^2}\,dx$

23. $\int \dfrac{x}{\sqrt{1 - x^2}}\,dx$

24. $\int \dfrac{x^3}{\sqrt{1 + x^4}}\,dx$

25. $\int \left(1 + \dfrac{1}{t}\right)^3 \left(\dfrac{1}{t^2}\right)\,dt$

26. $\int \left[x^2 + \dfrac{1}{(3x)^2}\right]\,dx$

27. $\int \dfrac{1}{\sqrt{2x}}\,dx$

28. $\int \dfrac{1}{2\sqrt{x}}\,dx$

29. $\int \dfrac{x^2 + 3x + 7}{\sqrt{x}}\,dx$

30. $\int \dfrac{t + 2t^2}{\sqrt{t}}\,dt$

31. $\int t^2 \left(t - \dfrac{2}{t}\right)\,dt$

32. $\int \left(\dfrac{t^3}{3} + \dfrac{1}{4t^2}\right)\,dt$

33. $\int (9 - y)\sqrt{y}\,dy$

34. $\int 2\pi y (8 - y^{3/2})\,dy$

In Exercises 35–38, solve the differential equation.

35. $\dfrac{dy}{dx} = 4x + \dfrac{4x}{\sqrt{16 - x^2}}$

36. $\dfrac{dy}{dx} = \dfrac{10x^2}{\sqrt{1 + x^3}}$

37. $\dfrac{dy}{dx} = \dfrac{x + 1}{(x^2 + 2x - 3)^2}$

38. $\dfrac{dy}{dx} = \dfrac{x - 4}{\sqrt{x^2 - 8x + 1}}$

 Slope Fields In Exercises 39–46, a differential equation, a point, and a slope field are given. A *slope field* consists of line segments with slopes given by the differential equation. These line segments give a visual perspective of the directions of the solutions of the differential equation. (a) Sketch two approximate solutions of the differential equation on the slope field, one of which passes through the given point. (To print an enlarged copy of the graph, go to the website *www.mathgraphs.com*.) (b) Use integration to find the particular solution of the differential equation and use a graphing utility to graph the solution. Compare the result with the sketches in part (a).

39. $\dfrac{dy}{dx} = x\sqrt{4 - x^2}$, $(2, 2)$

40. $\dfrac{dy}{dx} = x^2 (x^3 - 1)^2$, $(1, 0)$

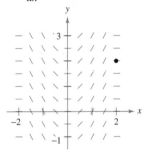

41. $\dfrac{dy}{dx} = x \cos x^2$, $(0, 1)$

42. $\dfrac{dy}{dx} = -2 \sec(2x) \tan(2x)$, $(0, -1)$

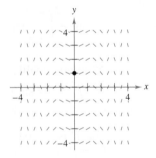

43. $\dfrac{dy}{dx} = 2e^{-x/2}$, $(0, 1)$

44. $\dfrac{dy}{dx} = xe^{-0.2x^2}$, $\left(0, -\dfrac{3}{2}\right)$

EXAMPLE 4   Completing the Square

Find $\displaystyle\int \frac{dx}{x^2 - 4x + 7}$.

Solution   You can write the denominator as the sum of two squares as shown.

$$x^2 - 4x + 7 = (x^2 - 4x + 4) - 4 + 7$$
$$= (x - 2)^2 + 3 = u^2 + a^2$$

Now, in this completed square form, let $u = x - 2$ and $a = \sqrt{3}$.

$$\int \frac{dx}{x^2 - 4x + 7} = \int \frac{dx}{(x-2)^2 + 3} = \frac{1}{\sqrt{3}} \arctan \frac{x-2}{\sqrt{3}} + C$$

If the leading coefficient is not 1, it helps to factor before completing the square. For instance, you can complete the square for $2x^2 - 8x + 10$ by factoring first.

$$2x^2 - 8x + 10 = 2(x^2 - 4x + 5)$$
$$= 2(x^2 - 4x + 4 - 4 + 5)$$
$$= 2[(x - 2)^2 + 1]$$

To complete the square when the coefficient of x^2 is negative, use the same factoring process shown above. For instance, you can complete the square for $3x - x^2$ as shown.

$$3x - x^2 = -(x^2 - 3x)$$
$$= -\left[x^2 - 3x + \left(\tfrac{3}{2}\right)^2 - \left(\tfrac{3}{2}\right)^2\right]$$
$$= \left(\tfrac{3}{2}\right)^2 - \left(x - \tfrac{3}{2}\right)^2$$

EXAMPLE 5   Completing the Square (Negative Leading Coefficient)

Find the area of the region bounded by the graph of

$$f(x) = \frac{1}{\sqrt{3x - x^2}}$$

the x-axis, and the lines $x = \tfrac{3}{2}$ and $x = \tfrac{9}{4}$.

Solution   From Figure 5.48, you can see that the area is given by

$$\text{Area} = \int_{3/2}^{9/4} \frac{1}{\sqrt{3x - x^2}}\, dx.$$

Using the completed square form derived above, you can integrate as shown.

$$\int_{3/2}^{9/4} \frac{dx}{\sqrt{3x - x^2}} = \int_{3/2}^{9/4} \frac{dx}{\sqrt{(3/2)^2 - [x - (3/2)]^2}}$$
$$= \arcsin \frac{x - (3/2)}{3/2} \bigg]_{3/2}^{9/4}$$
$$= \arcsin \frac{1}{2} - \arcsin 0$$
$$= \frac{\pi}{6}$$
$$\approx 0.524$$

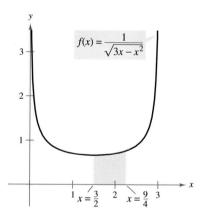

$$f(x) = \frac{1}{\sqrt{3x - x^2}}$$

The area of the region bounded by the graph of f, the x-axis, $x = \tfrac{3}{2}$, and $x = \tfrac{9}{4}$ is $\pi/6$.
Figure 5.48

TECHNOLOGY   With definite integrals such as the one given in Example 5, remember that you can resort to a numerical solution. For instance, applying Simpson's Rule (with $n = 12$) to the integral in the example, you obtain

$$\int_{3/2}^{9/4} \frac{1}{\sqrt{3x - x^2}}\, dx \approx 0.523599.$$

This differs from the exact value of the integral ($\pi/6 \approx 0.5235988$) by less than one millionth.

Review of Basic Integration Rules

You have now completed the introduction of the **basic integration rules.** To be efficient at applying these rules, you should have practiced enough so that each rule is committed to memory.

Basic Integration Rules ($a > 0$)

1. $\int kf(u)\,du = k\int f(u)\,du$

2. $\int [f(u) \pm g(u)]\,du = \int f(u)\,du \pm \int g(u)\,du$

3. $\int du = u + C$

4. $\int u^n\,du = \dfrac{u^{n+1}}{n+1} + C, \quad n \neq -1$

5. $\int \dfrac{du}{u} = \ln|u| + C$

6. $\int e^u\,du = e^u + C$

7. $\int a^u\,du = \left(\dfrac{1}{\ln a}\right)a^u + C$

8. $\int \sin u\,du = -\cos u + C$

9. $\int \cos u\,du = \sin u + C$

10. $\int \tan u\,du = -\ln|\cos u| + C$

11. $\int \cot u\,du = \ln|\sin u| + C$

12. $\int \sec u\,du = \ln|\sec u + \tan u| + C$

13. $\int \csc u\,du = -\ln|\csc u + \cot u| + C$

14. $\int \sec^2 u\,du = \tan u + C$

15. $\int \csc^2 u\,du = -\cot u + C$

16. $\int \sec u \tan u\,du = \sec u + C$

17. $\int \csc u \cot u\,du = -\csc u + C$

18. $\int \dfrac{du}{\sqrt{a^2 - u^2}} = \arcsin \dfrac{u}{a} + C$

19. $\int \dfrac{du}{a^2 + u^2} = \dfrac{1}{a}\arctan \dfrac{u}{a} + C$

20. $\int \dfrac{du}{u\sqrt{u^2 - a^2}} = \dfrac{1}{a}\text{arcsec} \dfrac{|u|}{a} + C$

You can learn a lot about the nature of integration by comparing this list with the summary of differentiation rules given in Section 3.6. For differentiation, you now have rules that allow you to differentiate *any* elementary function. For integration, this is far from true.

The integration rules listed above are primarily those that were happened on when developing differentiation rules. So far, you have not learned any rules or techniques for finding the antiderivative of a general product or quotient, the natural logarithmic function, or the inverse trigonometric functions. More importantly, you cannot apply any of the rules in this list unless you can create the proper *du* corresponding to the *u* in the formula. The point is that you need to work more on integration techniques, which you will do in Chapter 8. The next two examples should give you a better feeling for the integration problems that you *can* and *cannot* do with the techniques and rules you now know.

EXAMPLE 6   **Comparing Integration Problems**

Find as many of the following integrals as you can using the formulas and techniques you have studied so far in the text.

a. $\displaystyle\int \frac{dx}{x\sqrt{x^2 - 1}}$ b. $\displaystyle\int \frac{x\,dx}{\sqrt{x^2 - 1}}$ c. $\displaystyle\int \frac{dx}{\sqrt{x^2 - 1}}$

Solution

a. You *can* find this integral (it fits the Arcsecant Rule).

$$\int \frac{dx}{x\sqrt{x^2 - 1}} = \text{arcsec}|x| + C$$

b. You *can* find this integral (it fits the Power Rule).

$$\int \frac{x\,dx}{\sqrt{x^2 - 1}} = \frac{1}{2}\int (x^2 - 1)^{-1/2}(2x)\,dx$$

$$= \frac{1}{2}\left[\frac{(x^2 - 1)^{1/2}}{1/2}\right] + C$$

$$= \sqrt{x^2 - 1} + C$$

c. You *cannot* find this integral using present techniques. (You should scan the list of basic integration rules to verify this conclusion.)

EXAMPLE 7   **Comparing Integration Problems**

Find as many of the following integrals as you can using the formulas and techniques you have studied so far in the text.

a. $\displaystyle\int \frac{dx}{x\ln x}$ b. $\displaystyle\int \frac{\ln x\,dx}{x}$ c. $\displaystyle\int \ln x\,dx$

Solution

a. You *can* find this integral (it fits the Log Rule).

$$\int \frac{dx}{x\ln x} = \int \frac{1/x}{\ln x}\,dx$$

$$= \ln|\ln x| + C$$

b. You *can* find this integral (it fits the Power Rule).

$$\int \frac{\ln x\,dx}{x} = \int \left(\frac{1}{x}\right)(\ln x)^1\,dx$$

$$= \frac{(\ln x)^2}{2} + C$$

c. You *cannot* find this integral using present techniques. ▬▬▬▬

NOTE   Note in Examples 6 and 7 that the *simplest* functions are the ones that you cannot yet integrate.

In Exercises 1–20, find the integral.

1. $\displaystyle\int \frac{5}{\sqrt{9 - x^2}}\, dx$

2. $\displaystyle\int \frac{3}{\sqrt{1 - 4x^2}}\, dx$

3. $\displaystyle\int \frac{7}{16 + x^2}\, dx$

4. $\displaystyle\int \frac{4}{1 + 9x^2}\, dx$

5. $\displaystyle\int \frac{1}{x\sqrt{4x^2 - 1}}\, dx$

6. $\displaystyle\int \frac{1}{4 + (x - 1)^2}\, dx$

7. $\displaystyle\int \frac{x^3}{x^2 + 1}\, dx$

8. $\displaystyle\int \frac{x^4 - 1}{x^2 + 1}\, dx$

9. $\displaystyle\int \frac{1}{\sqrt{1 - (x + 1)^2}}\, dx$

10. $\displaystyle\int \frac{t}{t^4 + 16}\, dt$

11. $\displaystyle\int \frac{t}{\sqrt{1 - t^4}}\, dt$

12. $\displaystyle\int \frac{1}{x\sqrt{x^4 - 4}}\, dx$

13. $\displaystyle\int \frac{e^{2x}}{4 + e^{4x}}\, dx$

14. $\displaystyle\int \frac{1}{3 + (x - 2)^2}\, dx$

15. $\displaystyle\int \frac{1}{\sqrt{x}\sqrt{1 - x}}\, dx$

16. $\displaystyle\int \frac{3}{2\sqrt{x}(1 + x)}\, dx$

17. $\displaystyle\int \frac{x - 3}{x^2 + 1}\, dx$

18. $\displaystyle\int \frac{4x + 3}{\sqrt{1 - x^2}}\, dx$

19. $\displaystyle\int \frac{x + 5}{\sqrt{9 - (x - 3)^2}}\, dx$

20. $\displaystyle\int \frac{x - 2}{(x + 1)^2 + 4}\, dx$

In Exercises 21–30, evaluate the integral.

21. $\displaystyle\int_0^{1/6} \frac{1}{\sqrt{1 - 9x^2}}\, dx$

22. $\displaystyle\int_0^1 \frac{dx}{\sqrt{4 - x^2}}$

23. $\displaystyle\int_0^{\sqrt{3}/2} \frac{1}{1 + 4x^2}\, dx$

24. $\displaystyle\int_{\sqrt{3}}^3 \frac{1}{9 + x^2}\, dx$

25. $\displaystyle\int_0^{1/\sqrt{2}} \frac{\arcsin x}{\sqrt{1 - x^2}}\, dx$

26. $\displaystyle\int_0^{1/\sqrt{2}} \frac{\arccos x}{\sqrt{1 - x^2}}\, dx$

27. $\displaystyle\int_{-1/2}^0 \frac{x}{\sqrt{1 - x^2}}\, dx$

28. $\displaystyle\int_{-\sqrt{3}}^0 \frac{x}{1 + x^2}\, dx$

29. $\displaystyle\int_{\pi/2}^\pi \frac{\sin x}{1 + \cos^2 x}\, dx$

30. $\displaystyle\int_0^{\pi/2} \frac{\cos x}{1 + \sin^2 x}\, dx$

In Exercises 31–42, find or evaluate the integral. (Complete the square, if necessary.)

31. $\displaystyle\int_0^2 \frac{dx}{x^2 - 2x + 2}$

32. $\displaystyle\int_{-2}^2 \frac{dx}{x^2 + 4x + 13}$

33. $\displaystyle\int \frac{2x}{x^2 + 6x + 13}\, dx$

34. $\displaystyle\int \frac{2x - 5}{x^2 + 2x + 2}\, dx$

35. $\displaystyle\int \frac{1}{\sqrt{-x^2 - 4x}}\, dx$

36. $\displaystyle\int \frac{2}{\sqrt{-x^2 + 4x}}\, dx$

37. $\displaystyle\int \frac{x + 2}{\sqrt{-x^2 - 4x}}\, dx$

38. $\displaystyle\int \frac{x - 1}{\sqrt{x^2 - 2x}}\, dx$

39. $\displaystyle\int_2^3 \frac{2x - 3}{\sqrt{4x - x^2}}\, dx$

40. $\displaystyle\int \frac{1}{(x - 1)\sqrt{x^2 - 2x}}\, dx$

41. $\displaystyle\int \frac{x}{x^4 + 2x^2 + 2}\, dx$

42. $\displaystyle\int \frac{x}{\sqrt{9 + 8x^2 - x^4}}\, dx$

In Exercises 43–46, use the specified substitution to find or evaluate the integral.

43. $\displaystyle\int \sqrt{e^t - 3}\, dt$
$u = \sqrt{e^t - 3}$

44. $\displaystyle\int \frac{\sqrt{x - 2}}{x + 1}\, dx$
$u = \sqrt{x - 2}$

45. $\displaystyle\int_1^3 \frac{dx}{\sqrt{x}(1 + x)}$
$u = \sqrt{x}$

46. $\displaystyle\int_0^1 \frac{dx}{2\sqrt{3 - x}\sqrt{x + 1}}$
$u = \sqrt{x + 1}$

Writing About Concepts

In Exercises 47–50, determine which of the integrals can be found using the basic integration formulas you have studied so far in the text.

47. (a) $\displaystyle\int \frac{1}{\sqrt{1 - x^2}}\, dx$ (b) $\displaystyle\int \frac{x}{\sqrt{1 - x^2}}\, dx$ (c) $\displaystyle\int \frac{1}{x\sqrt{1 - x^2}}\, dx$

48. (a) $\displaystyle\int e^{x^2}\, dx$ (b) $\displaystyle\int xe^{x^2}\, dx$ (c) $\displaystyle\int \frac{1}{x^2}e^{1/x}\, dx$

49. (a) $\displaystyle\int \sqrt{x - 1}\, dx$ (b) $\displaystyle\int x\sqrt{x - 1}\, dx$ (c) $\displaystyle\int \frac{x}{\sqrt{x - 1}}\, dx$

50. (a) $\displaystyle\int \frac{1}{1 + x^4}\, dx$ (b) $\displaystyle\int \frac{x}{1 + x^4}\, dx$ (c) $\displaystyle\int \frac{x^3}{1 + x^4}\, dx$

51. Determine which value best approximates the area of the region between the x-axis and the function

$$f(x) = \frac{1}{\sqrt{1 - x^2}}$$

over the interval $[-0.5, 0.5]$. (Make your selection on the basis of a sketch of the region and *not* by performing any calculations.)

(a) 4 (b) -3 (c) 1 (d) 2 (e) 3

52. Decide whether you can find the integral

$$\int \frac{2\, dx}{\sqrt{x^2 + 4}}$$

using the formulas and techniques you have studied so far. Explain your reasoning.

Differential Equations **In Exercises 53 and 54, use the differential equation and the specified initial condition to find y.**

53. $\dfrac{dy}{dx} = \dfrac{1}{\sqrt{4 - x^2}}$
$y(0) = \pi$

54. $\dfrac{dy}{dx} = \dfrac{1}{4 + x^2}$
$y(2) = \pi$

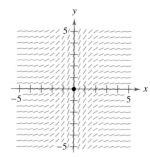 *Slope Fields* **In Exercises 55–58, a differential equation, a point, and a slope field are given. (a) Sketch two approximate solutions of the differential equation on the slope field, one of which passes through the given point. (b) Use integration to find the particular solution of the differential equation and use a graphing utility to graph the solution. Compare the result with the sketches in part (a). To print an enlarged copy of the graph, go to the website *www.mathgraphs.com*.**

55. $\dfrac{dy}{dx} = \dfrac{3}{1 + x^2}$, $(0, 0)$ **56.** $\dfrac{dy}{dx} = \dfrac{2}{9 + x^2}$, $(0, 2)$

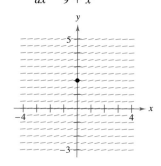

57. $\dfrac{dy}{dx} = \dfrac{1}{x\sqrt{x^2 - 4}}$, $(2, 1)$ **58.** $\dfrac{dy}{dx} = \dfrac{2}{\sqrt{25 - x^2}}$, $(5, \pi)$

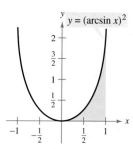 *Slope Fields* **In Exercises 59–62, use a computer algebra system to graph the slope field for the differential equation and graph the solution satisfying the specified initial condition.**

59. $\dfrac{dy}{dx} = \dfrac{10}{x\sqrt{x^2 - 1}}$

$y(3) = 0$

60. $\dfrac{dy}{dx} = \dfrac{1}{12 + x^2}$

$y(4) = 2$

61. $\dfrac{dy}{dx} = \dfrac{2y}{\sqrt{16 - x^2}}$

$y(0) = 2$

62. $\dfrac{dy}{dx} = \dfrac{\sqrt{y}}{1 + x^2}$

$y(0) = 4$

Area **In Exercises 63–68, find the area of the region.**

63. $y = \dfrac{1}{x^2 - 2x + 5}$ **64.** $y = \dfrac{2}{x^2 + 4x + 8}$

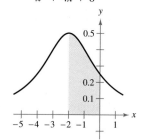

65. $y = \dfrac{1}{\sqrt{4 - x^2}}$ **66.** $y = \dfrac{1}{x\sqrt{x^2 - 1}}$

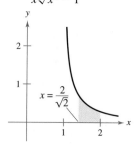

67. $y = \dfrac{3 \cos x}{1 + \sin^2 x}$ **68.** $y = \dfrac{e^x}{1 + e^{2x}}$

In Exercises 69 and 70, (a) verify the integration formula, and (b) use it to find the area of the region.

69. $\displaystyle\int \dfrac{\arctan x}{x^2}\, dx = \ln x - \dfrac{1}{2}\ln(1 + x^2) - \dfrac{\arctan x}{x} + C$

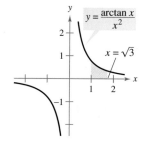

Figure for 69   Figure for 70

70. $\displaystyle\int (\arcsin x)^2\, dx$

$= x(\arcsin x)^2 - 2x + 2\sqrt{1 - x^2}\,\arcsin x + C$

71. (a) Sketch the region whose area is represented by

$$\int_0^1 \arcsin x \, dx.$$

(b) Use the integration capabilities of a graphing utility to approximate the area.

(c) Find the exact area analytically.

72. (a) Show that $\int_0^1 \dfrac{4}{1 + x^2} \, dx = \pi.$

(b) Approximate the number π using Simpson's Rule (with $n = 6$) and the integral in part (a).

(c) Approximate the number π by using the integration capabilities of a graphing utility.

73. *Investigation* Consider the function $F(x) = \dfrac{1}{2}\displaystyle\int_x^{x+2} \dfrac{2}{t^2 + 1} \, dt.$

(a) Write a short paragraph giving a geometric interpretation of the function $F(x)$ relative to the function $f(x) = \dfrac{2}{x^2 + 1}.$ Use what you have written to guess the value of x that will make F maximum.

(b) Perform the specified integration to find an alternative form of $F(x)$. Use calculus to locate the value of x that will make F maximum and compare the result with your guess in part (a).

74. Consider the integral $\displaystyle\int \dfrac{1}{\sqrt{6x - x^2}} \, dx.$

(a) Find the integral by completing the square of the radicand.

(b) Find the integral by making the substitution $u = \sqrt{x}.$

(c) The antiderivatives in parts (a) and (b) appear to be significantly different. Use a graphing utility to graph each antiderivative in the same viewing window and determine the relationship between them. Find the domain of each.

True or False? In Exercises 75–78, determine whether the statement is true or false. If it is false, explain why or give an example that shows it is false.

75. $\displaystyle\int \dfrac{dx}{3x\sqrt{9x^2 - 16}} = \dfrac{1}{4} \operatorname{arcsec} \dfrac{3x}{4} + C$

76. $\displaystyle\int \dfrac{dx}{25 + x^2} = \dfrac{1}{25} \arctan \dfrac{x}{25} + C$

77. $\displaystyle\int \dfrac{dx}{\sqrt{4 - x^2}} = -\arccos \dfrac{x}{2} + C$

78. One way to find $\displaystyle\int \dfrac{2e^{2x}}{\sqrt{9 - e^{2x}}} \, dx$ is to use the Arcsine Rule.

Verifying Integration Rules In Exercises 79–81, verify each rule by differentiating. Let $a > 0.$

79. $\displaystyle\int \dfrac{du}{\sqrt{a^2 - u^2}} = \arcsin \dfrac{u}{a} + C$

80. $\displaystyle\int \dfrac{du}{a^2 + u^2} = \dfrac{1}{a} \arctan \dfrac{u}{a} + C$

81. $\displaystyle\int \dfrac{du}{u\sqrt{u^2 - a^2}} = \dfrac{1}{a} \operatorname{arcsec} \dfrac{|u|}{a} + C$

82. *Numerical Integration* (a) Write an integral that represents the area of the region. (b) Then use the Trapezoidal Rule with $n = 8$ to estimate the area of the region. (c) Explain how you can use the results of parts (a) and (b) to estimate π.

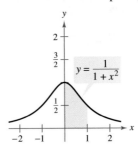

83. *Vertical Motion* An object is projected upward from ground level with an initial velocity of 500 feet per second. In this exercise, the goal is to analyze the motion of the object during its upward flight.

(a) If air resistance is neglected, find the velocity of the object as a function of time. Use a graphing utility to graph this function.

(b) Use the result in part (a) to find the position function and determine the maximum height attained by the object.

(c) If the air resistance is proportional to the square of the velocity, you obtain the equation

$$\dfrac{dv}{dt} = -(32 + kv^2)$$

where -32 feet per second per second is the acceleration due to gravity and k is a constant. Find the velocity as a function of time by solving the equation

$$\int \dfrac{dv}{32 + kv^2} = -\int dt.$$

(d) Use a graphing utility to graph the velocity function $v(t)$ in part (c) if $k = 0.001$. Use the graph to approximate the time t_0 at which the object reaches its maximum height.

(e) Use the integration capabilities of a graphing utility to approximate the integral

$$\int_0^{t_0} v(t) \, dt$$

where $v(t)$ and t_0 are those found in part (d). This is the approximation of the maximum height of the object.

(f) Explain the difference between the results in parts (b) and (e).

FOR FURTHER INFORMATION For more information on this topic, see "What Goes Up Must Come Down; Will Air Resistance Make It Return Sooner, or Later?" by John Lekner in *Mathematics Magazine*. To view this article, go to the website *www.matharticles.com*.

84. Graph $y_1 = \dfrac{x}{1 + x^2},$ $y_2 = \arctan x,$ and $y_3 = x$ on $[0, 10].$

Prove that $\dfrac{x}{1 + x^2} < \arctan x < x$ for $x > 0.$

Section 5.9

Hyperbolic Functions

- Develop properties of hyperbolic functions.
- Differentiate and integrate hyperbolic functions.
- Develop properties of inverse hyperbolic functions.
- Differentiate and integrate functions involving inverse hyperbolic functions.

Hyperbolic Functions

In this section you will look briefly at a special class of exponential functions called **hyperbolic functions.** The name *hyperbolic function* arose from comparison of the area of a semicircular region, as shown in Figure 5.49, with the area of a region under a hyperbola, as shown in Figure 5.50. The integral for the semicircular region involves an inverse trigonometric (circular) function:

$$\int_{-1}^{1} \sqrt{1 - x^2}\, dx = \frac{1}{2}\left[x\sqrt{1 - x^2} + \arcsin x \right]_{-1}^{1} = \frac{\pi}{2} \approx 1.571.$$

The integral for the hyperbolic region involves an inverse hyperbolic function:

$$\int_{-1}^{1} \sqrt{1 + x^2}\, dx = \frac{1}{2}\left[x\sqrt{1 + x^2} + \sinh^{-1} x \right]_{-1}^{1} \approx 2.296.$$

This is only one of many ways in which the hyperbolic functions are similar to the trigonometric functions.

JOHANN HEINRICH LAMBERT (1728–1777)

The first person to publish a comprehensive study on hyperbolic functions was Johann Heinrich Lambert, a Swiss-German mathematician and colleague of Euler.

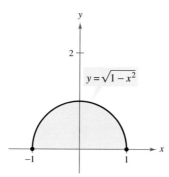

Circle: $x^2 + y^2 = 1$
Figure 5.49

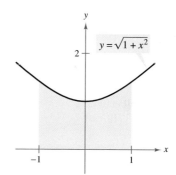

Hyperbola: $-x^2 + y^2 = 1$
Figure 5.50

FOR FURTHER INFORMATION For more information on the development of hyperbolic functions, see the article "An Introduction to Hyperbolic Functions in Elementary Calculus" by Jerome Rosenthal in *Mathematics Teacher*. To view this article, go to the website *www.matharticles.com*.

Definitions of the Hyperbolic Functions

$$\sinh x = \frac{e^x - e^{-x}}{2} \qquad\qquad \operatorname{csch} x = \frac{1}{\sinh x}, \quad x \neq 0$$

$$\cosh x = \frac{e^x + e^{-x}}{2} \qquad\qquad \operatorname{sech} x = \frac{1}{\cosh x}$$

$$\tanh x = \frac{\sinh x}{\cosh x} \qquad\qquad \coth x = \frac{1}{\tanh x}, \quad x \neq 0$$

NOTE   sinh x is read as "the hyperbolic sine of x," cosh x as "the hyperbolic cosine of x," and so on.

The graphs of the six hyperbolic functions and their domains and ranges are shown in Figure 5.51. Note that the graph of $\sinh x$ can be obtained by *addition of ordinates* using the exponential functions $f(x) = \frac{1}{2}e^x$ and $g(x) = -\frac{1}{2}e^{-x}$. Likewise, the graph of $\cosh x$ can be obtained by *addition of ordinates* using the exponential functions $f(x) = \frac{1}{2}e^x$ and $h(x) = \frac{1}{2}e^{-x}$.

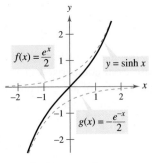

Domain: $(-\infty, \infty)$
Range: $(-\infty, \infty)$

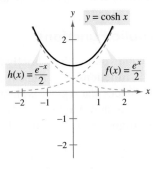

Domain: $(-\infty, \infty)$
Range: $[1, \infty)$

Domain: $(-\infty, \infty)$
Range: $(-1, 1)$

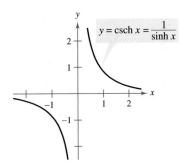

Domain: $(-\infty, 0) \cup (0, \infty)$
Range: $(-\infty, 0) \cup (0, \infty)$

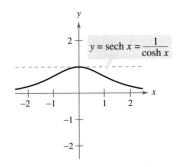

Domain: $(-\infty, \infty)$
Range: $(0, 1]$

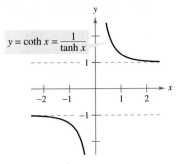

Domain: $(-\infty, 0) \cup (0, \infty)$
Range: $(-\infty, -1) \cup (1, \infty)$

Figure 5.51

Many of the trigonometric identities have corresponding *hyperbolic identities*. For instance,

$$
\begin{aligned}
\cosh^2 x - \sinh^2 x &= \left(\frac{e^x + e^{-x}}{2}\right)^2 - \left(\frac{e^x - e^{-x}}{2}\right)^2 \\
&= \frac{e^{2x} + 2 + e^{-2x}}{4} - \frac{e^{2x} - 2 + e^{-2x}}{4} \\
&= \frac{4}{4} \\
&= 1
\end{aligned}
$$

and

$$
\begin{aligned}
2 \sinh x \cosh x &= 2\left(\frac{e^x - e^{-x}}{2}\right)\left(\frac{e^x + e^{-x}}{2}\right) \\
&= \frac{e^{2x} - e^{-2x}}{2} \\
&= \sinh 2x.
\end{aligned}
$$

FOR FURTHER INFORMATION To understand geometrically the relationship between the hyperbolic and exponential functions, see the article "A Short Proof Linking the Hyperbolic and Exponential Functions" by Michael J. Seery in *The AMATYC Review*.

Hyperbolic Identities

$$\cosh^2 x - \sinh^2 x = 1 \qquad \sinh(x + y) = \sinh x \cosh y + \cosh x \sinh y$$

$$\tanh^2 x + \operatorname{sech}^2 x = 1 \qquad \sinh(x - y) = \sinh x \cosh y - \cosh x \sinh y$$

$$\coth^2 x - \operatorname{csch}^2 x = 1 \qquad \cosh(x + y) = \cosh x \cosh y + \sinh x \sinh y$$

$$\cosh(x - y) = \cosh x \cosh y - \sinh x \sinh y$$

$$\sinh^2 x = \frac{-1 + \cosh 2x}{2} \qquad \cosh^2 x = \frac{1 + \cosh 2x}{2}$$

$$\sinh 2x = 2 \sinh x \cosh x \qquad \cosh 2x = \cosh^2 x + \sinh^2 x$$

Differentiation and Integration of Hyperbolic Functions

Because the hyperbolic functions are written in terms of e^x and e^{-x}, you can easily derive rules for their derivatives. The following theorem lists these derivatives with the corresponding integration rules.

THEOREM 5.22   Derivatives and Integrals of Hyperbolic Functions

Let u be a differentiable function of x.

$$\frac{d}{dx}[\sinh u] = (\cosh u)u' \qquad \int \cosh u \, du = \sinh u + C$$

$$\frac{d}{dx}[\cosh u] = (\sinh u)u' \qquad \int \sinh u \, du = \cosh u + C$$

$$\frac{d}{dx}[\tanh u] = (\operatorname{sech}^2 u)u' \qquad \int \operatorname{sech}^2 u \, du = \tanh u + C$$

$$\frac{d}{dx}[\coth u] = -(\operatorname{csch}^2 u)u' \qquad \int \operatorname{csch}^2 u \, du = -\coth u + C$$

$$\frac{d}{dx}[\operatorname{sech} u] = -(\operatorname{sech} u \tanh u)u' \qquad \int \operatorname{sech} u \tanh u \, du = -\operatorname{sech} u + C$$

$$\frac{d}{dx}[\operatorname{csch} u] = -(\operatorname{csch} u \coth u)u' \qquad \int \operatorname{csch} u \coth u \, du = -\operatorname{csch} u + C$$

Proof

$$\frac{d}{dx}[\sinh x] = \frac{d}{dx}\left[\frac{e^x - e^{-x}}{2}\right]$$

$$= \frac{e^x + e^{-x}}{2} = \cosh x$$

$$\frac{d}{dx}[\tanh x] = \frac{d}{dx}\left[\frac{\sinh x}{\cosh x}\right]$$

$$= \frac{\cosh x(\cosh x) - \sinh x(\sinh x)}{\cosh^2 x}$$

$$= \frac{1}{\cosh^2 x}$$

$$= \operatorname{sech}^2 x$$

In Exercises 98 and 102, you are asked to prove some of the other differentiation rules.

EXAMPLE I Differentiation of Hyperbolic Functions

a. $\dfrac{d}{dx}\left[\sinh(x^2 - 3)\right] = 2x \cosh(x^2 - 3)$ **b.** $\dfrac{d}{dx}\left[\ln(\cosh x)\right] = \dfrac{\sinh x}{\cosh x} = \tanh x$

c. $\dfrac{d}{dx}\left[x \sinh x - \cosh x\right] = x \cosh x + \sinh x - \sinh x = x \cosh x$

EXAMPLE 2 Finding Relative Extrema

Find the relative extrema of $f(x) = (x - 1) \cosh x - \sinh x$.

Solution Begin by setting the first derivative of f equal to 0.

$$f'(x) = (x - 1) \sinh x + \cosh x - \cosh x = 0$$
$$(x - 1) \sinh x = 0$$

So, the critical numbers are $x = 1$ and $x = 0$. Using the Second Derivative Test, you can verify that the point $(0, -1)$ yields a relative maximum and the point $(1, -\sinh 1)$ yields a relative minimum, as shown in Figure 5.52. Try using a graphing utility to confirm this result. If your graphing utility does not have hyperbolic functions, you can use exponential functions as follows.

$$\begin{aligned} f(x) &= (x - 1)\left(\tfrac{1}{2}\right)(e^x + e^{-x}) - \tfrac{1}{2}(e^x - e^{-x}) \\ &= \tfrac{1}{2}(xe^x + xe^{-x} - e^x - e^{-x} - e^x + e^{-x}) \\ &= \tfrac{1}{2}(xe^x + xe^{-x} - 2e^x) \end{aligned}$$

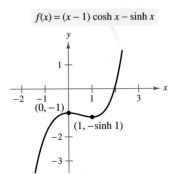

$f(x) = (x - 1) \cosh x - \sinh x$

$f''(0) < 0$, so $(0, -1)$ is a relative maximum. $f''(1) > 0$, so $(1, -\sinh 1)$ is a relative minimum.
Figure 5.52

When a uniform flexible cable, such as a telephone wire, is suspended from two points, it takes the shape of a *catenary*, as discussed in Example 3.

EXAMPLE 3 Hanging Power Cables

Power cables are suspended between two towers, forming the catenary shown in Figure 5.53. The equation for this catenary is

$$y = a \cosh \frac{x}{a}.$$

The distance between the two towers is $2b$. Find the slope of the catenary at the point where the cable meets the right-hand tower.

Solution Differentiating produces

$$y' = a\left(\frac{1}{a}\right) \sinh \frac{x}{a} = \sinh \frac{x}{a}.$$

At the point $(b, a \cosh(b/a))$, the slope (from the left) is given by $m = \sinh \dfrac{b}{a}$.

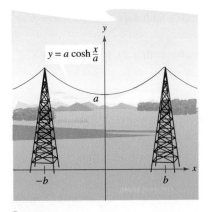

$y = a \cosh \dfrac{x}{a}$

Catenary
Figure 5.53

FOR FURTHER INFORMATION In Example 3, the cable is a catenary between two supports at the same height. To learn about the shape of a cable hanging between supports of different heights, see the article "Reexamining the Catenary" by Paul Cella in *The College Mathematics Journal*. To view this article, go to the website *www.matharticles.com*.

EXAMPLE 4   **Integrating a Hyperbolic Function**

Find $\displaystyle\int \cosh 2x \sinh^2 2x\, dx$.

Solution

$$\int \cosh 2x \sinh^2 2x\, dx = \frac{1}{2}\int (\sinh 2x)^2 (2 \cosh 2x)\, dx \qquad u = \sinh 2x$$

$$= \frac{1}{2}\left[\frac{(\sinh 2x)^3}{3} \right] + C$$

$$= \frac{\sinh^3 2x}{6} + C$$

Inverse Hyperbolic Functions

Unlike trigonometric functions, hyperbolic functions are not periodic. In fact, by looking back at Figure 5.51, you can see that four of the six hyperbolic functions are actually one-to-one (the hyperbolic sine, tangent, cosecant, and cotangent). So, you can conclude that these four functions have inverse functions. The other two (the hyperbolic cosine and secant) are one-to-one if their domains are restricted to the positive real numbers, and for this restricted domain they also have inverse functions. Because the hyperbolic functions are defined in terms of exponential functions, it is not surprising to find that the inverse hyperbolic functions can be written in terms of logarithmic functions, as shown in Theorem 5.23.

THEOREM 5.23   Inverse Hyperbolic Functions

Function	Domain
$\sinh^{-1} x = \ln(x + \sqrt{x^2 + 1})$	$(-\infty, \infty)$
$\cosh^{-1} x = \ln(x + \sqrt{x^2 - 1})$	$[1, \infty)$
$\tanh^{-1} x = \dfrac{1}{2}\ln\dfrac{1 + x}{1 - x}$	$(-1, 1)$
$\coth^{-1} x = \dfrac{1}{2}\ln\dfrac{x + 1}{x - 1}$	$(-\infty, -1) \cup (1, \infty)$
$\operatorname{sech}^{-1} x = \ln\dfrac{1 + \sqrt{1 - x^2}}{x}$	$(0, 1]$
$\operatorname{csch}^{-1} x = \ln\left(\dfrac{1}{x} + \dfrac{\sqrt{1 + x^2}}{\lvert x \rvert}\right)$	$(-\infty, 0) \cup (0, \infty)$

Proof   The proof of this theorem is a straightforward application of the properties of the exponential and logarithmic functions. For example, if

$$f(x) = \sinh x = \frac{e^x - e^{-x}}{2}$$

and

$$g(x) = \ln(x + \sqrt{x^2 + 1})$$

you can show that $f(g(x)) = x$ and $g(f(x)) = x$, which implies that g is the inverse function of f.

EXAMPLE 6   More About a Tractrix

For the tractrix given in Example 5, show that the boat is always pointing toward the person.

Solution   For a point (x, y) on a tractrix, the slope of the graph gives the direction of the boat, as shown in Figure 5.56.

$$y' = \frac{d}{dx}\left[20 \operatorname{sech}^{-1} \frac{x}{20} - \sqrt{20^2 - x^2} \right]$$

$$= -20\left(\frac{1}{20}\right)\left[\frac{1}{(x/20)\sqrt{1 - (x/20)^2}} \right] - \left(\frac{1}{2}\right)\left(\frac{-2x}{\sqrt{20^2 - x^2}} \right)$$

$$= \frac{-20^2}{x\sqrt{20^2 - x^2}} + \frac{x}{\sqrt{20^2 - x^2}}$$

$$= -\frac{\sqrt{20^2 - x^2}}{x}$$

However, from Figure 5.56, you can see that the slope of the line segment connecting the point $(0, y_1)$ with the point (x, y) is also

$$m = -\frac{\sqrt{20^2 - x^2}}{x}.$$

So, the boat is always pointing toward the person. (It is because of this property that a tractrix is called a *pursuit curve*.)

EXAMPLE 7   Integration Using Inverse Hyperbolic Functions

Find $\displaystyle\int \frac{dx}{x\sqrt{4 - 9x^2}}$.

Solution   Let $a = 2$ and $u = 3x$.

$$\int \frac{dx}{x\sqrt{4 - 9x^2}} = \int \frac{3\,dx}{(3x)\sqrt{4 - 9x^2}} \qquad\qquad \int \frac{du}{u\sqrt{a^2 - u^2}}$$

$$= -\frac{1}{2}\ln \frac{2 + \sqrt{4 - 9x^2}}{|3x|} + C \qquad -\frac{1}{a}\ln \frac{a + \sqrt{a^2 - u^2}}{|u|} + C$$

EXAMPLE 8   Integration Using Inverse Hyperbolic Functions

Find $\displaystyle\int \frac{dx}{5 - 4x^2}$.

Solution   Let $a = \sqrt{5}$ and $u = 2x$.

$$\int \frac{dx}{5 - 4x^2} = \frac{1}{2}\int \frac{2\,dx}{(\sqrt{5})^2 - (2x)^2} \qquad\qquad \int \frac{du}{a^2 - u^2}$$

$$= \frac{1}{2}\left(\frac{1}{2\sqrt{5}}\ln\left|\frac{\sqrt{5} + 2x}{\sqrt{5} - 2x}\right| \right) + C \qquad \frac{1}{2a}\ln\left|\frac{a + u}{a - u}\right| + C$$

$$= \frac{1}{4\sqrt{5}}\ln\left|\frac{\sqrt{5} + 2x}{\sqrt{5} - 2x}\right| + C$$

Exercises for Section 5.9

In Exercises 1–6, evaluate the function. If the value is not a rational number, round your answer to three decimal places.

1. (a) $\sinh 3$

   (b) $\tanh(-2)$

2. (a) $\cosh 0$

   (b) $\text{sech } 1$

3. (a) $\text{csch}(\ln 2)$

   (b) $\coth(\ln 5)$

4. (a) $\sinh^{-1} 0$

   (b) $\tanh^{-1} 0$

5. (a) $\cosh^{-1} 2$

   (b) $\text{sech}^{-1} \frac{2}{3}$

6. (a) $\text{csch}^{-1} 2$

   (b) $\coth^{-1} 3$

In Exercises 7–12, verify the identity.

7. $\tanh^2 x + \text{sech}^2 x = 1$

8. $\cosh^2 x = \dfrac{1 + \cosh 2x}{2}$

9. $\sinh(x + y) = \sinh x \cosh y + \cosh x \sinh y$

10. $\sinh 2x = 2 \sinh x \cosh x$

11. $\sinh 3x = 3 \sinh x + 4 \sinh^3 x$

12. $\cosh x + \cosh y = 2 \cosh \dfrac{x + y}{2} \cosh \dfrac{x - y}{2}$

In Exercises 13 and 14, use the value of the given hyperbolic function to find the values of the other hyperbolic functions at x.

13. $\sinh x = \dfrac{3}{2}$

14. $\tanh x = \dfrac{1}{2}$

In Exercises 15–24, find the derivative of the function.

15. $y = \text{sech}(x + 1)$

16. $y = \coth 3x$

17. $f(x) = \ln(\sinh x)$

18. $g(x) = \ln(\cosh x)$

19. $y = \ln\left(\tanh \dfrac{x}{2}\right)$

20. $y = x \cosh x - \sinh x$

21. $h(x) = \dfrac{1}{4} \sinh 2x - \dfrac{x}{2}$

22. $h(t) = t - \coth t$

23. $f(t) = \arctan(\sinh t)$

24. $g(x) = \text{sech}^2 3x$

In Exercises 25–28, find an equation of the tangent line to the graph of the function at the given point.

25. $y = \sinh(1 - x^2)$

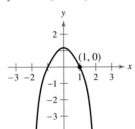

26. $y = x^{\cosh x}$

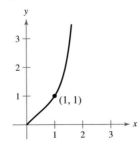

27. $y = (\cosh x - \sinh x)^2$

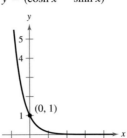

28. $y = e^{\sinh x}$

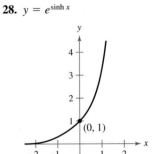

In Exercises 29–32, find any relative extrema of the function. Use a graphing utility to confirm your result.

29. $f(x) = \sin x \sinh x - \cos x \cosh x$, $\quad -4 \le x \le 4$

30. $f(x) = x \sinh(x - 1) - \cosh(x - 1)$

31. $g(x) = x \, \text{sech } x$

32. $h(x) = 2 \tanh x - x$

In Exercises 33 and 34, show that the function satisfies the differential equation.

Function	Differential Equation
33. $y = a \sinh x$	$y''' - y' = 0$
34. $y = a \cosh x$	$y'' - y = 0$

Linear and Quadratic Approximations **In Exercises 35 and 36, use a computer algebra system to find the linear approximation**

$$P_1(x) = f(a) + f'(a)(x - a)$$

and the quadratic approximation

$$P_2(x) = f(a) + f'(a)(x - a) + \tfrac{1}{2}f''(a)(x - a)^2$$

of the function f at $x = a$. Use a graphing utility to graph the function and its linear and quadratic approximations.

35. $f(x) = \tanh x$, $\quad a = 0$

36. $f(x) = \cosh x$, $\quad a = 0$

Catenary **In Exercises 37 and 38, a model for a power cable suspended between two towers is given. (a) Graph the model, (b) find the heights of the cable at the towers and at the midpoint between the towers, and (c) find the slope of the model at the point where the cable meets the right-hand tower.**

37. $y = 10 + 15 \cosh \dfrac{x}{15}$, $\quad -15 \le x \le 15$

38. $y = 18 + 25 \cosh \dfrac{x}{25}$, $\quad -25 \le x \le 25$

In Exercises 39–50, find the integral.

39. $\displaystyle \int \sinh(1 - 2x)\, dx$

40. $\displaystyle \int \dfrac{\cosh \sqrt{x}}{\sqrt{x}}\, dx$

41. $\displaystyle \int \cosh^2(x - 1) \sinh(x - 1)\, dx$

42. $\displaystyle \int \dfrac{\sinh x}{1 + \sinh^2 x}\, dx$

43. $\displaystyle\int \frac{\cosh x}{\sinh x}\,dx$

44. $\displaystyle\int \operatorname{sech}^2(2x - 1)\,dx$

45. $\displaystyle\int x\,\operatorname{csch}^2 \frac{x^2}{2}\,dx$

46. $\displaystyle\int \operatorname{sech}^3 x \tanh x\,dx$

47. $\displaystyle\int \frac{\operatorname{csch}(1/x)\coth(1/x)}{x^2}\,dx$

48. $\displaystyle\int \frac{\cosh x}{\sqrt{9 - \sinh^2 x}}\,dx$

49. $\displaystyle\int \frac{x}{x^4 + 1}\,dx$

50. $\displaystyle\int \frac{2}{x\sqrt{1 + 4x^2}}\,dx$

In Exercises 51–56, evaluate the integral.

51. $\displaystyle\int_0^{\ln 2} \tanh x\,dx$

52. $\displaystyle\int_0^1 \cosh^2 x\,dx$

53. $\displaystyle\int_0^4 \frac{1}{25 - x^2}\,dx$

54. $\displaystyle\int_0^4 \frac{1}{\sqrt{25 - x^2}}\,dx$

55. $\displaystyle\int_0^{\sqrt{2}/4} \frac{2}{\sqrt{1 - 4x^2}}\,dx$

56. $\displaystyle\int_0^{\ln 2} 2e^{-x}\cosh x\,dx$

In Exercises 57–64, find the derivative of the function.

57. $y = \cosh^{-1}(3x)$

58. $y = \tanh^{-1}\dfrac{x}{2}$

59. $y = \sinh^{-1}(\tan x)$

60. $y = \operatorname{sech}^{-1}(\cos 2x), \quad 0 < x < \pi/4$

61. $y = \tanh^{-1}(\sin 2x)$

62. $y = (\operatorname{csch}^{-1} x)^2$

63. $y = 2x\sinh^{-1}(2x) - \sqrt{1 + 4x^2}$

64. $y = x\tanh^{-1} x + \ln\sqrt{1 - x^2}$

Writing About Concepts

65. Discuss several ways in which the hyperbolic functions are similar to the trigonometric functions.

66. Sketch the graph of each hyperbolic function. Then identify the domain and range of each function.

Limits **In Exercises 67–72, find the limit.**

67. $\displaystyle\lim_{x\to\infty} \sinh x$

68. $\displaystyle\lim_{x\to\infty} \tanh x$

69. $\displaystyle\lim_{x\to\infty} \operatorname{sech} x$

70. $\displaystyle\lim_{x\to-\infty} \operatorname{csch} x$

71. $\displaystyle\lim_{x\to 0} \frac{\sinh x}{x}$

72. $\displaystyle\lim_{x\to 0^-} \coth x$

In Exercises 73–80, find the indefinite integral using the formulas of Theorem 5.24.

73. $\displaystyle\int \frac{1}{\sqrt{1 + e^{2x}}}\,dx$

74. $\displaystyle\int \frac{x}{9 - x^4}\,dx$

75. $\displaystyle\int \frac{1}{\sqrt{x}\sqrt{1 + x}}\,dx$

76. $\displaystyle\int \frac{\sqrt{x}}{\sqrt{1 + x^3}}\,dx$

77. $\displaystyle\int \frac{-1}{4x - x^2}\,dx$

78. $\displaystyle\int \frac{dx}{(x + 2)\sqrt{x^2 + 4x + 8}}$

79. $\displaystyle\int \frac{1}{1 - 4x - 2x^2}\,dx$

80. $\displaystyle\int \frac{dx}{(x + 1)\sqrt{2x^2 + 4x + 8}}$

In Exercises 81–84, solve the differential equation.

81. $\dfrac{dy}{dx} = \dfrac{1}{\sqrt{80 + 8x - 16x^2}}$

82. $\dfrac{dy}{dx} = \dfrac{1}{(x - 1)\sqrt{-4x^2 + 8x - 1}}$

83. $\dfrac{dy}{dx} = \dfrac{x^3 - 21x}{5 + 4x - x^2}$

84. $\dfrac{dy}{dx} = \dfrac{1 - 2x}{4x - x^2}$

Area **In Exercises 85–88, find the area of the region.**

85. $y = \operatorname{sech}\dfrac{x}{2}$

86. $y = \tanh 2x$

87. $y = \dfrac{5x}{\sqrt{x^4 + 1}}$

88. $y = \dfrac{6}{\sqrt{x^2 - 4}}$

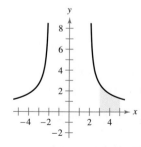

In Exercises 89 and 90, evaluate the integral in terms of (a) natural logarithms and (b) inverse hyperbolic functions.

89. $\displaystyle\int_0^{\sqrt{3}} \frac{dx}{\sqrt{x^2 + 1}}$

90. $\displaystyle\int_{-1/2}^{1/2} \frac{dx}{1 - x^2}$

91. *Chemical Reactions* Chemicals A and B combine in a 3-to-1 ratio to form a compound. The amount of compound x being produced at any time t is proportional to the unchanged amounts of A and B remaining in the solution. So, if 3 kilograms of A is mixed with 2 kilograms of B, you have

$$\frac{dx}{dt} = k\left(3 - \frac{3x}{4}\right)\left(2 - \frac{x}{4}\right) = \frac{3k}{16}(x^2 - 12x + 32).$$

One kilogram of the compound is formed after 10 minutes. Find the amount formed after 20 minutes by solving the equation

$$\int \frac{3k}{16}\,dt = \int \frac{dx}{x^2 - 12x + 32}.$$

92. *Vertical Motion* An object is dropped from a height of 400 feet.

(a) Find the velocity of the object as a function of time (neglect air resistance on the object).

(b) Use the result in part (a) to find the position function.

(c) If the air resistance is proportional to the square of the velocity, then $dv/dt = -32 + kv^2$, where -32 feet per second per second is the acceleration due to gravity and k is a constant. Show that the velocity v as a function of time t is

$$v(t) = -\sqrt{\frac{32}{k}} \tanh\left(\sqrt{32k}\, t\right)$$

by performing the following integration and simplifying the result.

$$\int \frac{dv}{32 - kv^2} = -\int dt$$

(d) Use the result in part (c) to find $\lim_{t \to \infty} v(t)$ and give its interpretation.

(e) Integrate the velocity function in part (c) and find the position s of the object as a function of t. Use a graphing utility to graph the position function when $k = 0.01$ and the position function in part (b) in the same viewing window. Estimate the additional time required for the object to reach ground level when air resistance is not neglected.

(f) Give a written description of what you believe would happen if k were increased. Then test your assertion with a particular value of k.

Tractrix **In Exercises 93 and 94, use the equation of the tractrix**

$$y = a \operatorname{sech}^{-1} \frac{x}{a} - \sqrt{a^2 - x^2}, \quad a > 0.$$

93. Find dy/dx.

94. Let L be the tangent line to the tractrix at the point P. If L intersects the y-axis at the point Q, show that the distance between P and Q is a.

95. Prove that $\tanh^{-1} x = \dfrac{1}{2} \ln\left(\dfrac{1 + x}{1 - x}\right)$, $\quad -1 < x < 1$.

96. Show that $\arctan(\sinh x) = \arcsin(\tanh x)$.

97. Let $x > 0$ and $b > 0$. Show that $\displaystyle\int_{-b}^{b} e^{xt}\, dt = \dfrac{2 \sinh bx}{x}$.

In Exercises 98–102, verify the differentiation formula.

98. $\dfrac{d}{dx}[\cosh x] = \sinh x$ **99.** $\dfrac{d}{dx}[\operatorname{sech}^{-1} x] = \dfrac{-1}{x\sqrt{1 - x^2}}$

100. $\dfrac{d}{dx}[\cosh^{-1} x] = \dfrac{1}{\sqrt{x^2 - 1}}$ **101.** $\dfrac{d}{dx}[\sinh^{-1} x] = \dfrac{1}{\sqrt{x^2 + 1}}$

102. $\dfrac{d}{dx}[\operatorname{sech} x] = -\operatorname{sech} x \tanh x$

Putnam Exam Challenge

103. From the vertex $(0, c)$ of the catenary $y = c \cosh(x/c)$ a line L is drawn perpendicular to the tangent to the catenary at a point P. Prove that the length of L intercepted by the axes is equal to the ordinate y of the point P.

104. Prove or disprove that there is at least one straight line normal to the graph of $y = \cosh x$ at a point $(a, \cosh a)$ and also normal to the graph of $y = \sinh x$ at a point $(c, \sinh c)$.

[At a point on a graph, the normal line is the perpendicular to the tangent at that point. Also, $\cosh x = (e^x + e^{-x})/2$ and $\sinh x = (e^x - e^{-x})/2$.]

These problems were composed by the Committee on the Putnam Prize Competition.
© The Mathematical Association of America. All rights reserved.

Section Project: St. Louis Arch

Roy Ooms/Masterfile

The Gateway Arch in St. Louis, Missouri was constructed using the hyperbolic cosine function. The equation used for construction was

$$y = 693.8597 - 68.7672 \cosh 0.0100333x,$$

$$-299.2239 \le x \le 299.2239$$

where x and y are measured in feet. Cross sections of the arch are equilateral triangles, and (x, y) traces the path of the centers of mass of the cross-sectional triangles. For each value of x, the area of the cross-sectional triangle is $A = 125.1406 \cosh 0.0100333x$.
(*Source:* Owner's Manual for the Gateway Arch, Saint Louis, MO, by William Thayer)

(a) How high above the ground is the center of the highest triangle? (At ground level, $y = 0$.)

(b) What is the height of the arch? (*Hint:* For an equilateral triangle, $A = \sqrt{3}c^2$, where c is one-half the base of the triangle, and the center of mass of the triangle is located at two-thirds the height of the triangle.)

(c) How wide is the arch at ground level?

See www.CalcChat.com for worked-out solutions to odd-numbered exercises.

Review Exercises for Chapter 5

In Exercises 1 and 2, use the graph of f' to sketch a graph of f. To print an enlarged copy of the graph, go to the website www.mathgraphs.com.

1.

2.

In Exercises 3–12, find the indefinite integral.

3. $\int (2x^2 + x - 1)\, dx$

4. $\int \frac{2}{\sqrt[3]{3x}}\, dx$

5. $\int \frac{x^3 + 1}{x^2}\, dx$

6. $\int \frac{x^3 - 2x^2 + 1}{x^2}\, dx$

7. $\int (4x - 3 \sin x)\, dx$

8. $\int (5 \cos x - 2 \sec^2 x)\, dx$

9. $\int (5 - e^x)\, dx$

10. $\int (t + e^t)\, dt$

11. $\int \frac{5}{x}\, dx$

12. $\int \frac{10}{x}\, dx$

13. Find the particular solution of the differential equation $f'(x) = -2x$ whose graph passes through the point $(-1, 1)$.

14. Find the particular solution of the differential equation $f''(x) = 2e^x$ whose graph passes through the point $(0, 1)$ and is tangent to the line $3x - y - 5 = 0$ at that point.

15. *Velocity and Acceleration* An airplane taking off from a runway travels 3600 feet before lifting off. If it starts from rest, moves with constant acceleration, and makes the run in 30 seconds, with what speed does it lift off?

16. *Velocity and Acceleration* The speed of a car traveling in a straight line is reduced from 45 to 30 miles per hour in a distance of 264 feet. Find the distance in which the car can be brought to rest from 30 miles per hour, assuming the same constant deceleration.

17. *Velocity and Acceleration* A ball is thrown vertically upward from ground level with an initial velocity of 96 feet per second.

(a) How long will it take the ball to rise to its maximum height?

(b) What is the maximum height?

(c) When is the velocity of the ball one-half the initial velocity?

(d) What is the height of the ball when its velocity is one-half the initial velocity?

18. *Velocity and Acceleration* Repeat Exercise 17 for an initial velocity of 40 meters per second.

19. Write in sigma notation (a) the sum of the first 10 positive odd integers, (b) the sum of the cubes of the first n positive integers, and (c) $6 + 10 + 14 + 18 + \cdots + 42$.

20. Evaluate each sum for $x_1 = 2$, $x_2 = -1$, $x_3 = 5$, $x_4 = 3$, and $x_5 = 7$.

(a) $\frac{1}{5}\sum_{i=1}^{5} x_i$

(b) $\sum_{i=1}^{5} \frac{1}{x_i}$

(c) $\sum_{i=1}^{5} (2x_i - x_i^2)$

(d) $\sum_{i=2}^{5} (x_i - x_{i-1})$

In Exercises 21 and 22, use upper and lower sums to approximate the area of the region using the indicated number of subintervals of equal width.

21. $y = \dfrac{10}{x^2 + 1}$

22. $y = 2^x$

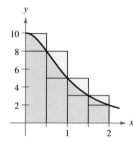

In Exercises 23–26, use the limit process to find the area of the region between the graph of the function and the x-axis over the indicated interval. Sketch the region.

Function	Interval
23. $y = 6 - x$	$[0, 4]$
24. $y = x^2 + 3$	$[0, 2]$
25. $y = 5 - x^2$	$[-2, 1]$
26. $y = \frac{1}{4}x^3$	$[2, 4]$

27. Use the limit process to find the area of the region bounded by $x = 5y - y^2$, $x = 0$, $y = 2$, and $y = 5$.

28. Consider the region bounded by $y = mx$, $y = 0$, $x = 0$, and $x = b$.

(a) Find the upper and lower sums to approximate the area of the region when $\Delta x = b/4$.

(b) Find the upper and lower sums to approximate the area of the region when $\Delta x = b/n$.

(c) Find the area of the region by letting n approach infinity in both sums in part (b). Show that in each case you obtain the formula for the area of a triangle.

In Exercises 29 and 30, express the limit as a definite integral on the interval $[a, b]$, where c_i is any point in the ith subinterval.

Limit	Interval
29. $\displaystyle\lim_{\|\Delta\|\to 0} \sum_{i=1}^{n} (2c_i - 3)\Delta x_i$	$[4, 6]$
30. $\displaystyle\lim_{\|\Delta\|\to 0} \sum_{i=1}^{n} 3c_i(9 - c_i^2)\Delta x_i$	$[1, 3]$

In Exercises 31 and 32, sketch the region whose area is given by the definite integral. Then use a geometric formula to evaluate the integral.

31. $\displaystyle\int_{0}^{5} (5 - |x - 5|)\, dx$

32. $\displaystyle\int_{-4}^{4} \sqrt{16 - x^2}\, dx$

In Exercises 33 and 34, use the given values to evaluate each definite integral.

33. If $\displaystyle\int_{2}^{6} f(x)\, dx = 10$ and $\displaystyle\int_{2}^{6} g(x)\, dx = 3$, find

(a) $\displaystyle\int_{2}^{6} [f(x) + g(x)]\, dx.$

(b) $\displaystyle\int_{2}^{6} [f(x) - g(x)]\, dx.$

(c) $\displaystyle\int_{2}^{6} [2f(x) - 3g(x)]\, dx.$

(d) $\displaystyle\int_{2}^{6} 5f(x)\, dx.$

34. If $\displaystyle\int_{0}^{3} f(x)\, dx = 4$ and $\displaystyle\int_{3}^{6} f(x)\, dx = -1$, find

(a) $\displaystyle\int_{0}^{6} f(x)\, dx.$

(b) $\displaystyle\int_{6}^{3} f(x)\, dx.$

(c) $\displaystyle\int_{4}^{4} f(x)\, dx.$

(d) $\displaystyle\int_{3}^{6} -10 f(x)\, dx.$

In Exercises 35 and 36, select the correct value of the definite integral.

35. $\displaystyle\int_{1}^{8} \left(\sqrt[3]{x} + 1\right) dx$

(a) $\frac{81}{4}$ (b) $\frac{331}{12}$

(c) $\frac{73}{4}$ (d) $\frac{355}{12}$

36. $\displaystyle\int_{1}^{3} \frac{12}{x^3}\, dx$

(a) $\frac{320}{9}$ (b) $-\frac{16}{3}$

(c) $-\frac{5}{9}$ (d) $\frac{16}{3}$

In Exercises 37–46, use the Fundamental Theorem of Calculus to evaluate the definite integral.

37. $\displaystyle\int_{0}^{4} (2 + x)\, dx$

38. $\displaystyle\int_{-1}^{1} (t^2 + 2)\, dt$

39. $\displaystyle\int_{-1}^{1} (4t^3 - 2t)\, dt$

40. $\displaystyle\int_{-2}^{2} (x^4 + 2x^2 - 5)\, dx$

41. $\displaystyle\int_{4}^{9} x\sqrt{x}\, dx$

42. $\displaystyle\int_{1}^{2} \left(\frac{1}{x^2} - \frac{1}{x^3}\right) dx$

43. $\displaystyle\int_{0}^{3\pi/4} \sin\theta\, d\theta$

44. $\displaystyle\int_{-\pi/4}^{\pi/4} \sec^2 t\, dt$

45. $\displaystyle\int_{0}^{2} (x + e^x)\, dx$

46. $\displaystyle\int_{1}^{6} \frac{3}{x}\, dx$

In Exercises 47–52, sketch the graph of the region whose area is given by the integral, and find the area.

47. $\displaystyle\int_{1}^{3} (2x - 1)\, dx$

48. $\displaystyle\int_{0}^{2} (x + 4)\, dx$

49. $\displaystyle\int_{3}^{4} (x^2 - 9)\, dx$

50. $\displaystyle\int_{-1}^{2} (-x^2 + x + 2)\, dx$

51. $\displaystyle\int_{0}^{1} (x - x^3)\, dx$

52. $\displaystyle\int_{0}^{1} \sqrt{x}\,(1 - x)\, dx$

In Exercises 53–56, sketch the region bounded by the graphs of the equations, and determine its area.

53. $y = \dfrac{4}{\sqrt{x}}, \quad y = 0, \quad x = 1, \quad x = 9$

54. $y = \sec^2 x, \quad y = 0, \quad x = 0, \quad x = \dfrac{\pi}{3}$

55. $y = \dfrac{2}{x}, \quad y = 0, \quad x = 1, \quad x = 3$

56. $y = 1 + e^x, \quad y = 0, \quad x = 0, \quad x = 2$

In Exercises 57 and 58, find the average value of the function over the interval. Find the values of x at which the function assumes its average value, and graph the function.

57. $f(x) = \dfrac{1}{\sqrt{x}}, \quad [4, 9]$

58. $f(x) = x^3, \quad [0, 2]$

In Exercises 59–62, use the Second Fundamental Theorem of Calculus to find $F'(x)$.

59. $F(x) = \displaystyle\int_{0}^{x} t^2 \sqrt{1 + t^3}\, dt$

60. $F(x) = \displaystyle\int_{1}^{x} \frac{1}{t^2}\, dt$

61. $F(x) = \displaystyle\int_{-3}^{x} (t^2 + 3t + 2)\, dt$

62. $F(x) = \displaystyle\int_{0}^{x} \csc^2 t\, dt$

In Exercises 63–80, find the indefinite integral.

63. $\displaystyle\int (x^2 + 1)^3\, dx$

64. $\displaystyle\int \left(x + \frac{1}{x}\right)^2 dx$

65. $\displaystyle\int \frac{x^2}{\sqrt{x^3 + 3}}\, dx$

66. $\displaystyle\int x^2\sqrt{x^3 + 3}\, dx$

67. $\displaystyle\int x(1 - 3x^2)^4\, dx$

68. $\displaystyle\int \frac{x + 3}{(x^2 + 6x - 5)^2}\, dx$

69. $\displaystyle\int \sin^3 x \cos x\, dx$

70. $\displaystyle\int x \sin 3x^2\, dx$

71. $\displaystyle\int \frac{\sin\theta}{\sqrt{1 - \cos\theta}}\, d\theta$

72. $\displaystyle\int \frac{\cos x}{\sqrt{\sin x}}\, dx$

73. $\displaystyle\int \tan^n x \sec^2 x\, dx, \quad n \neq -1$

74. $\displaystyle\int \sec 2x \tan 2x\, dx$

75. $\displaystyle\int (1 + \sec\pi x)^2 \sec\pi x \tan\pi x\, dx$

76. $\displaystyle\int \cot^4 \alpha \csc^2 \alpha\, d\alpha$

77. $\displaystyle\int xe^{-3x^2}\, dx$

78. $\displaystyle\int \frac{e^{1/x}}{x^2}\, dx$

79. $\displaystyle\int (x + 1)5^{(x+1)^2}\, dx$

80. $\displaystyle\int \frac{1}{t^2}(2^{-1/t})\, dt$

In Exercises 81–88, evaluate the definite integral. Use a graphing utility to verify your result.

81. $\int_{-1}^{2} x(x^2 - 4)\, dx$

82. $\int_{0}^{1} x^2(x^3 + 1)^3\, dx$

83. $\int_{0}^{3} \frac{1}{\sqrt{1 + x}}\, dx$

84. $\int_{3}^{6} \frac{x}{3\sqrt{x^2 - 8}}\, dx$

85. $2\pi \int_{0}^{1} (y + 1)\sqrt{1 - y}\, dy$

86. $2\pi \int_{-1}^{0} x^2\sqrt{x + 1}\, dx$

87. $\int_{0}^{\pi} \cos\frac{x}{2}\, dx$

88. $\int_{-\pi/4}^{\pi/4} \sin 2x\, dx$

89. *Fuel Cost* Suppose that gasoline is increasing in price according to the equation $p = 1.20 + 0.04t$, where p is the dollar price per gallon and t is the time in years, with $t = 0$ representing 1990. If an automobile is driven 15,000 miles a year and gets M miles per gallon, the annual fuel cost is

$$C = \frac{15{,}000}{M} \int_{t}^{t+1} p\, ds = \frac{15{,}000}{M} \int_{t}^{t+1} (1.20 + 0.04s)\, ds.$$

Estimate the annual fuel cost for the year (a) 2005 and (b) 2007.

90. *Respiratory Cycle* After exercising for a few minutes, a person has a respiratory cycle for which the rate of air intake is $v = 1.75 \sin(\pi t/2)$. Find the volume, in liters, of air inhaled during one cycle by integrating the function over the interval $[0, 2]$.

In Exercises 91–95, use the Trapezoidal Rule and Simpson's Rule with $n = 4$, and use the integration capabilities of a graphing utility, to approximate the definite integral.

91. $\int_{1}^{2} \frac{1}{1 + x^3}\, dx$

92. $\int_{0}^{1} \frac{x^{3/2}}{3 - x^2}\, dx$

93. $\int_{0}^{\pi/2} \sqrt{x} \cos x\, dx$

94. $\int_{0}^{\pi} \sqrt{1 + \sin^2 x}\, dx$

95. $\int_{-1}^{1} e^{-x^2}\, dx$

96. Let $I = \int_{0}^{4} f(x)\, dx$, where f is shown in the figure. Let $L(n)$ and $R(n)$ represent the Riemann sums using the left-hand endpoints and right-hand endpoints of n subintervals of equal width. (Assume n is even.) Let $T(n)$ and $S(n)$ be the corresponding values of the Trapezoidal Rule and Simpson's Rule.

(a) For any n, list $L(n)$, $R(n)$, $T(n)$, and I in increasing order.

(b) Approximate $S(4)$.

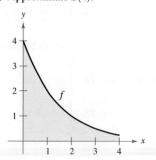

In Exercises 97–106, find or evaluate the integral.

97. $\int \frac{1}{7x - 2}\, dx$

98. $\int \frac{x}{x^2 - 1}\, dx$

99. $\int \frac{\sin x}{1 + \cos x}\, dx$

100. $\int \frac{\ln \sqrt{x}}{x}\, dx$

101. $\int_{1}^{4} \frac{x + 1}{x}\, dx$

102. $\int_{1}^{e} \frac{\ln x}{x}\, dx$

103. $\int_{0}^{\pi/3} \sec \theta\, d\theta$

104. $\int_{0}^{\pi/4} \tan\left(\frac{\pi}{4} - x\right) dx$

105. $\int \frac{e^{2x} - e^{-2x}}{e^{2x} + e^{-2x}}\, dx$

106. $\int \frac{e^{2x}}{e^{2x} + 1}\, dx$

In Exercises 107–114, find the indefinite integral.

107. $\int \frac{1}{e^{2x} + e^{-2x}}\, dx$

108. $\int \frac{1}{3 + 25x^2}\, dx$

109. $\int \frac{x}{\sqrt{1 - x^4}}\, dx$

110. $\int \frac{1}{16 + x^2}\, dx$

111. $\int \frac{x}{16 + x^2}\, dx$

112. $\int \frac{4 - x}{\sqrt{4 - x^2}}\, dx$

113. $\int \frac{\arctan(x/2)}{4 + x^2}\, dx$

114. $\int \frac{\arcsin x}{\sqrt{1 - x^2}}\, dx$

115. *Harmonic Motion* A weight of mass m is attached to a spring and oscillates with simple harmonic motion. By Hooke's Law, you can determine that

$$\int \frac{dy}{\sqrt{A^2 - y^2}} = \int \sqrt{\frac{k}{m}}\, dt$$

where A is the maximum displacement, t is the time, and k is a constant. Find y as a function of t, given that $y = 0$ when $t = 0$.

116. *Think About It* Sketch the region whose area is given by $\int_{0}^{1} \arcsin x\, dx$. Then find the area of the region. Explain how you arrived at your answer.

In Exercises 117 and 118, find the derivative of the function.

117. $y = 2x - \cosh \sqrt{x}$

118. $y = x \tanh^{-1} 2x$

In Exercises 119 and 120, find the indefinite integral.

119. $\int \frac{x}{\sqrt{x^4 - 1}}\, dx$

120. $\int x^2 \operatorname{sech}^2 x^3\, dx$

1. Let $L(x) = \int_1^x \frac{1}{t}\, dt, \ x > 0$.

  (a) Find $L(1)$.

  (b) Find $L'(x)$ and $L'(1)$.

   (c) Use a graphing utility to approximate the value of x (to three decimal places) for which $L(x) = 1$.

  (d) Prove that $L(x_1 x_2) = L(x_1) + L(x_2)$ for all positive values of x_1 and x_2.

2. Let $F(x) = \int_2^x \sin t^2\, dt$.

  (a) Use a graphing utility to complete the table.

x	0	1.0	1.5	1.9	2.0
$F(x)$					
x	2.1	2.5	3.0	4.0	5.0
$F(x)$					

  (b) Let $G(x) = \dfrac{1}{x-2} F(x) = \dfrac{1}{x-2} \displaystyle\int_2^x \sin t^2\, dt$. Use a graphing utility to complete the table and estimate $\displaystyle\lim_{x \to 2} G(x)$.

x	1.9	1.95	1.99	2.01	2.1
$G(x)$					

  (c) Use the definition of the derivative to find the exact value of $\displaystyle\lim_{x \to 2} G(x)$.

3. The **Fresnel function** S is defined by the integral

$$S(x) = \int_0^x \sin\left(\frac{\pi t^2}{2}\right) dt.$$

  (a) Graph the function $y = \sin\left(\dfrac{\pi x^2}{2}\right)$ on the interval $[0, 3]$.

  (b) Use the graph in part (a) to sketch the graph of S on the interval $[0, 3]$.

  (c) Locate all relative extrema of S on the interval $(0, 3)$.

  (d) Locate all points of inflection of S on the interval $(0, 3)$.

4. Galileo Galilei (1564–1642) stated the following proposition concerning falling objects:

   The time in which any space is traversed by a uniformly accelerating body is equal to the time in which that same space would be traversed by the same body moving at a uniform speed whose value is the mean of the highest speed of the accelerating body and the speed just before acceleration began.

  Use the techniques of this section to verify this proposition.

5. The graph of the function f consists of the three line segments joining the points $(0, 0)$, $(2, -2)$, $(6, 2)$, and $(8, 3)$. The function F is defined by the integral

$$F(x) = \int_0^x f(t)\, dt.$$

  (a) Sketch the graph of f.

  (b) Complete the table of values.

x	0	1	2	3	4	5	6	7	8
$F(x)$									

  (c) Find the extrema of F on the interval $[0, 8]$.

  (d) Determine all points of inflection of F on the interval $(0, 8)$.

6. A car is traveling in a straight line for 1 hour. Its velocity v in miles per hour at six-minute intervals is shown in the table.

t (hours)	0	0.1	0.2	0.3	0.4	0.5
v (mi/hr)	0	10	20	40	60	50

t (hours)	0.6	0.7	0.8	0.9	1.0
v (mi/hr)	40	35	40	50	65

  (a) Produce a reasonable graph of the velocity function v by graphing these points and connecting them with a smooth curve.

  (b) Find the open intervals over which the acceleration a is positive.

  (c) Find the average acceleration of the car (in miles per hour squared) over the interval $[0, 0.4]$.

  (d) What does the integral $\int_0^1 v(t)\, dt$ signify? Approximate this integral using the Trapezoidal Rule with five subintervals.

  (e) Approximate the acceleration at $t = 0.8$.

7. The **Two-Point Gaussian Quadrature Approximation** for f is

$$\int_{-1}^1 f(x)\, dx \approx f\left(-\frac{1}{\sqrt{3}}\right) + f\left(\frac{1}{\sqrt{3}}\right).$$

  (a) Use this formula to approximate

$$\int_{-1}^1 \cos x\, dx.$$

  Find the error of the approximation.

  (b) Use this formula to approximate

$$\int_{-1}^1 \frac{1}{1 + x^2}\, dx.$$

  (c) Prove that the Two-Point Gaussian Quadrature Approximation is exact for all polynomials of degree 3 or less.

8. Prove that $\displaystyle\int_0^x f(t)(x - t)\,dt = \int_0^x \left(\int_0^t f(v)\,dv \right) dt.$

9. Prove that $\displaystyle\int_a^b f(x)f'(x)\,dx = \tfrac{1}{2}([f(b)]^2 - [f(a)]^2).$

10. Use an appropriate Riemann sum to evaluate the limit

$$\lim_{n\to\infty} \frac{\sqrt{1} + \sqrt{2} + \sqrt{3} + \cdots + \sqrt{n}}{n^{3/2}}.$$

11. Use an appropriate Riemann sum to evaluate the limit

$$\lim_{n\to\infty} \frac{1^5 + 2^5 + 3^5 + \cdots + n^5}{n^6}.$$

12. Archimedes showed that the area of a parabolic arch is equal to $\frac{2}{3}$ the product of the base and the height, as indicated in the figure.

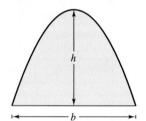

(a) Graph the parabolic arch bounded by $y = 9 - x^2$ and the x-axis. Use an appropriate integral to find the area A.

(b) Find the base and height of the arch and verify Archimedes' formula.

(c) Prove Archimedes' formula for a general parabola.

13. Suppose that f is integrable on $[a, b]$ and $0 < m \le f(x) \le M$ for all x in the interval $[a, b]$. Prove that

$$m(a - b) \le \int_a^b f(x)\,dx \le M(b - a).$$

Use this result to estimate $\displaystyle\int_0^1 \sqrt{1 + x^4}\,dx.$

14. Verify that

$$\sum_{i=1}^n i^2 = \frac{n(n + 1)(2n + 1)}{6}$$

by showing the following.

(a) $(1 + i)^3 - i^3 = 3i^2 + 3i + 1$

(b) $(n + 1)^3 = \displaystyle\sum_{i=1}^n (3i^2 + 3i + 1) + 1$

(c) $\displaystyle\sum_{i=1}^n i^2 = \frac{n(n + 1)(2n + 1)}{6}$

15. Prove that if f is a continuous function on a closed interval $[a, b]$, then

$$\left| \int_a^b f(x)\,dx \right| \le \int_a^b |f(x)|\,dx.$$

16. Consider the three regions A, B, and C determined by the graph of $f(x) = \arcsin x$, as indicated in the figure.

(a) Calculate the areas of regions A and B.

(b) Use your answer in part (a) to evaluate the integral

$$\int_{1/2}^{\sqrt{2}/2} \arcsin x\,dx.$$

(c) Use your answer in part (a) to evaluate the integral

$$\int_1^3 \ln x\,dx.$$

(d) Use your answer in part (a) to evaluate the integral

$$\int_1^{\sqrt{3}} \arctan x\,dx.$$

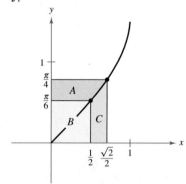

17. Use integration by substitution to find the area under the curve

$$y = \frac{1}{\sqrt{x} + x}$$

between $x = 1$ and $x = 4$.

18. Use integration by substitution to find the area under the curve

$$y = \frac{1}{\sin^2 x + 4\cos^2 x}$$

between $x = 0$ and $x = \pi/4$.

19. (a) Use a graphing utility to compare the graph of the function $y = e^x$ with the graphs of each of the given functions.

(i) $y_1 = 1 + \dfrac{x}{1!}$

(ii) $y_2 = 1 + \dfrac{x}{1!} + \dfrac{x^2}{2!}$

(iii) $y_3 = 1 + \dfrac{x}{1!} + \dfrac{x^2}{2!} + \dfrac{x^3}{3!}$

(b) Identify the pattern of successive polynomials in part (a) and extend the pattern one more term. Compare the graph of the resulting polynomial function with the graph of $y = e^x$.

(c) What do you think this pattern implies?

20. Let f be continuous on the interval $[0, b]$ such that $f(x) + f(b - x) \ne 0$. Show that

$$\int_0^b \frac{f(x)}{f(x) + f(b - x)}\,dx = \frac{b}{2}.$$

Use this result to evaluate $\displaystyle\int_0^1 \frac{\sin x}{\sin(1 - x) + \sin x}\,dx.$

6

Differential Equations

The SkyDome in Toronto is an entertainment center that has a retractable roof. How do you think the noise levels in the stadium compare when the roof is open and when the roof is closed? Explain your reasoning.

A function $y = f(x)$ is a solution of a differential equation if the equation is satisfied when y and its derivatives are replaced by $f(x)$ and its derivatives. One way to analyze a differential equation is to use slope fields, which show the general shape of all solutions of a differential equation. In Chapter 6, you will learn how to sketch slope fields and solve differential equations.

Image Gap/Alamy Images

Slope Fields and Euler's Method

- Use initial conditions to find particular solutions of differential equations.
- Use slope fields to approximate solutions of differential equations.
- Use Euler's Method to approximate solutions of differential equations.

General and Particular Solutions

In this text, you will learn that physical phenomena can be described by differential equations. In Section 6.2, you will see that problems involving radioactive decay, population growth, and Newton's Law of Cooling can be formulated in terms of differential equations.

A function $y = f(x)$ is called a **solution** of a differential equation if the equation is satisfied when y and its derivatives are replaced by $f(x)$ and its derivatives. For example, differentiation and substitution would show that $y = e^{-2x}$ is a solution of the differential equation $y' + 2y = 0$. It can be shown that every solution of this differential equation is of the form

$$y = Ce^{-2x} \qquad \text{General solution of } y' + 2y = 0$$

where C is any real number. This solution is called the **general solution.** Some differential equations have **singular solutions** that cannot be written as special cases of the general solution. However, such solutions are not considered in this text. The **order** of a differential equation is determined by the highest-order derivative in the equation. For instance, $y' = 4y$ is a first-order differential equation.

In Section 5.1, Example 8, you saw that the second-order differential equation $s''(t) = -32$ has the general solution

$$s(t) = -16t^2 + C_1 t + C_2 \qquad \text{General solution of } s''(t) = -32$$

which contains two arbitrary constants. It can be shown that a differential equation of order n has a general solution with n arbitrary constants.

NOTE First-order linear differential equations are discussed in Section 6.5.

EXAMPLE I   Verifying Solutions

Determine whether the function is a solution of the differential equation $y'' - y = 0$.

a. $y = \sin x$ **b.** $y = 4e^{-x}$ **c.** $y = Ce^x$

Solution

a. Because $y = \sin x$, $y' = \cos x$, and $y'' = -\sin x$, it follows that

$$y'' - y = -\sin x - \sin x = -2 \sin x \neq 0.$$

So, $y = \sin x$ is *not* a solution.

b. Because $y = 4e^{-x}$, $y' = -4e^{-x}$, and $y'' = 4e^{-x}$, it follows that

$$y'' - y = 4e^{-x} - 4e^{-x} = 0.$$

So, $y = 4e^{-x}$ is a solution.

c. Because $y = Ce^x$, $y' = Ce^x$, and $y'' = Ce^x$, it follows that

$$y'' - y = Ce^x - Ce^x = 0.$$

So, $y = Ce^x$ is a solution for any value of C.

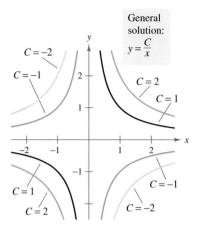

General
solution:
$y = \dfrac{C}{x}$

$C = -2$
$C = -1$
$C = 2$
$C = 1$
$C = 1$
$C = -1$
$C = 2$
$C = -2$

Solution curves for $xy' + y = 0$
Figure 6.1

Geometrically, the general solution of a first-order differential equation represents a family of curves known as **solution curves,** one for each value assigned to the arbitrary constant. For instance, you can verify that every function of the form

$$y = \frac{C}{x} \qquad \text{General solution of } xy' + y = 0$$

is a solution of the differential equation $xy' + y = 0$. Figure 6.1 shows four of the solution curves corresponding to different values of C.

As discussed in Section 5.1, **particular solutions** of a differential equation are obtained from **initial conditions** that give the value of the dependent variable or one of its derivatives for a particular value of the independent variable. The term "initial condition" stems from the fact that, often in problems involving time, the value of the dependent variable or one of its derivatives is known at the *initial* time $t = 0$. For instance, the second-order differential equation $s''(t) = -32$ having the general solution

$$s(t) = -16t^2 + C_1 t + C_2 \qquad \text{General solution of } s''(t) = -32$$

might have the following initial conditions.

$$s(0) = 80, \qquad s'(0) = 64 \qquad \text{Initial conditions}$$

In this case, the initial conditions yield the particular solution

$$s(t) = -16t^2 + 64t + 80. \qquad \text{Particular solution}$$

EXAMPLE 2   Finding a Particular Solution

For the differential equation $xy' - 3y = 0$, verify that $y = Cx^3$ is a solution, and find the particular solution determined by the initial condition $y = 2$ when $x = -3$.

Solution   You know that $y = Cx^3$ is a solution because $y' = 3Cx^2$ and

$$xy' - 3y = x(3Cx^2) - 3(Cx^3)$$
$$= 0.$$

Furthermore, the initial condition $y = 2$ when $x = -3$ yields

$$y = Cx^3 \qquad \text{General solution}$$
$$2 = C(-3)^3 \qquad \text{Substitute initial condition.}$$
$$-\frac{2}{27} = C \qquad \text{Solve for } C.$$

and you can conclude that the particular solution is

$$y = -\frac{2x^3}{27}. \qquad \text{Particular solution}$$

Try checking this solution by substituting for y and y' in the original differential equation.

NOTE   To determine a particular solution, the number of initial conditions must match the number of constants in the general solution.

indicates that in the HM mathSpace® *CD-ROM and the online* Eduspace® *system for this text, you will find an Open Exploration, which further explores this example using the computer algebra systems* Maple, Mathcad, Mathematica, *and* Derive.

Slope Fields

Solving a differential equation analytically can be difficult or even impossible. However, there is a graphical approach you can use to learn a lot about the solution of a differential equation. Consider a differential equation of the form

$$y' = F(x, y). \qquad \text{Differential equation}$$

At each point (x, y) in the xy-plane where F is defined, the differential equation determines the slope $y' = F(x, y)$ of the solution at that point. If you draw a short line segment with slope $F(x, y)$ at selected points (x, y) in the domain of F, then these line segments form a **slope field,** or a *direction field* for the differential equation $y' = F(x, y)$. Each line segment has the same slope as the solution curve through that point. A slope field shows the general shape of all the solutions.

EXAMPLE 3 Sketching a Slope Field

Sketch a slope field for the differential equation $y' = x - y$ for the points $(-1, 1)$, $(0, 1)$, and $(1, 1)$.

Solution

The slope of the solution curve at any point (x, y) is $F(x, y) = x - y$. So, the slope at $(-1, 1)$ is $y' = -1 - 1 = -2$, the slope at $(0, 1)$ is $y' = 0 - 1 = -1$, and the slope at $(1, 1)$ is $y' = 1 - 1 = 0$. Draw short line segments at the three points with their respective slopes, as shown in Figure 6.2.

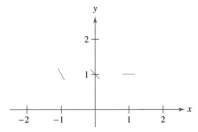

Figure 6.2

EXAMPLE 4 Identifying Slope Fields for Differential Equations

Match each slope field with its differential equation.

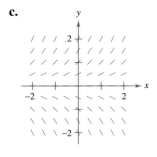

a.

b.

c.

Figure 6.3

i. $y' = x + y$ **ii.** $y' = x$ **iii.** $y' = y$

Solution

a. From Figure 6.3(a), you can see that the slope at any point along the y-axis is 0. The only equation that satisfies this condition is $y' = x$. So, the graph matches (ii).

b. From Figure 6.3(b), you can see that the slope at the point $(1, -1)$ is 0. The only equation that satisfies this condition is $y' = x + y$. So, the graph matches (i).

c. From Figure 6.3(c), you can see that the slope at any point along the x-axis is 0. The only equation that satisfies this condition is $y' = y$. So, the graph matches (iii).

A solution curve of a differential equation $y' = F(x, y)$ is simply a curve in the xy-plane whose tangent line at each point (x, y) has slope equal to $F(x, y)$. This is illustrated in Example 5.

EXAMPLE 5   Sketching a Solution Using a Slope Field

Sketch a slope field for the differential equation

$$y' = 2x + y.$$

Use the slope field to sketch the solution that passes through the point $(1, 1)$.

Solution

Make a table showing the slopes at several points. The table shown is a small sample. The slopes at many other points should be calculated to get a representative slope field.

x	-2	-2	-1	-1	0	0	1	1	2	2
y	-1	1	-1	1	-1	1	-1	1	-1	1
$y' = 2x + y$	-5	-3	-3	-1	-1	1	1	3	3	5

Next draw line segments at the points with their respective slopes, as shown in Figure 6.4.

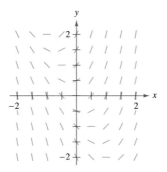

Slope field for $y' = 2x + y$
Figure 6.4

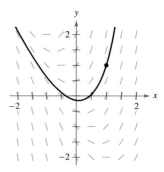

Particular solution for $y' = 2x + y$
passing through $(1, 1)$
Figure 6.5

After the slope field is drawn, start at the initial point $(1, 1)$ and move to the right in the direction of the line segment. Continue to draw the solution curve so that it moves parallel to the nearby line segments. Do the same to the left of $(1, 1)$. The resulting solution is shown in Figure 6.5.

From Example 5, note that the slope field shows that y' increases to infinity as x increases.

NOTE   Drawing a slope field by hand is tedious. In practice, slope fields are usually drawn using a graphing utility.

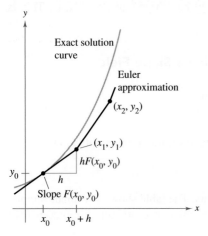

Figure 6.6

Euler's Method

Euler's Method is a numerical approach to approximating the particular solution of the differential equation

$$y' = F(x, y)$$

that passes through the point (x_0, y_0). From the given information, you know that the graph of the solution passes through the point (x_0, y_0) and has a slope of $F(x_0, y_0)$ at this point. This gives you a "starting point" for approximating the solution.

From this starting point, you can proceed in the direction indicated by the slope. Using a small step h, move along the tangent line until you arrive at the point (x_1, y_1), where

$$x_1 = x_0 + h \quad \text{and} \quad y_1 = y_0 + hF(x_0, y_0)$$

as shown in Figure 6.6. If you think of (x_1, y_1) as a new starting point, you can repeat the process to obtain a second point (x_2, y_2). The values of x_i and y_i are as follows.

$$x_1 = x_0 + h \qquad\qquad y_1 = y_0 + hF(x_0, y_0)$$
$$x_2 = x_1 + h \qquad\qquad y_2 = y_1 + hF(x_1, y_1)$$
$$\vdots \qquad\qquad\qquad\qquad \vdots$$
$$x_n = x_{n-1} + h \qquad\quad y_n = y_{n-1} + hF(x_{n-1}, y_{n-1})$$

NOTE   You can obtain better approximations of the exact solution by choosing smaller and smaller step sizes.

EXAMPLE 6   Approximating a Solution Using Euler's Method

Use Euler's Method to approximate the particular solution of the differential equation

$$y' = x - y$$

passing through the point $(0, 1)$. Use a step of $h = 0.1$.

Solution   Using $h = 0.1$, $x_0 = 0$, $y_0 = 1$, and $F(x, y) = x - y$, you have $x_0 = 0$, $x_1 = 0.1$, $x_2 = 0.2$, $x_3 = 0.3, \ldots,$ and

$$y_1 = y_0 + hF(x_0, y_0) = 1 + (0.1)(0 - 1) = 0.9$$
$$y_2 = y_1 + hF(x_1, y_1) = 0.9 + (0.1)(0.1 - 0.9) = 0.82$$
$$y_3 = y_2 + hF(x_2, y_2) = 0.82 + (0.1)(0.2 - 0.82) = 0.758.$$

The first 10 approximations are shown in the table. You can plot these values to see a graph of the approximate solution, as shown in Figure 6.7.

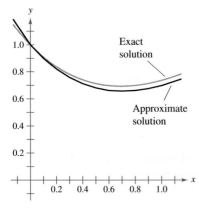

Figure 6.7

n	0	1	2	3	4	5	6	7	8	9	10
x_n	0	0.1	0.2	0.3	0.4	0.5	0.6	0.7	0.8	0.9	1.0
y_n	1	0.900	0.820	0.758	0.712	0.681	0.663	0.657	0.661	0.675	0.697

NOTE   For the differential equation in Example 6, you can verify the exact solution to be $y = x - 1 + 2e^{-x}$. Figure 6.7 compares this exact solution with the approximate solution obtained in Example 6.

Exercises for Section 6.1

See www.CalcChat.com for worked-out solutions to odd-numbered exercises.

In Exercises 1–8, verify the solution of the differential equation.

Solution	Differential Equation
1. $y = Ce^{4x}$	$y' = 4y$
2. $y = e^{-x}$	$3y' + 4y = e^{-x}$
3. $x^2 + y^2 = Cy$	$y' = 2xy/(x^2 - y^2)$
4. $y^2 - 2\ln y = x^2$	$\dfrac{dy}{dx} = \dfrac{xy}{y^2 - 1}$
5. $y = C_1 \cos x + C_2 \sin x$	$y'' + y = 0$
6. $y = C_1 e^{-x} \cos x + C_2 e^{-x} \sin x$	$y'' + 2y' + 2y = 0$
7. $y = -\cos x \ln|\sec x + \tan x|$	$y'' + y = \tan x$
8. $y = \frac{2}{3}(e^{-2x} + e^x)$	$y'' + 2y' = 2e^x$

In Exercises 9–12, verify the particular solution of the differential equation.

Solution	Differential Equation and Initial Condition
9. $y = \sin x \cos x - \cos^2 x$	$2y + y' = 2\sin(2x) - 1$ $y\left(\dfrac{\pi}{4}\right) = 0$
10. $y = \frac{1}{2}x^2 - 4\cos x + 2$	$y' = x + 4\sin x$ $y(0) = -2$
11. $y = 6e^{-2x^2}$	$y' = -4xy$ $y(0) = 6$
12. $y = e^{-\cos x}$	$y' = y \sin x$ $y\left(\dfrac{\pi}{2}\right) = 1$

In Exercises 13–18, determine whether the function is a solution of the differential equation $y^{(4)} - 16y = 0$.

13. $y = 3\cos x$

14. $y = 3\cos 2x$

15. $y = e^{-2x}$

16. $y = 5\ln x$

17. $y = C_1 e^{2x} + C_2 e^{-2x} + C_3 \sin 2x + C_4 \cos 2x$

18. $y = 3e^{2x} - 4\sin 2x$

In Exercises 19–24, determine whether the function is a solution of the differential equation $xy' - 2y = x^3 e^x$.

19. $y = x^2$

20. $y = x^2 e^x$

21. $y = x^2(2 + e^x)$

22. $y = \sin x$

23. $y = \ln x$

24. $y = x^2 e^x - 5x^2$

In Exercises 25–28, some of the curves corresponding to different values of C in the general solution of the differential equation are given. Find the particular solution that passes through the point shown on the graph.

Solution	Differential Equation
25. $y = Ce^{-x/2}$	$2y' + y = 0$
26. $y(x^2 + y) = C$	$2xy + (x^2 + 2y)y' = 0$
27. $y^2 = Cx^3$	$2xy' - 3y = 0$
28. $2x^2 - y^2 = C$	$yy' - 2x = 0$

Figure for 25

Figure for 26

Figure for 27   **Figure for 28**

In Exercises 29 and 30, the general solution of the differential equation is given. Use a graphing utility to graph the particular solutions for the given values of C.

29. $4yy' - x = 0$
$4y^2 - x^2 = C$
$C = 0, C = \pm 1, C = \pm 4$

30. $yy' + x = 0$
$x^2 + y^2 = C$
$C = 0, C = 1, C = 4$

In Exercises 31–36, verify that the general solution satisfies the differential equation. Then find the particular solution that satisfies the initial condition.

31. $y = Ce^{-2x}$
$y' + 2y = 0$
$y = 3$ when $x = 0$

32. $3x^2 + 2y^2 = C$
$3x + 2yy' = 0$
$y = 3$ when $x = 1$

33. $y = C_1 \sin 3x + C_2 \cos 3x$
$y'' + 9y = 0$
$y = 2$ when $x = \pi/6$
$y' = 1$ when $x = \pi/6$

34. $y = C_1 + C_2 \ln x$
$xy'' + y' = 0$
$y = 0$ when $x = 2$
$y' = \frac{1}{2}$ when $x = 2$

35. $y = C_1x + C_2x^3$
$x^2y'' - 3xy' + 3y = 0$
$y = 0$ when $x = 2$
$y' = 4$ when $x = 2$

36. $y = e^{2x/3}(C_1 + C_2x)$
$9y'' - 12y' + 4y = 0$
$y = 4$ when $x = 0$
$y = 0$ when $x = 3$

In Exercises 37–48, use integration to find a general solution of the differential equation.

37. $\dfrac{dy}{dx} = 3x^2$

38. $\dfrac{dy}{dx} = x^3 - 4x$

39. $\dfrac{dy}{dx} = \dfrac{x}{1 + x^2}$

40. $\dfrac{dy}{dx} = \dfrac{e^x}{1 + e^x}$

41. $\dfrac{dy}{dx} = \dfrac{x - 2}{x}$

42. $\dfrac{dy}{dx} = x \cos x^2$

43. $\dfrac{dy}{dx} = \sin 2x$

44. $\dfrac{dy}{dx} = \tan^2 x$

45. $\dfrac{dy}{dx} = x\sqrt{x - 3}$

46. $\dfrac{dy}{dx} = x\sqrt{5 - x}$

47. $\dfrac{dy}{dx} = xe^{x^2}$

48. $\dfrac{dy}{dx} = 5e^{-x/2}$

Slope Fields In Exercises 49–52, a differential equation and its slope field are given. Determine the slopes (if possible) in the slope field at the points given in the table.

x	-4	-2	0	2	4	8
y	2	0	4	4	6	8
dy/dx						

49. $\dfrac{dy}{dx} = \dfrac{x}{y}$

50. $\dfrac{dy}{dx} = x - y$

51. $\dfrac{dy}{dx} = x \cos \dfrac{\pi y}{8}$

52. $\dfrac{dy}{dx} = \tan\left(\dfrac{\pi y}{6}\right)$

In Exercises 53–56, match the differential equation with its slope field. [The slope fields are labeled (a), (b), (c), and (d).]

(a)

(b)

(c)

(d)

53. $\dfrac{dy}{dx} = \cos(2x)$

54. $\dfrac{dy}{dx} = \dfrac{1}{2}\sin x$

55. $\dfrac{dy}{dx} = e^{-2x}$

56. $\dfrac{dy}{dx} = \dfrac{1}{x}$

Slope Fields In Exercises 57–60, (a) sketch the slope field for the differential equation, (b) use the slope field to sketch the solution that passes through the given point, and (c) discuss the graph of the solution as $x \to \infty$ and $x \to -\infty$.

Differential Equation	*Point*
57. $y' = -x + 1$	$(2, 4)$
58. $y' = \frac{1}{3}x^2 - \frac{1}{2}x$	$(1, 1)$
59. $y' = y - 2x$	$(1, 1)$
60. $y' = y + xy$	$(0, 4)$

61. *Slope Field* Use the slope field for the differential equation $y' = 1/x$, where $x > 0$, to sketch the graph of the solution that satisfies each given initial condition. Then make a conjecture about the behavior of a particular solution of $y' = 1/x$ as $x \to \infty$. To print an enlarged copy of the graph, go to the website *www.mathgraphs.com*.

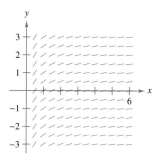

(a) $(1, 0)$ (b) $(2, -1)$

62. *Slope Field* Use the slope field for the differential equation $y' = 1/y$, where $y > 0$, to sketch the graph of the solution that satisfies each initial condition. Then make a conjecture about the behavior of a particular solution of $y' = 1/y$ as $x \to \infty$. To print an enlarged copy of the graph, go to the website *www.mathgraphs.com*.

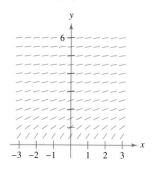

(a) $(0, 1)$ (b) $(1, 1)$

Slope Fields In Exercises 63–68, use a computer algebra system to (a) graph the slope field for the differential equation and (b) graph the solution satisfying the specified initial condition.

63. $\dfrac{dy}{dx} = 0.5y$, $y(0) = 6$

64. $\dfrac{dy}{dx} = 2 - y$, $y(0) = 4$

65. $\dfrac{dy}{dx} = 0.02y(10 - y)$, $y(0) = 2$

66. $\dfrac{dy}{dx} = 0.2x(2 - y)$, $y(0) = 9$

67. $\dfrac{dy}{dx} = 0.4y(3 - x)$, $y(0) = 1$

68. $\dfrac{dy}{dx} = \dfrac{1}{2}e^{-x/8} \sin \dfrac{\pi y}{4}$, $y(0) = 2$

Euler's Method **In Exercises 69–74, use Euler's Method to make a table of values for the approximate solution of the differential equation with the specified initial value. Use *n* steps of size *h*.**

69. $y' = x + y$, $y(0) = 2$, $n = 10$, $h = 0.1$

70. $y' = x + y$, $y(0) = 2$, $n = 20$, $h = 0.05$

71. $y' = 3x - 2y$, $y(0) = 3$, $n = 10$, $h = 0.05$

72. $y' = 0.5x(3 - y)$, $y(0) = 1$, $n = 5$, $h = 0.4$

73. $y' = e^{xy}$, $y(0) = 1$, $n = 10$, $h = 0.1$

74. $y' = \cos x + \sin y$, $y(0) = 5$, $n = 10$, $h = 0.1$

In Exercises 75–77, complete the table using the exact solution of the differential equation and two approximations obtained using Euler's Method to approximate the particular solution of the differential equation. Use $h = 0.2$ and 0.1 and compute each approximation to four decimal places.

x	0	0.2	0.4	0.6	0.8	1.0
$y(x)$ **(exact)**						
$y(x)$ **($h = 0.2$)**						
$y(x)$ **($h = 0.1$)**						

	Differential Equation	*Initial Condition*	*Exact Solution*
75.	$\dfrac{dy}{dx} = y$	$(0, 3)$	$y = 3e^x$
76.	$\dfrac{dy}{dx} = \dfrac{2x}{y}$	$(0, 2)$	$y = \sqrt{2x^2 + 4}$
77.	$\dfrac{dy}{dx} = y + \cos(x)$	$(0, 0)$	$y = \dfrac{1}{2}(\sin x - \cos x + e^x)$

78. Compare the values of the approximations in Exercises 75–77 with the values given by the exact solution. How does the error change as *h* increases?

79. *Temperature* At time $t = 0$ minutes, the temperature of an object is 140°F. The temperature of the object is changing at the rate given by the differential equation

$$\frac{dy}{dt} = -\frac{1}{2}(y - 72).$$

(a) Use a graphing utility and Euler's Method to approximate the particular solutions of this differential equation at $t = 1$, 2, and 3. Use a step size of $h = 0.1$. (A graphing utility program for Euler's Method is available on the website *college.hmco.com*.)

(b) Compare your results with the exact solution

$$y = 72 + 68e^{-t/2}.$$

80. *Temperature* Repeat Exercise 79 using a step size of $h = 0.05$. Compare the results.

Writing About Concepts

81. In your own words, describe the difference between a general solution of a differential equation and a particular solution.

82. Explain how to interpret a slope field.

83. Describe how to use Euler's Method to approximate the particular solution of a differential equation.

84. It is known that $y = Ce^{kx}$ is a solution of the differential equation $y' = 0.07y$. Is it possible to determine C or k from the information given? If so, find its value.

True or False? **In Exercises 85–88, determine whether the statement is true or false. If it is false, explain why or give an example that shows it is false.**

85. If $y = f(x)$ is a solution of a first-order differential equation, then $y = f(x) + C$ is also a solution.

86. The general solution of a differential equation is $y = -4.9x^2 + C_1x + C_2$. To find a particular solution, you must be given two initial conditions.

87. Slope fields represent the general solutions of differential equations.

88. A slope field shows that the slope at the point $(1, 1)$ is 6. This slope field represents the family of solutions for the differential equation $y' = 4x + 2y$.

89. ***Error and Euler's Method*** The exact solution of the differential equation

$$\frac{dy}{dx} = -2y$$

where $y(0) = 4$, is $y = 4e^{-2x}$.

(a) Use a graphing utility to complete the table, where y is the exact value of the solution, y_1 is the approximate solution using Euler's Method with $h = 0.1$, y_2 is the approximate solution using Euler's Method with $h = 0.2$, e_1 is the absolute error $|y - y_1|$, e_2 is the absolute error $|y - y_2|$, and r is the ratio e_1/e_2.

x	0	0.2	0.4	0.6	0.8	1
y						
y_1						
y_2						
e_1						
e_2						
r						

(b) What can you conclude about the ratio r as h changes?

(c) Predict the absolute error when $h = 0.05$.

90. ***Error and Euler's Method*** Repeat Exercise 89 where the exact solution of the differential equation

$$\frac{dy}{dx} = x - y$$

where $y(0) = 1$, is $y = x - 1 + 2e^{-x}$.

91. ***Electric Circuits*** The diagram shows a simple electric circuit consisting of a power source, a resistor, and an inductor.

A model of the current I, in amperes (A), at time t is given by the first-order differential equation

$$L\frac{dI}{dt} + RI = E(t)$$

where $E(t)$ is the voltage (V) produced by the power source, R is the resistance in ohms (Ω), and L is the inductance in henrys (H). Suppose the electric circuit consists of a 24-V power source, a 12-Ω resistor, and a 4-H inductor.

(a) Sketch a slope field for the differential equation.

(b) What is the limiting value of the current? Explain.

92. ***Think About It*** It is known that $y = e^{kt}$ is a solution of the differential equation $y'' - 16y = 0$. Find the values of k.

93. ***Think About It*** It is known that $y = A\sin \omega t$ is a solution of the differential equation $y'' + 16y = 0$. Find the values of ω.

Putnam Exam Challenge

94. Let f be a twice-differentiable real-valued function satisfying

$$f(x) + f''(x) = -xg(x)f'(x)$$

where $g(x) \geq 0$ for all real x. Prove that $|f(x)|$ is bounded.

95. Prove that if the family of integral curves of the differential equation

$$\frac{dy}{dx} + p(x)y = q(x), \qquad p(x) \cdot q(x) \neq 0$$

is cut by the line $x = k$, the tangents at the points of intersection are concurrent.

Differential Equations: Growth and Decay

- Use separation of variables to solve a simple differential equation.
- Use exponential functions to model growth and decay in applied problems.

Differential Equations

In the preceding section, you learned to analyze visually the solutions of differential equations using slope fields and to approximate solutions numerically using Euler's Method. Analytically, you have learned to solve only two types of differential equations—those of the forms

$$y' = f(x) \quad \text{and} \quad y'' = f(x).$$

In this section, you will learn how to solve a more general type of differential equation. The strategy is to rewrite the equation so that each variable occurs on only one side of the equation. This strategy is called *separation of variables*. (You will study this strategy in detail in Section 6.3.)

EXAMPLE 1   Solving a Differential Equation

NOTE   When you integrate both sides of the equation in Example 1, you don't need to add a constant of integration to both sides of the equation. If you did, you would obtain the same result as in Example 1.

$$\int y \, dy = \int 2x \, dx$$

$$\tfrac{1}{2}y^2 + C_2 = x^2 + C_3$$

$$\tfrac{1}{2}y^2 = x^2 + (C_3 - C_2)$$

$$\tfrac{1}{2}y^2 = x^2 + C_1$$

Solve the differential equation $y' = 2x/y$.

Solution

$$y' = \frac{2x}{y} \qquad \text{Write original equation.}$$

$$yy' = 2x \qquad \text{Multiply both sides by } y.$$

$$\int yy' \, dx = \int 2x \, dx \qquad \text{Integrate with respect to } x.$$

$$\int y \, dy = \int 2x \, dx \qquad dy = y' \, dx$$

$$\frac{1}{2}y^2 = x^2 + C_1 \qquad \text{Apply Power Rule.}$$

$$y^2 - 2x^2 = C \qquad \text{Rewrite, letting } C = 2C_1.$$

So, the general solution is given by

$$y^2 - 2x^2 = C.$$

You can use implicit differentiation to check this result.   ▬▬▬

In practice, most people prefer to use Leibniz notation and differentials when applying separation of variables. The solution of Example 1 is shown below using this notation.

$$\frac{dy}{dx} = \frac{2x}{y}$$

$$y \, dy = 2x \, dx$$

$$\int y \, dy = \int 2x \, dx$$

$$\frac{1}{2}y^2 = x^2 + C_1$$

$$y^2 - 2x^2 = C$$

EXPLORATION

In Example 1, the general solution of the differential equation is

$$y^2 - 2x^2 = C.$$

Use a graphing utility to sketch several particular solutions—those given by $C = \pm 2$, $C = \pm 1$, and $C = 0$. Describe the solutions graphically. Is the following statement true of each solution?

The slope of the graph at the point (x, y) is equal to twice the ratio of x and y.

Explain your reasoning. Are all curves for which this statement is true represented by the general solution?

First-Order Linear Differential Equations

- Solve a first-order linear differential equation.
- Solve a Bernoulli differential equation.
- Use linear differential equations to solve applied problems.

First-Order Linear Differential Equations

In this section, you will see how to solve a very important class of first-order differential equations—first-order linear differential equations.

Definition of First-Order Linear Differential Equation

A first-order linear differential equation is an equation of the form

$$\frac{dy}{dx} + P(x)y = Q(x)$$

where P and Q are continuous functions of x. This first-order linear differential equation is said to be in **standard form.**

NOTE It is instructive to see why the integrating factor helps solve a linear differential equation of the form $y' + P(x)y = Q(x)$. When both sides of the equation are multiplied by the integrating factor $u(x) = e^{\int P(x)\,dx}$, the left-hand side becomes the derivative of a product.

$$y'e^{\int P(x)\,dx} + P(x)ye^{\int P(x)\,dx} = Q(x)e^{\int P(x)\,dx}$$

$$\left[ye^{\int P(x)\,dx}\right]' = Q(x)e^{\int P(x)\,dx}$$

Integrating both sides of this second equation and dividing by $u(x)$ produces the general solution.

To solve a linear differential equation, write it in standard form to identify the functions $P(x)$ and $Q(x)$. Then integrate $P(x)$ and form the expression

$$u(x) = e^{\int P(x)\,dx} \qquad \text{Integrating factor}$$

which is called an **integrating factor.** The general solution of the equation is

$$y = \frac{1}{u(x)} \int Q(x)u(x)\,dx. \qquad \text{General solution}$$

EXAMPLE 1 **Solving a Linear Differential Equation**

Find the general solution of

$$y' + y = e^x.$$

Solution

For this equation, $P(x) = 1$ and $Q(x) = e^x$. So, the integrating factor is

$$u(x) = e^{\int P(x)\,dx} \qquad \text{Integrating factor}$$

$$= e^{\int dx}$$

$$= e^x.$$

This implies that the general solution is

$$y = \frac{1}{u(x)} \int Q(x)u(x)\,dx$$

$$= \frac{1}{e^x} \int e^x(e^x)\,dx$$

$$= e^{-x}\left(\frac{1}{2}e^{2x} + C\right)$$

$$= \frac{1}{2}e^x + Ce^{-x}. \qquad \text{General solution}$$

> **THEOREM 6.3   Solution of a First-Order Linear Differential Equation**
>
> An integrating factor for the first-order linear differential equation
>
> $$y' + P(x)y = Q(x)$$
>
> is $u(x) = e^{\int P(x)\,dx}$. The solution of the differential equation is
>
> $$ye^{\int P(x)\,dx} = \int Q(x)e^{\int P(x)\,dx}\,dx + C.$$

ANNA JOHNSON PELL WHEELER (1883–1966)

Anna Johnson Pell Wheeler was awarded a master's degree from the University of Iowa for her thesis *The Extension of Galois Theory to Linear Differential Equations* in 1904. Influenced by David Hilbert, she worked on integral equations while studying infinite linear spaces.

STUDY TIP   Rather than memorizing the formula in Theorem 6.3, just remember that multiplication by the integrating factor $e^{\int P(x)\,dx}$ converts the left side of the differential equation into the derivative of the product $ye^{\int P(x)\,dx}$.

EXAMPLE 2   Solving a First-Order Linear Differential Equation

Find the general solution of

$$xy' - 2y = x^2.$$

Solution   The standard form of the given equation is

$$y' + P(x)y = Q(x)$$

$$y' - \left(\frac{2}{x}\right)y = x. \qquad \text{Standard form}$$

So, $P(x) = -2/x$, and you have

$$\int P(x)\,dx = -\int \frac{2}{x}\,dx$$

$$= -\ln x^2$$

$$e^{\int P(x)\,dx} = e^{-\ln x^2}$$

$$= \frac{1}{e^{\ln x^2}}$$

$$= \frac{1}{x^2}. \qquad \text{Integrating factor}$$

So, multiplying each side of the standard form by $1/x^2$ yields

$$\frac{y'}{x^2} - \frac{2y}{x^3} = \frac{1}{x}$$

$$\frac{d}{dx}\left[\frac{y}{x^2}\right] = \frac{1}{x}$$

$$\frac{y}{x^2} = \int \frac{1}{x}\,dx$$

$$\frac{y}{x^2} = \ln|x| + C$$

$$y = x^2(\ln|x| + C). \qquad \text{General solution}$$

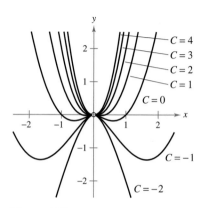

Figure 6.29

Several solution curves (for $C = -2, -1, 0, 1, 2, 3,$ and 4) are shown in Figure 6.29.

EXAMPLE 3   **Solving a First-Order Linear Differential Equation**

Find the general solution of

$$y' - y \tan t = 1, \quad -\frac{\pi}{2} < t < \frac{\pi}{2}.$$

Solution   The equation is already in the standard form $y' + P(t)y = Q(t)$. So, $P(t) = -\tan t$, and

$$\int P(t) \, dt = -\int \tan t \, dt = \ln|\cos t|$$

which implies that the integrating factor is

$$e^{\int P(t) \, dt} = e^{\ln|\cos t|}$$
$$= |\cos t|. \qquad \text{Integrating factor}$$

A quick check shows that $\cos t$ is also an integrating factor. So, multiplying $y' - y \tan t = 1$ by $\cos t$ produces

$$\frac{d}{dt}[y \cos t] = \cos t$$

$$y \cos t = \int \cos t \, dt$$

$$y \cos t = \sin t + C$$

$$y = \tan t + C \sec t. \qquad \text{General solution}$$

Several solution curves are shown in Figure 6.30.

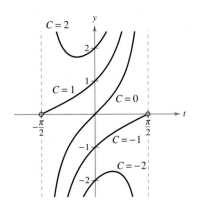

Figure 6.30

Bernoulli Equation

A well-known nonlinear equation that reduces to a linear one with an appropriate substitution is the **Bernoulli equation,** named after James Bernoulli (1654–1705).

$$y' + P(x)y = Q(x)y^n \qquad \text{Bernoulli equation}$$

This equation is linear if $n = 0$, and has separable variables if $n = 1$. So, in the following development, assume that $n \neq 0$ and $n \neq 1$. Begin by multiplying by y^{-n} and $(1 - n)$ to obtain

$$y^{-n}y' + P(x)y^{1-n} = Q(x)$$
$$(1 - n)y^{-n}y' + (1 - n)P(x)y^{1-n} = (1 - n)Q(x)$$
$$\frac{d}{dx}[y^{1-n}] + (1 - n)P(x)y^{1-n} = (1 - n)Q(x)$$

which is a linear equation in the variable y^{1-n}. Letting $z = y^{1-n}$ produces the linear equation

$$\frac{dz}{dx} + (1 - n)P(x)z = (1 - n)Q(x).$$

Finally, by Theorem 6.3, the *general solution of the Bernoulli equation* is

$$y^{1-n}e^{\int(1-n)P(x) \, dx} = \int (1 - n)Q(x)e^{\int(1-n)P(x) \, dx} \, dx + C.$$

EXAMPLE 4   Solving a Bernoulli Equation

Find the general solution of

$$y' + xy = xe^{-x^2}y^{-3}.$$

Solution   For this Bernoulli equation, let $n = -3$, and use the substitution

$z = y^4$ Let $z = y^{1-n} = y^{1-(-3)}$.

$z' = 4y^3y'.$ Differentiate.

Multiplying the original equation by $4y^3$ produces

$y' + xy = xe^{-x^2}y^{-3}$ Write original equation.

$4y^3y' + 4xy^4 = 4xe^{-x^2}$ Multiply each side by $4y^3$.

$z' + 4xz = 4xe^{-x^2}.$ Linear equation: $z' + P(x)z = Q(x)$

This equation is linear in z. Using $P(x) = 4x$ produces

$$\int P(x)\,dx = \int 4x\,dx$$

$$= 2x^2$$

which implies that e^{2x^2} is an integrating factor. Multiplying the linear equation by this factor produces

$z' + 4xz = 4xe^{-x^2}$ Linear equation

$z'e^{2x^2} + 4xze^{2x^2} = 4xe^{x^2}$ Multiply by integrating factor.

$$\frac{d}{dx}\left[ze^{2x^2}\right] = 4xe^{x^2}$$ Write left side as derivative.

$$ze^{2x^2} = \int 4xe^{x^2}\,dx$$ Integrate each side.

$$ze^{2x^2} = 2e^{x^2} + C$$

$$z = 2e^{-x^2} + Ce^{-2x^2}.$$ Divide each side by e^{2x^2}.

Finally, substituting $z = y^4$, the general solution is

$$y^4 = 2e^{-x^2} + Ce^{-2x^2}.$$ General solution

So far you have studied several types of first-order differential equations. Of these, the separable variables case is usually the simplest, and solution by an integrating factor is ordinarily used only as a last resort.

Summary of First-Order Differential Equations

Method	Form of Equation
1. Separable variables:	$M(x)\,dx + N(y)\,dy = 0$
2. Homogeneous:	$M(x, y)\,dx + N(x, y)\,dy = 0$, where M and N are nth-degree homogeneous
3. Linear:	$y' + P(x)y = Q(x)$
4. Bernoulli equation:	$y' + P(x)y = Q(x)y^n$

Applications

One type of problem that can be described in terms of a differential equation involves chemical mixtures, as illustrated in the next example.

4 gal/min

5 gal/min

Figure 6.31

EXAMPLE 5   A Mixture Problem

A tank contains 50 gallons of a solution composed of 90% water and 10% alcohol. A second solution containing 50% water and 50% alcohol is added to the tank at the rate of 4 gallons per minute. As the second solution is being added, the tank is being drained at a rate of 5 gallons per minute, as shown in Figure 6.31. Assuming the solution in the tank is stirred constantly, how much alcohol is in the tank after 10 minutes?

Solution   Let y be the number of gallons of alcohol in the tank at any time t. You know that $y = 5$ when $t = 0$. Because the number of gallons of solution in the tank at any time is $50 - t$, and the tank loses 5 gallons of solution per minute, it must lose

$$\left(\frac{5}{50 - t}\right)y$$

gallons of alcohol per minute. Furthermore, because the tank is gaining 2 gallons of alcohol per minute, the rate of change of alcohol in the tank is given by

$$\frac{dy}{dt} = 2 - \left(\frac{5}{50 - t}\right)y \quad \Longrightarrow \quad \frac{dy}{dt} + \left(\frac{5}{50 - t}\right)y = 2.$$

To solve this linear equation, let $P(t) = 5/(50 - t)$ and obtain

$$\int P(t)\,dt = \int \frac{5}{50 - t}\,dt = -5\ln|50 - t|.$$

Because $t < 50$, you can drop the absolute value signs and conclude that

$$e^{\int P(t)\,dt} = e^{-5\ln(50 - t)} = \frac{1}{(50 - t)^5}.$$

So, the general solution is

$$\frac{y}{(50 - t)^5} = \int \frac{2}{(50 - t)^5}\,dt = \frac{1}{2(50 - t)^4} + C$$

$$y = \frac{50 - t}{2} + C(50 - t)^5.$$

Because $y = 5$ when $t = 0$, you have

$$5 = \frac{50}{2} + C(50)^5 \quad \Longrightarrow \quad -\frac{20}{50^5} = C$$

which means that the particular solution is

$$y = \frac{50 - t}{2} - 20\left(\frac{50 - t}{50}\right)^5.$$

Finally, when $t = 10$, the amount of alcohol in the tank is

$$y = \frac{50 - 10}{2} - 20\left(\frac{50 - 10}{50}\right)^5 \approx 13.45 \text{ gal}$$

which represents a solution containing 33.6% alcohol.

In most falling-body problems discussed so far in the text, air resistance has been neglected. The next example includes this factor. In the example, the air resistance on the falling object is assumed to be proportional to its velocity v. If g is the gravitational constant, the downward force F on a falling object of mass m is given by the difference $mg - kv$. But by Newton's Second Law of Motion, you know that

$$F = ma$$
$$= m\frac{dv}{dt}$$

which yields the following differential equation.

$$m\frac{dv}{dt} = mg - kv \quad \Longrightarrow \quad \frac{dv}{dt} + \frac{k}{m}v = g$$

EXAMPLE 6   A Falling Object with Air Resistance

An object of mass m is dropped from a hovering helicopter. Find its velocity as a function of time t, assuming that the air resistance is proportional to the velocity of the object.

Solution   The velocity v satisfies the equation

$$\frac{dv}{dt} + \frac{kv}{m} = g$$

where g is the gravitational constant and k is the constant of proportionality. Letting $b = k/m$, you can *separate variables* to obtain

$$dv = (g - bv)\, dt$$
$$\int \frac{dv}{g - bv} = \int dt$$
$$-\frac{1}{b}\ln|g - bv| = t + C_1$$
$$\ln|g - bv| = -bt - bC_1$$
$$g - bv = Ce^{-bt}.$$

Because the object was dropped, $v = 0$ when $t = 0$; so $g = C$, and it follows that

$$-bv = -g + ge^{-bt} \quad \Longrightarrow \quad v = \frac{g - ge^{-bt}}{b} = \frac{mg}{k}(1 - e^{-kt/m}).$$

NOTE   Notice in Example 6 that the velocity approaches a limit of mg/k as a result of the air resistance. For falling-body problems in which air resistance is neglected, the velocity increases without bound.

A simple electric circuit consists of electric current I (in amperes), a resistance R (in ohms), an inductance L (in henrys), and a constant electromotive force E (in volts), as shown in Figure 6.32. According to Kirchhoff's Second Law, if the switch S is closed when $t = 0$, the applied electromotive force (voltage) is equal to the sum of the voltage drops in the rest of the circuit. This in turn means that the current I satisfies the differential equation

$$L\frac{dI}{dt} + RI = E.$$

Figure 6.32

EXAMPLE 7 An Electric Circuit Problem

Find the current I as a function of time t (in seconds), given that I satisfies the differential equation $L(dI/dt) + RI = \sin 2t$, where R and L are nonzero constants.

TECHNOLOGY The integral in Example 7 was found using a computer algebra system. If you have access to *Derive, Maple, Mathcad, Mathematica,* or the *TI-89,* try using it to integrate

$$\frac{1}{L}\int e^{(R/L)t} \sin 2t \, dt.$$

In Chapter 8 you will learn how to integrate functions of this type using integration by parts.

Solution In standard form, the given linear equation is

$$\frac{dI}{dt} + \frac{R}{L}I = \frac{1}{L}\sin 2t.$$

Let $P(t) = R/L$, so that $e^{\int P(t)\,dt} = e^{(R/L)t}$, and, by Theorem 6.3,

$$Ie^{(R/L)t} = \frac{1}{L}\int e^{(R/L)t}\sin 2t\,dt$$

$$= \frac{1}{4L^2 + R^2}e^{(R/L)t}(R\sin 2t - 2L\cos 2t) + C.$$

So the general solution is

$$I = e^{-(R/L)t}\left[\frac{1}{4L^2 + R^2}e^{(R/L)t}(R\sin 2t - 2L\cos 2t) + C\right]$$

$$I = \frac{1}{4L^2 + R^2}(R\sin 2t - 2L\cos 2t) + Ce^{-(R/L)t}.$$

Exercises for Section 6.5

See www.CalcChat.com for worked-out solutions to odd-numbered exercises.

In Exercises 1–4, determine whether the differential equation is linear. Explain your reasoning.

1. $x^3y' + xy = e^x + 1$

2. $2xy - y'\ln x = y$

3. $y' + y\cos x = xy^2$

4. $\dfrac{1 - y'}{y} = 3x$

In Exercises 5–14, solve the first-order linear differential equation.

5. $\dfrac{dy}{dx} + \left(\dfrac{1}{x}\right)y = 3x + 4$

6. $\dfrac{dy}{dx} + \left(\dfrac{2}{x}\right)y = 3x + 2$

7. $y' - y = 10$

8. $y' + 2xy = 4x$

9. $(y + 1)\cos x\,dx - dy = 0$

10. $(y - 1)\sin x\,dx - dy = 0$

11. $(x - 1)y' + y = x^2 - 1$

12. $y' + 3y = e^{3x}$

13. $y' - 3x^2y = e^{x^3}$

14. $y' - y = \cos x$

Slope Fields In Exercises 15 and 16, (a) sketch an approximate solution of the differential equation satisfying the initial condition by hand on the slope field, (b) find the particular solution that satisfies the initial condition, and (c) use a graphing utility to graph the particular solution. Compare the graph with the hand-drawn graph of part (a). To print an enlarged copy of the graph, go to the website *www.mathgraphs.com*.

Differential Equation	Initial Condition
15. $\dfrac{dy}{dx} = e^x - y$	$(0, 1)$
16. $y' + \left(\dfrac{1}{x}\right)y = \sin x^2$	$\left(\sqrt{\pi}, 0\right)$

Figure for 15   **Figure for 16**

In Exercises 17–24, find the particular solution of the differential equation that satisfies the boundary condition.

Differential Equation	Boundary Condition
17. $y'\cos^2 x + y - 1 = 0$	$y(0) = 5$
18. $x^3y' + 2y = e^{1/x^2}$	$y(1) = e$
19. $y' + y\tan x = \sec x + \cos x$	$y(0) = 1$
20. $y' + y\sec x = \sec x$	$y(0) = 4$
21. $y' + \left(\dfrac{1}{x}\right)y = 0$	$y(2) = 2$
22. $y' + (2x - 1)y = 0$	$y(1) = 2$
23. $x\,dy = (x + y + 2)\,dx$	$y(1) = 10$
24. $2xy' - y = x^3 - x$	$y(4) = 2$

In Exercises 25–30, solve the Bernoulli differential equation.

25. $y' + 3x^2y = x^2y^3$

26. $y' + xy = xy^{-1}$

27. $y' + \left(\dfrac{1}{x}\right)y = xy^2$

28. $y' + \left(\dfrac{1}{x}\right)y = x\sqrt{y}$

29. $y' - y = e^x\sqrt[3]{y}$

30. $yy' - 2y^2 = e^x$

Slope Fields   **In Exercises 31–34, (a) use a graphing utility to graph the slope field for the differential equation, (b) find the particular solutions of the differential equation passing through the given points, and (c) use a graphing utility to graph the particular solutions on the slope field.**

Differential Equation	Points
31. $\dfrac{dy}{dx} - \dfrac{1}{x}y = x^2$	$(-2, 4), \ (2, 8)$
32. $\dfrac{dy}{dx} + 4x^3y = x^3$	$\left(0, \dfrac{7}{2}\right), \ \left(0, -\dfrac{1}{2}\right)$
33. $\dfrac{dy}{dx} + (\cot x)y = 2$	$(1, 1), \ (3, -1)$
34. $\dfrac{dy}{dx} + 2xy = xy^2$	$(0, 3), \ (0, 1)$

35. *Intravenous Feeding*   Glucose is added intravenously to the bloodstream at the rate of q units per minute, and the body removes glucose from the bloodstream at a rate proportional to the amount present. Assume that $Q(t)$ is the amount of glucose in the bloodstream at time t.

(a) Determine the differential equation describing the rate of change of glucose in the bloodstream with respect to time.

(b) Solve the differential equation from part (a), letting $Q = Q_0$ when $t = 0$.

(c) Find the limit of $Q(t)$ as $t \to \infty$.

36. *Learning Curve*   The management at a certain factory has found that the maximum number of units a worker can produce in a day is 40. The rate of increase in the number of units N produced with respect to time t in days by a new employee is proportional to $40 - N$.

(a) Determine the differential equation describing the rate of change of performance with respect to time.

(b) Solve the differential equation from part (a).

(c) Find the particular solution for a new employee who produced 10 units on the first day at the factory and 19 units on the twentieth day.

Mixture   **In Exercises 37–42, consider a tank that at time $t = 0$ contains v_0 gallons of a solution that, by weight, contains q_0 pounds of soluble concentrate. Another solution containing q_1 pounds of the concentrate per gallon is running into the tank at the rate of r_1 gallons per minute. The solution in the tank is kept well stirred and is withdrawn at the rate of r_2 gallons per minute.**

37. If Q is the amount of concentrate in the solution at any time t, show that

$$\dfrac{dQ}{dt} + \dfrac{r_2 Q}{v_0 + (r_1 - r_2)t} = q_1 r_1.$$

38. If Q is the amount of concentrate in the solution at any time t, write the differential equation for the rate of change of Q with respect to t if $r_1 = r_2 = r$.

39. A 200-gallon tank is full of a solution containing 25 pounds of concentrate. Starting at time $t = 0$, distilled water is admitted to the tank at a rate of 10 gallons per minute, and the well-stirred solution is withdrawn at the same rate as shown in the figure.

10 gal/min

10 gal/min

(a) Find the amount of concentrate Q in the solution as a function of t.

(b) Find the time at which the amount of concentrate in the tank reaches 15 pounds.

(c) Find the quantity of the concentrate in the solution as $t \to \infty$.

40. Repeat Exercise 39, assuming that the solution entering the tank contains 0.04 pound of concentrate per gallon.

41. A 200-gallon tank is half full of distilled water. At time $t = 0$, a solution containing 0.5 pound of concentrate per gallon enters the tank at the rate of 5 gallons per minute, and the well-stirred mixture is withdrawn at the rate of 3 gallons per minute as shown in the figure.

5 gal/min

3 gal/min

(a) At what time will the tank be full?

(b) At the time the tank is full, how many pounds of concentrate will it contain?

42. Repeat Exercise 41, assuming that the solution entering the tank contains 1 pound of concentrate per gallon.

Falling Object   **In Exercises 43 and 44, consider an eight-pound object dropped from a height of 5000 feet, where the air resistance is proportional to the velocity.**

43. Write the velocity as a function of time if the object's velocity after 5 seconds is approximately -101 feet per second. What is the limiting value of the velocity function?

44. Use the result of Exercise 43 to write the position of the object as a function of time. Approximate the velocity of the object when it reaches ground level.

Electric Circuits   **In Exercises 45 and 46, use the differential equation for electric circuits given by**

$$L\frac{dI}{dt} + RI = E.$$

In this equation, I is the current, R is the resistance, L is the inductance, and E is the electromotive force (voltage).

45. Solve the differential equation given a constant voltage E_0.

46. Use the result of Exercise 45 to find the equation for the current if $I(0) = 0$, $E_0 = 120$ volts, $R = 600$ ohms, and $L = 4$ henrys. When does the current reach 90% of its limiting value?

Writing About Concepts

47. Give the standard form of a first-order linear differential equation. What is its integrating factor?

48. Give the standard form of the Bernoulli equation. Describe how one reduces it to a linear equation.

In Exercises 49–52, match the differential equation with its solution.

Differential Equation	*Solution*
49. $y' - 2x = 0$	(a) $y = Ce^{x^2}$
50. $y' - 2y = 0$	(b) $y = -\frac{1}{2} + Ce^{x^2}$
51. $y' - 2xy = 0$	(c) $y = x^2 + C$
52. $y' - 2xy = x$	(d) $y = Ce^{2x}$

In Exercises 53–64, solve the first-order differential equation by any appropriate method.

53. $\dfrac{dy}{dx} = \dfrac{e^{2x+y}}{e^{x-y}}$

54. $\dfrac{dy}{dx} = \dfrac{x+1}{y(y+2)}$

55. $y\cos x - \cos x + \dfrac{dy}{dx} = 0$

56. $y' = 2x\sqrt{1-y^2}$

57. $(3y^2 + 4xy)\,dx + (2xy + x^2)\,dy = 0$

58. $(x+y)\,dx - x\,dy = 0$

59. $(2y - e^x)\,dx + x\,dy = 0$

60. $(y^2 + xy)\,dx - x^2\,dy = 0$

61. $(x^2y^4 - 1)\,dx + x^3y^3\,dy = 0$

62. $y\,dx + (3x + 4y)\,dy = 0$

63. $3(y - 4x^2)\,dx + x\,dy = 0$

64. $x\,dx + (y + e^y)(x^2 + 1)\,dy = 0$

True or False?   **In Exercises 65 and 66, determine whether the statement is true or false. If it is false, explain why or give an example that shows it is false.**

65. $y' + x\sqrt{y} = x^2$ is a first-order linear differential equation.

66. $y' + xy = e^x y$ is a first-order linear differential equation.

Section Project: Weight Loss

A person's weight depends on both the number of calories consumed and the energy used. Moreover, the amount of energy used depends on a person's weight—the average amount of energy used by a person is 17.5 calories per pound per day. So, the more weight a person loses, the less energy a person uses (assuming that the person maintains a constant level of activity). An equation that can be used to model weight loss is

$$\frac{dw}{dt} = \frac{C}{3500} - \frac{17.5}{3500}w$$

where w is the person's weight (in pounds), t is the time in days, and C is the constant daily calorie consumption.

(a) Find the general solution of the differential equation.

(b) Consider a person who weighs 180 pounds and begins a diet of 2500 calories per day. How long will it take the person to lose 10 pounds? How long will it take the person to lose 35 pounds?

(c) Use a graphing utility to graph the solution. What is the "limiting" weight of the person?

(d) Repeat parts (b) and (c) for a person who weighs 200 pounds when the diet is started.

FOR FURTHER INFORMATION For more information on modeling weight loss, see the article "A Linear Diet Model" by Arthur C. Segal in *The College Mathematics Journal.*

Predator-Prey Differential Equations

- Analyze predator-prey differential equations.
- Analyze competing-species differential equations.

Predator–Prey Differential Equations

Although Alfred Lotka (1880-1949) and Vito Volterra (1860-1940) both worked on other problems, they are most known for their work on predator-prey equations. Lotka was also a statistician, and Volterra did work in the development of integral equations and functional analysis.

In the 1920s, mathematicians Alfred Lotka (1880–1949) and Vito Volterra (1860–1940) independently developed mathematical models to represent many of the different ways that two species can interact with each other. Two common ways that species interact with each other are as predator and prey, and as competing species.

Consider a predator-prey relationship involving foxes (predators) and rabbits (prey). Assume that the rabbits are the primary food source for the foxes, the rabbits have an unlimited food supply, and there is no threat to the rabbits other than from the foxes. Let x represent the number of rabbits, let y represent the number of foxes, and let t represent time. If there are no foxes, then the rabbit population will grow according to the exponential growth model $dx/dt = ax$, $a > 0$.

If there are foxes but no rabbits, the foxes have no food and their population will decay according to the exponential decay model $dy/dt = -my$, $m > 0$.

If both foxes and rabbits are present, there is an interaction rate of *decline* for the rabbit population given by $-bxy$, and an interaction rate of *increase* in the fox population given by nxy, where $b, n > 0$. So, the rates of change of each population can be modeled by the following predator-prey system of differential equations.

$$\frac{dx}{dt} = ax - bxy \qquad \text{Rate of change of prey}$$

$$\frac{dy}{dt} = -my + nxy \qquad \text{Rate of change of predators}$$

These equations are called **predator-prey equations** or **Lotka-Volterra equations**. The equations are **autonomous** because the rates of change do not depend explicitly on time t.

In general, it is not possible to solve the predator-prey equations explicitly for x and y. However, you can use techniques such as Euler's Method to approximate solutions. Also, you can discover properties of the solutions by analyzing the differential equations.

EXAMPLE 1   **Analyzing Predator-Prey Equations**

Write the predator-prey equations for $a = 0.04$, $b = 0.002$, $m = 0.08$, and $n = 0.0004$. Then find the values of x and y for which $dx/dt = dy/dt = 0$.

Solution   For $a = 0.04$, $b = 0.002$, $m = 0.08$, and $n = 0.0004$, the predator-prey equations are:

$$\frac{dx}{dt} = 0.04x - 0.002xy \qquad \text{Rate of change of prey}$$

$$\frac{dy}{dt} = -0.08y + 0.0004xy \qquad \text{Rate of change of predators}$$

Solving $dx/dt = x(0.04 - 0.002y) = 0$ and $dy/dt = y(-0.08 + 0.0004x) = 0$, you can see that $dx/dt = dy/dt = 0$ when $(x, y) = (0, 0)$ and when $(x, y) = (200, 20)$.

In general, for the predator-prey equations

$$\frac{dx}{dt} = ax - bxy \qquad \text{and} \qquad \frac{dy}{dt} = -my + nxy$$

$\dfrac{dx}{dt} = 0$ when $x = 0$ or $y = \dfrac{a}{b}$ and $\dfrac{dy}{dt} = 0$ when $y = 0$ or $x = \dfrac{m}{n}$. So, at the points

$(0, 0)$ and $\left(\dfrac{m}{n}, \dfrac{a}{b} \right)$, the prey and predator populations are constant. These points are called **critical points** or **equilibrium points** of the predator-prey equations.

EXAMPLE 2 Analyzing Predator-Prey Equations Graphically

Assume the predator-prey equations from Example 1

$$\frac{dx}{dt} = 0.04x - 0.002xy \qquad \text{and} \qquad \frac{dy}{dt} = -0.08y + 0.0004xy$$

model a predator-prey relationship involving foxes and rabbits where x is the number of rabbits and y is the number of foxes after t months. Use a graphing utility to graph the functions x and y when $0 \le t \le 240$ and the initial conditions are 200 rabbits and 10 foxes. What do you observe?

Solution The graphs of x and y are shown in Figure 6.33. Some observations are:

- The rabbit and fox populations oscillate periodically between their respective minimum and maximum values.
- The rabbit population oscillates from a minimum of about 125 rabbits to a maximum of about 300 rabbits.
- The fox population oscillates from about 10 foxes to about 35 foxes.
- About 20 months after the rabbit population peaks, the fox population peaks.
- The period of each population appears to be about 115 months.

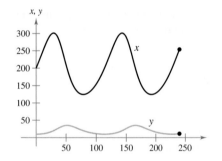

Figure 6.33

You have used slope fields (or direction fields) to analyze solutions of differential equations. In Example 2, the graph shows the curves plotted together with time t along the horizontal axis. You can also use the predator-prey equations dy/dt and dx/dt to graph a slope field. The slope field is graphed using the x-axis to represent the prey and the y-axis to represent the predators.

NOTE If you are using a graphing utility, you may need to rewrite the equations as a function of x:

$$\frac{dy}{dx} = \frac{dy/dt}{dx/dt} = \frac{-my + nxy}{ax - bxy}.$$

EXAMPLE 3 Predator-Prey Equations and Slope Fields

Use a graphing utility to graph the slope field of the predator-prey equations given in Example 2.

Solution The slope field is shown in Figure 6.34. The x-axis represents the rabbit population, and the y-axis represents the fox population.

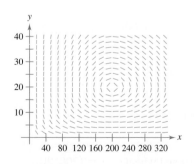

Figure 6.34

EXAMPLE 4   **Graphing a Solution Curve**

Use the predator-prey equations

$$\frac{dx}{dt} = 0.04x - 0.002xy \qquad \text{and} \qquad \frac{dy}{dt} = -0.08y + 0.0004xy$$

and the slope field from Example 3 to graph the solution curve using the initial conditions of 200 rabbits and 10 foxes. Describe the changes in the populations as you trace the solution curve.

Solution   The graph of the solution is a closed curve, as shown in Figures 6.35 and 6.36.

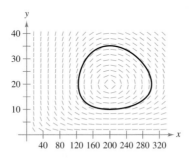

Figure 6.35   **Figure 6.36**

At $(200, 10)$, $dy/dt = 0$ and $dx/dt = 4$. So, the rabbit population is increasing at $(200, 10)$. This means that you should trace the curve counterclockwise as t increases. As you trace the curve, note the changes listed in Figure 6.37.

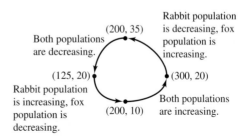

Figure 6.37

Although it is generally not possible to solve predator-prey equations explicitly for x and y, you can separate variables to derive an implicit solution. Begin by writing the equations dy/dt and dx/dt as a function of x.

$$\frac{dy}{dx} = \frac{y(-m + nx)}{x(a - by)} \qquad \text{Factor numerator and denominator.}$$

$$x(a - by)\,dy = y(-m + nx)\,dx \qquad \text{Differential form}$$

$$\frac{a - by}{y}\,dy = \frac{-m + nx}{x}\,dx \qquad \text{Separate variables.}$$

$$\int \frac{a - by}{y}\,dy = \int \frac{-m + nx}{x}\,dx \qquad \text{Integrate.}$$

$$a \ln y - by = -m \ln x + nx + C \qquad \text{Assume } x \text{ and } y \text{ are positive.}$$

$$a \ln y + m \ln x - by - nx = C \qquad \text{General solution}$$

The constant C is determined by the initial conditions.

NOTE   The general solution $a \ln y + m \ln x - by - nx = C$ can be rewritten as $\ln(y^a x^m) = C + by + nx$ or as $y^a x^m = C_1 e^{by + nx}$.

Competing Species

Consider two species that compete with each other for the food available in their common environment. Assume that their populations are given by x and y at time t. If there were no interaction or competition between the species, then the populations x and y would each have logistic growth and would satisfy the following differential equations.

$$\frac{dx}{dt} = ax - bx^2 \qquad \text{Rate of change of first species without interaction}$$

$$\frac{dy}{dt} = my - ny^2 \qquad \text{Rate of change of second species without interaction}$$

If the species interact, then their competition for resources causes a rate of decline in each population proportional to the product xy. Using a negative interaction factor leads to the following **competing-species equations** (where a, b, c, m, n, and p are positive constants).

$$\frac{dx}{dt} = ax - bx^2 - cxy \qquad \text{Rate of change of first species with interaction}$$

$$\frac{dy}{dt} = my - ny^2 - pxy \qquad \text{Rate of change of second species with interaction}$$

In this text it is assumed that competing-species equations have four critical points, as shown in Example 5.

EXAMPLE 5 Deriving the Critical Points

Show that the critical points of the competing-species equations

$$\frac{dx}{dt} = ax - bx^2 - cxy \qquad \text{and} \qquad \frac{dy}{dt} = my - ny^2 - pxy$$

are $(0, 0)$, $(0, m/n)$, $(a/b, 0)$, and $\left(\dfrac{an - mc}{bn - cp}, \dfrac{bm - ap}{bn - cp}\right)$.

Solution Set dx/dt and dy/dt equal to 0 and then factor to obtain the following system.

$$x(a - bx - cy) = 0 \qquad \text{Set } dx/dt \text{ equal to 0 and factor out } x.$$
$$y(m - ny - px) = 0 \qquad \text{Set } dy/dt \text{ equal to 0 and factor out } y.$$

If $x = 0$, then $y = 0$ or $y = m/n$. If $y = 0$, then $x = 0$ or $x = a/b$. So three of the critical points are $(0, 0)$, $(0, m/n)$, and $(a/b, 0)$.

At each of these critical points, one of the populations is 0. These points represent the possibility that both species cannot coexist. The fourth critical point is obtained by solving the system

$$a - bx - cy = 0$$
$$m - ny - px = 0.$$

The solution of this system is

$$(x, y) = \left(\frac{an - mc}{bn - cp}, \frac{bm - ap}{bn - cp}\right).$$

Assuming this point exists and lies in Quadrant I of the xy-plane, the point represents the possibility that both species can coexist.

EXAMPLE 6   Competing Species: One Species Survives

Consider the competing-species equations given by

$$\frac{dx}{dt} = 10x - x^2 - 2xy \qquad \text{and} \qquad \frac{dy}{dt} = 10y - y^2 - 2xy.$$

a. Find the critical points.

b. Use a graphing utility to graph the solution of the equations when $0 \le t \le 3$ and the initial conditions are $x(0) = 10$ and $y(0) = 15$. What do you observe?

Solution

a. Note that $a = 10$, $b = 1$, $c = 2$, $m = 10$, $n = 1$, and $p = 2$. So, the critical points are $(0,0)$, $(0, 10)$, $(10, 0)$, and $\left(\dfrac{10 - 20}{1 - 4}, \dfrac{10 - 20}{1 - 4}\right) = \left(\dfrac{10}{3}, \dfrac{10}{3}\right)$.

b. The solution of the competing-species equations is shown in Figure 6.38. From the graph, it appears that one species survives. The population of the surviving species, represented by the graph of y, appears to remain constant at 10.

Figure 6.38

EXAMPLE 7   Competing Species: Both Species Survive

Consider the competing-species equations given by

$$\frac{dx}{dt} = 10x - 3x^2 - xy \qquad \text{and} \qquad \frac{dy}{dt} = 14y - 3y^2 - xy.$$

a. Find the critical points.

b. Use a graphing utility to graph the solution of the equations when $0 \le t \le 15$ and the initial conditions are $x(0) = 10$ and $y(0) = 15$. What do you observe?

Solution

a. Note that $a = 10$, $b = 3$, $c = 1$, $m = 14$, $n = 3$, and $p = 1$. So, the critical points are $(0, 0)$, $\left(0, \dfrac{14}{3}\right)$, $\left(\dfrac{10}{3}, 0\right)$, and $\left(\dfrac{30 - 14}{9 - 1}, \dfrac{42 - 10}{9 - 1}\right) = (2, 4)$.

b. The solution of the competing-species equations is shown in Figure 6.39. From the graph, it appears that both species survive. The population represented by y appears to remain constant at 4. The population represented by x appears to remain constant at 2.

Figure 6.39

Examples 6 and 7 imply a general conclusion about competing-species equations that have precisely four critical points. In general, it can be shown that if $bn > cp$, both species survive. If $bn < cp$, then one species will survive and the other will not.

You can also use slope fields to analyze solutions of competing-species equations, as shown in Figures 6.40 (Example 6) and 6.41 (Example 7).

Figure 6.40

Figure 6.41

See www.CalcChat.com for worked-out solutions to odd-numbered exercises.

Exercises for Section 6.6

In Exercises 1–4, use the given values to write the predator-prey equations $dx/dt = ax - bxy$ and $dy/dt = -my + nxy$. Then find the values of x and y for which $dx/dt = dy/dt = 0$.

1. $a = 0.7, b = 0.05, m = 0.4, n = 0.007$

2. $a = 0.5, b = 0.004, m = 0.3$ $n = 0.01$

3. $a = 0.3, b = 0.006, m = 0.5, n = 0.009$

4. $a = 0.6, b = 0.02, m = 0.6, n = 0.01$

 Slope Fields In Exercises 5 and 6, predator-prey equations, a point, and a slope field are given. (a) Sketch a solution of the predator-prey equations on the slope field that passes through the given point. (b) Use a graphing utility to graph the solution. Compare the result with the sketch in part (a). To print an enlarged copy of the graph, go to the website *www.mathgraphs.com*.

5. $\dfrac{dx}{dt} = 0.04x - 0.002xy$

   $\dfrac{dy}{dt} = -0.08y + 0.0004xy$

   $(150, 30)$

6. $\dfrac{dx}{dt} = 0.03x - 0.006xy$

   $\dfrac{dy}{dt} = -0.04y + 0.004xy$

   $(15, 3)$

In Exercises 7 and 8, two graphs are given. The first is a graph of the functions x and y of a set of predator-prey equations where x is the number of prey and y is the number of predators at time t. The second graph is the corresponding slope field of the predator-prey equations. (a) Identify the initial conditions. (b) Sketch a solution of the predator-prey equations on the slope field that passes through the initial conditions. To print an enlarged copy of the graph, go to the website *www.mathgraphs.com*.

7.

8.

Rabbits and Foxes In Exercises 9–12, consider a predator-prey relationship involving foxes (predators) and rabbits (prey). Let x represent the number of rabbits, let y represent the number of foxes, and let t represent the time in months. Assume that the following predator-prey equations model the rates of change of each population.

$\dfrac{dx}{dt} = 0.8x - 0.04xy$ \quad Rate of change of prey population

$\dfrac{dy}{dt} = -0.3y + 0.006xy$ \quad Rate of change of predator population

When $t = 0, x = 55$ rabbits and $y = 10$ foxes.

9. Find the critical points of the predator-prey equations.

10. Use a graphing utility to graph the functions x and y when $0 \le t \le 36$. Describe the behavior of each solution as t increases.

11. Use a graphing utility to graph a slope field of the predator-prey equations when $0 \le x \le 150$ and $0 \le y \le 50$.

12. Use the predator-prey equations and the slope field in Exercise 11 to graph the solution curve using the initial conditions. Describe the changes in the rabbit and fox populations as you trace the solution curve.

Prairie Dogs and Black-Footed Ferrets In Exercises 13–16, consider a predator-prey relationship involving black-footed ferrets (predators) and prairie dogs (prey). Let x represent the number of prairie dogs, let y represent the number of black-footed ferrets, and let t represent the time in months. Assume that the following predator-prey equations model the rates of change of each population.

$\dfrac{dx}{dt} = 0.1x - 0.00008xy$ \quad Rate of change of prey population

$\dfrac{dy}{dt} = -0.4y + 0.00004xy$ \quad Rate of change of predator population

When $t = 0, x = 4000$ prairie dogs and $y = 1000$ black-footed ferrets.

13. Find the critical points of the predator-prey equations.

14. Use a graphing utility to graph the functions x and y when $0 \le t \le 240$. Describe the behavior of each solution as t increases.

15. Use a graphing utility to graph a slope field of the predator-prey equations when $0 \le x \le 25,000$ and $0 \le y \le 4000$.

16. Use the predator-prey equations and the slope field in Exercise 15 to graph the solution curve using the initial conditions. Describe the changes in the prairie dog and black-footed ferret populations as you trace the solution curve.

17. *Critical Point as the Initial Condition* In Exercise 9, you found the critical points of the predator-prey system. Assume that the critical point given by $(m/n, a/b)$ is the initial condition and repeat Exercises 10–12. Compare the results.

18. *Critical Point as the Initial Condition* In Exercise 13, you found the critical points of the predator-prey system. Assume that the critical point given by $(m/n, a/b)$ is the initial condition and repeat Exercises 14–16. Compare the results.

In Exercises 19–22, use the given values to write the competing-species equations $dx/dt = ax - bx^2 - cxy$ and $dy/dt = my - ny^2 - pxy$. Then find the values of x and y for which $dx/dt = dy/dt = 0$.

19. $a = 1, b = 2, c = 1, m = 1, n = 2, p = 1$

20. $a = 2, b = 1, c = 1, m = 5, n = 4, p = 1$

21. $a = 0.1, b = 0.4, c = 0.5, m = 0.1, n = 0.8, p = 0.3$

22. $a = 0.05, b = 0.2, c = 0.4, m = 0.06, n = 0.9, p = 0.2$

Bass and Trout In Exercises 23 and 24, consider a competing-species relationship involving bass and trout. Assume the bass and trout compete for the same resources. Let x represent the number of bass (in thousands), let y represent the number of trout (in thousands), and let t represent the time in months. Assume that the following competing-species equations model the rates of change of each population.

$$\frac{dx}{dt} = 0.8x - 0.4x^2 - 0.1xy \qquad \text{Rate of change of bass population}$$

$$\frac{dy}{dt} = 0.3y - 0.6y^2 - 0.1xy \qquad \text{Rate of change of trout population}$$

When $t = 0$, $x = 9$ and $y = 5$.

23. Find the critical points of the competing-species equations.

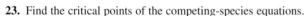 **24.** Use a graphing utility to graph the functions x and y when $0 \le t \le 36$. Describe the behavior of each solution as t increases.

Bass and Trout In Exercises 25 and 26, consider a competing-species relationship involving bass and trout. Assume the bass and trout compete for the same resources. Let x represent the number of bass (in thousands), let y represent the number of trout (in thousands), and let t represent the time in months. Assume that the following competing-species equations model the rates of change of each population.

$$\frac{dx}{dt} = 0.8x - 0.4x^2 - xy \qquad \text{Rate of change of bass population}$$

$$\frac{dy}{dt} = 0.3y - 0.6y^2 - xy \qquad \text{Rate of change of trout population}$$

When $t = 0$, $x = 7$ and $y = 6$.

25. Find the critical points of the competing-species equations.

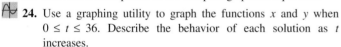 **26.** Use a graphing utility to graph the functions x and y when $0 \le t \le 36$. Describe the behavior of each solution as t increases.

27. *Critical Point as the Initial Condition* In Exercise 23, you found the critical points of the competing-species system. Assume the critical point given by $\left(\dfrac{an - mc}{bn - cp}, \dfrac{bm - ap}{bn - cp} \right)$ is the initial condition and repeat Exercise 24. Compare the results.

28. *Critical Point as the Initial Condition* In Exercise 23, you found the critical points of the competing-species system. Assume the critical point given by $(0, m/n)$ is the initial condition and repeat Exercise 24. Compare the results.

True or False? In Exercises 31–34, determine whether the statement is true or false. If it is false, explain why or give an example that shows it is false.

31. The predator-prey equations are separable differential equations.

32. The predator-prey equations are linear differential equations.

33. The competing-species equations are a special case of the predator-prey equations.

34. The predator-prey equations can always be solved explicitly for x and y.

35. *Revising the Predator-Prey Equations* Consider a predator-prey relationship with x prey and y predators at time t. Assume both predator and prey are present. Then the rates of change of each population can be modeled by the following revised predator-prey system of differential equations.

$$\frac{dx}{dt} = ax\left(1 - \frac{x}{L}\right) - bxy \qquad \text{Rate of change of prey population}$$

$$\frac{dy}{dt} = -my + nxy \qquad \text{Rate of change of predator population}$$

(a) If there are no predators, the prey population will grow according to what model?

(b) Write the revised predator-prey equations for $a = 0.4$, $L = 100$, $b = 0.01$, $m = 0.3$ and $n = 0.005$. Find the critical numbers.

(c) Use a graphing utility to graph the functions x and y of the revised predator-prey equations when $0 \le t \le 72$ and the initial conditions are $x(0) = 40$ and $y(0) = 80$. Describe the behavior of each solution as t increases.

(d) Use a graphing utility to graph a slope field of the revised predator-prey equations when $0 \le x \le 100$ and $0 \le y \le 80$.

(e) Use the predator-prey equations and the slope field in part (d) to graph the solution curve using the initial conditions in part (c). Describe the changes in the prey and predator populations as you trace the solution curve.

36. *Comparing Results* Repeat Exercise 35, parts (b)–(e), using the original predator-prey equations. Use $0 \le t \le 72$, $0 \le x \le 140$, and $0 \le y \le 100$ to graph the solutions and the slope field. Compare the results.

Review Exercises for Chapter 6

1. Determine whether the function $y = x^3$ is a solution of the differential equation $x^2y' + 3y = 6x^3$.

2. Determine whether the function $y = 2\sin 2x$ is a solution of the differential equation $y''' - 8y = 0$.

In Exercises 3–8, use integration to find a general solution of the differential equation.

3. $\dfrac{dy}{dx} = 2x^2 + 5$

4. $\dfrac{dy}{dx} = x^3 - 2x$

5. $\dfrac{dy}{dx} = \cos 2x$

6. $\dfrac{dy}{dx} = 2\sin x$

7. $\dfrac{dy}{dx} = 2x\sqrt{x - 7}$

8. $\dfrac{dy}{dx} = 3e^{-x/3}$

Slope Fields   **In Exercises 9 and 10, a differential equation and its slope field are given. Determine the slopes (if possible) in the slope field at the points given in the table.**

x	−4	−2	0	2	4	8
y	2	0	4	4	6	8
dy/dx						

9. $\dfrac{dy}{dx} = \dfrac{2x}{y}$

10. $\dfrac{dy}{dx} = x\sin\left(\dfrac{\pi y}{4}\right)$

Slope Fields   **In Exercises 11–16, (a) sketch the slope field for the differential equation, and (b) use the slope field to sketch the solution that passes through the given point.**

Differential Equation	*Point*
11. $y' = -x - 2$	$(-1, 1)$
12. $y' = 2x^2 - x$	$(0, 2)$
13. $y' = \dfrac{1}{4}x^2 - \dfrac{1}{3}x$	$(0, 3)$
14. $y' = y + 3x$	$(2, 1)$
15. $y' = \dfrac{xy}{x^2 + 4}$	$(0, 1)$
16. $y' = \dfrac{y}{x^2 + 1}$	$(0, -2)$

In Exercises 17–22, solve the differential equation.

17. $\dfrac{dy}{dx} = 6 - x$

18. $\dfrac{dy}{dx} = y + 6$

19. $\dfrac{dy}{dx} = (3 + y)^2$

20. $\dfrac{dy}{dx} = 4\sqrt{y}$

21. $(2 + x)y' - xy = 0$

22. $xy' - (x + 1)y = 0$

In Exercises 23–26, find the exponential function $y = Ce^{kt}$ that passes through the two points.

23.   **24.**

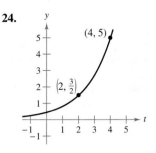

25. $(0, 5), \left(5, \dfrac{1}{6}\right)$ **26.** $(1, 9), (6, 2)$

27. *Air Pressure*   Under ideal conditions, air pressure decreases continuously with the height above sea level at a rate proportional to the pressure at that height. The barometer reads 30 inches at sea level and 15 inches at 18,000 feet. Find the barometric pressure at 35,000 feet.

28. *Radioactive Decay*   Radioactive radium has a half-life of approximately 1599 years. The initial quantity is 5 grams. How much remains after 600 years?

29. *Sales*   The sales S (in thousands of units) of a new product after it has been on the market for t years is given by

$$S = Ce^{k/t}.$$

(a) Find S as a function of t if 5000 units have been sold after 1 year and the saturation point for the market is 30,000 units (that is, $\lim\limits_{t\to\infty} S = 30$).

(b) How many units will have been sold after 5 years?

(c) Use a graphing utility to graph this sales function.

30. *Sales*   The sales S (in thousands of units) of a new product after it has been on the market for t years is given by

$$S = 25(1 - e^{kt}).$$

(a) Find S as a function of t if 4000 units have been sold after 1 year.

(b) How many units will saturate this market?

(c) How many units will have been sold after 5 years?

(d) Use a graphing utility to graph this sales function.

31. *Population Growth*   A population grows continuously at the rate of 1.5%. How long will it take the population to double?

32. *Fuel Economy* An automobile gets 28 miles per gallon of gasoline for speeds up to 50 miles per hour. Over 50 miles per hour, the number of miles per gallon drops at the rate of 12 percent for each 10 miles per hour.

(a) s is the speed and y is the number of miles per gallon. Find y as a function of s by solving the differential equation

$$\frac{dy}{ds} = -0.012y, \quad s > 50.$$

(b) Use the function in part (a) to complete the table.

Speed	50	55	60	65	70
Miles per Gallon					

In Exercises 33–38, solve the differential equation.

33. $\dfrac{dy}{dx} = \dfrac{x^2 + 3}{x}$

34. $\dfrac{dy}{dx} = \dfrac{e^{-2x}}{1 + e^{-2x}}$

35. $y' - 2xy = 0$

36. $y' - e^y \sin x = 0$

37. $\dfrac{dy}{dx} = \dfrac{x^2 + y^2}{2xy}$

38. $\dfrac{dy}{dx} = \dfrac{3(x + y)}{x}$

39. Verify that the general solution $y = C_1 x + C_2 x^3$ satisfies the differential equation $x^2 y'' - 3xy' + 3y = 0$. Then find the particular solution that satisfies the initial conditions $y = 0$ and $y' = 4$ when $x = 2$.

40. *Vertical Motion* A falling object encounters air resistance that is proportional to its velocity. The acceleration due to gravity is -9.8 meters per second per second. The net change in velocity is $dv/dt = kv - 9.8$.

(a) Find the velocity of the object as a function of time if the initial velocity is v_0.

(b) Use the result of part (a) to find the limit of the velocity as t approaches infinity.

(c) Integrate the velocity function found in part (a) to find the position function s.

Slope Fields **In Exercises 41 and 42, sketch a few solutions of the differential equation on the slope field and then find the general solution analytically. To print an enlarged copy of the graph, go to the website *www.mathgraphs.com*.**

41. $\dfrac{dy}{dx} = -\dfrac{4x}{y}$

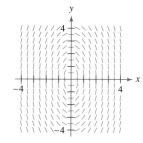

42. $\dfrac{dy}{dx} = 3 - 2y$

In Exercises 43 and 44, the logistic equation models the growth of a population. Use the equation to (a) find the value of k, (b) find the carrying capacity, (c) find the initial population, (d) determine when the population will reach 50% of its carrying capacity, and (e) write a logistic differential equation that has the solution $P(t)$.

43. $P(t) = \dfrac{7200}{1 + 44e^{-0.55t}}$

44. $P(t) = \dfrac{4800}{1 + 14e^{-0.15t}}$

45. *Environment* A conservation department releases 1200 brook trout into a lake. It is estimated that the carrying capacity of the lake for the species is 20,400. After the first year, there are 2000 brook trout in the lake.

(a) Write a logistic equation that models the number of brook trout in the lake.

(b) Find the number of brook trout in the lake after 8 years.

(c) When will the number of brook trout reach 10,000?

46. *Environment* Write a logistic differential equation that models the growth rate of the brook trout population in Exercise 45. Then repeat part (b) using Euler's Method with a step size of $h = 1$. Compare the approximation with the exact answers.

47. *Sales Growth* The rate of change in sales S (in thousands of units) of a new product is proportional to the difference between L and S (in thousands of units) at any time t. When $t = 0$, $S = 0$. Write and solve the differential equation for this sales model.

48. *Sales Growth* Use the result of Exercise 47 to write S as a function of t if (a) $L = 100$, $S = 25$ when $t = 2$, and (b) $L = 500$, $S = 50$ when $t = 1$.

Learning Theory **In Exercises 49 and 50, assume that the rate of change in the proportion P of correct responses after n trials is proportional to the product of P and $L - P$, where L is the limiting proportion of correct responses.**

49. Write and solve the differential equation for this learning theory model.

50. Use the solution of Exercise 49 to write P as a function of n, and then use a graphing utility to graph the solution.

(a) $L = 1.00$

$P = 0.50$ when $n = 0$

$P = 0.85$ when $n = 4$

(b) $L = 0.80$

$P = 0.25$ when $n = 0$

$P = 0.60$ when $n = 10$

7. The cylindrical water tank shown in the figure has a height of 18 feet. When the tank is full, a circular valve is opened at the bottom of the tank. After 30 minutes, the depth of the water is 12 feet.

(a) How long will it take for the tank to drain completely?

(b) What is the depth of the water in the tank after 1 hour?

8. Suppose the tank in Exercise 7 has a height of 20 feet, a radius of 8 feet, and the valve is circular with a radius of 2 inches. The tank is full when the valve is opened. How long will it take for the tank to drain completely?

9. In hilly areas, radio reception may be poor. Consider a situation where an FM transmitter is located at the point $(-1, 1)$ behind a hill modeled by the graph of

$$y = x - x^2$$

and a radio receiver is located on the opposite side of the hill. (Assume that the x-axis represents ground level at the base of the hill.)

(a) What is the closest position $(x, 0)$ the radio can be to the hill so that reception is unobstructed?

(b) Write the closest position $(x, 0)$ of the radio, with x represented as a function of h, if the transmitter is located at $(-1, h)$.

(c) Use a graphing utility to graph the function for x in part (b). Determine the vertical asymptote of the function and interpret the result.

10. Biomass is a measure of the amount of living matter in an ecosystem. Suppose the biomass $s(t)$ in a given ecosystem increases at a rate of about 3.5 tons per year, and decreases by about 1.9% per year. This situation can be modeled by the differential equation

$$\frac{ds}{dt} = 3.5 - 0.019s.$$

(a) Solve the differential equation.

(b) Use a graphing utility to graph the slope field for the differential equation. What do you notice?

(c) Explain what happens as $t \to \infty$.

In Exercises 11–13, a medical researcher wants to determine the concentration C (in moles per liter) of a tracer drug injected into a moving fluid. Solve this problem by considering a single-compartment dilution model (see figure). Assume that the fluid is continuously mixed and that the volume of the fluid in the compartment is constant.

Flow R (pure)

Volume V

Flow R (concentration C)

Figure for 11–13

11. If the tracer is injected instantaneously at time $t = 0$, then the concentration of the fluid in the compartment begins diluting according to the differential equation

$$\frac{dC}{dt} = \left(-\frac{R}{V}\right)C, \quad C = C_0 \text{ when } t = 0.$$

(a) Solve this differential equation to find the concentration C as a function of time t.

(b) Find the limit of C as $t \to \infty$.

12. Use the solution of the differential equation in Exercise 11 to find the concentration C as a function of time t, and use a graphing utility to graph the function.

(a) $V = 2$ liters, $R = 0.5$ liter per minute, and $C_0 = 0.6$ mole per liter

(b) $V = 2$ liters, $R = 1.5$ liters per minute, and $C_0 = 0.6$ mole per liter

13. In Exercises 11 and 12, it was assumed that there was a single initial injection of the tracer drug into the compartment. Now consider the case in which the tracer is continuously injected (beginning at $t = 0$) at the rate of Q moles per minute. Considering Q to be negligible compared with R, use the differential equation

$$\frac{dC}{dt} = \frac{Q}{V} - \left(\frac{R}{V}\right)C, \quad C = 0 \text{ when } t = 0.$$

(a) Solve this differential equation to find the concentration C as a function of time t.

(b) Find the limit of C as $t \to \infty$.

Applications of Integration

The Atomium, located in Belgium, represents an iron crystal molecule magnified 165 billion times. The structure contains nine spheres connected with cylindrical tubes. The central sphere has one tube passing directly through its center. Explain how to find the volume of the portion of the central sphere that does not include the tube.

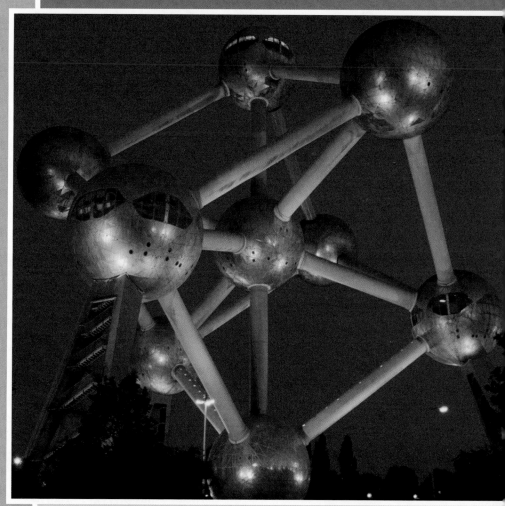

Andre Jenny/Alamy Images

The *disk method* is one method that is used to find the volume of a solid. This method requires finding the sum of the volumes of representative disks to approximate the volume of the solid. As you increase the number of disks, the approximation tends to become more accurate. In Section 7.2, you will use limits to write the exact volume of the solid as a definite integral.

Area of a Region Between Two Curves

- Find the area of a region between two curves using integration.
- Find the area of a region between intersecting curves using integration.
- Describe integration as an accumulation process.

Area of a Region Between Two Curves

With a few modifications you can extend the application of definite integrals from the area of a region *under* a curve to the area of a region *between* two curves. Consider two functions f and g that are continuous on the interval $[a, b]$. If, as in Figure 7.1, the graphs of both f and g lie above the x-axis, and the graph of g lies below the graph of f, you can geometrically interpret the area of the region between the graphs as the area of the region under the graph of g subtracted from the area of the region under the graph of f, as shown in Figure 7.2.

Figure 7.1

Area of region between f and g	=	Area of region under f	−	Area of region under g
$\displaystyle\int_a^b [f(x) - g(x)]\, dx$	=	$\displaystyle\int_a^b f(x)\, dx$	−	$\displaystyle\int_a^b g(x)\, dx$

Figure 7.2

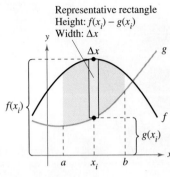

Representative rectangle
Height: $f(x_i) - g(x_i)$
Width: Δx

Figure 7.3

To verify the reasonableness of the result shown in Figure 7.2, you can partition the interval $[a, b]$ into n subintervals, each of width Δx. Then, as shown in Figure 7.3, sketch a **representative rectangle** of width Δx and height $f(x_i) - g(x_i)$, where x_i is in the ith interval. The area of this representative rectangle is

$$\Delta A_i = (\text{height})(\text{width}) = [f(x_i) - g(x_i)]\Delta x.$$

By adding the areas of the n rectangles and taking the limit as $\|\Delta\| \to 0$ $(n \to \infty)$, you obtain

$$\lim_{n\to\infty} \sum_{i=1}^{n} [f(x_i) - g(x_i)]\Delta x.$$

Because f and g are continuous on $[a, b]$, $f - g$ is also continuous on $[a, b]$ and the limit exists. So, the area of the given region is

$$\text{Area} = \lim_{n\to\infty} \sum_{i=1}^{n} [f(x_i) - g(x_i)]\Delta x$$

$$= \int_a^b [f(x) - g(x)]\, dx.$$

Area of a Region Between Two Curves

If f and g are continuous on $[a, b]$ and $g(x) \leq f(x)$ for all x in $[a, b]$, then the area of the region bounded by the graphs of f and g and the vertical lines $x = a$ and $x = b$ is

$$A = \int_a^b [f(x) - g(x)] \, dx.$$

In Figure 7.1, the graphs of f and g are shown above the x-axis. This, however, is not necessary. The same integrand $[f(x) - g(x)]$ can be used as long as f and g are continuous and $g(x) \leq f(x)$ for all x in the interval $[a, b]$. This result is summarized graphically in Figure 7.4.

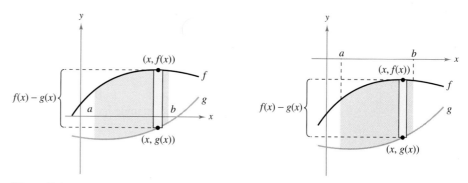

Figure 7.4

NOTE The height of a representative rectangle is $f(x) - g(x)$ regardless of the relative position of the x-axis, as shown in Figure 7.4.

Representative rectangles are used throughout this chapter in various applications of integration. A vertical rectangle (of width Δx) implies integration with respect to x, whereas a horizontal rectangle (of width Δy) implies integration with respect to y.

EXAMPLE 1   **Finding the Area of a Region Between Two Curves**

Find the area of the region bounded by the graphs of $y = x^2 + 2$, $y = -x$, $x = 0$, and $x = 1$.

Solution   Let $g(x) = -x$ and $f(x) = x^2 + 2$. Then $g(x) \leq f(x)$ for all x in $[0, 1]$, as shown in Figure 7.5. So, the area of the representative rectangle is

$$\Delta A = [f(x) - g(x)] \, \Delta x$$
$$= [(x^2 + 2) - (-x)] \, \Delta x$$

and the area of the region is

$$A = \int_a^b [f(x) - g(x)] \, dx = \int_0^1 [(x^2 + 2) - (-x)] \, dx$$
$$= \left[\frac{x^3}{3} + \frac{x^2}{2} + 2x \right]_0^1$$
$$= \frac{1}{3} + \frac{1}{2} + 2$$
$$= \frac{17}{6}.$$

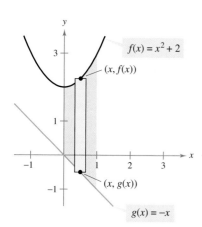

Region bounded by the graph of f, the graph of g, $x = 0$, and $x = 1$
Figure 7.5

Area of a Region Between Intersecting Curves

In Example 1, the graphs of $f(x) = x^2 + 2$ and $g(x) = -x$ do not intersect, and the values of a and b are given explicitly. A more common problem involves the area of a region bounded by two *intersecting* graphs, where the values of a and b must be calculated.

EXAMPLE 2   A Region Lying Between Two Intersecting Graphs

Find the area of the region bounded by the graphs of $f(x) = 2 - x^2$ and $g(x) = x$.

Solution   In Figure 7.6, notice that the graphs of f and g have two points of intersection. To find the x-coordinates of these points, set $f(x)$ and $g(x)$ equal to each other and solve for x.

$$2 - x^2 = x \qquad\qquad \text{Set } f(x) \text{ equal to } g(x).$$
$$-x^2 - x + 2 = 0 \qquad\qquad \text{Write in general form.}$$
$$-(x + 2)(x - 1) = 0 \qquad\qquad \text{Factor.}$$
$$x = -2 \text{ or } 1 \qquad\qquad \text{Solve for } x.$$

So, $a = -2$ and $b = 1$. Because $g(x) \le f(x)$ for all x in the interval $[-2, 1]$, the representative rectangle has an area of

$$\Delta A = [f(x) - g(x)]\,\Delta x$$
$$= [(2 - x^2) - x]\,\Delta x$$

and the area of the region is

$$A = \int_{-2}^{1} [(2 - x^2) - x]\,dx = \left[-\frac{x^3}{3} - \frac{x^2}{2} + 2x \right]_{-2}^{1}$$
$$= \frac{9}{2}.$$

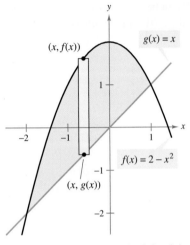

Region bounded by the graph of f and the graph of g
Figure 7.6

EXAMPLE 3   A Region Lying Between Two Intersecting Graphs

The sine and cosine curves intersect infinitely many times, bounding regions of equal areas, as shown in Figure 7.7. Find the area of one of these regions.

Solution

$$\sin x = \cos x \qquad\qquad \text{Set } f(x) \text{ equal to } g(x).$$
$$\frac{\sin x}{\cos x} = 1 \qquad\qquad \text{Divide each side by } \cos x.$$
$$\tan x = 1 \qquad\qquad \text{Trigonometric identity}$$
$$x = \frac{\pi}{4} \text{ or } \frac{5\pi}{4}, \qquad 0 \le x \le 2\pi \qquad \text{Solve for } x.$$

So, $a = \pi/4$ and $b = 5\pi/4$. Because $\sin x \ge \cos x$ for all x in the interval $[\pi/4, 5\pi/4]$, the area of the region is

$$A = \int_{\pi/4}^{5\pi/4} [\sin x - \cos x]\,dx = \left[-\cos x - \sin x \right]_{\pi/4}^{5\pi/4}$$
$$= 2\sqrt{2}.$$

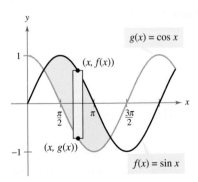

One of the regions bounded by the graphs of the sine and cosine functions
Figure 7.7

If two curves intersect at more than two points, then to find the area of the region between the curves, you must find all points of intersection and check to see which curve is above the other in each interval determined by these points.

 EXAMPLE 4   Curves That Intersect at More Than Two Points

Find the area of the region between the graphs of $f(x) = 3x^3 - x^2 - 10x$ and $g(x) = -x^2 + 2x$.

Solution   Begin by setting $f(x)$ and $g(x)$ equal to each other and solving for x. This yields the x-values at each point of intersection of the two graphs.

$$3x^3 - x^2 - 10x = -x^2 + 2x \qquad \text{Set } f(x) \text{ equal to } g(x).$$
$$3x^3 - 12x = 0 \qquad \text{Write in general form.}$$
$$3x(x - 2)(x + 2) = 0 \qquad \text{Factor.}$$
$$x = -2, 0, 2 \qquad \text{Solve for } x.$$

So, the two graphs intersect when $x = -2, 0,$ and 2. In Figure 7.8, notice that $g(x) \le f(x)$ on the interval $[-2, 0]$. However, the two graphs switch at the origin, and $f(x) \le g(x)$ on the interval $[0, 2]$. So, you need two integrals—one for the interval $[-2, 0]$ and one for the interval $[0, 2]$.

$$A = \int_{-2}^{0} [f(x) - g(x)]\, dx + \int_{0}^{2} [g(x) - f(x)]\, dx$$
$$= \int_{-2}^{0} (3x^3 - 12x)\, dx + \int_{0}^{2} (-3x^3 + 12x)\, dx$$
$$= \left[\frac{3x^4}{4} - 6x^2 \right]_{-2}^{0} + \left[\frac{-3x^4}{4} + 6x^2 \right]_{0}^{2}$$
$$= -(12 - 24) + (-12 + 24) = 24$$

NOTE   In Example 4, notice that you obtain an incorrect result if you integrate from -2 to 2. Such integration produces

$$\int_{-2}^{2} [f(x) - g(x)]\, dx = \int_{-2}^{2} (3x^3 - 12x)\, dx = 0.$$

If the graph of a function of y is a boundary of a region, it is often convenient to use representative rectangles that are *horizontal* and find the area by integrating with respect to y. In general, to determine the area between two curves, you can use

$$A = \int_{x_1}^{x_2} \underbrace{[(\text{top curve}) - (\text{bottom curve})]\, dx}_{\text{in variable } x} \qquad \text{Vertical rectangles}$$

$$A = \int_{y_1}^{y_2} \underbrace{[(\text{right curve}) - (\text{left curve})]\, dy}_{\text{in variable } y} \qquad \text{Horizontal rectangles}$$

where (x_1, y_1) and (x_2, y_2) are either adjacent points of intersection of the two curves involved or points on the specified boundary lines.

*Left margin figure:*

$g(x) \le f(x)$ $f(x) \le g(x)$

$(0, 0)$

$(2, 0)$

$(-2, -8)$

$g(x) = -x^2 + 2x$

$f(x) = 3x^3 - x^2 - 10x$

On $[-2, 0], g(x) \le f(x)$, and on $[0, 2]$, $f(x) \le g(x)$
Figure 7.8

---

indicates that in the HM mathSpace® CD-ROM and the online Eduspace® system for this text, you will find an Open Exploration, which further explores this example using the computer algebra systems Maple, Mathcad, Mathematica, and Derive.

EXAMPLE 5 **Horizontal Representative Rectangles**

Find the area of the region bounded by the graphs of $x = 3 - y^2$ and $x = y + 1$.

Solution Consider

$$g(y) = 3 - y^2 \quad \text{and} \quad f(y) = y + 1.$$

These two curves intersect when $y = -2$ and $y = 1$, as shown in Figure 7.9. Because $f(y) \le g(y)$ on this interval, you have

$$\Delta A = [g(y) - f(y)] \, \Delta y = [(3 - y^2) - (y + 1)] \, \Delta y.$$

So, the area is

$$\begin{aligned}
A &= \int_{-2}^{1} [(3 - y^2) - (y + 1)] \, dy \\
&= \int_{-2}^{1} (-y^2 - y + 2) \, dy \\
&= \left[\frac{-y^3}{3} - \frac{y^2}{2} + 2y \right]_{-2}^{1} \\
&= \left(-\frac{1}{3} - \frac{1}{2} + 2 \right) - \left(\frac{8}{3} - 2 - 4 \right) \\
&= \frac{9}{2}.
\end{aligned}$$

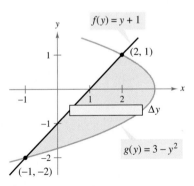

Horizontal rectangles (integration with respect to y)
Figure 7.9

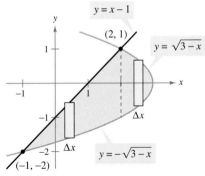

Vertical rectangles (integration with respect to x)
Figure 7.10

In Example 5, notice that by integrating with respect to y you need only one integral. If you had integrated with respect to x, you would have needed two integrals because the upper boundary would have changed at $x = 2$, as shown in Figure 7.10.

$$\begin{aligned}
A &= \int_{-1}^{2} \left[(x - 1) + \sqrt{3 - x} \right] dx + \int_{2}^{3} \left(\sqrt{3 - x} + \sqrt{3 - x} \right) dx \\
&= \int_{-1}^{2} \left[x - 1 + (3 - x)^{1/2} \right] dx + 2 \int_{2}^{3} (3 - x)^{1/2} \, dx \\
&= \left[\frac{x^2}{2} - x - \frac{(3 - x)^{3/2}}{3/2} \right]_{-1}^{2} - 2 \left[\frac{(3 - x)^{3/2}}{3/2} \right]_{2}^{3} \\
&= \left(2 - 2 - \frac{2}{3} \right) - \left(\frac{1}{2} + 1 - \frac{16}{3} \right) - 2(0) + 2 \left(\frac{2}{3} \right) \\
&= \frac{9}{2}
\end{aligned}$$

Integration as an Accumulation Process

In this section, the integration formula for the area between two curves was developed by using a rectangle as the *representative element*. For each new application in the remaining sections of this chapter, an appropriate representative element will be constructed using precalculus formulas you already know. Each integration formula will then be obtained by summing or accumulating these representative elements.

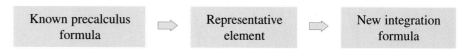

For example, in this section the area formula was developed as follows.

$$A = (\text{height})(\text{width}) \quad \Longrightarrow \quad \Delta A = \left[f(x) - g(x) \right] \Delta x \quad \Longrightarrow \quad A = \int_a^b \left[f(x) - g(x) \right] dx$$

EXAMPLE 6   Describing Integration as an Accumulation Process

Find the area of the region bounded by the graph of $y = 4 - x^2$ and the x-axis. Describe the integration as an accumulation process.

Solution   The area of the region is given by

$$A = \int_{-2}^{2} (4 - x^2)\, dx.$$

You can think of the integration as an accumulation of the areas of the rectangles formed as the representative rectangle slides from $x = -2$ to $x = 2$, as shown in Figure 7.11.

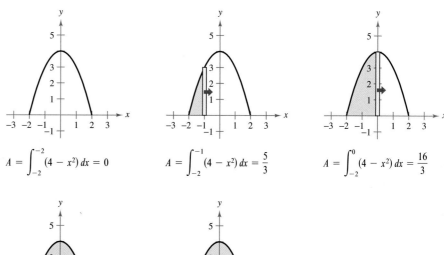

$$A = \int_{-2}^{-2} (4 - x^2)\, dx = 0 \qquad A = \int_{-2}^{-1} (4 - x^2)\, dx = \frac{5}{3} \qquad A = \int_{-2}^{0} (4 - x^2)\, dx = \frac{16}{3}$$

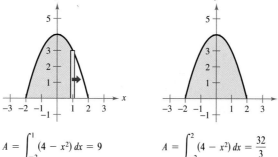

$$A = \int_{-2}^{1} (4 - x^2)\, dx = 9 \qquad A = \int_{-2}^{2} (4 - x^2)\, dx = \frac{32}{3}$$

Figure 7.11

In Exercises 1–6, set up the definite integral that gives the area of the region.

1. $f(x) = x^2 - 6x$
$g(x) = 0$

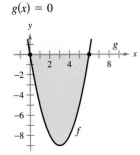

2. $f(x) = x^2 + 2x + 1$
$g(x) = 2x + 5$

3. $f(x) = x^2 - 4x + 3$
$g(x) = -x^2 + 2x + 3$

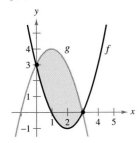

4. $f(x) = x^2$
$g(x) = x^3$

5. $f(x) = 3(x^3 - x)$
$g(x) = 0$

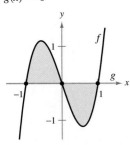

6. $f(x) = (x - 1)^3$
$g(x) = x - 1$

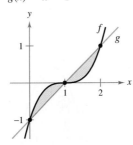

In Exercises 7–12, the integrand of the definite integral is a difference of two functions. Sketch the graph of each function and shade the region whose area is represented by the integral.

7. $\int_0^4 \left[(x + 1) - \frac{x}{2} \right] dx$

8. $\int_{-1}^1 \left[(1 - x^2) - (x^2 - 1) \right] dx$

9. $\int_0^6 \left[4(2^{-x/3}) - \frac{x}{6} \right] dx$

10. $\int_2^3 \left[\left(\frac{x^3}{3} - x \right) - \frac{x}{3} \right] dx$

11. $\int_{-\pi/3}^{\pi/3} (2 - \sec x) \, dx$

12. $\int_{-\pi/4}^{\pi/4} (\sec^2 x - \cos x) \, dx$

In Exercises 13 and 14, find the area of the region by integrating (a) with respect to x and (b) with respect to y.

13. $x = 4 - y^2$
$x = y - 2$

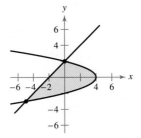

14. $y = x^2$
$y = 6 - x$

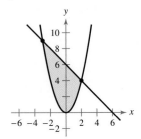

Think About It In Exercises 15 and 16, determine which value best approximates the area of the region bounded by the graphs of f and g. (Make your selection on the basis of a sketch of the region and not by performing any calculations.)

15. $f(x) = x + 1, \qquad g(x) = (x - 1)^2$
  (a) -2 (b) 2 (c) 10 (d) 4 (e) 8

16. $f(x) = 2 - \frac{1}{2}x, \qquad g(x) = 2 - \sqrt{x}$
  (a) 1 (b) 6 (c) -3 (d) 3 (e) 4

In Exercises 17–32, sketch the region bounded by the graphs of the algebraic functions and find the area of the region.

17. $y = \frac{1}{2}x^3 + 2, \; y = x + 1, \; x = 0, \; x = 2$

18. $y = -\frac{3}{8}x(x - 8), \; y = 10 - \frac{1}{2}x, \; x = 2, \; x = 8$

19. $f(x) = x^2 - 4x, \; g(x) = 0$

20. $f(x) = -x^2 + 4x + 1, \; g(x) = x + 1$

21. $f(x) = x^2 + 2x + 1, \; g(x) = 3x + 3$

22. $f(x) = -x^2 + 4x + 2, \; g(x) = x + 2$

23. $y = x, \; y = 2 - x, \; y = 0$

24. $y = \frac{1}{x^2}, \; y = 0, \; x = 1, \; x = 5$

25. $f(x) = \sqrt{3x} + 1, \; g(x) = x + 1$

26. $f(x) = \sqrt[3]{x - 1}, \; g(x) = x - 1$

27. $f(y) = y^2, \; g(y) = y + 2$

28. $f(y) = y(2 - y), \; g(y) = -y$

29. $f(y) = y^2 + 1, \; g(y) = 0, \; y = -1, \; y = 2$

30. $f(y) = \frac{y}{\sqrt{16 - y^2}}, \; g(y) = 0, \; y = 3$

31. $f(x) = \frac{10}{x}, \; x = 0, \; y = 2, \; y = 10$

32. $g(x) = \frac{4}{2 - x}, \; y = 4, \; x = 0$

In Exercises 33–42, (a) use a graphing utility to graph the region bounded by the graphs of the equations, (b) find the area of the region, and (c) use the integration capabilities of the graphing utility to verify your results.

33. $f(x) = x(x^2 - 3x + 3)$, $g(x) = x^2$

34. $f(x) = x^3 - 2x + 1$, $g(x) = -2x$, $x = 1$

35. $y = x^2 - 4x + 3$, $y = 3 + 4x - x^2$

36. $y = x^4 - 2x^2$, $y = 2x^2$

37. $f(x) = x^4 - 4x^2$, $g(x) = x^2 - 4$

38. $f(x) = x^4 - 4x^2$, $g(x) = x^3 - 4x$

39. $f(x) = 1/(1 + x^2)$, $g(x) = \frac{1}{2}x^2$

40. $f(x) = 6x/(x^2 + 1)$, $y = 0$, $0 \le x \le 3$

41. $y = \sqrt{1 + x^3}$, $y = \frac{1}{2}x + 2$, $x = 0$

42. $y = x\sqrt{\dfrac{4 - x}{4 + x}}$, $y = 0$, $x = 4$

In Exercises 43–48, sketch the region bounded by the graphs of the functions, and find the area of the region.

43. $f(x) = 2 \sin x$, $g(x) = \tan x$, $-\dfrac{\pi}{3} \le x \le \dfrac{\pi}{3}$

44. $f(x) = \sin x$, $g(x) = \cos 2x$, $-\dfrac{\pi}{2} \le x \le \dfrac{\pi}{6}$

45. $f(x) = \cos x$, $g(x) = 2 - \cos x$, $0 \le x \le 2\pi$

46. $f(x) = \sec \dfrac{\pi x}{4} \tan \dfrac{\pi x}{4}$, $g(x) = (\sqrt{2} - 4)x + 4$, $x = 0$

47. $f(x) = xe^{-x^2}$, $y = 0$, $0 \le x \le 1$

48. $f(x) = 3^x$, $g(x) = 2x + 1$

 In Exercises 49–52, (a) use a graphing utility to graph the region bounded by the graphs of the equations, (b) find the area of the region, and (c) use the integration capabilities of the graphing utility to verify your results.

49. $f(x) = 2 \sin x + \sin 2x$, $y = 0$, $0 \le x \le \pi$

50. $f(x) = 2 \sin x + \cos 2x$, $y = 0$, $0 < x \le \pi$

51. $f(x) = \dfrac{1}{x^2}e^{1/x}$, $y = 0$, $1 \le x \le 3$

52. $g(x) = \dfrac{4 \ln x}{x}$, $y = 0$, $x = 5$

 In Exercises 53–56, (a) use a graphing utility to graph the region bounded by the graphs of the equations, (b) explain why the area of the region is difficult to find by hand, and (c) use the integration capabilities of the graphing utility to approximate the area to four decimal places.

53. $y = \sqrt{\dfrac{x^3}{4 - x}}$, $y = 0$, $x = 3$

54. $y = \sqrt{x}\, e^x$, $y = 0$, $x = 0$, $x = 1$

55. $y = x^2$, $y = 4 \cos x$

56. $y = x^2$, $y = \sqrt{3 + x}$

In Exercises 57–60, find the accumulation function F. Then evaluate F at each value of the independent variable and graphically show the area given by each value of F.

57. $F(x) = \displaystyle\int_0^x \left(\tfrac{1}{2}t + 1\right) dt$ (a) $F(0)$ (b) $F(2)$ (c) $F(6)$

58. $F(x) = \displaystyle\int_0^x \left(\tfrac{1}{2}t^2 + 2\right) dt$ (a) $F(0)$ (b) $F(4)$ (c) $F(6)$

59. $F(\alpha) = \displaystyle\int_{-1}^{\alpha} \cos \dfrac{\pi\theta}{2}\, d\theta$ (a) $F(-1)$ (b) $F(0)$ (c) $F\!\left(\tfrac{1}{2}\right)$

60. $F(y) = \displaystyle\int_{-1}^{y} 4e^{x/2}\, dx$ (a) $F(-1)$ (b) $F(0)$ (c) $F(4)$

In Exercises 61–64, use integration to find the area of the figure having the given vertices.

61. $(2, -3), (4, 6), (6, 1)$ **62.** $(0, 0), (a, 0), (b, c)$

63. $(0, 2), (4, 2), (0, -2), (-4, -2)$

64. $(0, 0), (1, 2), (3, -2), (1, -3)$

65. *Numerical Integration*   Estimate the surface area of the golf green using (a) the Trapezoidal Rule and (b) Simpson's Rule.

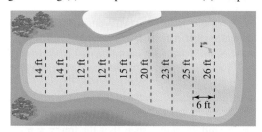

66. *Numerical Integration*   Estimate the surface area of the oil spill using (a) the Trapezoidal Rule and (b) Simpson's Rule.

In Exercises 67–70, set up and evaluate the definite integral that gives the area of the region bounded by the graph of the function and the tangent line to the graph at the given point.

67. $f(x) = x^3$, $(1, 1)$ **68.** $y = x^3 - 2x$, $(-1, 1)$

69. $f(x) = \dfrac{1}{x^2 + 1}$, $\left(1, \tfrac{1}{2}\right)$ **70.** $y = \dfrac{2}{1 + 4x^2}$, $\left(\tfrac{1}{2}, 1\right)$

Writing About Concepts

71. The graphs of $y = x^4 - 2x^2 + 1$ and $y = 1 - x^2$ intersect at three points. However, the area between the curves *can* be found by a single integral. Explain why this is so, and write an integral for this area.

72. The area of the region bounded by the graphs of $y = x^3$ and $y = x$ *cannot* be found by the single integral $\int_{-1}^{1}(x^3 - x)\,dx$. Explain why this is so. Use symmetry to write a single integral that does represent the area.

73. A college graduate has two job offers. The starting salary for each is $32,000, and after 8 years of service each will pay $54,000. The salary increase for each offer is shown in the figure. From a strictly monetary viewpoint, which is the better offer? Explain.

Figure for 73   **Figure for 74**

74. A state legislature is debating two proposals for eliminating the annual budget deficits by the year 2010. The rate of decrease of the deficits for each proposal is shown in the figure. From the viewpoint of minimizing the cumulative state deficit, which is the better proposal? Explain.

In Exercises 75 and 76, find b such that the line $y = b$ divides the region bounded by the graphs of the two equations into two regions of equal area.

75. $y = 9 - x^2$, $\quad y = 0$ **76.** $y = 9 - |x|$, $\quad y = 0$

In Exercises 77 and 78, find a such that the line $x = a$ divides the region bounded by the graphs of the equations into two regions of equal area.

77. $y = x$, $\quad y = 4$, $\quad x = 0$ **78.** $y^2 = 4 - x$, $\quad x = 0$

In Exercises 79 and 80, evaluate the limit and sketch the graph of the region whose area is represented by the limit.

79. $\displaystyle \lim_{\|\Delta\| \to 0} \sum_{i=1}^{n} (x_i - x_i^2)\,\Delta x$, where $x_i = i/n$ and $\Delta x = 1/n$

80. $\displaystyle \lim_{\|\Delta\| \to 0} \sum_{i=1}^{n} (4 - x_i^2)\,\Delta x$, where $x_i = -2 + (4i/n)$ and $\Delta x = 4/n$

Revenue **In Exercises 81 and 82, two models R_1 and R_2 are given for revenue (in billions of dollars per year) for a large corporation. The model R_1 gives projected annual revenues from 2000 to 2005, with $t = 0$ corresponding to 2000, and R_2 gives projected revenues if there is a decrease in the rate of growth of corporate sales over the period. Approximate the total reduction in revenue if corporate sales are actually closer to the model R_2.**

81. $R_1 = 7.21 + 0.58t$ **82.** $R_1 = 7.21 + 0.26t + 0.02t^2$
     $R_2 = 7.21 + 0.45t$ $R_2 = 7.21 + 0.1t + 0.01t^2$

 83. *Modeling Data* The table shows the total receipts R and total expenditures E for the Old-Age and Survivors Insurance Trust Fund (Social Security Trust Fund) in billions of dollars. The time t is given in years, with $t = 1$ corresponding to 1991. (*Source: Social Security Administration*)

t	1	2	3	4	5	6
R	299.3	311.2	323.3	328.3	342.8	363.7
E	245.6	259.9	273.1	284.1	297.8	308.2

t	7	8	9	10	11
R	397.2	424.8	457.0	490.5	518.1
E	322.1	332.3	339.9	358.3	377.5

(a) Use a graphing utility to fit an exponential model to the data for receipts. Plot the data and graph the model.

(b) Use a graphing utility to fit an exponential model to the data for expenditures. Plot the data and graph the model.

(c) If the models are assumed to be true for the years 2002 through 2007, use integration to approximate the surplus revenue generated during those years.

(d) Will the models found in parts (a) and (b) intersect? Explain. Based on your answer and news reports about the fund, will these models be accurate for long-term analysis?

84. *Lorenz Curve* Economists use *Lorenz curves* to illustrate the distribution of income in a country. A Lorenz curve, $y = f(x)$, represents the actual income distribution in the country. In this model, x represents percents of families in the country and y represents percents of total income. The model $y = x$ represents a country in which each family has the same income. The area between these two models, where $0 \leq x \leq 100$, indicates a country's "income inequality." The table lists percents of income y for selected percents of families x in a country.

x	10	20	30	40	50
y	3.35	6.07	9.17	13.39	19.45

x	60	70	80	90
y	28.03	39.77	55.28	75.12

(a) Use a graphing utility to find a quadratic model for the Lorenz curve.

(b) Plot the data and graph the model.

(c) Graph the model $y = x$. How does this model compare with the model in part (a)?

(d) Use the integration capabilities of a graphing utility to approximate the "income inequality."

85. Profit The chief financial officer of a company reports that profits for the past fiscal year were $893,000. The officer predicts that profits for the next 5 years will grow at a continuous annual rate somewhere between $3\frac{1}{2}\%$ and 5%. Estimate the cumulative difference in total profit over the 5 years based on the predicted range of growth rates.

86. Area The shaded region in the figure consists of all points whose distances from the center of the square are less than their distances from the edges of the square. Find the area of the region.

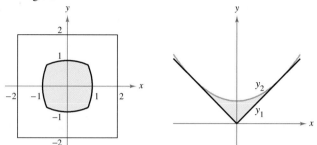

Figure for 86   **Figure for 87**

87. Mechanical Design The surface of a machine part is the region between the graphs of $y_1 = |x|$ and $y_2 = 0.08x^2 + k$ (see figure).

(a) Find k if the parabola is tangent to the graph of y_1.

(b) Find the area of the surface of the machine part.

88. Building Design Concrete sections for a new building have the dimensions (in meters) and shape shown in the figure.

(a) Find the area of the face of the section superimposed on the rectangular coordinate system.

(b) Find the volume of concrete in one of the sections by multiplying the area in part (a) by 2 meters.

(c) One cubic meter of concrete weighs 5000 pounds. Find the weight of the section.

89. Building Design To decrease the weight and to aid in the hardening process, the concrete sections in Exercise 88 often are not solid. Rework Exercise 88 to allow for cylindrical openings such as those shown in the figure.

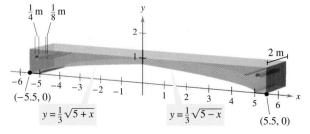

True or False? In Exercises 90–92, determine whether the statement is true or false. If it is false, explain why or give an example that shows it is false.

90. If the area of the region bounded by the graphs of f and g is 1, then the area of the region bounded by the graphs of $h(x) = f(x) + C$ and $k(x) = g(x) + C$ is also 1.

91. If $\displaystyle\int_a^b [f(x) - g(x)]\, dx = A$, then $\displaystyle\int_a^b [g(x) - f(x)]\, dx = -A$.

92. If the graphs of f and g intersect midway between $x = a$ and $x = b$, then

$$\int_a^b [f(x) - g(x)]\, dx = 0.$$

93. Area Find the area between the graph of $y = \sin x$ and the line segments joining the points $(0, 0)$ and $\left(\dfrac{7\pi}{6}, -\dfrac{1}{2}\right)$, as shown in the figure.

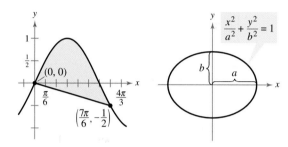

Figure for 93   **Figure for 94**

94. Area Let $a > 0$ and $b > 0$. Show that the area of the ellipse

$$\frac{x^2}{a^2} + \frac{y^2}{b^2} = 1$$

is πab (see figure).

Putnam Exam Challenge

95. The horizontal line $y = c$ intersects the curve $y = 2x - 3x^3$ in the first quadrant as shown in the figure. Find c so that the areas of the two shaded regions are equal.

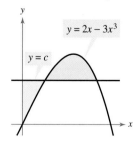

This problem was composed by the Committee on the Putnam Prize Competition.
© The Mathematical Association of America. All rights reserved.

Volume: The Disk Method

- Find the volume of a solid of revolution using the disk method.
- Find the volume of a solid of revolution using the washer method.
- Find the volume of a solid with known cross sections.

The Disk Method

In Chapter 5 we mentioned that area is only one of the *many* applications of the definite integral. Another important application is its use in finding the volume of a three-dimensional solid. In this section you will study a particular type of three-dimensional solid—one whose cross sections are similar. Solids of revolution are used commonly in engineering and manufacturing. Some examples are axles, funnels, pills, bottles, and pistons, as shown in Figure 7.12.

Rectangle

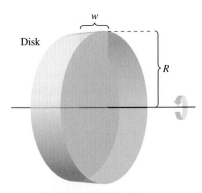

Disk

Volume of a disk: $\pi R^2 w$
Figure 7.13

Solids of revolution
Figure 7.12

If a region in the plane is revolved about a line, the resulting solid is a **solid of revolution,** and the line is called the **axis of revolution.** The simplest such solid is a right circular cylinder or **disk,** which is formed by revolving a rectangle about an axis adjacent to one side of the rectangle, as shown in Figure 7.13. The volume of such a disk is

Volume of disk = (area of disk)(width of disk)

$$= \pi R^2 w$$

where R is the radius of the disk and w is the width.

To see how to use the volume of a disk to find the volume of a general solid of revolution, consider a solid of revolution formed by revolving the plane region in Figure 7.14 about the indicated axis. To determine the volume of this solid, consider a representative rectangle in the plane region. When this rectangle is revolved about the axis of revolution, it generates a representative disk whose volume is

$$\Delta V = \pi R^2 \Delta x.$$

Approximating the volume of the solid by n such disks of width Δx and radius $R(x_i)$ produces

$$\text{Volume of solid} \approx \sum_{i=1}^{n} \pi [R(x_i)]^2 \, \Delta x$$

$$= \pi \sum_{i=1}^{n} [R(x_i)]^2 \, \Delta x.$$

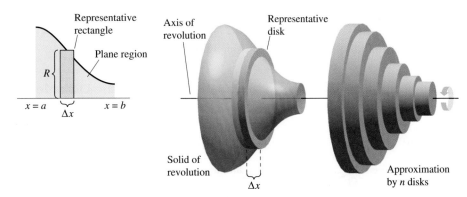

Disk method
Figure 7.14

This approximation appears to become better and better as $\|\Delta\| \to 0$ $(n \to \infty)$. So, you can define the volume of the solid as

$$\text{Volume of solid} = \lim_{\|\Delta\| \to 0} \pi \sum_{i=1}^{n} [R(x_i)]^2 \, \Delta x = \pi \int_a^b [R(x)]^2 \, dx.$$

Schematically, the disk method looks like this.

Known Precalculus Formula	*Representative Element*	*New Integration Formula*
Volume of disk $V = \pi R^2 w$	$\Delta V = \pi [R(x_i)]^2 \, \Delta x$	Solid of revolution $V = \pi \int_a^b [R(x)]^2 \, dx$

A similar formula can be derived if the axis of revolution is vertical.

The Disk Method

To find the volume of a solid of revolution with the **disk method,** use one of the following, as shown in Figure 7.15.

Horizontal Axis of Revolution	*Vertical Axis of Revolution*
Volume $= V = \pi \displaystyle\int_a^b [R(x)]^2 \, dx$	Volume $= V = \pi \displaystyle\int_c^d [R(y)]^2 \, dy$

NOTE   In Figure 7.15, note that you can determine the variable of integration by placing a representative rectangle in the plane region "perpendicular" to the axis of revolution. If the width of the rectangle is Δx, integrate with respect to x, and if the width of the rectangle is Δy, integrate with respect to y.

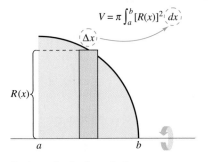

Horizontal axis of revolution
Figure 7.15

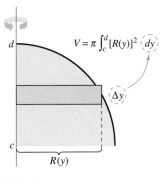

Vertical axis of revolution

The simplest application of the disk method involves a plane region bounded by the graph of f and the x-axis. If the axis of revolution is the x-axis, the radius $R(x)$ is simply $f(x)$.

EXAMPLE I   Using the Disk Method

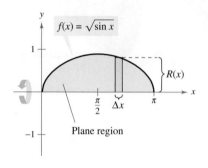

Find the volume of the solid formed by revolving the region bounded by the graph of

$$f(x) = \sqrt{\sin x}$$

and the x-axis $(0 \le x \le \pi)$ about the x-axis.

Solution   From the representative rectangle in the upper graph in Figure 7.16, you can see that the radius of this solid is

$$R(x) = f(x)$$
$$= \sqrt{\sin x}.$$

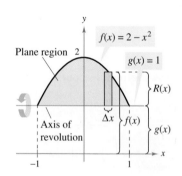

So, the volume of the solid of revolution is

$$V = \pi \int_a^b [R(x)]^2\, dx = \pi \int_0^\pi \left(\sqrt{\sin x} \right)^2 dx \qquad \text{Apply disk method.}$$

$$= \pi \int_0^\pi \sin x\, dx \qquad \text{Simplify.}$$

$$= \pi \left[-\cos x \right]_0^\pi \qquad \text{Integrate.}$$

$$= \pi(1 + 1)$$

$$= 2\pi.$$

Figure 7.16

EXAMPLE 2   Revolving About a Line That Is Not a Coordinate Axis

Find the volume of the solid formed by revolving the region bounded by

$$f(x) = 2 - x^2$$

and $g(x) = 1$ about the line $y = 1$, as shown in Figure 7.17.

Solution   By equating $f(x)$ and $g(x)$, you can determine that the two graphs intersect when $x = \pm 1$. To find the radius, subtract $g(x)$ from $f(x)$.

$$R(x) = f(x) - g(x)$$
$$= (2 - x^2) - 1$$
$$= 1 - x^2$$

Finally, integrate between -1 and 1 to find the volume.

$$V = \pi \int_a^b [R(x)]^2\, dx = \pi \int_{-1}^1 (1 - x^2)^2\, dx \qquad \text{Apply disk method.}$$

$$= \pi \int_{-1}^1 (1 - 2x^2 + x^4)\, dx \qquad \text{Simplify.}$$

$$= \pi \left[x - \frac{2x^3}{3} + \frac{x^5}{5} \right]_{-1}^1 \qquad \text{Integrate.}$$

$$= \frac{16\pi}{15}$$

Figure 7.17

Figure 7.18

The Washer Method

The disk method can be extended to cover solids of revolution with holes by replacing the representative disk with a representative **washer.** The washer is formed by revolving a rectangle about an axis, as shown in Figure 7.18. If r and R are the inner and outer radii of the washer and w is the width of the washer, the volume is given by

Volume of washer $= \pi(R^2 - r^2)w.$

To see how this concept can be used to find the volume of a solid of revolution, consider a region bounded by an **outer radius** $R(x)$ and an **inner radius** $r(x)$, as shown in Figure 7.19. If the region is revolved about its axis of revolution, the volume of the resulting solid is given by

$$V = \pi \int_a^b ([R(x)]^2 - [r(x)]^2)\, dx. \qquad \text{Washer method}$$

Note that the integral involving the inner radius represents the volume of the hole and is *subtracted* from the integral involving the outer radius.

Plane region

Solid of revolution with hole

Figure 7.19

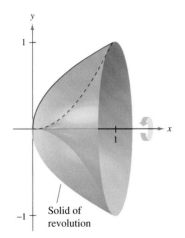

Solid of revolution
Figure 7.20

EXAMPLE 3   Using the Washer Method

Find the volume of the solid formed by revolving the region bounded by the graphs of $y = \sqrt{x}$ and $y = x^2$ about the x-axis, as shown in Figure 7.20.

Solution   In Figure 7.20, you can see that the outer and inner radii are as follows.

$R(x) = \sqrt{x}$ Outer radius

$r(x) = x^2$ Inner radius

Integrating between 0 and 1 produces

$$V = \pi \int_a^b ([R(x)]^2 - [r(x)]^2)\, dx \qquad \text{Apply washer method.}$$

$$= \pi \int_0^1 \left[(\sqrt{x})^2 - (x^2)^2 \right] dx$$

$$= \pi \int_0^1 (x - x^4)\, dx \qquad \text{Simplify.}$$

$$= \pi \left[\frac{x^2}{2} - \frac{x^5}{5} \right]_0^1 \qquad \text{Integrate.}$$

$$= \frac{3\pi}{10}.$$

In each example so far, the axis of revolution has been *horizontal* and you have integrated with respect to *x*. In the next example, the axis of revolution is *vertical* and you integrate with respect to *y*. In this example, you need two separate integrals to compute the volume.

EXAMPLE 4 Integrating with Respect to y, Two-Integral Case

Find the volume of the solid formed by revolving the region bounded by the graphs of $y = x^2 + 1$, $y = 0$, $x = 0$, and $x = 1$ about the *y*-axis, as shown in Figure 7.21.

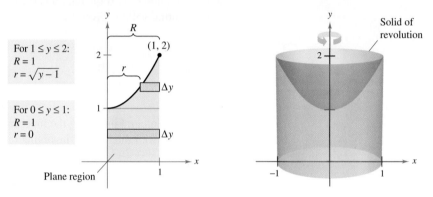

For $1 \leq y \leq 2$:
$R = 1$
$r = \sqrt{y - 1}$

For $0 \leq y \leq 1$:
$R = 1$
$r = 0$

Plane region

Solid of revolution

Figure 7.21

Solution For the region shown in Figure 7.21, the outer radius is simply $R = 1$. There is, however, no convenient formula that represents the inner radius. When $0 \leq y \leq 1$, $r = 0$, but when $1 \leq y \leq 2$, r is determined by the equation $y = x^2 + 1$, which implies that $r = \sqrt{y - 1}$.

$$r(y) = \begin{cases} 0, & 0 \leq y \leq 1 \\ \sqrt{y - 1}, & 1 \leq y \leq 2 \end{cases}$$

Using this definition of the inner radius, you can use two integrals to find the volume.

$$V = \pi \int_0^1 (1^2 - 0^2) \, dy + \pi \int_1^2 \left[1^2 - \left(\sqrt{y - 1} \right)^2 \right] dy \qquad \text{Apply washer method.}$$

$$= \pi \int_0^1 1 \, dy + \pi \int_1^2 (2 - y) \, dy \qquad \text{Simplify.}$$

$$= \pi \left[y \right]_0^1 + \pi \left[2y - \frac{y^2}{2} \right]_1^2 \qquad \text{Integrate.}$$

$$= \pi + \pi \left(4 - 2 - 2 + \frac{1}{2} \right) = \frac{3\pi}{2}$$

Note that the first integral $\pi \int_0^1 1 \, dy$ represents the volume of a right circular cylinder of radius 1 and height 1. This portion of the volume could have been determined without using calculus.

Generated by Mathematica

Figure 7.22

TECHNOLOGY Some graphing utilities have the capability to generate (or have built-in software capable of generating) a solid of revolution. If you have access to such a utility, use it to graph some of the solids of revolution described in this section. For instance, the solid in Example 4 might appear like that shown in Figure 7.22.

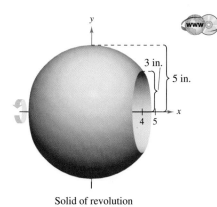

3 in.

5 in.

Solid of revolution

(a)

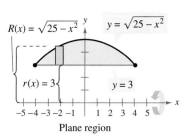

$R(x) = \sqrt{25 - x^2}$

$y = \sqrt{25 - x^2}$

$r(x) = 3$

$y = 3$

Plane region

(b)

Figure 7.23

www   *EXAMPLE 5*   **Manufacturing**

A manufacturer drills a hole through the center of a metal sphere of radius 5 inches, as shown in Figure 7.23(a). The hole has a radius of 3 inches. What is the volume of the resulting metal ring?

Solution   You can imagine the ring to be generated by a segment of the circle whose equation is $x^2 + y^2 = 25$, as shown in Figure 7.23(b). Because the radius of the hole is 3 inches, you can let $y = 3$ and solve the equation $x^2 + y^2 = 25$ to determine that the limits of integration are $x = \pm 4$. So, the inner and outer radii are $r(x) = 3$ and $R(x) = \sqrt{25 - x^2}$ and the volume is given by

$$V = \pi \int_a^b ([R(x)]^2 - [r(x)]^2)\, dx = \pi \int_{-4}^4 \left[\left(\sqrt{25 - x^2}\right)^2 - (3)^2\right] dx$$

$$= \pi \int_{-4}^4 (16 - x^2)\, dx$$

$$= \pi \left[16x - \frac{x^3}{3} \right]_{-4}^4$$

$$= \frac{256\pi}{3} \text{ cubic inches.}$$

Solids with Known Cross Sections

With the disk method, you can find the volume of a solid having a circular cross section whose area is $A = \pi R^2$. This method can be generalized to solids of any shape, as long as you know a formula for the area of an arbitrary cross section. Some common cross sections are squares, rectangles, triangles, semicircles, and trapezoids.

Volumes of Solids with Known Cross Sections

1. For cross sections of area $A(x)$ taken perpendicular to the x-axis,

$$\text{Volume} = \int_a^b A(x)\, dx. \qquad \text{See Figure 7.24(a).}$$

2. For cross sections of area $A(y)$ taken perpendicular to the y-axis,

$$\text{Volume} = \int_c^d A(y)\, dy. \qquad \text{See Figure 7.24(b).}$$

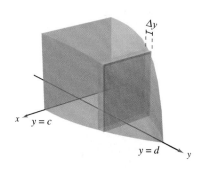

(a) Cross sections perpendicular to x-axis

(b) Cross sections perpendicular to y-axis

Figure 7.24

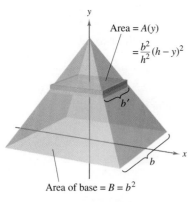

Cross sections are equilateral triangles.

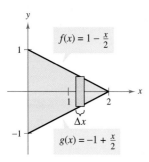

Triangular base in xy-plane
Figure 7.25

Figure 7.26

EXAMPLE 6 **Triangular Cross Sections**

Find the volume of the solid shown in Figure 7.25. The base of the solid is the region bounded by the lines

$$f(x) = 1 - \frac{x}{2}, \qquad g(x) = -1 + \frac{x}{2}, \qquad \text{and} \qquad x = 0.$$

The cross sections perpendicular to the x-axis are equilateral triangles.

Solution The base and area of each triangular cross section are as follows.

$$\text{Base} = \left(1 - \frac{x}{2}\right) - \left(-1 + \frac{x}{2}\right) = 2 - x \qquad \text{Length of base}$$

$$\text{Area} = \frac{\sqrt{3}}{4}(\text{base})^2 \qquad \text{Area of equilateral triangle}$$

$$A(x) = \frac{\sqrt{3}}{4}(2 - x)^2 \qquad \text{Area of cross section}$$

Because x ranges from 0 to 2, the volume of the solid is

$$V = \int_a^b A(x)\, dx = \int_0^2 \frac{\sqrt{3}}{4}(2 - x)^2\, dx$$

$$= -\frac{\sqrt{3}}{4}\left[\frac{(2 - x)^3}{3}\right]_0^2 = \frac{2\sqrt{3}}{3}.$$

EXAMPLE 7 **An Application to Geometry**

Prove that the volume of a pyramid with a square base is $V = \frac{1}{3}hB$, where h is the height of the pyramid and B is the area of the base.

Solution As shown in Figure 7.26, you can intersect the pyramid with a plane parallel to the base at height y to form a square cross section whose sides are of length b'. Using similar triangles, you can show that

$$\frac{b'}{b} = \frac{h - y}{h} \qquad \text{or} \qquad b' = \frac{b}{h}(h - y)$$

where b is the length of the sides of the base of the pyramid. So,

$$A(y) = (b')^2 = \frac{b^2}{h^2}(h - y)^2.$$

Integrating between 0 and h produces

$$V = \int_0^h A(y)\, dy = \int_0^h \frac{b^2}{h^2}(h - y)^2\, dy$$

$$= \frac{b^2}{h^2}\int_0^h (h - y)^2\, dy$$

$$= -\left(\frac{b^2}{h^2}\right)\left[\frac{(h - y)^3}{3}\right]_0^h$$

$$= \frac{b^2}{h^2}\left(\frac{h^3}{3}\right)$$

$$= \frac{1}{3}hB. \qquad B = b^2$$

Exercises for Section 7.2

In Exercises 1–6, set up and evaluate the integral that gives the volume of the solid formed by revolving the region about the x-axis.

1. $y = -x + 1$

2. $y = 4 - x^2$

3. $y = \sqrt{x}$

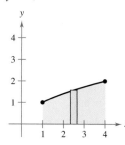

4. $y = \sqrt{9 - x^2}$

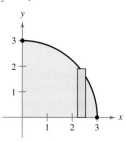

5. $y = x^2$, $y = x^3$

6. $y = 2$, $y = 4 - \dfrac{x^2}{4}$

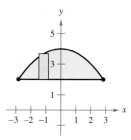

In Exercises 7–10, set up and evaluate the integral that gives the volume of the solid formed by revolving the region about the y-axis.

7. $y = x^2$

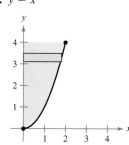

8. $y = \sqrt{16 - x^2}$

9. $y = x^{2/3}$

10. $x = -y^2 + 4y$

In Exercises 11–14, find the volume of the solid generated by revolving the region bounded by the graphs of the equations about the given lines.

11. $y = \sqrt{x}$, $y = 0$, $x = 4$

  (a) the x-axis   (b) the y-axis

  (c) the line $x = 4$ (d) the line $x = 6$

12. $y = 2x^2$, $y = 0$, $x = 2$

  (a) the y-axis   (b) the x-axis

  (c) the line $y = 8$ (d) the line $x = 2$

13. $y = x^2$, $y = 4x - x^2$

  (a) the x-axis   (b) the line $y = 6$

14. $y = 6 - 2x - x^2$, $y = x + 6$

  (a) the x-axis   (b) the line $y = 3$

In Exercises 15–18, find the volume of the solid generated by revolving the region bounded by the graphs of the equations about the line $y = 4$.

15. $y = x$, $y = 3$, $x = 0$ **16.** $y = \frac{1}{2}x^3$, $y = 4$, $x = 0$

17. $y = \dfrac{1}{1 + x}$, $y = 0$, $x = 0$, $x = 3$

18. $y = \sec x$, $y = 0$, $0 \le x \le \dfrac{\pi}{3}$

In Exercises 19–22, find the volume of the solid generated by revolving the region bounded by the graphs of the equations about the line $x = 6$.

19. $y = x$, $y = 0$, $y = 4$, $x = 6$

20. $y = 6 - x$, $y = 0$, $y = 4$, $x = 0$

21. $x = y^2$, $x = 4$

22. $xy = 6$, $y = 2$, $y = 6$, $x = 6$

In Exercises 23–30, find the volume of the solid generated by revolving the region bounded by the graphs of the equations about the x-axis.

23. $y = \dfrac{1}{\sqrt{x + 1}}$, $y = 0$, $x = 0$, $x = 3$

24. $y = x\sqrt{4 - x^2}$, $y = 0$

25. $y = \dfrac{1}{x}$, $y = 0$, $x = 1$, $x = 4$

26. $y = \dfrac{3}{x+1}$, $y = 0$, $x = 0$, $x = 8$

27. $y = e^{-x}$, $y = 0$, $x = 0$, $x = 1$

28. $y = e^{x/2}$, $y = 0$, $x = 0$, $x = 4$

29. $y = x^2 + 1$, $y = -x^2 + 2x + 5$, $x = 0$, $x = 3$

30. $y = \sqrt{x}$, $y = -\frac{1}{2}x + 4$, $x = 0$, $x = 8$

In Exercises 31 and 32, find the volume of the solid generated by revolving the region bounded by the graphs of the equations about the y-axis.

31. $y = 3(2 - x)$, $y = 0$, $x = 0$

32. $y = 9 - x^2$, $y = 0$, $x = 2$, $x = 3$

In Exercises 33–36, find the volume of the solid generated by revolving the region bounded by the graphs of the equations about the x-axis. Verify your results using the integration capabilities of a graphing utility.

33. $y = \sin x$, $y = 0$, $x = 0$, $x = \pi$

34. $y = \cos x$, $y = 0$, $x = 0$, $x = \dfrac{\pi}{2}$

35. $y = e^{x-1}$, $y = 0$, $x = 1$, $x = 2$

36. $y = e^{x/2} + e^{-x/2}$, $y = 0$, $x = -1$, $x = 2$

In Exercises 37–40, use the integration capabilities of a graphing utility to approximate the volume of the solid generated by revolving the region bounded by the graphs of the equations about the x-axis.

37. $y = e^{-x^2}$, $y = 0$, $x = 0$, $x = 2$

38. $y = \ln x$, $y = 0$, $x = 1$, $x = 3$

39. $y = 2 \arctan(0.2x)$, $y = 0$, $x = 0$, $x = 5$

40. $y = \sqrt{2x}$, $y = x^2$

Writing About Concepts

In Exercises 41 and 42, the integral represents the volume of a solid. Describe the solid.

41. $\pi \displaystyle\int_{0}^{\pi/2} \sin^2 x \, dx$ 42. $\pi \displaystyle\int_{2}^{4} y^4 \, dy$

Think About It **In Exercises 43 and 44, determine which value best approximates the volume of the solid generated by revolving the region bounded by the graphs of the equations about the x-axis. (Make your selection on the basis of a sketch of the solid and *not* by performing any calculations.)**

43. $y = e^{-x^2/2}$, $y = 0$, $x = 0$, $x = 2$

    (a) 3   (b) -5 (c) 10   (d) 7   (e) 20

44. $y = \arctan x$, $y = 0$, $x = 0$, $x = 1$

    (a) 10   (b) $\frac{3}{4}$ (c) 5   (d) -6 (e) 15

Writing About Concepts *(continued)*

45. A region bounded by the parabola $y = 4x - x^2$ and the x-axis is revolved about the x-axis. A second region bounded by the parabola $y = 4 - x^2$ and the x-axis is revolved about the x-axis. Without integrating, how do the volumes of the two solids compare? Explain.

46. The region in the figure is revolved about the indicated axes and line. Order the volumes of the resulting solids from least to greatest. Explain your reasoning.

    (a) x-axis   (b) y-axis   (c) $x = 8$

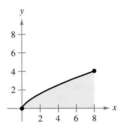

47. If the portion of the line $y = \frac{1}{2}x$ lying in the first quadrant is revolved about the x-axis, a cone is generated. Find the volume of the cone extending from $x = 0$ to $x = 6$.

48. Use the disk method to verify that the volume of a right circular cone is $\frac{1}{3}\pi r^2 h$, where r is the radius of the base and h is the height.

49. Use the disk method to verify that the volume of a sphere is $\frac{4}{3}\pi r^3$.

50. A sphere of radius r is cut by a plane h ($h < r$) units above the equator. Find the volume of the solid (spherical segment) above the plane.

51. A cone of height H with a base of radius r is cut by a plane parallel to and h units above the base. Find the volume of the solid (frustum of a cone) below the plane.

52. The region bounded by $y = \sqrt{x}$, $y = 0$, $x = 0$, and $x = 4$ is revolved about the x-axis.

    (a) Find the value of x in the interval $[0, 4]$ that divides the solid into two parts of equal volume.

    (b) Find the values of x in the interval $[0, 4]$ that divide the solid into three parts of equal volume.

53. ***Volume of a Fuel Tank*** A tank on the wing of a jet aircraft is formed by revolving the region bounded by the graph of $y = \frac{1}{8}x^2\sqrt{2 - x}$ and the x-axis about the x-axis (see figure), where x and y are measured in meters. Find the tank's volume.

54. *Volume of a Lab Glass*   A glass container can be modeled by revolving the graph of

$$y = \begin{cases} \sqrt{0.1x^3 - 2.2x^2 + 10.9x + 22.2}, & 0 \le x \le 11.5 \\ 2.95, & 11.5 < x \le 15 \end{cases}$$

about the *x*-axis, where *x* and *y* are measured in centimeters. Use a graphing utility to graph the function and find the volume of the container.

55. Find the volume of the solid generated if the upper half of the ellipse $9x^2 + 25y^2 = 225$ is revolved about (a) the *x*-axis to form a prolate spheroid (shaped like a football), and (b) the *y*-axis to form an oblate spheroid (shaped like half of a candy).

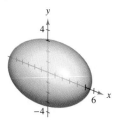

Figure for 55(a)	**Figure for 55(b)**

56. *Minimum Volume*   The arc of

$$y = 4 - \frac{x^2}{4}$$

on the interval $[0, 4]$ is revolved about the line $y = b$ (see figure).

(a) Find the volume of the resulting solid as a function of *b*.

(b) Use a graphing utility to graph the function in part (a), and use the graph to approximate the value of *b* that minimizes the volume of the solid.

(c) Use calculus to find the value of *b* that minimizes the volume of the solid, and compare the result with the answer to part (b).

Figure for 56	**Figure for 58**

57. *Water Depth in a Tank*   A tank on a water tower is a sphere of radius 50 feet. Determine the depths of the water when the tank is filled to one-fourth and three-fourths of its total capacity. (*Note:* Use the *zero* or *root* feature of a graphing utility after evaluating the definite integral.)

58. *Modeling Data*   A draftsman is asked to determine the amount of material required to produce a machine part (see figure in first column). The diameters *d* of the part at equally spaced points *x* are listed in the table. The measurements are listed in centimeters.

x	0	1	2	3	4	5
d	4.2	3.8	4.2	4.7	5.2	5.7

x	6	7	8	9	10
d	5.8	5.4	4.9	4.4	4.6

(a) Use these data with Simpson's Rule to approximate the volume of the part.

(b) Use the regression capabilities of a graphing utility to find a fourth-degree polynomial through the points representing the radius of the solid. Plot the data and graph the model.

(c) Use a graphing utility to approximate the definite integral yielding the volume of the part. Compare the result with the answer to part (a).

59. *Think About It*   Match each integral with the solid whose volume it represents, and give the dimensions of each solid.

(a) Right circular cylinder   (b) Ellipsoid

(c) Sphere   (d) Right circular cone   (e) Torus

(i) $\displaystyle \pi \int_0^h \left(\frac{rx}{h}\right)^2 dx$ (ii) $\displaystyle \pi \int_0^h r^2 \, dx$

(iii) $\displaystyle \pi \int_{-r}^r \left(\sqrt{r^2 - x^2}\right)^2 dx$ (iv) $\displaystyle \pi \int_{-b}^b \left(a\sqrt{1 - \frac{x^2}{b^2}}\right)^2 dx$

(v) $\displaystyle \pi \int_{-r}^r \left[(R + \sqrt{r^2 - x^2})^2 - (R - \sqrt{r^2 - x^2})^2\right] dx$

60. *Cavalieri's Theorem*   Prove that if two solids have equal altitudes and all plane sections parallel to their bases and at equal distances from their bases have equal areas, then the solids have the same volume (see figure).

Area of R_1 = area of R_2

61. Find the volume of the solid whose base is bounded by the graphs of $y = x + 1$ and $y = x^2 - 1$, with the indicated cross sections taken perpendicular to the *x*-axis.

(a) Squares   (b) Rectangles of height 1

62. Find the volume of the solid whose base is bounded by the circle

$$x^2 + y^2 = 4$$

with the indicated cross sections taken perpendicular to the x-axis.

(a) Squares   (b) Equilateral triangles

(c) Semicircles   (d) Isosceles right triangles

63. The base of a solid is bounded by $y = x^3$, $y = 0$, and $x = 1$. Find the volume of the solid for each of the following cross sections (taken perpendicular to the y-axis): (a) squares, (b) semicircles, (c) equilateral triangles, and (d) semiellipses whose heights are twice the lengths of their bases.

64. Find the volume of the solid of intersection (the solid common to both) of the two right circular cylinders of radius r whose axes meet at right angles (see figure).

Two intersecting cylinders   Solid of intersection

FOR FURTHER INFORMATION For more information on this problem, see the article "Estimating the Volumes of Solid Figures with Curved Surfaces" by Donald Cohen in *Mathematics Teacher*. To view this article, go to the website *www.matharticles.com*.

65. A manufacturer drills a hole through the center of a metal sphere of radius R. The hole has a radius r. Find the volume of the resulting ring.

66. For the metal sphere in Exercise 65, let $R = 5$. What value of r will produce a ring whose volume is exactly half the volume of the sphere?

In Exercises 67–74, find the volume generated by rotating the given region about the specified line.

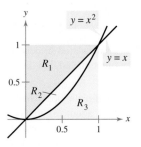

67. R_1 about $x = 0$ **68.** R_1 about $x = 1$

69. R_2 about $y = 0$ **70.** R_2 about $y = 1$

71. R_3 about $x = 0$ **72.** R_3 about $x = 1$

73. R_2 about $x = 0$ **74.** R_2 about $x = 1$

75. The solid shown in the figure has cross sections bounded by the graph of $|x|^a + |y|^a = 1$, where $1 \le a \le 2$.

(a) Describe the cross section when $a = 1$ and $a = 2$.

(b) Describe a procedure for approximating the volume of the solid.

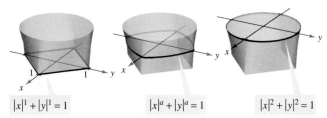

$|x|^1 + |y|^1 = 1$ $|x|^a + |y|^a = 1$ $|x|^2 + |y|^2 = 1$

76. Two planes cut a right circular cylinder to form a wedge. One plane is perpendicular to the axis of the cylinder and the second makes an angle of θ degrees with the first (see figure).

(a) Find the volume of the wedge if $\theta = 45°$.

(b) Find the volume of the wedge for an arbitrary angle θ. Assuming that the cylinder has sufficient length, how does the volume of the wedge change as θ increases from $0°$ to $90°$?

Figure for 76   **Figure for 77**

77. (a) Show that the volume of the torus shown is given by the integral $8\pi R \int_0^r \sqrt{r^2 - y^2} \, dy$, where $R > r > 0$.

(b) Find the volume of the torus.

Volume: The Shell Method

- Find the volume of a solid of revolution using the shell method.
- Compare the uses of the disk method and the shell method.

The Shell Method

Figure 7.27

In this section, you will study an alternative method for finding the volume of a solid of revolution. This method is called the **shell method** because it uses cylindrical shells. A comparison of the advantages of the disk and shell methods is given later in this section.

To begin, consider a representative rectangle as shown in Figure 7.27, where w is the width of the rectangle, h is the height of the rectangle, and p is the distance between the axis of revolution and the *center* of the rectangle. When this rectangle is revolved about its axis of revolution, it forms a cylindrical shell (or tube) of thickness w. To find the volume of this shell, consider two cylinders. The radius of the larger cylinder corresponds to the outer radius of the shell, and the radius of the smaller cylinder corresponds to the inner radius of the shell. Because p is the average radius of the shell, you know the outer radius is $p + (w/2)$ and the inner radius is $p - (w/2)$.

$$p + \frac{w}{2} \qquad \text{Outer radius}$$

$$p - \frac{w}{2} \qquad \text{Inner radius}$$

So, the volume of the shell is

Volume of shell = (volume of cylinder) − (volume of hole)

$$= \pi\left(p + \frac{w}{2}\right)^2 h - \pi\left(p - \frac{w}{2}\right)^2 h$$

$$= 2\pi p h w$$

$$= 2\pi(\text{average radius})(\text{height})(\text{thickness}).$$

You can use this formula to find the volume of a solid of revolution. Assume that the plane region in Figure 7.28 is revolved about a line to form the indicated solid. If you consider a horizontal rectangle of width Δy, then, as the plane region is revolved about a line parallel to the x-axis, the rectangle generates a representative shell whose volume is

$$\Delta V = 2\pi[p(y)h(y)]\,\Delta y.$$

Figure 7.28

You can approximate the volume of the solid by n such shells of thickness Δy, height $h(y_i)$, and average radius $p(y_i)$.

Volume of solid $\approx \displaystyle\sum_{i=1}^{n} 2\pi[p(y_i)h(y_i)]\,\Delta y = 2\pi\sum_{i=1}^{n}[p(y_i)h(y_i)]\,\Delta y$

This approximation appears to become better and better as $\|\Delta\| \to 0$ $(n \to \infty)$. So, the volume of the solid is

Volume of solid $= \displaystyle\lim_{\|\Delta\|\to 0} 2\pi\sum_{i=1}^{n}[p(y_i)h(y_i)]\,\Delta y$

$= 2\pi\displaystyle\int_{c}^{d}[p(y)h(y)]\,dy.$

The Shell Method

To find the volume of a solid of revolution with the **shell method,** use one of the following, as shown in Figure 7.29.

Horizontal Axis of Revolution	*Vertical Axis of Revolution*
$\text{Volume} = V = 2\pi \int_c^d p(y)h(y)\,dy$	$\text{Volume} = V = 2\pi \int_a^b p(x)h(x)\,dx$

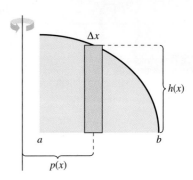

Horizontal axis of revolution
Figure 7.29

Vertical axis of revolution

EXAMPLE 1   Using the Shell Method to Find Volume

Find the volume of the solid of revolution formed by revolving the region bounded by

$$y = x - x^3$$

and the *x*-axis $(0 \le x \le 1)$ about the *y*-axis.

Solution   Because the axis of revolution is vertical, use a vertical representative rectangle, as shown in Figure 7.30. The width Δx indicates that *x* is the variable of integration. The distance from the center of the rectangle to the axis of revolution is $p(x) = x$, and the height of the rectangle is

$$h(x) = x - x^3.$$

Because *x* ranges from 0 to 1, the volume of the solid is

$$
\begin{aligned}
V = 2\pi \int_a^b p(x)h(x)\,dx &= 2\pi \int_0^1 x(x - x^3)\,dx && \text{Apply shell method.}\\
&= 2\pi \int_0^1 (-x^4 + x^2)\,dx && \text{Simplify.}\\
&= 2\pi \left[-\frac{x^5}{5} + \frac{x^3}{3} \right]_0^1 && \text{Integrate.}\\
&= 2\pi \left(-\frac{1}{5} + \frac{1}{3} \right)\\
&= \frac{4\pi}{15}.
\end{aligned}
$$

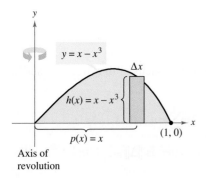

Axis of
revolution

Figure 7.30

EXAMPLE 2   Using the Shell Method to Find Volume

Find the volume of the solid of revolution formed by revolving the region bounded by the graph of

$$x = e^{-y^2}$$

and the y-axis $(0 \le y \le 1)$ about the x-axis.

Solution   Because the axis of revolution is horizontal, use a horizontal representative rectangle, as shown in Figure 7.31. The width Δy indicates that y is the variable of integration. The distance from the center of the rectangle to the axis of revolution is $p(y) = y$, and the height of the rectangle is $h(y) = e^{-y^2}$. Because y ranges from 0 to 1, the volume of the solid is

$$V = 2\pi \int_c^d p(y)h(y)\, dy = 2\pi \int_0^1 ye^{-y^2}\, dy \qquad \text{Apply shell method.}$$

$$= -\pi \left[e^{-y^2} \right]_0^1 \qquad \text{Integrate.}$$

$$= \pi \left(1 - \frac{1}{e} \right)$$

$$\approx 1.986.$$

Figure 7.31

NOTE   To see the advantage of using the shell method in Example 2, solve the equation $x = e^{-y^2}$ for y.

$$y = \begin{cases} 1, & 0 \le x \le 1/e \\ \sqrt{-\ln x}, & 1/e < x \le 1 \end{cases}$$

Then use this equation to find the volume using the disk method.

Comparison of Disk and Shell Methods

The disk and shell methods can be distinguished as follows. For the disk method, the representative rectangle is always *perpendicular* to the axis of revolution, whereas for the shell method, the representative rectangle is always *parallel* to the axis of revolution, as shown in Figure 7.32.

Vertical axis
of revolution

Horizontal axis
of revolution

Vertical axis
of revolution

Horizontal axis
of revolution

Disk method: Representative rectangle is perpendicular to the axis of revolution.
Figure 7.32

Shell method: Representative rectangle is parallel to the axis of revolution.

Often, one method is more convenient to use than the other. The following example illustrates a case in which the shell method is preferable.

EXAMPLE 3   Shell Method Preferable

Find the volume of the solid formed by revolving the region bounded by the graphs of

$$y = x^2 + 1, \quad y = 0, \quad x = 0, \quad \text{and} \quad x = 1$$

about the y-axis.

Solution   In Example 4 in the preceding section, you saw that the washer method requires two integrals to determine the volume of this solid. See Figure 7.33(a).

$$V = \pi \int_0^1 (1^2 - 0^2)\, dy + \pi \int_1^2 \left[1^2 - \left(\sqrt{y - 1} \right)^2 \right] dy \qquad \text{Apply washer method.}$$

$$= \pi \int_0^1 1\, dy + \pi \int_1^2 (2 - y)\, dy \qquad \text{Simplify.}$$

$$= \pi \Big[y \Big]_0^1 + \pi \left[2y - \frac{y^2}{2} \right]_1^2 \qquad \text{Integrate.}$$

$$= \pi + \pi \left(4 - 2 - 2 + \frac{1}{2} \right)$$

$$= \frac{3\pi}{2}$$

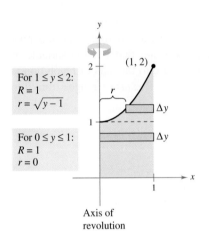

For $1 \le y \le 2$:
$R = 1$
$r = \sqrt{y - 1}$

For $0 \le y \le 1$:
$R = 1$
$r = 0$

Axis of revolution

(a) Disk method

In Figure 7.33(b), you can see that the shell method requires only one integral to find the volume.

$$V = 2\pi \int_a^b p(x)h(x)\, dx \qquad \text{Apply shell method.}$$

$$= 2\pi \int_0^1 x(x^2 + 1)\, dx$$

$$= 2\pi \left[\frac{x^4}{4} + \frac{x^2}{2} \right]_0^1 \qquad \text{Integrate.}$$

$$= 2\pi \left(\frac{3}{4} \right)$$

$$= \frac{3\pi}{2}$$

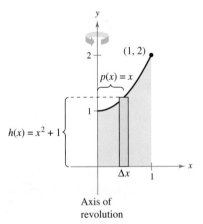

$p(x) = x$

$h(x) = x^2 + 1$

Axis of revolution

(b) Shell method

Figure 7.33

Suppose the region in Example 3 were revolved about the vertical line $x = 1$. Would the resulting solid of revolution have a greater volume or a smaller volume than the solid in Example 3? Without integrating, you should be able to reason that the resulting solid would have a smaller volume because "more" of the revolved region would be closer to the axis of revolution. To confirm this, try solving the following integral, which gives the volume of the solid.

$$V = 2\pi \int_0^1 (1 - x)(x^2 + 1)\, dx \qquad p(x) = 1 - x$$

FOR FURTHER INFORMATION   To learn more about the disk and shell methods, see the article "The Disk and Shell Method" by Charles A. Cable in *The American Mathematical Monthly*. To view this article, go to the website *www.matharticles.com*.

Figure 7.34

(a) Disk method

(b) Shell method
Figure 7.35

EXAMPLE 4   Volume of a Pontoon

A pontoon is to be made in the shape shown in Figure 7.34. The pontoon is designed by rotating the graph of

$$y = 1 - \frac{x^2}{16}, \qquad -4 \le x \le 4$$

about the x-axis, where x and y are measured in feet. Find the volume of the pontoon.

Solution   Refer to Figure 7.35(a) and use the disk method as follows.

$$V = \pi \int_{-4}^{4} \left(1 - \frac{x^2}{16}\right)^2 dx \qquad \text{Apply disk method.}$$

$$= \pi \int_{-4}^{4} \left(1 - \frac{x^2}{8} + \frac{x^4}{256}\right) dx \qquad \text{Simplify.}$$

$$= \pi \left[x - \frac{x^3}{24} + \frac{x^5}{1280}\right]_{-4}^{4} \qquad \text{Integrate.}$$

$$= \frac{64\pi}{15} \approx 13.4 \text{ cubic feet}$$

Try using Figure 7.35(b) to set up the integral for the volume using the shell method. Does the integral seem more complicated?

For the shell method in Example 4, you would have to solve for x in terms of y in the equation

$$y = 1 - (x^2/16).$$

Sometimes, solving for x is very difficult (or even impossible). In such cases you must use a vertical rectangle (of width Δx), thus making x the variable of integration. The position (horizontal or vertical) of the axis of revolution then determines the method to be used. This is shown in Example 5.

EXAMPLE 5   Shell Method Necessary

Find the volume of the solid formed by revolving the region bounded by the graphs of $y = x^3 + x + 1$, $y = 1$, and $x = 1$ about the line $x = 2$, as shown in Figure 7.36.

Solution   In the equation $y = x^3 + x + 1$, you cannot easily solve for x in terms of y. (See Section 3.8 on Newton's Method.) Therefore, the variable of integration must be x, and you should choose a vertical representative rectangle. Because the rectangle is parallel to the axis of revolution, use the shell method and obtain

$$V = 2\pi \int_{a}^{b} p(x)h(x)\, dx = 2\pi \int_{0}^{1} (2-x)(x^3 + x + 1 - 1)\, dx \qquad \text{Apply shell method.}$$

$$= 2\pi \int_{0}^{1} (-x^4 + 2x^3 - x^2 + 2x)\, dx \qquad \text{Simplify.}$$

$$= 2\pi \left[-\frac{x^5}{5} + \frac{x^4}{2} - \frac{x^3}{3} + x^2 \right]_{0}^{1} \qquad \text{Integrate.}$$

$$= 2\pi \left(-\frac{1}{5} + \frac{1}{2} - \frac{1}{3} + 1 \right)$$

$$= \frac{29\pi}{15}.$$

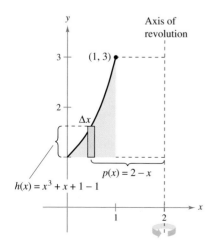

Figure 7.36

Exercises for Section 7.3

See www.CalcChat.com for worked-out solutions to odd-numbered exercises.

In Exercises 1–12, use the shell method to set up and evaluate the integral that gives the volume of the solid generated by revolving the plane region about the y-axis.

1. $y = x$

2. $y = 1 - x$

3. $y = \sqrt{x}$

4. $y = x^2 + 4$

5. $y = x^2$, $y = 0$, $x = 2$

6. $y = \frac{1}{2}x^2$, $y = 0$, $x = 6$

7. $y = x^2$, $y = 4x - x^2$

8. $y = 4 - x^2$, $y = 0$

9. $y = 4x - x^2$, $x = 0$, $y = 4$

10. $y = 2x$, $y = 4$, $x = 0$

11. $y = \dfrac{1}{\sqrt{2\pi}} e^{-x^2/2}$, $y = 0$, $x = 0$, $x = 1$

12. $y = \begin{cases} \dfrac{\sin x}{x}, & x > 0 \\ 1, & x = 0 \end{cases}$, $y = 0$, $x = 0$, $x = \pi$

In Exercises 13–20, use the shell method to set up and evaluate the integral that gives the volume of the solid generated by revolving the plane region about the x-axis.

13. $y = x$

14. $y = 2 - x$

15. $y = \dfrac{1}{x}$

16. $x + y^2 = 16$

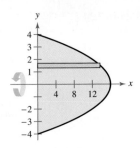

17. $y = x^3$, $x = 0$, $y = 8$ **18.** $y = x^2$, $x = 0$, $y = 9$

19. $x + y = 4$, $y = x$, $y = 0$

20. $y = \sqrt{x + 2}$, $y = x$, $y = 0$

In Exercises 21–24, use the shell method to find the volume of the solid generated by revolving the plane region about the given line.

21. $y = x^2$, $y = 4x - x^2$, about the line $x = 4$

22. $y = x^2$, $y = 4x - x^2$, about the line $x = 2$

23. $y = 4x - x^2$, $y = 0$, about the line $x = 5$

24. $y = \sqrt{x}$, $y = 0$, $x = 4$, about the line $x = 6$

In Exercises 25 and 26, decide whether it is more convenient to use the disk method or the shell method to find the volume of the solid of revolution. Explain your reasoning. (Do not find the volume.)

25. $(y - 2)^2 = 4 - x$

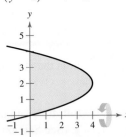

26. $y = 4 - e^x$

In Exercises 27–30, use the disk *or* the shell method to find the volume of the solid generated by revolving the region bounded by the graphs of the equations about each given line.

27. $y = x^3$, $y = 0$, $x = 2$

    (a) the x-axis   (b) the y-axis   (c) the line $x = 4$

28. $y = \dfrac{10}{x^2}$, $y = 0$, $x = 1$, $x = 5$

    (a) the x-axis   (b) the y-axis   (c) the line $y = 10$

29. $x^{1/2} + y^{1/2} = a^{1/2}$, $x = 0$, $y = 0$

    (a) the x-axis   (b) the y-axis   (c) the line $x = a$

30. $x^{2/3} + y^{2/3} = a^{2/3}$, $a > 0$ (hypocycloid)

    (a) the x-axis   (b) the y-axis

Writing About Concepts

31. Consider a solid that is generated by revolving a plane region about the y-axis. Describe the position of a representative rectangle when using (a) the shell method and (b) the disk method to find the volume of the solid.

32. The region in the figure is revolved about the indicated axes and line. Order the volumes of the resulting solids from least to greatest. Explain your reasoning.

    (a) x-axis   (b) y-axis   (c) $x = 5$

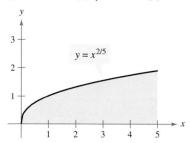

In Exercises 33 and 34, give a geometric argument that explains why the integrals have equal values.

33. $\pi \displaystyle\int_1^5 (x - 1)\, dx = 2\pi \int_0^2 y[5 - (y^2 + 1)]\, dy$

34. $\pi \displaystyle\int_0^2 [16 - (2y)^2]\, dy = 2\pi \int_0^4 x\left(\frac{x}{2}\right) dx$

In Exercises 35–38, (a) use a graphing utility to graph the plane region bounded by the graphs of the equations, and (b) use the integration capabilities of the graphing utility to approximate the volume of the solid generated by revolving the region about the y-axis.

35. $x^{4/3} + y^{4/3} = 1$, $x = 0$, $y = 0$, first quadrant

36. $y = \sqrt{1 - x^3}$, $y = 0$, $x = 0$

37. $y = \sqrt[3]{(x - 2)^2(x - 6)^2}$, $y = 0$, $x = 2$, $x = 6$

38. $y = \dfrac{2}{1 + e^{1/x}}$, $y = 0$, $x = 1$, $x = 3$

Think About It **In Exercises 39 and 40, determine which value best approximates the volume of the solid generated by revolving the region bounded by the graphs of the equations about the y-axis. (Make your selection on the basis of a sketch of the solid and *not* by performing any calculations.)**

39. $y = 2e^{-x}$, $y = 0$, $x = 0$, $x = 2$

    (a) $\frac{3}{2}$ (b) -2 (c) 4 (d) 7.5 (e) 15

40. $y = \tan x$, $y = 0$, $x = 0$, $x = \dfrac{\pi}{4}$

    (a) 3.5 (b) $-\frac{9}{4}$ (c) 8 (d) 10 (e) 1

41. *Machine Part* A solid is generated by revolving the region bounded by $y = \frac{1}{2}x^2$ and $y = 2$ about the y-axis. A hole, centered along the axis of revolution, is drilled through this solid so that one-fourth of the volume is removed. Find the diameter of the hole.

42. *Machine Part* A solid is generated by revolving the region bounded by $y = \sqrt{9 - x^2}$ and $y = 0$ about the y-axis. A hole, centered along the axis of revolution, is drilled through this solid so that one-third of the volume is removed. Find the diameter of the hole.

43. *Volume of a Torus* A torus is formed by revolving the region bounded by the circle $x^2 + y^2 = 1$ about the line $x = 2$ (see figure). Find the volume of this "doughnut-shaped" solid. (*Hint:* The integral $\int_{-1}^{1} \sqrt{1 - x^2}\, dx$ represents the area of a semicircle.)

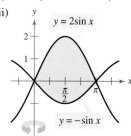

44. *Volume of a Torus* Repeat Exercise 43 for a torus formed by revolving the region bounded by the circle $x^2 + y^2 = r^2$ about the line $x = R$, where $r < R$.

45. (a) Use differentiation to verify that

$$\int x \sin x\, dx = \sin x - x \cos x + C.$$

    (b) Use the result of part (a) to find the volume of the solid generated by revolving each plane region about the y-axis.

    (i)   (ii)

46. (a) Use differentiation to verify that

$$\int x \cos x\, dx = \cos x + x \sin x + C.$$

    (b) Use the result of part (a) to find the volume of the solid generated by revolving each plane region about the y-axis. (*Hint:* Begin by approximating the points of intersection.)

    (i)   (ii)

In Exercises 47–50, the integral represents the volume of a solid of revolution. Identify (a) the plane region that is revolved and (b) the axis of revolution.

47. $2\pi \displaystyle\int_{0}^{2} x^3 \, dx$

48. $2\pi \displaystyle\int_{0}^{1} y - y^{3/2} \, dy$

49. $2\pi \displaystyle\int_{0}^{6} (y + 2)\sqrt{6 - y} \, dy$

50. $2\pi \displaystyle\int_{0}^{1} (4 - x)e^x \, dx$

51. *Volume of a Segment of a Sphere* Let a sphere of radius r be cut by a plane, thereby forming a segment of height h. Show that the volume of this segment is $\frac{1}{3}\pi h^2(3r - h)$.

52. *Volume of an Ellipsoid* Consider the plane region bounded by the graph of

$$\left(\frac{x}{a}\right)^2 + \left(\frac{y}{b}\right)^2 = 1$$

where $a > 0$ and $b > 0$. Show that the volume of the ellipsoid formed when this region revolves about the y-axis is $\dfrac{4\pi a^2 b}{3}$.

53. *Exploration* Consider the region bounded by the graphs of $y = ax^n$, $y = ab^n$, and $x = 0$ (see figure).

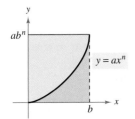

(a) Find the ratio $R_1(n)$ of the area of the region to the area of the circumscribed rectangle.

(b) Find $\lim\limits_{n \to \infty} R_1(n)$ and compare the result with the area of the circumscribed rectangle.

(c) Find the volume of the solid of revolution formed by revolving the region about the y-axis. Find the ratio $R_2(n)$ of this volume to the volume of the circumscribed right circular cylinder.

(d) Find $\lim\limits_{n \to \infty} R_2(n)$ and compare the result with the volume of the circumscribed cylinder.

(e) Use the results of parts (b) and (d) to make a conjecture about the shape of the graph of $y = ax^n$ $(0 \le x \le b)$ as $n \to \infty$.

54. *Think About It* Match each integral with the solid whose volume it represents, and give the dimensions of each solid.

(a) Right circular cone   (b) Torus   (c) Sphere

(d) Right circular cylinder   (e) Ellipsoid

(i) $2\pi \displaystyle\int_{0}^{r} hx \, dx$

(ii) $2\pi \displaystyle\int_{0}^{r} hx\left(1 - \frac{x}{r}\right) dx$

(iii) $2\pi \displaystyle\int_{0}^{r} 2x\sqrt{r^2 - x^2} \, dx$

(iv) $2\pi \displaystyle\int_{0}^{b} 2ax\sqrt{1 - \frac{x^2}{b^2}} \, dx$

(v) $2\pi \displaystyle\int_{-r}^{r} (R - x)\left(2\sqrt{r^2 - x^2}\right) dx$

55. *Volume of a Storage Shed* A storage shed has a circular base of diameter 80 feet (see figure). Starting at the center, the interior height is measured every 10 feet and recorded in the table.

x	0	10	20	30	40
Height	50	45	40	20	0

(a) Use Simpson's Rule to approximate the volume of the shed.

(b) Note that the roof line consists of two line segments. Find the equations of the line segments and use integration to find the volume of the shed.

Distance from center

56. *Modeling Data* A pond is approximately circular, with a diameter of 400 feet (see figure). Starting at the center, the depth of the water is measured every 25 feet and recorded in the table.

x	0	25	50	75	100	125	150	175	200
Depth	20	19	19	17	15	14	10	6	0

(a) Use Simpson's Rule to approximate the volume of water in the pond.

(b) Use the regression capabilities of a graphing utility to find a quadratic model for the depths recorded in the table. Use the graphing utility to plot the depths and graph the model.

(c) Use the integration capabilities of a graphing utility and the model in part (b) to approximate the volume of water in the pond.

(d) Use the result of part (c) to approximate the number of gallons of water in the pond if 1 cubic foot of water is approximately 7.48 gallons.

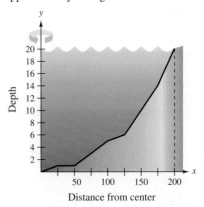

Distance from center

57. Consider the graph of $y^2 = x(4 - x)^2$ (see figure). Find the volumes of the solids that are generated when the loop of this graph is revolved around (a) the x-axis, (b) the y-axis, and (c) the line $x = 4$.

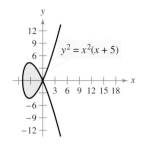

Figure for 57   **Figure for 58**

58. Consider the graph of $y^2 = x^2(x + 5)$ (see figure). Find the volume of the solid that is generated when the loop of this graph is revolved around (a) the x-axis, (b) the y-axis, and (c) the line $x = -5$.

59. Let V_1 and V_2 be the volumes of the solids that result when the plane region bounded by $y = 1/x$, $y = 0$, $x = \frac{1}{4}$, and $x = c \left(c > \frac{1}{4}\right)$ is revolved about the x-axis and y-axis, respectively. Find the value of c for which $V_1 = V_2$.

Section Project: Saturn

The Oblateness of Saturn   Saturn is the most oblate of the nine planets in our solar system. Its equatorial radius is 60,268 kilometers and its polar radius is 54,364 kilometers. The color enhanced photograph of Saturn was taken by Voyager 1. In the photograph, the oblateness of Saturn is clearly visible.

(a) Find the ratio of the volumes of the sphere and the oblate ellipsoid shown below.

(b) If a planet were spherical and had the same volume as Saturn, what would its radius be?

NSSDC

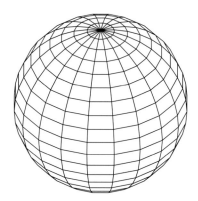

Computer model of "spherical Saturn," whose equatorial radius is equal to its polar radius. The equation of the cross section passing through the pole is

$$x^2 + y^2 = 60{,}268^2.$$

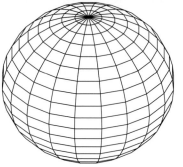

Computer model of "oblate Saturn," whose equatorial radius is greater than its polar radius. The equation of the cross section passing through the pole is

$$\frac{x^2}{60{,}268^2} + \frac{y^2}{54{,}364^2} = 1.$$

Arc Length and Surfaces of Revolution

- Find the arc length of a smooth curve.
- Find the area of a surface of revolution.

CHRISTIAN HUYGENS (1629–1695)

The Dutch mathematician Christian Huygens, who invented the pendulum clock, and James Gregory (1638–1675), a Scottish mathematician, both made early contributions to the problem of finding the length of a rectifiable curve.

Arc Length

In this section, definite integrals are used to find the arc lengths of curves and the areas of surfaces of revolution. In either case, an arc (a segment of a curve) is approximated by straight line segments whose lengths are given by the familiar Distance Formula

$$d = \sqrt{(x_2 - x_1)^2 + (y_2 - y_1)^2}.$$

A **rectifiable** curve is one that has a finite arc length. You will see that a sufficient condition for the graph of a function f to be rectifiable between $(a, f(a))$ and $(b, f(b))$ is that f' be continuous on $[a, b]$. Such a function is **continuously differentiable** on $[a, b]$, and its graph on the interval $[a, b]$ is a **smooth curve.**

Consider a function $y = f(x)$ that is continuously differentiable on the interval $[a, b]$. You can approximate the graph of f by n line segments whose endpoints are determined by the partition

$$a = x_0 < x_1 < x_2 < \cdots < x_n = b$$

as shown in Figure 7.37. By letting $\Delta x_i = x_i - x_{i-1}$ and $\Delta y_i = y_i - y_{i-1}$, you can approximate the length of the graph by

$$s \approx \sum_{i=1}^{n} \sqrt{(x_i - x_{i-1})^2 + (y_i - y_{i-1})^2}$$

$$= \sum_{i=1}^{n} \sqrt{(\Delta x_i)^2 + (\Delta y_i)^2}$$

$$= \sum_{i=1}^{n} \sqrt{(\Delta x_i)^2 + \left(\frac{\Delta y_i}{\Delta x_i}\right)^2 (\Delta x_i)^2}$$

$$= \sum_{i=1}^{n} \sqrt{1 + \left(\frac{\Delta y_i}{\Delta x_i}\right)^2} \, (\Delta x_i).$$

This approximation appears to become better and better as $\|\Delta\| \to 0 \ (n \to \infty)$. So, the length of the graph is

$$s = \lim_{\|\Delta\| \to 0} \sum_{i=1}^{n} \sqrt{1 + \left(\frac{\Delta y_i}{\Delta x_i}\right)^2} \, (\Delta x_i).$$

Because $f'(x)$ exists for each x in (x_{i-1}, x_i), the Mean Value Theorem guarantees the existence of c_i in (x_{i-1}, x_i) such that

$$f(x_i) - f(x_{i-1}) = f'(c_i)(x_i - x_{i-1})$$

$$\frac{\Delta y_i}{\Delta x_i} = f'(c_i).$$

Because f' is continuous on $[a, b]$, it follows that $\sqrt{1 + [f'(x)]^2}$ is also continuous (and therefore integrable) on $[a, b]$, which implies that

$$s = \lim_{\|\Delta\| \to 0} \sum_{i=1}^{n} \sqrt{1 + [f'(c_i)]^2} \, (\Delta x_i)$$

$$= \int_a^b \sqrt{1 + [f'(x)]^2} \, dx$$

where s is called the **arc length** of f between a and b.

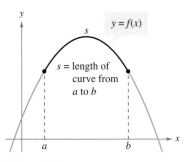

Figure 7.37

Definition of Arc Length

Let the function given by $y = f(x)$ represent a smooth curve on the interval $[a, b]$. The **arc length** of f between a and b is

$$s = \int_a^b \sqrt{1 + [f'(x)]^2}\, dx.$$

Similarly, for a smooth curve given by $x = g(y)$, the **arc length** of g between c and d is

$$s = \int_c^d \sqrt{1 + [g'(y)]^2}\, dy.$$

Because the definition of arc length can be applied to a linear function, you can check to see that this new definition agrees with the standard Distance Formula for the length of a line segment. This is shown in Example 1.

EXAMPLE I   **The Length of a Line Segment**

Find the arc length from (x_1, y_1) to (x_2, y_2) on the graph of $f(x) = mx + b$, as shown in Figure 7.38.

Solution   Because

$$m = f'(x) = \frac{y_2 - y_1}{x_2 - x_1}$$

it follows that

$$s = \int_{x_1}^{x_2} \sqrt{1 + [f'(x)]^2}\, dx \qquad \text{Formula for arc length}$$

$$= \int_{x_1}^{x_2} \sqrt{1 + \left(\frac{y_2 - y_1}{x_2 - x_1}\right)^2}\, dx$$

$$= \sqrt{\frac{(x_2 - x_1)^2 + (y_2 - y_1)^2}{(x_2 - x_1)^2}} \, (x) \Bigg]_{x_1}^{x_2} \qquad \text{Integrate and simplify.}$$

$$= \sqrt{\frac{(x_2 - x_1)^2 + (y_2 - y_1)^2}{(x_2 - x_1)^2}} \, (x_2 - x_1)$$

$$= \sqrt{(x_2 - x_1)^2 + (y_2 - y_1)^2}$$

which is the formula for the distance between two points in the plane.   ▬▬▬

TECHNOLOGY   Definite integrals representing arc length often are very difficult to evaluate. In this section, a few examples are presented. In the next chapter, with more advanced integration techniques, you will be able to tackle more difficult arc length problems. In the meantime, remember that you can always use a numerical integration program to approximate an arc length. For instance, use the *numerical integration* feature of a graphing utility to approximate the arc lengths in Examples 2 and 3.

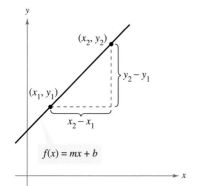

The arc length of the graph of f from (x_1, y_1) to (x_2, y_2) is the same as the standard Distance Formula.

Figure 7.38

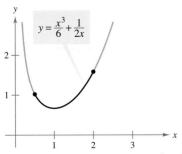

The arc length of the graph of y on $\left[\frac{1}{2}, 2\right]$
Figure 7.39

FOR FURTHER INFORMATION To see how arc length can be used to define trigonometric functions, see the article "Trigonometry Requires Calculus, Not Vice Versa" by Yves Nievergelt in *UMAP Modules*.

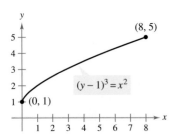

The arc length of the graph of y on $[0, 8]$
Figure 7.40

EXAMPLE 2 **Finding Arc Length**

Find the arc length of the graph of

$$y = \frac{x^3}{6} + \frac{1}{2x}$$

on the interval $\left[\frac{1}{2}, 2\right]$, as shown in Figure 7.39.

Solution Using

$$\frac{dy}{dx} = \frac{3x^2}{6} - \frac{1}{2x^2} = \frac{1}{2}\left(x^2 - \frac{1}{x^2}\right)$$

yields an arc length of

$$
\begin{aligned}
s &= \int_a^b \sqrt{1 + \left(\frac{dy}{dx}\right)^2}\, dx = \int_{1/2}^2 \sqrt{1 + \left[\frac{1}{2}\left(x^2 - \frac{1}{x^2}\right)\right]^2}\, dx \quad \text{Formula for arc length} \\
&= \int_{1/2}^2 \sqrt{\frac{1}{4}\left(x^4 + 2 + \frac{1}{x^4}\right)}\, dx \\
&= \int_{1/2}^2 \frac{1}{2}\left(x^2 + \frac{1}{x^2}\right)\, dx \qquad\qquad \text{Simplify.} \\
&= \frac{1}{2}\left[\frac{x^3}{3} - \frac{1}{x}\right]_{1/2}^2 \qquad\qquad\qquad \text{Integrate.} \\
&= \frac{1}{2}\left(\frac{13}{6} + \frac{47}{24}\right) \\
&= \frac{33}{16}.
\end{aligned}
$$

EXAMPLE 3 **Finding Arc Length**

Find the arc length of the graph of $(y - 1)^3 = x^2$ on the interval $[0, 8]$, as shown in Figure 7.40.

Solution Begin by solving for x in terms of y: $x = \pm(y - 1)^{3/2}$. Choosing the positive value of x produces

$$\frac{dx}{dy} = \frac{3}{2}(y - 1)^{1/2}.$$

The x-interval $[0, 8]$ corresponds to the y-interval $[1, 5]$, and the arc length is

$$
\begin{aligned}
s &= \int_c^d \sqrt{1 + \left(\frac{dx}{dy}\right)^2}\, dy = \int_1^5 \sqrt{1 + \left[\frac{3}{2}(y - 1)^{1/2}\right]^2}\, dy \quad \text{Formula for arc length} \\
&= \int_1^5 \sqrt{\frac{9}{4}y - \frac{5}{4}}\, dy \\
&= \frac{1}{2}\int_1^5 \sqrt{9y - 5}\, dy \qquad\qquad \text{Simplify.} \\
&= \frac{1}{18}\left[\frac{(9y - 5)^{3/2}}{3/2}\right]_1^5 \qquad\qquad \text{Integrate.} \\
&= \frac{1}{27}(40^{3/2} - 4^{3/2}) \\
&\approx 9.073.
\end{aligned}
$$

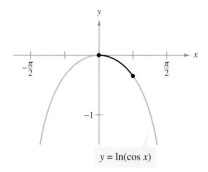

The arc length of the graph of y on $\left[0, \frac{\pi}{4}\right]$
Figure 7.41

www ⊙ ▷ *EXAMPLE 4*   **Finding Arc Length**

Find the arc length of the graph of $y = \ln(\cos x)$ from $x = 0$ to $x = \pi/4$, as shown in Figure 7.41.

Solution   Using

$$\frac{dy}{dx} = -\frac{\sin x}{\cos x} = -\tan x$$

yields an arc length of

$$s = \int_a^b \sqrt{1 + \left(\frac{dy}{dx}\right)^2}\, dx = \int_0^{\pi/4} \sqrt{1 + \tan^2 x}\, dx \qquad \text{Formula for arc length}$$

$$= \int_0^{\pi/4} \sqrt{\sec^2 x}\, dx \qquad \text{Trigonometric identity}$$

$$= \int_0^{\pi/4} \sec x\, dx \qquad \text{Simplify.}$$

$$= \Big[\ln|\sec x + \tan x|\Big]_0^{\pi/4} \qquad \text{Integrate.}$$

$$= \ln\left(\sqrt{2} + 1\right) - \ln 1$$

$$\approx 0.881.$$

EXAMPLE 5   **Length of a Cable**

An electric cable is hung between two towers that are 200 feet apart, as shown in Figure 7.42. The cable takes the shape of a catenary whose equation is

$$y = 75(e^{x/150} + e^{-x/150}) = 150 \cosh \frac{x}{150}.$$

Find the arc length of the cable between the two towers.

Figure 7.42

Solution   Because $y' = \frac{1}{2}(e^{x/150} - e^{-x/150})$, you can write

$$(y')^2 = \frac{1}{4}(e^{x/75} - 2 + e^{-x/75})$$

and

$$1 + (y')^2 = \frac{1}{4}(e^{x/75} + 2 + e^{-x/75}) = \left[\frac{1}{2}(e^{x/150} + e^{-x/150})\right]^2.$$

Therefore, the arc length of the cable is

$$s = \int_a^b \sqrt{1 + (y')^2}\, dx = \frac{1}{2}\int_{-100}^{100} (e^{x/150} + e^{-x/150})\, dx \qquad \text{Formula for arc length}$$

$$= 75\left[e^{x/150} - e^{-x/150}\right]_{-100}^{100} \qquad \text{Integrate.}$$

$$= 150(e^{2/3} - e^{-2/3})$$

$$\approx 215 \text{ feet.}$$

Area of a Surface of Revolution

In Sections 7.2 and 7.3, integration was used to calculate the volume of a solid of revolution. You will now look at a procedure for finding the area of a surface of revolution.

Definition of Surface of Revolution

If the graph of a continuous function is revolved about a line, the resulting surface is a **surface of revolution.**

Figure 7.43

The area of a surface of revolution is derived from the formula for the lateral surface area of the frustum of a right circular cone. Consider the line segment in Figure 7.43, where L is the length of the line segment, r_1 is the radius at the left end of the line segment, and r_2 is the radius at the right end of the line segment. When the line segment is revolved about its axis of revolution, it forms a frustum of a right circular cone, with

$$S = 2\pi r L \qquad \text{Lateral surface area of frustum}$$

where

$$r = \frac{1}{2}(r_1 + r_2). \qquad \text{Average radius of frustum}$$

(In Exercise 60, you are asked to verify the formula for S.)

Suppose the graph of a function f, having a continuous derivative on the interval $[a, b]$, is revolved about the x-axis to form a surface of revolution, as shown in Figure 7.44. Let Δ be a partition of $[a, b]$, with subintervals of width Δx_i. Then the line segment of length

$$\Delta L_i = \sqrt{\Delta x_i^2 + \Delta y_i^2}$$

generates a frustum of a cone. Let r_i be the average radius of this frustum. By the Intermediate Value Theorem, a point d_i exists (in the ith subinterval) such that $r_i = f(d_i)$. The lateral surface area ΔS_i of the frustum is

$$\begin{aligned}
\Delta S_i &= 2\pi r_i \Delta L_i \\
&= 2\pi f(d_i)\sqrt{\Delta x_i^2 + \Delta y_i^2} \\
&= 2\pi f(d_i)\sqrt{1 + \left(\frac{\Delta y_i}{\Delta x_i}\right)^2}\, \Delta x_i.
\end{aligned}$$

Figure 7.44

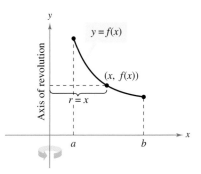

Figure 7.45

By the Mean Value Theorem, a point c_i exists in (x_{i-1}, x_i) such that

$$f'(c_i) = \frac{f(x_i) - f(x_{i-1})}{x_i - x_{i-1}}$$

$$= \frac{\Delta y_i}{\Delta x_i}.$$

So, $\Delta S_i = 2\pi f(d_i)\sqrt{1 + [f'(c_i)]^2}\,\Delta x_i$, and the total surface area can be approximated by

$$S \approx 2\pi \sum_{i=1}^{n} f(d_i)\sqrt{1 + [f'(c_i)]^2}\,\Delta x_i.$$

It can be shown that the limit of the right side as $\|\Delta\| \to 0$ $(n \to \infty)$ is

$$S = 2\pi \int_a^b f(x)\sqrt{1 + [f'(x)]^2}\,dx.$$

In a similar manner, if the graph of f is revolved about the y-axis, then S is

$$S = 2\pi \int_a^b x\sqrt{1 + [f'(x)]^2}\,dx.$$

In both formulas for S, you can regard the products $2\pi f(x)$ and $2\pi x$ as the circumference of the circle traced by a point (x, y) on the graph of f as it is revolved about the x- or y-axis (Figure 7.45). In one case the radius is $r = f(x)$, and in the other case the radius is $r = x$. Moreover, by appropriately adjusting r, you can generalize the formula for surface area to cover *any* horizontal or vertical axis of revolution, as indicated in the following definition.

Definition of the Area of a Surface of Revolution

Let $y = f(x)$ have a continuous derivative on the interval $[a, b]$. The area S of the surface of revolution formed by revolving the graph of f about a horizontal or vertical axis is

$$S = 2\pi \int_a^b r(x)\sqrt{1 + [f'(x)]^2}\,dx \qquad \text{y is a function of x.}$$

where $r(x)$ is the distance between the graph of f and the axis of revolution. If $x = g(y)$ on the interval $[c, d]$, then the surface area is

$$S = 2\pi \int_c^d r(y)\sqrt{1 + [g'(y)]^2}\,dy \qquad \text{x is a function of y.}$$

where $r(y)$ is the distance between the graph of g and the axis of revolution.

The formulas in this definition are sometimes written as

$$S = 2\pi \int_a^b r(x)\,ds \qquad \text{y is a function of x.}$$

and

$$S = 2\pi \int_c^d r(y)\,ds \qquad \text{x is a function of y.}$$

where $ds = \sqrt{1 + [f'(x)]^2}\,dx$ and $ds = \sqrt{1 + [g'(y)]^2}\,dy$, respectively.

EXAMPLE 6   The Area of a Surface of Revolution

Find the area of the surface formed by revolving the graph of

$$f(x) = x^3$$

on the interval $[0, 1]$ about the x-axis, as shown in Figure 7.46.

Solution   The distance between the x-axis and the graph of f is $r(x) = f(x)$, and because $f'(x) = 3x^2$, the surface area is

$$
\begin{aligned}
S &= 2\pi \int_a^b r(x) \sqrt{1 + [f'(x)]^2}\, dx &&\text{Formula for surface area} \\
&= 2\pi \int_0^1 x^3 \sqrt{1 + (3x^2)^2}\, dx \\
&= \frac{2\pi}{36} \int_0^1 (36x^3)(1 + 9x^4)^{1/2}\, dx &&\text{Simplify.} \\
&= \frac{\pi}{18} \left[\frac{(1 + 9x^4)^{3/2}}{3/2} \right]_0^1 &&\text{Integrate.} \\
&= \frac{\pi}{27}(10^{3/2} - 1) \\
&\approx 3.563.
\end{aligned}
$$

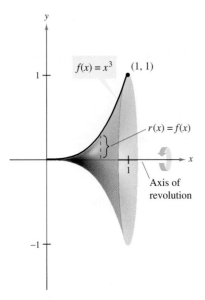

Figure 7.46

EXAMPLE 7   The Area of a Surface of Revolution

Find the area of the surface formed by revolving the graph of

$$f(x) = x^2$$

on the interval $\left[0, \sqrt{2}\right]$ about the y-axis, as shown in Figure 7.47.

Solution   In this case, the distance between the graph of f and the y-axis is $r(x) = x$. Using $f'(x) = 2x$, you can determine that the surface area is

$$
\begin{aligned}
S &= 2\pi \int_a^b r(x) \sqrt{1 + [f'(x)]^2}\, dx &&\text{Formula for surface area} \\
&= 2\pi \int_0^{\sqrt{2}} x \sqrt{1 + (2x)^2}\, dx \\
&= \frac{2\pi}{8} \int_0^{\sqrt{2}} (1 + 4x^2)^{1/2}(8x)\, dx &&\text{Simplify.} \\
&= \frac{\pi}{4} \left[\frac{(1 + 4x^2)^{3/2}}{3/2} \right]_0^{\sqrt{2}} &&\text{Integrate.} \\
&= \frac{\pi}{6}\left[(1 + 8)^{3/2} - 1\right] \\
&= \frac{13\pi}{3} \\
&\approx 13.614.
\end{aligned}
$$

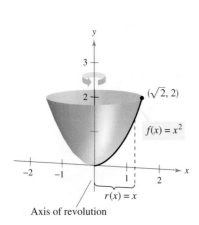

Figure 7.47

Exercises for Section 7.4

In Exercises 1 and 2, find the distance between the points using (a) the Distance Formula and (b) integration.

1. $(0, 0)$, $(5, 12)$

2. $(1, 2)$, $(7, 10)$

In Exercises 3–14, find the arc length of the graph of the function over the indicated interval.

3. $y = \dfrac{2}{3}x^{3/2} + 1$

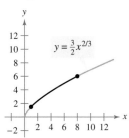

$y = \frac{2}{3}x^{3/2} + 1$

4. $y = 2x^{3/2} + 3$

$y = 2x^{3/2} + 3$

5. $y = \dfrac{3}{2}x^{2/3}$

$y = \frac{3}{2}x^{2/3}$

6. $y = \dfrac{x^4}{8} + \dfrac{1}{4x^2}$

$y = \frac{x^4}{8} + \frac{1}{4x^2}$

7. $y = \dfrac{x^5}{10} + \dfrac{1}{6x^3}$, $[1, 2]$

8. $y = \dfrac{3}{2}x^{2/3} + 4$, $[1, 27]$

9. $y = \ln(\sin x)$, $\left[\dfrac{\pi}{4}, \dfrac{3\pi}{4}\right]$

10. $y = \ln(\cos x)$, $\left[0, \dfrac{\pi}{3}\right]$

11. $y = \frac{1}{2}(e^x + e^{-x})$, $[0, 2]$

12. $y = \ln\left(\dfrac{e^x + 1}{e^x - 1}\right)$, $[\ln 2, \ln 3]$

13. $x = \frac{1}{3}(y^2 + 2)^{3/2}$, $0 \le y \le 4$

14. $x = \frac{1}{3}\sqrt{y}(y - 3)$, $1 \le y \le 4$

 In Exercises 15–24, (a) graph the function, highlighting the part indicated by the given interval, (b) find a definite integral that represents the arc length of the curve over the indicated interval and observe that the integral cannot be evaluated with the techniques studied so far, and (c) use the integration capabilities of a graphing utility to approximate the arc length.

15. $y = 4 - x^2$, $0 \le x \le 2$

16. $y = x^2 + x - 2$, $-2 \le x \le 1$

17. $y = \dfrac{1}{x}$, $1 \le x \le 3$

18. $y = \dfrac{1}{x + 1}$, $0 \le x \le 1$

19. $y = \sin x$, $0 \le x \le \pi$

20. $y = \cos x$, $-\dfrac{\pi}{2} \le x \le \dfrac{\pi}{2}$

21. $x = e^{-y}$, $0 \le y \le 2$

22. $y = \ln x$, $1 \le x \le 5$

23. $y = 2 \arctan x$, $0 \le x \le 1$

24. $x = \sqrt{36 - y^2}$, $0 \le y \le 3$

Approximation   In Exercises 25 and 26, determine which value best approximates the length of the arc represented by the integral. (Make your selection on the basis of a sketch of the arc and *not* by performing any calculations.)

25. $\displaystyle\int_0^2 \sqrt{1 + \left[\dfrac{d}{dx}\left(\dfrac{5}{x^2 + 1}\right)\right]^2}\, dx$

(a) 25   (b) 5   (c) 2   (d) -4 (e) 3

26. $\displaystyle\int_0^{\pi/4} \sqrt{1 + \left[\dfrac{d}{dx}(\tan x)\right]^2}\, dx$

(a) 3   (b) -2 (c) 4   (d) $\dfrac{4\pi}{3}$ (e) 1

 Approximation   In Exercises 27 and 28, approximate the arc length of the graph of the function over the interval $[0, 4]$ in four ways. (a) Use the Distance Formula to find the distance between the endpoints of the arc. (b) Use the Distance Formula to find the lengths of the four line segments connecting the points on the arc when $x = 0$, $x = 1$, $x = 2$, $x = 3$, and $x = 4$. Find the sum of the four lengths. (c) Use Simpson's Rule with $n = 10$ to approximate the integral yielding the indicated arc length. (d) Use the integration capabilities of a graphing utility to approximate the integral yielding the indicated arc length.

27. $f(x) = x^3$

28. $f(x) = (x^2 - 4)^2$

29. (a) Use a graphing utility to graph the function $f(x) = x^{2/3}$.

(b) Can you integrate with respect to x to find the arc length of the graph of f on the interval $[-1, 8]$? Explain.

(c) Find the arc length of the graph of f on the interval $[-1, 8]$.

30. *Astroid*   Find the total length of the graph of the astroid $x^{2/3} + y^{2/3} = 4$.

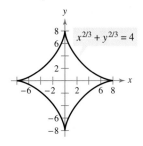

$x^{2/3} + y^{2/3} = 4$

31. *Think About It* The figure shows the graphs of the functions $y_1 = x$, $y_2 = \frac{1}{2}x^{3/2}$, $y_3 = \frac{1}{4}x^2$, and $y_4 = \frac{1}{8}x^{5/2}$ on the interval $[0, 4]$. To print an enlarged copy of the graph, go to the website *www.mathgraphs.com.*

(a) Label the functions.

(b) List the functions in order of increasing arc length.

(c) Verify your answer in part (b) by approximating each arc length accurate to three decimal places.

32. *Think About It* Explain why the two integrals are equal.

$$\int_1^e \sqrt{1 + \frac{1}{x^2}}\, dx = \int_0^1 \sqrt{1 + e^{2x}}\, dx$$

Use the integration capabilities of a graphing utility to verify that the integrals are equal.

33. *Length of Pursuit* A fleeing object leaves the origin and moves up the y-axis (see figure). At the same time, a pursuer leaves the point $(1, 0)$ and always moves toward the fleeing object. The pursuer's speed is twice that of the fleeing object. The equation of the path is modeled by

$$y = \frac{1}{3}(x^{3/2} - 3x^{1/2} + 2).$$

How far has the fleeing object traveled when it is caught? Show that the pursuer has traveled twice as far.

$$y = \frac{1}{3}(x^{3/2} - 3x^{1/2} + 2)$$

Figure for 33

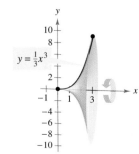

$$y = 31 - 10(e^{x/20} + e^{-x/20})$$

Figure for 34

34. *Roof Area* A barn is 100 feet long and 40 feet wide (see figure). A cross section of the roof is the inverted catenary $y = 31 - 10(e^{x/20} + e^{-x/20})$. Find the number of square feet of roofing on the barn.

35. *Length of a Catenary* Electrical wires suspended between two towers form a catenary (see figure) modeled by the equation

$$y = 20 \cosh \frac{x}{20}, \quad -20 \le x \le 20$$

where x and y are measured in meters. The towers are 40 meters apart. Find the length of the suspended cable.

Figure for 35

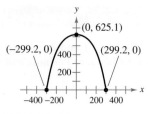

Figure for 36

36. *Length of Gateway Arch* The Gateway Arch in St. Louis, Missouri, is modeled by

$$y = 693.8597 - 68.7672 \cosh 0.0100333x,$$
$$-299.2239 \le x \le 299.2239.$$

(See Section 5.9, Section Project: St. Louis Arch.) Find the length of this curve (see figure).

37. Find the arc length from $(0, 3)$ clockwise to $\left(2, \sqrt{5}\right)$ along the circle $x^2 + y^2 = 9$.

38. Find the arc length from $(-3, 4)$ clockwise to $(4, 3)$ along the circle $x^2 + y^2 = 25$. Show that the result is one-fourth the circumference of the circle.

In Exercises 39–42, set up and evaluate the definite integral for the area of the surface generated by revolving the curve about the x-axis.

39. $y = \frac{1}{3}x^3$ **40.** $y = 2\sqrt{x}$

41. $y = \frac{x^3}{6} + \frac{1}{2x}$, $1 \le x \le 2$ **42.** $y = \frac{x}{2}$, $0 \le x \le 6$

In Exercises 43 and 44, set up and evaluate the definite integral for the area of the surface generated by revolving the curve about the y-axis.

43. $y = \sqrt[3]{x} + 2$ **44.** $y = 9 - x^2$

In Exercises 45 and 46, use the integration capabilities of a graphing utility to approximate the surface area of the solid of revolution.

Function	Interval
45. $y = \sin x$	$[0, \pi]$

revolved about the *x*-axis

46. $y = \ln x$ $[1, e]$

revolved about the *y*-axis

Writing About Concepts

47. Define a rectifiable curve.

48. What precalculus formula and representative element are used to develop the integration formula for arc length?

49. What precalculus formula and representative element are used to develop the integration formula for the area of a surface of revolution?

50. The graphs of the functions f_1 and f_2 on the interval $[a, b]$ are shown in the figure. The graph of each is revolved about the *x*-axis. Which surface of revolution has the greater surface area? Explain.

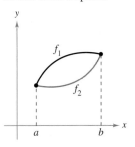

51. A right circular cone is generated by revolving the region bounded by $y = hx/r$, $y = h$, and $x = 0$ about the *y*-axis. Verify that the lateral surface area of the cone is

$$S = \pi r \sqrt{r^2 + h^2}.$$

52. A sphere of radius r is generated by revolving the graph of $y = \sqrt{r^2 - x^2}$ about the *x*-axis. Verify that the surface area of the sphere is $4\pi r^2$.

53. Find the area of the zone of a sphere formed by revolving the graph of $y = \sqrt{9 - x^2}$, $0 \le x \le 2$, about the *y*-axis.

54. Find the area of the zone of a sphere formed by revolving the graph of $y = \sqrt{r^2 - x^2}$, $0 \le x \le a$, about the *y*-axis. Assume that $a < r$.

55. *Bulb Design*   An ornamental light bulb is designed by revolving the graph of

$$y = \tfrac{1}{3}x^{1/2} - x^{3/2}, \quad 0 \le x \le \tfrac{1}{3}$$

about the *x*-axis, where *x* and *y* are measured in feet (see figure). Find the surface area of the bulb and use the result to approximate the amount of glass needed to make the bulb. (Assume that the glass is 0.015 inch thick.)

Figure for 55

 56. *Think About It*   Consider the equation $\dfrac{x^2}{9} + \dfrac{y^2}{4} = 1$.

(a) Use a graphing utility to graph the equation.

(b) Set up the definite integral for finding the first quadrant arc length of the graph in part (a).

(c) Compare the interval of integration in part (b) and the domain of the integrand. Is it possible to evaluate the definite integral? Is it possible to use Simpson's Rule to evaluate the definite integral? Explain. (You will learn how to evaluate this type of integral in Section 8.8.)

 57. *Modeling Data*   The circumference C (in inches) of a vase is measured at three-inch intervals starting at its base. The measurements are shown in the table, where *y* is the vertical distance in inches from the base.

y	0	3	6	9	12	15	18
C	50	65.5	70	66	58	51	48

(a) Use the data to approximate the volume of the vase by summing the volumes of approximating disks.

(b) Use the data to approximate the outside surface area (excluding the base) of the vase by summing the outside surface areas of approximating frustums of right circular cones.

(c) Use the regression capabilities of a graphing utility to find a cubic model for the points (y, r) where $r = C/(2\pi)$. Use the graphing utility to plot the points and graph the model.

(d) Use the model in part (c) and the integration capabilities of a graphing utility to approximate the volume and outside surface area of the vase. Compare the results with your answers in parts (a) and (b).

 58. *Modeling Data*   Property bounded by two perpendicular roads and a stream is shown in the figure on the next page. All distances are measured in feet.

(a) Use the regression capabilities of a graphing utility to fit a fourth-degree polynomial to the path of the stream.

(b) Use the model in part (a) to approximate the area of the property in acres.

(c) Use the integration capabilities of a graphing utility to find the length of the stream that bounds the property.

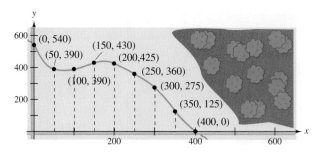

Figure for 58

59. Let R be the region bounded by $y = 1/x$, the x-axis, $x = 1$, and $x = b$, where $b > 1$. Let D be the solid formed when R is revolved about the x-axis.

(a) Find the volume V of D.

(b) Write the surface area S as an integral.

(c) Show that V approaches a finite limit as $b \to \infty$.

(d) Show that $S \to \infty$ as $b \to \infty$.

60. (a) Given a circular sector with radius L and central angle θ (see figure), show that the area of the sector is given by

$$S = \frac{1}{2} L^2 \theta.$$

(b) By joining the straight line edges of the sector in part (a), a right circular cone is formed (see figure) and the lateral surface area of the cone is the same as the area of the sector. Show that the area is $S = \pi r L$, where r is the radius of the base of the cone. (*Hint:* The arc length of the sector equals the circumference of the base of the cone.)

Figure for 60(a)   **Figure for 60(b)**

(c) Use the result of part (b) to verify that the formula for the lateral surface area of the frustum of a cone with slant height L and radii r_1 and r_2 (see figure) is $S = \pi(r_1 + r_2)L$. (*Note:* This formula was used to develop the integral for finding the surface area of a surface of revolution.)

Axis of revolution

61. *Individual Project* Select a solid of revolution from everyday life. Measure the radius of the solid at a minimum of seven points along its axis. Use the data to approximate the volume of the solid and the surface area of the lateral sides of the solid.

62. *Writing* Read the article "Arc Length, Area and the Arcsine Function" by Andrew M. Rockett in *Mathematics Magazine*. Then write a paragraph explaining how the arcsine function can be defined in terms of an arc length. (To view this article, go to the website *www.matharticles.com*.)

63. *Astroid* Find the area of the surface formed by revolving the portion in the first quadrant of the graph of $x^{2/3} + y^{2/3} = 4$, $0 \le y \le 8$ about the y-axis.

Figure for 63   **Figure for 64**

64. Consider the graph of $y^2 = \frac{1}{12}x(4 - x)^2$ (see figure). Find the area of the surface formed when the loop of this graph is revolved around the x-axis.

65. *Suspension Bridge* A cable for a suspension bridge has the shape of a parabola with equation $y = kx^2$. Let h represent the height of the cable from its lowest point to its highest point and let $2w$ represent the total span of the bridge (see figure). Show that the length C of the cable is given by

$$C = 2 \int_0^w \sqrt{1 + \frac{4h^2}{w^4}x^2}\, dx.$$

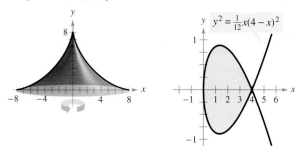

66. *Suspension Bridge* The Humber Bridge, located in the United Kingdom and opened in 1981, has a main span of about 1400 meters. Each of its towers has a height of about 155 meters. Use these dimensions, the integral in Exercise 65, and the integration capabilities of a graphing utility to approximate the length of a parabolic cable along the main span.

Putnam Exam Challenge

67. Find the length of the curve $y^2 = x^3$ from the origin to the point where the tangent makes an angle of $45°$ with the x-axis.

This problem was composed by the Committee on the Putnam Prize Competition.
© The Mathematical Association of America. All rights reserved.

Work

- **Find the work done by a constant force.**
- **Find the work done by a variable force.**

Work Done by a Constant Force

The concept of work is important to scientists and engineers for determining the energy needed to perform various jobs. For instance, it is useful to know the amount of work done when a crane lifts a steel girder, when a spring is compressed, when a rocket is propelled into the air, or when a truck pulls a load along a highway.

In general, **work** is done by a force when it moves an object. If the force applied to the object is *constant*, then the definition of work is as follows.

Definition of Work Done by a Constant Force

If an object is moved a distance D in the direction of an applied constant force F, then the **work** W done by the force is defined as $W = FD$.

There are many types of forces—centrifugal, electromotive, and gravitational, to name a few. A **force** can be thought of as a *push* or a *pull*; a force changes the state of rest or state of motion of a body. For gravitational forces on Earth, it is common to use units of measure corresponding to the weight of an object.

EXAMPLE 1   Lifting an Object

Determine the work done in lifting a 50-pound object 4 feet.

Solution   The magnitude of the required force F is the weight of the object, as shown in Figure 7.48. So, the work done in lifting the object 4 feet is

$W = FD$ Work = (force)(distance)

$\quad = 50(4)$ Force = 50 pounds, distance = 4 feet

$\quad = 200$ foot-pounds.

In the U.S. measurement system, work is typically expressed in foot-pounds (ft-lb), inch-pounds, or foot-tons. In the centimeter-gram-second (C-G-S) system, the basic unit of force is the **dyne**—the force required to produce an acceleration of 1 centimeter per second per second on a mass of 1 gram. In this system, work is typically expressed in dyne-centimeters (ergs) or newton-meters (joules), where 1 joule = 10^7 ergs.

The work done in lifting a 50-pound object 4 feet is 200 foot-pounds.
Figure 7.48

E X P L O R A T I O N

How Much Work?   In Example 1, 200 foot-pounds of work was needed to lift the 50-pound object 4 feet vertically off the ground. Suppose that once you lifted the object, you held it and walked a horizontal distance of 4 feet. Would this require an additional 200 foot-pounds of work? Explain your reasoning.

Work Done by a Variable Force

In Example 1, the force involved was *constant*. If a *variable* force is applied to an object, calculus is needed to determine the work done, because the amount of force changes as the object changes position. For instance, the force required to compress a spring increases as the spring is compressed.

Suppose that an object is moved along a straight line from $x = a$ to $x = b$ by a continuously varying force $F(x)$. Let Δ be a partition that divides the interval $[a, b]$ into n subintervals determined by

$$a = x_0 < x_1 < x_2 < \cdots < x_n = b$$

and let $\Delta x_i = x_i - x_{i-1}$. For each i, choose c_i such that $x_{i-1} \le c_i \le x_i$. Then at c_i the force is given by $F(c_i)$. Because F is continuous, you can approximate the work done in moving the object through the ith subinterval by the increment

$$\Delta W_i = F(c_i)\,\Delta x_i$$

as shown in Figure 7.49. So, the total work done as the object moves from a to b is approximated by

$$W \approx \sum_{i=1}^{n} \Delta W_i$$

$$= \sum_{i=1}^{n} F(c_i)\,\Delta x_i.$$

This approximation appears to become better and better as $\|\Delta\| \to 0$ ($n \to \infty$). So, the work done is

$$W = \lim_{\|\Delta\| \to 0} \sum_{i=1}^{n} F(c_i)\,\Delta x_i$$

$$= \int_a^b F(x)\,dx.$$

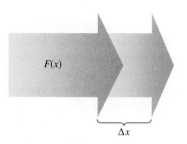

$F(x)$

Δx

The amount of force changes as an object changes position (Δx).
Figure 7.49

Definition of Work Done by a Variable Force

If an object is moved along a straight line by a continuously varying force $F(x)$, then the **work** W done by the force as the object is moved from $x = a$ to $x = b$ is

$$W = \lim_{\|\Delta\| \to 0} \sum_{i=1}^{n} \Delta W_i$$

$$= \int_a^b F(x)\,dx.$$

The remaining examples in this section use some well-known physical laws. The discoveries of many of these laws occurred during the same period in which calculus was being developed. In fact, during the seventeenth and eighteenth centuries, there was little difference between physicists and mathematicians. One such physicist-mathematician was Emilie de Breteuil. Breteuil was instrumental in synthesizing the work of many other scientists, including Newton, Leibniz, Huygens, Kepler, and Descartes. Her physics text *Institutions* was widely used for many years.

EMILIE DE BRETEUIL (1706–1749)

Another major work by de Breteuil was the translation of Newton's "Philosophiae Naturalis Principia Mathematica" into French. Her translation and commentary greatly contributed to the acceptance of Newtonian science in Europe.

The following three laws of physics were developed by Robert Hooke (1635–1703), Isaac Newton (1642–1727), and Charles Coulomb (1736–1806).

1. **Hooke's Law:** The force F required to compress or stretch a spring (within its elastic limits) is proportional to the distance d that the spring is compressed or stretched from its original length. That is,

$$F = kd$$

where the constant of proportionality k (the spring constant) depends on the specific nature of the spring.

2. **Newton's Law of Universal Gravitation:** The force F of attraction between two particles of masses m_1 and m_2 is proportional to the product of the masses and inversely proportional to the square of the distance d between the two particles. That is,

$$F = k\frac{m_1 m_2}{d^2}.$$

If m_1 and m_2 are given in grams and d in centimeters, F will be in dynes for a value of $k = 6.670 \times 10^{-8}$ cubic centimeter per gram-second squared.

3. **Coulomb's Law:** The force between two charges q_1 and q_2 in a vacuum is proportional to the product of the charges and inversely proportional to the square of the distance d between the two charges. That is,

$$F = k\frac{q_1 q_2}{d^2}.$$

If q_1 and q_2 are given in electrostatic units and d in centimeters, F will be in dynes for a value of $k = 1$.

EXPLORATION

The work done in compressing the spring in Example 2 from $x = 3$ inches to $x = 6$ inches is 3375 inch-pounds. Should the work done in compressing the spring from $x = 0$ inches to $x = 3$ inches be more than, the same as, or less than this? Explain.

 EXAMPLE 2   **Compressing a Spring**

A force of 750 pounds compresses a spring 3 inches from its natural length of 15 inches. Find the work done in compressing the spring an additional 3 inches.

Solution   By Hooke's Law, the force $F(x)$ required to compress the spring x units (from its natural length) is $F(x) = kx$. Using the given data, it follows that $F(3) = 750 = (k)(3)$ and so $k = 250$ and $F(x) = 250x$, as shown in Figure 7.50. To find the increment of work, assume that the force required to compress the spring over a small increment Δx is nearly constant. So, the increment of work is

$$\Delta W = (\text{force})(\text{distance increment}) = (250x)\,\Delta x.$$

Because the spring is compressed from $x = 3$ to $x = 6$ inches less than its natural length, the work required is

$$W = \int_a^b F(x)\,dx = \int_3^6 250x\,dx \qquad \text{Formula for work}$$

$$= 125x^2 \Big]_3^6 = 4500 - 1125 = 3375 \text{ inch-pounds.}$$

Note that you do *not* integrate from $x = 0$ to $x = 6$ because you were asked to determine the work done in compressing the spring an *additional* 3 inches (not including the first 3 inches).

Natural length ($F = 0$)

Compressed 3 inches ($F = 750$)

Compressed x inches ($F = 250x$)

Figure 7.50

Not drawn to scale

Figure 7.51

EXAMPLE 3 **Moving a Space Module into Orbit**

A space module weighs 15 metric tons on the surface of Earth. How much work is done in propelling the module to a height of 800 miles above Earth, as shown in Figure 7.51? (Use 4000 miles as the radius of Earth. Do not consider the effect of air resistance or the weight of the propellant.)

Solution Because the weight of a body varies inversely as the square of its distance from the center of Earth, the force $F(x)$ exerted by gravity is

$$F(x) = \frac{C}{x^2}.$$ *C* is the constant of proportionality.

Because the module weighs 15 metric tons on the surface of Earth and the radius of Earth is approximately 4000 miles, you have

$$15 = \frac{C}{(4000)^2}$$

$$240,000,000 = C.$$

So, the increment of work is

$$\Delta W = (\text{force})(\text{distance increment})$$

$$= \frac{240,000,000}{x^2} \Delta x.$$

Finally, because the module is propelled from $x = 4000$ to $x = 4800$ miles, the total work done is

$$W = \int_a^b F(x)\,dx = \int_{4000}^{4800} \frac{240,000,000}{x^2}\,dx \qquad \text{Formula for work}$$

$$= \frac{-240,000,000}{x} \Big]_{4000}^{4800} \qquad \text{Integrate.}$$

$$= -50,000 + 60,000$$

$$= 10,000 \text{ mile-tons}$$

$$\approx 1.164 \times 10^{11} \text{ foot-pounds.}$$

In the C-G-S system, using a conversion factor of 1 foot-pound ≈ 1.35582 joules, the work done is

$$W \approx 1.578 \times 10^{11} \text{ joules.}$$

The solutions to Examples 2 and 3 conform to our development of work as the summation of increments in the form

$$\Delta W = (\text{force})(\text{distance increment}) = (F)(\Delta x).$$

Another way to formulate the increment of work is

$$\Delta W = (\text{force increment})(\text{distance}) = (\Delta F)(x).$$

This second interpretation of ΔW is useful in problems involving the movement of nonrigid substances such as fluids and chains.

EXAMPLE 4   **Emptying a Tank of Oil**

A spherical tank of radius 8 feet is half full of oil that weighs 50 pounds per cubic foot. Find the work required to pump oil out through a hole in the top of the tank.

Solution   Consider the oil to be subdivided into disks of thickness Δy and radius x, as shown in Figure 7.52. Because the increment of force for each disk is given by its weight, you have

$$\Delta F = \text{weight}$$
$$= \left(\frac{50 \text{ pounds}}{\text{cubic foot}}\right)(\text{volume})$$
$$= 50(\pi x^2 \Delta y) \text{ pounds.}$$

For a circle of radius 8 and center at $(0, 8)$, you have

$$x^2 + (y - 8)^2 = 8^2$$
$$x^2 = 16y - y^2$$

and you can write the force increment as

$$\Delta F = 50(\pi x^2 \Delta y)$$
$$= 50\pi(16y - y^2)\,\Delta y.$$

In Figure 7.52, note that a disk y feet from the bottom of the tank must be moved a distance of $(16 - y)$ feet. So, the increment of work is

$$\Delta W = \Delta F(16 - y)$$
$$= 50\pi(16y - y^2)\,\Delta y(16 - y)$$
$$= 50\pi(256y - 32y^2 + y^3)\,\Delta y.$$

Because the tank is half full, y ranges from 0 to 8, and the work required to empty the tank is

$$W = \int_0^8 50\pi(256y - 32y^2 + y^3)\,dy$$
$$= 50\pi\left[128y^2 - \frac{32}{3}y^3 + \frac{y^4}{4}\right]_0^8$$
$$= 50\pi\left(\frac{11{,}264}{3}\right)$$
$$\approx 589{,}782 \text{ foot-pounds.}$$

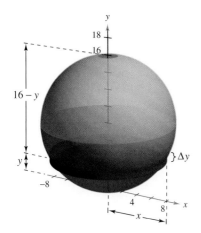

Figure 7.52

To estimate the reasonableness of the result in Example 4, consider that the weight of the oil in the tank is

$$\left(\frac{1}{2}\right)(\text{volume})(\text{density}) = \frac{1}{2}\left(\frac{4}{3}\pi 8^3\right)(50)$$
$$\approx 53{,}616.5 \text{ pounds.}$$

Lifting the entire half-tank of oil 8 feet would involve work of $8(53{,}616.5) \approx 428{,}932$ foot-pounds. Because the oil is actually lifted between 8 and 16 feet, it seems reasonable that the work done is 589,782 foot-pounds.

Work required to raise one end of the chain
Figure 7.53

Work done by expanding gas
Figure 7.54

EXAMPLE 5 **Lifting a Chain**

A 20-foot chain weighing 5 pounds per foot is lying coiled on the ground. How much work is required to raise one end of the chain to a height of 20 feet so that it is fully extended, as shown in Figure 7.53?

Solution   Imagine that the chain is divided into small sections, each of length Δy. Then the weight of each section is the increment of force

$$\Delta F = (\text{weight}) = \left(\frac{5 \text{ pounds}}{\text{foot}}\right)(\text{length}) = 5\Delta y.$$

Because a typical section (initially on the ground) is raised to a height of y, the increment of work is

$$\Delta W = (\text{force increment})(\text{distance}) = (5 \, \Delta y)y = 5y \, \Delta y.$$

Because y ranges from 0 to 20, the total work is

$$W = \int_0^{20} 5y \, dy = \frac{5y^2}{2}\Bigg]_0^{20} = \frac{5(400)}{2} = 1000 \text{ foot-pounds.}$$

In the next example you will consider a piston of radius r in a cylindrical casing, as shown in Figure 7.54. As the gas in the cylinder expands, the piston moves and work is done. If p represents the pressure of the gas (in pounds per square foot) against the piston head and V represents the volume of the gas (in cubic feet), the work increment involved in moving the piston Δx feet is

$$\Delta W = (\text{force})(\text{distance increment}) = F(\Delta x) = p(\pi r^2) \, \Delta x = p \, \Delta V.$$

So, as the volume of the gas expands from V_0 to V_1, the work done in moving the piston is

$$W = \int_{V_0}^{V_1} p \, dV.$$

Assuming the pressure of the gas to be inversely proportional to its volume, you have $p = k/V$ and the integral for work becomes

$$W = \int_{V_0}^{V_1} \frac{k}{V} \, dV.$$

EXAMPLE 6 **Work Done by an Expanding Gas**

A quantity of gas with an initial volume of 1 cubic foot and a pressure of 500 pounds per square foot expands to a volume of 2 cubic feet. Find the work done by the gas. (Assume that the pressure is inversely proportional to the volume.)

Solution   Because $p = k/V$ and $p = 500$ when $V = 1$, you have $k = 500$. So, the work is

$$W = \int_{V_0}^{V_1} \frac{k}{V} \, dV$$

$$= \int_1^2 \frac{500}{V} \, dV$$

$$= 500 \ln|V| \Bigg]_1^2 \approx 346.6 \text{ foot-pounds.}$$

Exercises for Section 7.5

See www.CalcChat.com for worked-out solutions to odd-numbered exercises.

Constant Force   In Exercises 1–4, determine the work done by the constant force.

1. A 100-pound bag of sugar is lifted 10 feet.

2. An electric hoist lifts a 2800-pound car 4 feet.

3. A force of 112 newtons is required to slide a cement block 4 meters in a construction project.

4. The locomotive of a freight train pulls its cars with a constant force of 9 tons a distance of one-half mile.

Writing About Concepts

5. State the definition of work done by a constant force.

6. State the definition of work done by a variable force.

7. The graphs show the force F_i (in pounds) required to move an object 9 feet along the *x*-axis. Order the force functions from the one that yields the least work to the one that yields the most work without doing any calculations. Explain your reasoning.

(a)

(b)

(c)

(d)
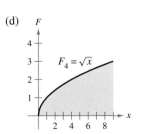

8. Verify your answer to Exercise 7 by calculating the work for each force function.

Hooke's Law   In Exercises 9–16, use Hooke's Law to determine the variable force in the spring problem.

9. A force of 5 pounds compresses a 15-inch spring a total of 4 inches. How much work is done in compressing the spring 7 inches?

10. How much work is done in compressing the spring in Exercise 9 from a length of 10 inches to a length of 6 inches?

11. A force of 250 newtons stretches a spring 30 centimeters. How much work is done in stretching the spring from 20 centimeters to 50 centimeters?

12. A force of 800 newtons stretches a spring 70 centimeters on a mechanical device for driving fence posts. Find the work done in stretching the spring the required 70 centimeters.

13. A force of 20 pounds stretches a spring 9 inches in an exercise machine. Find the work done in stretching the spring 1 foot from its natural position.

14. An overhead garage door has two springs, one on each side of the door. A force of 15 pounds is required to stretch each spring 1 foot. Because of the pulley system, the springs stretch only one-half the distance the door travels. The door moves a total of 8 feet and the springs are at their natural length when the door is open. Find the work done by the pair of springs.

15. Eighteen foot-pounds of work is required to stretch a spring 4 inches from its natural length. Find the work required to stretch the spring an additional 3 inches.

16. Seven and one-half foot-pounds of work is required to compress a spring 2 inches from its natural length. Find the work required to compress the spring an additional one-half inch.

17. *Propulsion*   Neglecting air resistance and the weight of the propellant, determine the work done in propelling a five-ton satellite to a height of

    (a) 100 miles above Earth.

    (b) 300 miles above Earth.

18. *Propulsion*   Use the information in Exercise 17 to write the work *W* of the propulsion system as a function of the height *h* of the satellite above Earth. Find the limit (if it exists) of *W* as *h* approaches infinity.

19. *Propulsion*   Neglecting air resistance and the weight of the propellant, determine the work done in propelling a 10-ton satellite to a height of

    (a) 11,000 miles above Earth.

    (b) 22,000 miles above Earth.

20. *Propulsion*   A lunar module weighs 12 tons on the surface of Earth. How much work is done in propelling the module from the surface of the moon to a height of 50 miles? Consider the radius of the moon to be 1100 miles and its force of gravity to be one-sixth that of Earth.

21. *Pumping Water*   A rectangular tank with a base 4 feet by 5 feet and a height of 4 feet is full of water (see figure). The water weighs 62.4 pounds per cubic foot. How much work is done in pumping water out over the top edge in order to empty (a) half of the tank? (b) all of the tank?

22. *Think About It* Explain why the answer in part (b) of Exercise 21 is not twice the answer in part (a).

23. *Pumping Water* A cylindrical water tank 4 meters high with a radius of 2 meters is buried so that the top of the tank is 1 meter below ground level (see figure). How much work is done in pumping a full tank of water up to ground level? (The water weighs 9800 newtons per cubic meter.)

Figure for 23   **Figure for 24**

24. *Pumping Water* Suppose the tank in Exercise 23 is located on a tower so that the bottom of the tank is 10 meters above the level of a stream (see figure). How much work is done in filling the tank half full of water through a hole in the bottom, using water from the stream?

25. *Pumping Water* An open tank has the shape of a right circular cone (see figure). The tank is 8 feet across the top and 6 feet high. How much work is done in emptying the tank by pumping the water over the top edge?

Figure for 25   **Figure for 28**

26. *Pumping Water* Water is pumped in through the bottom of the tank in Exercise 25. How much work is done to fill the tank

(a) to a depth of 2 feet?

(b) from a depth of 4 feet to a depth of 6 feet?

27. *Pumping Water* A hemispherical tank of radius 6 feet is positioned so that its base is circular. How much work is required to fill the tank with water through a hole in the base if the water source is at the base?

28. *Pumping Diesel Fuel* The fuel tank on a truck has trapezoidal cross sections with dimensions (in feet) shown in the figure. Assume that an engine is approximately 3 feet above the top of the fuel tank and that diesel fuel weighs approximately 53.1 pounds per cubic foot. Find the work done by the fuel pump in raising a full tank of fuel to the level of the engine.

Pumping Gasoline In Exercises 29 and 30, find the work done in pumping gasoline that weighs 42 pounds per cubic foot. (*Hint:* Evaluate one integral by a geometric formula and the other by observing that the integrand is an odd function.)

29. A cylindrical gasoline tank 3 feet in diameter and 4 feet long is carried on the back of a truck and is used to fuel tractors. The axis of the tank is horizontal. The opening on the tractor tank is 5 feet above the top of the tank in the truck. Find the work done in pumping the entire contents of the fuel tank into a tractor.

30. The top of a cylindrical storage tank for gasoline at a service station is 4 feet below ground level. The axis of the tank is horizontal and its diameter and length are 5 feet and 12 feet, respectively. Find the work done in pumping the entire contents of the full tank to a height of 3 feet above ground level.

Lifting a Chain In Exercises 31–34, consider a 15-foot chain that weighs 3 pounds per foot hanging from a winch 15 feet above ground level. Find the work done by the winch in winding up the specified amount of chain.

31. Wind up the entire chain.

32. Wind up one-third of the chain.

33. Run the winch until the bottom of the chain is at the 10-foot level.

34. Wind up the entire chain with a 500-pound load attached to it.

Lifting a Chain In Exercises 35 and 36, consider a 15-foot hanging chain that weighs 3 pounds per foot. Find the work done in lifting the chain vertically to the indicated position.

35. Take the bottom of the chain and raise it to the 15-foot level, leaving the chain doubled and still hanging vertically (see figure).

36. Repeat Exercise 35 raising the bottom of the chain to the 12-foot level.

Demolition Crane In Exercises 37 and 38, consider a demolition crane with a 500-pound ball suspended from a 40-foot cable that weighs 1 pound per foot.

37. Find the work required to wind up 15 feet of the apparatus.

38. Find the work required to wind up all 40 feet of the apparatus.

Boyle's Law   In Exercises 39 and 40, find the work done by the gas for the given volume and pressure. Assume that the pressure is inversely proportional to the volume. (See Example 6.)

39. A quantity of gas with an initial volume of 2 cubic feet and a pressure of 1000 pounds per square foot expands to a volume of 3 cubic feet.

40. A quantity of gas with an initial volume of 1 cubic foot and a pressure of 2500 pounds per square foot expands to a volume of 3 cubic feet.

41. ***Electric Force***   Two electrons repel each other with a force that varies inversely as the square of the distance between them. One electron is fixed at the point $(2, 4)$. Find the work done in moving the second electron from $(-2, 4)$ to $(1, 4)$.

 42. ***Modeling Data***   The hydraulic cylinder on a woodsplitter has a four-inch bore (diameter) and a stroke of 2 feet. The hydraulic pump creates a maximum pressure of 2000 pounds per square inch. Therefore, the maximum force created by the cylinder is $2000(\pi 2^2) = 8000\pi$ pounds.

(a) Find the work done through one extension of the cylinder given that the maximum force is required.

(b) The force exerted in splitting a piece of wood is variable. Measurements of the force obtained when a piece of wood was split are shown in the table. The variable x measures the extension of the cylinder in feet, and F is the force in pounds. Use Simpson's Rule to approximate the work done in splitting the piece of wood.

x	0	$\frac{1}{3}$	$\frac{2}{3}$	1	$\frac{4}{3}$	$\frac{5}{3}$	2
$F(x)$	0	20,000	22,000	15,000	10,000	5000	0

Table for 42(b)

(c) Use the regression capabilities of a graphing utility to find a fourth-degree polynomial model for the data. Plot the data and graph the model.

(d) Use the model in part (c) to approximate the extension of the cylinder when the force is maximum.

(e) Use the model in part (c) to approximate the work done in splitting the piece of wood.

 Hydraulic Press   In Exercises 43–46, use the integration capabilities of a graphing utility to approximate the work done by a press in a manufacturing process. A model for the variable force F (in pounds) and the distance x (in feet) the press moves is given.

Force	Interval
43. $F(x) = 1000[1.8 - \ln(x + 1)]$	$0 \leq x \leq 5$
44. $F(x) = \dfrac{e^{x^2} - 1}{100}$	$0 \leq x \leq 4$
45. $F(x) = 100x\sqrt{125 - x^3}$	$0 \leq x \leq 5$
46. $F(x) = 1000 \sinh x$	$0 \leq x \leq 2$

Section Project: Tidal Energy

Tidal power plants use "tidal energy" to produce electrical energy. To construct a tidal power plant, a dam is built to separate a basin from the sea. Electrical energy is produced as the water flows back and forth between the basin and the sea. The amount of "natural energy" produced depends on the volume of the basin and the tidal range—the vertical distance between high and low tides. (Several natural basin have tidal ranges in excess of 15 feet; the Bay of Fundy in Nova Scotia has a tidal range of 53 feet.)

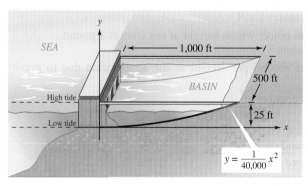

(a) Consider a basin with a rectangular base, as shown in the figure. The basin has a tidal range of 25 feet, with low tide corresponding to $y = 0$. How much water does the basin hold at high tide?

(b) The amount of energy produced during the filling (or the emptying) of the basin is proportional to the amount of work required to fill (or empty) the basin. How much work is required to fill the basin with seawater? (Use a seawater density of 64 pounds per cubic foot.)

The Bay of Fundy in Nova Scotia has an extreme tidal range, as displayed in the greatly contrasting photos above.

FOR FURTHER INFORMATION   For more information on tidal power, see the article "LaRance: Six Years of Operating a Tidal Power Plant in France" by J. Cotillon in *Water Power Magazine*.

Mass of a Planar Lamina

section you have assumed the total mass of a system to be distributed at
s in a plane or on a line. Now consider a thin, flat plate of material of
ity called a **planar lamina** (see Figure 7.60). **Density** is a measure of
of volume, such as grams per cubic centimeter. For planar laminas,
ity is considered to be a measure of mass per unit of area. Density is
the lowercase Greek letter rho.

an irregularly shaped planar lamina of uniform density ρ, bounded by
$y = f(x)$, $y = g(x)$, and $a \leq x \leq b$, as shown in Figure 7.61. The mass
s given by

sity)(area)

$$\int [f(x) - g(x)]\, dx$$

where A is the area of the region. To find the center of mass of this lamina, partition
the interval $[a, b]$ into n subintervals of equal width Δx. Let x_i be the center of the ith
subinterval. You can approximate the portion of the lamina lying in the ith subinterval
by a rectangle whose height is $h = f(x_i) - g(x_i)$. Because the density of the rectangle
is ρ, its mass is

$$m_i = (\text{density})(\text{area})$$
$$= \rho \underbrace{[f(x_i) - g(x_i)]}_{\text{Density} \quad \text{Height}} \underbrace{\Delta x}_{\text{Width}}.$$

Now, considering this mass to be located at the center (x_i, y_i) of the rectangle, the
directed distance from the x-axis to (x_i, y_i) is $y_i = [f(x_i) + g(x_i)]/2$. So, the moment
of m_i about the x-axis is

$$\text{Moment} = (\text{mass})(\text{distance})$$
$$= m_i y_i$$
$$= \rho[f(x_i) - g(x_i)] \Delta x \left[\frac{f(x_i) + g(x_i)}{2} \right].$$

Summing the moments and taking the limit as $n \to \infty$ suggest the definitions below.

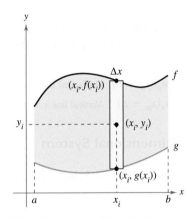

Planar lamina of uniform density ρ
Figure 7.61

Moments and Center of Mass of a Planar Lamina

Let f and g be continuous functions such that $f(x) \geq g(x)$ on $[a, b]$, and
consider the planar lamina of uniform density ρ bounded by the graphs of
$y = f(x)$, $y = g(x)$, and $a \leq x \leq b$.

1. The **moments about the x- and y-axes** are

$$M_x = \rho \int_a^b \left[\frac{f(x) + g(x)}{2} \right] [f(x) - g(x)]\, dx$$

$$M_y = \rho \int_a^b x[f(x) - g(x)]\, dx.$$

2. The **center of mass** $(\bar{x}, \bar{y})$ is given by $\bar{x} = \dfrac{M_y}{m}$ and $\bar{y} = \dfrac{M_x}{m}$, where
$m = \rho \int_a^b [f(x) - g(x)]\, dx$ is the mass of the lamina.

EXAMPLE 4   The Center of Mass of a Planar Lamina

Find the center of mass of the lamina of uniform density ρ bounded by the graph of $f(x) = 4 - x^2$ and the x-axis.

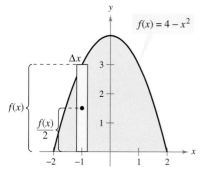

Figure 7.62

Solution   Because the center of mass lies on the axis of symmetry, you know that $\bar{x} = 0$. Moreover, the mass of the lamina is

$$m = \rho \int_{-2}^{2} (4 - x^2)\, dx$$

$$= \rho \left[4x - \frac{x^3}{3} \right]_{-2}^{2}$$

$$= \frac{32\rho}{3}.$$

To find the moment about the x-axis, place a representative rectangle in the region, as shown in Figure 7.62. The distance from the x-axis to the center of this rectangle is

$$y_i = \frac{f(x)}{2} = \frac{4 - x^2}{2}.$$

Because the mass of the representative rectangle is

$$\rho f(x)\, \Delta x = \rho(4 - x^2)\, \Delta x$$

you have

$$M_x = \rho \int_{-2}^{2} \frac{4 - x^2}{2} (4 - x^2)\, dx$$

$$= \frac{\rho}{2} \int_{-2}^{2} (16 - 8x^2 + x^4)\, dx$$

$$= \frac{\rho}{2} \left[16x - \frac{8x^3}{3} + \frac{x^5}{5} \right]_{-2}^{2}$$

$$= \frac{256\rho}{15}$$

and $\bar{y}$ is given by

$$\bar{y} = \frac{M_x}{m} = \frac{256\rho/15}{32\rho/3} = \frac{8}{5}.$$

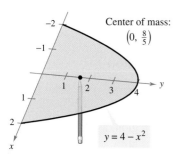

Center of mass:

$\left(0, \frac{8}{5}\right)$

$y = 4 - x^2$

The center of mass is the balancing point.
Figure 7.63

So, the center of mass (the balancing point) of the lamina is $\left(0, \frac{8}{5}\right)$, as shown in Figure 7.63.

The density ρ in Example 4 is a common factor of both the moments and the mass, and as such divides out of the quotients representing the coordinates of the center of mass. So, the center of mass of a lamina of *uniform* density depends only on the shape of the lamina and not on its density. For this reason, the point

$$(\bar{x}, \bar{y}) \qquad \text{Center of mass or centroid}$$

is sometimes called the center of mass of a *region* in the plane, or the **centroid** of the region. In other words, to find the centroid of a region in the plane, you simply assume that the region has a constant density of $\rho = 1$ and compute the corresponding center of mass.

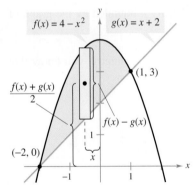

$f(x) = 4 - x^2$ $g(x) = x + 2$

$\dfrac{f(x) + g(x)}{2}$

(1, 3)

$f(x) - g(x)$

(−2, 0)

Figure 7.64

EXPLORATION

Cut an irregular shape from a piece of cardboard.

a. Hold a pencil vertically and move the object on the pencil point until the centroid is located.

b. Divide the object into representative elements. Make the necessary measurements and numerically approximate the centroid. Compare your result with the result in part (a).

EXAMPLE 5 The Centroid of a Plane Region

Find the centroid of the region bounded by the graphs of $f(x) = 4 - x^2$ and $g(x) = x + 2$.

Solution The two graphs intersect at the points $(-2, 0)$ and $(1, 3)$, as shown in Figure 7.64. So, the area of the region is

$$A = \int_{-2}^{1} [f(x) - g(x)]\, dx = \int_{-2}^{1} (2 - x - x^2)\, dx = \frac{9}{2}.$$

The centroid $(\bar{x}, \bar{y})$ of the region has the following coordinates.

$$\bar{x} = \frac{1}{A} \int_{-2}^{1} x[(4 - x^2) - (x + 2)]\, dx = \frac{2}{9} \int_{-2}^{1} (-x^3 - x^2 + 2x)\, dx$$

$$= \frac{2}{9} \left[-\frac{x^4}{4} - \frac{x^3}{3} + x^2 \right]_{-2}^{1} = -\frac{1}{2}$$

$$\bar{y} = \frac{1}{A} \int_{-2}^{1} \left[\frac{(4 - x^2) + (x + 2)}{2} \right][(4 - x^2) - (x + 2)]\, dx$$

$$= \frac{2}{9} \left(\frac{1}{2} \right) \int_{-2}^{1} (-x^2 + x + 6)(-x^2 - x + 2)\, dx$$

$$= \frac{1}{9} \int_{-2}^{1} (x^4 - 9x^2 - 4x + 12)\, dx$$

$$= \frac{1}{9} \left[\frac{x^5}{5} - 3x^3 - 2x^2 + 12x \right]_{-2}^{1} = \frac{12}{5}.$$

So, the centroid of the region is $(\bar{x}, \bar{y}) = \left(-\frac{1}{2}, \frac{12}{5} \right)$.

For simple plane regions, you may be able to find the centroids without resorting to integration.

EXAMPLE 6 The Centroid of a Simple Plane Region

Find the centroid of the region shown in Figure 7.65(a).

Solution By superimposing a coordinate system on the region, as shown in Figure 7.65(b), you can locate the centroids of the three rectangles at

$$\left(\frac{1}{2}, \frac{3}{2} \right), \quad \left(\frac{5}{2}, \frac{1}{2} \right), \quad \text{and} \quad (5, 1).$$

Using these three points, you can find the centroid of the region.

$$A = \text{area of region} = 3 + 3 + 4 = 10$$

$$\bar{x} = \frac{(1/2)(3) + (5/2)(3) + (5)(4)}{10} = \frac{29}{10} = 2.9$$

$$\bar{y} = \frac{(3/2)(3) + (1/2)(3) + (1)(4)}{10} = \frac{10}{10} = 1$$

So, the centroid of the region is $(2.9, 1)$.

NOTE In Example 6, notice that $(2.9, 1)$ is not the "average" of $\left(\frac{1}{2}, \frac{3}{2} \right)$, $\left(\frac{5}{2}, \frac{1}{2} \right)$, and $(5, 1)$.

(a) Original region

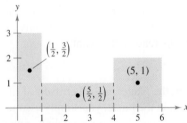

$\left(\frac{1}{2}, \frac{3}{2} \right)$

(5, 1)

$\left(\frac{5}{2}, \frac{1}{2} \right)$

(b) The centroids of the three rectangles

Figure 7.65

Theorem of Pappus

The final topic in this section is a useful theorem credited to Pappus of Alexandria (ca. 300 A.D.), a Greek mathematician whose eight-volume *Mathematical Collection* is a record of much of classical Greek mathematics. The proof of this theorem is given in Section 14.4.

Centroid of R

The volume V is $2\pi\,rA$, where A is the area of region R.

Figure 7.66

> **THEOREM 7.1   The Theorem of Pappus**
>
> Let R be a region in a plane and let L be a line in the same plane such that L does not intersect the interior of R, as shown in Figure 7.66. If r is the distance between the centroid of R and the line, then the volume V of the solid of revolution formed by revolving R about the line is
>
> $$V = 2\pi\,rA$$
>
> where A is the area of R. (Note that $2\pi r$ is the distance traveled by the centroid as the region is revolved about the line.)

The Theorem of Pappus can be used to find the volume of a torus, as shown in the following example. Recall that a torus is a doughnut-shaped solid formed by revolving a circular region about a line that lies in the same plane as the circle (but does not intersect the circle).

EXAMPLE 7   Finding Volume by the Theorem of Pappus

Find the volume of the torus shown in Figure 7.67(a), which was formed by revolving the circular region bounded by

$$(x - 2)^2 + y^2 = 1$$

about the y-axis, as shown in Figure 7.67(b).

Torus

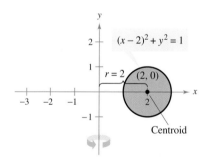

(a)   **(b)**

Figure 7.67

Solution   In Figure 7.67(b), you can see that the centroid of the circular region is $(2, 0)$. So, the distance between the centroid and the axis of revolution is $r = 2$. Because the area of the circular region is $A = \pi$, the volume of the torus is

$$\begin{aligned} V &= 2\pi\,rA \\ &= 2\pi(2)(\pi) \\ &= 4\pi^2 \\ &\approx 39.5. \end{aligned}$$

See www.CalcChat.com for worked-out solutions to odd-numbered exercises.

In Exercises 1–4, find the center of mass of the point masses lying on the *x*-axis.

1. $m_1 = 6, m_2 = 3, m_3 = 5$
  $x_1 = -5, x_2 = 1, x_3 = 3$

2. $m_1 = 7, m_2 = 4, m_3 = 3, m_4 = 8$
  $x_1 = -3, x_2 = -2, x_3 = 5, x_4 = 6$

3. $m_1 = 1, m_2 = 1, m_3 = 1, m_4 = 1, m_5 = 1$
  $x_1 = 7, x_2 = 8, x_3 = 12, x_4 = 15, x_5 = 18$

4. $m_1 = 12, m_2 = 1, m_3 = 6, m_4 = 3, m_5 = 11$
  $x_1 = -6, x_2 = -4, x_3 = -2, x_4 = 0, x_5 = 8$

5. *Graphical Reasoning*

  (a) Translate each point mass in Exercise 3 to the right five units and determine the resulting center of mass.

  (b) Translate each point mass in Exercise 4 to the left three units and determine the resulting center of mass.

6. *Conjecture* Use the result of Exercise 5 to make a conjecture about the change in the center of mass that results when each point mass is translated *k* units horizontally.

Statics Problems **In Exercises 7 and 8, consider a beam of length *L* with a fulcrum *x* feet from one end (see figure). There are objects with weights W_1 and W_2 placed on opposite ends of the beam. Find *x* such that the system is in equilibrium.**

7. Two children weighing 50 pounds and 75 pounds are going to play on a seesaw that is 10 feet long.

8. In order to move a 550-pound rock, a person weighing 200 pounds wants to balance it on a beam that is 5 feet long.

In Exercise 9–12, find the center of mass of the given system of point masses.

9.

m_i	5	1	3
(x_1, y_1)	$(2, 2)$	$(-3, 1)$	$(1, -4)$

10.

m_i	10	2	5
(x_1, y_1)	$(1, -1)$	$(5, 5)$	$(-4, 0)$

11.

m_i	3	4
(x_1, y_1)	$(-2, -3)$	$(5, 5)$

m_i	2	1	6
(x_1, y_1)	$(7, 1)$	$(0, 0)$	$(-3, 0)$

12.

m_i	12	6	$\frac{15}{2}$	15
(x_1, y_1)	$(2, 3)$	$(-1, 5)$	$(6, 8)$	$(2, -2)$

In Exercises 13–24, find $M_x, M_y,$ and $(\bar{x}, \bar{y})$ for the laminas of uniform density ρ bounded by the graphs of the equations.

13. $y = \sqrt{x}, y = 0, x = 4$

14. $y = \frac{1}{2}x^2, y = 0, x = 2$

15. $y = x^2, y = x^3$

16. $y = \sqrt{x}, y = x$

17. $y = -x^2 + 4x + 2, y = x + 2$

18. $y = \sqrt{x} + 1, y = \frac{1}{3}x + 1$

19. $y = x^{2/3}, y = 0, x = 8$

20. $y = x^{2/3}, y = 4$

21. $x = 4 - y^2, x = 0$

22. $x = 2y - y^2, x = 0$

23. $x = -y, x = 2y - y^2$

24. $x = y + 2, x = y^2$

In Exercises 25–28, set up and evaluate the integrals for finding the area and moments about the *x*- and *y*-axes for the region bounded by the graphs of the equations. (Assume $\rho = 1$.)

25. $y = x^2, y = x$

26. $y = \frac{1}{x}, y = 0, 1 \le x \le 4$

27. $y = 2x + 4, y = 0, 0 \le x \le 3$

28. $y = x^2 - 4, y = 0$

In Exercises 29–32, use a graphing utility to graph the region bounded by the graphs of the equations. Use the integration capabilities of the graphing utility to approximate the centroid of the region.

29. $y = 10x\sqrt{125 - x^3}, y = 0$

30. $y = xe^{-x/2}, y = 0, x = 0, x = 4$

31. *Prefabricated End Section of a Building*
  $y = 5\sqrt[3]{400 - x^2}, y = 0$

32. *Witch of Agnesi*
  $y = 8/(x^2 + 4), y = 0, x = -2, x = 2$

In Exercises 33–38, find and/or verify the centroid of the common region used in engineering.

33. *Triangle* Show that the centroid of the triangle with vertices $(-a, 0)$, $(a, 0)$, and (b, c) is the point of intersection of the medians (see figure).

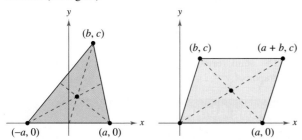

Figure for 33   **Figure for 34**

34. *Parallelogram* Show that the centroid of the parallelogram with vertices $(0, 0)$, $(a, 0)$, (b, c), and $(a + b, c)$ is the point of intersection of the diagonals (see figure).

35. *Trapezoid* Find the centroid of the trapezoid with vertices $(0, 0)$, $(0, a)$, (c, b), and $(c, 0)$. Show that it is the intersection of the line connecting the midpoints of the parallel sides and the line connecting the extended parallel sides, as shown in the figure.

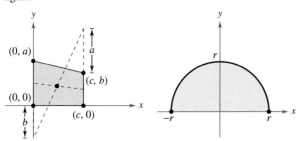

Figure for 35   **Figure for 36**

36. *Semicircle* Find the centroid of the region bounded by the graphs of $y = \sqrt{r^2 - x^2}$ and $y = 0$ (see figure).

37. *Semiellipse* Find the centroid of the region bounded by the graphs of $y = \dfrac{b}{a}\sqrt{a^2 - x^2}$ and $y = 0$ (see figure).

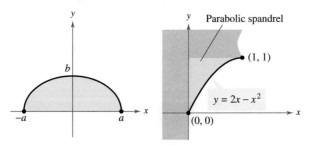

Figure for 37   **Figure for 38**

38. *Parabolic Spandrel* Find the centroid of the **parabolic spandrel** shown in the figure.

39. *Graphical Reasoning* Consider the region bounded by the graphs of $y = x^2$ and $y = b$, where $b > 0$.

(a) Sketch a graph of the region.

(b) Use the graph in part (a) to determine $\bar{x}$. Explain.

(c) Set up the integral for finding M_y. Because of the form of the integrand, the value of the integral can be obtained without integrating. What is the form of the integrand and what is the value of the integral? Compare with the result in part (b).

(d) Use the graph in part (a) to determine whether $\bar{y} > \dfrac{b}{2}$ or $\bar{y} < \dfrac{b}{2}$. Explain.

(e) Use integration to verify your answer in part (d).

40. *Graphical and Numerical Reasoning* Consider the region bounded by the graphs of $y = x^{2n}$ and $y = b$, where $b > 0$ and n is a positive integer.

(a) Set up the integral for finding M_y. Because of the form of the integrand, the value of the integral can be obtained without integrating. What is the form of the integrand and what is the value of the integral? Compare with the result in part (b).

(b) Is $\bar{y} > \dfrac{b}{2}$ or $\bar{y} < \dfrac{b}{2}$? Explain.

(c) Use integration to find $\bar{y}$ as a function of n.

(d) Use the result of part (c) to complete the table.

n	1	2	3	4
$\bar{y}$				

(e) Find $\lim\limits_{n \to \infty} \bar{y}$.

(f) Give a geometric explanation of the result in part (e).

41. *Modeling Data* The manufacturer of glass for a window in a conversion van needs to approximate its center of mass. A coordinate system is superimposed on a prototype of the glass (see figure). The measurements (in centimeters) for the right half of the symmetric piece of glass are shown in the table.

x	0	10	20	30	40
y	30	29	26	20	0

(a) Use Simpson's Rule to approximate the center of mass of the glass.

(b) Use the regression capabilities of a graphing utility to find a fourth-degree polynomial model for the data.

(c) Use the integration capabilities of a graphing utility and the model to approximate the center of mass of the glass. Compare with the result in part (a).

42. *Modeling Data* The manufacturer of a boat needs to approximate the center of mass of a section of the hull. A coordinate system is superimposed on a prototype (see figure). The measurements (in feet) for the right half of the symmetric prototype are listed in the table.

x	0	0.5	1.0	1.5	2
l	1.50	1.45	1.30	0.99	0
d	0.50	0.48	0.43	0.33	0

(a) Use Simpson's Rule to approximate the center of mass of the hull section.

(b) Use the regression capabilities of a graphing utility to find fourth-degree polynomial models for both curves shown in the figure. Plot the data and graph the models.

(c) Use the integration capabilities of a graphing utility and the model to approximate the center of mass of the hull section. Compare with the result in part (a).

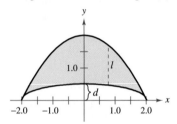

In Exercises 43–46, introduce an appropriate coordinate system and find the coordinates of the center of mass of the planar lamina. (The answer depends on the position of the coordinate system.)

43.

44.

45.

46.

47. Find the center of mass of the lamina in Exercise 43 if the circular portion of the lamina has twice the density of the square portion of the lamina.

48. Find the center of mass of the lamina in Exercise 43 if the square portion of the lamina has twice the density of the circular portion of the lamina.

In Exercises 49–52, use the Theorem of Pappus to find the volume of the solid of revolution.

49. The torus formed by revolving the circle $(x - 5)^2 + y^2 = 16$ about the y-axis

50. The torus formed by revolving the circle $x^2 + (y - 3)^2 = 4$ about the x-axis

51. The solid formed by revolving the region bounded by the graphs of $y = x$, $y = 4$, and $x = 0$ about the x-axis

52. The solid formed by revolving the region bounded by the graphs of $y = 2\sqrt{x - 2}$, $y = 0$, and $x = 6$ about the y-axis

Writing About Concepts

53. Let the point masses $m_1, m_2, \ldots, m_n$ be located at (x_1, y_1), $(x_2, y_2), \ldots, (x_n, y_n)$. Define the center of mass $(\bar{x}, \bar{y})$.

54. What is a planar lamina? Describe what is meant by the center of mass $(\bar{x}, \bar{y})$ of a planar lamina.

55. The centroid of the plane region bounded by the graphs of $y = f(x)$, $y = 0$, $x = 0$, and $x = 1$ is $\left(\frac{5}{6}, \frac{5}{18}\right)$. Is it possible to find the centroid of each of the regions bounded by the graphs of the following sets of equations? If so, identify the centroid and explain your answer.

(a) $y = f(x) + 2$, $y = 2$, $x = 0$, and $x = 1$

(b) $y = f(x - 2)$, $y = 0$, $x = 2$, and $x = 3$

(c) $y = -f(x)$, $y = 0$, $x = 0$, and $x = 1$

(d) $y = f(x)$, $y = 0$, $x = -1$, and $x = 1$

56. State the Theorem of Pappus.

In Exercises 57 and 58, use the *Second Theorem of Pappus*, which is stated as follows. If a segment of a plane curve C is revolved about an axis that does not intersect the curve (except possibly at its endpoints), the area S of the resulting surface of revolution is given by the product of the length of C times the distance d traveled by the centroid of C.

57. A sphere is formed by revolving the graph of $y = \sqrt{r^2 - x^2}$ about the x-axis. Use the formula for surface area, $S = 4\pi r^2$, to find the centroid of the semicircle $y = \sqrt{r^2 - x^2}$.

58. A torus is formed by revolving the graph of $(x - 1)^2 + y^2 = 1$ about the y-axis. Find the surface area of the torus.

59. Let $n \geq 1$ be constant, and consider the region bounded by $f(x) = x^n$, the x-axis, and $x = 1$. Find the centroid of this region. As $n \to \infty$, what does the region look like, and where is its centroid?

Putnam Exam Challenge

60. Let V be the region in the cartesian plane consisting of all points (x, y) satisfying the simultaneous conditions

$$|x| \leq y \leq |x| + 3 \quad \text{and} \quad y \leq 4.$$

Find the centroid $(\bar{x}, \bar{y})$ of V.

Fluid Pressure and Fluid Force

- **Find fluid pressure and fluid force.**

Fluid Pressure and Fluid Force

Swimmers know that the deeper an object is submerged in a fluid, the greater the pressure on the object. **Pressure** is defined as the force per unit of area over the surface of a body. For example, because a column of water that is 10 feet in height and 1 inch square weighs 4.3 pounds, the *fluid pressure* at a depth of 10 feet of water is 4.3 pounds per square inch.* At 20 feet, this would increase to 8.6 pounds per square inch, and in general the pressure is proportional to the depth of the object in the fluid.

BLAISE PASCAL (1623–1662)

Pascal is well known for his work in many areas of mathematics and physics, and also for his influence on Leibniz. Although much of Pascal's work in calculus was intuitive and lacked the rigor of modern mathematics, he nevertheless anticipated many important results.

Definition of Fluid Pressure

The **pressure** on an object at depth h in a liquid is

$$\text{Pressure} = P = wh$$

where w is the weight-density of the liquid per unit of volume.

Below are some common weight-densities of fluids in pounds per cubic foot.

Ethyl alcohol	49.4
Gasoline	41.0–43.0
Glycerin	78.6
Kerosene	51.2
Mercury	849.0
Seawater	64.0
Water	62.4

When calculating fluid pressure, you can use an important (and rather surprising) physical law called **Pascal's Principle,** named after the French mathematician Blaise Pascal. Pascal's Principle states that the pressure exerted by a fluid at a depth h is transmitted equally *in all directions*. For example, in Figure 7.68, the pressure at the indicated depth is the same for all three objects. Because fluid pressure is given in terms of force per unit area $(P = F/A)$, the fluid force on a *submerged horizontal* surface of area A is

$$\text{Fluid force} = F = PA = (\text{pressure})(\text{area}).$$

The pressure at h is the same for all three objects.
Figure 7.68

---

* The total pressure on an object in 10 feet of water would also include the pressure due to Earth's atmosphere. At sea level, atmospheric pressure is approximately 14.7 pounds per square inch.

The fluid force on a horizontal metal sheet is equal to the fluid pressure times the area.
Figure 7.69

EXAMPLE 1 Fluid Force on a Submerged Sheet

Find the fluid force on a rectangular metal sheet measuring 3 feet by 4 feet that is submerged in 6 feet of water, as shown in Figure 7.69.

Solution Because the weight-density of water is 62.4 pounds per cubic foot and the sheet is submerged in 6 feet of water, the fluid pressure is

$$P = (62.4)(6) \qquad P = wh$$
$$= 374.4 \text{ pounds per square foot.}$$

Because the total area of the sheet is $A = (3)(4) = 12$ square feet, the fluid force is

$$F = PA = \left(374.4 \frac{\text{pounds}}{\text{square foot}}\right)(12 \text{ square feet})$$
$$= 4492.8 \text{ pounds.}$$

This result is independent of the size of the body of water. The fluid force would be the same in a swimming pool or lake.

In Example 1, the fact that the sheet is rectangular and horizontal means that you do not need the methods of calculus to solve the problem. Consider a surface that is submerged vertically in a fluid. This problem is more difficult because the pressure is not constant over the surface.

Suppose a vertical plate is submerged in a fluid of weight-density w (per unit of volume), as shown in Figure 7.70. To determine the total force against *one side* of the region from depth c to depth d, you can subdivide the interval $[c, d]$ into n subintervals, each of width Δy. Next, consider the representative rectangle of width Δy and length $L(y_i)$, where y_i is in the ith subinterval. The force against this representative rectangle is

$$\Delta F_i = w(\text{depth})(\text{area})$$
$$= wh(y_i)L(y_i)\,\Delta y.$$

The force against n such rectangles is

$$\sum_{i=1}^{n} \Delta F_i = w \sum_{i=1}^{n} h(y_i)L(y_i)\,\Delta y.$$

Note that w is considered to be constant and is factored out of the summation. Therefore, taking the limit as $\|\Delta\| \to 0$ ($n \to \infty$) suggests the following definition.

Calculus methods must be used to find the fluid force on a vertical metal plate.
Figure 7.70

Definition of Force Exerted by a Fluid

The **force F exerted by a fluid** of constant weight-density w (per unit of volume) against a submerged vertical plane region from $y = c$ to $y = d$ is

$$F = w \lim_{\|\Delta\| \to 0} \sum_{i=1}^{n} h(y_i)L(y_i)\,\Delta y$$
$$= w \int_{c}^{d} h(y)L(y)\,dy$$

where $h(y)$ is the depth of the fluid at y and $L(y)$ is the horizontal length of the region at y.

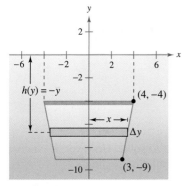

(a) Water gate in a dam

(b) The fluid force against the gate
Figure 7.71

EXAMPLE 2   **Fluid Force on a Vertical Surface**

A vertical gate in a dam has the shape of an isosceles trapezoid 8 feet across the top and 6 feet across the bottom, with a height of 5 feet, as shown in Figure 7.71(a). What is the fluid force on the gate when the top of the gate is 4 feet below the surface of the water?

Solution   In setting up a mathematical model for this problem, you are at liberty to locate the *x*- and *y*-axes in several different ways. A convenient approach is to let the *y*-axis bisect the gate and place the *x*-axis at the surface of the water, as shown in Figure 7.71(b). So, the depth of the water at *y* in feet is

$$\text{Depth } = h(y) = -y.$$

To find the length $L(y)$ of the region at *y*, find the equation of the line forming the right side of the gate. Because this line passes through the points $(3, -9)$ and $(4, -4)$, its equation is

$$y - (-9) = \frac{-4 - (-9)}{4 - 3}(x - 3)$$

$$y + 9 = 5(x - 3)$$

$$y = 5x - 24$$

$$x = \frac{y + 24}{5}.$$

In Figure 7.71(b) you can see that the length of the region at *y* is

$$\text{Length} = 2x$$

$$= \frac{2}{5}(y + 24)$$

$$= L(y).$$

Finally, by integrating from $y = -9$ to $y = -4$, you can calculate the fluid force to be

$$F = w\int_{c}^{d} h(y)L(y)\,dy$$

$$= 62.4\int_{-9}^{-4}(-y)\left(\frac{2}{5}\right)(y + 24)\,dy$$

$$= -62.4\left(\frac{2}{5}\right)\int_{-9}^{-4}(y^2 + 24y)\,dy$$

$$= -62.4\left(\frac{2}{5}\right)\left[\frac{y^3}{3} + 12y^2\right]_{-9}^{-4}$$

$$= -62.4\left(\frac{2}{5}\right)\left(\frac{-1675}{3}\right)$$

$$= 13{,}936 \text{ pounds.}$$

NOTE   In Example 2, the *x*-axis coincided with the surface of the water. This was convenient, but arbitrary. In choosing a coordinate system to represent a physical situation, you should consider various possibilities. Often you can simplify the calculations in a problem by locating the coordinate system to take advantage of special characteristics of the problem, such as symmetry.

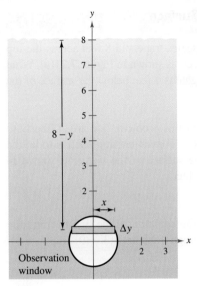

The fluid force on the window
Figure 7.72

EXAMPLE 3   Fluid Force on a Vertical Surface

A circular observation window on a marine science ship has a radius of 1 foot, and the center of the window is 8 feet below water level, as shown in Figure 7.72. What is the fluid force on the window?

Solution   To take advantage of symmetry, locate a coordinate system such that the origin coincides with the center of the window, as shown in Figure 7.72. The depth at y is then

$$\text{Depth } = h(y) = 8 - y.$$

The horizontal length of the window is $2x$, and you can use the equation for the circle, $x^2 + y^2 = 1$, to solve for x as follows.

$$\text{Length } = 2x$$
$$= 2\sqrt{1 - y^2} = L(y)$$

Finally, because y ranges from -1 to 1, and using 64 pounds per cubic foot as the weight-density of seawater, you have

$$F = w \int_c^d h(y)L(y)\,dy$$
$$= 64 \int_{-1}^1 (8 - y)(2)\sqrt{1 - y^2}\,dy.$$

Initially it looks as if this integral would be difficult to solve. However, if you break the integral into two parts and apply symmetry, the solution is simple.

$$F = 64(16) \int_{-1}^1 \sqrt{1 - y^2}\,dy - 64(2) \int_{-1}^1 y\sqrt{1 - y^2}\,dy$$

The second integral is 0 (because the integrand is odd and the limits of integration are symmetric to the origin). Moreover, by recognizing that the first integral represents the area of a semicircle of radius 1, you obtain

$$F = 64(16)\left(\frac{\pi}{2}\right) - 64(2)(0)$$
$$= 512\pi$$
$$\approx 1608.5 \text{ pounds.}$$

So, the fluid force on the window is 1608.5 pounds.

TECHNOLOGY   To confirm the result obtained in Example 3, you might have considered using Simpson's Rule to approximate the value of

$$128 \int_{-1}^1 (8 - x)\sqrt{1 - x^2}\,dx.$$

From the graph of

$$f(x) = (8 - x)\sqrt{1 - x^2}$$

however, you can see that f is not differentiable when $x = \pm 1$ (see Figure 7.73). This means that you cannot apply Theorem 5.19 from Section 5.6 to determine the potential error in Simpson's Rule. Without knowing the potential error, the approximation is of little value. Use a graphing utility to approximate the integral.

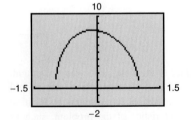

f is not differentiable at $x = \pm 1$.
Figure 7.73

Exercises for Section 7.7

See www.CalcChat.com for worked-out solutions to odd-numbered exercises.

Force on a Submerged Sheet   In Exercises 1 and 2, the area of the top side of a piece of sheet metal is given. The sheet metal is submerged horizontally in 5 feet of water. Find the fluid force on the top side.

1. 3 square feet

2. 16 square feet

Buoyant Force   In Exercises 3 and 4, find the buoyant force of a rectangular solid of the given dimensions submerged in water so that the top side is parallel to the surface of the water. The buoyant force is the difference between the fluid forces on the top and bottom sides of the solid.

3.

4.

Fluid Force on a Tank Wall   In Exercises 5–10, find the fluid force on the vertical side of the tank, where the dimensions are given in feet. Assume that the tank is full of water.

5. Rectangle

6. Triangle

7. Trapezoid

8. Semicircle

9. Parabola, $y = x^2$

10. Semiellipse,
$$y = -\frac{1}{2}\sqrt{36 - 9x^2}$$

Fluid Force of Water   In Exercises 11–14, find the fluid force on the vertical plate submerged in water, where the dimensions are given in meters and the weight-density of water is 9800 newtons per cubic meter.

11. Square

12. Square

13. Triangle

14. Rectangle

Force on a Concrete Form   In Exercises 15–18, the figure is the vertical side of a form for poured concrete that weighs 140.7 pounds per cubic foot. Determine the force on this part of the concrete form.

15. Rectangle

16. Semiellipse,
$$y = -\frac{3}{4}\sqrt{16 - x^2}$$

17. Rectangle

18. Triangle

19. ***Fluid Force of Gasoline***   A cylindrical gasoline tank is placed so that the axis of the cylinder is horizontal. Find the fluid force on a circular end of the tank if the tank is half full, assuming that the diameter is 3 feet and the gasoline weighs 42 pounds per cubic foot.

20. Fluid Force of Gasoline Repeat Exercise 19 for a tank that is full. (Evaluate one integral by a geometric formula and the other by observing that the integrand is an odd function.)

21. Fluid Force on a Circular Plate A circular plate of radius r feet is submerged vertically in a tank of fluid that weighs w pounds per cubic foot. The center of the circle is k ($k > r$) feet below the surface of the fluid. Show that the fluid force on the surface of the plate is

$$F = wk(\pi r^2).$$

(Evaluate one integral by a geometric formula and the other by observing that the integrand is an odd function.)

22. Fluid Force on a Circular Plate Use the result of Exercise 21 to find the fluid force on the circular plate shown in each figure. Assume the plates are in the wall of a tank filled with water and the measurements are given in feet.

(a)

(b)

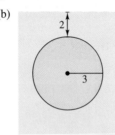

23. Fluid Force on a Rectangular Plate A rectangular plate of height h feet and base b feet is submerged vertically in a tank of fluid that weighs w pounds per cubic foot. The center is k feet below the surface of the fluid, where $h \le k/2$. Show that the fluid force on the surface of the plate is

$$F = wkhb.$$

24. Fluid Force on a Rectangular Plate Use the result of Exercise 23 to find the fluid force on the rectangular plate shown in each figure. Assume the plates are in the wall of a tank filled with water and the measurements are given in feet.

(a)

(b)

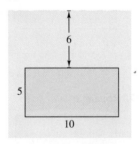

25. Submarine Porthole A porthole on a vertical side of a submarine (submerged in seawater) is 1 square foot. Find the fluid force on the porthole, assuming that the center of the square is 15 feet below the surface.

26. Submarine Porthole Repeat Exercise 25 for a circular porthole that has a diameter of 1 foot. The center is 15 feet below the surface.

27. Modeling Data The vertical stern of a boat with a superimposed coordinate system is shown in the figure. The table shows the width w of the stern at indicated values of y. Find the fluid force against the stern if the measurements are given in feet.

y	0	$\frac{1}{2}$	1	$\frac{3}{2}$	2	$\frac{5}{2}$	3	$\frac{7}{2}$	4
w	0	3	5	8	9	10	10.25	10.5	10.5

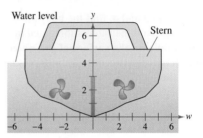

28. Irrigation Canal Gate The vertical cross section of an irrigation canal is modeled by

$$f(x) = \frac{5x^2}{x^2 + 4}$$

where x is measured in feet and $x = 0$ corresponds to the center of the canal. Use the integration capabilities of a graphing utility to approximate the fluid force against a vertical gate used to stop the flow of water if the water is 3 feet deep.

In Exercises 29 and 30, use the integration capabilities of a graphing utility to approximate the fluid force on the vertical plate bounded by the x-axis and the top half of the graph of the equation. Assume that the base of the plate is 12 feet beneath the surface of the water.

29. $x^{2/3} + y^{2/3} = 4^{2/3}$ **30.** $\dfrac{x^2}{28} + \dfrac{y^2}{16} = 1$

31. Think About It

(a) Approximate the depth of the water in the tank in Exercise 5 if the fluid force is one-half as great as when the tank is full.

(b) Explain why the answer in part (a) is not $\frac{3}{2}$.

Writing About Concepts

32. Define fluid pressure.

33. Define fluid force against a submerged vertical plane region.

34. Two identical semicircular windows are placed at the same depth in the vertical wall of an aquarium (see figure). Which has the greater fluid force? Explain.

Review Exercises for Chapter 7

See www.CalcChat.com for worked-out solutions to odd-numbered exercises.

In Exercises 1–10, sketch the region bounded by the graphs of the equations, and determine the area of the region.

1. $y = \dfrac{1}{x^2},\ y = 0,\ x = 1,\ x = 5$

2. $y = \dfrac{1}{x^2},\ y = 4,\ x = 5$

3. $y = \dfrac{1}{x^2 + 1},\ y = 0,\ x = -1,\ x = 1$

4. $x = y^2 - 2y,\ x = -1,\ y = 0$

5. $y = x,\ y = x^3$

6. $x = y^2 + 1,\ x = y + 3$

7. $y = e^x,\ y = e^2,\ x = 0$

8. $y = \csc x,\ y = 2$ (one region)

9. $y = \sin x,\ y = \cos x,\ \dfrac{\pi}{4} \le x \le \dfrac{5\pi}{4}$

10. $x = \cos y,\ x = \dfrac{1}{2},\ \dfrac{\pi}{3} \le y \le \dfrac{7\pi}{3}$

In Exercises 11–14, use a graphing utility to graph the region bounded by the graphs of the functions, and use the integration capabilities of the graphing utility to find the area of the region.

11. $y = x^2 - 8x + 3,\ y = 3 + 8x - x^2$

12. $y = x^2 - 4x + 3,\ y = x^3,\ x = 0$

13. $\sqrt{x} + \sqrt{y} = 1,\ y = 0,\ x = 0$

14. $y = x^4 - 2x^2,\ y = 2x^2$

In Exercises 15–18, use vertical and horizontal representative rectangles to set up integrals for finding the area of the region bounded by the graphs of the equations. Find the area of the region by evaluating the easier of the two integrals.

15. $x = y^2 - 2y,\ x = 0$

16. $y = \sqrt{x - 1},\ y = \dfrac{x - 1}{2}$

17. $y = 1 - \dfrac{x}{2},\ y = x - 2,\ y = 1$

18. $y = \sqrt{x - 1},\ y = 2,\ y = 0,\ x = 0$

19. **Think About It** A person has two job offers. The starting salary for each is $30,000, and after 10 years of service each will pay $56,000. The salary increases for each offer are shown in the figure. From a strictly monetary viewpoint, which is the better offer? Explain.

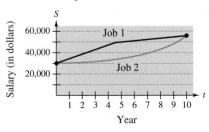

20. **Modeling Data** The table shows the annual service revenue R_1 in billions of dollars for the cellular telephone industry for the years 1995 through 2001. *(Source: Cellular Telecommunications & Internet Association)*

Year	1995	1996	1997	1998	1999	2000	2001
R_1	19.1	23.6	27.5	33.1	40.0	52.5	65.0

(a) Use the regression capabilities of a graphing utility to find an exponential model for the data. Let t represent the year, with $t = 5$ corresponding to 1995. Use the graphing utility to plot the data and graph the model in the same viewing window.

(b) A financial consultant believes that a model for service revenue for the years 2005 through 2010 is

$$R_2 = 5 + 6.83e^{0.2t}.$$

What is the difference in total service revenue between the two models for the years 2005 through 2010?

In Exercises 21–28, find the volume of the solid generated by revolving the plane region bounded by the equations about the indicated line(s).

21. $y = x,\ y = 0,\ x = 4$
   (a) the x-axis   (b) the y-axis
   (c) the line $x = 4$ (d) the line $x = 6$

22. $y = \sqrt{x},\ y = 2,\ x = 0$
   (a) the x-axis   (b) the line $y = 2$
   (c) the y-axis   (d) the line $x = -1$

23. $\dfrac{x^2}{16} + \dfrac{y^2}{9} = 1$ (a) the y-axis (oblate spheroid)
                        (b) the x-axis (prolate spheroid)

24. $\dfrac{x^2}{a^2} + \dfrac{y^2}{b^2} = 1$ (a) the y-axis (oblate spheroid)
                        (b) the x-axis (prolate spheroid)

25. $y = \dfrac{1}{x^4 + 1},\ y = 0,\ x = 0,\ x = 1$
revolved about the y-axis

26. $y = \dfrac{1}{\sqrt{1 + x^2}},\ y = 0,\ x = -1,\ x = 1$
revolved about the x-axis

27. $y = 1/(1 + \sqrt{x - 2}),\ y = 0,\ x = 2,\ x = 6$
revolved about the y-axis

28. $y = e^{-x},\ y = 0,\ x = 0,\ x = 1$
revolved about the x-axis

In Exercises 29 and 30, consider the region bounded by the graphs of the equations $y = x\sqrt{x + 1}$ and $y = 0$.

29. **Area** Find the area of the region.

30. **Volume** Find the volume of the solid generated by revolving the region about (a) the x-axis and (b) the y-axis.

31. ***Depth of Gasoline in a Tank*** A gasoline tank is an oblate spheroid generated by revolving the region bounded by the graph of $(x^2/16) + (y^2/9) = 1$ about the y-axis, where x and y are measured in feet. Find the depth of the gasoline in the tank when it is filled to one-fourth its capacity.

32. ***Magnitude of a Base*** The base of a solid is a circle of radius a, and its vertical cross sections are equilateral triangles. The volume of the solid is 10 cubic meters. Find the radius of the circle.

In Exercises 33 and 34, find the arc length of the graph of the function over the given interval.

33. $f(x) = \dfrac{4}{5}x^{5/4}$, $[0, 4]$ **34.** $y = \dfrac{1}{6}x^3 + \dfrac{1}{2x}$, $[1, 3]$

35. ***Length of a Catenary*** A cable of a suspension bridge forms a catenary modeled by the equation

$$y = 300 \cosh\left(\frac{x}{2000}\right) - 280, \quad -2000 \le x \le 2000$$

where x and y are measured in feet. Use a graphing utility to approximate the length of the cable.

36. ***Approximation*** Determine which value best approximates the length of the arc represented by the integral

$$\int_0^{\pi/4} \sqrt{1 + (\sec^2 x)^2}\, dx.$$

(Make your selection on the basis of a sketch of the arc and *not* by performing any calculations.)

(a) -2 (b) 1 (c) π (d) 4 (e) 3

37. ***Surface Area*** Use integration to find the lateral surface area of a right circular cone of height 4 and radius 3.

38. ***Surface Area*** The region bounded by the graphs of $y = 2\sqrt{x}$, $y = 0$, and $x = 3$ is revolved about the x-axis. Find the surface area of the solid generated.

39. ***Work*** A force of 4 pounds is needed to stretch a spring 1 inch from its natural position. Find the work done in stretching the spring from its natural length of 10 inches to a length of 15 inches.

40. ***Work*** The force required to stretch a spring is 50 pounds. Find the work done in stretching the spring from its natural length of 9 inches to double that length.

41. ***Work*** A water well has an eight-inch casing (diameter) and is 175 feet deep. The water is 25 feet from the top of the well. Determine the amount of work done in pumping the well dry, assuming that no water enters it while it is being pumped.

42. ***Work*** Repeat Exercise 41, assuming that water enters the well at a rate of 4 gallons per minute and the pump works at a rate of 12 gallons per minute. How many gallons are pumped in this case?

43. ***Work*** A chain 10 feet long weighs 5 pounds per foot and is hung from a platform 20 feet above the ground. How much work is required to raise the entire chain to the 20-foot level?

44. ***Work*** A windlass, 200 feet above ground level on the top of a building, uses a cable weighing 4 pounds per foot. Find the work done in winding up the cable if

(a) one end is at ground level.

(b) there is a 300-pound load attached to the end of the cable.

45. ***Work*** The work done by a variable force in a press is 80 foot-pounds. The press moves a distance of 4 feet and the force is a quadratic of the form $F = ax^2$. Find a.

46. ***Work*** Find the work done by the force F shown in the figure.

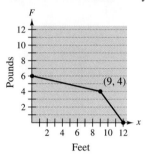

In Exercises 47–50, find the centroid of the region bounded by the graphs of the equations.

47. $\sqrt{x} + \sqrt{y} = \sqrt{a}$, $x = 0$, $y = 0$

48. $y = x^2$, $y = 2x + 3$

49. $y = a^2 - x^2$, $y = 0$

50. $y = x^{2/3}$, $y = \dfrac{1}{2}x$

51. ***Centroid*** A blade on an industrial fan has the configuration of a semicircle attached to a trapezoid (see figure). Find the centroid of the blade.

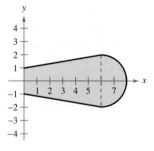

52. ***Fluid Force*** A swimming pool is 5 feet deep at one end and 10 feet deep at the other, and the bottom is an inclined plane. The length and width of the pool are 40 feet and 20 feet. If the pool is full of water, what is the fluid force on each of the vertical walls?

53. ***Fluid Force*** Show that the fluid force against any vertical region in a liquid is the product of the weight per cubic volume of the liquid, the area of the region, and the depth of the centroid of the region.

54. ***Fluid Force*** Using the result of Exercise 53, find the fluid force on one side of a vertical circular plate of radius 4 feet that is submerged in water so that its center is 5 feet below the surface.

1. Let R be the area of the region in the first quadrant bounded by the parabola $y = x^2$ and the line $y = cx$, $c > 0$. Let T be the area of the triangle AOB. Calculate the limit

$$\lim_{c \to 0^+} \frac{T}{R}.$$

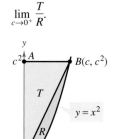

2. Let R be the region bounded by the parabola $y = x - x^2$ and the x-axis. Find the equation of the line $y = mx$ that divides this region into two regions of equal area.

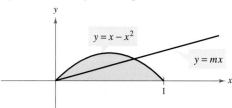

3. (a) A torus is formed by revolving the region bounded by the circle

$$(x - 2)^2 + y^2 = 1$$

about the y-axis (see figure). Use the disk method to calculate the volume of the torus.

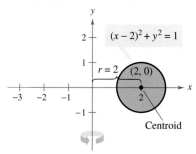

(b) Use the disk method to find the volume of the general torus if the circle has radius r and its center is R units from the axis of rotation.

4. Graph the curve

$$8y^2 = x^2(1 - x^2).$$

Use a computer algebra system to find the surface area of the solid of revolution obtained by revolving the curve about the x-axis.

5. A hole is cut through the center of a sphere of radius r (see figure). The height of the remaining spherical ring is h. Find the volume of the ring and show that it is independent of the radius of the sphere.

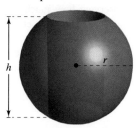

6. A rectangle R of length l and width w is revolved about the line L (see figure). Find the volume of the resulting solid of revolution.

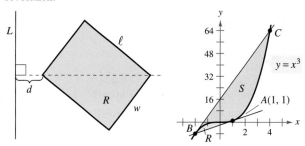

Figure for 6   **Figure for 7**

7. (a) The tangent line to the curve $y = x^3$ at the point $A(1, 1)$ intersects the curve at another point B. Let R be the area of the region bounded by the curve and the tangent line. The tangent line at B intersects the curve at another point C (see figure). Let S be the area of the region bounded by the curve and this second tangent line. How are the areas R and S related?

(b) Repeat the construction in part (a) by selecting an arbitrary point A on the curve $y = x^3$. Show that the two areas R and S are always related in the same way.

8. The graph of $y = f(x)$ passes through the origin. The arc length of the curve from $(0, 0)$ to $(x, f(x))$ is given by

$$s(x) = \int_0^x \sqrt{1 + e^t}\, dt.$$

Identify the function f.

9. Let f be rectifiable on the interval $[a, b]$, and let

$$s(x) = \int_a^x \sqrt{1 + [f'(t)]^2}\, dt.$$

(a) Find $\dfrac{ds}{dx}$.

(b) Find ds and $(ds)^2$.

(c) If $f(t) = t^{3/2}$, find $s(x)$ on $[1, 3]$.

(d) Calculate $s(2)$ and describe what it signifies.

10. The **Archimedes Principle** states that the upward or buoyant force on an object within a fluid is equal to the weight of the fluid that the object displaces. For a partially submerged object, you can obtain information about the relative densities of the floating object and the fluid by observing how much of the object is above and below the surface. You can also determine the size of a floating object if you know the amount that is above the surface and the relative densities. You can see the top of a floating iceberg (see figure). The density of ocean water is 1.03×10^3 kilograms per cubic meter, and that of ice is 0.92×10^3 kilograms per cubic meter. What percent of the total iceberg is below the surface?

11. Sketch the region bounded on the left by $x = 1$, bounded above by $y = 1/x^3$, and bounded below by $y = -1/x^3$.

   (a) Find the centroid of the region for $1 \le x \le 6$.

   (b) Find the centroid of the region for $1 \le x \le b$.

   (c) Where is the centroid as $b \to \infty$?

12. Sketch the region to the right of the y-axis, bounded above by $y = 1/x^4$ and bounded below by $y = -1/x^4$.

   (a) Find the centroid of the region for $1 \le x \le 6$.

   (b) Find the centroid of the region for $1 \le x \le b$.

   (c) Where is the centroid as $b \to \infty$?

13. Find the work done by each force F.

14. Estimate the surface area of the pond using (a) the Trapezoidal Rule and (b) Simpson's Rule.

In Exercises 15 and 16, find the consumer surplus and producer surplus for the given demand $[p_1(x)]$ and supply $[p_2(x)]$ curves. The consumer surplus and producer surplus are represented by the areas shown in the figure.

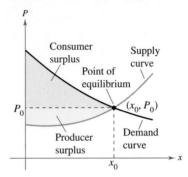

15. $p_1(x) = 50 - 0.5x, \quad p_2(x) = 0.125x$

16. $p_1(x) = 1000 - 0.4x^2, \quad p_2(x) = 42x$

17. A swimming pool is 20 feet wide, 40 feet long, 4 feet deep at one end, and 8 feet deep at the other end (see figure). The bottom is an inclined plane. Find the fluid force on each vertical wall.

18. (a) Find at least two continuous functions f that satisfy each condition.

   (i) $f(x) \ge 0$ on $[0, 1]$ (ii) $f(0) = 0$ and $f(1) = 0$

   (iii) The area bounded by the graph of f and the x-axis for $0 \le x \le 1$ equals 1.

   (b) For each function found in part (a), approximate the arc length of the graph of the function on the interval $[0, 1]$. (Use a graphing utility if necessary.)

   (c) Can you find a function f that satisfies the conditions in part (a) and whose graph has an arc length of less than 3 on the interval $[0, 1]$?

Integration Techniques, L'Hôpital's Rule, and Improper Integrals

$$\int_0^1 \frac{1}{\sqrt{x}}\,dx = 2$$

$$\int_1^4 \frac{1}{\sqrt{x}}\,dx = 2$$

$$\int_4^\infty \frac{1}{\sqrt{x}}\,dx = \infty$$

The NASA Hubble Space Telescope image of a planetary nebula nicknamed the "Cat's Eye Nebula" gives just a glimpse of the kinds of things you might see if you could travel through space. Would it be possible to propel a spacecraft an unlimited distance away from Earth's surface? Why?

P. Harrington and K.J. Borkowski (University of Maryland), and NASA

From your studies of calculus thus far, you know that a definite integral has finite limits of integration and a continuous integrand. In Chapter 8, you will study *improper integrals*. Improper integrals have at least one infinite limit of integration or have an integrand with an infinite discontinuity. You will see that improper integrals either converge or diverge.

Basic Integration Rules

- Review procedures for fitting an integrand to one of the basic integration rules.

Fitting Integrands to Basic Rules

In this chapter, you will study several integration techniques that greatly expand the set of integrals to which the basic integration rules can be applied. These rules are reviewed on page 520. A major step in solving any integration problem is recognizing which basic integration rule to use. As shown in Example 1, slight differences in the integrand can lead to very different solution techniques.

 EXAMPLE 1   **A Comparison of Three Similar Integrals**

Find each integral.

a. $\displaystyle\int \frac{4}{x^2 + 9}\, dx$ **b.** $\displaystyle\int \frac{4x}{x^2 + 9}\, dx$ **c.** $\displaystyle\int \frac{4x^2}{x^2 + 9}\, dx$

Solution

a. Use the Arctangent Rule and let $u = x$ and $a = 3$.

$$\int \frac{4}{x^2 + 9}\, dx = 4 \int \frac{1}{x^2 + 3^2}\, dx \qquad \text{Constant Multiple Rule}$$

$$= 4\left(\frac{1}{3} \arctan \frac{x}{3}\right) + C \qquad \text{Arctangent Rule}$$

$$= \frac{4}{3} \arctan \frac{x}{3} + C \qquad \text{Simplify.}$$

b. Here the Arctangent Rule does not apply because the numerator contains a factor of x. Consider the Log Rule and let $u = x^2 + 9$. Then $du = 2x\, dx$, and you have

$$\int \frac{4x}{x^2 + 9}\, dx = 2 \int \frac{2x\, dx}{x^2 + 9} \qquad \text{Constant Multiple Rule}$$

$$= 2 \int \frac{du}{u} \qquad \text{Substitution: } u = x^2 + 9$$

$$= 2 \ln|u| + C = 2 \ln(x^2 + 9) + C. \qquad \text{Log Rule}$$

c. Because the degree of the numerator is equal to the degree of the denominator, you should first use division to rewrite the improper rational function as the sum of a polynomial and a proper rational function.

$$\int \frac{4x^2}{x^2 + 9}\, dx = \int \left(4 - \frac{36}{x^2 + 9}\right) dx \qquad \text{Rewrite using long division.}$$

$$= \int 4\, dx - 36 \int \frac{1}{x^2 + 9}\, dx \qquad \text{Write as two integrals.}$$

$$= 4x - 36\left(\frac{1}{3} \arctan \frac{x}{3}\right) + C \qquad \text{Integrate.}$$

$$= 4x - 12 \arctan \frac{x}{3} + C \qquad \text{Simplify.}$$

🌐 *indicates that in the* HM mathSpace® *CD-ROM and the online* Eduspace® *system for this text, you will find an Open Exploration, which further explores this example using the computer algebra systems* Maple, Mathcad, Mathematica, *and* Derive.

EXPLORATION

A Comparison of Three Similar Integrals   Which, if any, of the following integrals can be evaluated using the 20 basic integration rules? For any that can be evaluated, do so. For any that can't, explain why.

a. $\displaystyle\int \frac{3}{\sqrt{1 - x^2}}\, dx$

b. $\displaystyle\int \frac{3x}{\sqrt{1 - x^2}}\, dx$

c. $\displaystyle\int \frac{3x^2}{\sqrt{1 - x^2}}\, dx$

NOTE   Notice in Example 1(c) that some preliminary algebra is required before applying the rules for integration, and that subsequently more than one rule is needed to evaluate the resulting integral.

EXAMPLE 2   Using Two Basic Rules to Solve a Single Integral

Evaluate $\displaystyle\int_0^1 \frac{x+3}{\sqrt{4-x^2}}\,dx$.

Solution   Begin by writing the integral as the sum of two integrals. Then apply the Power Rule and the Arcsine Rule as follows.

$$\int_0^1 \frac{x+3}{\sqrt{4-x^2}}\,dx = \int_0^1 \frac{x}{\sqrt{4-x^2}}\,dx + \int_0^1 \frac{3}{\sqrt{4-x^2}}\,dx$$

$$= -\frac{1}{2}\int_0^1 (4-x^2)^{-1/2}(-2x)\,dx + 3\int_0^1 \frac{1}{\sqrt{2^2-x^2}}\,dx$$

$$= \left[-(4-x^2)^{1/2} + 3\arcsin\frac{x}{2} \right]_0^1$$

$$= \left(-\sqrt{3} + \frac{\pi}{2} \right) - (-2+0)$$

$$\approx 1.839$$

See Figure 8.1.

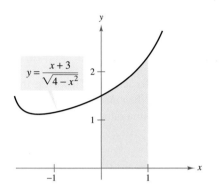

$y = \dfrac{x+3}{\sqrt{4-x^2}}$

The area of the region is approximately 1.839.

Figure 8.1

TECHNOLOGY   Simpson's Rule can be used to give a good approximation of the value of the integral in Example 2 (for $n = 10$, the approximation is 1.839). When using numerical integration, however, you should be aware that Simpson's Rule does not always give good approximations when one or both of the limits of integration are near a vertical asymptote. For instance, using the Fundamental Theorem of Calculus, you can obtain

$$\int_0^{1.99} \frac{x+3}{\sqrt{4-x^2}}\,dx \approx 6.213.$$

Applying Simpson's Rule (with $n = 10$) to this integral produces an approximation of 6.889.

EXAMPLE 3   A Substitution Involving $a^2 - u^2$

Find $\displaystyle\int \frac{x^2}{\sqrt{16-x^6}}\,dx$.

Solution   Because the radical in the denominator can be written in the form

$$\sqrt{a^2 - u^2} = \sqrt{4^2 - (x^3)^2}$$

you can try the substitution $u = x^3$. Then $du = 3x^2\,dx$, and you have

$$\int \frac{x^2}{\sqrt{16-x^6}}\,dx = \frac{1}{3}\int \frac{3x^2\,dx}{\sqrt{16-(x^3)^2}} \qquad \text{Rewrite integral.}$$

$$= \frac{1}{3}\int \frac{du}{\sqrt{4^2 - u^2}} \qquad \text{Substitution: } u = x^3$$

$$= \frac{1}{3}\arcsin\frac{u}{4} + C \qquad \text{Arcsine Rule}$$

$$= \frac{1}{3}\arcsin\frac{x^3}{4} + C. \qquad \text{Rewrite as a function of } x.$$

STUDY TIP   Rules 18, 19, and 20 of the basic integration rules on the next page all have expressions involving the sum or difference of two squares:

$$a^2 - u^2$$

$$a^2 + u^2$$

$$u^2 - a^2$$

With such an expression, consider the substitution $u = f(x)$, as in Example 3.

Surprisingly, two of the most commonly overlooked integration rules are the Log Rule and the Power Rule. Notice in the next two examples how these two integration rules can be disguised.

EXAMPLE 4 A Disguised Form of the Log Rule

Find $\displaystyle \int \frac{1}{1 + e^x}\,dx$.

Solution The integral does not appear to fit any of the basic rules. However, the quotient form suggests the Log Rule. If you let $u = 1 + e^x$, then $du = e^x\,dx$. You can obtain the required du by adding and subtracting e^x in the numerator, as follows.

$$\int \frac{1}{1 + e^x}\,dx = \int \frac{1 + e^x - e^x}{1 + e^x}\,dx \qquad \text{Add and subtract } e^x \text{ in numerator.}$$

$$= \int \left(\frac{1 + e^x}{1 + e^x} - \frac{e^x}{1 + e^x} \right) dx \qquad \text{Rewrite as two fractions.}$$

$$= \int dx - \int \frac{e^x\,dx}{1 + e^x} \qquad \text{Rewrite as two integrals.}$$

$$= x - \ln(1 + e^x) + C \qquad \text{Integrate.} \qquad \blacksquare$$

NOTE There is usually more than one way to solve an integration problem. For instance, in Example 4, try integrating by multiplying the numerator and denominator by e^{-x} to obtain an integral of the form $-\int du/u$. See if you can get the same answer by this procedure. (Be careful: the answer will appear in a different form.)

EXAMPLE 5 A Disguised Form of the Power Rule

Find $\displaystyle \int (\cot x)[\ln(\sin x)]\,dx$.

Solution Again, the integral does not appear to fit any of the basic rules. However, considering the two primary choices for u [$u = \cot x$ and $u = \ln(\sin x)$], you can see that the second choice is the appropriate one because

$$u = \ln(\sin x) \qquad \text{and} \qquad du = \frac{\cos x}{\sin x}\,dx = \cot x\,dx.$$

So,

$$\int (\cot x)[\ln(\sin x)]\,dx = \int u\,du \qquad \text{Substitution: } u = \ln(\sin x)$$

$$= \frac{u^2}{2} + C \qquad \text{Integrate.}$$

$$= \frac{1}{2}[\ln(\sin x)]^2 + C. \qquad \text{Rewrite as a function of } x.$$

$\blacksquare$

NOTE In Example 5, try *checking* that the derivative of

$$\frac{1}{2}[\ln(\sin x)]^2 + C$$

is the integrand of the original integral.

Review of Basic Integration Rules ($a > 0$)

1. $\displaystyle \int kf(u)\,du = k\int f(u)\,du$

2. $\displaystyle \int [f(u) \pm g(u)]\,du =$
   $\displaystyle \int f(u)\,du \pm \int g(u)\,du$

3. $\displaystyle \int du = u + C$

4. $\displaystyle \int u^n\,du = \frac{u^{n+1}}{n+1} + C,\ n \neq -1$

5. $\displaystyle \int \frac{du}{u} = \ln|u| + C$

6. $\displaystyle \int e^u\,du = e^u + C$

7. $\displaystyle \int a^u\,du = \left(\frac{1}{\ln a}\right)a^u + C$

8. $\displaystyle \int \sin u\,du = -\cos u + C$

9. $\displaystyle \int \cos u\,du = \sin u + C$

10. $\displaystyle \int \tan u\,du = -\ln|\cos u| + C$

11. $\displaystyle \int \cot u\,du = \ln|\sin u| + C$

12. $\displaystyle \int \sec u\,du =$
    $\ln|\sec u + \tan u| + C$

13. $\displaystyle \int \csc u\,du =$
    $-\ln|\csc u + \cot u| + C$

14. $\displaystyle \int \sec^2 u\,du = \tan u + C$

15. $\displaystyle \int \csc^2 u\,du = -\cot u + C$

16. $\displaystyle \int \sec u \tan u\,du = \sec u + C$

17. $\displaystyle \int \csc u \cot u\,du = -\csc u + C$

18. $\displaystyle \int \frac{du}{\sqrt{a^2 - u^2}} = \arcsin \frac{u}{a} + C$

19. $\displaystyle \int \frac{du}{a^2 + u^2} = \frac{1}{a}\arctan \frac{u}{a} + C$

20. $\displaystyle \int \frac{du}{u\sqrt{u^2 - a^2}} = \frac{1}{a}\text{arcsec} \frac{|u|}{a} + C$

Trigonometric identities can often be used to fit integrals to one of the basic integration rules.

EXAMPLE 6   Using Trigonometric Identities

Find $\displaystyle\int \tan^2 2x \, dx$.

TECHNOLOGY   If you have access to a computer algebra system, try using it to evaluate the integrals in this section. Compare the *form* of the antiderivative given by the software with the form obtained by hand. Sometimes the forms will be the same, but often they will differ. For instance, why is the antiderivative $\ln 2x + C$ equivalent to the antiderivative $\ln x + C$?

Solution   Note that $\tan^2 u$ is not in the list of basic integration rules. However, $\sec^2 u$ is in the list. This suggests the trigonometric identity $\tan^2 u = \sec^2 u - 1$. If you let $u = 2x$, then $du = 2 \, dx$ and

$$
\begin{aligned}
\int \tan^2 2x \, dx &= \frac{1}{2}\int \tan^2 u \, du && \text{Substitution: } u = 2x \\
&= \frac{1}{2}\int (\sec^2 u - 1) \, du && \text{Trigonometric identity} \\
&= \frac{1}{2}\int \sec^2 u \, du - \frac{1}{2}\int du && \text{Rewrite as two integrals.} \\
&= \frac{1}{2}\tan u - \frac{u}{2} + C && \text{Integrate.} \\
&= \frac{1}{2}\tan 2x - x + C. && \text{Rewrite as a function of } x.
\end{aligned}
$$

This section concludes with a summary of the common procedures for fitting integrands to the basic integration rules.

Procedures for Fitting Integrands to Basic Rules

Technique	Example
Expand (numerator).	$(1 + e^x)^2 = 1 + 2e^x + e^{2x}$
Separate numerator.	$\dfrac{1+x}{x^2+1} = \dfrac{1}{x^2+1} + \dfrac{x}{x^2+1}$
Complete the square.	$\dfrac{1}{\sqrt{2x - x^2}} = \dfrac{1}{\sqrt{1 - (x-1)^2}}$
Divide improper rational function.	$\dfrac{x^2}{x^2+1} = 1 - \dfrac{1}{x^2+1}$
Add and subtract terms in numerator.	$\dfrac{2x}{x^2 + 2x + 1} = \dfrac{2x + 2 - 2}{x^2 + 2x + 1} = \dfrac{2x + 2}{x^2 + 2x + 1} - \dfrac{2}{(x+1)^2}$
Use trigonometric identities.	$\cot^2 x = \csc^2 x - 1$
Multiply and divide by Pythagorean conjugate.	$\dfrac{1}{1 + \sin x} = \left(\dfrac{1}{1 + \sin x}\right)\left(\dfrac{1 - \sin x}{1 - \sin x}\right) = \dfrac{1 - \sin x}{1 - \sin^2 x}$
	$= \dfrac{1 - \sin x}{\cos^2 x} = \sec^2 x - \dfrac{\sin x}{\cos^2 x}$

NOTE   Remember that you can separate numerators but not denominators. Watch out for this common error when fitting integrands to basic rules.

$$\frac{1}{x^2 + 1} \neq \frac{1}{x^2} + \frac{1}{1} \qquad \text{Do not separate denominators.}$$

In Exercises 1–4, select the correct antiderivative.

1. $\dfrac{dy}{dx} = \dfrac{x}{\sqrt{x^2 + 1}}$

   (a) $2\sqrt{x^2 + 1} + C$ (b) $\sqrt{x^2 + 1} + C$

   (c) $\frac{1}{2}\sqrt{x^2 + 1} + C$ (d) $\ln(x^2 + 1) + C$

2. $\dfrac{dy}{dx} = \dfrac{x}{x^2 + 1}$

   (a) $\ln\sqrt{x^2 + 1} + C$ (b) $\dfrac{2x}{(x^2 + 1)^2} + C$

   (c) $\arctan x + C$ (d) $\ln(x^2 + 1) + C$

3. $\dfrac{dy}{dx} = \dfrac{1}{x^2 + 1}$

   (a) $\ln\sqrt{x^2 + 1} + C$ (b) $\dfrac{2x}{(x^2 + 1)^2} + C$

   (c) $\arctan x + C$ (d) $\ln(x^2 + 1) + C$

4. $\dfrac{dy}{dx} = x\cos(x^2 + 1)$

   (a) $2x\sin(x^2 + 1) + C$ (b) $-\frac{1}{2}\sin(x^2 + 1) + C$

   (c) $\frac{1}{2}\sin(x^2 + 1) + C$ (d) $-2x\sin(x^2 + 1) + C$

In Exercises 5–14, select the basic integration formula you can use to find the integral, and identify u and a when appropriate.

5. $\displaystyle\int (3x - 2)^4 \, dx$

6. $\displaystyle\int \dfrac{2t - 1}{t^2 - t + 2} \, dt$

7. $\displaystyle\int \dfrac{1}{\sqrt{x}\left(1 - 2\sqrt{x}\right)} \, dx$

8. $\displaystyle\int \dfrac{2}{(2t - 1)^2 + 4} \, dt$

9. $\displaystyle\int \dfrac{3}{\sqrt{1 - t^2}} \, dt$

10. $\displaystyle\int \dfrac{-2x}{\sqrt{x^2 - 4}} \, dx$

11. $\displaystyle\int t\sin t^2 \, dt$

12. $\displaystyle\int \sec 3x \tan 3x \, dx$

13. $\displaystyle\int (\cos x)e^{\sin x} \, dx$

14. $\displaystyle\int \dfrac{1}{x\sqrt{x^2 - 4}} \, dx$

In Exercises 15–50, find the indefinite integral.

15. $\displaystyle\int 6(x - 4)^5 \, dx$

16. $\displaystyle\int \dfrac{2}{(t - 9)^2} \, dt$

17. $\displaystyle\int \dfrac{5}{(z - 4)^5} \, dz$

18. $\displaystyle\int t^2 \sqrt[3]{t^3 - 1} \, dt$

19. $\displaystyle\int \left[v + \dfrac{1}{(3v - 1)^3}\right] dv$

20. $\displaystyle\int \left[x - \dfrac{3}{(2x + 3)^2}\right] dx$

21. $\displaystyle\int \dfrac{t^2 - 3}{-t^3 + 9t + 1} \, dt$

22. $\displaystyle\int \dfrac{x + 1}{\sqrt{x^2 + 2x - 4}} \, dx$

23. $\displaystyle\int \dfrac{x^2}{x - 1} \, dx$

24. $\displaystyle\int \dfrac{2x}{x - 4} \, dx$

25. $\displaystyle\int \dfrac{e^x}{1 + e^x} \, dx$

26. $\displaystyle\int \left(\dfrac{1}{3x - 1} - \dfrac{1}{3x + 1}\right) dx$

27. $\displaystyle\int (1 + 2x^2)^2 \, dx$

28. $\displaystyle\int x\left(1 + \dfrac{1}{x}\right)^3 dx$

29. $\displaystyle\int x\cos 2\pi x^2 \, dx$

30. $\displaystyle\int \sec 4x \, dx$

31. $\displaystyle\int \csc \pi x \cot \pi x \, dx$

32. $\displaystyle\int \dfrac{\sin x}{\sqrt{\cos x}} \, dx$

33. $\displaystyle\int e^{5x} \, dx$

34. $\displaystyle\int \csc^2 xe^{\cot x} \, dx$

35. $\displaystyle\int \dfrac{2}{e^{-x} + 1} \, dx$

36. $\displaystyle\int \dfrac{5}{3e^x - 2} \, dx$

37. $\displaystyle\int \dfrac{\ln x^2}{x} \, dx$

38. $\displaystyle\int (\tan x)[\ln(\cos x)] \, dx$

39. $\displaystyle\int \dfrac{1 + \sin x}{\cos x} \, dx$

40. $\displaystyle\int \dfrac{1 + \cos \alpha}{\sin \alpha} \, d\alpha$

41. $\displaystyle\int \dfrac{1}{\cos \theta - 1} \, d\theta$

42. $\displaystyle\int \dfrac{2}{3(\sec x - 1)} \, dx$

43. $\displaystyle\int \dfrac{-1}{\sqrt{1 - (2t - 1)^2}} \, dt$

44. $\displaystyle\int \dfrac{1}{4 + 3x^2} \, dx$

45. $\displaystyle\int \dfrac{\tan(2/t)}{t^2} \, dt$

46. $\displaystyle\int \dfrac{e^{1/t}}{t^2} \, dt$

47. $\displaystyle\int \dfrac{3}{\sqrt{6x - x^2}} \, dx$

48. $\displaystyle\int \dfrac{1}{(x - 1)\sqrt{4x^2 - 8x + 3}} \, dx$

49. $\displaystyle\int \dfrac{4}{4x^2 + 4x + 65} \, dx$

50. $\displaystyle\int \dfrac{1}{\sqrt{1 - 4x - x^2}} \, dx$

 Slope Fields **In Exercises 51–54, a differential equation, a point, and a slope field are given. (a) Sketch two approximate solutions of the differential equation on the slope field, one of which passes through the given point. (b) Use integration to find the particular solution of the differential equation and use a graphing utility to graph the solution. Compare the result with the sketches in part (a). To print an enlarged copy of the graph, go to the website *www.mathgraphs.com*.**

51. $\dfrac{ds}{dt} = \dfrac{t}{\sqrt{1 - t^4}}$

   $\left(0, -\dfrac{1}{2}\right)$

52. $\dfrac{dy}{dx} = \tan^2(2x)$

   $(0, 0)$

53. $\dfrac{dy}{dx} = (\sec x + \tan x)^2$

$(0, 1)$

54. $\dfrac{dy}{dx} = \dfrac{1}{\sqrt{4x - x^2}}$

$\left(2, \dfrac{1}{2}\right)$

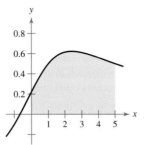

71. $y = \dfrac{3x + 2}{x^2 + 9}$

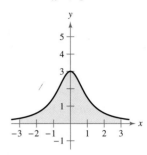

72. $y = \dfrac{3}{x^2 + 1}$

 Slope Fields **In Exercises 55 and 56, use a computer algebra system to graph the slope field for the differential equation and graph the solution through the specified initial condition.**

55. $\dfrac{dy}{dx} = 0.2y,\ y(0) = 3$

56. $\dfrac{dy}{dx} = 5 - y,\ y(0) = 1$

In Exercises 57–60, solve the differential equation.

57. $\dfrac{dy}{dx} = (1 + e^x)^2$

58. $\dfrac{dr}{dt} = \dfrac{(1 + e^t)^2}{e^t}$

59. $(4 + \tan^2 x)\,y' = \sec^2 x$

60. $y' = \dfrac{1}{x\sqrt{4x^2 - 1}}$

In Exercises 61–68, evaluate the definite integral. Use the integration capabilities of a graphing utility to verify your result.

61. $\displaystyle\int_0^{\pi/4} \cos 2x\, dx$

62. $\displaystyle\int_0^{\pi} \sin^2 t \cos t\, dt$

63. $\displaystyle\int_0^1 x e^{-x^2}\, dx$

64. $\displaystyle\int_1^e \dfrac{1 - \ln x}{x}\, dx$

65. $\displaystyle\int_0^4 \dfrac{2x}{\sqrt{x^2 + 9}}\, dx$

66. $\displaystyle\int_1^2 \dfrac{x - 2}{x}\, dx$

67. $\displaystyle\int_0^{2/\sqrt{3}} \dfrac{1}{4 + 9x^2}\, dx$

68. $\displaystyle\int_0^4 \dfrac{1}{\sqrt{25 - x^2}}\, dx$

Area **In Exercises 69–74, find the area of the region.**

69. $y = (-2x + 5)^{3/2}$

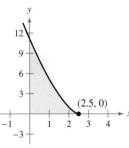

70. $y = x\sqrt{8 - 2x^2}$

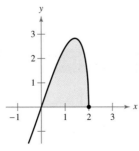

73. $y^2 = x^2(1 - x^2)$

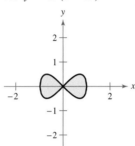

74. $y = \sin 2x$

In Exercises 75–78, use a computer algebra system to find the integral. Use the computer algebra system to graph two antiderivatives. Describe the relationship between the two graphs of the antiderivatives.

75. $\displaystyle\int \dfrac{1}{x^2 + 4x + 13}\, dx$

76. $\displaystyle\int \dfrac{x - 2}{x^2 + 4x + 13}\, dx$

77. $\displaystyle\int \dfrac{1}{1 + \sin \theta}\, d\theta$

78. $\displaystyle\int \left(\dfrac{e^x + e^{-x}}{2}\right)^3 dx$

Writing About Concepts

In Exercises 79–82, state the integration formula you would use to perform the integration. Explain why you chose that formula. Do not integrate.

79. $\displaystyle\int x(x^2 + 1)^3\, dx$

80. $\displaystyle\int x \sec(x^2 + 1)\tan(x^2 + 1)\, dx$

81. $\displaystyle\int \dfrac{x}{x^2 + 1}\, dx$

82. $\displaystyle\int \dfrac{1}{x^2 + 1}\, dx$

83. Explain why the antiderivative $y_1 = e^{x + C_1}$ is equivalent to the antiderivative $y_2 = Ce^x$.

84. Explain why the antiderivative $y_1 = \sec^2 x + C_1$ is equivalent to the antiderivative $y_2 = \tan^2 x + C$.

85. Determine the constants a and b such that

$$\sin x + \cos x = a \sin(x + b).$$

Use this result to integrate $\displaystyle\int \frac{dx}{\sin x + \cos x}$.

86. *Area* The graphs of $f(x) = x$ and $g(x) = ax^2$ intersect at the points $(0, 0)$ and $(1/a, 1/a)$. Find a $(a > 0)$ such that the area of the region bounded by the graphs of these two functions is $\frac{2}{3}$.

87. *Think About It* Use a graphing utility to graph the function $f(x) = \frac{1}{5}(x^3 - 7x^2 + 10x)$. Use the graph to determine whether

$$\int_0^5 f(x)\, dx$$ is positive or negative. Explain.

88. *Think About It* When evaluating

$$\int_{-1}^1 x^2\, dx$$

is it appropriate to substitute $u = x^2$, $x = \sqrt{u}$, and $dx = \dfrac{du}{2\sqrt{u}}$

to obtain

$$\frac{1}{2}\int_1^1 \sqrt{u}\, du = 0?$$

Explain.

Approximation **In Exercises 89 and 90, determine which value best approximates the area of the region between the x-axis and the function over the given interval. (Make your selection on the basis of a sketch of the region and *not* by integrating.)**

89. $f(x) = \dfrac{4x}{x^2 + 1}$, $[0, 2]$

  (a) 3 (b) 1 (c) -8 (d) 8 (e) 10

90. $f(x) = \dfrac{4}{x^2 + 1}$, $[0, 2]$

  (a) 3 (b) 1 (c) -4 (d) 4 (e) 10

Interpreting Integrals **In Exercises 91 and 92, (a) sketch the region whose area is given by the integral, (b) sketch the solid whose volume is given by the integral if the disk method is used, and (c) sketch the solid whose volume is given by the integral if the shell method is used. (There is more than one correct answer for each part.)**

91. $\displaystyle\int_0^2 2\pi x^2\, dx$

92. $\displaystyle\int_0^4 \pi y\, dy$

93. *Volume* The region bounded by $y = e^{-x^2}$, $y = 0$, $x = 0$, and $x = b$ $(b > 0)$ is revolved about the y-axis.

  (a) Find the volume of the solid generated if $b = 1$.

  (b) Find b such that the volume of the generated solid is $\frac{4}{3}$ cubic units.

94. *Arc Length* Find the arc length of the graph of $y = \ln(\sin x)$ from $x = \pi/4$ to $x = \pi/2$.

95. *Surface Area* Find the area of the surface formed by revolving the graph of $y = 2\sqrt{x}$ on the interval $[0, 9]$ about the x-axis.

96. *Centroid* Find the x-coordinate of the centroid of the region bounded by the graphs of

$$y = \frac{5}{\sqrt{25 - x^2}}, \quad y = 0, \quad x = 0, \quad \text{and} \quad x = 4.$$

In Exercises 97 and 98, find the average value of the function over the given interval.

97. $f(x) = \dfrac{1}{1 + x^2}$, $\quad -3 \le x \le 3$

98. $f(x) = \sin nx$, $\quad 0 \le x \le \pi/n$, n is a positive integer.

 Arc Length **In Exercises 99 and 100, use the integration capabilities of a graphing utility to approximate the arc length of the curve over the given interval.**

99. $y = \tan \pi x$, $\left[0, \frac{1}{4}\right]$ **100.** $y = x^{2/3}$, $[1, 8]$

101. *Finding a Pattern*

  (a) Find $\displaystyle\int \cos^3 x\, dx$.

  (b) Find $\displaystyle\int \cos^5 x\, dx$.

  (c) Find $\displaystyle\int \cos^7 x\, dx$.

  (d) Explain how to find $\int \cos^{15} x\, dx$ without actually integrating.

102. *Finding a Pattern*

  (a) Write $\int \tan^3 x\, dx$ in terms of $\int \tan x\, dx$. Then find $\int \tan^3 x\, dx$.

  (b) Write $\int \tan^5 x\, dx$ in terms of $\int \tan^3 x\, dx$.

  (c) Write $\int \tan^{2k+1} x\, dx$, where k is a positive integer, in terms of $\int \tan^{2k-1} x\, dx$.

  (d) Explain how to find $\int \tan^{15} x\, dx$ without actually integrating.

103. *Methods of Integration* Show that the following results are equivalent.

Integration by tables:

$$\int \sqrt{x^2 + 1}\, dx = \frac{1}{2}\left(x\sqrt{x^2 + 1} + \ln\left|x + \sqrt{x^2 + 1}\right|\right) + C$$

Integration by computer algebra system:

$$\int \sqrt{x^2 + 1}\, dx = \frac{1}{2}\left(x\sqrt{x^2 + 1} + \operatorname{arcsinh}(x)\right) + C$$

Putnam Exam Challenge

104. Evaluate $\displaystyle\int_2^4 \frac{\sqrt{\ln(9 - x)}\, dx}{\sqrt{\ln(9 - x)} + \sqrt{\ln(x + 3)}}$.

This problem was composed by the Committee on the Putnam Prize Competition.
© The Mathematical Association of America. All rights reserved.

Section 8.2

Integration by Parts

- Find an antiderivative using integration by parts.
- Use a tabular method to perform integration by parts.

Integration by Parts

In this section you will study an important integration technique called **integration by parts.** This technique can be applied to a wide variety of functions and is particularly useful for integrands involving *products* of algebraic and transcendental functions. For instance, integration by parts works well with integrals such as

$$\int x \ln x \, dx, \quad \int x^2 e^x \, dx, \quad \text{and} \quad \int e^x \sin x \, dx.$$

Integration by parts is based on the formula for the derivative of a product

$$\frac{d}{dx}[uv] = u\frac{dv}{dx} + v\frac{du}{dx}$$

$$= uv' + vu'$$

where both u and v are differentiable functions of x. If u' and v' are continuous, you can integrate both sides of this equation to obtain

$$uv = \int uv' \, dx + \int vu' \, dx$$

$$= \int u \, dv + \int v \, du.$$

By rewriting this equation, you obtain the following theorem.

THEOREM 8.1   Integration by Parts

If u and v are functions of x and have continuous derivatives, then

$$\int u \, dv = uv - \int v \, du.$$

This formula expresses the original integral in terms of another integral. Depending on the choices of u and dv, it may be easier to evaluate the second integral than the original one. Because the choices of u and dv are critical in the integration by parts process, the following guidelines are provided.

Guidelines for Integration by Parts

1. Try letting dv be the most complicated portion of the integrand that fits a basic integration rule. Then u will be the remaining factor(s) of the integrand.

2. Try letting u be the portion of the integrand whose derivative is a function simpler than u. Then dv will be the remaining factor(s) of the integrand.

EXPLORATION

Proof Without Words   Here is a different approach to proving the formula for integration by parts. Exercise taken from "Proof Without Words: Integration by Parts" by Roger B. Nelsen, *Mathematics Magazine*, April 1991, by permission of the author.

Area ▢ + Area ▢ = $qs - pr$

$$\int_r^s u \, dv + \int_q^p v \, du = \left[uv\right]_{(p,r)}^{(q,s)}$$

$$\int_r^s u \, dv = \left[uv\right]_{(p,r)}^{(q,s)} - \int_q^p v \, du$$

Explain how this graph proves the theorem. Which notation in this proof is unfamiliar? What do you think it means?

EXAMPLE 1 Integration by Parts

Find $\int xe^x \, dx$.

Solution To apply integration by parts, you need to write the integral in the form $\int u \, dv$. There are several ways to do this.

$$\int \underbrace{(x)}_{u} \underbrace{(e^x \, dx)}_{dv}, \quad \int \underbrace{(e^x)}_{u} \underbrace{(x \, dx)}_{dv}, \quad \int \underbrace{(1)}_{u} \underbrace{(xe^x \, dx)}_{dv}, \quad \int \underbrace{(xe^x)}_{u} \underbrace{(dx)}_{dv}$$

The guidelines on page 525 suggest choosing the first option because the derivative of $u = x$ is simpler than x, and $dv = e^x \, dx$ is the most complicated portion of the integrand that fits a basic integration formula.

$$dv = e^x \, dx \quad \Longrightarrow \quad v = \int dv = \int e^x \, dx = e^x$$

$$u = x \quad \Longrightarrow \quad du = dx$$

NOTE In Example 1, note that it is not necessary to include a constant of integration when solving

$$v = \int e^x \, dx = e^x + C_1.$$

To illustrate this, replace $v = e^x$ by $v = e^x + C_1$ and apply integration by parts to see that you obtain the same result.

Now, integration by parts produces

$$\int u \, dv = uv - \int v \, du \qquad \text{Integration by parts formula}$$

$$\int xe^x \, dx = xe^x - \int e^x \, dx \qquad \text{Substitute.}$$

$$= xe^x - e^x + C. \qquad \text{Integrate.}$$

To check this, differentiate $xe^x - e^x + C$ to see that you obtain the original integrand.

EXAMPLE 2 Integration by Parts

Find $\int x^2 \ln x \, dx$.

Solution In this case, x^2 is more easily integrated than $\ln x$. Furthermore, the derivative of $\ln x$ is simpler than $\ln x$. So, you should let $dv = x^2 \, dx$.

$$dv = x^2 \, dx \quad \Longrightarrow \quad v = \int x^2 \, dx = \frac{x^3}{3}$$

$$u = \ln x \quad \Longrightarrow \quad du = \frac{1}{x} \, dx$$

FOR FURTHER INFORMATION To see how integration by parts is used to prove Stirling's approximation

$$\ln(n!) = n \ln n - n$$

see the article "The Validity of Stirling's Approximation: A Physical Chemistry Project" by A. S. Wallner and K. A. Brandt in *Journal of Chemical Education.*

Integration by parts produces

$$\int u \, dv = uv - \int v \, du \qquad \text{Integration by parts formula}$$

$$\int x^2 \ln x \, dx = \frac{x^3}{3} \ln x - \int \left(\frac{x^3}{3}\right)\left(\frac{1}{x}\right) dx \qquad \text{Substitute.}$$

$$= \frac{x^3}{3} \ln x - \frac{1}{3} \int x^2 \, dx \qquad \text{Simplify.}$$

$$= \frac{x^3}{3} \ln x - \frac{x^3}{9} + C. \qquad \text{Integrate.}$$

You can check this result by differentiating.

$$\frac{d}{dx}\left[\frac{x^3}{3} \ln x - \frac{x^3}{9}\right] = \frac{x^3}{3}\left(\frac{1}{x}\right) + (\ln x)(x^2) - \frac{x^2}{3} = x^2 \ln x$$

TECHNOLOGY Try graphing

$$\int x^2 \ln x \, dx \quad \text{and} \quad \frac{x^3}{3} \ln x - \frac{x^3}{9}$$

on your graphing utility. Do you get the same graph? (This will take a while, so be patient.)

One surprising application of integration by parts involves integrands consisting of a single term, such as $\int \ln x \, dx$ or $\int \arcsin x \, dx$. In these cases, try letting $dv = dx$, as shown in the next example.

EXAMPLE 3   An Integrand with a Single Term

Evaluate $\displaystyle\int_0^1 \arcsin x \, dx$.

Solution   Let $dv = dx$.

$$dv = dx \qquad \Longrightarrow \qquad v = \int dx = x$$

$$u = \arcsin x \qquad \Longrightarrow \qquad du = \frac{1}{\sqrt{1 - x^2}} \, dx$$

Integration by parts now produces

$$\int u \, dv = uv - \int v \, du \qquad \text{Integration by parts formula}$$

$$\int \arcsin x \, dx = x \arcsin x - \int \frac{x}{\sqrt{1 - x^2}} \, dx \qquad \text{Substitute.}$$

$$= x \arcsin x + \frac{1}{2} \int (1 - x^2)^{-1/2} \, (-2x) \, dx \qquad \text{Rewrite.}$$

$$= x \arcsin x + \sqrt{1 - x^2} + C. \qquad \text{Integrate.}$$

Using this antiderivative, you can evaluate the definite integral as follows.

$$\int_0^1 \arcsin x \, dx = \left[x \arcsin x + \sqrt{1 - x^2} \right]_0^1$$

$$= \frac{\pi}{2} - 1$$

$$\approx 0.571$$

The area represented by this definite integral is shown in Figure 8.2.

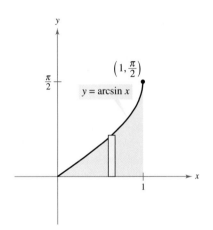

$\left(1, \dfrac{\pi}{2}\right)$

$y = \arcsin x$

The area of the region is approximately 0.571.

Figure 8.2

TECHNOLOGY   Remember that there are two ways to use technology to evaluate a definite integral: (1) you can use a numerical approximation such as the Trapezoidal Rule or Simpson's Rule, or (2) you can use a computer algebra system to find the antiderivative and then apply the Fundamental Theorem of Calculus. Both methods have shortcomings. To find the possible error when using a numerical method, the integrand must have a second derivative (Trapezoidal Rule) or a fourth derivative (Simpson's Rule) in the interval of integration: the integrand in Example 3 fails to meet either of these requirements. To apply the Fundamental Theorem of Calculus, the symbolic integration utility must be able to find the antiderivative.

Which method would you use to evaluate

$$\int_0^1 \arctan x \, dx?$$

Which method would you use to evaluate

$$\int_0^1 \arctan x^2 \, dx?$$

Some integrals require repeated use of the integration by parts formula.

EXAMPLE 4 Repeated Use of Integration by Parts

Find $\int x^2 \sin x \, dx$.

Solution   The factors x^2 and $\sin x$ are equally easy to integrate. However, the derivative of x^2 becomes simpler, whereas the derivative of $\sin x$ does not. So, you should let $u = x^2$.

$$dv = \sin x \, dx \quad \Longrightarrow \quad v = \int \sin x \, dx = -\cos x$$

$$u = x^2 \quad \Longrightarrow \quad du = 2x \, dx$$

Now, integration by parts produces

$$\int x^2 \sin x \, dx = -x^2 \cos x + \int 2x \cos x \, dx. \qquad \text{First use of integration by parts}$$

This first use of integration by parts has succeeded in simplifying the original integral, but the integral on the right still doesn't fit a basic integration rule. To evaluate that integral, you can apply integration by parts again. This time, let $u = 2x$.

$$dv = \cos x \, dx \quad \Longrightarrow \quad v = \int \cos x \, dx = \sin x$$

$$u = 2x \quad \Longrightarrow \quad du = 2 \, dx$$

Now, integration by parts produces

$$\int 2x \cos x \, dx = 2x \sin x - \int 2 \sin x \, dx \qquad \text{Second use of integration by parts}$$

$$= 2x \sin x + 2 \cos x + C.$$

Combining these two results, you can write

$$\int x^2 \sin x \, dx = -x^2 \cos x + 2x \sin x + 2 \cos x + C.$$

When making repeated applications of integration by parts, you need to be careful not to interchange the substitutions in successive applications. For instance, in Example 4, the first substitution was $u = x^2$ and $dv = \sin x \, dx$. If, in the second application, you had switched the substitution to $u = \cos x$ and $dv = 2x$, you would have obtained

$$\int x^2 \sin x \, dx = -x^2 \cos x + \int 2x \cos x \, dx$$

$$= -x^2 \cos x + x^2 \cos x + \int x^2 \sin x \, dx$$

$$= \int x^2 \sin x \, dx$$

EXPLORATION

Try to find

$$\int e^x \cos 2x \, dx$$

by letting $u = \cos 2x$ and $dv = e^x \, dx$ in the first substitution. For the second substitution, let $u = \sin 2x$ and $dv = e^x \, dx$.

thereby undoing the previous integration and returning to the *original* integral. When making repeated applications of integration by parts, you should also watch for the appearance of a *constant multiple* of the original integral. For instance, this occurs when you use integration by parts to evaluate $\int e^x \cos 2x \, dx$, and also occurs in the next example.

EXAMPLE 5   Integration by Parts

NOTE The integral in Example 5 is an important one. In Section 8.4 (Example 5), you will see that it is used to find the arc length of a parabolic segment.

Find $\displaystyle\int \sec^3 x \, dx$.

Solution   The most complicated portion of the integrand that can be easily integrated is $\sec^2 x$, so you should let $dv = \sec^2 x \, dx$ and $u = \sec x$.

$$dv = \sec^2 x \, dx \quad \Longrightarrow \quad v = \int \sec^2 x \, dx = \tan x$$

$$u = \sec x \quad \Longrightarrow \quad du = \sec x \tan x \, dx$$

Integration by parts produces

$$\int u \, dv = uv - \int v \, du \qquad \text{Integration by parts formula}$$

$$\int \sec^3 x \, dx = \sec x \tan x - \int \sec x \tan^2 x \, dx \qquad \text{Substitute.}$$

$$\int \sec^3 x \, dx = \sec x \tan x - \int \sec x (\sec^2 x - 1) \, dx \qquad \text{Trigonometric identity}$$

$$\int \sec^3 x \, dx = \sec x \tan x - \int \sec^3 x \, dx + \int \sec x \, dx \qquad \text{Rewrite.}$$

$$2\int \sec^3 x \, dx = \sec x \tan x + \int \sec x \, dx \qquad \text{Collect like integrals.}$$

$$\int \sec^3 x \, dx = \frac{1}{2} \sec x \tan x + \frac{1}{2} \ln|\sec x + \tan x| + C. \qquad \text{Integrate and divide by 2.}$$

STUDY TIP The trigonometric identities

$$\sin^2 x = \frac{1 - \cos 2x}{2}$$

$$\cos^2 x = \frac{1 + \cos 2x}{2}$$

play an important role in this chapter.

EXAMPLE 6   Finding a Centroid

A machine part is modeled by the region bounded by the graph of $y = \sin x$ and the x-axis, $0 \le x \le \pi/2$, as shown in Figure 8.3. Find the centroid of this region.

Solution   Begin by finding the area of the region.

$$A = \int_0^{\pi/2} \sin x \, dx = \left[-\cos x \right]_0^{\pi/2} = 1$$

Now, you can find the coordinates of the centroid as follows.

$$\bar{y} = \frac{1}{A} \int_0^{\pi/2} \frac{\sin x}{2} (\sin x) \, dx = \frac{1}{4} \int_0^{\pi/2} (1 - \cos 2x) \, dx = \frac{1}{4}\left[x - \frac{\sin 2x}{2} \right]_0^{\pi/2} = \frac{\pi}{8}$$

You can evaluate the integral for $\bar{x}$, $(1/A)\int_0^{\pi/2} x \sin x \, dx$, with integration by parts. To do this, let $dv = \sin x \, dx$ and $u = x$. This produces $v = -\cos x$ and $du = dx$, and you can write

$$\int x \sin x \, dx = -x \cos x + \int \cos x \, dx$$

$$= -x \cos x + \sin x + C.$$

Finally, you can determine $\bar{x}$ to be

$$\bar{x} = \frac{1}{A} \int_0^{\pi/2} x \sin x \, dx = \left[-x \cos x + \sin x \right]_0^{\pi/2} = 1.$$

So, the centroid of the region is $(1, \pi/8)$.

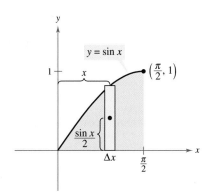

Figure 8.3

As you gain experience in using integration by parts, your skill in determining u and dv will increase. The following summary lists several common integrals with suggestions for the choices of u and dv.

Summary of Common Integrals Using Integration by Parts

1. For integrals of the form

$$\int x^n e^{ax}\, dx, \qquad \int x^n \sin ax\, dx, \qquad \text{or} \qquad \int x^n \cos ax\, dx$$

let $u = x^n$ and let $dv = e^{ax}\, dx$, $\sin ax\, dx$, or $\cos ax\, dx$.

2. For integrals of the form

$$\int x^n \ln x\, dx, \qquad \int x^n \arcsin ax\, dx, \qquad \text{or} \qquad \int x^n \arctan ax\, dx$$

let $u = \ln x$, $\arcsin ax$, or $\arctan ax$ and let $dv = x^n\, dx$.

3. For integrals of the form

$$\int e^{ax} \sin bx\, dx \qquad \text{or} \qquad \int e^{ax} \cos bx\, dx$$

let $u = \sin bx$ or $\cos bx$ and let $dv = e^{ax}\, dx$.

Tabular Method

In problems involving repeated applications of integration by parts, a tabular method, illustrated in Example 7, can help to organize the work. This method works well for integrals of the form $\int x^n \sin ax\, dx$, $\int x^n \cos ax\, dx$, and $\int x^n e^{ax}\, dx$.

 EXAMPLE 7   Using the Tabular Method

Find $\displaystyle\int x^2 \sin 4x\, dx$.

Solution   Begin as usual by letting $u = x^2$ and $dv = v'\, dx = \sin 4x\, dx$. Next, create a table consisting of three columns, as shown.

Alternate Signs	u and Its Derivatives	v' and Its Antiderivatives
$+$	x^2	$\sin 4x$
$-$	$2x$	$-\frac{1}{4}\cos 4x$
$+$	2	$-\frac{1}{16}\sin 4x$
$-$	0	$\frac{1}{64}\cos 4x$

Differentiate until you obtain 0 as a derivative.

The solution is obtained by adding the signed products of the diagonal entries:

$$\int x^2 \sin 4x\, dx = -\frac{1}{4}x^2 \cos 4x + \frac{1}{8}x \sin 4x + \frac{1}{32}\cos 4x + C.$$

In Exercises 1–4, match the antiderivative with the correct integral. [Integrals are labeled (a), (b), (c), and (d).]

(a) $\int \ln x \, dx$ (b) $\int x \sin x \, dx$

(c) $\int x^2 e^x \, dx$ (d) $\int x^2 \cos x \, dx$

1. $y = \sin x - x \cos x$

2. $y = x^2 \sin x + 2x \cos x - 2 \sin x$

3. $y = x^2 e^x - 2xe^x + 2e^x$

4. $y = -x + x \ln x$

In Exercises 5–10, identify u and dv for finding the integral using integration by parts. (Do not evaluate the integral.)

5. $\int xe^{2x} \, dx$ 6. $\int x^2 e^{2x} \, dx$

7. $\int (\ln x)^2 \, dx$ 8. $\int \ln 3x \, dx$

9. $\int x \sec^2 x \, dx$ 10. $\int x^2 \cos x \, dx$

In Exercises 11–36, find the integral. (*Note:* Solve by the simplest method—not all require integration by parts.)

11. $\int xe^{-2x} \, dx$ 12. $\int \dfrac{2x}{e^x} \, dx$

13. $\int x^3 e^x \, dx$ 14. $\int \dfrac{e^{1/t}}{t^2} \, dt$

15. $\int x^2 e^{x^3} \, dx$ 16. $\int x^4 \ln x \, dx$

17. $\int t \ln(t + 1) \, dt$ 18. $\int \dfrac{1}{x(\ln x)^3} \, dx$

19. $\int \dfrac{(\ln x)^2}{x} \, dx$ 20. $\int \dfrac{\ln x}{x^2} \, dx$

21. $\int \dfrac{xe^{2x}}{(2x + 1)^2} \, dx$ 22. $\int \dfrac{x^3 e^{x^2}}{(x^2 + 1)^2} \, dx$

23. $\int (x^2 - 1)e^x \, dx$ 24. $\int \dfrac{\ln 2x}{x^2} \, dx$

25. $\int x\sqrt{x - 1} \, dx$ 26. $\int \dfrac{x}{\sqrt{2 + 3x}} \, dx$

27. $\int x \cos x \, dx$ 28. $\int x \sin x \, dx$

29. $\int x^3 \sin x \, dx$ 30. $\int x^2 \cos x \, dx$

31. $\int t \csc t \cot t \, dt$ 32. $\int \theta \sec \theta \tan \theta \, d\theta$

33. $\int \arctan x \, dx$ 34. $\int 4 \arccos x \, dx$

35. $\int e^{2x} \sin x \, dx$ 36. $\int e^x \cos 2x \, dx$

In Exercises 37–42, solve the differential equation.

37. $y' = xe^{x^2}$ 38. $y' = \ln x$

39. $\dfrac{dy}{dt} = \dfrac{t^2}{\sqrt{2 + 3t}}$ 40. $\dfrac{dy}{dx} = x^2 \sqrt{x - 1}$

41. $(\cos y)y' = 2x$ 42. $y' = \arctan \dfrac{x}{2}$

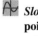 *Slope Fields*   **In Exercises 43 and 44, a differential equation, a point, and a slope field are given. (a) Sketch two approximate solutions of the differential equation on the slope field, one of which passes through the given point. (b) Use integration to find the particular solution of the differential equation and use a graphing utility to graph the solution. Compare the result with the sketches in part (a). To print an enlarged copy of the graph, go to the website *www.mathgraphs.com*.**

43. $\dfrac{dy}{dx} = x\sqrt{y} \cos x, \ (0, 4)$ 44. $\dfrac{dy}{dx} = e^{-x/3} \sin 2x, \ \left(0, -\dfrac{18}{37}\right)$

 Slope Fields   **In Exercises 45 and 46, use a computer algebra system to graph the slope field for the differential equation and graph the solution through the specified initial condition.**

45. $\dfrac{dy}{dx} = \dfrac{x}{y}e^{x/8}$ 46. $\dfrac{dy}{dx} = \dfrac{x}{y} \sin x$
    $y(0) = 2$ $y(0) = 4$

In Exercises 47–58, evaluate the definite integral. Use a graphing utility to confirm your result.

47. $\displaystyle\int_0^4 xe^{-x/2} \, dx$ 48. $\displaystyle\int_0^1 x^2 e^x \, dx$

49. $\displaystyle\int_0^{\pi/2} x \cos x \, dx$ 50. $\displaystyle\int_0^{\pi} x \sin 2x \, dx$

51. $\displaystyle\int_0^{1/2} \arccos x \, dx$ 52. $\displaystyle\int_0^1 x \arcsin x^2 \, dx$

53. $\displaystyle\int_0^1 e^x \sin x \, dx$ 54. $\displaystyle\int_0^2 e^{-x} \cos x \, dx$

55. $\displaystyle\int_1^2 x^2 \ln x \, dx$ 56. $\displaystyle\int_0^1 \ln(1 + x^2) \, dx$

57. $\displaystyle\int_2^4 x \, \text{arcsec} \, x \, dx$ 58. $\displaystyle\int_0^{\pi/4} x \sec^2 x \, dx$

In Exercises 59–64, use the tabular method to find the integral.

59. $\int x^2 e^{2x}\,dx$ **60.** $\int x^3 e^{-2x}\,dx$

61. $\int x^3 \sin x\,dx$ **62.** $\int x^3 \cos 2x\,dx$

63. $\int x \sec^2 x\,dx$ **64.** $\int x^2(x-2)^{3/2}\,dx$

In Exercises 65–70, find or evaluate the integral using substitution first, then using integration by parts.

65. $\int \sin \sqrt{x}\,dx$ **66.** $\int 2x^3 \cos x^2\,dx$

67. $\int_0^4 x\sqrt{4-x}\,dx$ **68.** $\int_0^2 e^{\sqrt{2x}}\,dx$

69. $\int \cos(\ln x)\,dx$ **70.** $\int \ln(x^2+1)\,dx$

Writing About Concepts

71. Integration by parts is based on what differentiation rule? Explain.

72. In your own words, state guidelines for integration by parts.

In Exercises 73–78, state whether you would use integration by parts to evaluate the integral. If so, identify what you would use for u and dv. Explain your reasoning.

73. $\int \dfrac{\ln x}{x}\,dx$ **74.** $\int x \ln x\,dx$

75. $\int x^2 e^{2x}\,dx$ **76.** $\int 2x\,e^{x^2}\,dx$

77. $\int \dfrac{x}{\sqrt{x+1}}\,dx$ **78.** $\int \dfrac{x}{\sqrt{x^2+1}}\,dx$

In Exercises 79–82, use a computer algebra system to (a) find or evaluate the integral and (b) graph two antiderivatives. (c) Describe the relationship between the graphs of the antiderivatives.

79. $\int t^3 e^{-4t}\,dt$ **80.** $\int \alpha^4 \sin \pi\alpha\,d\alpha$

81. $\int_0^{\pi/2} e^{-2x} \sin 3x\,dx$ **82.** $\int_0^5 x^4(25-x^2)^{3/2}\,dx$

83. Integrate $\int 2x\sqrt{2x-3}\,dx$

(a) by parts, letting $dv = \sqrt{2x-3}\,dx$.

(b) by substitution, letting $u = 2x-3$.

84. Integrate $\int x\sqrt{4+x}\,dx$

(a) by parts, letting $dv = \sqrt{4+x}\,dx$.

(b) by substitution, letting $u = 4+x$.

85. Integrate $\int \dfrac{x^3}{\sqrt{4+x^2}}\,dx$

(a) by parts, letting $dv = \left(x/\sqrt{4+x^2}\right)dx$.

(b) by substitution, letting $u = 4+x^2$.

86. Integrate $\int x\sqrt{4-x}\,dx$

(a) by parts, letting $dv = \sqrt{4-x}\,dx$.

(b) by substitution, letting $u = 4-x$.

In Exercises 87 and 88, use a computer algebra system to find the integral for $n = 0, 1, 2,$ and 3. Use the result to obtain a general rule for the integral for any positive integer n and test your results for $n = 4$.

87. $\int x^n \ln x\,dx$ **88.** $\int x^n e^x\,dx$

In Exercises 89–94, use integration by parts to verify the formula. (For Exercises 89–92, assume that n is a positive integer.)

89. $\int x^n \sin x\,dx = -x^n \cos x + n\int x^{n-1} \cos x\,dx$

90. $\int x^n \cos x\,dx = x^n \sin x - n\int x^{n-1} \sin x\,dx$

91. $\int x^n \ln x\,dx = \dfrac{x^{n+1}}{(n+1)^2}\big[-1 + (n+1)\ln x\big] + C$

92. $\int x^n e^{ax}\,dx = \dfrac{x^n e^{ax}}{a} - \dfrac{n}{a}\int x^{n-1} e^{ax}\,dx$

93. $\int e^{ax} \sin bx\,dx = \dfrac{e^{ax}(a \sin bx - b \cos bx)}{a^2+b^2} + C$

94. $\int e^{ax} \cos bx\,dx = \dfrac{e^{ax}(a \cos bx + b \sin bx)}{a^2+b^2} + C$

In Exercises 95–98, find the integral by using the appropriate formula from Exercises 89–94.

95. $\int x^3 \ln x\,dx$

96. $\int x^2 \cos x\,dx$

97. $\int e^{2x} \cos 3x\,dx$

98. $\int x^3 e^{2x}\,dx$

Area In Exercises 99–102, use a graphing utility to graph the region bounded by the graphs of the equations, and find the area of the region.

99. $y = xe^{-x}, y = 0, x = 4$

100. $y = \frac{1}{9} xe^{-x/3}, y = 0, x = 0, x = 3$

101. $y = e^{-x} \sin \pi x, y = 0, x = 0, x = 1$

102. $y = x \sin x, y = 0, x = 0, x = \pi$

103. *Area, Volume, and Centroid*   Given the region bounded by the graphs of $y = \ln x$, $y = 0$, and $x = e$, find

(a) the area of the region.

(b) the volume of the solid generated by revolving the region about the x-axis.

(c) the volume of the solid generated by revolving the region about the y-axis.

(d) the centroid of the region.

104. *Volume and Centroid*   Given the region bounded by the graphs of $y = x \sin x$, $y = 0$, $x = 0$, and $x = \pi$, find

(a) the volume of the solid generated by revolving the region about the x-axis.

(b) the volume of the solid generated by revolving the region about the y-axis.

(c) the centroid of the region.

105. *Centroid*   Find the centroid of the region bounded by the graphs of $y = \arcsin x$, $x = 0$, and $y = \pi/2$. How is this problem related to Example 6 in this section?

106. *Centroid*   Find the centroid of the region bounded by the graphs of $f(x) = x^2$, $g(x) = 2^x$, $x = 2$, and $x = 4$.

107. *Average Displacement*   A damping force affects the vibration of a spring so that the displacement of the spring is given by $y = e^{-4t}(\cos 2t + 5 \sin 2t)$. Find the average value of y on the interval from $t = 0$ to $t = \pi$.

108. *Memory Model*   A model for the ability M of a child to memorize, measured on a scale from 0 to 10, is given by $M = 1 + 1.6t \ln t$, $0 < t \le 4$, where t is the child's age in years. Find the average value of this model

(a) between the child's first and second birthdays.

(b) between the child's third and fourth birthdays.

Present Value   **In Exercises 109 and 110, find the present value P of a continuous income flow of $c(t)$ dollars per year if**

$$P = \int_0^{t_1} c(t) e^{-rt} \, dt$$

where t_1 is the time in years and r is the annual interest rate compounded continuously.

109. $c(t) = 100{,}000 + 4000t$, $r = 5\%$, $t_1 = 10$

110. $c(t) = 30{,}000 + 500t$, $r = 7\%$, $t_1 = 5$

Integrals Used to Find Fourier Coefficients   **In Exercises 111 and 112, verify the value of the definite integral, where n is a positive integer.**

111. $\displaystyle\int_{-\pi}^{\pi} x \sin nx \, dx = \begin{cases} \dfrac{2\pi}{n}, & n \text{ is odd} \\[3mm] -\dfrac{2\pi}{n}, & n \text{ is even} \end{cases}$

112. $\displaystyle\int_{-\pi}^{\pi} x^2 \cos nx \, dx = \dfrac{(-1)^n 4\pi}{n^2}$

113. *Vibrating String*   A string stretched between the two points $(0, 0)$ and $(2, 0)$ is plucked by displacing the string h units at its midpoint. The motion of the string is modeled by a **Fourier Sine Series** whose coefficients are given by

$$b_n = h \int_0^1 x \sin \frac{n\pi x}{2} \, dx + h \int_1^2 (-x + 2) \sin \frac{n\pi x}{2} \, dx.$$

Find b_n.

114. Find the fallacy in the following argument that $0 = 1$.

$$dv = dx \quad \Longrightarrow \quad v = \int dx = x$$

$$u = \frac{1}{x} \quad \Longrightarrow \quad du = -\frac{1}{x^2} \, dx$$

$$0 + \int \frac{dx}{x} = \left(\frac{1}{x}\right)(x) - \int \left(-\frac{1}{x^2}\right)(x) \, dx = 1 + \int \frac{dx}{x}$$

So, $0 = 1$.

115. Let $y = f(x)$ be positive and strictly increasing on the interval $0 < a \le x \le b$. Consider the region R bounded by the graphs of $y = f(x)$, $y = 0$, $x = a$, and $x = b$. If R is revolved about the y-axis, show that the disk method and shell method yield the same volume.

116. *Euler's Method*   Consider the differential equation $f'(x) = xe^{-x}$ with the initial condition $f(0) = 0$.

(a) Use integration to solve the differential equation.

(b) Use a graphing utility to graph the solution of the differential equation.

(c) Use Euler's Method with $h = 0.05$, and the recursive capabilities of a graphing utility, to generate the first 80 points of the graph of the approximate solution. Use the graphing utility to plot the points. Compare the result with the graph in part (b).

(d) Repeat part (c) using $h = 0.1$ and generate the first 40 points.

(e) Why is the result in part (c) a better approximation of the solution than the result in part (d)?

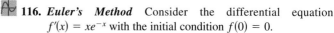

Euler's Method   **In Exercises 117 and 118, consider the differential equation and repeat parts (a)–(d) of Exercise 116.**

117. $f'(x) = 3x \sin(2x)$
$\quad\; f(0) = 0$

118. $f'(x) = \cos \sqrt{x}$
$\quad\;\; f(0) = 1$

119. *Think About It*   Give a geometric explanation to explain why

$$\int_0^{\pi/2} x \sin x \, dx \le \int_0^{\pi/2} x \, dx.$$

Verify the inequality by evaluating the integrals.

120. *Finding a Pattern*   Find the area bounded by the graphs of $y = x \sin x$ and $y = 0$ over each interval.

(a) $[0, \pi]$ (b) $[\pi, 2\pi]$ (c) $[2\pi, 3\pi]$

Describe any patterns that you notice. What is the area between the graphs of $y = x \sin x$ and $y = 0$ over the interval $[n\pi, (n + 1)\pi]$, where n is any nonnegative integer? Explain.

| Section 8.3 | **Trigonometric Integrals** |

- Solve trigonometric integrals involving powers of sine and cosine.
- Solve trigonometric integrals involving powers of secant and tangent.
- Solve trigonometric integrals involving sine-cosine products with different angles.

Integrals Involving Powers of Sine and Cosine

SHEILA SCOTT MACINTYRE (1910–1960)

Sheila Scott Macintyre published her first paper on the asymptotic periods of integral functions in 1935. She completed her doctorate work at Aberdeen University, where she taught. In 1958 she accepted a visiting research fellowship at the University of Cincinnati.

In this section you will study techniques for evaluating integrals of the form

$$\int \sin^m x \cos^n x \, dx \qquad \text{and} \qquad \int \sec^m x \tan^n x \, dx$$

where either m or n is a positive integer. To find antiderivatives for these forms, try to break them into combinations of trigonometric integrals to which you can apply the Power Rule.

For instance, you can evaluate $\int \sin^5 x \cos x \, dx$ with the Power Rule by letting $u = \sin x$. Then, $du = \cos x \, dx$ and you have

$$\int \sin^5 x \cos x \, dx = \int u^5 \, du = \frac{u^6}{6} + C = \frac{\sin^6 x}{6} + C.$$

To break up $\int \sin^m x \cos^n x \, dx$ into forms to which you can apply the Power Rule, use the following identities.

$$\sin^2 x + \cos^2 x = 1 \qquad \text{Pythagorean identity}$$

$$\sin^2 x = \frac{1 - \cos 2x}{2} \qquad \text{Half-angle identity for } \sin^2 x$$

$$\cos^2 x = \frac{1 + \cos 2x}{2} \qquad \text{Half-angle identity for } \cos^2 x$$

Guidelines for Evaluating Integrals Involving Sine and Cosine

1. If the power of the sine is odd and positive, save one sine factor and convert the remaining factors to cosines. Then, expand and integrate.

$$\int \underbrace{\sin^{2k+1} x}_{\text{Odd}} \cos^n x \, dx = \int \underbrace{(\sin^2 x)^k}_{\text{Convert to cosines}} \cos^n x \underbrace{\sin x \, dx}_{\text{Save for } du} = \int (1 - \cos^2 x)^k \cos^n x \sin x \, dx$$

2. If the power of the cosine is odd and positive, save one cosine factor and convert the remaining factors to sines. Then, expand and integrate.

$$\int \sin^m x \underbrace{\cos^{2k+1} x}_{\text{Odd}} \, dx = \int \sin^m x \underbrace{(\cos^2 x)^k}_{\text{Convert to sines}} \underbrace{\cos x \, dx}_{\text{Save for } du} = \int \sin^m x \, (1 - \sin^2 x)^k \cos x \, dx$$

3. If the powers of both the sine and cosine are even and nonnegative, make repeated use of the identities

$$\sin^2 x = \frac{1 - \cos 2x}{2} \qquad \text{and} \qquad \cos^2 x = \frac{1 + \cos 2x}{2}$$

to convert the integrand to odd powers of the cosine. Then proceed as in guideline 2.

TECHNOLOGY Use a computer algebra system to find the integral in Example 1. You should obtain

$$\int \sin^3 x \cos^4 x \, dx =$$

$$-\cos^5 x \left(\frac{1}{7} \sin^2 x + \frac{2}{35} \right) + C.$$

Is this equivalent to the result obtained in Example 1?

EXAMPLE 1 **Power of Sine Is Odd and Positive**

Find $\displaystyle\int \sin^3 x \cos^4 x \, dx$.

Solution Because you expect to use the Power Rule with $u = \cos x$, *save one sine factor* to form du and convert the remaining sine factors to cosines.

$$\int \sin^3 x \cos^4 x \, dx = \int \sin^2 x \cos^4 x (\sin x) \, dx \qquad \text{Rewrite.}$$

$$= \int (1 - \cos^2 x) \cos^4 x \sin x \, dx \qquad \text{Trigonometric identity}$$

$$= \int (\cos^4 x - \cos^6 x) \sin x \, dx \qquad \text{Multiply.}$$

$$= \int \cos^4 x \sin x \, dx - \int \cos^6 x \sin x \, dx \qquad \text{Rewrite.}$$

$$= -\int \cos^4 x (-\sin x) \, dx + \int \cos^6 x (-\sin x) \, dx$$

$$= -\frac{\cos^5 x}{5} + \frac{\cos^7 x}{7} + C \qquad \text{Integrate.}$$

In Example 1, *both* of the powers m and n happened to be positive integers. However, the same strategy will work as long as either m or n is odd and positive. For instance, in the next example the power of the cosine is 3, but the power of the sine is $-\frac{1}{2}$.

EXAMPLE 2 **Power of Cosine Is Odd and Positive**

Evaluate $\displaystyle\int_{\pi/6}^{\pi/3} \frac{\cos^3 x}{\sqrt{\sin x}} \, dx$.

Solution Because you expect to use the Power Rule with $u = \sin x$, *save one cosine factor* to form du and convert the remaining cosine factors to sines.

$$\int_{\pi/6}^{\pi/3} \frac{\cos^3 x}{\sqrt{\sin x}} \, dx = \int_{\pi/6}^{\pi/3} \frac{\cos^2 x \cos x}{\sqrt{\sin x}} \, dx$$

$$= \int_{\pi/6}^{\pi/3} \frac{(1 - \sin^2 x)(\cos x)}{\sqrt{\sin x}} \, dx$$

$$= \int_{\pi/6}^{\pi/3} [(\sin x)^{-1/2} \cos x - (\sin x)^{3/2} \cos x] \, dx$$

$$= \left[\frac{(\sin x)^{1/2}}{1/2} - \frac{(\sin x)^{5/2}}{5/2} \right]_{\pi/6}^{\pi/3}$$

$$= 2 \left(\frac{\sqrt{3}}{2} \right)^{1/2} - \frac{2}{5} \left(\frac{\sqrt{3}}{2} \right)^{5/2} - \sqrt{2} + \frac{\sqrt{32}}{80}$$

$$\approx 0.239$$

Figure 8.4 shows the region whose area is represented by this integral.

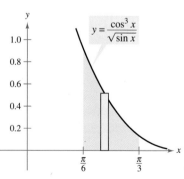

$$y = \frac{\cos^3 x}{\sqrt{\sin x}}$$

The area of the region is approximately 0.239.

Figure 8.4

EXAMPLE 3 **Power of Cosine Is Even and Nonnegative**

Find $\int \cos^4 x \, dx$.

Solution Because m and n are both even and nonnegative ($m = 0$), you can replace $\cos^4 x$ by $[(1 + \cos 2x)/2]^2$.

$$
\begin{aligned}
\int \cos^4 x \, dx &= \int \left(\frac{1 + \cos 2x}{2} \right)^2 dx \\
&= \int \left(\frac{1}{4} + \frac{\cos 2x}{2} + \frac{\cos^2 2x}{4} \right) dx \\
&= \int \left[\frac{1}{4} + \frac{\cos 2x}{2} + \frac{1}{4} \left(\frac{1 + \cos 4x}{2} \right) \right] dx \\
&= \frac{3}{8} \int dx + \frac{1}{4} \int 2 \cos 2x \, dx + \frac{1}{32} \int 4 \cos 4x \, dx \\
&= \frac{3x}{8} + \frac{\sin 2x}{4} + \frac{\sin 4x}{32} + C
\end{aligned}
$$

Use a symbolic differentiation utility to verify this. Can you simplify the derivative to obtain the original integrand? ▬▬▬▬▬

In Example 3, if you were to evaluate the definite integral from 0 to $\pi/2$, you would obtain

$$
\begin{aligned}
\int_0^{\pi/2} \cos^4 x \, dx &= \left[\frac{3x}{8} + \frac{\sin 2x}{4} + \frac{\sin 4x}{32} \right]_0^{\pi/2} \\
&= \left(\frac{3\pi}{16} + 0 + 0 \right) - (0 + 0 + 0) \\
&= \frac{3\pi}{16}.
\end{aligned}
$$

Note that the only term that contributes to the solution is $3x/8$. This observation is generalized in the following formulas developed by John Wallis.

JOHN WALLIS (1616–1703)

Wallis did much of his work in calculus prior to Newton and Leibniz, and he influenced the thinking of both of these men. Wallis is also credited with introducing the present symbol (∞) for infinity.

Wallis's Formulas

1. If n is odd ($n \geq 3$), then

$$
\int_0^{\pi/2} \cos^n x \, dx = \left(\frac{2}{3} \right) \left(\frac{4}{5} \right) \left(\frac{6}{7} \right) \cdots \left(\frac{n-1}{n} \right).
$$

2. If n is even ($n \geq 2$), then

$$
\int_0^{\pi/2} \cos^n x \, dx = \left(\frac{1}{2} \right) \left(\frac{3}{4} \right) \left(\frac{5}{6} \right) \cdots \left(\frac{n-1}{n} \right) \left(\frac{\pi}{2} \right).
$$

These formulas are also valid if $\cos^n x$ is replaced by $\sin^n x$. (You are asked to prove both formulas in Exercise 104.)

Integrals Involving Powers of Secant and Tangent

The following guidelines can help you evaluate integrals of the form

$$\int \sec^m x \tan^n x \, dx.$$

Guidelines for Evaluating Integrals Involving Secant and Tangent

1. If the power of the secant is even and positive, save a secant-squared factor and convert the remaining factors to tangents. Then expand and integrate.

$$\int \underbrace{\sec^{2k} x}_{\text{Even}} \tan^n x \, dx = \int \overbrace{(\sec^2 x)^{k-1}}^{\text{Convert to tangents}} \tan^n x \, \overbrace{\sec^2 x}^{\text{Save for } du} \, dx = \int (1 + \tan^2 x)^{k-1} \tan^n x \sec^2 x \, dx$$

2. If the power of the tangent is odd and positive, save a secant-tangent factor and convert the remaining factors to secants. Then expand and integrate.

$$\int \sec^m x \underbrace{\tan^{2k+1} x}_{\text{Odd}} \, dx = \int \sec^{m-1} x \overbrace{(\tan^2 x)^k}^{\text{Convert to secants}} \overbrace{\sec x \tan x}^{\text{Save for } du} \, dx = \int \sec^{m-1} x (\sec^2 x - 1)^k \sec x \tan x \, dx$$

3. If there are no secant factors and the power of the tangent is even and positive, convert a tangent-squared factor to a secant-squared factor, then expand and repeat if necessary.

$$\int \tan^n x \, dx = \int \tan^{n-2} x \overbrace{(\tan^2 x)}^{\text{Convert to secants}} \, dx = \int \tan^{n-2} x (\sec^2 x - 1) \, dx$$

4. If the integral is of the form $\int \sec^m x \, dx$, where m is odd and positive, use integration by parts, as illustrated in Example 5 in the preceding section.

5. If none of the first four guidelines applies, try converting to sines and cosines.

EXAMPLE 4   Power of Tangent Is Odd and Positive

Find $\displaystyle\int \frac{\tan^3 x}{\sqrt{\sec x}} \, dx.$

Solution   Because you expect to use the Power Rule with $u = \sec x$, *save a factor of* $(\sec x \tan x)$ to form du and convert the remaining tangent factors to secants.

$$\int \frac{\tan^3 x}{\sqrt{\sec x}} \, dx = \int (\sec x)^{-1/2} \tan^3 x \, dx$$

$$= \int (\sec x)^{-3/2} (\tan^2 x)(\sec x \tan x) \, dx$$

$$= \int (\sec x)^{-3/2} (\sec^2 x - 1)(\sec x \tan x) \, dx$$

$$= \int [(\sec x)^{1/2} - (\sec x)^{-3/2}](\sec x \tan x) \, dx$$

$$= \frac{2}{3}(\sec x)^{3/2} + 2(\sec x)^{-1/2} + C$$

NOTE   In Example 5, the power of the tangent is odd and positive. So, you could also find the integral using the procedure described in guideline 2 on page 537. In Exercise 85, you are asked to show that the results obtained by these two procedures differ only by a constant.

EXAMPLE 5   Power of Secant Is Even and Positive

Find $\displaystyle\int \sec^4 3x \tan^3 3x \, dx$.

Solution   Let $u = \tan 3x$, then $du = 3 \sec^2 3x \, dx$ and you can write

$$\int \sec^4 3x \tan^3 3x \, dx = \int \sec^2 3x \tan^3 3x(\sec^2 3x) \, dx$$

$$= \int (1 + \tan^2 3x) \tan^3 3x(\sec^2 3x) \, dx$$

$$= \frac{1}{3}\int (\tan^3 3x + \tan^5 3x)(3 \sec^2 3x) \, dx$$

$$= \frac{1}{3}\left(\frac{\tan^4 3x}{4} + \frac{\tan^6 3x}{6}\right) + C$$

$$= \frac{\tan^4 3x}{12} + \frac{\tan^6 3x}{18} + C.$$

EXAMPLE 6   Power of Tangent Is Even

Evaluate $\displaystyle\int_0^{\pi/4} \tan^4 x \, dx$.

Solution   Because there are no secant factors, you can begin by converting a tangent-squared factor to a secant-squared factor.

$$\int \tan^4 x \, dx = \int \tan^2 x(\tan^2 x) \, dx$$

$$= \int \tan^2 x(\sec^2 x - 1) \, dx$$

$$= \int \tan^2 x \sec^2 x \, dx - \int \tan^2 x \, dx$$

$$= \int \tan^2 x \sec^2 x \, dx - \int (\sec^2 x - 1) \, dx$$

$$= \frac{\tan^3 x}{3} - \tan x + x + C$$

You can evaluate the definite integral as follows.

$$\int_0^{\pi/4} \tan^4 x \, dx = \left[\frac{\tan^3 x}{3} - \tan x + x\right]_0^{\pi/4}$$

$$= \frac{\pi}{4} - \frac{2}{3}$$

$$\approx 0.119$$

The area represented by the definite integral is shown in Figure 8.5. Try using Simpson's Rule to approximate this integral. With $n = 18$, you should obtain an approximation that is within 0.00001 of the actual value.

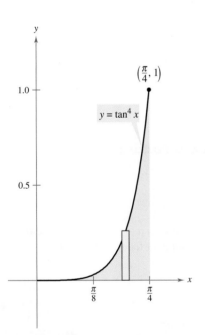

The area of the region is approximately 0.119.

Figure 8.5

• Use trigonometric substitution to solve an integral.
• Use integrals to model and solve real-life applications.

Trigonometric Substitution

Now that you can evaluate integrals involving powers of trigonometric functions, you can use **trigonometric substitution** to evaluate integrals involving the radicals

$$\sqrt{a^2 - u^2}, \qquad \sqrt{a^2 + u^2}, \qquad \text{and} \qquad \sqrt{u^2 - a^2}.$$

The objective with trigonometric substitution is to eliminate the radical in the integrand. You do this with the Pythagorean identities

$$\cos^2 \theta = 1 - \sin^2 \theta, \quad \sec^2 \theta = 1 + \tan^2 \theta, \quad \text{and} \quad \tan^2 \theta = \sec^2 \theta - 1.$$

For example, if $a > 0$, let $u = a \sin \theta$, where $-\pi/2 \leq \theta \leq \pi/2$. Then

$$\begin{aligned} \sqrt{a^2 - u^2} &= \sqrt{a^2 - a^2 \sin^2 \theta} \\ &= \sqrt{a^2(1 - \sin^2 \theta)} \\ &= \sqrt{a^2 \cos^2 \theta} \\ &= a \cos \theta. \end{aligned}$$

Note that $\cos \theta \geq 0$, because $-\pi/2 \leq \theta \leq \pi/2$.

Trigonometric Substitution ($a > 0$)

1. For integrals involving $\sqrt{a^2 - u^2}$, let

$$u = a \sin \theta.$$

Then $\sqrt{a^2 - u^2} = a \cos \theta$, where $-\pi/2 \leq \theta \leq \pi/2$.

2. For integrals involving $\sqrt{a^2 + u^2}$, let

$$u = a \tan \theta.$$

Then $\sqrt{a^2 + u^2} = a \sec \theta$, where $-\pi/2 < \theta < \pi/2$.

3. For integrals involving $\sqrt{u^2 - a^2}$, let

$$u = a \sec \theta.$$

Then $\sqrt{u^2 - a^2} = \pm a \tan \theta$, where $0 \leq \theta < \pi/2$ or $\pi/2 < \theta \leq \pi$. Use the positive value if $u > a$ and the negative value if $u < -a$.

NOTE   The restrictions on θ ensure that the function that defines the substitution is one-to-one. In fact, these are the same intervals over which the arcsine, arctangent, and arcsecant are defined.

EXPLORATION

Integrating a Radical Function
Up to this point in the text, you have not evaluated the following integral.

$$\int_{-1}^{1} \sqrt{1 - x^2} \, dx$$

From geometry, you should be able to find the exact value of this integral— what is it? Using numerical integration, with Simpson's Rule or the Trapezoidal Rule, you can't be sure of the accuracy of the approximation. Why?

Try finding the exact value using the substitution

$$x = \sin \theta \quad \text{and} \quad dx = \cos \theta \, d\theta.$$

Does your answer agree with the value you obtained using geometry?

$$\sin \theta = \frac{x}{3}, \cot \theta = \frac{\sqrt{9 - x^2}}{x}$$

Figure 8.6

EXAMPLE 1   Trigonometric Substitution: $u = a \sin \theta$

Find $\displaystyle\int \frac{dx}{x^2 \sqrt{9 - x^2}}$.

Solution   First, note that none of the basic integration rules applies. To use trigonometric substitution, you should observe that $\sqrt{9 - x^2}$ is of the form $\sqrt{a^2 - u^2}$. So, you can use the substitution

$$x = a \sin \theta = 3 \sin \theta.$$

Using differentiation and the triangle shown in Figure 8.6, you obtain

$$dx = 3 \cos \theta \, d\theta, \qquad \sqrt{9 - x^2} = 3 \cos \theta, \qquad \text{and} \qquad x^2 = 9 \sin^2 \theta.$$

So, trigonometric substitution yields

$$\int \frac{dx}{x^2 \sqrt{9 - x^2}} = \int \frac{3 \cos \theta \, d\theta}{(9 \sin^2 \theta)(3 \cos \theta)} \qquad \text{Substitute.}$$

$$= \frac{1}{9} \int \frac{d\theta}{\sin^2 \theta} \qquad \text{Simplify.}$$

$$= \frac{1}{9} \int \csc^2 \theta \, d\theta \qquad \text{Trigonometric identity}$$

$$= -\frac{1}{9} \cot \theta + C \qquad \text{Apply Cosecant Rule.}$$

$$= -\frac{1}{9} \left(\frac{\sqrt{9 - x^2}}{x} \right) + C \qquad \text{Substitute for } \cot \theta.$$

$$= -\frac{\sqrt{9 - x^2}}{9x} + C.$$

Note that the triangle in Figure 8.6 can be used to convert the θ's back to x's as follows.

$$\cot \theta = \frac{\text{adj.}}{\text{opp.}}$$

$$= \frac{\sqrt{9 - x^2}}{x}$$

TECHNOLOGY   Use a computer algebra system to find each indefinite integral.

$$\int \frac{dx}{\sqrt{9 - x^2}} \qquad \int \frac{dx}{x \sqrt{9 - x^2}} \qquad \int \frac{dx}{x^2 \sqrt{9 - x^2}} \qquad \int \frac{dx}{x^3 \sqrt{9 - x^2}}$$

Then use trigonometric substitution to duplicate the results obtained with the computer algebra system.

In an earlier chapter, you saw how the inverse hyperbolic functions can be used to evaluate the integrals

$$\int \frac{du}{\sqrt{u^2 \pm a^2}}, \qquad \int \frac{du}{a^2 - u^2}, \qquad \text{and} \qquad \int \frac{du}{u \sqrt{a^2 \pm u^2}}.$$

You can also evaluate these integrals using trigonometric substitution. This is shown in the next example.

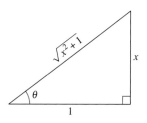

$\tan \theta = 2x, \sec \theta = \sqrt{4x^2 + 1}$

Figure 8.7

EXAMPLE 2   **Trigonometric Substitution: $u = a \tan \theta$**

Find $\displaystyle\int \frac{dx}{\sqrt{4x^2 + 1}}$.

Solution   Let $u = 2x$, $a = 1$, and $2x = \tan \theta$, as shown in Figure 8.7. Then,

$$dx = \frac{1}{2} \sec^2 \theta \, d\theta \qquad \text{and} \qquad \sqrt{4x^2 + 1} = \sec \theta.$$

Trigonometric substitution produces

$$\int \frac{1}{\sqrt{4x^2 + 1}} \, dx = \frac{1}{2} \int \frac{\sec^2 \theta \, d\theta}{\sec \theta} \qquad\qquad \text{Substitute.}$$

$$= \frac{1}{2} \int \sec \theta \, d\theta \qquad\qquad \text{Simplify.}$$

$$= \frac{1}{2} \ln|\sec \theta + \tan \theta| + C \qquad\qquad \text{Apply Secant Rule.}$$

$$= \frac{1}{2} \ln\left|\sqrt{4x^2 + 1} + 2x\right| + C. \qquad\qquad \text{Back-substitute.}$$

Try checking this result with a computer algebra system. Is the result given in this form or in the form of an inverse hyperbolic function?

You can extend the use of trigonometric substitution to cover integrals involving expressions such as $(a^2 - u^2)^{n/2}$ by writing the expression as

$$(a^2 - u^2)^{n/2} = \left(\sqrt{a^2 - u^2}\right)^n.$$

  EXAMPLE 3   **Trigonometric Substitution: Rational Powers**

Find $\displaystyle\int \frac{dx}{(x^2 + 1)^{3/2}}$.

Solution   Begin by writing $(x^2 + 1)^{3/2}$ as $\left(\sqrt{x^2 + 1}\right)^3$. Then, let $a = 1$ and $u = x = \tan \theta$, as shown in Figure 8.8. Using

$$dx = \sec^2 \theta \, d\theta \qquad \text{and} \qquad \sqrt{x^2 + 1} = \sec \theta$$

you can apply trigonometric substitution as follows.

$$\int \frac{dx}{(x^2 + 1)^{3/2}} = \int \frac{dx}{\left(\sqrt{x^2 + 1}\right)^3} \qquad\qquad \text{Rewrite denominator.}$$

$$= \int \frac{\sec^2 \theta \, d\theta}{\sec^3 \theta} \qquad\qquad \text{Substitute.}$$

$$= \int \frac{d\theta}{\sec \theta} \qquad\qquad \text{Simplify.}$$

$$= \int \cos \theta \, d\theta \qquad\qquad \text{Trigonometric identity}$$

$$= \sin \theta + C \qquad\qquad \text{Apply Cosine Rule.}$$

$$= \frac{x}{\sqrt{x^2 + 1}} + C \qquad\qquad \text{Back-substitute.}$$

$\tan \theta = x, \sin \theta = \dfrac{x}{\sqrt{x^2 + 1}}$

Figure 8.8

For definite integrals, it is often convenient to determine the integration limits for θ that avoid converting back to x. You might want to review this procedure in Section 5.5, Examples 8 and 9.

EXAMPLE 4 Converting the Limits of Integration

Evaluate $\displaystyle\int_{\sqrt{3}}^{2} \frac{\sqrt{x^2 - 3}}{x}\, dx$.

Solution Because $\sqrt{x^2 - 3}$ has the form $\sqrt{u^2 - a^2}$, you can consider

$$u = x, \quad a = \sqrt{3}, \quad \text{and} \quad x = \sqrt{3}\sec\theta$$

as shown in Figure 8.9. Then,

$$dx = \sqrt{3}\sec\theta\tan\theta\, d\theta \quad \text{and} \quad \sqrt{x^2 - 3} = \sqrt{3}\tan\theta.$$

To determine the upper and lower limits of integration, use the substitution $x = \sqrt{3}\sec\theta$, as follows.

Lower Limit	Upper Limit
When $x = \sqrt{3}$, $\sec\theta = 1$ and $\theta = 0$.	When $x = 2$, $\sec\theta = \dfrac{2}{\sqrt{3}}$ and $\theta = \dfrac{\pi}{6}$.

So, you have

$$\int_{\sqrt{3}}^{2} \frac{\sqrt{x^2 - 3}}{x}\, dx = \int_{0}^{\pi/6} \frac{\left(\sqrt{3}\tan\theta\right)\left(\sqrt{3}\sec\theta\tan\theta\right) d\theta}{\sqrt{3}\sec\theta}$$

$$= \int_{0}^{\pi/6} \sqrt{3}\tan^2\theta\, d\theta$$

$$= \sqrt{3}\int_{0}^{\pi/6} (\sec^2\theta - 1)\, d\theta$$

$$= \sqrt{3}\left[\tan\theta - \theta\right]_{0}^{\pi/6}$$

$$= \sqrt{3}\left(\frac{1}{\sqrt{3}} - \frac{\pi}{6}\right)$$

$$= 1 - \frac{\sqrt{3}\pi}{6}$$

$$\approx 0.0931.$$

In Example 4, try converting back to the variable x and evaluating the antiderivative at the original limits of integration. You should obtain

$$\int_{\sqrt{3}}^{2} \frac{\sqrt{x^2 - 3}}{x}\, dx = \sqrt{3}\left.\frac{\sqrt{x^2 - 3}}{\sqrt{3}} - \text{arcsec}\frac{x}{\sqrt{3}}\right]_{\sqrt{3}}^{2}.$$

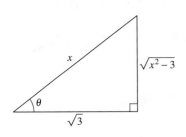

$$\sec\theta = \frac{x}{\sqrt{3}}, \tan\theta = \frac{\sqrt{x^2 - 3}}{\sqrt{3}}$$

Figure 8.9

When using trigonometric substitution to evaluate definite integrals, you must be careful to check that the values of θ lie in the intervals discussed at the beginning of this section. For instance, if in Example 4 you had been asked to evaluate the definite integral

$$\int_{-2}^{-\sqrt{3}} \frac{\sqrt{x^2 - 3}}{x} \, dx$$

then using $u = x$ and $a = \sqrt{3}$ in the interval $\left[-2, -\sqrt{3}\right]$ would imply that $u < -a$. So, when determining the upper and lower limits of integration, you would have to choose θ such that $\pi/2 < \theta \le \pi$. In this case the integral would be evaluated as follows.

$$
\begin{aligned}
\int_{-2}^{-\sqrt{3}} \frac{\sqrt{x^2 - 3}}{x} \, dx &= \int_{5\pi/6}^{\pi} \frac{\left(-\sqrt{3} \tan \theta\right)\left(\sqrt{3} \sec \theta \tan \theta\right) d\theta}{\sqrt{3} \sec \theta} \\
&= \int_{5\pi/6}^{\pi} -\sqrt{3} \tan^2 \theta \, d\theta \\
&= -\sqrt{3} \int_{5\pi/6}^{\pi} (\sec^2 \theta - 1) \, d\theta \\
&= -\sqrt{3} \left[\tan \theta - \theta \right]_{5\pi/6}^{\pi} \\
&= -\sqrt{3} \left[(0 - \pi) - \left(-\frac{1}{\sqrt{3}} - \frac{5\pi}{6} \right) \right] \\
&= -1 + \frac{\sqrt{3}\pi}{6} \\
&\approx -0.0931
\end{aligned}
$$

Trigonometric substitution can be used with completing the square. For instance, try evaluating the following integral.

$$\int \sqrt{x^2 - 2x} \, dx$$

To begin, you could complete the square and write the integral as

$$\int \sqrt{(x - 1)^2 - 1^2} \, dx.$$

Trigonometric substitution can be used to evaluate the three integrals listed in the following theorem. These integrals will be encountered several times in the remainder of the text. When this happens, we will simply refer to this theorem. (In Exercise 85, you are asked to verify the formulas given in the theorem.)

THEOREM 8.2   Special Integration Formulas ($a > 0$)

1. $\displaystyle\int \sqrt{a^2 - u^2} \, du = \frac{1}{2}\left(a^2 \arcsin \frac{u}{a} + u\sqrt{a^2 - u^2} \right) + C$

2. $\displaystyle\int \sqrt{u^2 - a^2} \, du = \frac{1}{2}\left(u\sqrt{u^2 - a^2} - a^2 \ln\left|u + \sqrt{u^2 - a^2}\right| \right) + C, \quad u > a$

3. $\displaystyle\int \sqrt{u^2 + a^2} \, du = \frac{1}{2}\left(u\sqrt{u^2 + a^2} + a^2 \ln\left|u + \sqrt{u^2 + a^2}\right| \right) + C$

Applications

EXAMPLE 5 Finding Arc Length

Find the arc length of the graph of $f(x) = \frac{1}{2}x^2$ from $x = 0$ to $x = 1$ (see Figure 8.10).

Solution Refer to the arc length formula in Section 7.4.

$$
\begin{aligned}
s &= \int_0^1 \sqrt{1 + [f'(x)]^2}\, dx && \text{Formula for arc length} \\
&= \int_0^1 \sqrt{1 + x^2}\, dx && f'(x) = x \\
&= \int_0^{\pi/4} \sec^3 \theta\, d\theta && \text{Let } a = 1 \text{ and } x = \tan\theta. \\
&= \frac{1}{2}\Big[\sec\theta \tan\theta + \ln|\sec\theta + \tan\theta| \Big]_0^{\pi/4} && \text{Example 5, Section 8.2} \\
&= \frac{1}{2}\Big[\sqrt{2} + \ln\big(\sqrt{2} + 1\big) \Big] \approx 1.148
\end{aligned}
$$

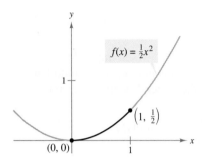

$f(x) = \frac{1}{2}x^2$

$\left(1, \tfrac{1}{2}\right)$

$(0, 0)$

The arc length of the curve from $(0, 0)$ to $\left(1, \frac{1}{2}\right)$

Figure 8.10

EXAMPLE 6 Comparing Two Fluid Forces

A sealed barrel of oil (weighing 48 pounds per cubic foot) is floating in seawater (weighing 64 pounds per cubic foot), as shown in Figures 8.11 and 8.12. (The barrel is not completely full of oil—on its side, the top 0.2 foot of the barrel is empty.) Compare the fluid forces against one end of the barrel from the inside and from the outside.

Solution In Figure 8.12, locate the coordinate system with the origin at the center of the circle given by $x^2 + y^2 = 1$. To find the fluid force against an end of the barrel *from the inside*, integrate between -1 and 0.8 (using a weight of $w = 48$).

$$
F = w \int_c^d h(y)L(y)\, dy \qquad \text{General equation (see Section 7.7)}
$$

$$
\begin{aligned}
F_{\text{inside}} &= 48 \int_{-1}^{0.8} (0.8 - y)(2)\sqrt{1 - y^2}\, dy \\
&= 76.8 \int_{-1}^{0.8} \sqrt{1 - y^2}\, dy - 96 \int_{-1}^{0.8} y\sqrt{1 - y^2}\, dy
\end{aligned}
$$

To find the fluid force *from the outside*, integrate between -1 and 0.4 (using a weight of $w = 64$).

$$
\begin{aligned}
F_{\text{outside}} &= 64 \int_{-1}^{0.4} (0.4 - y)(2)\sqrt{1 - y^2}\, dy \\
&= 51.2 \int_{-1}^{0.4} \sqrt{1 - y^2}\, dy - 128 \int_{-1}^{0.4} y\sqrt{1 - y^2}\, dy
\end{aligned}
$$

The details of integration are left for you to complete in Exercise 84. Intuitively, would you say that the force from the oil (the inside) or the force from the seawater (the outside) is greater? By evaluating these two integrals, you can determine that

$$
F_{\text{inside}} \approx 121.3 \text{ pounds} \qquad \text{and} \qquad F_{\text{outside}} \approx 93.0 \text{ pounds.}
$$

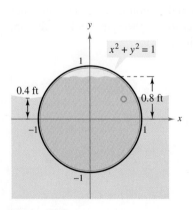

The barrel is not quite full of oil—the top 0.2 foot of the barrel is empty.

Figure 8.11

$x^2 + y^2 = 1$

0.4 ft

0.8 ft

Figure 8.12

In Exercises 1–4, use differentiation to match the antiderivative with the correct integral. [Integrals are labeled (a), (b), (c), and (d).]

(a) $\int \dfrac{x^2}{\sqrt{16-x^2}}\,dx$

(b) $\int \dfrac{\sqrt{x^2+16}}{x}\,dx$

(c) $\int \sqrt{7+6x-x^2}\,dx$

(d) $\int \dfrac{x^2}{\sqrt{x^2-16}}\,dx$

1. $4\ln\left|\dfrac{\sqrt{x^2+16}-4}{x}\right| + \sqrt{x^2+16} + C$

2. $8\ln\left|\sqrt{x^2-16}+x\right| + \dfrac{x\sqrt{x^2-16}}{2} + C$

3. $8\arcsin\dfrac{x}{4} - \dfrac{x\sqrt{16-x^2}}{2} + C$

4. $8\arcsin\dfrac{x-3}{4} + \dfrac{(x-3)\sqrt{7+6x-x^2}}{2} + C$

In Exercises 5–8, find the indefinite integral using the substitution $x = 5\sin\theta$.

5. $\int \dfrac{1}{(25-x^2)^{3/2}}\,dx$

6. $\int \dfrac{10}{x^2\sqrt{25-x^2}}\,dx$

7. $\int \dfrac{\sqrt{25-x^2}}{x}\,dx$

8. $\int \dfrac{x^2}{\sqrt{25-x^2}}\,dx$

In Exercises 9–12, find the indefinite integral using the substitution $x = 2\sec\theta$.

9. $\int \dfrac{1}{\sqrt{x^2-4}}\,dx$

10. $\int \dfrac{\sqrt{x^2-4}}{x}\,dx$

11. $\int x^3\sqrt{x^2-4}\,dx$

12. $\int \dfrac{x^3}{\sqrt{x^2-4}}\,dx$

In Exercises 13–16, find the indefinite integral using the substitution $x = \tan\theta$.

13. $\int x\sqrt{1+x^2}\,dx$

14. $\int \dfrac{9x^3}{\sqrt{1+x^2}}\,dx$

15. $\int \dfrac{1}{(1+x^2)^2}\,dx$

16. $\int \dfrac{x^2}{(1+x^2)^2}\,dx$

In Exercises 17–20, use the Special Integration Formulas (Theorem 8.2) to find the integral.

17. $\int \sqrt{4+9x^2}\,dx$

18. $\int \sqrt{1+x^2}\,dx$

19. $\int \sqrt{25-4x^2}\,dx$

20. $\int \sqrt{2x^2-1}\,dx$

In Exercises 21–42, find the integral.

21. $\int \dfrac{x}{\sqrt{x^2+9}}\,dx$

22. $\int \dfrac{x}{\sqrt{9-x^2}}\,dx$

23. $\int \dfrac{1}{\sqrt{16-x^2}}\,dx$

24. $\int \dfrac{1}{\sqrt{25-x^2}}\,dx$

25. $\int \sqrt{16-4x^2}\,dx$

26. $\int x\sqrt{16-4x^2}\,dx$

27. $\int \dfrac{1}{\sqrt{x^2-9}}\,dx$

28. $\int \dfrac{t}{(1-t^2)^{3/2}}\,dt$

29. $\int \dfrac{\sqrt{1-x^2}}{x^4}\,dx$

30. $\int \dfrac{\sqrt{4x^2+9}}{x^4}\,dx$

31. $\int \dfrac{1}{x\sqrt{4x^2+9}}\,dx$

32. $\int \dfrac{1}{x\sqrt{4x^2+16}}\,dx$

33. $\int \dfrac{-5x}{(x^2+5)^{3/2}}\,dx$

34. $\int \dfrac{1}{(x^2+3)^{3/2}}\,dx$

35. $\int e^{2x}\sqrt{1+e^{2x}}\,dx$

36. $\int (x+1)\sqrt{x^2+2x+2}\,dx$

37. $\int e^x\sqrt{1-e^{2x}}\,dx$

38. $\int \dfrac{\sqrt{1-x}}{\sqrt{x}}\,dx$

39. $\int \dfrac{1}{4+4x^2+x^4}\,dx$

40. $\int \dfrac{x^3+x+1}{x^4+2x^2+1}\,dx$

41. $\int \operatorname{arcsec} 2x\,dx, \quad x > \tfrac{1}{2}$

42. $\int x\arcsin x\,dx$

In Exercises 43–46, complete the square and find the integral.

43. $\int \dfrac{1}{\sqrt{4x-x^2}}\,dx$

44. $\int \dfrac{x^2}{\sqrt{2x-x^2}}\,dx$

45. $\int \dfrac{x}{\sqrt{x^2+4x+8}}\,dx$

46. $\int \dfrac{x}{\sqrt{x^2-6x+5}}\,dx$

In Exercises 47–52, evaluate the integral using (a) the given integration limits and (b) the limits obtained by trigonometric substitution.

47. $\int_0^{\sqrt{3}/2} \dfrac{t^2}{(1-t^2)^{3/2}}\,dt$

48. $\int_0^{\sqrt{3}/2} \dfrac{1}{(1-t^2)^{5/2}}\,dt$

49. $\int_0^3 \dfrac{x^3}{\sqrt{x^2+9}}\,dx$

50. $\int_0^{3/5} \sqrt{9-25x^2}\,dx$

51. $\int_4^6 \dfrac{x^2}{\sqrt{x^2-9}}\,dx$

52. $\int_3^6 \dfrac{\sqrt{x^2-9}}{x^2}\,dx$

In Exercises 53 and 54, find the particular solution of the differential equation.

53. $x\dfrac{dy}{dx} = \sqrt{x^2-9}, \quad x \geq 3, \quad y(3) = 1$

54. $\sqrt{x^2+4}\dfrac{dy}{dx} = 1, \quad x \geq -2, \quad y(0) = 4$

In Exercises 55–58, use a computer algebra system to find the integral. Verify the result by differentiation.

55. $\displaystyle\int \frac{x^2}{\sqrt{x^2 + 10x + 9}}\, dx$ **56.** $\displaystyle\int (x^2 + 2x + 11)^{3/2}\, dx$

57. $\displaystyle\int \frac{x^2}{\sqrt{x^2 - 1}}\, dx$ **58.** $\displaystyle\int x^2 \sqrt{x^2 - 4}\, dx$

Writing About Concepts

59. State the substitution you would make if you used trigonometric substitution and the integral involving the given radical, where $a > 0$. Explain your reasoning.

(a) $\sqrt{a^2 - u^2}$ (b) $\sqrt{a^2 + u^2}$ (c) $\sqrt{u^2 - a^2}$

60. State the method of integration you would use to perform each integration. Explain why you chose that method. Do not integrate.

(a) $\displaystyle\int x\sqrt{x^2 + 1}\, dx$ (b) $\displaystyle\int x^2\sqrt{x^2 - 1}\, dx$

61. Evaluate the integral $\displaystyle\int \frac{x}{x^2 + 9}\, dx$ using (a) u-substitution and (b) trigonometric substitution. Discuss the results.

62. Evaluate the integral $\displaystyle\int \frac{x^2}{x^2 + 9}\, dx$ (a) algebraically using $x^2 = (x^2 + 9) - 9$ and (b) using trigonometric substitution. Discuss the results.

True or False? **In Exercises 63–66, determine whether the statement is true or false. If it is false, explain why or give an example that shows it is false.**

63. If $x = \sin\theta$, then $\displaystyle\int \frac{dx}{\sqrt{1 - x^2}} = \int d\theta$.

64. If $x = \sec\theta$, then $\displaystyle\int \frac{\sqrt{x^2 - 1}}{x}\, dx = \int \sec\theta \tan\theta\, d\theta$.

65. If $x = \tan\theta$, then $\displaystyle\int_0^{\sqrt{3}} \frac{dx}{(1 + x^2)^{3/2}} = \int_0^{4\pi/3} \cos\theta\, d\theta$.

66. If $x = \sin\theta$, then $\displaystyle\int_{-1}^{1} x^2\sqrt{1 - x^2}\, dx = 2\int_0^{\pi/2} \sin^2\theta \cos^2\theta\, d\theta$.

67. ***Area*** Find the area enclosed by the ellipse shown in the figure.

$$\frac{x^2}{a^2} + \frac{y^2}{b^2} = 1$$

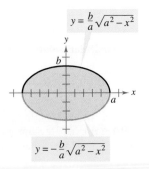

$y = \dfrac{b}{a}\sqrt{a^2 - x^2}$

$y = -\dfrac{b}{a}\sqrt{a^2 - x^2}$

68. ***Area*** Find the area of the shaded region of the circle of radius a, if the chord is h units $(0 < h < a)$ from the center of the circle (see figure).

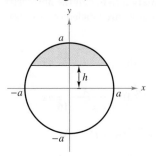

69. ***Mechanical Design*** The surface of a machine part is the region between the graphs of $y = |x|$ and $x^2 + (y - k)^2 = 25$ (see figure).

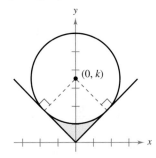

(a) Find k if the circle is tangent to the graph of $y = |x|$.

(b) Find the area of the surface of the machine part.

(c) Find the area of the surface of the machine part as a function of the radius r of the circle.

 70. ***Volume*** The axis of a storage tank in the form of a right circular cylinder is horizontal (see figure). The radius and length of the tank are 1 meter and 3 meters, respectively.

(a) Determine the volume of fluid in the tank as a function of its depth d.

(b) Use a graphing utility to graph the function in part (a).

(c) Design a dip stick for the tank with markings of $\frac{1}{4}$, $\frac{1}{2}$, and $\frac{3}{4}$.

(d) Fluid is entering the tank at a rate of $\frac{1}{4}$ cubic meter per minute. Determine the rate of change of the depth of the fluid as a function of its depth d.

(e) Use a graphing utility to graph the function in part (d). When will the rate of change of the depth be minimum? Does this agree with your intuition? Explain.

Volume of a Torus   In Exercises 71 and 72, find the volume of the torus generated by revolving the region bounded by the graph of the circle about the *y*-axis.

71. $(x - 3)^2 + y^2 = 1$ (see figure)

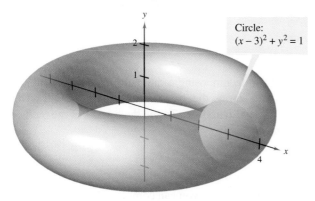

Circle:
$(x - 3)^2 + y^2 = 1$

72. $(x - h)^2 + y^2 = r^2, \ h > r$

Arc Length   In Exercises 73 and 74, find the arc length of the curve over the given interval.

73. $y = \ln x, \quad [1, 5]$

74. $y = \frac{1}{2}x^2, \quad [0, 4]$

75. Arc Length   Show that the length of one arch of the sine curve is equal to the length of one arch of the cosine curve.

76. Conjecture

(a) Find formulas for the distance between $(0, 0)$ and (a, a^2) along the line between these points and along the parabola $y = x^2$.

(b) Use the formulas from part (a) to find the distances for $a = 1$ and $a = 10$.

(c) Make a conjecture about the difference between the two distances as *a* increases.

Projectile Motion   In Exercises 77 and 78, (a) use a graphing utility to graph the path of a projectile that follows the path given by the graph of the equation, (b) determine the range of the projectile, and (c) use the integration capabilities of a graphing utility to determine the distance the projectile travels.

77. $y = x - 0.005x^2$

78. $y = x - \dfrac{x^2}{72}$

Centroid   In Exercises 79 and 80, find the centroid of the region determined by the graphs of the inequalities.

79. $y \leq 3/\sqrt{x^2 + 9}, \ y \geq 0, \ x \geq -4, \ x \leq 4$

80. $y \leq \frac{1}{4}x^2, \ (x - 4)^2 + y^2 \leq 16, \ y \geq 0$

81. Surface Area   Find the surface area of the solid generated by revolving the region bounded by the graphs of $y = x^2$, $y = 0$, $x = 0$, and $x = \sqrt{2}$ about the *x*-axis.

82. Field Strength   The field strength *H* of a magnet of length 2*L* on a particle *r* units from the center of the magnet is

$$H = \frac{2mL}{(r^2 + L^2)^{3/2}}$$

where $\pm m$ are the poles of the magnet (see figure). Find the average field strength as the particle moves from 0 to *R* units from the center by evaluating the integral

$$\frac{1}{R} \int_0^R \frac{2mL}{(r^2 + L^2)^{3/2}} \, dr.$$

Figure for 82   **Figure for 83**

83. Fluid Force   Find the fluid force on a circular observation window of radius 1 foot in a vertical wall of a large water-filled tank at a fish hatchery when the center of the window is (a) 3 feet and (b) *d* feet $(d > 1)$ below the water's surface (see figure). Use trigonometric substitution to evaluate the one integral. (Recall that in Section 7.7 in a similar problem, you evaluated one integral by a geometric formula and the other by observing that the integrand was odd.)

84. Fluid Force   Evaluate the following two integrals, which yield the fluid forces given in Example 6.

(a) $F_{\text{inside}} = 48 \displaystyle\int_{-1}^{0.8} (0.8 - y)(2)\sqrt{1 - y^2} \, dy$

(b) $F_{\text{outside}} = 64 \displaystyle\int_{-1}^{0.4} (0.4 - y)(2)\sqrt{1 - y^2} \, dy$

85. Use trigonometric substitution to verify the integration formulas given in Theorem 8.2.

86. Arc Length   Show that the arc length of the graph of $y = \sin x$ on the interval $[0, 2\pi]$ is equal to the circumference of the ellipse $x^2 + 2y^2 = 2$ (see figure).

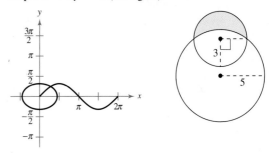

Figure for 86   **Figure for 87**

87. Area of a Lune   The crescent-shaped region bounded by two circles forms a *lune* (see figure). Find the area of the lune given that the radius of the smaller circle is 3 and the radius of the larger circle is 5.

Section 8.5

Partial Fractions

- Understand the concept of a partial fraction decomposition.
- Use partial fraction decomposition with linear factors to integrate rational functions.
- Use partial fraction decomposition with quadratic factors to integrate rational functions.

Partial Fractions

This section examines a procedure for decomposing a rational function into simpler rational functions to which you can apply the basic integration formulas. This procedure is called the **method of partial fractions.** To see the benefit of the method of partial fractions, consider the integral

$$\int \frac{1}{x^2 - 5x + 6} \, dx.$$

To evaluate this integral *without* partial fractions, you can complete the square and use trigonometric substitution (see Figure 8.13) to obtain

$$\int \frac{1}{x^2 - 5x + 6} \, dx = \int \frac{dx}{(x - 5/2)^2 - (1/2)^2} \qquad a = \tfrac{1}{2}, x - \tfrac{5}{2} = \tfrac{1}{2} \sec \theta$$

$$= \int \frac{(1/2) \sec \theta \tan \theta \, d\theta}{(1/4) \tan^2 \theta} \qquad dx = \tfrac{1}{2} \sec \theta \tan \theta \, d\theta$$

$$= 2 \int \csc \theta \, d\theta$$

$$= 2 \ln|\csc \theta - \cot \theta| + C$$

$$= 2 \ln\left| \frac{2x - 5}{2\sqrt{x^2 - 5x + 6}} - \frac{1}{2\sqrt{x^2 - 5x + 6}} \right| + C$$

$$= 2 \ln\left| \frac{x - 3}{\sqrt{x^2 - 5x + 6}} \right| + C$$

$$= 2 \ln\left| \frac{\sqrt{x - 3}}{\sqrt{x - 2}} \right| + C$$

$$= \ln\left| \frac{x - 3}{x - 2} \right| + C$$

$$= \ln|x - 3| - \ln|x - 2| + C.$$

Now, suppose you had observed that

$$\frac{1}{x^2 - 5x + 6} = \frac{1}{x - 3} - \frac{1}{x - 2}. \qquad \text{Partial fraction decomposition}$$

Then you could evaluate the integral easily, as follows.

$$\int \frac{1}{x^2 - 5x + 6} \, dx = \int \left(\frac{1}{x - 3} - \frac{1}{x - 2} \right) dx$$

$$= \ln|x - 3| - \ln|x - 2| + C$$

This method is clearly preferable to trigonometric substitution. However, its use depends on the ability to factor the denominator, $x^2 - 5x + 6$, and to find the **partial fractions**

$$\frac{1}{x - 3} \qquad \text{and} \qquad -\frac{1}{x - 2}.$$

In this section, you will study techniques for finding partial fraction decompositions.

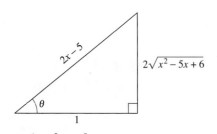

$\sec \theta = 2x - 5$
Figure 8.13

Mary Evans Picture Library

JOHN BERNOULLI (1667–1748)

The method of partial fractions was introduced by John Bernoulli, a Swiss mathematician who was instrumental in the early development of calculus. John Bernoulli was a professor at the University of Basel and taught many outstanding students, the most famous of whom was Leonhard Euler.

STUDY TIP   In precalculus you learned
how to combine functions such as

$$\frac{1}{x-2} + \frac{-1}{x+3} = \frac{5}{(x-2)(x+3)}.$$

The method of partial fractions shows
you how to reverse this process.

$$\frac{5}{(x-2)(x+3)} = \frac{?}{x-2} + \frac{?}{x+3}$$

Recall from algebra that every polynomial with real coefficients can be factored
into linear and irreducible quadratic factors.* For instance, the polynomial

$$x^5 + x^4 - x - 1$$

can be written as

$$
\begin{aligned}
x^5 + x^4 - x - 1 &= x^4(x+1) - (x+1) \\
&= (x^4 - 1)(x+1) \\
&= (x^2 + 1)(x^2 - 1)(x+1) \\
&= (x^2 + 1)(x+1)(x-1)(x+1) \\
&= (x-1)(x+1)^2(x^2+1)
\end{aligned}
$$

where $(x-1)$ is a linear factor, $(x+1)^2$ is a repeated linear factor, and (x^2+1) is an
irreducible quadratic factor. Using this factorization, you can write the partial fraction
decomposition of the rational expression

$$\frac{N(x)}{x^5 + x^4 - x - 1}$$

where $N(x)$ is a polynomial of degree less than 5, as follows.

$$\frac{N(x)}{(x-1)(x+1)^2(x^2+1)} = \frac{A}{x-1} + \frac{B}{x+1} + \frac{C}{(x+1)^2} + \frac{Dx+E}{x^2+1}$$

Decomposition of N(x)/D(x) into Partial Fractions

1. **Divide if improper:** If $N(x)/D(x)$ is an improper fraction (that is, if the
    degree of the numerator is greater than or equal to the degree of the denomi-
    nator), divide the denominator into the numerator to obtain

    $$\frac{N(x)}{D(x)} = (\text{a polynomial}) + \frac{N_1(x)}{D(x)}$$

    where the degree of $N_1(x)$ is less than the degree of $D(x)$. Then apply Steps 2,
    3, and 4 to the proper rational expression $N_1(x)/D(x)$.

2. **Factor denominator:** Completely factor the denominator into factors of the
    form

    $$(px + q)^m \qquad \text{and} \qquad (ax^2 + bx + c)^n$$

    where $ax^2 + bx + c$ is irreducible.

3. **Linear factors:** For each factor of the form $(px + q)^m$, the partial fraction
    decomposition must include the following sum of m fractions.

    $$\frac{A_1}{(px+q)} + \frac{A_2}{(px+q)^2} + \cdots + \frac{A_m}{(px+q)^m}$$

4. **Quadratic factors:** For each factor of the form $(ax^2 + bx + c)^n$, the partial
    fraction decomposition must include the following sum of n fractions.

    $$\frac{B_1x + C_1}{ax^2 + bx + c} + \frac{B_2x + C_2}{(ax^2 + bx + c)^2} + \cdots + \frac{B_nx + C_n}{(ax^2 + bx + c)^n}$$

For a review of factorization techniques, see Precalculus, *6th edition, by Larson and Hostetler
or* Precalculus: A Graphing Approach, *4th edition, by Larson, Hostetler, and Edwards (Boston,
Massachusetts: Houghton Mifflin, 2004 and 2005, respectively).*

Linear Factors

Algebraic techniques for determining the constants in the numerators of a partial decomposition with linear or repeated linear factors are shown in Examples 1 and 2.

EXAMPLE I Distinct Linear Factors

Write the partial fraction decomposition for $\dfrac{1}{x^2 - 5x + 6}$.

Solution Because $x^2 - 5x + 6 = (x - 3)(x - 2)$, you should include one partial fraction for each factor and write

$$\frac{1}{x^2 - 5x + 6} = \frac{A}{x - 3} + \frac{B}{x - 2}$$

where A and B are to be determined. Multiplying this equation by the least common denominator $(x - 3)(x - 2)$ yields the **basic equation**

$$1 = A(x - 2) + B(x - 3). \qquad \text{Basic equation}$$

Because this equation is to be true for all x, you can substitute any *convenient* values for x to obtain equations in A and B. The most convenient values are the ones that make particular factors equal to 0.

NOTE Note that the substitutions for x in Example 1 are chosen for their convenience in determining values for A and B; $x = 2$ is chosen to eliminate the term $A(x - 2)$, and $x = 3$ is chosen to eliminate the term $B(x - 3)$. The goal is to make *convenient* substitutions whenever possible.

To solve for A, let $x = 3$ and obtain

$$1 = A(3 - 2) + B(3 - 3) \qquad \text{Let } x = 3 \text{ in basic equation.}$$
$$1 = A(1) + B(0)$$
$$A = 1.$$

To solve for B, let $x = 2$ and obtain

$$1 = A(2 - 2) + B(2 - 3) \qquad \text{Let } x = 2 \text{ in basic equation.}$$
$$1 = A(0) + B(-1)$$
$$B = -1.$$

So, the decomposition is

$$\frac{1}{x^2 - 5x + 6} = \frac{1}{x - 3} - \frac{1}{x - 2}$$

as shown at the beginning of this section.

FOR FURTHER INFORMATION To learn a different method for finding the partial fraction decomposition, called the Heavyside Method, see the article "Calculus to Algebra Connections in Partial Fraction Decomposition" by Joseph Wiener and Will Watkins in *The AMATYC Review*.

Be sure you see that the method of partial fractions is practical only for integrals of rational functions whose denominators factor "nicely." For instance, if the denominator in Example 1 were changed to $x^2 - 5x + 5$, its factorization as

$$x^2 - 5x + 5 = \left[x + \frac{5 + \sqrt{5}}{2} \right]\left[x - \frac{5 - \sqrt{5}}{2} \right]$$

would be too cumbersome to use with partial fractions. In such cases, you should use completing the square or a computer algebra system to perform the integration. If you do this, you should obtain

$$\int \frac{1}{x^2 - 5x + 5}\, dx = \frac{\sqrt{5}}{5} \ln\left| 2x - \sqrt{5} - 5 \right| - \frac{\sqrt{5}}{5} \ln\left| 2x + \sqrt{5} - 5 \right| + C.$$

EXAMPLE 2   Repeated Linear Factors

Find $\displaystyle\int \frac{5x^2 + 20x + 6}{x^3 + 2x^2 + x}\,dx$.

Solution   Because

$$x^3 + 2x^2 + x = x(x^2 + 2x + 1)$$
$$= x(x + 1)^2$$

you should include one fraction for *each power* of x and $(x + 1)$ and write

$$\frac{5x^2 + 20x + 6}{x(x + 1)^2} = \frac{A}{x} + \frac{B}{x + 1} + \frac{C}{(x + 1)^2}.$$

Multiplying by the least common denominator $x(x + 1)^2$ yields the *basic equation*

$$5x^2 + 20x + 6 = A(x + 1)^2 + Bx(x + 1) + Cx. \qquad \text{Basic equation}$$

To solve for A, let $x = 0$. This eliminates the B and C terms and yields

$$6 = A(1) + 0 + 0$$
$$A = 6.$$

To solve for C, let $x = -1$. This eliminates the A and B terms and yields

$$5 - 20 + 6 = 0 + 0 - C$$
$$C = 9.$$

The most convenient choices for x have been used, so to find the value of B, you can use *any other value* of x along with the calculated values of A and C. Using $x = 1$, $A = 6$, and $C = 9$ produces

$$5 + 20 + 6 = A(4) + B(2) + C$$
$$31 = 6(4) + 2B + 9$$
$$-2 = 2B$$
$$B = -1.$$

So, it follows that

$$\int \frac{5x^2 + 20x + 6}{x(x + 1)^2}\,dx = \int \left(\frac{6}{x} - \frac{1}{x + 1} + \frac{9}{(x + 1)^2} \right) dx$$
$$= 6\ln|x| - \ln|x + 1| + 9\frac{(x + 1)^{-1}}{-1} + C$$
$$= \ln\left| \frac{x^6}{x + 1} \right| - \frac{9}{x + 1} + C.$$

Try checking this result by differentiating. Include algebra in your check, simplifying the derivative until you have obtained the original integrand.   ■

NOTE   It is necessary to make as many substitutions for x as there are unknowns ($A, B, C, \ldots$) to be determined. For instance, in Example 2, three substitutions ($x = 0$, $x = -1$, and $x = 1$) were made to solve for A, B, and C.

FOR FURTHER INFORMATION For an alternative approach to using partial fractions, see the article " A Shortcut in Partial Fractions" by Xun-Cheng Huang in *The College Mathematics Journal*.

TECHNOLOGY   Most computer algebra systems, such as *Derive, Maple, Mathcad, Mathematica,* and the *TI-89*, can be used to convert a rational function to its partial fraction decomposition. For instance, using *Maple*, you obtain the following.

> convert$\left(\dfrac{5x^2 + 20x + 6}{x^3 + 2x^2 + x}, \text{parfrac}, x \right)$

$$\frac{6}{x} + \frac{9}{(x + 1)^2} - \frac{1}{x + 1}$$

Quadratic Factors

When using the method of partial fractions with *linear* factors, a convenient choice of x immediately yields a value for one of the coefficients. With *quadratic* factors, a system of linear equations usually has to be solved, regardless of the choice of x.

 EXAMPLE 3   **Distinct Linear and Quadratic Factors**

Find $\displaystyle\int \frac{2x^3 - 4x - 8}{(x^2 - x)(x^2 + 4)}\, dx.$

Solution   Because

$$(x^2 - x)(x^2 + 4) = x(x - 1)(x^2 + 4)$$

you should include one partial fraction for each factor and write

$$\frac{2x^3 - 4x - 8}{x(x - 1)(x^2 + 4)} = \frac{A}{x} + \frac{B}{x - 1} + \frac{Cx + D}{x^2 + 4}.$$

Multiplying by the least common denominator $x(x - 1)(x^2 + 4)$ yields the *basic equation*

$$2x^3 - 4x - 8 = A(x - 1)(x^2 + 4) + Bx(x^2 + 4) + (Cx + D)(x)(x - 1).$$

To solve for A, let $x = 0$ and obtain

$$-8 = A(-1)(4) + 0 + 0 \quad \Longrightarrow \quad 2 = A.$$

To solve for B, let $x = 1$ and obtain

$$-10 = 0 + B(5) + 0 \quad \Longrightarrow \quad -2 = B.$$

At this point, C and D are yet to be determined. You can find these remaining constants by choosing two other values for x and solving the resulting system of linear equations. If $x = -1$, then, using $A = 2$ and $B = -2$, you can write

$$-6 = (2)(-2)(5) + (-2)(-1)(5) + (-C + D)(-1)(-2)$$
$$2 = -C + D.$$

If $x = 2$, you have

$$0 = (2)(1)(8) + (-2)(2)(8) + (2C + D)(2)(1)$$
$$8 = 2C + D.$$

Solving the linear system by subtracting the first equation from the second

$$-C + D = 2$$
$$2C + D = 8$$

yields $C = 2$. Consequently, $D = 4$, and it follows that

$$\int \frac{2x^3 - 4x - 8}{x(x - 1)(x^2 + 4)}\, dx = \int \left(\frac{2}{x} - \frac{2}{x - 1} + \frac{2x}{x^2 + 4} + \frac{4}{x^2 + 4} \right) dx$$

$$= 2 \ln|x| - 2 \ln|x - 1| + \ln(x^2 + 4) + 2 \arctan \frac{x}{2} + C.$$

In Examples 1, 2, and 3, the solution of the basic equation began with substituting values of x that made the linear factors equal to 0. This method works well when the partial fraction decomposition involves linear factors. However, if the decomposition involves only quadratic factors, an alternative procedure is often more convenient.

EXAMPLE 4   Repeated Quadratic Factors

Find $\displaystyle\int \frac{8x^3 + 13x}{(x^2 + 2)^2}\, dx.$

Solution   Include one partial fraction for each power of $(x^2 + 2)$ and write

$$\frac{8x^3 + 13x}{(x^2 + 2)^2} = \frac{Ax + B}{x^2 + 2} + \frac{Cx + D}{(x^2 + 2)^2}.$$

Multiplying by the least common denominator $(x^2 + 2)^2$ yields the *basic equation*

$$8x^3 + 13x = (Ax + B)(x^2 + 2) + Cx + D.$$

Expanding the basic equation and collecting like terms produces

$$8x^3 + 13x = Ax^3 + 2Ax + Bx^2 + 2B + Cx + D$$
$$8x^3 + 13x = Ax^3 + Bx^2 + (2A + C)x + (2B + D).$$

Now, you can equate the coefficients of like terms on opposite sides of the equation.

$$8 = A \qquad\qquad 0 = 2B + D$$

$$8x^3 + 0x^2 + 13x + 0 = Ax^3 + Bx^2 + (2A + C)x + (2B + D)$$

$$0 = B$$

$$13 = 2A + C$$

Using the known values $A = 8$ and $B = 0$, you can write

$$13 = 2A + C = 2(8) + C \implies C = -3$$
$$0 = 2B + D = 2(0) + D \implies D = 0.$$

Finally, you can conclude that

$$\int \frac{8x^3 + 13x}{(x^2 + 2)^2}\, dx = \int \left(\frac{8x}{x^2 + 2} + \frac{-3x}{(x^2 + 2)^2} \right) dx$$

$$= 4 \ln(x^2 + 2) + \frac{3}{2(x^2 + 2)} + C.$$

TECHNOLOGY   Use a computer algebra system to evaluate the integral in Example 4—you might find that the form of the antiderivative is different. For instance, when you use a computer algebra system to work Example 4, you obtain

$$\int \frac{8x^3 + 13x}{(x^2 + 2)^2}\, dx = \ln(x^8 + 8x^6 + 24x^4 + 32x^2 + 16) + \frac{3}{2(x^2 + 2)} + C.$$

Is this result equivalent to that obtained in Example 4?

Writing About Concepts (continued)

56. Determine which value best approximates the area of the region between the x-axis and the graph of $f(x) = 10/[x(x^2 + 1)]$ over the interval $[1, 3]$. Make your selection on the basis of a sketch of the region and not by performing any calculations. Explain your reasoning.

(a) -6 (b) 6 (c) 3 (d) 5 (e) 8

57. *Area*   Find the area of the region bounded by the graphs of $y = 12/(x^2 + 5x + 6)$, $y = 0$, $x = 0$, and $x = 1$.

58. *Area*   Find the area of the region bounded by the graphs of $y = 7/(16 - x^2)$ and $y = 1$.

59. *Modeling Data*   The predicted cost C (in hundreds of thousands of dollars) for a company to remove $p\%$ of a chemical from its waste water is shown in the table.

p	0	10	20	30	40	50	60	70	80	90
C	0	0.7	1.0	1.3	1.7	2.0	2.7	3.6	5.5	11.2

A model for the data is given by

$$C = \frac{124p}{(10 + p)(100 - p)}, \quad 0 \le p < 100.$$

Use the model to find the average cost for removing between 75% and 80% of the chemical.

60. *Logistic Growth*   In Chapter 6, the exponential growth equation was derived from the assumption that the rate of growth was proportional to the existing quantity. In practice, there often exists some upper limit L past which growth cannot occur. In such cases, you assume the rate of growth to be proportional not only to the existing quantity, but also to the difference between the existing quantity y and the upper limit L. That is, $dy/dt = ky(L - y)$. In integral form, you can write this relationship as

$$\int \frac{dy}{y(L - y)} = \int k \, dt.$$

(a) A slope field for the differential equation $dy/dt = y(3 - y)$ is shown. Draw a possible solution to the differential equation if $y(0) = 5$, and another if $y(0) = \frac{1}{2}$. To print an enlarged copy of the graph, go to the website *www.mathgraphs.com*.

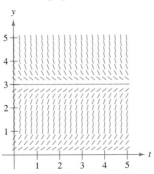

(b) Where $y(0)$ is greater than 3, what is the sign of the slope of the solution?

(c) For $y > 0$, find $\lim_{t \to \infty} y(t)$.

(d) Evaluate the two given integrals and solve for y as a function of t, where y_0 is the initial quantity.

(e) Use the result of part (d) to find and graph the solutions in part (a). Use a graphing utility to graph the solutions and compare the results with the solutions in part (a).

(f) The graph of the function y is a **logistic curve**. Show that the rate of growth is maximum at the point of inflection, and that this occurs when $y = L/2$.

61. *Volume and Centroid*   Consider the region bounded by the graphs of $y = 2x/(x^2 + 1)$, $y = 0$, $x = 0$, and $x = 3$. Find the volume of the solid generated by revolving the region about the x-axis. Find the centroid of the region.

62. *Volume*   Consider the region bounded by the graph of $y^2 = (2 - x)^2/(1 + x)^2$ on the interval $[0, 1]$. Find the volume of the solid generated by revolving this region about the x-axis.

63. *Epidemic Model*   A single infected individual enters a community of n susceptible individuals. Let x be the number of newly infected individuals at time t. The common epidemic model assumes that the disease spreads at a rate proportional to the product of the total number infected and the number not yet infected. So, $dx/dt = k(x + 1)(n - x)$ and you obtain

$$\int \frac{1}{(x + 1)(n - x)} \, dx = \int k \, dt.$$

Solve for x as a function of t.

64. *Chemical Reactions*   In a chemical reaction, one unit of compound Y and one unit of compound Z are converted into a single unit of compound X. x is the amount of compound X formed, and the rate of formation of X is proportional to the product of the amounts of unconverted compounds Y and Z. So, $dx/dt = k(y_0 - x)(z_0 - x)$, where y_0 and z_0 are the initial amounts of compounds Y and Z. From this equation you obtain

$$\int \frac{1}{(y_0 - x)(z_0 - x)} \, dx = \int k \, dt.$$

(a) Perform the two integrations and solve for x in terms of t.

(b) Use the result of part (a) to find x as $t \to \infty$ if (1) $y_0 < z_0$, (2) $y_0 > z_0$, and (3) $y_0 = z_0$.

65. Evaluate

$$\int_0^1 \frac{x}{1 + x^4} \, dx$$

in two different ways, one of which is partial fractions.

Putnam Exam Challenge

66. Prove $\dfrac{22}{7} - \pi = \displaystyle\int_0^1 \frac{x^4(1 - x)^4}{1 + x^2} \, dx$.

- **Evaluate an indefinite integral using a table of integrals.**
- **Evaluate an indefinite integral using reduction formulas.**
- **Evaluate an indefinite integral involving rational functions of sine and cosine.**

Integration by Tables

So far in this chapter you have studied several integration techniques that can be used with the basic integration rules. But merely knowing *how* to use the various techniques is not enough. You also need to know *when* to use them. Integration is first and foremost a problem of recognition. That is, you must recognize which rule or technique to apply to obtain an antiderivative. Frequently, a slight alteration of an integrand will require a different integration technique (or produce a function whose antiderivative is not an elementary function), as shown below.

$$\int x \ln x \, dx = \frac{x^2}{2} \ln x - \frac{x^2}{4} + C \qquad \text{Integration by parts}$$

$$\int \frac{\ln x}{x} \, dx = \frac{(\ln x)^2}{2} + C \qquad \text{Power Rule}$$

$$\int \frac{1}{x \ln x} \, dx = \ln|\ln x| + C \qquad \text{Log Rule}$$

$$\int \frac{x}{\ln x} \, dx = ? \qquad \text{Not an elementary function}$$

TECHNOLOGY   A computer algebra system consists, in part, of a database of integration formulas. The primary difference between using a computer algebra system and using tables of integrals is that with a computer algebra system the computer searches through the database to find a fit. With integration tables, *you* must do the searching.

Many people find tables of integrals to be a valuable supplement to the integration techniques discussed in this chapter. Tables of common integrals can be found in Appendix B. **Integration by tables** is not a "cure-all" for all of the difficulties that can accompany integration—using tables of integrals requires considerable thought and insight and often involves substitution.

Each integration formula in Appendix B can be developed using one or more of the techniques in this chapter. You should try to verify several of the formulas. For instance, Formula 4

$$\int \frac{u}{(a + bu)^2} \, du = \frac{1}{b^2} \left(\frac{a}{a + bu} + \ln|a + bu| \right) + C \qquad \text{Formula 4}$$

can be verified using the method of partial fractions, and Formula 19

$$\int \frac{\sqrt{a + bu}}{u} \, du = 2\sqrt{a + bu} + a \int \frac{du}{u\sqrt{a + bu}} \qquad \text{Formula 19}$$

can be verified using integration by parts. Note that the integrals in Appendix B are classified according to forms involving the following.

u^n	$(a + bu)$
$(a + bu + cu^2)$	$\sqrt{a + bu}$
$(a^2 \pm u^2)$	$\sqrt{u^2 \pm a^2}$
$\sqrt{a^2 - u^2}$	Trigonometric functions
Inverse trigonometric functions	Exponential functions
Logarithmic functions	

EXAMPLE 1   Integration by Tables

Find $\displaystyle\int \frac{dx}{x\sqrt{x - 1}}$.

Solution   Because the expression inside the radical is linear, you should consider forms involving $\sqrt{a + bu}$.

$$\int \frac{du}{u\sqrt{a + bu}} = \frac{2}{\sqrt{-a}} \arctan \sqrt{\frac{a + bu}{-a}} + C \qquad \text{Formula 17 } (a < 0)$$

Let $a = -1$, $b = 1$, and $u = x$. Then $du = dx$, and you can write

$$\int \frac{dx}{x\sqrt{x - 1}} = 2 \arctan \sqrt{x - 1} + C.$$

 ### EXAMPLE 2   Integration by Tables

Find $\displaystyle\int x\sqrt{x^4 - 9}\, dx$.

Solution   Because the radical has the form $\sqrt{u^2 - a^2}$, you should consider Formula 26.

$$\int \sqrt{u^2 - a^2}\, du = \frac{1}{2}\left(u\sqrt{u^2 - a^2} - a^2 \ln\left|u + \sqrt{u^2 - a^2}\right|\right) + C$$

Let $u = x^2$ and $a = 3$. Then $du = 2x\, dx$, and you have

$$\int x\sqrt{x^4 - 9}\, dx = \frac{1}{2}\int \sqrt{(x^2)^2 - 3^2}\,(2x)\, dx$$

$$= \frac{1}{4}\left(x^2\sqrt{x^4 - 9} - 9 \ln\left|x^2 + \sqrt{x^4 - 9}\right|\right) + C.$$

EXAMPLE 3   Integration by Tables

Find $\displaystyle\int \frac{x}{1 + e^{-x^2}}\, dx$.

Solution   Of the forms involving e^u, consider the following formula.

$$\int \frac{du}{1 + e^u} = u - \ln(1 + e^u) + C \qquad \text{Formula 84}$$

Let $u = -x^2$. Then $du = -2x\, dx$, and you have

$$\int \frac{x}{1 + e^{-x^2}}\, dx = -\frac{1}{2}\int \frac{-2x\, dx}{1 + e^{-x^2}}$$

$$= -\frac{1}{2}\left[-x^2 - \ln(1 + e^{-x^2})\right] + C$$

$$= \frac{1}{2}\left[x^2 + \ln(1 + e^{-x^2})\right] + C.$$

TECHNOLOGY   Example 3 shows the importance of having several solution techniques at your disposal. This integral is not difficult to solve with a table, but when it was entered into a well-known computer algebra system, the utility was unable to find the antiderivative.

Reduction Formulas

Several of the integrals in the integration tables have the form $\int f(x)\, dx = g(x) + \int h(x)\, dx$. Such integration formulas are called **reduction formulas** because they reduce a given integral to the sum of a function and a simpler integral.

EXAMPLE 4   Using a Reduction Formula

Find $\displaystyle\int x^3 \sin x\, dx$.

Solution   Consider the following three formulas.

$$\int u \sin u\, du = \sin u - u \cos u + C \qquad\qquad \text{Formula 52}$$

$$\int u^n \sin u\, du = -u^n \cos u + n \int u^{n-1} \cos u\, du \qquad \text{Formula 54}$$

$$\int u^n \cos u\, du = u^n \sin u - n \int u^{n-1} \sin u\, du \qquad \text{Formula 55}$$

Using Formula 54, Formula 55, and then Formula 52 produces

$$\int x^3 \sin x\, dx = -x^3 \cos x + 3 \int x^2 \cos x\, dx$$

$$= -x^3 \cos x + 3 \left(x^2 \sin x - 2 \int x \sin x\, dx \right)$$

$$= -x^3 \cos x + 3x^2 \sin x + 6x \cos x - 6 \sin x + C.$$

EXAMPLE 5   Using a Reduction Formula

Find $\displaystyle\int \frac{\sqrt{3 - 5x}}{2x}\, dx$.

Solution   Consider the following two formulas.

$$\int \frac{du}{u\sqrt{a + bu}} = \frac{1}{\sqrt{a}} \ln \left| \frac{\sqrt{a + bu} - \sqrt{a}}{\sqrt{a + bu} + \sqrt{a}} \right| + C \qquad \text{Formula 17 } (a > 0)$$

$$\int \frac{\sqrt{a + bu}}{u}\, du = 2\sqrt{a + bu} + a \int \frac{du}{u\sqrt{a + bu}} \qquad \text{Formula 19}$$

Using Formula 19, with $a = 3$, $b = -5$, and $u = x$, produces

$$\frac{1}{2} \int \frac{\sqrt{3 - 5x}}{x}\, dx = \frac{1}{2} \left(2\sqrt{3 - 5x} + 3 \int \frac{dx}{x\sqrt{3 - 5x}} \right)$$

$$= \sqrt{3 - 5x} + \frac{3}{2} \int \frac{dx}{x\sqrt{3 - 5x}}.$$

Using Formula 17, with $a = 3$, $b = -5$, and $u = x$, produces

$$\int \frac{\sqrt{3 - 5x}}{2x}\, dx = \sqrt{3 - 5x} + \frac{3}{2} \left(\frac{1}{\sqrt{3}} \ln \left| \frac{\sqrt{3 - 5x} - \sqrt{3}}{\sqrt{3 - 5x} + \sqrt{3}} \right| \right) + C$$

$$= \sqrt{3 - 5x} + \frac{\sqrt{3}}{2} \ln \left| \frac{\sqrt{3 - 5x} - \sqrt{3}}{\sqrt{3 - 5x} + \sqrt{3}} \right| + C.$$

TECHNOLOGY   Sometimes when you use computer algebra systems you obtain results that look very different, but are actually equivalent. Here is how several different systems evaluated the integral in Example 5.

Maple

$$\sqrt{3 - 5x} - \sqrt{3} \operatorname{arctanh}\left(\tfrac{1}{3}\sqrt{3 - 5x}\sqrt{3} \right)$$

Derive

$$\sqrt{3} \ln \left[\frac{\sqrt{(3 - 5x)} - \sqrt{3}}{\sqrt{x}} \right] + \sqrt{(3 - 5x)}$$

Mathematica

$$\operatorname{Sqrt}[3 - 5x] - \operatorname{Sqrt}[3] \operatorname{ArcTanh}\left[\frac{\operatorname{Sqrt}[3 - 5x]}{\operatorname{Sqrt}[3]} \right]$$

Mathcad

$$\sqrt{3 - 5x} + \tfrac{1}{2}\sqrt{3} \ln \left[-\tfrac{1}{5} \frac{(-6 + 5x + 2\sqrt{3}\sqrt{3 - 5x})}{x} \right]$$

Notice that computer algebra systems do not include a constant of integration.

Rational Functions of Sine and Cosine

EXAMPLE 6   Integration by Tables

Find $\displaystyle\int \frac{\sin 2x}{2 + \cos x}\,dx$.

Solution   Substituting $2 \sin x \cos x$ for $\sin 2x$ produces

$$\int \frac{\sin 2x}{2 + \cos x}\,dx = 2 \int \frac{\sin x \cos x}{2 + \cos x}\,dx.$$

A check of the forms involving $\sin u$ or $\cos u$ in Appendix B shows that none of those listed applies. So, you can consider forms involving $a + bu$. For example,

$$\int \frac{u\,du}{a + bu} = \frac{1}{b^2}\left(bu - a \ln|a + bu|\right) + C. \qquad \text{Formula 3}$$

Let $a = 2$, $b = 1$, and $u = \cos x$. Then $du = -\sin x\,dx$, and you have

$$
\begin{aligned}
2 \int \frac{\sin x \cos x}{2 + \cos x}\,dx &= -2 \int \frac{\cos x(-\sin x\,dx)}{2 + \cos x} \\
&= -2(\cos x - 2 \ln|2 + \cos x|) + C \\
&= -2 \cos x + 4 \ln|2 + \cos x| + C.
\end{aligned}
$$

Example 6 involves a rational expression of $\sin x$ and $\cos x$. If you are unable to find an integral of this form in the integration tables, try using the following special substitution to convert the trigonometric expression to a standard rational expression.

Substitution for Rational Functions of Sine and Cosine

For integrals involving rational functions of sine and cosine, the substitution

$$u = \frac{\sin x}{1 + \cos x} = \tan \frac{x}{2}$$

yields

$$\cos x = \frac{1 - u^2}{1 + u^2}, \quad \sin x = \frac{2u}{1 + u^2}, \quad \text{and} \quad dx = \frac{2\,du}{1 + u^2}.$$

Proof   From the substitution for u, it follows that

$$u^2 = \frac{\sin^2 x}{(1 + \cos x)^2} = \frac{1 - \cos^2 x}{(1 + \cos x)^2} = \frac{1 - \cos x}{1 + \cos x}.$$

Solving for $\cos x$ produces $\cos x = (1 - u^2)/(1 + u^2)$. To find $\sin x$, write $u = \sin x/(1 + \cos x)$ as

$$\sin x = u(1 + \cos x) = u\left(1 + \frac{1 - u^2}{1 + u^2}\right) = \frac{2u}{1 + u^2}.$$

Finally, to find dx, consider $u = \tan(x/2)$. Then you have $\arctan u = x/2$ and $dx = (2\,du)/(1 + u^2)$.

Exercises for Section 8.6

In Exercises 1 and 2, use a table of integrals with forms involving $a + bu$ to find the integral.

1. $\displaystyle\int \frac{x^2}{1+x}\,dx$

2. $\displaystyle\int \frac{2}{3x^2(2x-5)^2}\,dx$

In Exercises 3 and 4, use a table of integrals with forms involving $\sqrt{u^2 \pm a^2}$ to find the integral.

3. $\displaystyle\int e^x\sqrt{1+e^{2x}}\,dx$

4. $\displaystyle\int \frac{\sqrt{x^2-9}}{3x}\,dx$

In Exercises 5 and 6, use a table of integrals with forms involving $\sqrt{a^2 - u^2}$ to find the integral.

5. $\displaystyle\int \frac{1}{x^2\sqrt{1-x^2}}\,dx$

6. $\displaystyle\int \frac{x}{\sqrt{9-x^4}}\,dx$

In Exercises 7–10, use a table of integrals with forms involving the trigonometric functions to find the integral.

7. $\displaystyle\int \sin^4 2x\,dx$

8. $\displaystyle\int \frac{\cos^3\sqrt{x}}{\sqrt{x}}\,dx$

9. $\displaystyle\int \frac{1}{\sqrt{x}\left(1-\cos\sqrt{x}\right)}\,dx$

10. $\displaystyle\int \frac{1}{1-\tan 5x}\,dx$

In Exercises 11 and 12, use a table of integrals with forms involving e^u to find the integral.

11. $\displaystyle\int \frac{1}{1+e^{2x}}\,dx$

12. $\displaystyle\int e^{-x/2}\sin 2x\,dx$

In Exercises 13 and 14, use a table of integrals with forms involving $\ln u$ to find the integral.

13. $\displaystyle\int x^3\ln x\,dx$

14. $\displaystyle\int (\ln x)^3\,dx$

In Exercises 15–18, find the indefinite integral (a) using integration tables and (b) using the given method.

Integral	Method
15. $\displaystyle\int x^2 e^x\,dx$	Integration by parts
16. $\displaystyle\int x^4\ln x\,dx$	Integration by parts
17. $\displaystyle\int \frac{1}{x^2(x+1)}\,dx$	Partial fractions
18. $\displaystyle\int \frac{1}{x^2-75}\,dx$	Partial fractions

In Exercises 19–42, use integration tables to find the integral.

19. $\displaystyle\int x\,\mathrm{arcsec}(x^2+1)\,dx$

20. $\displaystyle\int \mathrm{arcsec}\,2x\,dx$

21. $\displaystyle\int \frac{1}{x^2\sqrt{x^2-4}}\,dx$

22. $\displaystyle\int \frac{1}{x^2+2x+2}\,dx$

23. $\displaystyle\int \frac{2x}{(1-3x)^2}\,dx$

24. $\displaystyle\int \frac{\theta^2}{1-\sin\theta^3}\,d\theta$

25. $\displaystyle\int e^x \arccos e^x\,dx$

26. $\displaystyle\int \frac{e^x}{1-\tan e^x}\,dx$

27. $\displaystyle\int \frac{x}{1-\sec x^2}\,dx$

28. $\displaystyle\int \frac{1}{t[1+(\ln t)^2]}\,dt$

29. $\displaystyle\int \frac{\cos\theta}{3+2\sin\theta+\sin^2\theta}\,d\theta$

30. $\displaystyle\int x^2\sqrt{2+9x^2}\,dx$

31. $\displaystyle\int \frac{1}{x^2\sqrt{2+9x^2}}\,dx$

32. $\displaystyle\int \sqrt{x}\,\arctan x^{3/2}\,dx$

33. $\displaystyle\int \frac{\ln x}{x(3+2\ln x)}\,dx$

34. $\displaystyle\int \frac{e^x}{(1-e^{2x})^{3/2}}\,dx$

35. $\displaystyle\int \frac{x}{(x^2-6x+10)^2}\,dx$

36. $\displaystyle\int (2x-3)^2\sqrt{(2x-3)^2+4}\,dx$

37. $\displaystyle\int \frac{x}{\sqrt{x^4-6x^2+5}}\,dx$

38. $\displaystyle\int \frac{\cos x}{\sqrt{\sin^2 x+1}}\,dx$

39. $\displaystyle\int \frac{x^3}{\sqrt{4-x^2}}\,dx$

40. $\displaystyle\int \sqrt{\frac{3-x}{3+x}}\,dx$

41. $\displaystyle\int \frac{e^{3x}}{(1+e^x)^3}\,dx$

42. $\displaystyle\int \tan^3\theta\,d\theta$

In Exercises 43–50, use integration tables to evaluate the integral.

43. $\displaystyle\int_0^1 xe^{x^2}\,dx$

44. $\displaystyle\int_0^3 \frac{x}{\sqrt{1+x}}\,dx$

45. $\displaystyle\int_1^3 x^2\ln x\,dx$

46. $\displaystyle\int_0^\pi x\sin x\,dx$

47. $\displaystyle\int_{-\pi/2}^{\pi/2} \frac{\cos x}{1+\sin^2 x}\,dx$

48. $\displaystyle\int_2^4 \frac{x^2}{(3x-5)^2}\,dx$

49. $\displaystyle\int_0^{\pi/2} t^3\cos t\,dt$

50. $\displaystyle\int_0^1 \sqrt{3+x^2}\,dx$

In Exercises 51–56, verify the integration formula.

51. $\displaystyle\int \frac{u^2}{(a+bu)^2}\,du = \frac{1}{b^3}\left(bu - \frac{a^2}{a+bu} - 2a\ln|a+bu|\right) + C$

52. $\displaystyle\int \frac{u^n}{\sqrt{a+bu}}\,du = \frac{2}{(2n+1)b}\left(u^n\sqrt{a+bu} - na\int \frac{u^{n-1}}{\sqrt{a+bu}}\,du\right)$

53. $\displaystyle\int \frac{1}{(u^2\pm a^2)^{3/2}}\,du = \frac{\pm u}{a^2\sqrt{u^2\pm a^2}} + C$

54. $\displaystyle\int u^n\cos u\,du = u^n\sin u - n\int u^{n-1}\sin u\,du$

55. $\displaystyle\int \arctan u\,du = u\arctan u - \ln\sqrt{1+u^2} + C$

56. $\int (\ln u)^n \, du = u(\ln u)^n - n \int (\ln u)^{n-1} \, du$

 In Exercises 57–62, use a computer algebra system to determine the antiderivative that passes through the given point. Use the system to graph the resulting antiderivative.

57. $\int \dfrac{1}{x^{3/2}\sqrt{1-x}} \, dx, \; \left(\frac{1}{2}, 5\right)$ **58.** $\int x\sqrt{x^2 + 2x} \, dx, \; (0, 0)$

59. $\int \dfrac{1}{(x^2 - 6x + 10)^2} \, dx, \; (3, 0)$

60. $\int \dfrac{\sqrt{2 - 2x - x^2}}{x + 1} \, dx, \; (0, \sqrt{2})$

61. $\int \dfrac{1}{\sin \theta \tan \theta} \, d\theta, \; \left(\frac{\pi}{4}, 2\right)$

62. $\int \dfrac{\sin \theta}{(\cos \theta)(1 + \sin \theta)} \, d\theta, \; (0, 1)$

In Exercises 63–70, find or evaluate the integral.

63. $\int \dfrac{1}{2 - 3\sin \theta} \, d\theta$ **64.** $\int \dfrac{\sin \theta}{1 + \cos^2 \theta} \, d\theta$

65. $\int_0^{\pi/2} \dfrac{1}{1 + \sin \theta + \cos \theta} \, d\theta$ **66.** $\int_0^{\pi/2} \dfrac{1}{3 - 2\cos \theta} \, d\theta$

67. $\int \dfrac{\sin \theta}{3 - 2\cos \theta} \, d\theta$ **68.** $\int \dfrac{\cos \theta}{1 + \cos \theta} \, d\theta$

69. $\int \dfrac{\cos \sqrt{\theta}}{\sqrt{\theta}} \, d\theta$ **70.** $\int \dfrac{1}{\sec \theta - \tan \theta} \, d\theta$

Area **In Exercises 71 and 72, find the area of the region bounded by the graphs of the equations.**

71. $y = \dfrac{x}{\sqrt{x + 1}}, \; y = 0, \; x = 8$ **72.** $y = \dfrac{x}{1 + e^{x^2}}, \; y = 0, \; x = 2$

Writing About Concepts

In Exercises 73–78, state (if possible) the method or integration formula you would use to find the antiderivative. Explain why you chose that method or formula. Do not integrate.

73. $\int \dfrac{e^x}{e^{2x} + 1} \, dx$ **74.** $\int \dfrac{e^x}{e^x + 1} \, dx$

75. $\int x \, e^{x^2} \, dx$ **76.** $\int x \, e^x \, dx$

77. $\int e^{x^2} \, dx$ **78.** $\int e^{2x} \sqrt{e^{2x} + 1} \, dx$

79. (a) Evaluate $\int x^n \ln x \, dx$ for $n = 1, 2$, and 3. Describe any patterns you notice.

(b) Write a general rule for evaluating the integral in part (a), for an integer $n \geq 1$.

80. Describe what is meant by a reduction formula. Give an example.

True or False? **In Exercises 81 and 82, determine whether the statement is true or false. If it is false, explain why or give an example that shows it is false.**

81. To use a table of integrals, the integral you are evaluating must appear in the table.

82. When using a table of integrals, you may have to make substitutions to rewrite your integral in the form in which it appears in the table.

83. *Work* A hydraulic cylinder on an industrial machine pushes a steel block a distance of x feet $(0 \leq x \leq 5)$, where the variable force required is $F(x) = 2000xe^{-x}$ pounds. Find the work done in pushing the block the full 5 feet through the machine.

84. *Work* Repeat Exercise 83, using $F(x) = \dfrac{500x}{\sqrt{26 - x^2}}$ pounds.

85. *Building Design* The cross section of a precast concrete beam for a building is bounded by the graphs of the equations

$$x = \dfrac{2}{\sqrt{1 + y^2}}, \; x = \dfrac{-2}{\sqrt{1 + y^2}}, \; y = 0, \text{ and } y = 3$$

where x and y are measured in feet. The length of the beam is 20 feet (see figure). (a) Find the volume V and the weight W of the beam. Assume the concrete weighs 148 pounds per cubic foot. (b) Then find the centroid of a cross section of the beam.

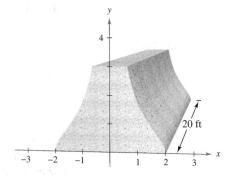

86. *Population* A population is growing according to the logistic model $N = \dfrac{5000}{1 + e^{4.8 - 1.9t}}$ where t is the time in days. Find the average population over the interval $[0, 2]$.

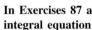 **In Exercises 87 and 88, use a graphing utility to (a) solve the integral equation for the constant k and (b) graph the region whose area is given by the integral.**

87. $\int_0^4 \dfrac{k}{2 + 3x} \, dx = 10$ **88.** $\int_0^k 6x^2 \, e^{-x/2} \, dx = 50$

Putnam Exam Challenge

89. Evaluate $\displaystyle\int_0^{\pi/2} \dfrac{dx}{1 + (\tan x)^{\sqrt{2}}}$.

This problem was composed by the Committee on the Putnam Prize Competition.
© The Mathematical Association of America. All rights reserved.

| Section 8.7 | Indeterminate Forms and L'Hôpital's Rule |

- Recognize limits that produce indeterminate forms.
- Apply L'Hôpital's Rule to evaluate a limit.

Indeterminate Forms

Recall from Chapters 2 and 4 that the forms $0/0$ and ∞/∞ are called *indeterminate* because they do not guarantee that a limit exists, nor do they indicate what the limit is, if one does exist. When you encountered one of these indeterminate forms earlier in the text, you attempted to rewrite the expression by using various algebraic techniques.

Indeterminate Form	*Limit*	*Algebraic Technique*
$\dfrac{0}{0}$	$\displaystyle\lim_{x\to-1}\frac{2x^2-2}{x+1}=\lim_{x\to-1}2(x-1)$ $=-4$	Divide numerator and denominator by $(x+1)$.
$\dfrac{\infty}{\infty}$	$\displaystyle\lim_{x\to\infty}\frac{3x^2-1}{2x^2+1}=\lim_{x\to\infty}\frac{3-(1/x^2)}{2+(1/x^2)}$ $=\dfrac{3}{2}$	Divide numerator and denominator by x^2.

Occasionally, you can extend these algebraic techniques to find limits of transcendental functions. For instance, the limit

$$\lim_{x\to0}\frac{e^{2x}-1}{e^x-1}$$

produces the indeterminate form $0/0$. Factoring and then dividing produces

$$\lim_{x\to0}\frac{e^{2x}-1}{e^x-1}=\lim_{x\to0}\frac{(e^x+1)(e^x-1)}{e^x-1}=\lim_{x\to0}(e^x+1)=2.$$

However, not all indeterminate forms can be evaluated by algebraic manipulation. This is often true when *both* algebraic and transcendental functions are involved. For instance, the limit

$$\lim_{x\to0}\frac{e^{2x}-1}{x}$$

produces the indeterminate form $0/0$. Rewriting the expression to obtain

$$\lim_{x\to0}\left(\frac{e^{2x}}{x}-\frac{1}{x}\right)$$

merely produces another indeterminate form, $\infty-\infty$. Of course, you could use technology to estimate the limit, as shown in the table and in Figure 8.14. From the table and the graph, the limit appears to be 2. (This limit will be verified in Example 1.)

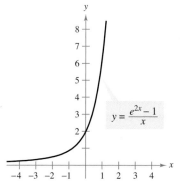

The limit as x approaches 0 appears to be 2.
Figure 8.14

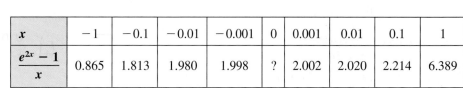

x	-1	-0.1	-0.01	-0.001	0	0.001	0.01	0.1	1
$\dfrac{e^{2x}-1}{x}$	0.865	1.813	1.980	1.998	?	2.002	2.020	2.214	6.389

GUILLAUME L'HÔPITAL (1661–1704)

L'Hôpital's Rule is named after the French mathematician Guillaume François Antoine de L'Hôpital. L'Hôpital is credited with writing the first text on differential calculus (in 1696) in which the rule publicly appeared. It was recently discovered that the rule and its proof were written in a letter from John Bernoulli to L'Hôpital. "... I acknowledge that I owe very much to the bright minds of the Bernoulli brothers. ... I have made free use of their discoveries ...," said L'Hôpital.

L'Hôpital's Rule

To find the limit illustrated in Figure 8.14, you can use a theorem called **L'Hôpital's Rule.** This theorem states that under certain conditions the limit of the quotient $f(x)/g(x)$ is determined by the limit of the quotient of the derivatives

$$\frac{f'(x)}{g'(x)}.$$

To prove this theorem, you can use a more general result called the **Extended Mean Value Theorem.**

THEOREM 8.3   The Extended Mean Value Theorem

If f and g are differentiable on an open interval (a, b) and continuous on $[a, b]$ such that $g'(x) \neq 0$ for any x in (a, b), then there exists a point c in (a, b) such that

$$\frac{f'(c)}{g'(c)} = \frac{f(b) - f(a)}{g(b) - g(a)}.$$

NOTE   To see why this is called the Extended Mean Value Theorem, consider the special case in which $g(x) = x$. For this case, you obtain the "standard" Mean Value Theorem as presented in Section 4.2.

The Extended Mean Value Theorem and L'Hôpital's Rule are both proved in Appendix A.

THEOREM 8.4   L'Hôpital's Rule

Let f and g be functions that are differentiable on an open interval (a, b) containing c, except possibly at c itself. Assume that $g'(x) \neq 0$ for all x in (a, b), except possibly at c itself. If the limit of $f(x)/g(x)$ as x approaches c produces the indeterminate form $0/0$, then

$$\lim_{x \to c} \frac{f(x)}{g(x)} = \lim_{x \to c} \frac{f'(x)}{g'(x)}$$

provided the limit on the right exists (or is infinite). This result also applies if the limit of $f(x)/g(x)$ as x approaches c produces any one of the indeterminate forms ∞/∞, $(-\infty)/\infty$, $\infty/(-\infty)$, or $(-\infty)/(-\infty)$.

NOTE   People occasionally use L'Hôpital's Rule incorrectly by applying the Quotient Rule to $f(x)/g(x)$. Be sure you see that the rule involves $f'(x)/g'(x)$, not the derivative of $f(x)/g(x)$.

L'Hôpital's Rule can also be applied to one-sided limits. For instance, if the limit of $f(x)/g(x)$ as x approaches c *from the right* produces the indeterminate form $0/0$, then

$$\lim_{x \to c^+} \frac{f(x)}{g(x)} = \lim_{x \to c^+} \frac{f'(x)}{g'(x)}$$

provided the limit exists (or is infinite).

FOR FURTHER INFORMATION To enhance your understanding of the necessity of the restriction that $g'(x)$ be nonzero for all x in (a, b), except possibly at c, see the article "Counterexamples to L'Hôpital's Rule" by R. P. Boas in *The American Mathematical Monthly*. To view this article, go to the website *www.matharticles.com*.

TECHNOLOGY   *Numerical and Graphical Approaches*   Use a numerical or a graphical approach to approximate each limit.

a. $\lim\limits_{x\to 0} \dfrac{2^{2x}-1}{x}$

b. $\lim\limits_{x\to 0} \dfrac{3^{2x}-1}{x}$

c. $\lim\limits_{x\to 0} \dfrac{4^{2x}-1}{x}$

d. $\lim\limits_{x\to 0} \dfrac{5^{2x}-1}{x}$

What pattern do you observe? Does an analytic approach have an advantage for these limits? If so, explain your reasoning.

EXAMPLE 1   **Indeterminate Form 0/0**

Evaluate $\lim\limits_{x\to 0} \dfrac{e^{2x}-1}{x}$.

Solution   Because direct substitution results in the indeterminate form 0/0

$$\lim_{x\to 0} \frac{e^{2x}-1}{x} \quad \overset{\displaystyle \lim_{x\to 0}(e^{2x}-1)=0}{\underset{\displaystyle \lim_{x\to 0} x = 0}{\diagup\diagdown}}$$

you can apply L'Hôpital's Rule as shown below.

$$\lim_{x\to 0} \frac{e^{2x}-1}{x} = \lim_{x\to 0} \frac{\dfrac{d}{dx}[e^{2x}-1]}{\dfrac{d}{dx}[x]} \qquad \text{Apply L'Hôpital's Rule.}$$

$$= \lim_{x\to 0} \frac{2e^{2x}}{1} \qquad \text{Differentiate numerator and denominator.}$$

$$= 2 \qquad \text{Evaluate the limit.} \qquad \rule[0.5ex]{2em}{0.4pt}$$

NOTE   In writing the string of equations in Example 1, you actually do not know that the first limit is equal to the second until you have shown that the second limit exists. In other words, if the second limit had not existed, it would not have been permissible to apply L'Hôpital's Rule.

Another form of L'Hôpital's Rule states that if the limit of $f(x)/g(x)$ as x approaches ∞ (or $-\infty$) produces the indeterminate form 0/0 or ∞/∞, then

$$\lim_{x\to\infty} \frac{f(x)}{g(x)} = \lim_{x\to\infty} \frac{f'(x)}{g'(x)}$$

provided the limit on the right exists.

EXAMPLE 2   **Indeterminate Form ∞/∞**

Evaluate $\lim\limits_{x\to\infty} \dfrac{\ln x}{x}$.

Solution   Because direct substitution results in the indeterminate form ∞/∞, you can apply L'Hôpital's Rule to obtain

$$\lim_{x\to\infty} \frac{\ln x}{x} = \lim_{x\to\infty} \frac{\dfrac{d}{dx}[\ln x]}{\dfrac{d}{dx}[x]} \qquad \text{Apply L'Hôpital's Rule.}$$

$$= \lim_{x\to\infty} \frac{1}{x} \qquad \text{Differentiate numerator and denominator.}$$

$$= 0. \qquad \text{Evaluate the limit.} \qquad \rule[0.5ex]{2em}{0.4pt}$$

NOTE   Try graphing $y_1 = \ln x$ and $y_2 = x$ in the same viewing window. Which function grows faster as x approaches ∞? How is this observation related to Example 2?

–3

The li
infini
Figu

Exercises for Section 8.7   See www.CalcChat.com for worked-out solutions to odd-numbered exercises.

Numerical and Graphical Analysis   In Exercises 1–4, complete the table and use the result to estimate the limit. Use a graphing utility to graph the function to support your result.

1. $\displaystyle\lim_{x\to 0}\frac{\sin 5x}{\sin 2x}$

x	-0.1	-0.01	-0.001	0.001	0.01	0.1
$f(x)$						

2. $\displaystyle\lim_{x\to 0}\frac{1-e^x}{x}$

x	-0.1	-0.01	-0.001	0.001	0.01	0.1
$f(x)$						

3. $\displaystyle\lim_{x\to\infty} x^5\, e^{-x/100}$

x	1	10	10^2	10^3	10^4	10^5
$f(x)$						

4. $\displaystyle\lim_{x\to\infty}\frac{6x}{\sqrt{3x^2-2x}}$

x	1	10	10^2	10^3	10^4	10^5
$f(x)$						

In Exercises 5–10, evaluate the limit (a) using techniques from Chapters 2 and 4 and (b) using L'Hôpital's Rule.

5. $\displaystyle\lim_{x\to 3}\frac{2(x-3)}{x^2-9}$

6. $\displaystyle\lim_{x\to -1}\frac{2x^2-x-3}{x+1}$

7. $\displaystyle\lim_{x\to 3}\frac{\sqrt{x+1}-2}{x-3}$

8. $\displaystyle\lim_{x\to 0}\frac{\sin 4x}{2x}$

9. $\displaystyle\lim_{x\to\infty}\frac{5x^2-3x+1}{3x^2-5}$

10. $\displaystyle\lim_{x\to\infty}\frac{2x+1}{4x^2+x}$

In Exercises 11–36, evaluate the limit, using L'Hôpital's Rule if necessary. (In Exercise 18, n is a positive integer.)

11. $\displaystyle\lim_{x\to 2}\frac{x^2-x-2}{x-2}$

12. $\displaystyle\lim_{x\to -1}\frac{x^2-x-2}{x+1}$

13. $\displaystyle\lim_{x\to 0}\frac{\sqrt{4-x^2}-2}{x}$

14. $\displaystyle\lim_{x\to 2^-}\frac{\sqrt{4-x^2}}{x-2}$

15. $\displaystyle\lim_{x\to 0}\frac{e^x-(1-x)}{x}$

16. $\displaystyle\lim_{x\to 1}\frac{\ln x^2}{x^2-1}$

17. $\displaystyle\lim_{x\to 0^+}\frac{e^x-(1+x)}{x^3}$

18. $\displaystyle\lim_{x\to 0^+}\frac{e^x-(1+x)}{x^n}$

19. $\displaystyle\lim_{x\to 0}\frac{\sin 2x}{\sin 3x}$

20. $\displaystyle\lim_{x\to 0}\frac{\sin ax}{\sin bx}$

21. $\displaystyle\lim_{x\to 0}\frac{\arcsin x}{x}$

22. $\displaystyle\lim_{x\to 1}\frac{\arctan x-(\pi/4)}{x-1}$

23. $\displaystyle\lim_{x\to\infty}\frac{3x^2-2x+1}{2x^2+3}$

24. $\displaystyle\lim_{x\to\infty}\frac{x-1}{x^2+2x+3}$

25. $\displaystyle\lim_{x\to\infty}\frac{x^2+2x+3}{x-1}$

26. $\displaystyle\lim_{x\to\infty}\frac{x^3}{x+2}$

27. $\displaystyle\lim_{x\to\infty}\frac{x^3}{e^{x/2}}$

28. $\displaystyle\lim_{x\to\infty}\frac{x^2}{e^x}$

29. $\displaystyle\lim_{x\to\infty}\frac{x}{\sqrt{x^2+1}}$

30. $\displaystyle\lim_{x\to\infty}\frac{x^2}{\sqrt{x^2+1}}$

31. $\displaystyle\lim_{x\to\infty}\frac{\cos x}{x}$

32. $\displaystyle\lim_{x\to\infty}\frac{\sin x}{x-\pi}$

33. $\displaystyle\lim_{x\to\infty}\frac{\ln x}{x^2}$

34. $\displaystyle\lim_{x\to\infty}\frac{\ln x^4}{x^3}$

35. $\displaystyle\lim_{x\to\infty}\frac{e^x}{x^2}$

36. $\displaystyle\lim_{x\to\infty}\frac{e^{x/2}}{x}$

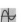 In Exercises 37–54, (a) describe the type of indeterminate form (if any) that is obtained by direct substitution. (b) Evaluate the limit, using L'Hôpital's Rule if necessary. (c) Use a graphing utility to graph the function and verify the result in part (b).

37. $\displaystyle\lim_{x\to\infty} x\ln x$

38. $\displaystyle\lim_{x\to 0^+} x^3\cot x$

39. $\displaystyle\lim_{x\to\infty}\left(x\sin\frac{1}{x}\right)$

40. $\displaystyle\lim_{x\to\infty} x\tan\frac{1}{x}$

41. $\displaystyle\lim_{x\to 0^+} x^{1/x}$

42. $\displaystyle\lim_{x\to 0^+}(e^x+x)^{2/x}$

43. $\displaystyle\lim_{x\to\infty} x^{1/x}$

44. $\displaystyle\lim_{x\to\infty}\left(1+\frac{1}{x}\right)^x$

45. $\displaystyle\lim_{x\to 0^+}(1+x)^{1/x}$

46. $\displaystyle\lim_{x\to\infty}(1+x)^{1/x}$

47. $\displaystyle\lim_{x\to 0^+}[3(x)^{x/2}]$

48. $\displaystyle\lim_{x\to 4^+}[3(x-4)]^{x-4}$

49. $\displaystyle\lim_{x\to 1^+}(\ln x)^{x-1}$

50. $\displaystyle\lim_{x\to 0^+}\left[\cos\left(\frac{\pi}{2}-x\right)\right]^x$

51. $\displaystyle\lim_{x\to 2^+}\left(\frac{8}{x^2-4}-\frac{x}{x-2}\right)$

52. $\displaystyle\lim_{x\to 2^+}\left(\frac{1}{x^2-4}-\frac{\sqrt{x-1}}{x^2-4}\right)$

53. $\displaystyle\lim_{x\to 1^+}\left(\frac{3}{\ln x}-\frac{2}{x-1}\right)$

54. $\displaystyle\lim_{x\to 0^+}\left(\frac{10}{x}-\frac{3}{x^2}\right)$

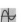 In Exercises 55–58, use a graphing utility to (a) graph the function and (b) find the required limit (if it exists).

55. $\displaystyle\lim_{x\to 3}\frac{x-3}{\ln(2x-5)}$

56. $\displaystyle\lim_{x\to 0^+}(\sin x)^x$

57. $\displaystyle\lim_{x\to\infty}\left(\sqrt{x^2+5x+2}-x\right)$

58. $\displaystyle\lim_{x\to\infty}\frac{x^3}{e^{2x}}$

Exercises for Section 8.8

See www.CalcChat.com for worked-out solutions to odd-numbered exercises.

In Exercises 1–4, decide whether the integral is improper. Explain your reasoning.

1. $\displaystyle\int_0^1 \frac{dx}{3x-2}$

2. $\displaystyle\int_1^3 \frac{dx}{x^2}$

3. $\displaystyle\int_0^1 \frac{2x-5}{x^2-5x+6}\,dx$

4. $\displaystyle\int_1^\infty \ln(x^2)\,dx$

In Exercises 5–10, explain why the integral is improper and determine whether it diverges or converges. Evaluate the integral if it converges.

5. $\displaystyle\int_0^4 \frac{1}{\sqrt{x}}\,dx$

6. $\displaystyle\int_3^4 \frac{1}{(x-3)^{3/2}}\,dx$

7. $\displaystyle\int_0^2 \frac{1}{(x-1)^2}\,dx$

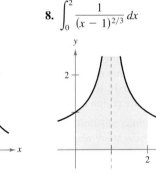

8. $\displaystyle\int_0^2 \frac{1}{(x-1)^{2/3}}\,dx$

9. $\displaystyle\int_0^\infty e^{-x}\,dx$

10. $\displaystyle\int_{-\infty}^0 e^{2x}\,dx$

Writing   In Exercises 11–14, explain why the evaluation of the integral is *incorrect*. Use the integration capabilities of a graphing utility to attempt to evaluate the integral. Determine whether the utility gives the correct answer.

11. $\displaystyle\int_{-1}^1 \frac{1}{x^2}\,dx = -2$

12. $\displaystyle\int_{-2}^2 \frac{-2}{(x-1)^3}\,dx = \frac{8}{9}$

13. $\displaystyle\int_0^\infty e^{-x}\,dx = 0$

14. $\displaystyle\int_0^\pi \sec x\,dx = 0$

In Exercises 15–32, determine whether the improper integral diverges or converges. Evaluate the integral if it converges.

15. $\displaystyle\int_1^\infty \frac{1}{x^2}\,dx$

16. $\displaystyle\int_1^\infty \frac{5}{x^3}\,dx$

17. $\displaystyle\int_1^\infty \frac{3}{\sqrt[3]{x}}\,dx$

18. $\displaystyle\int_1^\infty \frac{4}{\sqrt[4]{x}}\,dx$

19. $\displaystyle\int_{-\infty}^0 xe^{-2x}\,dx$

20. $\displaystyle\int_0^\infty xe^{-x/2}\,dx$

21. $\displaystyle\int_0^\infty x^2 e^{-x}\,dx$

22. $\displaystyle\int_0^\infty (x-1)e^{-x}\,dx$

23. $\displaystyle\int_0^\infty e^{-x}\cos x\,dx$

24. $\displaystyle\int_0^\infty e^{-ax}\sin bx\,dx, \quad a>0$

25. $\displaystyle\int_4^\infty \frac{1}{x(\ln x)^3}\,dx$

26. $\displaystyle\int_1^\infty \frac{\ln x}{x}\,dx$

27. $\displaystyle\int_{-\infty}^\infty \frac{2}{4+x^2}\,dx$

28. $\displaystyle\int_0^\infty \frac{x^3}{(x^2+1)^2}\,dx$

29. $\displaystyle\int_0^\infty \frac{1}{e^x+e^{-x}}\,dx$

30. $\displaystyle\int_0^\infty \frac{e^x}{1+e^x}\,dx$

31. $\displaystyle\int_0^\infty \cos \pi x\,dx$

32. $\displaystyle\int_0^\infty \sin \frac{x}{2}\,dx$

In Exercises 33–48, determine whether the improper integral diverges or converges. Evaluate the integral if it converges, and check your results with the results obtained by using the integration capabilities of a graphing utility.

33. $\displaystyle\int_0^1 \frac{1}{x^2}\,dx$

34. $\displaystyle\int_0^4 \frac{8}{x}\,dx$

35. $\displaystyle\int_0^8 \frac{1}{\sqrt[3]{8-x}}\,dx$

36. $\displaystyle\int_0^6 \frac{4}{\sqrt{6-x}}\,dx$

37. $\displaystyle\int_0^1 x \ln x\,dx$

38. $\displaystyle\int_0^e \ln x^2\,dx$

39. $\displaystyle\int_0^{\pi/2} \tan \theta\,d\theta$

40. $\displaystyle\int_0^{\pi/2} \sec \theta\,d\theta$

41. $\displaystyle\int_2^4 \frac{2}{x\sqrt{x^2-4}}\,dx$

42. $\displaystyle\int_0^2 \frac{1}{\sqrt{4-x^2}}\,dx$

43. $\displaystyle\int_2^4 \frac{1}{\sqrt{x^2-4}}\,dx$

44. $\displaystyle\int_0^2 \frac{1}{4-x^2}\,dx$

45. $\displaystyle\int_0^2 \frac{1}{\sqrt[3]{x-1}}\,dx$

46. $\displaystyle\int_1^3 \frac{2}{(x-2)^{8/3}}\,dx$

47. $\displaystyle\int_0^\infty \frac{4}{\sqrt{x}(x+6)}\,dx$

48. $\displaystyle\int_1^\infty \frac{1}{x \ln x}\,dx$

91. *The Gamma Function* The Gamma Function $\Gamma(n)$ is defined by

$$\Gamma(n) = \int_0^\infty x^{n-1}e^{-x}\,dx, \quad n > 0.$$

(a) Find $\Gamma(1)$, $\Gamma(2)$, and $\Gamma(3)$.

(b) Use integration by parts to show that $\Gamma(n + 1) = n\Gamma(n)$.

(c) Write $\Gamma(n)$ using factorial notation where n is a positive integer.

92. Prove that $I_n = \left(\dfrac{n - 1}{n + 2}\right)I_{n-1}$, where

$$I_n = \int_0^\infty \frac{x^{2n-1}}{(x^2 + 1)^{n+3}}\,dx, \quad n \geq 1.$$

Then evaluate each integral.

(a) $\displaystyle\int_0^\infty \frac{x}{(x^2 + 1)^4}\,dx$

(b) $\displaystyle\int_0^\infty \frac{x^3}{(x^2 + 1)^5}\,dx$

(c) $\displaystyle\int_0^\infty \frac{x^5}{(x^2 + 1)^6}\,dx$

Laplace Transforms Let $f(t)$ be a function defined for all positive values of t. The Laplace Transform of $f(t)$ is defined by

$$F(s) = \int_0^\infty e^{-st}f(t)\,dt$$

if the improper integral exists. Laplace Transforms are used to solve differential equations. In Exercises 93–100, find the Laplace Transform of the function.

93. $f(t) = 1$

94. $f(t) = t$

95. $f(t) = t^2$

96. $f(t) = e^{at}$

97. $f(t) = \cos at$

98. $f(t) = \sin at$

99. $f(t) = \cosh at$

100. $f(t) = \sinh at$

101. *Normal Probability* The mean height of American men between 18 and 24 years old is 70 inches, and the standard deviation is 3 inches. An 18- to 24-year-old man is chosen at random from the population. The probability that he is 6 feet tall or taller is

$$P(72 \leq x < \infty) = \int_{72}^\infty \frac{1}{3\sqrt{2\pi}}e^{-(x-70)^2/18}\,dx.$$

(Source: National Center for Health Statistics)

(a) Use a graphing utility to graph the integrand. Use the graphing utility to convince yourself that the area between the x-axis and the integrand is 1.

(b) Use a graphing utility to approximate $P(72 \leq x < \infty)$.

(c) Approximate $0.5 - P(70 \leq x \leq 72)$ using a graphing utility. Use the graph in part (a) to explain why this result is the same as the answer in part (b).

102. (a) Sketch the semicircle $y = \sqrt{4 - x^2}$.

(b) Explain why

$$\int_{-2}^2 \frac{2\,dx}{\sqrt{4 - x^2}} = \int_{-2}^2 \sqrt{4 - x^2}\,dx$$

without evaluating either integral.

103. For what value of c does the integral

$$\int_0^\infty \left(\frac{1}{\sqrt{x^2 + 1}} - \frac{c}{x + 1}\right)dx$$

converge? Evaluate the integral for this value of c.

104. For what value of c does the integral

$$\int_1^\infty \left(\frac{cx}{x^2 + 2} - \frac{1}{3x}\right)dx$$

converge? Evaluate the integral for this value of c.

105. *Volume* Find the volume of the solid generated by revolving the region bounded by the graph of f about the x-axis.

$$f(x) = \begin{cases} x \ln x, & 0 < x \leq 2 \\ 0, & x = 0 \end{cases}$$

106. *Volume* Find the volume of the solid generated by revolving the unbounded region lying between $y = -\ln x$ and the y-axis $(y \geq 0)$ about the x-axis.

u-Substitution In Exercises 107 and 108, rewrite the improper integral as a proper integral using the given u-substitution. Then use the Trapezoidal Rule with $n = 5$ to approximate the integral.

107. $\displaystyle\int_0^1 \frac{\sin x}{\sqrt{x}}\,dx, \quad u = \sqrt{x}$

108. $\displaystyle\int_0^1 \frac{\cos x}{\sqrt{1 - x}}\,dx, \quad u = \sqrt{1 - x}$

109. (a) Use a graphing utility to graph the function $y = e^{-x^2}$.

(b) Show that $\displaystyle\int_0^\infty e^{-x^2}\,dx = \int_0^1 \sqrt{-\ln y}\,dy$.

110. Let $\displaystyle\int_{-\infty}^\infty f(x)\,dx$ be convergent and let a and b be real numbers where $a \neq b$. Show that

$$\int_{-\infty}^a f(x)\,dx + \int_a^\infty f(x)\,dx = \int_{-\infty}^b f(x)\,dx + \int_b^\infty f(x)\,dx.$$

In Exercises 1–8, use the basic integration rules to find or evaluate the integral.

1. $\int x\sqrt{x^2 - 1}\, dx$

2. $\int xe^{x^2 - 1}\, dx$

3. $\int \dfrac{x}{x^2 - 1}\, dx$

4. $\int \dfrac{x}{\sqrt{1 - x^2}}\, dx$

5. $\int_1^e \dfrac{\ln(2x)}{x}\, dx$

6. $\int_{3/2}^2 2x\sqrt{2x - 3}\, dx$

7. $\int \dfrac{16}{\sqrt{16 - x^2}}\, dx$

8. $\int \dfrac{x^4 + 2x^2 + x + 1}{(x^2 + 1)^2}\, dx$

In Exercises 9–16, use integration by parts to find the integral.

9. $\int e^{2x} \sin 3x\, dx$

10. $\int (x^2 - 1)e^x\, dx$

11. $\int x\sqrt{x - 5}\, dx$

12. $\int \arctan 2x\, dx$

13. $\int x^2 \sin 2x\, dx$

14. $\int \ln\sqrt{x^2 - 1}\, dx$

15. $\int x \arcsin 2x\, dx$

16. $\int e^x \arctan e^x\, dx$

In Exercises 17–22, find the trigonometric integral.

17. $\int \cos^3(\pi x - 1)\, dx$

18. $\int \sin^2 \dfrac{\pi x}{2}\, dx$

19. $\int \sec^4 \dfrac{x}{2}\, dx$

20. $\int \tan \theta \sec^4 \theta\, d\theta$

21. $\int \dfrac{1}{1 - \sin \theta}\, d\theta$

22. $\int \cos 2\theta(\sin \theta + \cos \theta)^2\, d\theta$

Area   **In Exercises 23 and 24, find the area of the region.**

23. $y = \sin^4 x$

24. $y = \cos(3x) \cos x$

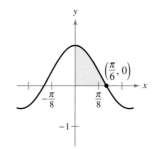

In Exercises 25–30, use trigonometric substitution to find or evaluate the integral.

25. $\int \dfrac{-12}{x^2\sqrt{4 - x^2}}\, dx$

26. $\int \dfrac{\sqrt{x^2 - 9}}{x}\, dx, \quad x > 3$

27. $\int \dfrac{x^3}{\sqrt{4 + x^2}}\, dx$

28. $\int \sqrt{9 - 4x^2}\, dx$

29. $\int_{-2}^0 \sqrt{4 - x^2}\, dx$

30. $\int_0^{\pi/2} \dfrac{\sin \theta}{1 + 2\cos^2 \theta}\, d\theta$

In Exercises 31 and 32, find the integral using each method.

31. $\int \dfrac{x^3}{\sqrt{4 + x^2}}\, dx$

  (a) Trigonometric substitution

  (b) Substitution: $u^2 = 4 + x^2$

  (c) Integration by parts: $dv = \left(x/\sqrt{4 + x^2}\right) dx$

32. $\int x\sqrt{4 + x}\, dx$

  (a) Trigonometric substitution

  (b) Substitution: $u^2 = 4 + x$

  (c) Substitution: $u = 4 + x$

  (d) Integration by parts: $dv = \sqrt{4 + x}\, dx$

In Exercises 33–38, use partial fractions to find the integral.

33. $\int \dfrac{x - 28}{x^2 - x - 6}\, dx$

34. $\int \dfrac{2x^3 - 5x^2 + 4x - 4}{x^2 - x}\, dx$

35. $\int \dfrac{x^2 + 2x}{x^3 - x^2 + x - 1}\, dx$

36. $\int \dfrac{4x - 2}{3(x - 1)^2}\, dx$

37. $\int \dfrac{x^2}{x^2 + 2x - 15}\, dx$

38. $\int \dfrac{\sec^2 \theta}{\tan \theta(\tan \theta - 1)}\, d\theta$

In Exercises 39–46, use integration tables to find or evaluate the integral.

39. $\int \dfrac{x}{(2 + 3x)^2}\, dx$

40. $\int \dfrac{x}{\sqrt{2 + 3x}}\, dx$

41. $\int_0^{\sqrt{\pi}/2} \dfrac{x}{1 + \sin x^2}\, dx$

42. $\int_0^1 \dfrac{x}{1 + e^{x^2}}\, dx$

43. $\int \dfrac{x}{x^2 + 4x + 8}\, dx$

44. $\int \dfrac{3}{2x\sqrt{9x^2 - 1}}\, dx, \quad x > \dfrac{1}{3}$

45. $\int \dfrac{1}{\sin \pi x \cos \pi x}\, dx$

46. $\int \dfrac{1}{1 + \tan \pi x}\, dx$

47. Verify the reduction formula

$$\int (\ln x)^n\, dx = x(\ln x)^n - n \int (\ln x)^{n-1}\, dx.$$

48. Verify the reduction formula

$$\int \tan^n x\, dx = \dfrac{1}{n - 1} \tan^{n-1} x - \int \tan^{n-2} x\, dx.$$

In Exercises 49–56, find the integral using any method.

49. $\int \theta \sin \theta \cos \theta \, d\theta$

50. $\int \dfrac{\csc \sqrt{2x}}{\sqrt{x}} \, dx$

51. $\int \dfrac{x^{1/4}}{1 + x^{1/2}} \, dx$

52. $\int \sqrt{1 + \sqrt{x}} \, dx$

53. $\int \sqrt{1 + \cos x} \, dx$

54. $\int \dfrac{3x^3 + 4x}{(x^2 + 1)^2} \, dx$

55. $\int \cos x \ln(\sin x) \, dx$

56. $\int (\sin \theta + \cos \theta)^2 \, d\theta$

In Exercises 57–60, solve the differential equation using any method.

57. $\dfrac{dy}{dx} = \dfrac{9}{x^2 - 9}$

58. $\dfrac{dy}{dx} = \dfrac{\sqrt{4 - x^2}}{2x}$

59. $y' = \ln(x^2 + x)$

60. $y' = \sqrt{1 - \cos \theta}$

In Exercises 61–66, evaluate the definite integral using any method. Use a graphing utility to verify your result.

61. $\displaystyle\int_2^{\sqrt{5}} x(x^2 - 4)^{3/2} \, dx$

62. $\displaystyle\int_0^1 \dfrac{x}{(x - 2)(x - 4)} \, dx$

63. $\displaystyle\int_1^4 \dfrac{\ln x}{x} \, dx$

64. $\displaystyle\int_0^2 xe^{3x} \, dx$

65. $\displaystyle\int_0^{\pi} x \sin x \, dx$

66. $\displaystyle\int_0^3 \dfrac{x}{\sqrt{1 + x}} \, dx$

Area   **In Exercises 67 and 68, find the area of the region.**

67. $y = x\sqrt{4 - x}$

68. $y = \dfrac{1}{25 - x^2}$

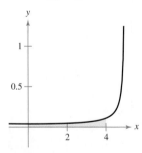

Centroid   **In Exercises 69 and 70, find the centroid of the region bounded by the graphs of the equations.**

69. $y = \sqrt{1 - x^2}, \quad y = 0$

70. $(x - 1)^2 + y^2 = 1, \quad (x - 4)^2 + y^2 = 4$

Arc Length   **In Exercises 71 and 72, approximate to two decimal places the arc length of the curve over the given interval.**

Function	Interval
71. $y = \sin x$	$[0, \pi]$
72. $y = \sin^2 x$	$[0, \pi]$

In Exercises 73–80, use L'Hôpital's Rule to evaluate the limit.

73. $\displaystyle\lim_{x \to 1} \dfrac{(\ln x)^2}{x - 1}$

74. $\displaystyle\lim_{x \to 0} \dfrac{\sin \pi x}{\sin 2\pi x}$

75. $\displaystyle\lim_{x \to \infty} \dfrac{e^{2x}}{x^2}$

76. $\displaystyle\lim_{x \to \infty} xe^{-x^2}$

77. $\displaystyle\lim_{x \to \infty} (\ln x)^{2/x}$

78. $\displaystyle\lim_{x \to 1^+} (x - 1)^{\ln x}$

79. $\displaystyle\lim_{n \to \infty} 1000\left(1 + \dfrac{0.09}{n}\right)^n$

80. $\displaystyle\lim_{x \to 1^+} \left(\dfrac{2}{\ln x} - \dfrac{2}{x - 1}\right)$

In Exercises 81–86, determine whether the improper integral diverges or converges. Evaluate the integral if it converges.

81. $\displaystyle\int_0^{16} \dfrac{1}{\sqrt[4]{x}} \, dx$

82. $\displaystyle\int_0^1 \dfrac{6}{x - 1} \, dx$

83. $\displaystyle\int_1^{\infty} x^2 \ln x \, dx$

84. $\displaystyle\int_0^{\infty} \dfrac{e^{-1/x}}{x^2} \, dx$

85. $\displaystyle\int_1^{\infty} \dfrac{\ln x}{x^2} \, dx$

86. $\displaystyle\int_1^{\infty} \dfrac{1}{\sqrt[4]{x}} \, dx$

87. ***Present Value***   The board of directors of a corporation is calculating the price to pay for a business that is forecast to yield a continuous flow of profit of \$500,000 per year. If money will earn a nominal rate of 5% per year compounded continuously, what is the present value of the business

(a) for 20 years?

(b) forever (in perpetuity)?

(*Note:* The present value for t_0 years is $\int_0^{t_0} 500{,}000e^{-0.05t} \, dt$.)

88. ***Volume***   Find the volume of the solid generated by revolving the region bounded by the graphs of $y = xe^{-x}$, $y = 0$, and $x = 0$ about the x-axis.

89. ***Probability***   The average lengths (from beak to tail) of different species of warblers in the eastern United States are approximately normally distributed with a mean of 12.9 centimeters and a standard deviation of 0.95 centimeter (see figure). The probability that a randomly selected warbler has a length between a and b centimeters is

$$P(a \le x \le b) = \dfrac{1}{0.95\sqrt{2\pi}} \int_a^b e^{-(x - 12.9)^2/2(0.95)^2} \, dx.$$

Use a graphing utility to approximate the probability that a randomly selected warbler has a length of (a) 13 centimeters or greater and (b) 15 centimeters or greater.   (*Source: Peterson's Field Guide: Eastern Birds*)

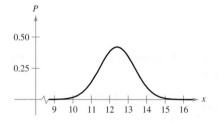

1. (a) Evaluate the integrals

$$\int_{-1}^{1} (1 - x^2) \, dx \quad \text{and} \quad \int_{-1}^{1} (1 - x^2)^2 \, dx.$$

(b) Use Wallis's Formulas to prove that

$$\int_{-1}^{1} (1 - x^2)^n \, dx = \frac{2^{2n+1}(n!)^2}{(2n + 1)!}$$

for all positive integers n.

2. (a) Evaluate the integrals $\int_{0}^{1} \ln x \, dx$ and $\int_{0}^{1} (\ln x)^2 \, dx$.

(b) Prove that

$$\int_{0}^{1} (\ln x)^n \, dx = (-1)^n \, n!$$

for all positive integers n.

3. Find the value of the positive constant c such that

$$\lim_{x \to \infty} \left(\frac{x + c}{x - c} \right)^x = 9.$$

4. Find the value of the positive constant c such that

$$\lim_{x \to \infty} \left(\frac{x - c}{x + c} \right)^x = \frac{1}{4}.$$

5. In the figure, the line $x = 1$ is tangent to the unit circle at A. The length of segment QA equals the length of the circular arc $\overset{\frown}{PA}$. Show that the length of segment OR approaches 2 as P approaches A.

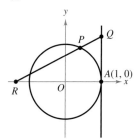

6. In the figure, the segment BD is the height of $\triangle OAB$. Let R be the ratio of the area of $\triangle DAB$ to that of the shaded region formed by deleting $\triangle OAB$ from the circular sector subtended by angle θ. Find $\lim\limits_{\theta \to 0^+} R$.

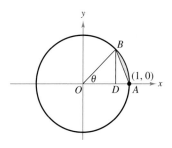

7. Consider the problem of finding the area of the region bounded by the x-axis, the line $x = 4$, and the curve

$$y = \frac{x^2}{(x^2 + 9)^{3/2}}.$$

(a) Use a graphing utility to graph the region and approximate its area.

(b) Use an appropriate trigonometric substitution to find the exact area.

(c) Use the substitution $x = 3 \sinh u$ to find the exact area and verify that you obtain the same answer as in part (b).

8. Use the substitution $u = \tan\dfrac{x}{2}$ to find the area of the shaded region under the graph of $y = \dfrac{1}{2 + \cos x}$, $0 \le x \le \pi/2$ (see figure).

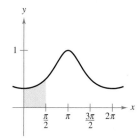

9. Find the arc length of the graph of the function $y = \ln(1 - x^2)$ on the interval $0 \le x \le \frac{1}{2}$ (see figure).

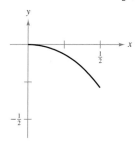

10. Find the centroid of the region above the x-axis and bounded above by the curve $y = e^{-c^2 x^2}$, where c is a positive constant (see figure).

$$\left(\textit{Hint:} \text{ Show that } \int_{0}^{\infty} e^{-c^2 x^2} \, dx = \frac{1}{c} \int_{0}^{\infty} e^{-x^2} \, dx. \right)$$

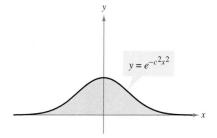

11. Some elementary functions, such as $f(x) = \sin(x^2)$, do not have antiderivatives that are elementary functions. Joseph Liouville proved that

$$\int \frac{e^x}{x} \, dx$$

does not have an elementary antiderivative. Use this fact to prove that

$$\int \frac{1}{\ln x} \, dx$$

is not elementary.

12. (a) Let $y = f^{-1}(x)$ be the inverse function of f. Use integration by parts to derive the formula

$$\int f^{-1}(x) \, dx = xf^{-1}(x) - \int f(y) \, dy.$$

(b) Use the formula in part (a) to find the integral

$$\int \arcsin x \, dx.$$

(c) Use the formula in part (a) to find the area under the graph of $y = \ln x$, $1 \le x \le e$ (see figure).

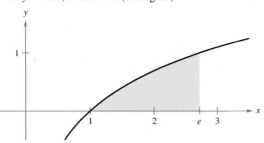

13. Factor the polynomial $p(x) = x^4 + 1$ and then find the area under the graph of $y = \dfrac{1}{x^4 + 1}$, $0 \le x \le 1$ (see figure).

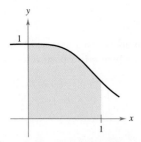

14. (a) Use the substitution $u = \dfrac{\pi}{2} - x$ to evaluate the integral

$$\int_0^{\pi/2} \frac{\sin x}{\cos x + \sin x} \, dx.$$

(b) Let n be a positive integer. Evaluate the integral

$$\int_0^{\pi/2} \frac{\sin^n x}{\cos^n x + \sin^n x} \, dx.$$

15. Use a graphing utility to estimate each limit. Then calculate each limit using L'Hôpital's Rule. What can you conclude about the indeterminate form $0 \cdot \infty$?

(a) $\displaystyle\lim_{x \to 0^+} \left(\cot x + \frac{1}{x} \right)$ (b) $\displaystyle\lim_{x \to 0^+} \left(\cot x - \frac{1}{x} \right)$

(c) $\displaystyle\lim_{x \to 0^+} \left[\left(\cot x + \frac{1}{x} \right)\left(\cot x - \frac{1}{x} \right) \right]$

16. Suppose the denominator of a rational function can be factored into distinct linear factors

$$D(x) = (x - c_1)(x - c_2) \cdots (x - c_n)$$

for a positive integer n and distinct real numbers $c_1, c_2, \ldots, c_n$. If N is a polynomial of degree less than n, show that

$$\frac{N(x)}{D(x)} = \frac{P_1}{x - c_1} + \frac{P_2}{x - c_2} + \cdots + \frac{P_n}{x - c_n}$$

where $P_k = N(c_k)/D'(c_k)$ for $k = 1, 2, \ldots, n$. Note that this is the partial fraction decomposition of $N(x)/D(x)$.

17. Use the results of Exercise 16 to find the partial fraction decomposition of

$$\frac{x^3 - 3x^2 + 1}{x^4 - 13x^2 + 12x}.$$

18. The velocity v (in feet per second) of a rocket whose initial mass (including fuel) is m is given by

$$v = gt + u \ln \frac{m}{m - rt}, \quad t < \frac{m}{r}$$

where u is the expulsion speed of the fuel, r is the rate at which the fuel is consumed, and $g = -32$ feet per second per second is the acceleration due to gravity. Find the position equation for a rocket for which $m = 50{,}000$ pounds, $u = 12{,}000$ feet per second, and $r = 400$ pounds per second. What is the height of the rocket when $t = 100$ seconds? (Assume that the rocket was fired from ground level and is moving straight upward.)

19. Suppose that $f(a) = f(b) = g(a) = g(b) = 0$ and the second derivatives of f and g are continuous on the closed interval $[a, b]$. Prove that

$$\int_a^b f(x)g''(x) \, dx = \int_a^b f''(x)g(x) \, dx.$$

20. Suppose that $f(a) = f(b) = 0$ and the second derivatives of f exist on the closed interval $[a, b]$. Prove that

$$\int_a^b (x - a)(x - b)f''(x) \, dx = 2\int_a^b f(x) \, dx.$$

21. Using the inequality

$$\frac{1}{x^5} + \frac{1}{x^{10}} + \frac{1}{x^{15}} < \frac{1}{x^5 - 1} < \frac{1}{x^5} + \frac{1}{x^{10}} + \frac{2}{x^{15}}$$

for $x \ge 2$, approximate $\displaystyle\int_2^\infty \frac{1}{x^5 - 1} \, dx.$

9 Infinite Series

A *ballista* was used as a portable rock-throwing machine. The propulsion mechanism was similar in appearance to a crossbow. Skilled artillerymen aimed and fired the ballista entirely by eye. What type of projectile path do you think these artillerymen preferred–a high, arching trajectory or a low, relatively level trajectory? Why?

Maclaurin *polynomials* approximate a given function in an interval around $x = 0$. As you add terms to the Maclaurin polynomial, it becomes a better approximation of the given function near $x = 0$. In Section 9.10, you will see that a Maclaurin *series* is equivalent to the given function (under suitable conditions).

Charles & Josette Lenars/Corbis

593

Sequences

- List the terms of a sequence.
- Determine whether a sequence converges or diverges.
- Write a formula for the nth term of a sequence.
- Use properties of monotonic sequences and bounded sequences.

Sequences

In mathematics, the word "sequence" is used in much the same way as in ordinary English. To say that a collection of objects or events is *in sequence* usually means that the collection is ordered so that it has an identified first member, second member, third member, and so on.

Mathematically, a **sequence** is defined as a function whose domain is the set of positive integers. Although a sequence is a function, it is common to represent sequences by subscript notation rather than by the standard function notation. For instance, in the sequence

$$1, \quad 2, \quad 3, \quad 4, \quad \ldots, \quad n, \quad \ldots$$
$$\downarrow \quad \downarrow \quad \downarrow \quad \downarrow \qquad \downarrow \qquad \downarrow \qquad \downarrow \qquad \text{Sequence}$$
$$a_1, \quad a_2, \quad a_3, \quad a_4, \quad \ldots, \quad a_n, \quad \ldots$$

1 is mapped onto a_1, 2 is mapped onto a_2, and so on. The numbers $a_1, a_2, a_3, \ldots, a_n$, $\ldots$ are the **terms** of the sequence. The number a_n is the **nth term** of the sequence, and the entire sequence is denoted by $\{a_n\}$.

EXAMPLE 1 Listing the Terms of a Sequence

a. The terms of the sequence $\{a_n\} = \{3 + (-1)^n\}$ are

$$3 + (-1)^1, \ 3 + (-1)^2, \ 3 + (-1)^3, \ 3 + (-1)^4, \ldots$$
$$2, \qquad 4, \qquad 2, \qquad 4, \qquad \ldots.$$

b. The terms of the sequence $\{b_n\} = \left\{\dfrac{n}{1 - 2n}\right\}$ are

$$\frac{1}{1 - 2 \cdot 1}, \ \frac{2}{1 - 2 \cdot 2}, \ \frac{3}{1 - 2 \cdot 3}, \ \frac{4}{1 - 2 \cdot 4}, \ \cdots$$
$$-1, \qquad -\frac{2}{3}, \qquad -\frac{3}{5}, \qquad -\frac{4}{7}, \qquad \cdots.$$

c. The terms of the sequence $\{c_n\} = \left\{\dfrac{n^2}{2^n - 1}\right\}$ are

$$\frac{1^2}{2^1 - 1}, \ \frac{2^2}{2^2 - 1}, \ \frac{3^2}{2^3 - 1}, \ \frac{4^2}{2^4 - 1}, \ \cdots$$
$$\frac{1}{1}, \qquad \frac{4}{3}, \qquad \frac{9}{7}, \qquad \frac{16}{15}, \qquad \cdots.$$

d. The terms of the **recursively defined** sequence $\{d_n\}$, where $d_1 = 25$ and $d_{n+1} = d_n - 5$ are

$$25, \quad 25 - 5 = 20, \quad 20 - 5 = 15, \quad 15 - 5 = 10, \ldots.$$

EXPLORATION

Finding Patterns Describe a pattern for each of the following sequences. Then use your description to write a formula for the nth term of each sequence. As n increases, do the terms appear to be approaching a limit? Explain your reasoning.

a. $1, \frac{1}{2}, \frac{1}{4}, \frac{1}{8}, \frac{1}{16}, \ldots$

b. $1, \frac{1}{2}, \frac{1}{6}, \frac{1}{24}, \frac{1}{120}, \ldots$

c. $10, \frac{10}{3}, \frac{10}{6}, \frac{10}{10}, \frac{10}{15}, \ldots$

d. $\frac{1}{4}, \frac{4}{9}, \frac{9}{16}, \frac{16}{25}, \frac{25}{36}, \ldots$

e. $\frac{3}{7}, \frac{5}{10}, \frac{7}{13}, \frac{9}{16}, \frac{11}{19}, \ldots$

NOTE Occasionally, it is convenient to begin a sequence with a_0, so that the terms of the sequence become

$$a_0, a_1, a_2, a_3, \ldots, a_n, \ldots$$

STUDY TIP Some sequences are defined recursively. To define a sequence recursively, you need to be given one or more of the first few terms. All other terms of the sequence are then defined using previous terms, as shown in Example 1(d).

Limit of a Sequence

The primary focus of this chapter concerns sequences whose terms approach limiting values. Such sequences are said to **converge.** For instance, the sequence $\{1/2^n\}$

$$\frac{1}{2}, \frac{1}{4}, \frac{1}{8}, \frac{1}{16}, \frac{1}{32}, \cdots$$

converges to 0, as indicated in the following definition.

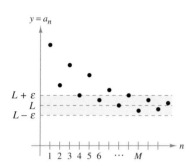

For $n > M$, the terms of the sequence all lie within ε units of L.
Figure 9.1

Definition of the Limit of a Sequence

Let L be a real number. The **limit** of a sequence $\{a_n\}$ is L, written as

$$\lim_{n \to \infty} a_n = L$$

if for each $\varepsilon > 0$, there exists $M > 0$ such that $|a_n - L| < \varepsilon$ whenever $n > M$. If the limit L of a sequence exists, then the sequence **converges** to L. If the limit of a sequence does not exist, then the sequence **diverges.**

Graphically, this definition says that eventually (for $n > M$ and $\varepsilon > 0$) the terms of a sequence that converges to L will lie within the band between the lines $y = L + \varepsilon$ and $y = L - \varepsilon$, as shown in Figure 9.1.

If a sequence $\{a_n\}$ agrees with a function f at every positive integer, and if $f(x)$ approaches a limit L as $x \to \infty$, the sequence must converge to the same limit L.

THEOREM 9.1   Limit of a Sequence

Let L be a real number. Let f be a function of a real variable such that

$$\lim_{x \to \infty} f(x) = L.$$

If $\{a_n\}$ is a sequence such that $f(n) = a_n$ for every positive integer n, then

$$\lim_{n \to \infty} a_n = L.$$

NOTE   There are different ways in which a sequence can fail to have a limit. One way is that the terms of the sequence increase without bound or decrease without bound. These cases are written symbolically as follows.

Terms increase without bound:

$$\lim_{n \to \infty} a_n = \infty$$

Terms decrease without bound:

$$\lim_{n \to \infty} a_n = -\infty$$

EXAMPLE 2   **Finding the Limit of a Sequence**

Find the limit of the sequence whose nth term is

$$a_n = \left(1 + \frac{1}{n}\right)^n.$$

Solution   In Theorem 5.15, you learned that

$$\lim_{x \to \infty} \left(1 + \frac{1}{x}\right)^x = e.$$

So, you can apply Theorem 9.1 to conclude that

$$\lim_{n \to \infty} a_n = \lim_{n \to \infty} \left(1 + \frac{1}{n}\right)^n$$

$$= e.$$

The following properties of limits of sequences parallel those given for limits of functions of a real variable in Section 2.3.

THEOREM 9.2   Properties of Limits of Sequences

Let $\lim\limits_{n\to\infty} a_n = L$ and $\lim\limits_{n\to\infty} b_n = K$.

1. $\lim\limits_{n\to\infty} (a_n \pm b_n) = L \pm K$ **2.** $\lim\limits_{n\to\infty} ca_n = cL$, c is any real number

3. $\lim\limits_{n\to\infty} (a_n b_n) = LK$ **4.** $\lim\limits_{n\to\infty} \dfrac{a_n}{b_n} = \dfrac{L}{K}$, $b_n \neq 0$ and $K \neq 0$

 EXAMPLE 3   Determining Convergence or Divergence

a. Because the sequence $\{a_n\} = \{3 + (-1)^n\}$ has terms

$$2, 4, 2, 4, \ldots$$ See Example 1(a), page 594.

that alternate between 2 and 4, the limit

$$\lim_{n\to\infty} a_n$$

does not exist. So, the sequence diverges.

b. For $\{b_n\} = \left\{ \dfrac{n}{1 - 2n} \right\}$, divide the numerator and denominator by n to obtain

$$\lim_{n\to\infty} \frac{n}{1 - 2n} = \lim_{n\to\infty} \frac{1}{(1/n) - 2} = -\frac{1}{2}$$ See Example 1(b), page 594.

which implies that the sequence converges to $-\frac{1}{2}$.

EXAMPLE 4   Using L'Hôpital's Rule to Determine Convergence

Show that the sequence whose nth term is $a_n = \dfrac{n^2}{2^n - 1}$ converges.

Solution   Consider the function of a real variable

$$f(x) = \frac{x^2}{2^x - 1}.$$

Applying L'Hôpital's Rule twice produces

$$\lim_{x\to\infty} \frac{x^2}{2^x - 1} = \lim_{x\to\infty} \frac{2x}{(\ln 2)2^x} = \lim_{x\to\infty} \frac{2}{(\ln 2)^2 2^x} = 0.$$

Because $f(n) = a_n$ for every positive integer, you can apply Theorem 9.1 to conclude that

$$\lim_{n\to\infty} \frac{n^2}{2^n - 1} = 0.$$ See Example 1(c), page 594.

So, the sequence converges to 0.

TECHNOLOGY   Use a graphing utility to graph the function in Example 4. Notice that as x approaches infinity, the value of the function gets closer and closer to 0. If you have access to a graphing utility that can generate terms of a sequence, try using it to calculate the first 20 terms of the sequence in Example 4. Then view the terms to observe numerically that the sequence converges to 0.

---

 indicates that in the HM mathSpace® *CD-ROM and the online* Eduspace® *system for this text, you will find an Open Exploration, which further explores this example using the computer algebra systems* Maple, Mathcad, Mathematica, *and* Derive.

The symbol $n!$ (read "n factorial") is used to simplify some of the formulas developed in this chapter. Let n be a positive integer; then **n factorial** is defined as

$$n! = 1 \cdot 2 \cdot 3 \cdot 4 \cdots (n - 1) \cdot n.$$

As a special case, **zero factorial** is defined as $0! = 1$. From this definition, you can see that $1! = 1$, $2! = 1 \cdot 2 = 2$, $3! = 1 \cdot 2 \cdot 3 = 6$, and so on. Factorials follow the same conventions for order of operations as exponents. That is, just as $2x^3$ and $(2x)^3$ imply different orders of operations, $2n!$ and $(2n)!$ imply the following orders.

$$2n! = 2(n!) = 2(1 \cdot 2 \cdot 3 \cdot 4 \cdots n)$$

and

$$(2n)! = 1 \cdot 2 \cdot 3 \cdot 4 \cdots n \cdot (n + 1) \cdots 2n$$

Another useful limit theorem that can be rewritten for sequences is the Squeeze Theorem from Section 2.3.

THEOREM 9.3   Squeeze Theorem for Sequences

If

$$\lim_{n \to \infty} a_n = L = \lim_{n \to \infty} b_n$$

and there exists an integer N such that $a_n \le c_n \le b_n$ for all $n > N$, then

$$\lim_{n \to \infty} c_n = L.$$

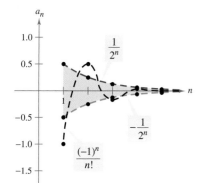

For $n \ge 4$, $(-1)^n / n!$ is squeezed between $-1/2^n$ and $1/2^n$.
Figure 9.2

EXAMPLE 5   **Using the Squeeze Theorem**

Show that the sequence $\{c_n\} = \left\{ (-1)^n \dfrac{1}{n!} \right\}$ converges, and find its limit.

Solution   To apply the Squeeze Theorem, you must find two convergent sequences that can be related to the given sequence. Two possibilities are $a_n = -1/2^n$ and $b_n = 1/2^n$, both of which converge to 0. By comparing the term $n!$ with 2^n, you can see that

$$n! = 1 \cdot 2 \cdot 3 \cdot 4 \cdot 5 \cdot 6 \cdots n = 24 \cdot \underbrace{5 \cdot 6 \cdots n}_{n - 4 \text{ factors}} \qquad (n \ge 4)$$

and

$$2^n = 2 \cdot 2 \cdot 2 \cdot 2 \cdot 2 \cdot 2 \cdots 2 = 16 \cdot \underbrace{2 \cdot 2 \cdots 2}_{n - 4 \text{ factors}}. \qquad (n \ge 4)$$

This implies that for $n \ge 4$, $2^n < n!$, and you have

$$\frac{-1}{2^n} \le (-1)^n \frac{1}{n!} \le \frac{1}{2^n}, \quad n \ge 4$$

as shown in Figure 9.2. So, by the Squeeze Theorem it follows that

$$\lim_{n \to \infty} (-1)^n \frac{1}{n!} = 0.$$

NOTE   Example 5 suggests something about the rate at which $n!$ increases as $n \to \infty$. As Figure 9.2 suggests, both $1/2^n$ and $1/n!$ approach 0 as $n \to \infty$. Yet $1/n!$ approaches 0 so much faster than $1/2^n$ does that

$$\lim_{n \to \infty} \frac{1/n!}{1/2^n} = \lim_{n \to \infty} \frac{2^n}{n!} = 0.$$

In fact, it can be shown that for any fixed number k,

$$\lim_{n \to \infty} \frac{k^n}{n!} = 0.$$

This means that *the factorial function grows faster than any exponential function.*

In Example 5, the sequence $\{c_n\}$ has both positive and negative terms. For this sequence, it happens that the sequence of absolute values, $\{|c_n|\}$, also converges to 0. You can show this by the Squeeze Theorem using the inequality

$$0 \le \frac{1}{n!} \le \frac{1}{2^n}, \quad n \ge 4.$$

In such cases, it is often convenient to consider the sequence of absolute values—and then apply Theorem 9.4, which states that if the absolute value sequence converges to 0, the original signed sequence also converges to 0.

THEOREM 9.4 **Absolute Value Theorem**

For the sequence $\{a_n\}$, if

$$\lim_{n\to\infty} |a_n| = 0 \qquad \text{then} \qquad \lim_{n\to\infty} a_n = 0.$$

Proof Consider the two sequences $\{|a_n|\}$ and $\{-|a_n|\}$. Because both of these sequences converge to 0 and

$$-|a_n| \le a_n \le |a_n|$$

you can use the Squeeze Theorem to conclude that $\{a_n\}$ converges to 0. ▪

Pattern Recognition for Sequences

Sometimes the terms of a sequence are generated by some rule that does not explicitly identify the nth term of the sequence. In such cases, you may be required to discover a *pattern* in the sequence and to describe the nth term. Once the nth term has been specified, you can investigate the convergence or divergence of the sequence.

EXAMPLE 6 **Finding the nth Term of a Sequence**

Find a sequence $\{a_n\}$ whose first five terms are

$$\frac{2}{1}, \frac{4}{3}, \frac{8}{5}, \frac{16}{7}, \frac{32}{9}, \ldots$$

and then determine whether the particular sequence you have chosen converges or diverges.

Solution First, note that the numerators are successive powers of 2, and the denominators form the sequence of positive odd integers. By comparing a_n with n, you have the following pattern.

$$\frac{2^1}{1}, \frac{2^2}{3}, \frac{2^3}{5}, \frac{2^4}{7}, \frac{2^5}{9}, \ldots, \frac{2^n}{2n-1}$$

Using L'Hôpital's Rule to evaluate the limit of $f(x) = 2^x/(2x-1)$, you obtain

$$\lim_{x\to\infty} \frac{2^x}{2x-1} = \lim_{x\to\infty} \frac{2^x(\ln 2)}{2} = \infty \quad \Longrightarrow \quad \lim_{n\to\infty} \frac{2^n}{2n-1} = \infty.$$

So, the sequence diverges. ▪

Without a specific rule for generating the terms of a sequence or some knowledge of the context in which the terms of the sequence are obtained, it is not possible to determine the convergence or divergence of the sequence merely from its first several terms. For instance, although the first three terms of the following four sequences are identical, the first two sequences converge to 0, the third sequence converges to $\frac{1}{9}$, and the fourth sequence diverges.

$$\{a_n\} : \frac{1}{2}, \frac{1}{4}, \frac{1}{8}, \frac{1}{16}, \cdots, \frac{1}{2^n}, \cdots$$

$$\{b_n\} : \frac{1}{2}, \frac{1}{4}, \frac{1}{8}, \frac{1}{15}, \cdots, \frac{6}{(n+1)(n^2-n+6)}, \cdots$$

$$\{c_n\} : \frac{1}{2}, \frac{1}{4}, \frac{1}{8}, \frac{7}{62}, \cdots, \frac{n^2-3n+3}{9n^2-25n+18}, \cdots$$

$$\{d_n\} : \frac{1}{2}, \frac{1}{4}, \frac{1}{8}, 0, \cdots, \frac{-n(n+1)(n-4)}{6(n^2+3n-2)}, \cdots$$

The process of determining an nth term from the pattern observed in the first several terms of a sequence is an example of *inductive reasoning*.

EXAMPLE 7   Finding the nth Term of a Sequence

Determine an nth term for a sequence whose first five terms are

$$-\frac{2}{1}, \frac{8}{2}, -\frac{26}{6}, \frac{80}{24}, -\frac{242}{120}, \cdots$$

and then decide whether the sequence converges or diverges.

Solution   Note that the numerators are 1 less than 3^n. So, you can reason that the numerators are given by the rule $3^n - 1$. Factoring the denominators produces

$$1 = 1$$
$$2 = 1 \cdot 2$$
$$6 = 1 \cdot 2 \cdot 3$$
$$24 = 1 \cdot 2 \cdot 3 \cdot 4$$
$$120 = 1 \cdot 2 \cdot 3 \cdot 4 \cdot 5 \cdots.$$

This suggests that the denominators are represented by $n!$. Finally, because the signs alternate, you can write the nth term as

$$a_n = (-1)^n \left(\frac{3^n - 1}{n!} \right).$$

From the discussion about the growth of $n!$, it follows that

$$\lim_{n \to \infty} |a_n| = \lim_{n \to \infty} \frac{3^n - 1}{n!} = 0.$$

Applying Theorem 9.4, you can conclude that

$$\lim_{n \to \infty} a_n = 0.$$

So, the sequence $\{a_n\}$ converges to 0.

Monotonic Sequences and Bounded Sequences

So far you have determined the convergence of a sequence by finding its limit. Even if you cannot determine the limit of a particular sequence, it still may be useful to know whether the sequence converges. Theorem 9.5 provides a test for convergence of sequences without determining the limit. First, some preliminary definitions are given.

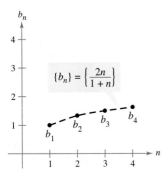

(a) Not monotonic

Definition of a Monotonic Sequence

A sequence $\{a_n\}$ is **monotonic** if its terms are nondecreasing

$$a_1 \leq a_2 \leq a_3 \leq \cdots \leq a_n \leq \cdots$$

or if its terms are nonincreasing

$$a_1 \geq a_2 \geq a_3 \geq \cdots \geq a_n \geq \cdots .$$

EXAMPLE 8 Determining Whether a Sequence Is Monotonic

Determine whether each sequence having the given nth term is monotonic.

a. $a_n = 3 + (-1)^n$ **b.** $b_n = \dfrac{2n}{1+n}$ **c.** $c_n = \dfrac{n^2}{2^n - 1}$

Solution

a. This sequence alternates between 2 and 4. So, it is not monotonic.

b. This sequence is monotonic because each successive term is larger than its predecessor. To see this, compare the terms b_n and b_{n+1}. [Note that, because n is positive, you can multiply each side of the inequality by $(1 + n)$ and $(2 + n)$ without reversing the inequality sign.]

$$b_n = \frac{2n}{1+n} \overset{?}{<} \frac{2(n+1)}{1+(n+1)} = b_{n+1}$$
$$2n(2+n) \overset{?}{<} (1+n)(2n+2)$$
$$4n + 2n^2 \overset{?}{<} 2 + 4n + 2n^2$$
$$0 < 2$$

Starting with the final inequality, which is valid, you can reverse the steps to conclude that the original inequality is also valid.

c. This sequence is not monotonic, because the second term is larger than the first term, and larger than the third. (Note that if you drop the first term, the remaining sequence $c_2, c_3, c_4, \ldots$ is monotonic.)

Figure 9.3 graphically illustrates these three sequences.

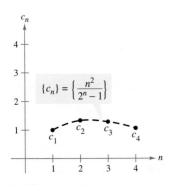

(b) Monotonic

(c) Not monotonic

Figure 9.3

NOTE   In Example 8(b), another way to see that the sequence is monotonic is to argue that the derivative of the corresponding differentiable function $f(x) = 2x/(1+x)$ is positive for all x. This implies that f is increasing, which in turn implies that $\{a_n\}$ is increasing.

NOTE   All three sequences shown in Figure 9.3 are bounded. To see this, consider the following.

$$2 \le a_n \le 4$$

$$1 \le b_n \le 2$$

$$0 \le c_n \le \frac{4}{3}$$

Definition of a Bounded Sequence

1. A sequence $\{a_n\}$ is **bounded above** if there is a real number M such that $a_n \le M$ for all n. The number M is called an **upper bound** of the sequence.

2. A sequence $\{a_n\}$ is **bounded below** if there is a real number N such that $N \le a_n$ for all n. The number N is called a **lower bound** of the sequence.

3. A sequence $\{a_n\}$ is **bounded** if it is bounded above and bounded below.

One important property of the real numbers is that they are **complete.** Informally, this means that there are no holes or gaps on the real number line. (The set of rational numbers does not have the completeness property.) The completeness axiom for real numbers can be used to conclude that if a sequence has an upper bound, it must have a **least upper bound** (an upper bound that is smaller than all other upper bounds for the sequence). For example, the least upper bound of the sequence $\{a_n\} = \{n/(n + 1)\}$,

$$\frac{1}{2}, \frac{2}{3}, \frac{3}{4}, \frac{4}{5}, \cdots, \frac{n}{n + 1}, \cdots$$

is 1. The completeness axiom is used in the proof of Theorem 9.5.

THEOREM 9.5   Bounded Monotonic Sequences

If a sequence $\{a_n\}$ is bounded and monotonic, then it converges.

Every bounded nondecreasing sequence converges.

Figure 9.4

Proof   Assume that the sequence is nondecreasing, as shown in Figure 9.4. For the sake of simplicity, also assume that each term in the sequence is positive. Because the sequence is bounded, there must exist an upper bound M such that

$$a_1 \le a_2 \le a_3 \le \cdots \le a_n \le \cdots \le M.$$

From the completeness axiom, it follows that there is a least upper bound L such that

$$a_1 \le a_2 \le a_3 \le \cdots \le a_n \le \cdots \le L.$$

For $\varepsilon > 0$, it follows that $L - \varepsilon < L$, and therefore $L - \varepsilon$ cannot be an upper bound for the sequence. Consequently, at least one term of $\{a_n\}$ is greater than $L - \varepsilon$. That is, $L - \varepsilon < a_N$ for some positive integer N. Because the terms of $\{a_n\}$ are nondecreasing, it follows that $a_N \le a_n$ for $n > N$. You now know that $L - \varepsilon < a_N \le a_n \le L < L + \varepsilon$, for every $n > N$. It follows that $|a_n - L| < \varepsilon$ for $n > N$, which by definition means that $\{a_n\}$ converges to L. The proof for a nonincreasing sequence is similar.

EXAMPLE 9   Bounded and Monotonic Sequences

a. The sequence $\{a_n\} = \{1/n\}$ is both bounded and monotonic and so, by Theorem 9.5, must converge.

b. The divergent sequence $\{b_n\} = \{n^2/(n + 1)\}$ is monotonic, but not bounded. (It is bounded below.)

c. The divergent sequence $\{c_n\} = \{(-1)^n\}$ is bounded, but not monotonic.

Exercises for Section 9.1

See www.CalcChat.com for worked-out solutions to odd-numbered exercises.

In Exercises 1–10, write the first five terms of the sequence.

1. $a_n = 2^n$

2. $a_n = \dfrac{3^n}{n!}$

3. $a_n = \left(-\dfrac{1}{2}\right)^n$

4. $a_n = \left(-\dfrac{2}{3}\right)^n$

5. $a_n = \sin\dfrac{n\pi}{2}$

6. $a_n = \dfrac{2n}{n+3}$

7. $a_n = \dfrac{(-1)^{n(n+1)/2}}{n^2}$

8. $a_n = (-1)^{n+1}\left(\dfrac{2}{n}\right)$

9. $a_n = 5 - \dfrac{1}{n} + \dfrac{1}{n^2}$

10. $a_n = 10 + \dfrac{2}{n} + \dfrac{6}{n^2}$

In Exercises 11–14, write the first five terms of the recursively defined sequence.

11. $a_1 = 3,\ a_{k+1} = 2(a_k - 1)$

12. $a_1 = 4,\ a_{k+1} = \left(\dfrac{k+1}{2}\right)a_k$

13. $a_1 = 32,\ a_{k+1} = \tfrac{1}{2}a_k$

14. $a_1 = 6,\ a_{k+1} = \tfrac{1}{3}a_k^2$

In Exercises 15–20, match the sequence with its graph. [The graphs are labeled (a), (b), (c), (d), (e), and (f).]

(a)

(b)

(c)

(d)

(e)

(f)
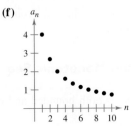

15. $a_n = \dfrac{8}{n+1}$

16. $a_n = \dfrac{8n}{n+1}$

17. $a_n = 4(0.5)^{n-1}$

18. $a_n = \dfrac{4^n}{n!}$

19. $a_n = (-1)^n$

20. $a_n = \dfrac{(-1)^n}{n}$

 In Exercises 21–24, use a graphing utility to graph the first 10 terms of the sequence.

21. $a_n = \dfrac{2}{3}n$

22. $a_n = 2 - \dfrac{4}{n}$

23. $a_n = 16(-0.5)^{n-1}$

24. $a_n = \dfrac{2n}{n+1}$

In Exercises 25–30, write the next two apparent terms of the sequence. Describe the pattern you used to find these terms.

25. $2, 5, 8, 11, \ldots$

26. $\dfrac{7}{2}, 4, \dfrac{9}{2}, 5, \ldots$

27. $5, 10, 20, 40, \ldots$

28. $1, -\dfrac{1}{2}, \dfrac{1}{4}, -\dfrac{1}{8}, \ldots$

29. $3, -\dfrac{3}{2}, \dfrac{3}{4}, -\dfrac{3}{8}, \ldots$

30. $1, -\dfrac{3}{2}, \dfrac{9}{4}, -\dfrac{27}{8}, \ldots$

In Exercises 31–36, simplify the ratio of factorials.

31. $\dfrac{10!}{8!}$

32. $\dfrac{25!}{23!}$

33. $\dfrac{(n+1)!}{n!}$

34. $\dfrac{(n+2)!}{n!}$

35. $\dfrac{(2n-1)!}{(2n+1)!}$

36. $\dfrac{(2n+2)!}{(2n)!}$

In Exercises 37–42, find the limit (if possible) of the sequence.

37. $a_n = \dfrac{5n^2}{n^2+2}$

38. $a_n = 5 - \dfrac{1}{n^2}$

39. $a_n = \dfrac{2n}{\sqrt{n^2+1}}$

40. $a_n = \dfrac{5n}{\sqrt{n^2+4}}$

41. $a_n = \sin\dfrac{1}{n}$

42. $a_n = \cos\dfrac{2}{n}$

 In Exercises 43–46, use a graphing utility to graph the first 10 terms of the sequence. Use the graph to make an inference about the convergence or divergence of the sequence. Verify your inference analytically and, if the sequence converges, find its limit.

43. $a_n = \dfrac{n+1}{n}$

44. $a_n = \dfrac{1}{n^{3/2}}$

45. $a_n = \cos\dfrac{n\pi}{2}$

46. $a_n = 3 - \dfrac{1}{2^n}$

In Exercises 47–68, determine the convergence or divergence of the sequence with the given nth term. If the sequence converges, find its limit.

47. $a_n = (-1)^n\left(\dfrac{n}{n+1}\right)$

48. $a_n = 1 + (-1)^n$

49. $a_n = \dfrac{3n^2 - n + 4}{2n^2 + 1}$

50. $a_n = \dfrac{\sqrt[3]{n}}{\sqrt[3]{n}+1}$

51. $a_n = \dfrac{1 \cdot 3 \cdot 5 \cdot \cdots \cdot (2n-1)}{(2n)^n}$

52. $a_n = \dfrac{1 \cdot 3 \cdot 5 \cdot \cdots \cdot (2n-1)}{n!}$

53. $a_n = \dfrac{1 + (-1)^n}{n}$

54. $a_n = \dfrac{1 + (-1)^n}{n^2}$

55. $a_n = \dfrac{\ln(n^3)}{2n}$

56. $a_n = \dfrac{\ln \sqrt{n}}{n}$

57. $a_n = \dfrac{3^n}{4^n}$

58. $a_n = (0.5)^n$

59. $a_n = \dfrac{(n + 1)!}{n!}$

60. $a_n = \dfrac{(n - 2)!}{n!}$

61. $a_n = \dfrac{n - 1}{n} - \dfrac{n}{n - 1}, \quad n \geq 2$

62. $a_n = \dfrac{n^2}{2n + 1} - \dfrac{n^2}{2n - 1}$

63. $a_n = \dfrac{n^p}{e^n}, \quad p > 0$

64. $a_n = n \sin \dfrac{1}{n}$

65. $a_n = \left(1 + \dfrac{k}{n}\right)^n$

66. $a_n = 2^{1/n}$

67. $a_n = \dfrac{\sin n}{n}$

68. $a_n = \dfrac{\cos \pi n}{n^2}$

In Exercises 69–82, write an expression for the nth term of the sequence. (There is more than one correct answer.)

69. $1, 4, 7, 10, \ldots$

70. $3, 7, 11, 15, \ldots$

71. $-1, 2, 7, 14, 23, \ldots$

72. $1, -\frac{1}{4}, \frac{1}{9}, -\frac{1}{16}, \ldots$

73. $\frac{2}{3}, \frac{3}{4}, \frac{4}{5}, \frac{5}{6}, \ldots$

74. $2, -1, \frac{1}{2}, -\frac{1}{4}, \frac{1}{8}, \ldots$

75. $2, 1 + \frac{1}{2}, 1 + \frac{1}{3}, 1 + \frac{1}{4}, 1 + \frac{1}{5}, \ldots$

76. $1 + \frac{1}{2}, 1 + \frac{3}{4}, 1 + \frac{7}{8}, 1 + \frac{15}{16}, 1 + \frac{31}{32}, \ldots$

77. $\dfrac{1}{2 \cdot 3}, \dfrac{2}{3 \cdot 4}, \dfrac{3}{4 \cdot 5}, \dfrac{4}{5 \cdot 6}, \cdots$

78. $1, \frac{1}{2}, \frac{1}{6}, \frac{1}{24}, \frac{1}{120}, \ldots$

79. $1, -\dfrac{1}{1 \cdot 3}, \dfrac{1}{1 \cdot 3 \cdot 5}, -\dfrac{1}{1 \cdot 3 \cdot 5 \cdot 7}, \cdots$

80. $1, x, \dfrac{x^2}{2}, \dfrac{x^3}{6}, \dfrac{x^4}{24}, \dfrac{x^5}{120}, \cdots$

81. $2, 24, 720, 40{,}320, 3{,}628{,}800, \ldots$

82. $1, 6, 120, 5040, 362{,}880, \ldots$

In Exercises 83–94, determine whether the sequence with the given nth term is monotonic. Discuss the boundedness of the sequence. Use a graphing utility to confirm your results.

83. $a_n = 4 - \dfrac{1}{n}$

84. $a_n = \dfrac{3n}{n + 2}$

85. $a_n = \dfrac{n}{2^{n+2}}$

86. $a_n = ne^{-n/2}$

87. $a_n = (-1)^n \left(\dfrac{1}{n}\right)$

88. $a_n = \left(-\dfrac{2}{3}\right)^n$

89. $a_n = \left(\dfrac{2}{3}\right)^n$

90. $a_n = \left(\dfrac{3}{2}\right)^n$

91. $a_n = \sin \dfrac{n\pi}{6}$

92. $a_n = \cos\left(\dfrac{n\pi}{2}\right)$

93. $a_n = \dfrac{\cos n}{n}$

94. $a_n = \dfrac{\sin \sqrt{n}}{n}$

In Exercises 95–98, (a) use Theorem 9.5 to show that the sequence with the given nth term converges and (b) use a graphing utility to graph the first 10 terms of the sequence and find its limit.

95. $a_n = 5 + \dfrac{1}{n}$

96. $a_n = 4 - \dfrac{3}{n}$

97. $a_n = \dfrac{1}{3}\left(1 - \dfrac{1}{3^n}\right)$

98. $a_n = 4 + \dfrac{1}{2^n}$

99. Let $\{a_n\}$ be an increasing sequence such that $2 \leq a_n \leq 4$. Explain why $\{a_n\}$ has a limit. What can you conclude about the limit?

100. Let $\{a_n\}$ be a monotonic sequence such that $a_n \leq 1$. Discuss the convergence of $\{a_n\}$. If $\{a_n\}$ converges, what can you conclude about its limit?

101. **Compound Interest** Consider the sequence $\{A_n\}$ whose nth term is given by

$$A_n = P\left(1 + \dfrac{r}{12}\right)^n$$

where P is the principal, A_n is the account balance after n months, and r is the interest rate compounded annually.

(a) Is $\{A_n\}$ a convergent sequence? Explain.

(b) Find the first 10 terms of the sequence if $P = \$9000$ and $r = 0.055$.

102. **Compound Interest** A deposit of $100 is made at the beginning of each month in an account at an annual interest rate of 3% compounded monthly. The balance in the account after n months is $A_n = 100(401)(1.0025^n - 1)$.

(a) Compute the first six terms of the sequence $\{A_n\}$.

(b) Find the balance in the account after 5 years by computing the 60th term of the sequence.

(c) Find the balance in the account after 20 years by computing the 240th term of the sequence.

Writing About Concepts

103. In your own words, define each of the following.

(a) Sequence   (b) Convergence of a sequence

(c) Monotonic sequence (d) Bounded sequence

104. The graphs of two sequences are shown in the figures. Which graph represents the sequence with alternating signs? Explain.

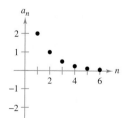

In Exercises 105–108, give an example of a sequence satisfying the condition or explain why no such sequence exists. (Examples are not unique.)

105. A monotonically increasing sequence that converges to 10

106. A monotonically increasing bounded sequence that does not converge

107. A sequence that converges to $\frac{3}{4}$

108. An unbounded sequence that converges to 100

109. *Government Expenditures* A government program that currently costs taxpayers $2.5 billion per year is cut back by 20 percent per year.

(a) Write an expression for the amount budgeted for this program after n years.

(b) Compute the budgets for the first 4 years.

(c) Determine the convergence or divergence of the sequence of reduced budgets. If the sequence converges, find its limit.

110. *Inflation* If the rate of inflation is $4\frac{1}{2}\%$ per year and the average price of a car is currently $16,000, the average price after n years is

$$P_n = \$16{,}000(1.045)^n.$$

Compute the average prices for the next 5 years.

111. *Modeling Data* The number a_n of endangered and threatened species in the United States from 1996 through 2002 is shown in the table, where n represents the year, with $n = 6$ corresponding to 1996. *(Source: U.S. Fish and Wildlife Service)*

n	6	7	8	9	10	11	12
a_n	1053	1132	1194	1205	1244	1254	1263

(a) Use the regression capabilities of a graphing utility to find a model of the form

$$a_n = bn^2 + cn + d, \qquad n = 6, 7, \ldots, 12$$

for the data. Use the graphing utility to plot the points and graph the model.

(b) Use the model to predict the number of endangered and threatened species in the year 2008.

112. *Modeling Data* The annual sales a_n (in millions of dollars) for Avon Products, Inc. from 1993 through 2002 are given below as ordered pairs of the form (n, a_n), where n represents the year, with $n = 3$ corresponding to 1993. *(Source: 2002 Avon Products, Inc. Annual Report)*

(3, 3844), (4, 4267), (5, 4492), (6, 4814), (7, 5079),

(8, 5213), (9, 5289), (10, 5682), (11, 5958), (12, 6171)

(a) Use the regression capabilities of a graphing utility to find a model of the form

$$a_n = bn + c, \quad n = 3, 4, \ldots, 12$$

for the data. Graphically compare the points and the model.

(b) Use the model to predict sales in the year 2008.

113. *Comparing Exponential and Factorial Growth* Consider the sequence $a_n = 10^n/n!$.

(a) Find two consecutive terms that are equal in magnitude.

(b) Are the terms following those found in part (a) increasing or decreasing?

(c) In Section 8.7, Exercises 65–70, it was shown that for "large" values of the independent variable an exponential function increases more rapidly than a polynomial function. From the result in part (b), what inference can you make about the rate of growth of an exponential function versus a factorial function for "large" integer values of n?

114. Compute the first six terms of the sequence

$$\{a_n\} = \left\{ \left(1 + \frac{1}{n}\right)^n \right\}.$$

If the sequence converges, find its limit.

115. Compute the first six terms of the sequence $\{a_n\} = \{\sqrt[n]{n}\}$. If the sequence converges, find its limit.

116. Prove that if $\{s_n\}$ converges to L and $L > 0$, then there exists a number N such that $s_n > 0$ for $n > N$.

True or False? In Exercises 117–120, determine whether the statement is true or false. If it is false, explain why or give an example that shows it is false.

117. If $\{a_n\}$ converges to 3 and $\{b_n\}$ converges to 2, then $\{a_n + b_n\}$ converges to 5.

118. If $\{a_n\}$ converges, then $\lim_{n\to\infty} (a_n - a_{n+1}) = 0$.

119. If $n > 1$, then $n! = n(n - 1)!$.

120. If $\{a_n\}$ converges, then $\{a_n/n\}$ converges to 0.

121. *Fibonacci Sequence* In a study of the progeny of rabbits, Fibonacci (ca. 1170–ca. 1240) encountered the sequence now bearing his name. It is defined recursively by

$$a_{n+2} = a_n + a_{n+1}, \quad \text{where} \quad a_1 = 1 \text{ and } a_2 = 1.$$

(a) Write the first 12 terms of the sequence.

(b) Write the first 10 terms of the sequence defined by

$$b_n = \frac{a_{n+1}}{a_n}, \quad n \geq 1.$$

(c) Using the definition in part (b), show that

$$b_n = 1 + \frac{1}{b_{n-1}}.$$

(d) The **golden ratio** ρ can be defined by $\lim_{n\to\infty} b_n = \rho$. Show that $\rho = 1 + 1/\rho$ and solve this equation for ρ.

122. *Conjecture* Let $x_0 = 1$ and consider the sequence x_n given by the formula

$$x_n = \frac{1}{2}x_{n-1} + \frac{1}{x_{n-1}}, \qquad n = 1, 2, \ldots.$$

Use a graphing utility to compute the first 10 terms of the sequence and make a conjecture about the limit of the sequence.

123. Consider the sequence

$$\sqrt{2}, \sqrt{2 + \sqrt{2}}, \sqrt{2 + \sqrt{2 + \sqrt{2}}}, \ldots$$

(a) Compute the first five terms of this sequence.

(b) Write a recursion formula for a_n, for $n \geq 2$.

(c) Find $\lim\limits_{n \to \infty} a_n$.

124. Consider the sequence

$$\sqrt{6}, \sqrt{6 + \sqrt{6}}, \sqrt{6 + \sqrt{6 + \sqrt{6}}}, \ldots$$

(a) Compute the first five terms of this sequence.

(b) Write a recursion formula for a_n, for $n \geq 2$.

(c) Find $\lim\limits_{n \to \infty} a_n$.

125. Consider the sequence $\{a_n\}$ where $a_1 = \sqrt{k}$, $a_{n+1} = \sqrt{k + a_n}$, and $k > 0$.

(a) Show that $\{a_n\}$ is increasing and bounded.

(b) Prove that $\lim\limits_{n \to \infty} a_n$ exists.

(c) Find $\lim\limits_{n \to \infty} a_n$.

126. *Arithmetic-Geometric Mean* Let $a_0 > b_0 > 0$. Let a_1 be the arithmetic mean of a_0 and b_0 and let b_1 be the geometric mean of a_0 and b_0.

$$a_1 = \frac{a_0 + b_0}{2} \qquad \text{Arithmetic mean}$$

$$b_1 = \sqrt{a_0 b_0} \qquad \text{Geometric mean}$$

Now define the sequences $\{a_n\}$ and $\{b_n\}$ as follows.

$$a_n = \frac{a_{n-1} + b_{n-1}}{2} \qquad b_n = \sqrt{a_{n-1}b_{n-1}}$$

(a) Let $a_0 = 10$ and $b_0 = 3$. Write out the first five terms of $\{a_n\}$ and $\{b_n\}$. Compare the terms of $\{b_n\}$. Compare a_n and b_n. What do you notice?

(b) Use induction to show that $a_n > a_{n+1} > b_{n+1} > b_n$, for $a_0 > b_0 > 0$.

(c) Explain why $\{a_n\}$ and $\{b_n\}$ are both convergent.

(d) Show that $\lim\limits_{n \to \infty} a_n = \lim\limits_{n \to \infty} b_n$.

127. (a) Let $f(x) = \sin x$ and $a_n = n \sin 1/n$. Show that

$$\lim_{n \to \infty} a_n - f'(0) = 1.$$

(b) Let $f(x)$ be differentiable on the interval $[0, 1]$ and $f(0) = 0$. Consider the sequence $\{a_n\}$, where $a_n = n f(1/n)$. Show that $\lim\limits_{n \to \infty} a_n = f'(0)$.

128. Consider the sequence $\{a_n\} = \{nr^n\}$. Decide whether $\{a_n\}$ converges for each value of r.

(a) $r = \frac{1}{2}$　　(b) $r = 1$　　(c) $r = \frac{3}{2}$

(d) For what values or r does the sequence $\{nr^n\}$ converge?

129. (a) Show that $\int_1^n \ln x \, dx < \ln(n!)$ for $n \geq 2$.

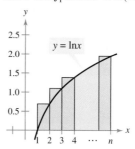

(b) Draw a graph similar to the one above that shows

$$\ln(n!) < \int_1^{n+1} \ln x \, dx.$$

(c) Use the results of parts (a) and (b) to show that

$$\frac{n^n}{e^{n-1}} < n! < \frac{(n+1)^{n+1}}{e^n}, \text{ for } n > 1.$$

(d) Use the Squeeze Theorem for Sequences and the result of part (c) to show that

$$\lim_{n \to \infty} \frac{\sqrt[n]{n!}}{n} = \frac{1}{e}.$$

(e) Test the result of part (d) for $n = 20$, 50, and 100.

130. Consider the sequence $\{a_n\} = \left\{ \dfrac{1}{n} \displaystyle\sum_{k=1}^{n} \dfrac{1}{1 + (k/n)} \right\}$.

(a) Write the first five terms of $\{a_n\}$.

(b) Show that $\lim\limits_{n \to \infty} a_n = \ln 2$ by interpreting a_n as a Riemann sum of a definite integral.

131. Prove, using the definition of the limit of a sequence, that

$$\lim_{n \to \infty} \frac{1}{n^3} = 0.$$

132. Prove, using the definition of the limit of a sequence, that

$$\lim_{n \to \infty} r^n = 0 \text{ for } -1 < r < 1.$$

133. Complete the proof of Theorem 9.5.

Putnam Exam Challenge

134. Let $\{x_n\}$, $n \geq 0$, be a sequence of nonzero real numbers such that $x_n^2 - x_{n-1}x_{n+1} = 1$ for $n = 1, 2, 3, \ldots$. Prove that there exists a real number a such that $x_{n+1} = ax_n - x_{n-1}$, for all $n \geq 1$.

135. Let $T_0 = 2$, $T_1 = 3$, $T_2 = 6$, and, for $n \geq 3$,

$$T_n = (n + 4)T_{n-1} - 4nT_{n-2} + (4n - 8)T_{n-3}.$$

The first 10 terms of the sequence are

2, 3, 6, 14, 40, 152, 784, 5168, 40,576, 363,392.

Find, with proof, a formula for T_n of the form $T_n = A_n + B_n$, where $\{A_n\}$ and $\{B_n\}$ are well-known sequences.

Series and Convergence

- Understand the definition of a convergent infinite series.
- Use properties of infinite geometric series.
- Use the nth-Term Test for Divergence of an infinite series.

Infinite Series

One important application of infinite sequences is in representing "infinite summations." Informally, if $\{a_n\}$ is an infinite sequence, then

$$\sum_{n=1}^{\infty} a_n = a_1 + a_2 + a_3 + \cdots + a_n + \cdots \qquad \text{Infinite series}$$

is an **infinite series** (or simply a **series**). The numbers a_1, a_2, a_3, are the **terms** of the series. For some series it is convenient to begin the index at $n = 0$ (or some other integer). As a typesetting convention, it is common to represent an infinite series as simply $\Sigma \, a_n$. In such cases, the starting value for the index must be taken from the context of the statement.

To find the sum of an infinite series, consider the following **sequence of partial sums.**

$$S_1 = a_1$$
$$S_2 = a_1 + a_2$$
$$S_3 = a_1 + a_2 + a_3$$
$$\vdots$$
$$S_n = a_1 + a_2 + a_3 + \cdots + a_n$$

If this sequence of partial sums converges, the series is said to converge and has the sum indicated in the following definition.

Definitions of Convergent and Divergent Series

For the infinite series $\displaystyle\sum_{n=1}^{\infty} a_n$, the **$n$th partial sum** is given by

$$S_n = a_1 + a_2 + \cdots + a_n.$$

If the sequence of partial sums $\{S_n\}$ converges to S, then the series $\displaystyle\sum_{n=1}^{\infty} a_n$ **converges.** The limit S is called the **sum of the series.**

$$S = a_1 + a_2 + \cdots + a_n + \cdots$$

If $\{S_n\}$ diverges, then the series **diverges.**

STUDY TIP As you study this chapter, you will see that there are two basic questions involving infinite series. Does a series converge or does it diverge? If a series converges, what is its sum? These questions are not always easy to answer, especially the second one.

EXPLORATION

Finding the Sum of an Infinite Series Find the sum of each infinite series. Explain your reasoning.

a. $0.1 + 0.01 + 0.001 + 0.0001 + \cdots$

b. $\frac{3}{10} + \frac{3}{100} + \frac{3}{1000} + \frac{3}{10,000} + \cdots$

c. $1 + \frac{1}{2} + \frac{1}{4} + \frac{1}{8} + \frac{1}{16} + \cdots$

d. $\frac{15}{100} + \frac{15}{10,000} + \frac{15}{1,000,000} + \cdots$

TECHNOLOGY   Figure 9.5 shows the first 15 partial sums of the infinite series in Example 1(a). Notice how the values appear to approach the line $y = 1$.

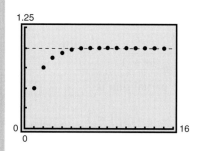

Figure 9.5

NOTE   You can geometrically determine the partial sums of the series in Example 1(a) using Figure 9.6.

Figure 9.6

FOR FURTHER INFORMATION   To learn more about the partial sums of infinite series, see the article "Six Ways to Sum a Series" by Dan Kalman in *The College Mathematics Journal*. To view this article, go to the website *www.matharticles.com*.

EXAMPLE 1   Convergent and Divergent Series

a. The series

$$\sum_{n=1}^{\infty} \frac{1}{2^n} = \frac{1}{2} + \frac{1}{4} + \frac{1}{8} + \frac{1}{16} + \cdots$$

has the following partial sums.

$$S_1 = \frac{1}{2}$$

$$S_2 = \frac{1}{2} + \frac{1}{4} = \frac{3}{4}$$

$$S_3 = \frac{1}{2} + \frac{1}{4} + \frac{1}{8} = \frac{7}{8}$$

$$\vdots$$

$$S_n = \frac{1}{2} + \frac{1}{4} + \frac{1}{8} + \cdots + \frac{1}{2^n} = \frac{2^n - 1}{2^n}$$

Because

$$\lim_{n \to \infty} \frac{2^n - 1}{2^n} = 1$$

it follows that the series converges and its sum is 1.

b. The nth partial sum of the series

$$\sum_{n=1}^{\infty} \left(\frac{1}{n} - \frac{1}{n+1} \right) = \left(1 - \frac{1}{2} \right) + \left(\frac{1}{2} - \frac{1}{3} \right) + \left(\frac{1}{3} - \frac{1}{4} \right) + \cdots$$

is given by

$$S_n = 1 - \frac{1}{n+1}.$$

Because the limit of S_n is 1, the series converges and its sum is 1.

c. The series

$$\sum_{n=1}^{\infty} 1 = 1 + 1 + 1 + 1 + \cdots$$

diverges because $S_n = n$ and the sequence of partial sums diverges.

The series in Example 1(b) is a **telescoping series** of the form

$$(b_1 - b_2) + (b_2 - b_3) + (b_3 - b_4) + (b_4 - b_5) + \cdots. \qquad \text{Telescoping series}$$

Note that b_2 is canceled by the second term, b_3 is canceled by the third term, and so on. Because the nth partial sum of this series is

$$S_n = b_1 - b_{n+1}$$

it follows that a telescoping series will converge if and only if b_n approaches a finite number as $n \to \infty$. Moreover, if the series converges, its sum is

$$S = b_1 - \lim_{n \to \infty} b_{n+1}.$$

EXAMPLE 2 Writing a Series in Telescoping Form

Find the sum of the series $\displaystyle\sum_{n=1}^{\infty} \frac{2}{4n^2 - 1}$.

Solution

Using partial fractions, you can write

$$a_n = \frac{2}{4n^2 - 1} = \frac{2}{(2n - 1)(2n + 1)} = \frac{1}{2n - 1} - \frac{1}{2n + 1}.$$

From this telescoping form, you can see that the nth partial sum is

$$S_n = \left(\frac{1}{1} - \frac{1}{3}\right) + \left(\frac{1}{3} - \frac{1}{5}\right) + \cdots + \left(\frac{1}{2n - 1} - \frac{1}{2n + 1}\right) = 1 - \frac{1}{2n + 1}.$$

So, the series converges and its sum is 1. That is,

$$\sum_{n=1}^{\infty} \frac{2}{4n^2 - 1} = \lim_{n \to \infty} S_n = \lim_{n \to \infty} \left(1 - \frac{1}{2n + 1}\right) = 1.$$

Geometric Series

The series given in Example 1(a) is a **geometric series.** In general, the series given by

$$\sum_{n=0}^{\infty} ar^n = a + ar + ar^2 + \cdots + ar^n + \cdots, \quad a \neq 0 \qquad \text{Geometric series}$$

is a **geometric series** with ratio r.

THEOREM 9.6 Convergence of a Geometric Series

A geometric series with ratio r diverges if $|r| \geq 1$. If $0 < |r| < 1$, then the series converges to the sum

$$\sum_{n=0}^{\infty} ar^n = \frac{a}{1 - r}, \quad 0 < |r| < 1.$$

Proof It is easy to see that the series diverges if $r = \pm 1$. If $r \neq \pm 1$, then $S_n = a + ar + ar^2 + \cdots + ar^{n-1}$. Multiplication by r yields

$$rS_n = ar + ar^2 + ar^3 + \cdots + ar^n.$$

Subtracting the second equation from the first produces $S_n - rS_n = a - ar^n$. Therefore, $S_n(1 - r) = a(1 - r^n)$, and the nth partial sum is

$$S_n = \frac{a}{1 - r}(1 - r^n).$$

If $0 < |r| < 1$, it follows that $r^n \to 0$ as $n \to \infty$, and you obtain

$$\lim_{n \to \infty} S_n = \lim_{n \to \infty} \left[\frac{a}{1 - r}(1 - r^n)\right] = \frac{a}{1 - r}\left[\lim_{n \to \infty} (1 - r^n)\right] = \frac{a}{1 - r}$$

which means that the series *converges* and its sum is $a/(1 - r)$. It is left to you to show that the series diverges if $|r| > 1$.

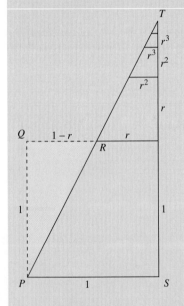

37. $\displaystyle\sum_{n=1}^{\infty} \frac{8}{(n+1)(n+2)}$

38. $\displaystyle\sum_{n=1}^{\infty} \frac{1}{(2n+1)(2n+3)}$

39. $\displaystyle\sum_{n=0}^{\infty} \left(\frac{1}{2}\right)^n$

40. $\displaystyle\sum_{n=0}^{\infty} 6\left(\frac{4}{5}\right)^n$

41. $\displaystyle\sum_{n=0}^{\infty} \left(-\frac{1}{2}\right)^n$

42. $\displaystyle\sum_{n=0}^{\infty} 2\left(-\frac{2}{3}\right)^n$

43. $1 + 0.1 + 0.01 + 0.001 + \cdots$

44. $8 + 6 + \frac{9}{2} + \frac{27}{8} + \cdots$

45. $3 - 1 + \frac{1}{3} - \frac{1}{9} + \cdots$

46. $4 - 2 + 1 - \frac{1}{2} + \cdots$

47. $\displaystyle\sum_{n=0}^{\infty} \left(\frac{1}{2^n} - \frac{1}{3^n}\right)$

48. $\displaystyle\sum_{n=1}^{\infty} [(0.7)^n + (0.9)^n]$

49. $\displaystyle\sum_{n=1}^{\infty} (\sin 1)^n$

50. $\displaystyle\sum_{n=1}^{\infty} \frac{1}{9n^2 + 3n - 2}$

In Exercises 51–56, (a) write the repeating decimal as a geometric series and (b) write its sum as the ratio of two integers.

51. $0.\overline{4}$

52. $0.\overline{9}$

53. $0.8\overline{1}$

54. $0.0\overline{1}$

55. $0.0\overline{75}$

56. $0.2\overline{15}$

In Exercises 57–72, determine the convergence or divergence of the series.

57. $\displaystyle\sum_{n=1}^{\infty} \frac{n+10}{10n+1}$

58. $\displaystyle\sum_{n=1}^{\infty} \frac{n+1}{2n-1}$

59. $\displaystyle\sum_{n=1}^{\infty} \left(\frac{1}{n} - \frac{1}{n+2}\right)$

60. $\displaystyle\sum_{n=1}^{\infty} \frac{1}{n(n+3)}$

61. $\displaystyle\sum_{n=1}^{\infty} \frac{3n-1}{2n+1}$

62. $\displaystyle\sum_{n=1}^{\infty} \frac{3^n}{n^3}$

63. $\displaystyle\sum_{n=0}^{\infty} \frac{4}{2^n}$

64. $\displaystyle\sum_{n=0}^{\infty} \frac{1}{4^n}$

65. $\displaystyle\sum_{n=0}^{\infty} (1.075)^n$

66. $\displaystyle\sum_{n=1}^{\infty} \frac{2^n}{100}$

67. $\displaystyle\sum_{n=2}^{\infty} \frac{n}{\ln n}$

68. $\displaystyle\sum_{n=1}^{\infty} \ln\frac{1}{n}$

69. $\displaystyle\sum_{n=1}^{\infty} \left(1 + \frac{k}{n}\right)^n$

70. $\displaystyle\sum_{n=1}^{\infty} e^{-n}$

71. $\displaystyle\sum_{n=1}^{\infty} \arctan n$

72. $\displaystyle\sum_{n=1}^{\infty} \ln\left(\frac{n+1}{n}\right)$

Writing About Concepts

73. State the definitions of convergent and divergent series.

74. Describe the difference between $\displaystyle\lim_{n\to\infty} a_n = 5$ and

$\displaystyle\sum_{n=1}^{\infty} a_n = 5.$

75. Define a geometric series, state when it converges, and give the formula for the sum of a convergent geometric series.

Writing About Concepts *(continued)*

76. State the nth-Term Test for Divergence.

77. Let $a_n = \dfrac{n+1}{n}$. Discuss the convergence of $\{a_n\}$ and

$\displaystyle\sum_{n=1}^{\infty} a_n.$

78. Explain any differences among the following series.

(a) $\displaystyle\sum_{n=1}^{\infty} a_n$ (b) $\displaystyle\sum_{k=1}^{\infty} a_k$ (c) $\displaystyle\sum_{n=1}^{\infty} a_k$

In Exercises 79–86, find all values of x for which the series converges. For these values of x, write the sum of the series as a function of x.

79. $\displaystyle\sum_{n=1}^{\infty} \frac{x^n}{2^n}$

80. $\displaystyle\sum_{n=1}^{\infty} (3x)^n$

81. $\displaystyle\sum_{n=1}^{\infty} (x-1)^n$

82. $\displaystyle\sum_{n=0}^{\infty} 4\left(\frac{x-3}{4}\right)^n$

83. $\displaystyle\sum_{n=0}^{\infty} (-1)^n x^n$

84. $\displaystyle\sum_{n=0}^{\infty} (-1)^n x^{2n}$

85. $\displaystyle\sum_{n=0}^{\infty} \left(\frac{1}{x}\right)^n$

86. $\displaystyle\sum_{n=1}^{\infty} \left(\frac{x^2}{x^2+4}\right)^n$

87. (a) You delete a finite number of terms from a divergent series. Will the new series still diverge? Explain your reasoning.

(b) You add a finite number of terms to a convergent series. Will the new series still converge? Explain your reasoning.

88. *Think About It* Consider the formula

$$\frac{1}{x-1} = 1 + x + x^2 + x^3 + \cdots.$$

Given $x = -1$ and $x = 2$, can you conclude that either of the following statements is true? Explain your reasoning.

(a) $\dfrac{1}{2} = 1 - 1 + 1 - 1 + \cdots$

(b) $-1 = 1 + 2 + 4 + 8 + \cdots$

 In Exercises 89 and 90, (a) find the common ratio of the geometric series, (b) write the function that gives the sum of the series, and (c) use a graphing utility to graph the function and the partial sums S_3 and S_5. What do you notice?

89. $1 + x + x^2 + x^3 + \cdots$

90. $1 - \dfrac{x}{2} + \dfrac{x^2}{4} - \dfrac{x^3}{8} + \cdots$

In Exercises 91 and 92, use a graphing utility to graph the function. Identify the horizontal asymptote of the graph and determine its relationship to the sum of the series.

Function	Series
91. $f(x) = 3\left[\dfrac{1-(0.5)^x}{1-0.5}\right]$	$\displaystyle\sum_{n=0}^{\infty} 3\left(\frac{1}{2}\right)^n$
92. $f(x) = 2\left[\dfrac{1-(0.8)^x}{1-0.8}\right]$	$\displaystyle\sum_{n=0}^{\infty} 2\left(\frac{4}{5}\right)^n$

Writing In Exercises 93 and 94, use a graphing utility to determine the first term that is less than 0.0001 in each of the convergent series. Note that the answers are very different. Explain how this will affect the rate at which the series converges.

93. $\displaystyle\sum_{n=1}^{\infty} \frac{1}{n(n+1)}$, $\displaystyle\sum_{n=1}^{\infty} \left(\frac{1}{8}\right)^n$ **94.** $\displaystyle\sum_{n=1}^{\infty} \frac{1}{2^n}$, $\displaystyle\sum_{n=1}^{\infty} (0.01)^n$

95. Marketing An electronic games manufacturer producing a new product estimates the annual sales to be 8000 units. Each year 10% of the units that have been sold will become inoperative. So, 8000 units will be in use after 1 year, $[8000 + 0.9(8000)]$ units will be in use after 2 years, and so on. How many units will be in use after n years?

96. Depreciation A company buys a machine for $225,000 that depreciates at a rate of 30% per year. Find a formula for the value of the machine after n years. What is its value after 5 years?

97. Multiplier Effect The annual spending by tourists in a resort city is $100 million. Approximately 75% of that revenue is again spent in the resort city, and of that amount approximately 75% is again spent in the same city, and so on. Write the geometric series that gives the total amount of spending generated by the $100 million and find the sum of the series.

98. Multiplier Effect Repeat Exercise 97 if the percent of the revenue that is spent again in the city decreases to 60%.

99. Distance A ball is dropped from a height of 16 feet. Each time it drops h feet, it rebounds $0.81h$ feet. Find the total distance traveled by the ball.

100. Time The ball in Exercise 99 takes the following times for each fall.

$s_1 = -16t^2 + 16$, $s_1 = 0$ if $t = 1$

$s_2 = -16t^2 + 16(0.81)$, $s_2 = 0$ if $t = 0.9$

$s_3 = -16t^2 + 16(0.81)^2$, $s_3 = 0$ if $t = (0.9)^2$

$s_4 = -16t^2 + 16(0.81)^3$, $s_4 = 0$ if $t = (0.9)^3$

$\vdots$ $\vdots$

$s_n = -16t^2 + 16(0.81)^{n-1}$, $s_n = 0$ if $t = (0.9)^{n-1}$

Beginning with s_2, the ball takes the same amount of time to bounce up as it does to fall, and so the total time elapsed before it comes to rest is given by

$$t = 1 + 2\sum_{n=1}^{\infty} (0.9)^n.$$

Find this total time.

Probability In Exercises 101 and 102, the random variable n represents the number of units of a product sold per day in a store. The probability distribution of n is given by $P(n)$. Find the probability that two units are sold in a given day $[P(2)]$ and show that $P(1) + P(2) + P(3) + \cdots = 1$.

101. $P(n) = \dfrac{1}{2}\left(\dfrac{1}{2}\right)^n$ **102.** $P(n) = \dfrac{1}{3}\left(\dfrac{2}{3}\right)^n$

103. Probability A fair coin is tossed repeatedly. The probability that the first head occurs on the nth toss is given by $P(n) = \left(\frac{1}{2}\right)^n$, where $n \geq 1$.

(a) Show that $\displaystyle\sum_{n=1}^{\infty} \left(\frac{1}{2}\right)^n = 1$.

(b) The expected number of tosses required until the first head occurs in the experiment is given by

$$\sum_{n=1}^{\infty} n\left(\frac{1}{2}\right)^n.$$

Is this series geometric?

(c) Use a computer algebra system to find the sum in part (b).

104. Probability In an experiment, three people toss a fair coin one at a time until one of them tosses a head. Determine, for each person, the probability that he or she tosses the first head. Verify that the sum of the three probabilities is 1.

105. Area The sides of a square are 16 inches in length. A new square is formed by connecting the midpoints of the sides of the original square, and two of the triangles outside the second square are shaded (see figure). Determine the area of the shaded regions (a) if this process is continued five more times and (b) if this pattern of shading is continued infinitely.

Figure for 105   **Figure for 106**

106. Length A right triangle XYZ is shown above where $|XY| = z$ and $\angle X = \theta$. Line segments are continually drawn to be perpendicular to the triangle, as shown in the figure.

(a) Find the total length of the perpendicular line segments $|Yy_1| + |x_1y_1| + |x_1y_2| + \cdots$ in terms of z and θ.

(b) If $z = 1$ and $\theta = \pi/6$, find the total length of the perpendicular line segments.

In Exercises 107–110, use the formula for the nth partial sum of a geometric series

$$\sum_{i=0}^{n-1} ar^i = \frac{a(1-r^n)}{1-r}.$$

107. Present Value The winner of a $1,000,000 sweepstakes will be paid $50,000 per year for 20 years. The money earns 6% interest per year. The present value of the winnings is

$$\sum_{n=1}^{20} 50{,}000\left(\frac{1}{1.06}\right)^n.$$

Compute the present value and interpret its meaning.

108. *Sphereflake* A sphereflake shown below is a computer-generated fractal that was created by Eric Haines. The radius of the large sphere is 1. To the large sphere, nine spheres of radius $\frac{1}{3}$ are attached. To each of these, nine spheres of radius $\frac{1}{9}$ are attached. This process is continued infinitely. Prove that the sphereflake has an infinite surface area.

Eric Haines

109. *Salary* You go to work at a company that pays $0.01 for the first day, $0.02 for the second day, $0.04 for the third day, and so on. If the daily wage keeps doubling, what would your total income be for working (a) 29 days, (b) 30 days, and (c) 31 days?

110. *Annuities* When an employee receives a paycheck at the end of each month, P dollars is invested in a retirement account. These deposits are made each month for t years and the account earns interest at the annual percentage rate r. If the interest is compounded monthly, the amount A in the account at the end of t years is

$$A = P + P\left(1 + \frac{r}{12}\right) + \cdots + P\left(1 + \frac{r}{12}\right)^{12t-1}$$

$$= P\left(\frac{12}{r}\right)\left[\left(1 + \frac{r}{12}\right)^{12t} - 1\right].$$

If the interest is compounded continuously, the amount A in the account after t years is

$$A = P + Pe^{r/12} + Pe^{2r/12} + Pe^{(12t-1)r/12}$$

$$= \frac{P(e^{rt} - 1)}{e^{r/12} - 1}.$$

Verify the formulas for the sums given above.

Annuities **In Exercises 111–114, consider making monthly deposits of P dollars in a savings account at an annual interest rate r. Use the results of Exercise 110 to find the balance A after t years if the interest is compounded (a) monthly and (b) continuously.**

111. $P = \$50$, $r = 3\%$, $t = 20$ years

112. $P = \$75$, $r = 5\%$, $t = 25$ years

113. $P = \$100$, $r = 4\%$, $t = 40$ years

114. $P = \$20$, $r = 6\%$, $t = 50$ years

115. *Modeling Data* The annual sales a_n (in millions of dollars) for Avon Products, Inc. from 1993 through 2002 are given below as ordered pairs of the form (n, a_n), where n represents the year, with $n = 3$ corresponding to 1993. *(Source: 2002 Avon Products, Inc. Annual Report)*

(3, 3844), (4, 4267), (5, 4492), (6, 4814), (7, 5079), (8, 5213), (9, 5289), (10, 5682), (11, 5958), (12, 6171)

(a) Use the regression capabilities of a graphing utility to find a model of the form

$$a_n = ce^{kn}, \quad n = 3, 4, 5, \ldots, 12$$

for the data. Graphically compare the points and the model.

(b) Use the data to find the total sales for the 10-year period.

(c) Approximate the total sales for the 10-year period using the formula for the sum of a geometric series. Compare the result with that in part (b).

116. *Salary* You accept a job that pays a salary of $40,000 for the first year. During the next 39 years you receive a 4% raise each year. What would be your total compensation over the 40-year period?

True or False? **In Exercises 117–122, determine whether the statement is true or false. If it is false, explain why or give an example that shows it is false.**

117. If $\lim\limits_{n \to \infty} a_n = 0$, then $\sum\limits_{n=1}^{\infty} a_n$ converges.

118. If $\sum\limits_{n=1}^{\infty} a_n = L$, then $\sum\limits_{n=0}^{\infty} a_n = L + a_0$.

119. If $|r| < 1$, then $\sum\limits_{n=1}^{\infty} ar^n = \dfrac{a}{(1 - r)}$.

120. The series $\sum\limits_{n=1}^{\infty} \dfrac{n}{1000(n + 1)}$ diverges.

121. $0.75 = 0.749999\ldots$.

122. Every decimal with a repeating pattern of digits is a rational number.

123. Show that the series $\sum\limits_{n=1}^{\infty} a_n$ can be written in the telescoping form

$$\sum_{n=1}^{\infty} [(c - S_{n-1}) - (c - S_n)]$$

where $S_0 = 0$ and S_n is the nth partial sum.

124. Let Σa_n be a convergent series, and let

$$R_N = a_{N+1} + a_{N+2} + \cdots$$

be the remainder of the series after the first N terms. Prove that $\lim\limits_{N \to \infty} R_N = 0$.

125. Find two divergent series Σa_n and Σb_n such that $\Sigma(a_n + b_n)$ converges.

126. Given two infinite series Σa_n and Σb_n such that Σa_n converges and Σb_n diverges, prove that $\Sigma(a_n + b_n)$ diverges.

127. Suppose that Σa_n diverges and c is a nonzero constant. Prove that Σca_n diverges.

128. If $\displaystyle\sum_{n=1}^{\infty} a_n$ converges where a_n is nonzero, show that $\displaystyle\sum_{n=1}^{\infty} \frac{1}{a_n}$ diverges.

129. The Fibonacci sequence is defined recursively by $a_{n+2} = a_n + a_{n+1}$, where $a_1 = 1$ and $a_2 = 1$.

(a) Show that $\dfrac{1}{a_{n+1} a_{n+3}} = \dfrac{1}{a_{n+1} a_{n+2}} - \dfrac{1}{a_{n+2} a_{n+3}}$.

(b) Show that $\displaystyle\sum_{n=0}^{\infty} \frac{1}{a_{n+1} a_{n+3}} = 1$.

130. Find the values of x for which the infinite series

$$1 + 2x + x^2 + 2x^3 + x^4 + 2x^5 + x^6 + \cdots$$

converges. What is the sum when the series converges?

131. Prove that $\dfrac{1}{r} + \dfrac{1}{r^2} + \dfrac{1}{r^3} + \cdots = \dfrac{1}{r-1}$, for $|r| > 1$.

132. *Writing* The figure below represents an informal way of showing that $\displaystyle\sum_{n=1}^{\infty} \frac{1}{n^2} < 2$. Explain how the figure implies this conclusion.

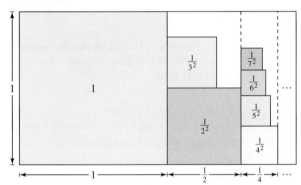

FOR FURTHER INFORMATION For more on this exercise, see the article "Convergence with Pictures" by P.J. Rippon in *American Mathematical Monthly*.

133. *Writing* Read the article "The Exponential-Decay Law Applied to Medical Dosages" by Gerald M. Armstrong and Calvin P. Midgley in *Mathematics Teacher*. (To view this article, go to the website *www.matharticles.com*.) Then write a paragraph on how a geometric sequence can be used to find the total amount of a drug that remains in a patient's system after n equal doses have been administered (at equal time intervals).

Putnam Exam Challenge

134. Write $\displaystyle\sum_{k=1}^{\infty} \frac{6^k}{(3^{k+1} - 2^{k+1})(3^k - 2^k)}$ as a rational number.

135. Let $f(n)$ be the sum of the first n terms of the sequence 0, 1, 1, 2, 2, 3, 3, 4, . . . , where the nth term is given by

$$a_n = \begin{cases} n/2, & \text{if } n \text{ is even} \\ (n-1)/2, & \text{if } n \text{ is odd} \end{cases}.$$

Show that if x and y are positive integers and $x > y$ then $xy = f(x + y) - f(x - y)$.

Section Project: Cantor's Disappearing Table

The following procedure shows how to make a table disappear by removing only half of the table!

(a) Original table has a length of L.

(b) Remove $\frac{1}{4}$ of the table centered at the midpoint. Each remaining piece has a length that is less than $\frac{1}{2}L$.

(c) Remove $\frac{1}{8}$ of the table by taking sections of length $\frac{1}{16}L$ from the centers of each of the two remaining pieces. Now, you have removed $\frac{1}{4} + \frac{1}{8}$ of the table. Each remaining piece has a length that is less than $\frac{1}{4}L$.

(d) Remove $\frac{1}{16}$ of the table by taking sections of length $\frac{1}{64}L$ from the centers of each of the four remaining pieces. Now, you have removed $\frac{1}{4} + \frac{1}{8} + \frac{1}{16}$ of the table. Each remaining piece has a length that is less than $\frac{1}{8}L$.

Will continuing this process cause the table to disappear, even though you have only removed half of the table? Why?

FOR FURTHER INFORMATION Read the article "Cantor's Disappearing Table" by Larry E. Knop in *The College Mathematics Journal*. To view this article, go to the website *www.matharticles.com*.

The Integral Test and *p*-Series

- Use the Integral Test to determine whether an infinite series converges or diverges.
- Use properties of *p*-series and harmonic series.

The Integral Test

In this and the following section, you will study several convergence tests that apply to series with *positive* terms.

THEOREM 9.10   The Integral Test

If f is positive, continuous, and decreasing for $x \geq 1$ and $a_n = f(n)$, then

$$\sum_{n=1}^{\infty} a_n \quad \text{and} \quad \int_1^{\infty} f(x)\, dx$$

either both converge or both diverge.

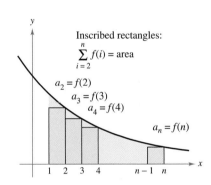

Inscribed rectangles:
$$\sum_{i=2}^{n} f(i) = \text{area}$$
$a_2 = f(2)$
$a_3 = f(3)$
$a_4 = f(4)$
$a_n = f(n)$

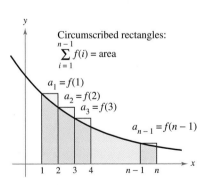

Circumscribed rectangles:
$$\sum_{i=1}^{n-1} f(i) = \text{area}$$
$a_1 = f(1)$
$a_2 = f(2)$
$a_3 = f(3)$
$a_{n-1} = f(n-1)$

Figure 9.8

Proof   Begin by partitioning the interval $[1, n]$ into $n - 1$ unit intervals, as shown in Figure 9.8. The total areas of the inscribed rectangles and the circumscribed rectangles are as follows.

$$\sum_{i=2}^{n} f(i) = f(2) + f(3) + \cdots + f(n) \qquad \text{Inscribed area}$$

$$\sum_{i=1}^{n-1} f(i) = f(1) + f(2) + \cdots + f(n-1) \qquad \text{Circumscribed area}$$

The exact area under the graph of f from $x = 1$ to $x = n$ lies between the inscribed and circumscribed areas.

$$\sum_{i=2}^{n} f(i) \leq \int_1^{n} f(x)\, dx \leq \sum_{i=1}^{n-1} f(i)$$

Using the nth partial sum, $S_n = f(1) + f(2) + \cdots + f(n)$, you can write this inequality as

$$S_n - f(1) \leq \int_1^{n} f(x)\, dx \leq S_{n-1}.$$

Now, assuming that $\int_1^{\infty} f(x)\, dx$ converges to L, it follows that for $n \geq 1$

$$S_n - f(1) \leq L \quad \Longrightarrow \quad S_n \leq L + f(1).$$

Consequently, $\{S_n\}$ is bounded and monotonic, and by Theorem 9.5 it converges. So, $\Sigma\, a_n$ converges. For the other direction of the proof, assume that the improper integral diverges. Then $\int_1^{n} f(x)\, dx$ approaches infinity as $n \to \infty$, and the inequality $S_{n-1} \geq \int_1^{n} f(x)\, dx$ implies that $\{S_n\}$ diverges. So, $\Sigma\, a_n$ diverges.

NOTE   Remember that the convergence or divergence of $\Sigma\, a_n$ is not affected by deleting the first N terms. Similarly, if the conditions for the Integral Test are satisfied for all $x \geq N > 1$, you can simply use the integral $\int_N^{\infty} f(x)\, dx$ to test for convergence or divergence. (This is illustrated in Example 4.)

EXAMPLE 1   **Using the Integral Test**

Apply the Integral Test to the series $\displaystyle\sum_{n=1}^{\infty} \frac{n}{n^2 + 1}$.

Solution   The function $f(x) = x/(x^2 + 1)$ is positive and continuous for $x \geq 1$. To determine whether f is decreasing, find the derivative.

$$f'(x) = \frac{(x^2 + 1)(1) - x(2x)}{(x^2 + 1)^2} = \frac{-x^2 + 1}{(x^2 + 1)^2}$$

So, $f'(x) < 0$ for $x > 1$ and it follows that f satisfies the conditions for the Integral Test. You can integrate to obtain

$$\int_1^{\infty} \frac{x}{x^2 + 1}\, dx = \frac{1}{2} \int_1^{\infty} \frac{2x}{x^2 + 1}\, dx$$

$$= \frac{1}{2} \lim_{b \to \infty} \int_1^b \frac{2x}{x^2 + 1}\, dx$$

$$= \frac{1}{2} \lim_{b \to \infty} \Big[\ln(x^2 + 1) \Big]_1^b$$

$$= \frac{1}{2} \lim_{b \to \infty} \big[\ln(b^2 + 1) - \ln 2 \big]$$

$$= \infty.$$

So, the series *diverges*.

EXAMPLE 2   **Using the Integral Test**

Apply the Integral Test to the series $\displaystyle\sum_{n=1}^{\infty} \frac{1}{n^2 + 1}$.

Solution   Because $f(x) = 1/(x^2 + 1)$ satisfies the conditions for the Integral Test (check this), you can integrate to obtain

$$\int_1^{\infty} \frac{1}{x^2 + 1}\, dx = \lim_{b \to \infty} \int_1^b \frac{1}{x^2 + 1}\, dx$$

$$= \lim_{b \to \infty} \Big[\arctan x \Big]_1^b$$

$$= \lim_{b \to \infty} (\arctan b - \arctan 1)$$

$$= \frac{\pi}{2} - \frac{\pi}{4} = \frac{\pi}{4}.$$

So, the series *converges* (see Figure 9.9).

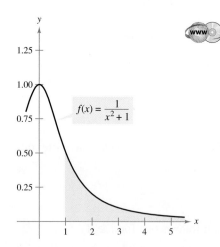

Because the improper integral converges, the infinite series also converges.
Figure 9.9

TECHNOLOGY

In Example 2, the fact that the improper integral converges to $\pi/4$ does not imply that the infinite series converges to $\pi/4$. To approximate the sum of the series, you can use the inequality

$$\sum_{n=1}^{N} \frac{1}{n^2 + 1} \leq \sum_{n=1}^{\infty} \frac{1}{n^2 + 1} \leq \sum_{n=1}^{N} \frac{1}{n^2 + 1} + \int_N^{\infty} \frac{1}{x^2 + 1}\, dx.$$

(See Exercise 60.) The larger the value of N, the better the approximation. For instance, using $N = 200$ produces $1.072 \leq \sum 1/(n^2 + 1) \leq 1.077$.

p-Series and Harmonic Series

HARMONIC SERIES

Pythagoras and his students paid close attention to the development of music as an abstract science. This led to the discovery of the relationship between the tone and the length of the vibrating string. It was observed that the most beautiful musical harmonies corresponded to the simplest ratios of whole numbers. Later mathematicians developed this idea into the harmonic series, where the terms in the harmonic series correspond to the nodes on a vibrating string that produce multiples of the fundamental frequency. For example, $\frac{1}{2}$ is twice the fundamental frequency, $\frac{1}{3}$ is three times the fundamental frequency, and so on.

In the remainder of this section, you will investigate a second type of series that has a simple arithmetic test for convergence or divergence. A series of the form

$$\sum_{n=1}^{\infty} \frac{1}{n^p} = \frac{1}{1^p} + \frac{1}{2^p} + \frac{1}{3^p} + \cdots \qquad \textit{p-series}$$

is a ***p*-series,** where *p* is a positive constant. For $p = 1$, the series

$$\sum_{n=1}^{\infty} \frac{1}{n} = 1 + \frac{1}{2} + \frac{1}{3} + \cdots \qquad \text{Harmonic series}$$

is the **harmonic** series. A **general harmonic series** is of the form $\Sigma 1/(an + b)$. In music, strings of the same material, diameter, and tension, whose lengths form a harmonic series, produce harmonic tones.

The Integral Test is convenient for establishing the convergence or divergence of *p*-series. This is shown in the proof of Theorem 9.11.

THEOREM 9.11   Convergence of *p*-Series

The *p*-series

$$\sum_{n=1}^{\infty} \frac{1}{n^p} = \frac{1}{1^p} + \frac{1}{2^p} + \frac{1}{3^p} + \frac{1}{4^p} + \cdots$$

1. converges if $p > 1$, and

2. diverges if $0 < p \le 1$.

Proof   The proof follows from the Integral Test and from Theorem 8.5, which states that

$$\int_{1}^{\infty} \frac{1}{x^p}\, dx$$

converges if $p > 1$ and diverges if $0 < p \le 1$.   ▬▬▬

EXAMPLE 3   Convergent and Divergent *p*-Series

Discuss the convergence or divergence of (a) the harmonic series and (b) the *p*-series with $p = 2$.

NOTE   The sum of the series in Example 3(b) can be shown to be $\pi^2/6$. (This was proved by Leonhard Euler, but the proof is too difficult to present here.) Be sure you see that the Integral Test does not tell you that the sum of the series is equal to the value of the integral. For instance, the sum of the series in Example 3(b) is

$$\sum_{n=1}^{\infty} \frac{1}{n^2} = \frac{\pi^2}{6} \approx 1.645$$

but the value of the corresponding improper integral is

$$\int_{1}^{\infty} \frac{1}{x^2}\, dx = 1.$$

Solution

a. From Theorem 9.11, it follows that the harmonic series

$$\sum_{n=1}^{\infty} \frac{1}{n} = \frac{1}{1} + \frac{1}{2} + \frac{1}{3} + \cdots \qquad p = 1$$

diverges.

b. From Theorem 9.11, it follows that the *p*-series

$$\sum_{n=1}^{\infty} \frac{1}{n^2} = \frac{1}{1^2} + \frac{1}{2^2} + \frac{1}{3^2} + \cdots \qquad p = 2$$

converges.   ▬▬▬

EXAMPLE 4 **Testing a Series for Convergence**

Determine whether the following series converges or diverges.

$$\sum_{n=2}^{\infty} \frac{1}{n \ln n}$$

Solution This series is similar to the divergent harmonic series. If its terms were larger than those of the harmonic series, you would expect it to diverge. However, because its terms are smaller, you are not sure what to expect. The function $f(x) = 1/(x \ln x)$ is positive and continuous for $x \geq 2$. To determine whether f is decreasing, first rewrite f as $f(x) = (x \ln x)^{-1}$ and then find its derivative.

$$f'(x) = (-1)(x \ln x)^{-2}(1 + \ln x) = -\frac{1 + \ln x}{x^2(\ln x)^2}$$

So, $f'(x) < 0$ for $x > 2$ and it follows that f satisfies the conditions for the Integral Test.

$$\int_2^{\infty} \frac{1}{x \ln x}\, dx = \int_2^{\infty} \frac{1/x}{\ln x}\, dx$$
$$= \lim_{b \to \infty} \left[\ln(\ln x)\right]_2^b$$
$$= \lim_{b \to \infty} \left[\ln(\ln b) - \ln(\ln 2)\right] = \infty$$

The series diverges.

NOTE The infinite series in Example 4 diverges very slowly. For instance, the sum of the first 10 terms is approximately 1.6878196, whereas the sum of the first 100 terms is just slightly larger: 2.3250871. In fact, the sum of the first 10,000 terms is approximately 3.015021704. You can see that although the infinite series "adds up to infinity," it does so very slowly.

Exercises for Section 9.3   See www.CalcChat.com for worked-out solutions to odd-numbered exercises.

In Exercises 1–18, use the Integral Test to determine the convergence or divergence of the series.

1. $\displaystyle\sum_{n=1}^{\infty} \frac{1}{n+1}$

2. $\displaystyle\sum_{n=1}^{\infty} \frac{2}{3n+5}$

3. $\displaystyle\sum_{n=1}^{\infty} e^{-n}$

4. $\displaystyle\sum_{n=1}^{\infty} ne^{-n/2}$

5. $\dfrac{1}{2} + \dfrac{1}{5} + \dfrac{1}{10} + \dfrac{1}{17} + \dfrac{1}{26} + \cdots$

6. $\dfrac{1}{3} + \dfrac{1}{5} + \dfrac{1}{7} + \dfrac{1}{9} + \dfrac{1}{11} + \cdots$

7. $\dfrac{\ln 2}{2} + \dfrac{\ln 3}{3} + \dfrac{\ln 4}{4} + \dfrac{\ln 5}{5} + \dfrac{\ln 6}{6} + \cdots$

8. $\dfrac{\ln 2}{\sqrt{2}} + \dfrac{\ln 3}{\sqrt{3}} + \dfrac{\ln 4}{\sqrt{4}} + \dfrac{\ln 5}{\sqrt{5}} + \dfrac{\ln 6}{\sqrt{6}} + \cdots$

9. $\dfrac{1}{\sqrt{1}(\sqrt{1}+1)} + \dfrac{1}{\sqrt{2}(\sqrt{2}+1)} + \dfrac{1}{\sqrt{3}(\sqrt{3}+1)}$

$+ \cdots + \dfrac{1}{\sqrt{n}(\sqrt{n}+1)} + \cdots$

10. $\dfrac{1}{4} + \dfrac{2}{7} + \dfrac{3}{12} + \cdots + \dfrac{n}{n^2+3} + \cdots$

11. $\displaystyle\sum_{n=1}^{\infty} \frac{1}{\sqrt{n+1}}$

12. $\displaystyle\sum_{n=2}^{\infty} \frac{\ln n}{n^3}$

13. $\displaystyle\sum_{n=1}^{\infty} \frac{\ln n}{n^2}$

14. $\displaystyle\sum_{n=2}^{\infty} \frac{1}{n\sqrt{\ln n}}$

15. $\displaystyle\sum_{n=1}^{\infty} \frac{\arctan n}{n^2+1}$

16. $\displaystyle\sum_{n=3}^{\infty} \frac{1}{n \ln n \ln(\ln n)}$

17. $\displaystyle\sum_{n=1}^{\infty} \frac{2n}{n^2+1}$

18. $\displaystyle\sum_{n=1}^{\infty} \frac{n}{n^4+1}$

In Exercises 19 and 20, use the Integral Test to determine the convergence or divergence of the series, where k is a positive integer.

19. $\displaystyle\sum_{n=1}^{\infty} \frac{n^{k-1}}{n^k+c}$

20. $\displaystyle\sum_{n=1}^{\infty} n^k e^{-n}$

In Exercises 21–24, explain why the Integral Test does not apply to the series.

21. $\displaystyle\sum_{n=1}^{\infty} \frac{(-1)^n}{n}$

22. $\displaystyle\sum_{n=1}^{\infty} e^{-n} \cos n$

23. $\displaystyle\sum_{n=1}^{\infty} \frac{2 + \sin n}{n}$

24. $\displaystyle\sum_{n=1}^{\infty} \left(\frac{\sin n}{n}\right)^2$

In Exercises 25–28, use the Integral Test to determine the convergence or divergence of the *p*-series.

25. $\displaystyle\sum_{n=1}^{\infty} \frac{1}{n^3}$

26. $\displaystyle\sum_{n=1}^{\infty} \frac{1}{n^{1/3}}$

27. $\displaystyle\sum_{n=1}^{\infty} \frac{1}{\sqrt{n}}$

28. $\displaystyle\sum_{n=1}^{\infty} \frac{1}{n^2}$

In Exercises 29–36, use Theorem 9.11 to determine the convergence or divergence of the *p*-series.

29. $\displaystyle\sum_{n=1}^{\infty} \frac{1}{\sqrt[5]{n}}$

30. $\displaystyle\sum_{n=1}^{\infty} \frac{3}{n^{5/3}}$

31. $1 + \dfrac{1}{\sqrt{2}} + \dfrac{1}{\sqrt{3}} + \dfrac{1}{\sqrt{4}} + \cdots$

32. $1 + \dfrac{1}{4} + \dfrac{1}{9} + \dfrac{1}{16} + \dfrac{1}{25} + \cdots$

33. $1 + \dfrac{1}{2\sqrt{2}} + \dfrac{1}{3\sqrt{3}} + \dfrac{1}{4\sqrt{4}} + \dfrac{1}{5\sqrt{5}} + \cdots$

34. $1 + \dfrac{1}{\sqrt[3]{4}} + \dfrac{1}{\sqrt[3]{9}} + \dfrac{1}{\sqrt[3]{16}} + \dfrac{1}{\sqrt[3]{25}} + \cdots$

35. $\displaystyle\sum_{n=1}^{\infty} \frac{1}{n^{1.04}}$

36. $\displaystyle\sum_{n=1}^{\infty} \frac{1}{n^{\pi}}$

In Exercises 37–42, match the series with the graph of its sequence of partial sums. [The graphs are labeled (a), (b), (c), (d), (e), and (f).] Determine the convergence or divergence of the series.

(a)

(b)

(c)

(d)

(e)

(f)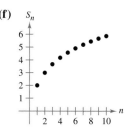

37. $\displaystyle\sum_{n=1}^{\infty} \frac{2}{\sqrt[4]{n^3}}$

38. $\displaystyle\sum_{n=1}^{\infty} \frac{2}{n}$

39. $\displaystyle\sum_{n=1}^{\infty} \frac{2}{\sqrt{n^{\pi}}}$

40. $\displaystyle\sum_{n=1}^{\infty} \frac{2}{\sqrt[5]{n^2}}$

41. $\displaystyle\sum_{n=1}^{\infty} \frac{2}{n\sqrt{n}}$

42. $\displaystyle\sum_{n=1}^{\infty} \frac{2}{n^2}$

 43. ***Numerical and Graphical Analysis*** Use a graphing utility to find the indicated partial sum S_n and complete the table. Then use a graphing utility to graph the first 10 terms of the sequence of partial sums. Compare the rate at which the sequence of partial sums approaches the sum of the series for each series.

n	5	10	20	50	100
S_n					

(a) $\displaystyle\sum_{n=1}^{\infty} 3\left(\frac{1}{5}\right)^{n-1} = \frac{15}{4}$ (b) $\displaystyle\sum_{n=1}^{\infty} \frac{1}{n^2} = \frac{\pi^2}{6}$

44. ***Numerical Reasoning*** Because the harmonic series diverges, it follows that for any positive real number M there exists a positive integer N such that the partial sum

$$\sum_{n=1}^{N} \frac{1}{n} > M.$$

(a) Use a graphing utility to complete the table.

M	2	4	6	8
N				

(b) As the real number M increases in equal increments, does the number N increase in equal increments? Explain.

Writing About Concepts

45. State the Integral Test and give an example of its use.

46. Define a *p*-series and state the requirements for its convergence.

47. A friend in your calculus class tells you that the following series converges because the terms are very small and approach 0 rapidly. Is your friend correct? Explain.

$$\frac{1}{10,000} + \frac{1}{10,001} + \frac{1}{10,002} + \cdots$$

Writing About Concepts *(continued)*

48. In Exercises 37–42, $\lim\limits_{n\to\infty} a_n = 0$ for each series but they do not all converge. Is this a contradiction of Theorem 9.9? Why do you think some converge and others diverge? Explain.

49. Use a graph to show that

$$\sum_{n=1}^{\infty} \frac{1}{\sqrt{n}} > \int_{1}^{\infty} \frac{1}{\sqrt{x}}\, dx.$$

What can you conclude about the convergence or divergence of the series? Explain.

50. Let f be a positive, continuous, and decreasing function for $x \geq 1$, such that $a_n = f(n)$. Use a graph to rank the following quantities in decreasing order. Explain your reasoning.

   (a) $\displaystyle\sum_{n=2}^{7} a_n$ (b) $\displaystyle\int_{1}^{7} f(x)\, dx$ (c) $\displaystyle\sum_{n=1}^{6} a_n$

In Exercises 51–54, find the positive values of *p* for which the series converges.

51. $\displaystyle\sum_{n=2}^{\infty} \frac{1}{n(\ln n)^p}$ **52.** $\displaystyle\sum_{n=2}^{\infty} \frac{\ln n}{n^p}$

53. $\displaystyle\sum_{n=1}^{\infty} \frac{n}{(1 + n^2)^p}$ **54.** $\displaystyle\sum_{n=1}^{\infty} n(1 + n^2)^p$

In Exercises 55–58, use the result of Exercise 51 to determine the convergence or divergence of the series.

55. $\displaystyle\sum_{n=2}^{\infty} \frac{1}{n \ln n}$ **56.** $\displaystyle\sum_{n=2}^{\infty} \frac{1}{n \sqrt[3]{(\ln n)^2}}$

57. $\displaystyle\sum_{n=2}^{\infty} \frac{1}{n(\ln n)^2}$ **58.** $\displaystyle\sum_{n=2}^{\infty} \frac{1}{n \ln(n^2)}$

59. Let f be a positive, continuous, and decreasing function for $x \geq 1$, such that $a_n = f(n)$. Prove that if the series

$$\sum_{n=1}^{\infty} a_n$$

converges to S, then the remainder $R_N = S - S_N$ is bounded by

$$0 \leq R_N \leq \int_{N}^{\infty} f(x)\, dx.$$

60. Show that the result of Exercise 59 can be written as

$$\sum_{n=1}^{N} a_n \leq \sum_{n=1}^{\infty} a_n \leq \sum_{n=1}^{N} a_n + \int_{N}^{\infty} f(x)\, dx.$$

In Exercises 61–66, use the result of Exercise 59 to approximate the sum of the convergent series using the indicated number of terms. Include an estimate of the maximum error for your approximation.

61. $\displaystyle\sum_{n=1}^{\infty} \frac{1}{n^4}$, six terms

62. $\displaystyle\sum_{n=1}^{\infty} \frac{1}{n^5}$, four terms

63. $\displaystyle\sum_{n=1}^{\infty} \frac{1}{n^2 + 1}$, ten terms

64. $\displaystyle\sum_{n=1}^{\infty} \frac{1}{(n + 1)[\ln(n + 1)]^3}$, ten terms

65. $\displaystyle\sum_{n=1}^{\infty} ne^{-n^2}$, four terms

66. $\displaystyle\sum_{n=1}^{\infty} e^{-n}$, four terms

In Exercises 67–72, use the result of Exercise 59 to find *N* such that $R_N \leq 0.001$ for the convergent series.

67. $\displaystyle\sum_{n=1}^{\infty} \frac{1}{n^4}$ **68.** $\displaystyle\sum_{n=1}^{\infty} \frac{1}{n^{3/2}}$

69. $\displaystyle\sum_{n=1}^{\infty} e^{-5n}$ **70.** $\displaystyle\sum_{n=1}^{\infty} e^{-n/2}$

71. $\displaystyle\sum_{n=1}^{\infty} \frac{1}{n^2 + 1}$ **72.** $\displaystyle\sum_{n=1}^{\infty} \frac{2}{n^2 + 5}$

73. (a) Show that $\displaystyle\sum_{n=2}^{\infty} \frac{1}{n^{1.1}}$ converges and $\displaystyle\sum_{n=2}^{\infty} \frac{1}{n \ln n}$ diverges.

   (b) Compare the first five terms of each series in part (a).

   (c) Find $n > 3$ such that

   $$\frac{1}{n^{1.1}} < \frac{1}{n \ln n}.$$

74. Ten terms are used to approximate a convergent *p*-series. Therefore, the remainder is a function of p and is

$$0 \leq R_{10}(p) \leq \int_{10}^{\infty} \frac{1}{x^p}\, dx, \quad p > 1.$$

   (a) Perform the integration in the inequality.

   (b) Use a graphing utility to represent the inequality graphically.

   (c) Identify any asymptotes of the error function and interpret their meaning.

75. *Euler's Constant* Let

$$S_n = \sum_{k=1}^{n} \frac{1}{k} = 1 + \frac{1}{2} + \cdots + \frac{1}{n}.$$

(a) Show that $\ln(n + 1) \le S_n \le 1 + \ln n$.

(b) Show that the sequence $\{a_n\} = \{S_n - \ln n\}$ is bounded.

(c) Show that the sequence $\{a_n\}$ is decreasing.

(d) Show that a_n converges to a limit γ (called Euler's constant).

(e) Approximate γ using a_{100}.

76. Find the sum of the series $\displaystyle\sum_{n=2}^{\infty} \ln\left(1 - \frac{1}{n^2}\right)$.

77. Consider the series

$$\sum_{n=2}^{\infty} x^{\ln n}.$$

(a) Determine the convergence or divergence of the series for $x = 1$.

(b) Determine the convergence or divergence of the series for $x = 1/e$.

(c) Find the positive values of x for which the series converges.

78. The **Riemann zeta function** for real numbers is defined for all x for which the series

$$\zeta(x) = \sum_{n=1}^{\infty} n^{-x}$$

converges. Find the domain of the function.

Review In Exercises 79–90, determine the convergence or divergence of the series.

79. $\displaystyle\sum_{n=1}^{\infty} \frac{1}{2n - 1}$

80. $\displaystyle\sum_{n=2}^{\infty} \frac{1}{n\sqrt{n^2 - 1}}$

81. $\displaystyle\sum_{n=1}^{\infty} \frac{1}{n\sqrt[4]{n}}$

82. $\displaystyle 3\sum_{n=1}^{\infty} \frac{1}{n^{0.95}}$

83. $\displaystyle\sum_{n=0}^{\infty} \left(\frac{2}{3}\right)^n$

84. $\displaystyle\sum_{n=0}^{\infty} (1.075)^n$

85. $\displaystyle\sum_{n=1}^{\infty} \frac{n}{\sqrt{n^2 + 1}}$

86. $\displaystyle\sum_{n=1}^{\infty} \left(\frac{1}{n^2} - \frac{1}{n^3}\right)$

87. $\displaystyle\sum_{n=1}^{\infty} \left(1 + \frac{1}{n}\right)^n$

88. $\displaystyle\sum_{n=2}^{\infty} \ln n$

89. $\displaystyle\sum_{n=2}^{\infty} \frac{1}{n(\ln n)^3}$

90. $\displaystyle\sum_{n=2}^{\infty} \frac{\ln n}{n^3}$

Section Project: The Harmonic Series

The harmonic series

$$\sum_{n=1}^{\infty} \frac{1}{n} = 1 + \frac{1}{2} + \frac{1}{3} + \frac{1}{4} + \cdots + \frac{1}{n} + \cdots$$

is one of the most important series in this chapter. Even though its terms tend to zero as *n* increases,

$$\lim_{n\to\infty} \frac{1}{n} = 0$$

the harmonic series diverges. In other words, even though the terms are getting smaller and smaller, the sum "adds up to infinity."

(a) One way to show that the harmonic series diverges is attributed to Jakob Bernoulli. He grouped the terms of the harmonic series as follows:

$$1 + \frac{1}{2} + \underbrace{\frac{1}{3} + \frac{1}{4}}_{> \frac{1}{2}} + \underbrace{\frac{1}{5} + \cdots + \frac{1}{8}}_{> \frac{1}{2}} + \underbrace{\frac{1}{9} + \cdots + \frac{1}{16}}_{> \frac{1}{2}} +$$

$$\underbrace{\frac{1}{17} + \cdots + \frac{1}{32}}_{> \frac{1}{2}} + \cdots$$

Write a short paragraph explaining how you can use this grouping to show that the harmonic series diverges.

(b) Use the proof of the Integral Test, Theorem 9.10, to show that

$$\ln(n + 1) \le 1 + \frac{1}{2} + \frac{1}{3} + \frac{1}{4} + \cdots + \frac{1}{n} \le 1 + \ln n.$$

(c) Use part (b) to determine how many terms *M* you would need so that

$$\sum_{n=1}^{M} \frac{1}{n} > 50.$$

(d) Show that the sum of the first million terms of the harmonic series is less than 15.

(e) Show that the following inequalities are valid.

$$\ln \frac{21}{10} \le \frac{1}{10} + \frac{1}{11} + \cdots + \frac{1}{20} \le \ln \frac{20}{9}$$

$$\ln \frac{201}{100} \le \frac{1}{100} + \frac{1}{101} + \cdots + \frac{1}{200} \le \ln \frac{200}{99}$$

(f) Use the ideas in part (e) to find the limit

$$\lim_{m\to\infty} \sum_{n=m}^{2m} \frac{1}{n}.$$

Comparisons of Series

- Use the Direct Comparison Test to determine whether a series converges or diverges.
- Use the Limit Comparison Test to determine whether a series converges or diverges.

Direct Comparison Test

For the convergence tests developed so far, the terms of the series have to be fairly simple and the series must have special characteristics in order for the convergence tests to be applied. A slight deviation from these special characteristics can make a test nonapplicable. For example, in the following pairs, the second series cannot be tested by the same convergence test as the first series even though it is similar to the first.

1. $\displaystyle\sum_{n=0}^{\infty} \frac{1}{2^n}$ is geometric, but $\displaystyle\sum_{n=0}^{\infty} \frac{n}{2^n}$ is not.

2. $\displaystyle\sum_{n=1}^{\infty} \frac{1}{n^3}$ is a p-series, but $\displaystyle\sum_{n=1}^{\infty} \frac{1}{n^3 + 1}$ is not.

3. $a_n = \dfrac{n}{(n^2 + 3)^2}$ is easily integrated, but $b_n = \dfrac{n^2}{(n^2 + 3)^2}$ is not.

In this section you will study two additional tests for positive-term series. These two tests greatly expand the variety of series you are able to test for convergence or divergence. They allow you to *compare* a series having complicated terms with a simpler series whose convergence or divergence is known.

THEOREM 9.12   Direct Comparison Test

Let $0 < a_n \le b_n$ for all n.

1. If $\displaystyle\sum_{n=1}^{\infty} b_n$ converges, then $\displaystyle\sum_{n=1}^{\infty} a_n$ converges.

2. If $\displaystyle\sum_{n=1}^{\infty} a_n$ diverges, then $\displaystyle\sum_{n=1}^{\infty} b_n$ diverges.

Proof   To prove the first property, let $L = \displaystyle\sum_{n=1}^{\infty} b_n$ and let

$$S_n = a_1 + a_2 + \cdots + a_n.$$

Because $0 < a_n \le b_n$, the sequence $S_1, S_2, S_3, \ldots$ is nondecreasing and bounded above by L; so, it must converge. Because

$$\lim_{n \to \infty} S_n = \sum_{n=1}^{\infty} a_n$$

it follows that $\Sigma\, a_n$ converges. The second property is logically equivalent to the first.

NOTE   As stated, the Direct Comparison Test requires that $0 < a_n \le b_n$ for all n. Because the convergence of a series is not dependent on its first several terms, you could modify the test to require only that $0 < a_n \le b_n$ for all n greater than some integer N.

EXAMPLE 1   **Using the Direct Comparison Test**

Determine the convergence or divergence of

$$\sum_{n=1}^{\infty} \frac{1}{2 + 3^n}.$$

Solution   This series resembles

$$\sum_{n=1}^{\infty} \frac{1}{3^n}. \qquad \text{Convergent geometric series}$$

Term-by-term comparison yields

$$a_n = \frac{1}{2 + 3^n} < \frac{1}{3^n} = b_n, \quad n \geq 1.$$

So, by the Direct Comparison Test, the series converges.

EXAMPLE 2   **Using the Direct Comparison Test**

Determine the convergence or divergence of

$$\sum_{n=1}^{\infty} \frac{1}{2 + \sqrt{n}}.$$

Solution   This series resembles

$$\sum_{n=1}^{\infty} \frac{1}{n^{1/2}}. \qquad \text{Divergent } p\text{-series}$$

Term-by-term comparison yields

$$\frac{1}{2 + \sqrt{n}} \leq \frac{1}{\sqrt{n}}, \quad n \geq 1$$

which *does not* meet the requirements for divergence. (Remember that if term-by-term comparison reveals a series that is *smaller* than a divergent series, the Direct Comparison Test tells you nothing.) Still expecting the series to diverge, you can compare the given series with

$$\sum_{n=1}^{\infty} \frac{1}{n}. \qquad \text{Divergent harmonic series}$$

In this case, term-by-term comparison yields

NOTE   To verify the last inequality in Example 2, try showing that $2 + \sqrt{n} \leq n$ whenever $n \geq 4$.

$$a_n = \frac{1}{n} \leq \frac{1}{2 + \sqrt{n}} = b_n, \quad n \geq 4$$

and, by the Direct Comparison Test, the given series diverges.

Remember that both parts of the Direct Comparison Test require that $0 < a_n \leq b_n$. Informally, the test says the following about the two series with nonnegative terms.

1. If the "larger" series converges, the "smaller" series must also converge.

2. If the "smaller" series diverges, the "larger" series must also diverge.

Limit Comparison Test

Often a given series closely resembles a *p*-series or a geometric series, yet you cannot establish the term-by-term comparison necessary to apply the Direct Comparison Test. Under these circumstances you may be able to apply a second comparison test, called the **Limit Comparison Test.**

THEOREM 9.13   Limit Comparison Test

Suppose that $a_n > 0$, $b_n > 0$, and

$$\lim_{n \to \infty} \left(\frac{a_n}{b_n} \right) = L$$

where L is *finite and positive*. Then the two series $\Sigma \, a_n$ and $\Sigma \, b_n$ either both converge or both diverge.

NOTE As with the Direct Comparison Test, the Limit Comparison Test could be modified to require only that a_n and b_n be positive for all n greater than some integer N.

Proof   Because $a_n > 0$, $b_n > 0$, and

$$\lim_{n \to \infty} \left(\frac{a_n}{b_n} \right) = L$$

there exists $N > 0$ such that

$$0 < \frac{a_n}{b_n} < L + 1, \quad \text{for } n \geq N.$$

This implies that

$$0 < a_n < (L + 1)b_n.$$

So, by the Direct Comparison Test, the convergence of $\Sigma \, b_n$ implies the convergence of $\Sigma \, a_n$. Similarly, the fact that

$$\lim_{n \to \infty} \left(\frac{b_n}{a_n} \right) = \frac{1}{L}$$

can be used to show that the convergence of $\Sigma \, a_n$ implies the convergence of $\Sigma \, b_n$.

EXAMPLE 3   Using the Limit Comparison Test

Show that the following general harmonic series diverges.

$$\sum_{n=1}^{\infty} \frac{1}{an + b}, \quad a > 0, \quad b > 0$$

Solution   By comparison with

$$\sum_{n=1}^{\infty} \frac{1}{n} \qquad \text{Divergent harmonic series}$$

you have

$$\lim_{n \to \infty} \frac{1/(an + b)}{1/n} = \lim_{n \to \infty} \frac{n}{an + b} = \frac{1}{a}.$$

Because this limit is greater than 0, you can conclude from the Limit Comparison Test that the given series diverges.

The Limit Comparison Test works well for comparing a "messy" algebraic series with a *p*-series. In choosing an appropriate *p*-series, you must choose one with an *n*th term of the same magnitude as the *n*th term of the given series.

Given Series	*Comparison Series*	*Conclusion*
$\displaystyle\sum_{n=1}^{\infty} \frac{1}{3n^2 - 4n + 5}$	$\displaystyle\sum_{n=1}^{\infty} \frac{1}{n^2}$	Both series converge.
$\displaystyle\sum_{n=1}^{\infty} \frac{1}{\sqrt{3n - 2}}$	$\displaystyle\sum_{n=1}^{\infty} \frac{1}{\sqrt{n}}$	Both series diverge.
$\displaystyle\sum_{n=1}^{\infty} \frac{n^2 - 10}{4n^5 + n^3}$	$\displaystyle\sum_{n=1}^{\infty} \frac{n^2}{n^5} = \sum_{n=1}^{\infty} \frac{1}{n^3}$	Both series converge.

In other words, when choosing a series for comparison, you can disregard all but the *highest powers of n* in both the numerator and the denominator.

EXAMPLE 4   Using the Limit Comparison Test

Determine the convergence or divergence of

$$\sum_{n=1}^{\infty} \frac{\sqrt{n}}{n^2 + 1}.$$

Solution   Disregarding all but the highest powers of *n* in the numerator and the denominator, you can compare the series with

$$\sum_{n=1}^{\infty} \frac{\sqrt{n}}{n^2} = \sum_{n=1}^{\infty} \frac{1}{n^{3/2}}. \qquad \text{Convergent } p\text{-series}$$

Because

$$\lim_{n \to \infty} \frac{a_n}{b_n} = \lim_{n \to \infty} \left(\frac{\sqrt{n}}{n^2 + 1} \right) \left(\frac{n^{3/2}}{1} \right)$$

$$= \lim_{n \to \infty} \frac{n^2}{n^2 + 1} = 1$$

you can conclude by the Limit Comparison Test that the given series converges.

EXAMPLE 5   Using the Limit Comparison Test

Determine the convergence or divergence of

$$\sum_{n=1}^{\infty} \frac{n2^n}{4n^3 + 1}.$$

Solution   A reasonable comparison would be with the series

$$\sum_{n=1}^{\infty} \frac{2^n}{n^2}. \qquad \text{Divergent series}$$

Note that this series diverges by the *n*th-Term Test. From the limit

$$\lim_{n \to \infty} \frac{a_n}{b_n} = \lim_{n \to \infty} \left(\frac{n2^n}{4n^3 + 1} \right) \left(\frac{n^2}{2^n} \right)$$

$$= \lim_{n \to \infty} \frac{1}{4 + (1/n^3)} = \frac{1}{4}$$

you can conclude that the given series diverges.

See www.CalcChat.com for worked-out solutions to odd-numbered exercises.

1. Graphical Analysis The figures show the graphs of the first 10 terms, and the graphs of the first 10 terms of the sequence of partial sums, of each series.

$$\sum_{n=1}^{\infty} \frac{6}{n^{3/2}}, \quad \sum_{n=1}^{\infty} \frac{6}{n^{3/2} + 3}, \quad \text{and} \quad \sum_{n=1}^{\infty} \frac{6}{n\sqrt{n^2 + 0.5}}$$

(a) Identify the series in each figure.

(b) Which series is a *p*-series? Does it converge or diverge?

(c) For the series that are not *p*-series, how do the magnitudes of the terms compare with the magnitudes of the terms of the *p*-series? What conclusion can you draw about the convergence or divergence of the series?

(d) Explain the relationship between the magnitudes of the terms of the series and the magnitudes of the terms of the partial sums.

Graphs of terms   Graphs of partial sums

2. Graphical Analysis The figures show the graphs of the first 10 terms, and the graphs of the first 10 terms of the sequence of partial sums, of each series.

$$\sum_{n=1}^{\infty} \frac{2}{\sqrt{n}}, \quad \sum_{n=1}^{\infty} \frac{2}{\sqrt{n} - 0.5}, \quad \text{and} \quad \sum_{n=1}^{\infty} \frac{4}{\sqrt{n} + 0.5}$$

(a) Identify the series in each figure.

(b) Which series is a *p*-series? Does it converge or diverge?

(c) For the series that are not *p*-series, how do the magnitudes of the terms compare with the magnitudes of the terms of the *p*-series? What conclusion can you draw about the convergence or divergence of the series?

(d) Explain the relationship between the magnitudes of the terms of the series and the magnitudes of the terms of the partial sums.

Graphs of terms   Graphs of partial sums

In Exercises 3–14, use the Direct Comparison Test to determine the convergence or divergence of the series.

3. $\displaystyle\sum_{n=1}^{\infty} \frac{1}{n^2 + 1}$

4. $\displaystyle\sum_{n=1}^{\infty} \frac{1}{3n^2 + 2}$

5. $\displaystyle\sum_{n=2}^{\infty} \frac{1}{n - 1}$

6. $\displaystyle\sum_{n=2}^{\infty} \frac{1}{\sqrt{n} - 1}$

7. $\displaystyle\sum_{n=0}^{\infty} \frac{1}{3^n + 1}$

8. $\displaystyle\sum_{n=0}^{\infty} \frac{3^n}{4^n + 5}$

9. $\displaystyle\sum_{n=2}^{\infty} \frac{\ln n}{n + 1}$

10. $\displaystyle\sum_{n=1}^{\infty} \frac{1}{\sqrt{n^3 + 1}}$

11. $\displaystyle\sum_{n=0}^{\infty} \frac{1}{n!}$

12. $\displaystyle\sum_{n=1}^{\infty} \frac{1}{4\sqrt[3]{n} - 1}$

13. $\displaystyle\sum_{n=0}^{\infty} e^{-n^2}$

14. $\displaystyle\sum_{n=1}^{\infty} \frac{4^n}{3^n - 1}$

In Exercises 15–28, use the Limit Comparison Test to determine the convergence or divergence of the series.

15. $\displaystyle\sum_{n=1}^{\infty} \frac{n}{n^2 + 1}$

16. $\displaystyle\sum_{n=1}^{\infty} \frac{2}{3^n - 5}$

17. $\displaystyle\sum_{n=0}^{\infty} \frac{1}{\sqrt{n^2 + 1}}$

18. $\displaystyle\sum_{n=3}^{\infty} \frac{3}{\sqrt{n^2 - 4}}$

19. $\displaystyle\sum_{n=1}^{\infty} \frac{2n^2 - 1}{3n^5 + 2n + 1}$

20. $\displaystyle\sum_{n=1}^{\infty} \frac{5n - 3}{n^2 - 2n + 5}$

21. $\displaystyle\sum_{n=1}^{\infty} \frac{n + 3}{n(n + 2)}$

22. $\displaystyle\sum_{n=1}^{\infty} \frac{1}{n(n^2 + 1)}$

23. $\displaystyle\sum_{n=1}^{\infty} \frac{1}{n\sqrt{n^2 + 1}}$

24. $\displaystyle\sum_{n=1}^{\infty} \frac{n}{(n + 1)2^{n-1}}$

25. $\displaystyle\sum_{n=1}^{\infty} \frac{n^{k-1}}{n^k + 1}, \quad k > 2$

26. $\displaystyle\sum_{n=1}^{\infty} \frac{5}{n + \sqrt{n^2 + 4}}$

27. $\displaystyle\sum_{n=1}^{\infty} \sin\frac{1}{n}$

28. $\displaystyle\sum_{n=1}^{\infty} \tan\frac{1}{n}$

In Exercises 29–36, test for convergence or divergence, using each test at least once. Identify which test was used.

(a) *n*th-Term Test   (b) **Geometric Series Test**

(c) *p*-Series Test   (d) **Telescoping Series Test**

(e) **Integral Test**   (f) **Direct Comparison Test**

(g) **Limit Comparison Test**

29. $\displaystyle\sum_{n=1}^{\infty} \frac{\sqrt{n}}{n}$

30. $\displaystyle\sum_{n=0}^{\infty} 5\left(-\frac{1}{5}\right)^n$

31. $\displaystyle\sum_{n=1}^{\infty} \frac{1}{3^n + 2}$

32. $\displaystyle\sum_{n=4}^{\infty} \frac{1}{3n^2 - 2n - 15}$

33. $\displaystyle\sum_{n=1}^{\infty} \frac{n}{2n + 3}$

34. $\displaystyle\sum_{n=1}^{\infty} \left(\frac{1}{n + 1} - \frac{1}{n + 2}\right)$

35. $\displaystyle\sum_{n=1}^{\infty} \frac{n}{(n^2 + 1)^2}$

36. $\displaystyle\sum_{n=1}^{\infty} \frac{3}{n(n + 3)}$

37. Use the Limit Comparison Test with the harmonic series to show that the series $\Sigma\, a_n$ (where $0 < a_n < a_{n-1}$) diverges if $\lim_{n \to \infty} na_n$ is finite and nonzero.

38. Prove that, if $P(n)$ and $Q(n)$ are polynomials of degree j and k, respectively, then the series

$$\sum_{n=1}^{\infty} \frac{P(n)}{Q(n)}$$

converges if $j < k - 1$ and diverges if $j \geq k - 1$.

In Exercises 39–42, use the polynomial test given in Exercise 38 to determine whether the series converges or diverges.

39. $\frac{1}{2} + \frac{2}{5} + \frac{3}{10} + \frac{4}{17} + \frac{5}{26} + \cdots$

40. $\frac{1}{3} + \frac{1}{8} + \frac{1}{15} + \frac{1}{24} + \frac{1}{35} + \cdots$

41. $\displaystyle\sum_{n=1}^{\infty} \frac{1}{n^3 + 1}$ **42.** $\displaystyle\sum_{n=1}^{\infty} \frac{n^2}{n^3 + 1}$

In Exercises 43 and 44, use the divergence test given in Exercise 37 to show that the series diverges.

43. $\displaystyle\sum_{n=1}^{\infty} \frac{n^3}{5n^4 + 3}$ **44.** $\displaystyle\sum_{n=2}^{\infty} \frac{1}{\ln n}$

In Exercises 45–48, determine the convergence or divergence of the series.

45. $\frac{1}{200} + \frac{1}{400} + \frac{1}{600} + \frac{1}{800} + \cdots$

46. $\frac{1}{200} + \frac{1}{210} + \frac{1}{220} + \frac{1}{230} + \cdots$

47. $\frac{1}{201} + \frac{1}{204} + \frac{1}{209} + \frac{1}{216} + \cdots$

48. $\frac{1}{201} + \frac{1}{208} + \frac{1}{227} + \frac{1}{264} + \cdots$

Writing About Concepts

49. Review the results of Exercises 45–48. Explain why careful analysis is required to determine the convergence or divergence of a series and why only considering the magnitudes of the terms of a series could be misleading.

50. State the Direct Comparison Test and give an example of its use.

51. State the Limit Comparison Test and give an example of its use.

52. It appears that the terms of the series

$$\frac{1}{1000} + \frac{1}{1001} + \frac{1}{1002} + \frac{1}{1003} + \cdots$$

are less than the corresponding terms of the convergent series

$$1 + \frac{1}{4} + \frac{1}{9} + \frac{1}{16} + \cdots.$$

If the statement above is correct, the first series converges. Is this correct? Why or why not? Make a statement about how the divergence or convergence of a series is affected by inclusion or exclusion of the first finite number of terms.

Writing About Concepts (continued)

53. The figure shows the first 20 terms of the convergent series $\displaystyle\sum_{n=1}^{\infty} a_n$ and the first 20 terms of the series $\displaystyle\sum_{n=1}^{\infty} a_n^2$. Identify the two series and explain your reasoning in making the selection.

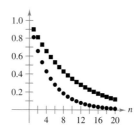

54. Consider the series $\displaystyle\sum_{n=1}^{\infty} \frac{1}{(2n - 1)^2}$.

(a) Verify that the series converges.

(b) Use a graphing utility to complete the table.

n	5	10	20	50	100
S_n					

(c) The sum of the series is $\pi^2/8$. Find the sum of the series

$$\sum_{n=3}^{\infty} \frac{1}{(2n - 1)^2}.$$

(d) Use a graphing utility to find the sum of the series

$$\sum_{n=10}^{\infty} \frac{1}{(2n - 1)^2}.$$

True or False? **In Exercises 55–60, determine whether the statement is true or false. If it is false, explain why or give an example that shows it is false.**

55. If $0 < a_n \leq b_n$ and $\displaystyle\sum_{n=1}^{\infty} a_n$ converges, then $\displaystyle\sum_{n=1}^{\infty} b_n$ diverges.

56. If $0 < a_{n+10} \leq b_n$ and $\displaystyle\sum_{n=1}^{\infty} b_n$ converges, then $\displaystyle\sum_{n=1}^{\infty} a_n$ converges.

57. If $a_n + b_n \leq c_n$ and $\displaystyle\sum_{n=1}^{\infty} c_n$ converges, then the series $\displaystyle\sum_{n=1}^{\infty} a_n$ and $\displaystyle\sum_{n=1}^{\infty} b_n$ both converge. (Assume that the terms of all three series are positive.)

58. If $a_n \leq b_n + c_n$ and $\displaystyle\sum_{n=1}^{\infty} a_n$ diverges, then the series $\displaystyle\sum_{n=1}^{\infty} b_n$ and $\displaystyle\sum_{n=1}^{\infty} c_n$ both diverge. (Assume that the terms of all three series are positive.)

59. If $0 < a_n \leq b_n$ and $\displaystyle\sum_{n=1}^{\infty} a_n$ diverges, then $\displaystyle\sum_{n=1}^{\infty} b_n$ diverges.

EXAMPLE 7 **Rearrangement of a Series**

The alternating harmonic series converges to ln 2. That is,

$$\sum_{n=1}^{\infty} (-1)^{n+1} \frac{1}{n} = \frac{1}{1} - \frac{1}{2} + \frac{1}{3} - \frac{1}{4} + \cdots = \ln 2. \qquad \text{(See Exercise 49, Section 9.10.)}$$

Rearrange the series to produce a different sum.

Solution   Consider the following rearrangement.

$$1 - \frac{1}{2} - \frac{1}{4} + \frac{1}{3} - \frac{1}{6} - \frac{1}{8} + \frac{1}{5} - \frac{1}{10} - \frac{1}{12} + \frac{1}{7} - \frac{1}{14} - \cdots$$

$$= \left(1 - \frac{1}{2}\right) - \frac{1}{4} + \left(\frac{1}{3} - \frac{1}{6}\right) - \frac{1}{8} + \left(\frac{1}{5} - \frac{1}{10}\right) - \frac{1}{12} + \left(\frac{1}{7} - \frac{1}{14}\right) - \cdots$$

$$= \frac{1}{2} - \frac{1}{4} + \frac{1}{6} - \frac{1}{8} + \frac{1}{10} - \frac{1}{12} + \frac{1}{14} - \cdots$$

$$= \frac{1}{2}\left(1 - \frac{1}{2} + \frac{1}{3} - \frac{1}{4} + \frac{1}{5} - \frac{1}{6} + \frac{1}{7} - \cdots\right) = \frac{1}{2}(\ln 2)$$

By rearranging the terms, you obtain a sum that is half the original sum. ▬▬▬

Exercises for Section 9.5

**In Exercises 1–6, match the series with the graph of its sequence
of partial sums. [The graphs are labeled (a), (b), (c), (d), (e), and
(f).]**

(a)

(b)

(c)

(d)

(e)

(f)

1. $\sum_{n=1}^{\infty} \frac{6}{n^2}$

2. $\sum_{n=1}^{\infty} \frac{(-1)^{n-1}6}{n^2}$

3. $\sum_{n=1}^{\infty} \frac{3}{n!}$

4. $\sum_{n=1}^{\infty} \frac{(-1)^{n-1}3}{n!}$

5. $\sum_{n=1}^{\infty} \frac{10}{n2^n}$

6. $\sum_{n=1}^{\infty} \frac{(-1)^{n-1}10}{n2^n}$

 Numerical and Graphical Analysis   **In Exercises 7–10, explore
the Alternating Series Remainder.**

**(a) Use a graphing utility to find the indicated partial sum S_n
and complete the table.**

n	1	2	3	4	5	6	7	8	9	10
S_n										

**(b) Use a graphing utility to graph the first 10 terms of the
sequence of partial sums and a horizontal line representing
the sum.**

**(c) What pattern exists between the plot of the successive points
in part (b) relative to the horizontal line representing the
sum of the series? Do the distances between the successive
points and the horizontal line increase or decrease?**

**(d) Discuss the relationship between the answers in part (c) and
the Alternating Series Remainder as given in Theorem 9.15.**

7. $\sum_{n=1}^{\infty} \frac{(-1)^{n-1}}{2n-1} = \frac{\pi}{4}$

8. $\sum_{n=1}^{\infty} \frac{(-1)^{n-1}}{(n-1)!} = \frac{1}{e}$

9. $\sum_{n=1}^{\infty} \frac{(-1)^{n-1}}{n^2} = \frac{\pi^2}{12}$

10. $\sum_{n=1}^{\infty} \frac{(-1)^{n-1}}{(2n-1)!} = \sin 1$

In Exercises 11–32, determine the convergence or divergence of the series.

11. $\displaystyle\sum_{n=1}^{\infty} \frac{(-1)^{n+1}}{n}$

12. $\displaystyle\sum_{n=1}^{\infty} \frac{(-1)^{n+1}n}{2n-1}$

13. $\displaystyle\sum_{n=1}^{\infty} \frac{(-1)^{n+1}}{2n-1}$

14. $\displaystyle\sum_{n=1}^{\infty} \frac{(-1)^{n}}{\ln(n+1)}$

15. $\displaystyle\sum_{n=1}^{\infty} \frac{(-1)^{n}\,n^2}{n^2+1}$

16. $\displaystyle\sum_{n=1}^{\infty} \frac{(-1)^{n+1}\,n}{n^2+1}$

17. $\displaystyle\sum_{n=1}^{\infty} \frac{(-1)^{n}}{\sqrt{n}}$

18. $\displaystyle\sum_{n=1}^{\infty} \frac{(-1)^{n+1}\,n^2}{n^2+5}$

19. $\displaystyle\sum_{n=1}^{\infty} \frac{(-1)^{n+1}\,(n+1)}{\ln(n+1)}$

20. $\displaystyle\sum_{n=1}^{\infty} \frac{(-1)^{n+1}\ln(n+1)}{n+1}$

21. $\displaystyle\sum_{n=1}^{\infty} \sin\frac{(2n-1)\pi}{2}$

22. $\displaystyle\sum_{n=1}^{\infty} \frac{1}{n}\sin\frac{(2n-1)\pi}{2}$

23. $\displaystyle\sum_{n=1}^{\infty} \cos n\pi$

24. $\displaystyle\sum_{n=1}^{\infty} \frac{1}{n}\cos n\pi$

25. $\displaystyle\sum_{n=0}^{\infty} \frac{(-1)^{n}}{n!}$

26. $\displaystyle\sum_{n=0}^{\infty} \frac{(-1)^{n}}{(2n+1)!}$

27. $\displaystyle\sum_{n=1}^{\infty} \frac{(-1)^{n+1}\sqrt{n}}{n+2}$

28. $\displaystyle\sum_{n=1}^{\infty} \frac{(-1)^{n+1}\sqrt{n}}{\sqrt[3]{n}}$

29. $\displaystyle\sum_{n=1}^{\infty} \frac{(-1)^{n+1}\,n!}{1\cdot3\cdot5\cdots(2n-1)}$

30. $\displaystyle\sum_{n=1}^{\infty} (-1)^{n+1}\frac{1\cdot3\cdot5\cdots(2n-1)}{1\cdot4\cdot7\cdots(3n-2)}$

31. $\displaystyle\sum_{n=1}^{\infty} \frac{2(-1)^{n+1}}{e^n-e^{-n}} = \sum_{n=1}^{\infty}(-1)^{n+1}\operatorname{csch} n$

32. $\displaystyle\sum_{n=1}^{\infty} \frac{2(-1)^{n+1}}{e^n+e^{-n}} = \sum_{n=1}^{\infty}(-1)^{n+1}\operatorname{sech} n$

In Exercises 33–36, approximate the sum of the series by using the first six terms. (See Example 4.)

33. $\displaystyle\sum_{n=1}^{\infty} \frac{(-1)^{n+1}\,3}{n^2}$

34. $\displaystyle\sum_{n=1}^{\infty} \frac{(-1)^{n+1}\,4}{\ln(n+1)}$

35. $\displaystyle\sum_{n=0}^{\infty} \frac{(-1)^{n}\,2}{n!}$

36. $\displaystyle\sum_{n=1}^{\infty} \frac{(-1)^{n+1}\,n}{2^n}$

In Exercises 37–42, (a) use Theorem 9.15 to determine the number of terms required to approximate the sum of the convergent series with an error of less than 0.001, and (b) use a graphing utility to approximate the sum of the series with an error of less than 0.001.

37. $\displaystyle\sum_{n=0}^{\infty} \frac{(-1)^{n}}{n!} = \frac{1}{e}$

38. $\displaystyle\sum_{n=0}^{\infty} \frac{(-1)^{n}}{2^n\,n!} = \frac{1}{\sqrt{e}}$

39. $\displaystyle\sum_{n=0}^{\infty} \frac{(-1)^{n}}{(2n+1)!} = \sin 1$

40. $\displaystyle\sum_{n=0}^{\infty} \frac{(-1)^{n}}{(2n)!} = \cos 1$

41. $\displaystyle\sum_{n=1}^{\infty} \frac{(-1)^{n+1}}{n} = \ln 2$

42. $\displaystyle\sum_{n=1}^{\infty} \frac{(-1)^{n+1}}{n4^n} = \ln\frac{5}{4}$

In Exercises 43–46, use Theorem 9.15 to determine the number of terms required to approximate the sum of the series with an error of less than 0.001.

43. $\displaystyle\sum_{n=1}^{\infty} \frac{(-1)^{n+1}}{n^3}$

44. $\displaystyle\sum_{n=1}^{\infty} \frac{(-1)^{n+1}}{n^2}$

45. $\displaystyle\sum_{n=1}^{\infty} \frac{(-1)^{n+1}}{2n^3-1}$

46. $\displaystyle\sum_{n=1}^{\infty} \frac{(-1)^{n+1}}{n^4}$

In Exercises 47–62, determine whether the series converges conditionally or absolutely, or diverges.

47. $\displaystyle\sum_{n=1}^{\infty} \frac{(-1)^{n+1}}{(n+1)^2}$

48. $\displaystyle\sum_{n=1}^{\infty} \frac{(-1)^{n+1}}{n+1}$

49. $\displaystyle\sum_{n=1}^{\infty} \frac{(-1)^{n+1}}{\sqrt{n}}$

50. $\displaystyle\sum_{n=1}^{\infty} \frac{(-1)^{n+1}}{n\sqrt{n}}$

51. $\displaystyle\sum_{n=1}^{\infty} \frac{(-1)^{n+1}\,n^2}{(n+1)^2}$

52. $\displaystyle\sum_{n=1}^{\infty} \frac{(-1)^{n+1}(2n+3)}{n+10}$

53. $\displaystyle\sum_{n=2}^{\infty} \frac{(-1)^{n}}{\ln n}$

54. $\displaystyle\sum_{n=0}^{\infty} (-1)^{n}\,e^{-n^2}$

55. $\displaystyle\sum_{n=2}^{\infty} \frac{(-1)^{n}\,n}{n^3-1}$

56. $\displaystyle\sum_{n=1}^{\infty} \frac{(-1)^{n+1}}{n^{1.5}}$

57. $\displaystyle\sum_{n=0}^{\infty} \frac{(-1)^{n}}{(2n+1)!}$

58. $\displaystyle\sum_{n=0}^{\infty} \frac{(-1)^{n}}{\sqrt{n+4}}$

59. $\displaystyle\sum_{n=0}^{\infty} \frac{\cos n\pi}{n+1}$

60. $\displaystyle\sum_{n=1}^{\infty} (-1)^{n+1}\arctan n$

61. $\displaystyle\sum_{n=1}^{\infty} \frac{\cos n\pi}{n^2}$

62. $\displaystyle\sum_{n=1}^{\infty} \frac{\sin[(2n-1)\pi/2]}{n}$

Writing About Concepts

63. Define an alternating series and state the Alternating Series Test.

64. Give the remainder after N terms of a convergent alternating series.

65. In your own words, state the difference between absolute and conditional convergence of an alternating series.

66. The graphs of the sequences of partial sums of two series are shown in the figures. Which graph represents the partial sums of an alternating series? Explain.

(a) S_n

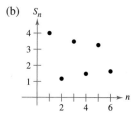

(b) S_n

True or False? **In Exercises 67–70, determine whether the statement is true or false. If it is false, explain why or give an example that shows it is false.**

67. If both $\Sigma\, a_n$ and $\Sigma\,(-a_n)$ converge, then $\Sigma\,|a_n|$ converges.

68. If $\Sigma\, a_n$ diverges, then $\Sigma\,|a_n|$ diverges.

69. For the alternating series $\displaystyle\sum_{n=1}^{\infty} \frac{(-1)^n}{n}$, the partial sum S_{100} is an overestimate of the sum of the series.

70. If $\Sigma\, a_n$ and $\Sigma\, b_n$ both converge, then $\Sigma\, a_n b_n$ converges.

In Exercises 71 and 72, find the values of p for which the series converges.

71. $\displaystyle\sum_{n=1}^{\infty} (-1)^n\left(\frac{1}{n^p}\right)$ **72.** $\displaystyle\sum_{n=1}^{\infty} (-1)^n\left(\frac{1}{n+p}\right)$

73. Prove that if $\Sigma\,|a_n|$ converges, then $\Sigma\, a_n^2$ converges. Is the converse true? If not, give an example that shows it is false.

74. Use the result of Exercise 71 to give an example of an alternating p-series that converges, but whose corresponding p-series diverges.

75. Give an example of a series that demonstrates the statement you proved in Exercise 73.

76. Find all values of x for which the series $\Sigma\,(x^n/n)$ (a) converges absolutely and (b) converges conditionally.

77. Consider the following series.

$$\frac{1}{2} - \frac{1}{3} + \frac{1}{4} - \frac{1}{9} + \frac{1}{8} - \frac{1}{27} + \cdots + \frac{1}{2^n} - \frac{1}{3^n} + \cdots$$

(a) Does the series meet the conditions of Theorem 9.14? Explain why or why not.

(b) Does the series converge? If so, what is the sum?

78. Consider the following series.

$$\sum_{n=1}^{\infty} (-1)^{n+1} a_n, \quad a_n = \begin{cases} \dfrac{1}{\sqrt{n}}, & \text{if } n \text{ is odd} \\[2mm] \dfrac{1}{n^3}, & \text{if } n \text{ is even} \end{cases}$$

(a) Does the series meet the conditions of Theorem 9.14? Explain why or why not.

(b) Does the series converge? If so, what is the sum?

Review **In Exercises 79–88, test for convergence or divergence and identify the test used.**

79. $\displaystyle\sum_{n=1}^{\infty} \frac{10}{n^{3/2}}$ **80.** $\displaystyle\sum_{n=1}^{\infty} \frac{3}{n^2+5}$

81. $\displaystyle\sum_{n=1}^{\infty} \frac{3^n}{n^2}$ **82.** $\displaystyle\sum_{n=1}^{\infty} \frac{1}{2^n+1}$

83. $\displaystyle\sum_{n=0}^{\infty} 5\left(\frac{7}{8}\right)^n$ **84.** $\displaystyle\sum_{n=1}^{\infty} \frac{3n^2}{2n^2+1}$

85. $\displaystyle\sum_{n=1}^{\infty} 100e^{-n/2}$

86. $\displaystyle\sum_{n=0}^{\infty} \frac{(-1)^n}{n+4}$

87. $\displaystyle\sum_{n=1}^{\infty} \frac{(-1)^{n+1}4}{3n^2-1}$

88. $\displaystyle\sum_{n=2}^{\infty} \frac{\ln n}{n}$

89. The following argument, that $0 = 1$, is *incorrect*. Describe the error.

$$\begin{aligned} 0 &= 0 + 0 + 0 + \cdots \\ &= (1-1) + (1-1) + (1-1) + \cdots \\ &= 1 + (-1+1) + (-1+1) + \cdots \\ &= 1 + 0 + 0 + \cdots \\ &= 1 \end{aligned}$$

90. The following argument, $2 = 1$, is *incorrect*. Describe the error. Multiply each side of the alternating harmonic series

$$S = 1 - \frac{1}{2} + \frac{1}{3} - \frac{1}{4} + \frac{1}{5} - \frac{1}{6} + \frac{1}{7} - \frac{1}{8} + \frac{1}{9} - \frac{1}{10} + \cdots$$

by 2 to get

$$2S = 2 - 1 + \frac{2}{3} - \frac{1}{2} + \frac{2}{5} - \frac{1}{3} + \frac{2}{7} - \frac{1}{4} + \frac{2}{9} - \frac{1}{5} + \cdots$$

Now collect terms with like denominators (as indicated by the arrows) to get

$$2S = 1 - \frac{1}{2} + \frac{1}{3} - \frac{1}{4} + \frac{1}{5} + \cdots$$

The resulting series is the same one that you started with. So, $2S = S$ and divide each side by S to get $2 = 1$.

FOR FURTHER INFORMATION For more on this exercise, see the article "Riemann's Rearrangement Theorem" by Stewart Galanor in *Mathematics Teacher*. To view this article, go to the website *www.matharticles.com*.

Putnam Exam Challenge

91. Assume as known the (true) fact that the alternating harmonic series

(1) $1 - \frac{1}{2} + \frac{1}{3} - \frac{1}{4} + \frac{1}{5} - \frac{1}{6} + \frac{1}{7} - \frac{1}{8} + \cdots$

is convergent, and denote its sum by s. Rearrange the series (1) as follows:

(2) $1 + \frac{1}{3} - \frac{1}{2} + \frac{1}{5} + \frac{1}{7} - \frac{1}{4} + \frac{1}{9} + \frac{1}{11} - \frac{1}{6} + \cdots$

Assume as known the (true) fact that the series (2) is also convergent, and denote its sum by S. Denote by s_k, S_k the kth partial sum of the series (1) and (2), respectively. Prove each statement.

(i) $S_{3n} = s_{4n} + \frac{1}{2}s_{2n}$,   (ii) $S \neq s$

The Ratio and Root Tests

- Use the Ratio Test to determine whether a series converges or diverges.
- Use the Root Test to determine whether a series converges or diverges.
- Review the tests for convergence and divergence of an infinite series.

The Ratio Test

This section begins with a test for absolute convergence—the **Ratio Test.**

THEOREM 9.17   Ratio Test

Let $\Sigma\, a_n$ be a series with nonzero terms.

1. $\Sigma\, a_n$ converges absolutely if $\lim\limits_{n \to \infty} \left| \dfrac{a_{n+1}}{a_n} \right| < 1.$

2. $\Sigma\, a_n$ diverges if $\lim\limits_{n \to \infty} \left| \dfrac{a_{n+1}}{a_n} \right| > 1$ or $\lim\limits_{n \to \infty} \left| \dfrac{a_{n+1}}{a_n} \right| = \infty.$

3. The Ratio Test is inconclusive if $\lim\limits_{n \to \infty} \left| \dfrac{a_{n+1}}{a_n} \right| = 1.$

Proof   To prove Property 1, assume that

$$\lim_{n \to \infty} \left| \frac{a_{n+1}}{a_n} \right| = r < 1$$

and choose R such that $0 \le r < R < 1$. By the definition of the limit of a sequence, there exists some $N > 0$ such that $|a_{n+1}/a_n| < R$ for all $n > N$. Therefore, you can write the following inequalities.

$$|a_{N+1}| < |a_N|R$$
$$|a_{N+2}| < |a_{N+1}|R < |a_N|R^2$$
$$|a_{N+3}| < |a_{N+2}|R < |a_{N+1}|R^2 < |a_N|R^3$$
$$\vdots$$

The geometric series $\Sigma\, |a_N|R^n = |a_N|R + |a_N|R^2 + \cdots + |a_N|R^n + \cdots$ converges, and so, by the Direct Comparison Test, the series

$$\sum_{n=1}^{\infty} |a_{N+n}| = |a_{N+1}| + |a_{N+2}| + \cdots + |a_{N+n}| + \cdots$$

also converges. This in turn implies that the series $\Sigma\, |a_n|$ converges, because discarding a finite number of terms $(n = N - 1)$ does not affect convergence. Consequently, by Theorem 9.16, the series $\Sigma\, a_n$ converges absolutely. The proof of Property 2 is similar and is left as an exercise (see Exercise 98).   ▬▬▬

NOTE   The fact that the Ratio Test is inconclusive when $|a_{n+1}/a_n| \to 1$ can be seen by comparing the two series $\Sigma\, (1/n)$ and $\Sigma\, (1/n^2)$. The first series diverges and the second one converges, but in both cases

$$\lim_{n \to \infty} \left| \frac{a_{n+1}}{a_n} \right| = 1.$$

Although the Ratio Test is not a cure for all ills related to tests for convergence, it is particularly useful for series that *converge rapidly*. Series involving factorials or exponentials are frequently of this type.

EXAMPLE 1   Using the Ratio Test

Determine the convergence or divergence of

$$\sum_{n=0}^{\infty} \frac{2^n}{n!}.$$

Solution   Because $a_n = 2^n/n!$, you can write the following.

$$\lim_{n \to \infty} \left| \frac{a_{n+1}}{a_n} \right| = \lim_{n \to \infty} \left[\frac{2^{n+1}}{(n+1)!} \div \frac{2^n}{n!} \right]$$

$$= \lim_{n \to \infty} \left[\frac{2^{n+1}}{(n+1)!} \cdot \frac{n!}{2^n} \right]$$

$$= \lim_{n \to \infty} \frac{2}{n+1}$$

$$= 0$$

Therefore, the series converges.

STUDY TIP   A step frequently used in applications of the Ratio Test involves simplifying quotients of factorials. In Example 1, for instance, notice that

$$\frac{n!}{(n+1)!} = \frac{n!}{(n+1)n!} = \frac{1}{n+1}.$$

EXAMPLE 2   Using the Ratio Test

Determine whether each series converges or diverges.

a. $\displaystyle\sum_{n=0}^{\infty} \frac{n^2 2^{n+1}}{3^n}$ **b.** $\displaystyle\sum_{n=1}^{\infty} \frac{n^n}{n!}$

Solution

a. This series converges because the limit of $|a_{n+1}/a_n|$ is less than 1.

$$\lim_{n \to \infty} \left| \frac{a_{n+1}}{a_n} \right| = \lim_{n \to \infty} \left[(n+1)^2 \left(\frac{2^{n+2}}{3^{n+1}} \right) \left(\frac{3^n}{n^2 2^{n+1}} \right) \right]$$

$$= \lim_{n \to \infty} \frac{2(n+1)^2}{3n^2}$$

$$= \frac{2}{3} < 1$$

b. This series diverges because the limit of $|a_{n+1}/a_n|$ is greater than 1.

$$\lim_{n \to \infty} \left| \frac{a_{n+1}}{a_n} \right| = \lim_{n \to \infty} \left[\frac{(n+1)^{n+1}}{(n+1)!} \left(\frac{n!}{n^n} \right) \right]$$

$$= \lim_{n \to \infty} \left[\frac{(n+1)^{n+1}}{(n+1)} \left(\frac{1}{n^n} \right) \right]$$

$$= \lim_{n \to \infty} \frac{(n+1)^n}{n^n}$$

$$= \lim_{n \to \infty} \left(1 + \frac{1}{n} \right)^n$$

$$= e > 1$$

EXAMPLE 3   A Failure of the Ratio Test

Determine the convergence or divergence of $\displaystyle\sum_{n=1}^{\infty} (-1)^n \frac{\sqrt{n}}{n+1}$.

Solution   The limit of $|a_{n+1}/a_n|$ is equal to 1.

$$\lim_{n\to\infty} \left|\frac{a_{n+1}}{a_n}\right| = \lim_{n\to\infty} \left[\left(\frac{\sqrt{n+1}}{n+2}\right)\left(\frac{n+1}{\sqrt{n}}\right)\right]$$

$$= \lim_{n\to\infty} \left[\sqrt{\frac{n+1}{n}}\left(\frac{n+1}{n+2}\right)\right]$$

$$= \sqrt{1}\,(1)$$

$$= 1$$

NOTE   The Ratio Test is also inconclusive for any *p*-series.

So, the Ratio Test is inconclusive. To determine whether the series converges, you need to try a different test. In this case, you can apply the Alternating Series Test. To show that $a_{n+1} \leq a_n$, let

$$f(x) = \frac{\sqrt{x}}{x+1}.$$

Then the derivative is

$$f'(x) = \frac{-x+1}{2\sqrt{x}(x+1)^2}.$$

Because the derivative is negative for $x > 1$, you know that f is a decreasing function. Also, by L'Hôpital's Rule,

$$\lim_{x\to\infty} \frac{\sqrt{x}}{x+1} = \lim_{x\to\infty} \frac{1/(2\sqrt{x})}{1}$$

$$= \lim_{x\to\infty} \frac{1}{2\sqrt{x}}$$

$$= 0.$$

Therefore, by the Alternating Series Test, the series converges.

The series in Example 3 is *conditionally convergent*. This follows from the fact that the series

$$\sum_{n=1}^{\infty} |a_n|$$

diverges $\left(\text{by the Limit Comparison Test with } \Sigma\, 1/\sqrt{n}\right)$, but the series

$$\sum_{n=1}^{\infty} a_n$$

converges.

TECHNOLOGY   A computer or programmable calculator can reinforce the conclusion that the series in Example 3 converges *conditionally*. By adding the first 100 terms of the series, you obtain a sum of about -0.2. (The sum of the first 100 terms of the series $\Sigma\, |a_n|$ is about 17.)

The Root Test

The next test for convergence or divergence of series works especially well for series involving nth powers. The proof of this theorem is similar to that given for the Ratio Test, and is left as an exercise (see Exercise 99).

THEOREM 9.18   Root Test

Let $\Sigma\, a_n$ be a series.

1. $\Sigma\, a_n$ converges absolutely if $\displaystyle\lim_{n\to\infty} \sqrt[n]{|a_n|} < 1$.

2. $\Sigma\, a_n$ diverges if $\displaystyle\lim_{n\to\infty} \sqrt[n]{|a_n|} > 1$ or $\displaystyle\lim_{n\to\infty} \sqrt[n]{|a_n|} = \infty$.

3. The Root Test is inconclusive if $\displaystyle\lim_{n\to\infty} \sqrt[n]{|a_n|} = 1$.

EXAMPLE 4   Using the Root Test

Determine the convergence or divergence of

$$\sum_{n=1}^{\infty} \frac{e^{2n}}{n^n}.$$

Solution   You can apply the Root Test as follows.

$$
\begin{aligned}
\lim_{n\to\infty} \sqrt[n]{|a_n|} &= \lim_{n\to\infty} \sqrt[n]{\frac{e^{2n}}{n^n}} \\
&= \lim_{n\to\infty} \frac{e^{2n/n}}{n^{n/n}} \\
&= \lim_{n\to\infty} \frac{e^2}{n} \\
&= 0 < 1
\end{aligned}
$$

NOTE   The Root Test is always inconclusive for any p-series.

Because this limit is less than 1, you can conclude that the series converges absolutely (and therefore converges).

FOR FURTHER INFORMATION   For more information on the usefulness of the Root Test, see the article *"N! and the Root Test"* by Charles C. Mumma II in *The American Mathematical Monthly*. To view this article, go to the website *www.matharticles.com*.

To see the usefulness of the Root Test for the series in Example 4, try applying the Ratio Test to that series. When you do this, you obtain the following.

$$
\begin{aligned}
\lim_{n\to\infty} \left| \frac{a_{n+1}}{a_n} \right| &= \lim_{n\to\infty} \left[\frac{e^{2(n+1)}}{(n+1)^{n+1}} \div \frac{e^{2n}}{n^n} \right] \\
&= \lim_{n\to\infty} e^2 \frac{n^n}{(n+1)^{n+1}} \\
&= \lim_{n\to\infty} e^2 \left(\frac{n}{n+1} \right)^n \left(\frac{1}{n+1} \right) \\
&= 0
\end{aligned}
$$

Note that this limit is not as easily evaluated as the limit obtained by the Root Test in Example 4.

Strategies for Testing Series

You have now studied 10 tests for determining the convergence or divergence of an infinite series. (See the summary in the table on page 644.) Skill in choosing and applying the various tests will come only with practice. Below is a set of guidelines for choosing an appropriate test.

> **Guidelines for Testing a Series for Convergence or Divergence**
>
> **1.** Does the nth term approach 0? If not, the series diverges.
> **2.** Is the series one of the special types—geometric, p-series, telescoping, or alternating?
> **3.** Can the Integral Test, the Root Test, or the Ratio Test be applied?
> **4.** Can the series be compared favorably to one of the special types?

In some instances, more than one test is applicable. However, your objective should be to learn to choose the most efficient test.

EXAMPLE 5   Applying the Strategies for Testing Series

Determine the convergence or divergence of each series.

a. $\displaystyle\sum_{n=1}^{\infty} \frac{n+1}{3n+1}$ **b.** $\displaystyle\sum_{n=1}^{\infty} \left(\frac{\pi}{6}\right)^n$ **c.** $\displaystyle\sum_{n=1}^{\infty} ne^{-n^2}$

d. $\displaystyle\sum_{n=1}^{\infty} \frac{1}{3n+1}$ **e.** $\displaystyle\sum_{n=1}^{\infty} (-1)^n \frac{3}{4n+1}$ **f.** $\displaystyle\sum_{n=1}^{\infty} \frac{n!}{10^n}$

g. $\displaystyle\sum_{n=1}^{\infty} \left(\frac{n+1}{2n+1}\right)^n$

Solution

a. For this series, the limit of the nth term is not 0 $\left(a_n \to \frac{1}{3} \text{ as } n \to \infty\right)$. So, by the nth-Term Test, the series diverges.

b. This series is geometric. Moreover, because the ratio $r = \pi/6$ of the terms is less than 1 in absolute value, you can conclude that the series converges.

c. Because the function $f(x) = xe^{-x^2}$ is easily integrated, you can use the Integral Test to conclude that the series converges.

d. The nth term of this series can be compared to the nth term of the harmonic series. After using the Limit Comparison Test, you can conclude that the series diverges.

e. This is an alternating series whose nth term approaches 0. Because $a_{n+1} \leq a_n$, you can use the Alternating Series Test to conclude that the series converges.

f. The nth term of this series involves a factorial, which indicates that the Ratio Test may work well. After applying the Ratio Test, you can conclude that the series diverges.

g. The nth term of this series involves a variable that is raised to the nth power, which indicates that the Root Test may work well. After applying the Root Test, you can conclude that the series converges.

Summary of Tests for Series

Test	Series	Condition(s) of Convergence	Condition(s) of Divergence	Comment
nth-Term	$\displaystyle\sum_{n=1}^{\infty} a_n$		$\displaystyle\lim_{n\to\infty} a_n \neq 0$	This test cannot be used to show convergence.
Geometric Series	$\displaystyle\sum_{n=0}^{\infty} ar^n$	$\lvert r \rvert < 1$	$\lvert r \rvert \geq 1$	Sum: $S = \dfrac{a}{1-r}$
Telescoping Series	$\displaystyle\sum_{n=1}^{\infty} (b_n - b_{n+1})$	$\displaystyle\lim_{n\to\infty} b_n = L$		Sum: $S = b_1 - L$
p-Series	$\displaystyle\sum_{n=1}^{\infty} \dfrac{1}{n^p}$	$p > 1$	$p \leq 1$	
Alternating Series	$\displaystyle\sum_{n=1}^{\infty} (-1)^{n-1} a_n$	$0 < a_{n+1} \leq a_n$ and $\displaystyle\lim_{n\to\infty} a_n = 0$		Remainder: $\lvert R_N \rvert \leq a_{N+1}$
Integral (f is continuous, positive, and decreasing)	$\displaystyle\sum_{n=1}^{\infty} a_n,$ $a_n = f(n) \geq 0$	$\displaystyle\int_1^{\infty} f(x)\,dx$ converges	$\displaystyle\int_1^{\infty} f(x)\,dx$ diverges	Remainder: $0 < R_N < \displaystyle\int_N^{\infty} f(x)\,dx$
Root	$\displaystyle\sum_{n=1}^{\infty} a_n$	$\displaystyle\lim_{n\to\infty} \sqrt[n]{\lvert a_n \rvert} < 1$	$\displaystyle\lim_{n\to\infty} \sqrt[n]{\lvert a_n \rvert} > 1$	Test is inconclusive if $\displaystyle\lim_{n\to\infty} \sqrt[n]{\lvert a_n \rvert} = 1$.
Ratio	$\displaystyle\sum_{n=1}^{\infty} a_n$	$\displaystyle\lim_{n\to\infty} \left\lvert \dfrac{a_{n+1}}{a_n} \right\rvert < 1$	$\displaystyle\lim_{n\to\infty} \left\lvert \dfrac{a_{n+1}}{a_n} \right\rvert > 1$	Test is inconclusive if $\displaystyle\lim_{n\to\infty} \left\lvert \dfrac{a_{n+1}}{a_n} \right\rvert = 1$.
Direct Comparison ($a_n, b_n > 0$)	$\displaystyle\sum_{n=1}^{\infty} a_n$	$0 < a_n \leq b_n$ and $\displaystyle\sum_{n=1}^{\infty} b_n$ converges	$0 < b_n \leq a_n$ and $\displaystyle\sum_{n=1}^{\infty} b_n$ diverges	
Limit Comparison ($a_n, b_n > 0$)	$\displaystyle\sum_{n=1}^{\infty} a_n$	$\displaystyle\lim_{n\to\infty} \dfrac{a_n}{b_n} = L > 0$ and $\displaystyle\sum_{n=1}^{\infty} b_n$ converges	$\displaystyle\lim_{n\to\infty} \dfrac{a_n}{b_n} = L > 0$ and $\displaystyle\sum_{n=1}^{\infty} b_n$ diverges	

In Exercises 1–4, verify the formula.

1. $\dfrac{(n + 1)!}{(n - 2)!} = (n + 1)(n)(n - 1)$

2. $\dfrac{(2k - 2)!}{(2k)!} = \dfrac{1}{(2k)(2k - 1)}$

3. $1 \cdot 3 \cdot 5 \cdots (2k - 1) = \dfrac{(2k)!}{2^k k!}$

4. $\dfrac{1}{1 \cdot 3 \cdot 5 \cdots (2k - 5)} = \dfrac{2^k k!(2k - 3)(2k - 1)}{(2k)!}, \quad k \geq 3$

In Exercises 5–10, match the series with the graph of its sequence of partial sums. [The graphs are labeled (a), (b), (c), (d), (e), and (f).]

(a)

(b)

(c)

(d)

(e)

(f)

5. $\displaystyle\sum_{n=1}^{\infty} n\left(\dfrac{3}{4}\right)^n$

6. $\displaystyle\sum_{n=1}^{\infty} \left(\dfrac{3}{4}\right)^n\left(\dfrac{1}{n!}\right)$

7. $\displaystyle\sum_{n=1}^{\infty} \dfrac{(-3)^{n+1}}{n!}$

8. $\displaystyle\sum_{n=1}^{\infty} \dfrac{(-1)^{n-1}4}{(2n)!}$

9. $\displaystyle\sum_{n=1}^{\infty} \left(\dfrac{4n}{5n - 3}\right)^n$

10. $\displaystyle\sum_{n=0}^{\infty} 4e^{-n}$

Numerical, Graphical, and Analytic Analysis   **In Exercises 11 and 12, (a) verify that the series converges. (b) Use a graphing utility to find the indicated partial sum S_n and complete the table. (c) Use a graphing utility to graph the first 10 terms of the sequence of partial sums. (d) Use the table to estimate the sum of the series. (e) Explain the relationship between the magnitudes of the terms of the series and the rate at which the sequence of partial sums approaches the sum of the series.**

n	5	10	15	20	25
S_n					

11. $\displaystyle\sum_{n=1}^{\infty} n^2\left(\dfrac{5}{8}\right)^n$

12. $\displaystyle\sum_{n=1}^{\infty} \dfrac{n^2 + 1}{n!}$

In Exercises 13–32, use the Ratio Test to determine the convergence or divergence of the series.

13. $\displaystyle\sum_{n=0}^{\infty} \dfrac{n!}{3^n}$

14. $\displaystyle\sum_{n=0}^{\infty} \dfrac{3^n}{n!}$

15. $\displaystyle\sum_{n=1}^{\infty} n\left(\dfrac{3}{4}\right)^n$

16. $\displaystyle\sum_{n=1}^{\infty} n\left(\dfrac{3}{2}\right)^n$

17. $\displaystyle\sum_{n=1}^{\infty} \dfrac{n}{2^n}$

18. $\displaystyle\sum_{n=1}^{\infty} \dfrac{n^3}{2^n}$

19. $\displaystyle\sum_{n=1}^{\infty} \dfrac{2^n}{n^2}$

20. $\displaystyle\sum_{n=1}^{\infty} \dfrac{(-1)^{n+1}(n + 2)}{n(n + 1)}$

21. $\displaystyle\sum_{n=0}^{\infty} \dfrac{(-1)^n 2^n}{n!}$

22. $\displaystyle\sum_{n=1}^{\infty} \dfrac{(-1)^{n-1}(3/2)^n}{n^2}$

23. $\displaystyle\sum_{n=1}^{\infty} \dfrac{n!}{n3^n}$

24. $\displaystyle\sum_{n=1}^{\infty} \dfrac{(2n)!}{n^5}$

25. $\displaystyle\sum_{n=0}^{\infty} \dfrac{4^n}{n!}$

26. $\displaystyle\sum_{n=1}^{\infty} \dfrac{n^n}{n!}$

27. $\displaystyle\sum_{n=0}^{\infty} \dfrac{3^n}{(n + 1)^n}$

28. $\displaystyle\sum_{n=0}^{\infty} \dfrac{(n!)^2}{(3n)!}$

29. $\displaystyle\sum_{n=0}^{\infty} \dfrac{4^n}{3^n + 1}$

30. $\displaystyle\sum_{n=0}^{\infty} \dfrac{(-1)^n 2^{4n}}{(2n + 1)!}$

31. $\displaystyle\sum_{n=0}^{\infty} \dfrac{(-1)^{n+1}n!}{1 \cdot 3 \cdot 5 \cdots (2n + 1)}$

32. $\displaystyle\sum_{n=1}^{\infty} \dfrac{(-1)^n[2 \cdot 4 \cdot 6 \cdots (2n)]}{2 \cdot 5 \cdot 8 \cdots (3n - 1)}$

In Exercises 33–36, verify that the Ratio Test is inconclusive for the p-series.

33. $\displaystyle\sum_{n=1}^{\infty} \dfrac{1}{n^{3/2}}$

34. $\displaystyle\sum_{n=1}^{\infty} \dfrac{1}{n^{1/2}}$

35. $\displaystyle\sum_{n=1}^{\infty} \dfrac{1}{n^4}$

36. $\displaystyle\sum_{n=1}^{\infty} \dfrac{1}{n^p}$

In Exercises 37–50, use the Root Test to determine the convergence or divergence of the series.

37. $\displaystyle\sum_{n=1}^{\infty} \left(\frac{n}{2n+1}\right)^n$

38. $\displaystyle\sum_{n=1}^{\infty} \left(\frac{2n}{n+1}\right)^n$

39. $\displaystyle\sum_{n=2}^{\infty} \left(\frac{2n+1}{n-1}\right)^n$

40. $\displaystyle\sum_{n=1}^{\infty} \left(\frac{4n+3}{2n-1}\right)^n$

41. $\displaystyle\sum_{n=2}^{\infty} \frac{(-1)^n}{(\ln n)^n}$

42. $\displaystyle\sum_{n=1}^{\infty} \left(\frac{-3n}{2n+1}\right)^{3n}$

43. $\displaystyle\sum_{n=1}^{\infty} \left(2\sqrt[n]{n}+1\right)^n$

44. $\displaystyle\sum_{n=0}^{\infty} e^{-n}$

45. $\displaystyle\sum_{n=1}^{\infty} \frac{n}{4^n}$

46. $\displaystyle\sum_{n=1}^{\infty} \left(\frac{n}{500}\right)^n$

47. $\displaystyle\sum_{n=1}^{\infty} \left(\frac{1}{n}-\frac{1}{n^2}\right)^n$

48. $\displaystyle\sum_{n=1}^{\infty} \left(\frac{\ln n}{n}\right)^n$

49. $\displaystyle\sum_{n=2}^{\infty} \frac{n}{(\ln n)^n}$

50. $\displaystyle\sum_{n=1}^{\infty} \frac{(n!)^n}{(n^n)^2}$

In Exercises 51–68, determine the convergence or divergence of the series using any appropriate test from this chapter. Identify the test used.

51. $\displaystyle\sum_{n=1}^{\infty} \frac{(-1)^{n+1}5}{n}$

52. $\displaystyle\sum_{n=1}^{\infty} \frac{5}{n}$

53. $\displaystyle\sum_{n=1}^{\infty} \frac{3}{n\sqrt{n}}$

54. $\displaystyle\sum_{n=1}^{\infty} \left(\frac{\pi}{4}\right)^n$

55. $\displaystyle\sum_{n=1}^{\infty} \frac{2n}{n+1}$

56. $\displaystyle\sum_{n=1}^{\infty} \frac{n}{2n^2+1}$

57. $\displaystyle\sum_{n=1}^{\infty} \frac{(-1)^n 3^{n-2}}{2^n}$

58. $\displaystyle\sum_{n=1}^{\infty} \frac{10}{3\sqrt[3]{n}}$

59. $\displaystyle\sum_{n=1}^{\infty} \frac{10n+3}{n2^n}$

60. $\displaystyle\sum_{n=1}^{\infty} \frac{2^n}{4n^2-1}$

61. $\displaystyle\sum_{n=1}^{\infty} \frac{\cos n}{2^n}$

62. $\displaystyle\sum_{n=2}^{\infty} \frac{(-1)^n}{n\ln n}$

63. $\displaystyle\sum_{n=1}^{\infty} \frac{n7^n}{n!}$

64. $\displaystyle\sum_{n=1}^{\infty} \frac{\ln n}{n^2}$

65. $\displaystyle\sum_{n=1}^{\infty} \frac{(-1)^n 3^{n-1}}{n!}$

66. $\displaystyle\sum_{n=1}^{\infty} \frac{(-1)^n 3^n}{n2^n}$

67. $\displaystyle\sum_{n=1}^{\infty} \frac{(-3)^n}{3\cdot5\cdot7\cdots(2n+1)}$

68. $\displaystyle\sum_{n=1}^{\infty} \frac{3\cdot5\cdot7\cdots(2n+1)}{18^n(2n-1)n!}$

In Exercises 69–72, identify the two series that are the same.

69. (a) $\displaystyle\sum_{n=1}^{\infty} \frac{n5^n}{n!}$

(b) $\displaystyle\sum_{n=0}^{\infty} \frac{n5^n}{(n+1)!}$

(c) $\displaystyle\sum_{n=0}^{\infty} \frac{(n+1)5^{n+1}}{(n+1)!}$

70. (a) $\displaystyle\sum_{n=4}^{\infty} n\left(\frac{3}{4}\right)^n$

(b) $\displaystyle\sum_{n=0}^{\infty} (n+1)\left(\frac{3}{4}\right)^n$

(c) $\displaystyle\sum_{n=1}^{\infty} n\left(\frac{3}{4}\right)^{n-1}$

71. (a) $\displaystyle\sum_{n=0}^{\infty} \frac{(-1)^n}{(2n+1)!}$

(b) $\displaystyle\sum_{n=1}^{\infty} \frac{(-1)^{n-1}}{(2n-1)!}$

(c) $\displaystyle\sum_{n=1}^{\infty} \frac{(-1)^{n-1}}{(2n+1)!}$

72. (a) $\displaystyle\sum_{n=2}^{\infty} \frac{(-1)^n}{(n-1)2^{n-1}}$

(b) $\displaystyle\sum_{n=1}^{\infty} \frac{(-1)^{n+1}}{n2^n}$

(c) $\displaystyle\sum_{n=0}^{\infty} \frac{(-1)^{n+1}}{(n+1)2^n}$

In Exercises 73 and 74, write an equivalent series with the index of summation beginning at $n=0$.

73. $\displaystyle\sum_{n=1}^{\infty} \frac{n}{4^n}$

74. $\displaystyle\sum_{n=2}^{\infty} \frac{2^n}{(n-2)!}$

In Exercises 75 and 76, (a) determine the number of terms required to approximate the sum of the series with an error less than 0.0001, and (b) use a graphing utility to approximate the sum of the series with an error less than 0.0001.

75. $\displaystyle\sum_{k=1}^{\infty} \frac{(-3)^k}{2^k k!}$

76. $\displaystyle\sum_{k=0}^{\infty} \frac{(-3)^k}{1\cdot3\cdot5\cdots(2k+1)}$

In Exercises 77–82, the terms of a series $\displaystyle\sum_{n=1}^{\infty} a_n$ are defined recursively. Determine the convergence or divergence of the series. Explain your reasoning.

77. $a_1 = \dfrac{1}{2}, \; a_{n+1} = \dfrac{4n-1}{3n+2}a_n$

78. $a_1 = 2, \; a_{n+1} = \dfrac{2n+1}{5n-4}a_n$

79. $a_1 = 1, \; a_{n+1} = \dfrac{\sin n + 1}{\sqrt{n}}a_n$

80. $a_1 = \dfrac{1}{5}, \; a_{n+1} = \dfrac{\cos n + 1}{n}a_n$

81. $a_1 = \dfrac{1}{3}, \; a_{n+1} = \left(1+\dfrac{1}{n}\right)a_n$

82. $a_1 = \dfrac{1}{4}, \; a_{n+1} = \sqrt[n]{a_n}$

In Exercises 83–86, use the Ratio Test or the Root Test to determine the convergence or divergence of the series.

83. $1 + \dfrac{1\cdot2}{1\cdot3} + \dfrac{1\cdot2\cdot3}{1\cdot3\cdot5} + \dfrac{1\cdot2\cdot3\cdot4}{1\cdot3\cdot5\cdot7} + \cdots$

84. $1 + \dfrac{2}{3} + \dfrac{3}{3^2} + \dfrac{4}{3^3} + \dfrac{5}{3^4} + \dfrac{6}{3^5} + \cdots$

85. $\dfrac{1}{(\ln 3)^3} + \dfrac{1}{(\ln 4)^4} + \dfrac{1}{(\ln 5)^5} + \dfrac{1}{(\ln 6)^6} + \cdots$

86. $1 + \dfrac{1\cdot3}{1\cdot2\cdot3} + \dfrac{1\cdot3\cdot5}{1\cdot2\cdot3\cdot4\cdot5}$
$+ \dfrac{1\cdot3\cdot5\cdot7}{1\cdot2\cdot3\cdot4\cdot5\cdot6\cdot7} + \cdots$

In Exercises 87–92, find the values of x for which the series converges.

87. $\displaystyle\sum_{n=0}^{\infty} 2\left(\frac{x}{3}\right)^n$

88. $\displaystyle\sum_{n=0}^{\infty} \left(\frac{x+1}{4}\right)^n$

89. $\displaystyle\sum_{n=1}^{\infty} \frac{(-1)^n(x+1)^n}{n}$

90. $\displaystyle\sum_{n=0}^{\infty} 2(x-1)^n$

91. $\displaystyle\sum_{n=0}^{\infty} n!\left(\frac{x}{2}\right)^n$

92. $\displaystyle\sum_{n=0}^{\infty} \frac{(x+1)^n}{n!}$

Writing About Concepts

93. State the Ratio Test.

94. State the Root Test.

95. You are told that the terms of a positive series appear to approach zero rapidly as n approaches infinity. In fact, $a_7 \le 0.0001$. Given no other information, does this imply that the series converges? Support your conclusion with examples.

96. The graph shows the first 10 terms of the sequence of partial sums of the convergent series

$$\sum_{n=1}^{\infty} \left(\frac{2n}{3n+2}\right)^n.$$

Find a series such that the terms of its sequence of partial sums are less than the corresponding terms of the sequence in the figure, but such that the series diverges. Explain your reasoning.

97. Using the Ratio Test, it is determined that an alternating series converges. Does the series converge conditionally or absolutely? Explain.

98. Prove Property 2 of Theorem 9.17.

99. Prove Theorem 9.18. (*Hint for Property 1:* If the limit equals $r < 1$, choose a real number R such that $r < R < 1$. By the definitions of the limit, there exists some $N > 0$ such that $\sqrt[n]{|a_n|} < R$ for $n > N$.)

100. Show that the Root Test is inconclusive for the p-series

$$\sum_{n=1}^{\infty} \frac{1}{n^p}.$$

101. Show that the Ratio Test and the Root Test are both inconclusive for the logarithmic p-series

$$\sum_{n=2}^{\infty} \frac{1}{n(\ln n)^p}.$$

102. Determine the convergence or divergence of the series

$$\sum_{n=1}^{\infty} \frac{(n!)^2}{(xn)!}$$

when (a) $x = 1$, (b) $x = 2$, (c) $x = 3$, and (d) x is a positive integer.

103. Show that if $\displaystyle\sum_{n=1}^{\infty} a_n$ is absolutely convergent, then

$$\left|\sum_{n=1}^{\infty} a_n\right| \le \sum_{n=1}^{\infty} |a_n|.$$

104. *Writing* Read the article "A Differentiation Test for Absolute Convergence" by Yaser S. Abu-Mostafa in *Mathematics Magazine*. Then write a paragraph that describes the test. Include examples of series that converge and examples of series that diverge.

Putnam Exam Challenge

105. Is the following series convergent or divergent?

$$1 + \frac{1}{2}\cdot\frac{19}{7} + \frac{2!}{3^2}\left(\frac{19}{7}\right)^2 + \frac{3!}{4^3}\left(\frac{19}{7}\right)^3 + \frac{4!}{5^4}\left(\frac{19}{7}\right)^4 + \cdots$$

106. Show that if the series

$$a_1 + a_2 + a_3 + \cdots + a_n + \cdots$$

converges, then the series

$$a_1 + \frac{a_2}{2} + \frac{a_3}{3} + \cdots + \frac{a_n}{n} + \cdots$$

converges also.

In Exercises 1–4, match the Taylor polynomial approximation of the function $f(x) = e^{-x^2/2}$ with the correct graph. [The graphs are labeled (a), (b), (c), and (d).]

(a)

(b)

(c)

(d)

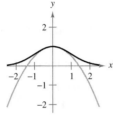

1. $g(x) = -\frac{1}{2}x^2 + 1$
2. $g(x) = \frac{1}{8}x^4 - \frac{1}{2}x^2 + 1$
3. $g(x) = e^{-1/2}[(x + 1) + 1]$
4. $g(x) = e^{-1/2}\left[\frac{1}{3}(x - 1)^3 - (x - 1) + 1\right]$

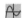 **In Exercises 5–8, find a first-degree polynomial function P_1 whose value and slope agree with the value and slope of f at $x = c$. Use a graphing utility to graph f and P_1. What is P_1 called?**

5. $f(x) = \dfrac{4}{\sqrt{x}}, \quad c = 1$

6. $f(x) = \dfrac{4}{\sqrt[3]{x}}, \quad c = 8$

7. $f(x) = \sec x, \quad c = \dfrac{\pi}{4}$

8. $f(x) = \tan x, \quad c = \dfrac{\pi}{4}$

 Graphical and Numerical Analysis   **In Exercises 9 and 10, use a graphing utility to graph f and its second-degree polynomial approximation P_2 at $x = c$. Complete the table comparing the values of f and P_2.**

9. $f(x) = \dfrac{4}{\sqrt{x}}, \quad c = 1$

$P_2(x) = 4 - 2(x - 1) + \frac{3}{2}(x - 1)^2$

x	0	0.8	0.9	1	1.1	1.2	2
$f(x)$							
$P_2(x)$							

10. $f(x) = \sec x, \quad c = \dfrac{\pi}{4}$

$$P_2(x) = \sqrt{2} + \sqrt{2}\left(x - \frac{\pi}{4}\right) + \frac{3}{2}\sqrt{2}\left(x - \frac{\pi}{4}\right)^2$$

x	-2.15	0.585	0.685	$\dfrac{\pi}{4}$	0.885	0.985	1.785
$f(x)$							
$P_2(x)$							

11. ***Conjecture***   Consider the function $f(x) = \cos x$ and its Maclaurin polynomials P_2, P_4, and P_6 (see Example 5).

   (a) Use a graphing utility to graph f and the indicated polynomial approximations.

   (b) Evaluate and compare the values of $f^{(n)}(0)$ and $P_n^{(n)}(0)$ for $n = 2, 4,$ and 6.

   (c) Use the results in part (b) to make a conjecture about $f^{(n)}(0)$ and $P_n^{(n)}(0)$.

12. ***Conjecture***   Consider the function $f(x) = x^2 e^x$.

   (a) Find the Maclaurin polynomials P_2, P_3, and P_4 for f.

   (b) Use a graphing utility to graph f, P_2, P_3, and P_4.

   (c) Evaluate and compare the values of $f^{(n)}(0)$ and $P_n^{(n)}(0)$ for $n = 2, 3,$ and 4.

   (d) Use the results in part (c) to make a conjecture about $f^{(n)}(0)$ and $P_n^{(n)}(0)$.

In Exercises 13–24, find the Maclaurin polynomial of degree n for the function.

13. $f(x) = e^{-x}, \quad n = 3$

14. $f(x) = e^{-x}, \quad n = 5$

15. $f(x) = e^{2x}, \quad n = 4$

16. $f(x) = e^{3x}, \quad n = 4$

17. $f(x) = \sin x, \quad n = 5$

18. $f(x) = \sin \pi x, \quad n = 3$

19. $f(x) = xe^x, \quad n = 4$

20. $f(x) = x^2 e^{-x}, \quad n = 4$

21. $f(x) = \dfrac{1}{x + 1}, \quad n = 4$

22. $f(x) = \dfrac{x}{x + 1}, \quad n = 4$

23. $f(x) = \sec x, \quad n = 2$

24. $f(x) = \tan x, \quad n = 3$

In Exercises 25–30, find the nth Taylor polynomial centered at c.

25. $f(x) = \dfrac{1}{x}, \quad n = 4, \quad c = 1$

26. $f(x) = \dfrac{2}{x^2}, \quad n = 4, \quad c = 2$

27. $f(x) = \sqrt{x}, \quad n = 4, \quad c = 1$

28. $f(x) = \sqrt[3]{x}, \quad n = 3, \quad c = 8$

29. $f(x) = \ln x, \quad n = 4, \quad c = 1$

30. $f(x) = x^2 \cos x, \quad n = 2, \quad c = \pi$

In Exercises 31 and 32, use a computer algebra system to find the indicated Taylor polynomials for the function f. Graph the function and the Taylor polynomials.

31. $f(x) = \tan x$

    (a) $n = 3$, $c = 0$

    (b) $n = 3$, $c = \pi/4$

32. $f(x) = \dfrac{1}{x^2 + 1}$

    (a) $n = 4$, $c = 0$

    (b) $n = 4$, $c = 1$

33. *Numerical and Graphical Approximations*

    (a) Use the Maclaurin polynomials $P_1(x)$, $P_3(x)$, and $P_5(x)$ for $f(x) = \sin x$ to complete the table.

x	0	0.25	0.50	0.75	1.00
$\sin x$	0	0.2474	0.4794	0.6816	0.8415
$P_1(x)$					
$P_3(x)$					
$P_5(x)$					

    (b) Use a graphing utility to graph $f(x) = \sin x$ and the Maclaurin polynomials in part (a).

    (c) Describe the change in accuracy of a polynomial approximation as the distance from the point where the polynomial is centered increases.

34. *Numerical and Graphical Approximations*

    (a) Use the Taylor polynomials $P_1(x)$, $P_2(x)$, and $P_4(x)$ for $f(x) = \ln x$ centered at $c = 1$ to complete the table.

x	1.00	1.25	1.50	1.75	2.00
$\ln x$	0	0.2231	0.4055	0.5596	0.6931
$P_1(x)$					
$P_2(x)$					
$P_4(x)$					

    (b) Use a graphing utility to graph $f(x) = \ln x$ and the Taylor polynomials in part (a).

    (c) Describe the change in accuracy of polynomial approximations as the degree increases.

Numerical and Graphical Approximations **In Exercises 35 and 36, (a) find the Maclaurin polynomial $P_3(x)$ for $f(x)$, (b) complete the table for $f(x)$ and $P_3(x)$, and (c) sketch the graphs of $f(x)$ and $P_3(x)$ on the same set of coordinate axes.**

x	-0.75	-0.50	-0.25	0	0.25	0.50	0.75
$f(x)$							
$P_3(x)$							

35. $f(x) = \arcsin x$ **36.** $f(x) = \arctan x$

In Exercises 37–40, the graph of $y = f(x)$ is shown with four of its Maclaurin polynomials. Identify the Maclaurin polynomials and use a graphing utility to confirm your results.

37.

38.

39.

40.

In Exercises 41–44, approximate the function at the given value of x, using the polynomial found in the indicated exercise.

41. $f(x) = e^{-x}$, $f\!\left(\frac{1}{2}\right)$, Exercise 13

42. $f(x) = x^2 e^{-x}$, $f\!\left(\frac{1}{5}\right)$, Exercise 20

43. $f(x) = \ln x$, $f(1.2)$, Exercise 29

44. $f(x) = x^2 \cos x$, $f\!\left(\dfrac{7\pi}{8}\right)$, Exercise 30

In Exercises 45–48, use Taylor's Theorem to obtain an upper bound for the error of the approximation. Then calculate the exact value of the error.

45. $\cos(0.3) \approx 1 - \dfrac{(0.3)^2}{2!} + \dfrac{(0.3)^4}{4!}$

46. $e \approx 1 + 1 + \dfrac{1^2}{2!} + \dfrac{1^3}{3!} + \dfrac{1^4}{4!} + \dfrac{1^5}{5!}$

47. $\arcsin(0.4) \approx 0.4 + \dfrac{(0.4)^3}{2 \cdot 3}$ **48.** $\arctan(0.4) \approx 0.4 - \dfrac{(0.4)^3}{3}$

In Exercises 49–52, determine the degree of the Maclaurin polynomial required for the error in the approximation of the function at the indicated value of x to be less than 0.001.

49. $\sin(0.3)$ **50.** $\cos(0.1)$

51. $e^{0.6}$ **52.** $e^{0.3}$

 In Exercises 53–56, determine the degree of the Maclaurin polynomial required for the error in the approximation of the function at the indicated value of x to be less than 0.0001. Use a computer algebra system to obtain and evaluate the required derivatives.

53. $f(x) = \ln(x + 1)$, approximate $f(0.5)$.

54. $f(x) = \cos(\pi x^2)$, approximate $f(0.6)$.

55. $f(x) = e^{-\pi x}$, approximate $f(1.3)$.

56. $f(x) = e^{-x}$, approximate $f(1)$.

In Exercises 57–60, determine the values of x for which the function can be replaced by the Taylor polynomial if the error cannot exceed 0.001.

57. $f(x) = e^x \approx 1 + x + \dfrac{x^2}{2!} + \dfrac{x^3}{3!}, \qquad x < 0$

58. $f(x) = \sin x \approx x - \dfrac{x^3}{3!}$

59. $f(x) = \cos x \approx 1 - \dfrac{x^2}{2!} + \dfrac{x^4}{4!}$

60. $f(x) = e^{-2x} \approx 1 - 2x + 2x^2 - \dfrac{4}{3}x^3$

Writing About Concepts

61. An elementary function is approximated by a polynomial. In your own words, describe what is meant by saying that the polynomial is *expanded about c* or *centered at c*.

62. When an elementary function f is approximated by a second-degree polynomial P_2 centered at c, what is known about f and P_2 at c? Explain your reasoning.

63. State the definition of an nth-degree Taylor polynomial of f centered at c.

64. Describe the accuracy of the nth-degree Taylor polynomial of f centered at c as the distance between c and x increases.

65. In general, how does the accuracy of a Taylor polynomial change as the degree of the polynomial is increased? Explain your reasoning.

66. The graphs show first-, second-, and third-degree polynomial approximations P_1, P_2, and P_3 of a function f. Label the graphs of P_1, P_2, and P_3. To print an enlarged copy of the graph, go to the website *www.mathgraphs.com*.

67. *Comparing Maclaurin Polynomials*

  (a) Compare the Maclaurin polynomials of degree 4 and degree 5, respectively, for the functions $f(x) = e^x$ and $g(x) = xe^x$. What is the relationship between them?

  (b) Use the result in part (a) and the Maclaurin polynomial of degree 5 for $f(x) = \sin x$ to find a Maclaurin polynomial of degree 6 for the function $g(x) = x \sin x$.

  (c) Use the result in part (a) and the Maclaurin polynomial of degree 5 for $f(x) = \sin x$ to find a Maclaurin polynomial of degree 4 for the function $g(x) = (\sin x)/x$.

68. *Differentiating Maclaurin Polynomials*

  (a) Differentiate the Maclaurin polynomial of degree 5 for $f(x) = \sin x$ and compare the result with the Maclaurin polynomial of degree 4 for $g(x) = \cos x$.

  (b) Differentiate the Maclaurin polynomial of degree 6 for $f(x) = \cos x$ and compare the result with the Maclaurin polynomial of degree 5 for $g(x) = \sin x$.

  (c) Differentiate the Maclaurin polynomial of degree 4 for $f(x) = e^x$. Describe the relationship between the two series.

69. *Graphical Reasoning*   The figure shows the graph of the function

$$f(x) = \sin\left(\frac{\pi x}{4}\right)$$

and the second-degree Taylor polynomial

$$P_2(x) = 1 - \frac{\pi^2}{32}(x - 2)^2$$

centered at $x = 2$.

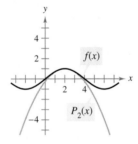

  (a) Use the symmetry of the graph of f to write the second-degree Taylor polynomial for f centered at $x = -2$.

  (b) Use a horizontal translation of the result in part (a) to find the second-degree Taylor polynomial for f centered at $x = 6$.

  (c) Is it possible to use a horizontal translation of the result in part (a) to write a second-degree Taylor polynomial for f centered at $x = 4$? Explain.

70. Prove that if f is an odd function, then its nth Maclaurin polynomial contains only terms with odd powers of x.

71. Prove that if f is an even function, then its nth Maclaurin polynomial contains only terms with even powers of x.

72. Let $P_n(x)$ be the nth Taylor polynomial for f at c. Prove that $P_n(c) = f(c)$ and $P^{(k)}(c) = f^{(k)}(c)$ for $1 \le k \le n$. (See Exercises 9 and 10.)

73. *Writing*   The proof in Exercise 72 guarantees that the Taylor polynomial and its derivatives agree with the function and its derivatives at $x = c$. Use the graphs and tables in Exercises 33–36 to discuss what happens to the accuracy of the Taylor polynomial as you move away from $x = c$.

Power Series

- Understand the definition of a power series.
- Find the radius and interval of convergence of a power series.
- Determine the endpoint convergence of a power series.
- Differentiate and integrate a power series.

Power Series

In Section 9.7, you were introduced to the concept of approximating functions by Taylor polynomials. For instance, the function $f(x) = e^x$ can be *approximated* by its Maclaurin polynomials as follows.

$e^x \approx 1 + x$ 1st-degree polynomial

$e^x \approx 1 + x + \dfrac{x^2}{2!}$ 2nd-degree polynomial

$e^x \approx 1 + x + \dfrac{x^2}{2!} + \dfrac{x^3}{3!}$ 3rd-degree polynomial

$e^x \approx 1 + x + \dfrac{x^2}{2!} + \dfrac{x^3}{3!} + \dfrac{x^4}{4!}$ 4th-degree polynomial

$e^x \approx 1 + x + \dfrac{x^2}{2!} + \dfrac{x^3}{3!} + \dfrac{x^4}{4!} + \dfrac{x^5}{5!}$ 5th-degree polynomial

In that section, you saw that the higher the degree of the approximating polynomial, the better the approximation becomes.

In this and the next two sections, you will see that several important types of functions, including

$$f(x) = e^x$$

can be represented *exactly* by an infinite series called a **power series.** For example, the power series representation for e^x is

$$e^x = 1 + x + \frac{x^2}{2!} + \frac{x^3}{3!} + \cdots + \frac{x^n}{n!} + \cdots.$$

For each real number x, it can be shown that the infinite series on the right converges to the number e^x. Before doing this, however, some preliminary results dealing with power series will be discussed—beginning with the following definition.

EXPLORATION

Graphical Reasoning Use a graphing utility to approximate the graph of each power series near $x = 0$. (Use the first several terms of each series.) Each series represents a well-known function. What is the function?

a. $\displaystyle\sum_{n=0}^{\infty} \frac{(-1)^n x^n}{n!}$

b. $\displaystyle\sum_{n=0}^{\infty} \frac{(-1)^n x^{2n}}{(2n)!}$

c. $\displaystyle\sum_{n=0}^{\infty} \frac{(-1)^n x^{2n+1}}{(2n+1)!}$

d. $\displaystyle\sum_{n=0}^{\infty} \frac{(-1)^n x^{2n+1}}{2n+1}$

e. $\displaystyle\sum_{n=0}^{\infty} \frac{2^n x^n}{n!}$

Definition of Power Series

If x is a variable, then an infinite series of the form

$$\sum_{n=0}^{\infty} a_n x^n = a_0 + a_1 x + a_2 x^2 + a_3 x^3 + \cdots + a_n x^n + \cdots$$

is called a **power series.** More generally, an infinite series of the form

$$\sum_{n=0}^{\infty} a_n (x - c)^n = a_0 + a_1(x - c) + a_2(x - c)^2 + \cdots + a_n(x - c)^n + \cdots$$

is called a **power series centered at *c*,** where c is a constant.

NOTE   To simplify the notation for power series, we agree that $(x - c)^0 = 1$, even if $x = c$.

Exercises for Section 9.8

See www.CalcChat.com for worked-out solutions to odd-numbered exercises.

In Exercises 1–4, state where the power series is centered.

1. $\sum_{n=0}^{\infty} n x^n$

2. $\sum_{n=1}^{\infty} \frac{(-1)^n 1 \cdot 3 \cdots (2n-1)}{2^n n!} x^n$

3. $\sum_{n=1}^{\infty} \frac{(x-2)^n}{n^3}$

4. $\sum_{n=0}^{\infty} \frac{(-1)^n (x - \pi)^{2n}}{(2n)!}$

In Exercises 5–10, find the radius of convergence of the power series.

5. $\sum_{n=0}^{\infty} (-1)^n \frac{x^n}{n+1}$

6. $\sum_{n=0}^{\infty} (2x)^n$

7. $\sum_{n=1}^{\infty} \frac{(2x)^n}{n^2}$

8. $\sum_{n=0}^{\infty} \frac{(-1)^n x^n}{2^n}$

9. $\sum_{n=0}^{\infty} \frac{(2x)^{2n}}{(2n)!}$

10. $\sum_{n=0}^{\infty} \frac{(2n)! x^{2n}}{n!}$

In Exercises 11–34, find the interval of convergence of the power series. (Be sure to include a check for convergence at the endpoints of the interval.)

11. $\sum_{n=0}^{\infty} \left(\frac{x}{2}\right)^n$

12. $\sum_{n=0}^{\infty} \left(\frac{x}{5}\right)^n$

13. $\sum_{n=1}^{\infty} \frac{(-1)^n x^n}{n}$

14. $\sum_{n=0}^{\infty} (-1)^{n+1}(n+1)x^n$

15. $\sum_{n=0}^{\infty} \frac{x^n}{n!}$

16. $\sum_{n=0}^{\infty} \frac{(3x)^n}{(2n)!}$

17. $\sum_{n=0}^{\infty} (2n)! \left(\frac{x}{2}\right)^n$

18. $\sum_{n=0}^{\infty} \frac{(-1)^n x^n}{(n+1)(n+2)}$

19. $\sum_{n=1}^{\infty} \frac{(-1)^{n+1} x^n}{4^n}$

20. $\sum_{n=0}^{\infty} \frac{(-1)^n n!(x-4)^n}{3^n}$

21. $\sum_{n=1}^{\infty} \frac{(-1)^{n+1}(x-5)^n}{n5^n}$

22. $\sum_{n=0}^{\infty} \frac{(x-2)^{n+1}}{(n+1)4^{n+1}}$

23. $\sum_{n=0}^{\infty} \frac{(-1)^{n+1}(x-1)^{n+1}}{n+1}$

24. $\sum_{n=1}^{\infty} \frac{(-1)^{n+1}(x-2)^n}{n2^n}$

25. $\sum_{n=1}^{\infty} \frac{(x-3)^{n-1}}{3^{n-1}}$

26. $\sum_{n=0}^{\infty} \frac{(-1)^n x^{2n+1}}{2n+1}$

27. $\sum_{n=1}^{\infty} \frac{n}{n+1}(-2x)^{n-1}$

28. $\sum_{n=0}^{\infty} \frac{(-1)^n x^{2n}}{n!}$

29. $\sum_{n=0}^{\infty} \frac{x^{2n+1}}{(2n+1)!}$

30. $\sum_{n=1}^{\infty} \frac{n! x^n}{(2n)!}$

31. $\sum_{n=1}^{\infty} \frac{2 \cdot 3 \cdot 4 \cdots (n+1)x^n}{n!}$

32. $\sum_{n=1}^{\infty} \left[\frac{2 \cdot 4 \cdot 6 \cdots 2n}{3 \cdot 5 \cdot 7 \cdots (2n+1)} \right] x^{2n+1}$

33. $\sum_{n=1}^{\infty} \frac{(-1)^{n+1} 3 \cdot 7 \cdot 11 \cdots (4n-1)(x-3)^n}{4^n}$

34. $\sum_{n=1}^{\infty} \frac{n!(x+1)^n}{1 \cdot 3 \cdot 5 \cdots (2n-1)}$

In Exercises 35 and 36, find the radius of convergence of the power series, where $c > 0$ and k is a positive integer.

35. $\sum_{n=1}^{\infty} \frac{(x-c)^{n-1}}{c^{n-1}}$

36. $\sum_{n=0}^{\infty} \frac{(n!)^k x^n}{(kn)!}$

In Exercises 37–40, find the interval of convergence of the power series. (Be sure to include a check for convergence at the endpoints of the interval.)

37. $\sum_{n=0}^{\infty} \left(\frac{x}{k}\right)^n, \quad k > 0$

38. $\sum_{n=1}^{\infty} \frac{(-1)^{n+1}(x-c)^n}{nc^n}$

39. $\sum_{n=1}^{\infty} \frac{k(k+1)(k+2)\cdots(k+n-1)x^n}{n!}, \quad k \geq 1$

40. $\sum_{n=1}^{\infty} \frac{n!(x-c)^n}{1 \cdot 3 \cdot 5 \cdots (2n-1)}$

In Exercises 41–44, write an equivalent series with the index of summation beginning at $n = 1$.

41. $\sum_{n=0}^{\infty} \frac{x^n}{n!}$

42. $\sum_{n=0}^{\infty} (-1)^{n+1}(n+1)x^n$

43. $\sum_{n=0}^{\infty} \frac{x^{2n+1}}{(2n+1)!}$

44. $\sum_{n=0}^{\infty} \frac{(-1)^n x^{2n+1}}{2n+1}$

In Exercises 45–48, find the intervals of convergence of (a) $f(x)$, (b) $f'(x)$, (c) $f''(x)$, and (d) $\int f(x)\, dx$. Include a check for convergence at the endpoints of the interval.

45. $f(x) = \sum_{n=0}^{\infty} \left(\frac{x}{2}\right)^n$

46. $f(x) = \sum_{n=1}^{\infty} \frac{(-1)^{n+1}(x-5)^n}{n5^n}$

47. $f(x) = \sum_{n=0}^{\infty} \frac{(-1)^{n+1}(x-1)^{n+1}}{n+1}$

48. $f(x) = \sum_{n=1}^{\infty} \frac{(-1)^{n+1}(x-2)^n}{n}$

Writing **In Exercises 49–52, match the graph of the first 10 terms of the sequence of partial sums of the series**

$$g(x) = \sum_{n=0}^{\infty} \left(\frac{x}{3}\right)^n$$

with the indicated value of the function. [The graphs are labeled (a), (b), (c), and (d).] Explain how you made your choice.

(a)

(b)

(c) S_n

(d) S_n

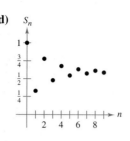

49. $g(1)$

50. $g(2)$

51. $g(3.1)$

52. $g(-2)$

Writing In Exercises 53–56, match the graph of the first 10 terms of the sequence of partial sums of the series

$$g(x) = \sum_{n=0}^{\infty} (2x)^n$$

with the indicated value of the function. [The graphs are labeled (a), (b), (c), and (d).] Explain how you made your choice.

(a) S_n

(b) S_n

(c) S_n

(d) S_n

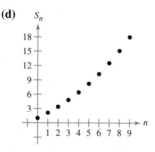

53. $g\left(\dfrac{1}{8}\right)$

54. $g\left(-\dfrac{1}{8}\right)$

55. $g\left(\dfrac{9}{16}\right)$

56. $g\left(\dfrac{3}{8}\right)$

Writing About Concepts

57. Define a power series centered at c.

58. Describe the radius of convergence of a power series. Describe the interval of convergence of a power series.

59. Describe the three basic forms of the domain of a power series.

Writing About Concepts *(continued)*

60. Describe how to differentiate and integrate a power series with a radius of convergence R. Will the series resulting from the operations of differentiation and integration have a different radius of convergence? Explain.

61. Give examples that show that the convergence of a power series at an endpoint of its interval of convergence may be either conditional or absolute. Explain your reasoning.

62. Write a power series that has the indicated interval of convergence. Explain your reasoning.

(a) $(-2, 2)$ (b) $(-1, 1]$ (c) $(-1, 0)$ (d) $[-2, 6)$

63. Let $f(x) = \displaystyle\sum_{n=0}^{\infty} \frac{(-1)^n x^{2n+1}}{(2n+1)!}$ and $g(x) = \displaystyle\sum_{n=0}^{\infty} \frac{(-1)^n x^{2n}}{(2n)!}$.

(a) Find the intervals of convergence of f and g.

(b) Show that $f'(x) = g(x)$.

(c) Show that $g'(x) = -f(x)$.

(d) Identify the functions f and g.

64. Let $f(x) = \displaystyle\sum_{n=0}^{\infty} \frac{x^n}{n!}$.

(a) Find the interval of convergence of f.

(b) Show that $f'(x) = f(x)$.

(c) Show that $f(0) = 1$.

(d) Identify the function f.

In Exercises 65–70, show that the function represented by the power series is a solution of the differential equation.

65. $y = \displaystyle\sum_{n=0}^{\infty} \frac{(-1)^n x^{2n+1}}{(2n+1)!}$, $y'' + y = 0$

66. $y = \displaystyle\sum_{n=0}^{\infty} \frac{(-1)^n x^{2n}}{(2n)!}$, $y'' + y = 0$

67. $y = \displaystyle\sum_{n=0}^{\infty} \frac{x^{2n+1}}{(2n+1)!}$, $y'' - y = 0$

68. $y = \displaystyle\sum_{n=0}^{\infty} \frac{x^{2n}}{(2n)!}$, $y'' - y = 0$

69. $y = \displaystyle\sum_{n=0}^{\infty} \frac{x^{2n}}{2^n n!}$, $y'' - xy' - y = 0$

70. $y = 1 + \displaystyle\sum_{n=1}^{\infty} \frac{(-1)^n x^{4n}}{2^{2n} n! \cdot 3 \cdot 7 \cdot 11 \cdots (4n-1)}$, $y'' + x^2 y = 0$

71. *Bessel Function* The Bessel function of order 0 is

$$J_0(x) = \sum_{k=0}^{\infty} \frac{(-1)^k x^{2k}}{2^{2k} (k!)^2}.$$

(a) Show that the series converges for all x.

(b) Show that the series is a solution of the differential equation $x^2 J_0'' + x J_0' + x^2 J_0 = 0$.

(c) Use a graphing utility to graph the polynomial composed of the first four terms of J_0.

(d) Approximate $\int_0^1 J_0 \, dx$ accurate to two decimal places.

72. Bessel Function The Bessel function of order 1 is

$$J_1(x) = x \sum_{k=0}^{\infty} \frac{(-1)^k x^{2k}}{2^{2k+1} k!(k+1)!}.$$

(a) Show that the series converges for all x.

(b) Show that the series is a solution of the differential equation $x^2 J_1'' + x J_1' + (x^2 - 1) J_1 = 0$.

(c) Use a graphing utility to graph the polynomial composed of the first four terms of J_1.

(d) Show that $J_0'(x) = -J_1(x)$.

In Exercises 73–76, the series represents a well-known function. Use a computer algebra system to graph the partial sum S_{10} and identify the function from the graph.

73. $f(x) = \sum_{n=0}^{\infty} (-1)^n \frac{x^{2n}}{(2n)!}$

74. $f(x) = \sum_{n=0}^{\infty} (-1)^n \frac{x^{2n+1}}{(2n+1)!}$

75. $f(x) = \sum_{n=0}^{\infty} (-1)^n x^n, \quad -1 < x < 1$

76. $f(x) = \sum_{n=0}^{\infty} (-1)^n \frac{x^{2n+1}}{2n+1}, \quad -1 \le x \le 1$

77. Investigation In Exercise 11 you found that the interval of convergence of the geometric series $\sum_{n=0}^{\infty} \left(\frac{x}{2}\right)^n$ is $(-2, 2)$.

(a) Find the sum of the series when $x = \frac{3}{4}$. Use a graphing utility to graph the first six terms of the sequence of partial sums and the horizontal line representing the sum of the series.

(b) Repeat part (a) for $x = -\frac{3}{4}$.

(c) Write a short paragraph comparing the rate of convergence of the partial sums with the sum of the series in parts (a) and (b). How do the plots of the partial sums differ as they converge toward the sum of the series?

(d) Given any positive real number M, there exists a positive integer N such that the partial sum

$$\sum_{n=0}^{N} \left(\frac{3}{2}\right)^n > M.$$

Use a graphing utility to complete the table.

M	10	100	1000	10,000
N				

78. Investigation The interval of convergence of the series $\sum_{n=0}^{\infty} (3x)^n$ is $\left(-\frac{1}{3}, \frac{1}{3}\right)$.

(a) Find the sum of the series when $x = \frac{1}{6}$. Use a graphing utility to graph the first six terms of the sequence of partial sums and the horizontal line representing the sum of the series.

(b) Repeat part (a) for $x = -\frac{1}{6}$.

(c) Write a short paragraph comparing the rate of convergence of the partial sums with the sum of the series in parts (a) and (b). How do the plots of the partial sums differ as they converge toward the sum of the series?

(d) Given any positive real number M, there exists a positive integer N such that the partial sum

$$\sum_{n=0}^{N} \left(3 \cdot \frac{2}{3}\right)^n > M.$$

Use a graphing utility to complete the table.

M	10	100	1000	10,000
N				

True or False? In Exercises 79–82, determine whether the statement is true or false. If it is false, explain why or give an example that shows it is false.

79. If the power series $\sum_{n=0}^{\infty} a_n x^n$ converges for $x = 2$, then it also converges for $x = -2$.

80. If the power series $\sum_{n=0}^{\infty} a_n x^n$ converges for $x = 2$, then it also converges for $x = -1$.

81. If the interval of convergence for $\sum_{n=0}^{\infty} a_n x^n$ is $(-1, 1)$, then the interval of convergence for $\sum_{n=0}^{\infty} a_n (x-1)^n$ is $(0, 2)$.

82. If $f(x) = \sum_{n=0}^{\infty} a_n x^n$ converges for $|x| < 2$, then

$$\int_0^1 f(x)\, dx = \sum_{n=0}^{\infty} \frac{a_n}{n+1}.$$

83. Prove that the power series

$$\sum_{n=0}^{\infty} \frac{(n+p)!}{n!(n+q)!} x^n$$

has a radius of convergence of $R = \infty$ if p and q are positive integers.

84. Let $g(x) = 1 + 2x + x^2 + 2x^3 + x^4 + \cdots$, where the coefficients are $c_{2n} = 1$ and $c_{2n+1} = 2$ for $n \ge 0$.

(a) Find the interval of convergence of the series.

(b) Find an explicit formula for $g(x)$.

85. Let $f(x) = \sum_{n=0}^{\infty} c_n x^n$, where $c_{n+3} = c_n$ for $n \ge 0$.

(a) Find the interval of convergence of the series.

(b) Find an explicit formula for $f(x)$.

86. Prove that if the power series $\sum_{n=0}^{\infty} c_n x^n$ has a radius of convergence of R, then $\sum_{n=0}^{\infty} c_n x^{2n}$ has a radius of convergence of $\sqrt{R}$.

87. For $n > 0$, let $R > 0$ and $c_n > 0$. Prove that if the interval of convergence of the series $\sum_{n=0}^{\infty} c_n (x - x_0)^n$ is $(x_0 - R, x_0 + R]$, then the series converges conditionally at $x_0 + R$.

Representation of Functions by Power Series

- Find a geometric power series that represents a function.
- Construct a power series using series operations.

Geometric Power Series

In this section and the next, you will study several techniques for finding a power series that represents a given function.

Consider the function given by $f(x) = 1/(1 - x)$. The form of f closely resembles the sum of a geometric series

$$\sum_{n=0}^{\infty} ar^n = \frac{a}{1 - r}, \quad |r| < 1.$$

In other words, if you let $a = 1$ and $r = x$, a power series representation for $1/(1 - x)$, centered at 0, is

$$\frac{1}{1 - x} = \sum_{n=0}^{\infty} x^n$$
$$= 1 + x + x^2 + x^3 + \cdots, \quad |x| < 1.$$

Of course, this series represents $f(x) = 1/(1 - x)$ only on the interval $(-1, 1)$, whereas f is defined for all $x \neq 1$, as shown in Figure 9.22. To represent f in another interval, you must develop a different series. For instance, to obtain the power series centered at -1, you could write

$$\frac{1}{1 - x} = \frac{1}{2 - (x + 1)} = \frac{1/2}{1 - [(x + 1)/2]} = \frac{a}{1 - r}$$

which implies that $a = \frac{1}{2}$ and $r = (x + 1)/2$. So, for $|x + 1| < 2$, you have

$$\frac{1}{1 - x} = \sum_{n=0}^{\infty} \left(\frac{1}{2}\right)\left(\frac{x + 1}{2}\right)^n$$
$$= \frac{1}{2}\left[1 + \frac{(x + 1)}{2} + \frac{(x + 1)^2}{4} + \frac{(x + 1)^3}{8} + \cdots\right], \quad |x + 1| < 2$$

which converges on the interval $(-3, 1)$.

JOSEPH FOURIER (1768–1830)

Some of the early work in representing functions by power series was done by the French mathematician Joseph Fourier. Fourier's work is important in the history of calculus, partly because it forced eighteenth century mathematicians to question the then-prevailing narrow concept of a function. Both Cauchy and Dirichlet were motivated by Fourier's work with series, and in 1837 Dirichlet published the general definition of a function that is used today.

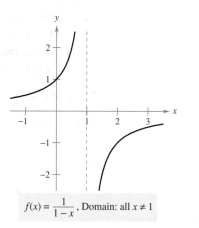

$f(x) = \dfrac{1}{1 - x}$, Domain: all $x \neq 1$

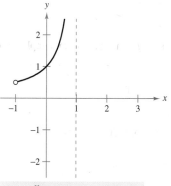

$f(x) = \sum_{n=0}^{\infty} x^n$, Domain: $-1 < x < 1$

Figure 9.22

EXAMPLE 1 **Finding a Geometric Power Series Centered at 0**

Find a power series for $f(x) = \dfrac{4}{x + 2}$, centered at 0.

Solution Writing $f(x)$ in the form $a/(1 - r)$ produces

$$\frac{4}{2 + x} = \frac{2}{1 - (-x/2)} = \frac{a}{1 - r}$$

which implies that $a = 2$ and $r = -x/2$. So, the power series for $f(x)$ is

$$\frac{4}{x + 2} = \sum_{n=0}^{\infty} ar^n$$

$$= \sum_{n=0}^{\infty} 2\left(-\frac{x}{2}\right)^n$$

$$= 2\left(1 - \frac{x}{2} + \frac{x^2}{4} - \frac{x^3}{8} + \cdots\right).$$

This power series converges when

$$\left|-\frac{x}{2}\right| < 1$$

which implies that the interval of convergence is $(-2, 2)$.

Another way to determine a power series for a rational function such as the one in Example 1 is to use long division. For instance, by dividing $2 + x$ into 4, you obtain the result shown at the left.

Long Division

$$2 - x + \tfrac{1}{2}x^2 - \tfrac{1}{4}x^3 + \cdots$$

$$\begin{array}{r} 2 + x \overline{)\,4} \\ \underline{4 + 2x} \\ -2x \\ \underline{-2x - x^2} \\ x^2 \\ x^2 + \tfrac{1}{2}x^3 \\ \underline{-\tfrac{1}{2}x^3} \\ -\tfrac{1}{2}x^3 - \tfrac{1}{4}x^4 \end{array}$$

EXAMPLE 2 **Finding a Geometric Power Series Centered at 1**

Find a power series for $f(x) = \dfrac{1}{x}$, centered at 1.

Solution Writing $f(x)$ in the form $a/(1 - r)$ produces

$$\frac{1}{x} = \frac{1}{1 - (-x + 1)} = \frac{a}{1 - r}$$

which implies that $a = 1$ and $r = 1 - x = -(x - 1)$. So, the power series for $f(x)$ is

$$\frac{1}{x} = \sum_{n=0}^{\infty} ar^n$$

$$= \sum_{n=0}^{\infty} [-(x - 1)]^n$$

$$= \sum_{n=0}^{\infty} (-1)^n (x - 1)^n$$

$$= 1 - (x - 1) + (x - 1)^2 - (x - 1)^3 + \cdots.$$

This power series converges when

$$|x - 1| < 1$$

which implies that the interval of convergence is $(0, 2)$.

Operations with Power Series

The versatility of geometric power series will be shown later in this section, following a discussion of power series operations. These operations, used with differentiation and integration, provide a means of developing power series for a variety of elementary functions. (For simplicity, the following properties are stated for a series centered at 0.)

Operations with Power Series

Let $f(x) = \Sigma\, a_n x^n$ and $g(x) = \Sigma\, b_n x^n$.

1. $f(kx) = \displaystyle\sum_{n=0}^{\infty} a_n k^n x^n$

2. $f(x^N) = \displaystyle\sum_{n=0}^{\infty} a_n x^{nN}$

3. $f(x) \pm g(x) = \displaystyle\sum_{n=0}^{\infty} (a_n \pm b_n) x^n$

The operations described above can change the interval of convergence for the resulting series. For example, in the following addition, the interval of convergence for the sum is the *intersection* of the intervals of convergence of the two original series.

$$\underbrace{\sum_{n=0}^{\infty} x^n}_{(-1,\,1)} + \underbrace{\sum_{n=0}^{\infty} \left(\frac{x}{2}\right)^n}_{(-2,\,2)} = \underbrace{\sum_{n=0}^{\infty} \left(1 + \frac{1}{2^n}\right)x^n}_{(-1,\,1)}$$

$$(-1, 1) \cap (-2, 2) = (-1, 1)$$

EXAMPLE 3   Adding Two Power Series

Find a power series, centered at 0, for $f(x) = \dfrac{3x - 1}{x^2 - 1}$.

Solution   Using partial fractions, you can write $f(x)$ as

$$\frac{3x - 1}{x^2 - 1} = \frac{2}{x + 1} + \frac{1}{x - 1}.$$

By adding the two geometric power series

$$\frac{2}{x + 1} = \frac{2}{1 - (-x)} = \sum_{n=0}^{\infty} 2(-1)^n x^n, \quad |x| < 1$$

and

$$\frac{1}{x - 1} = \frac{-1}{1 - x} = -\sum_{n=0}^{\infty} x^n, \quad |x| < 1$$

you obtain the following power series.

$$\frac{3x - 1}{x^2 - 1} = \sum_{n=0}^{\infty} [2(-1)^n - 1]x^n = 1 - 3x + x^2 - 3x^3 + x^4 - \cdots$$

The interval of convergence for this power series is $(-1, 1)$.

EXAMPLE 4   **Finding a Power Series by Integration**

Find a power series for $f(x) = \ln x$, centered at 1.

Solution   From Example 2, you know that

$$\frac{1}{x} = \sum_{n=0}^{\infty} (-1)^n (x-1)^n.$$ Interval of convergence: (0, 2)

Integrating this series produces

$$\ln x = \int \frac{1}{x}\, dx + C$$

$$= C + \sum_{n=0}^{\infty} (-1)^n \frac{(x-1)^{n+1}}{n+1}.$$

By letting $x = 1$, you can conclude that $C = 0$. Therefore,

$$\ln x = \sum_{n=0}^{\infty} (-1)^n \frac{(x-1)^{n+1}}{n+1}$$

$$= \frac{(x-1)}{1} - \frac{(x-1)^2}{2} + \frac{(x-1)^3}{3} - \frac{(x-1)^4}{4} + \cdots.$$ Interval of convergence: (0, 2]

Note that the series converges at $x = 2$. This is consistent with the observation in the preceding section that integration of a power series may alter the convergence at the endpoints of the interval of convergence.

TECHNOLOGY   In Section 9.7, the fourth-degree Taylor polynomial for the natural logarithmic function

$$\ln x \approx (x-1) - \frac{(x-1)^2}{2} + \frac{(x-1)^3}{3} - \frac{(x-1)^4}{4}$$

was used to approximate $\ln(1.1)$.

$$\ln(1.1) \approx (0.1) - \frac{1}{2}(0.1)^2 + \frac{1}{3}(0.1)^3 - \frac{1}{4}(0.1)^4$$

$$\approx 0.0953083$$

You now know from Example 4 that this polynomial represents the first four terms of the power series for $\ln x$. Moreover, using the Alternating Series Remainder, you can determine that the error in this approximation is less than

$$|R_4| \leq |a_5|$$

$$= \frac{1}{5}(0.1)^5$$

$$= 0.000002.$$

During the seventeenth and eighteenth centuries, mathematical tables for logarithms and values of other transcendental functions were computed in this manner. Such numerical techniques are far from outdated, because it is precisely by such means that many modern calculating devices are programmed to evaluate transcendental functions.

EXAMPLE 5   **Finding a Power Series by Integration**

Find a power series for $g(x) = \arctan x$, centered at 0.

Solution   Because $D_x[\arctan x] = 1/(1 + x^2)$, you can use the series

$$f(x) = \frac{1}{1 + x} = \sum_{n=0}^{\infty} (-1)^n x^n.$$ Interval of convergence: $(-1, 1)$

Substituting x^2 for x produces

$$f(x^2) = \frac{1}{1 + x^2} = \sum_{n=0}^{\infty} (-1)^n x^{2n}.$$

Finally, by integrating, you obtain

$$\arctan x = \int \frac{1}{1 + x^2} \, dx + C$$

$$= C + \sum_{n=0}^{\infty} (-1)^n \frac{x^{2n+1}}{2n + 1}$$

$$= \sum_{n=0}^{\infty} (-1)^n \frac{x^{2n+1}}{2n + 1}$$ Let $x = 0$, then $C = 0$.

$$= x - \frac{x^3}{3} + \frac{x^5}{5} - \frac{x^7}{7} + \cdots.$$ Interval of convergence: $(-1, 1)$

It can be shown that the power series developed for $\arctan x$ in Example 5 also converges (to $\arctan x$) for $x = \pm 1$. For instance, when $x = 1$, you can write

$$\arctan 1 = 1 - \frac{1}{3} + \frac{1}{5} - \frac{1}{7} + \cdots$$

$$= \frac{\pi}{4}.$$

However, this series (developed by James Gregory in 1671) does not give us a practical way of approximating π because it converges so slowly that hundreds of terms would have to be used to obtain reasonable accuracy. Example 6 shows how to use *two* different arctangent series to obtain a very good approximation of π using only a few terms. This approximation was developed by John Machin in 1706.

EXAMPLE 6   **Approximating π with a Series**

Use the trigonometric identity

$$4 \arctan \frac{1}{5} - \arctan \frac{1}{239} = \frac{\pi}{4}$$

to approximate the number π [see Exercise 50(b)].

Solution   By using only five terms from each of the series for $\arctan(1/5)$ and $\arctan(1/239)$, you obtain

$$4 \left(4 \arctan \frac{1}{5} - \arctan \frac{1}{239} \right) \approx 3.1415926$$

which agrees with the exact value of π with an error of less than 0.0000001.

The Granger Collection

SRINIVASA RAMANUJAN (1887–1920)

Series that can be used to approximate π have interested mathematicians for the past 300 years. An amazing series for approximating $1/\pi$ was discovered by the Indian mathematician Srinivasa Ramanujan in 1914 (see Exercise 64). Each successive term of Ramanujan's series adds roughly eight more correct digits to the value of $1/\pi$. For more information about Ramanujan's work, see the article "Ramanujan and Pi" by Jonathan M. Borwein and Peter B. Borwein in *Scientific American*.

Exercises for Section 9.9

See www.CalcChat.com for worked-out solutions to odd-numbered exercises.

In Exercises 1–4, find a geometric power series for the function, centered at 0, (a) by the technique shown in Examples 1 and 2 and (b) by long division.

1. $f(x) = \dfrac{1}{2 - x}$

2. $f(x) = \dfrac{4}{5 - x}$

3. $f(x) = \dfrac{1}{2 + x}$

4. $f(x) = \dfrac{1}{1 + x}$

In Exercises 5–16, find a power series for the function, centered at c, and determine the interval of convergence.

5. $f(x) = \dfrac{1}{2 - x}, \quad c = 5$

6. $f(x) = \dfrac{4}{5 - x}, \quad c = -2$

7. $f(x) = \dfrac{3}{2x - 1}, \quad c = 0$

8. $f(x) = \dfrac{3}{2x - 1}, \quad c = 2$

9. $g(x) = \dfrac{1}{2x - 5}, \quad c = -3$

10. $h(x) = \dfrac{1}{2x - 5}, \quad c = 0$

11. $f(x) = \dfrac{3}{x + 2}, \quad c = 0$

12. $f(x) = \dfrac{4}{3x + 2}, \quad c = 2$

13. $g(x) = \dfrac{3x}{x^2 + x - 2}, \quad c = 0$

14. $g(x) = \dfrac{4x - 7}{2x^2 + 3x - 2}, \quad c = 0$

15. $f(x) = \dfrac{2}{1 - x^2}, \quad c = 0$

16. $f(x) = \dfrac{4}{4 + x^2}, \quad c = 0$

In Exercises 17–26, use the power series

$$\frac{1}{1 + x} = \sum_{n=0}^{\infty} (-1)^n x^n$$

to determine a power series, centered at 0, for the function. Identify the interval of convergence.

17. $h(x) = \dfrac{-2}{x^2 - 1} = \dfrac{1}{1 + x} + \dfrac{1}{1 - x}$

18. $h(x) = \dfrac{x}{x^2 - 1} = \dfrac{1}{2(1 + x)} - \dfrac{1}{2(1 - x)}$

19. $f(x) = -\dfrac{1}{(x + 1)^2} = \dfrac{d}{dx}\left[\dfrac{1}{x + 1}\right]$

20. $f(x) = \dfrac{2}{(x + 1)^3} = \dfrac{d^2}{dx^2}\left[\dfrac{1}{x + 1}\right]$

21. $f(x) = \ln(x + 1) = \displaystyle\int \dfrac{1}{x + 1}\, dx$

22. $f(x) = \ln(1 - x^2) = \displaystyle\int \dfrac{1}{1 + x}\, dx - \int \dfrac{1}{1 - x}\, dx$

23. $g(x) = \dfrac{1}{x^2 + 1}$

24. $f(x) = \ln(x^2 + 1)$

25. $h(x) = \dfrac{1}{4x^2 + 1}$

26. $f(x) = \arctan 2x$

 Graphical and Numerical Analysis   In Exercises 27 and 28, let

$$S_n = x - \frac{x^2}{2} + \frac{x^3}{3} - \frac{x^4}{4} + \cdots \pm \frac{x^n}{n}.$$

Use a graphing utility to confirm the inequality graphically. Then complete the table to confirm the inequality numerically.

x	0.0	0.2	0.4	0.6	0.8	1.0
S_n						
$\ln(x + 1)$						
S_{n+1}						

27. $S_2 \leq \ln(x + 1) \leq S_3$

28. $S_4 \leq \ln(x + 1) \leq S_5$

In Exercises 29 and 30, (a) graph several partial sums of the series, (b) find the sum of the series and its radius of convergence, (c) use 50 terms of the series to approximate the sum when $x = 0.5$, and (d) determine what the approximation represents and how good the approximation is.

29. $\displaystyle\sum_{n=1}^{\infty} \frac{(-1)^{n+1}(x - 1)^n}{n}$

30. $\displaystyle\sum_{n=0}^{\infty} \frac{(-1)^n x^{2n+1}}{(2n + 1)!}$

In Exercises 31–34, match the polynomial approximation of the function $f(x) = \arctan x$ with the correct graph. [The graphs are labeled (a), (b), (c), and (d).]

(a)

(b)

(c)

(d)

31. $g(x) = x$

32. $g(x) = x - \dfrac{x^3}{3}$

33. $g(x) = x - \dfrac{x^3}{3} + \dfrac{x^5}{5}$

34. $g(x) = x - \dfrac{x^3}{3} + \dfrac{x^5}{5} - \dfrac{x^7}{7}$

In Exercises 35–38, use the series for $f(x) = \arctan x$ to approximate the value, using $R_N \leq 0.001$.

35. $\arctan \dfrac{1}{4}$

36. $\displaystyle\int_0^{3/4} \arctan x^2 \, dx$

37. $\displaystyle\int_0^{1/2} \dfrac{\arctan x^2}{x} \, dx$

38. $\displaystyle\int_0^{1/2} x^2 \arctan x \, dx$

In Exercises 39–42, use the power series

$$\frac{1}{1-x} = \sum_{n=0}^{\infty} x^n, \quad |x| < 1.$$

Find the series representation of the function and determine its interval of convergence.

39. $f(x) = \dfrac{1}{(1-x)^2}$

40. $f(x) = \dfrac{x}{(1-x)^2}$

41. $f(x) = \dfrac{1+x}{(1-x)^2}$

42. $f(x) = \dfrac{x(1+x)}{(1-x)^2}$

43. *Probability*   A fair coin is tossed repeatedly. The probability that the first head occurs on the nth toss is $P(n) = \left(\frac{1}{2}\right)^n$. When this game is repeated many times, the average number of tosses required until the first head occurs is

$$E(n) = \sum_{n=1}^{\infty} nP(n).$$

(This value is called the *expected value of n.*) Use the results of Exercises 39–42 to find $E(n)$. Is the answer what you expected? Why or why not?

44. Use the results of Exercises 39–42 to find the sum of each series.

(a) $\dfrac{1}{3}\displaystyle\sum_{n=1}^{\infty} n\left(\dfrac{2}{3}\right)^n$

(b) $\dfrac{1}{10}\displaystyle\sum_{n=1}^{\infty} n\left(\dfrac{9}{10}\right)^n$

Writing   **In Exercises 45–48, explain how to use the geometric series**

$$g(x) = \frac{1}{1-x} = \sum_{n=0}^{\infty} x^n, \quad |x| < 1$$

to find the series for the function. Do not find the series.

45. $f(x) = \dfrac{1}{1+x}$

46. $f(x) = \dfrac{1}{1-x^2}$

47. $f(x) = \dfrac{5}{1+x}$

48. $f(x) = \ln(1-x)$

49. Prove that $\arctan x + \arctan y = \arctan \dfrac{x+y}{1-xy}$ for $xy \neq 1$ provided the value of the left side of the equation is between $-\pi/2$ and $\pi/2$.

50. Use the result of Exercise 49 to verify each identity.

(a) $\arctan \dfrac{120}{119} - \arctan \dfrac{1}{239} = \dfrac{\pi}{4}$

(b) $4\arctan \dfrac{1}{5} - \arctan \dfrac{1}{239} = \dfrac{\pi}{4}$

[*Hint:* Use Exercise 49 twice to find $4\arctan \frac{1}{5}$. Then use part (a).]

In Exercises 51 and 52, (a) verify the given equation and (b) use the equation and the series for the arctangent to approximate π to two-decimal-place accuracy.

51. $2\arctan \dfrac{1}{2} - \arctan \dfrac{1}{7} = \dfrac{\pi}{4}$

52. $\arctan \dfrac{1}{2} + \arctan \dfrac{1}{3} = \dfrac{\pi}{4}$

In Exercises 53–58, find the sum of the convergent series by using a well-known function. Identify the function and explain how you obtained the sum.

53. $\displaystyle\sum_{n=1}^{\infty} (-1)^{n+1} \dfrac{1}{2^n n}$

54. $\displaystyle\sum_{n=1}^{\infty} (-1)^{n+1} \dfrac{1}{3^n n}$

55. $\displaystyle\sum_{n=1}^{\infty} (-1)^{n+1} \dfrac{2^n}{5^n n}$

56. $\displaystyle\sum_{n=0}^{\infty} (-1)^n \dfrac{1}{2n+1}$

57. $\displaystyle\sum_{n=0}^{\infty} (-1)^n \dfrac{1}{2^{2n+1}(2n+1)}$

58. $\displaystyle\sum_{n=1}^{\infty} (-1)^{n+1} \dfrac{1}{3^{2n-1}(2n-1)}$

Writing About Concepts

59. Use the results of Exercises 31–34 to make a geometric argument for why the series approximations of $f(x) = \arctan x$ have only odd powers of x.

60. Use the results of Exercises 31–34 to make a conjecture about the degrees of series approximations of $f(x) = \arctan x$ that have relative extrema.

61. One of the series in Exercises 53–58 converges to its sum at a much lower rate than the other five series. Which is it? Explain why this series converges so slowly. Use a graphing utility to illustrate the rate of convergence.

62. The radius of convergence of the power series $\displaystyle\sum_{n=0}^{\infty} a_n x^n$ is 3. What is the radius of convergence of the series $\displaystyle\sum_{n=1}^{\infty} n a_n x^{n-1}$? Explain.

63. The power series $\displaystyle\sum_{n=0}^{\infty} a_n x^n$ converges for $|x+1| < 4$. What can you conclude about the series $\displaystyle\sum_{n=0}^{\infty} a_n \dfrac{x^{n+1}}{n+1}$? Explain.

---

 64. Use a graphing utility to show that

$$\frac{\sqrt{8}}{9801} \sum_{n=0}^{\infty} \frac{(4n)!(1103 + 26{,}390n)}{(n!)^4 396^{4n}} = \frac{1}{\pi}.$$

(*Note:* This series was discovered by the Indian mathematician Srinivasa Ramanujan in 1914.)

In Exercises 65 and 66, find the sum of the series.

65. $\displaystyle\sum_{n=0}^{\infty} \dfrac{(-1)^n}{3^n(2n+1)}$

66. $\displaystyle\sum_{n=0}^{\infty} \dfrac{(-1)^n \pi^{2n+1}}{3^{2n+1}(2n+1)!}$

Taylor and Maclaurin Series

- Find a Taylor or Maclaurin series for a function.
- Find a binomial series.
- Use a basic list of Taylor series to find other Taylor series.

Taylor Series and Maclaurin Series

In Section 9.9, you derived power series for several functions using geometric series with term-by-term differentiation or integration. In this section you will study a *general* procedure for deriving the power series for a function that has derivatives of all orders. The following theorem gives the form that *every* convergent power series must take.

THEOREM 9.22   The Form of a Convergent Power Series

If f is represented by a power series $f(x) = \Sigma\, a_n(x - c)^n$ for all x in an open interval I containing c, then $a_n = f^{(n)}(c)/n!$ and

$$f(x) = f(c) + f'(c)(x - c) + \frac{f''(c)}{2!}(x - c)^2 + \cdots + \frac{f^{(n)}(c)}{n!}(x - c)^n + \cdots.$$

Proof   Suppose the power series $\Sigma\, a_n(x - c)^n$ has a radius of convergence R. Then, by Theorem 9.21, you know that the nth derivative of f exists for $|x - c| < R$, and by successive differentiation you obtain the following.

$$f^{(0)}(x) = a_0 + a_1(x - c) + a_2(x - c)^2 + a_3(x - c)^3 + a_4(x - c)^4 + \cdots$$
$$f^{(1)}(x) = a_1 + 2a_2(x - c) + 3a_3(x - c)^2 + 4a_4(x - c)^3 + \cdots$$
$$f^{(2)}(x) = 2a_2 + 3!a_3(x - c) + 4 \cdot 3a_4(x - c)^2 + \cdots$$
$$f^{(3)}(x) = 3!a_3 + 4!a_4(x - c) + \cdots$$
$$\vdots$$
$$f^{(n)}(x) = n!a_n + (n + 1)!a_{n+1}(x - c) + \cdots$$

Evaluating each of these derivatives at $x = c$ yields

$$f^{(0)}(c) = 0!a_0$$
$$f^{(1)}(c) = 1!a_1$$
$$f^{(2)}(c) = 2!a_2$$
$$f^{(3)}(c) = 3!a_3$$

and, in general, $f^{(n)}(c) = n!a_n$. By solving for a_n, you find that the coefficients of the power series representation of $f(x)$ are

$$a_n = \frac{f^{(n)}(c)}{n!}.$$

Notice that the coefficients of the power series in Theorem 9.22 are precisely the coefficients of the Taylor polynomials for $f(x)$ at c as defined in Section 9.7. For this reason, the series is called the **Taylor series** for $f(x)$ at c.

COLIN MACLAURIN (1698–1746)

The development of power series to represent functions is credited to the combined work of many seventeenth and eighteenth century mathematicians. Gregory, Newton, John and James Bernoulli, Leibniz, Euler, Lagrange, Wallis, and Fourier all contributed to this work. However, the two names that are most commonly associated with power series are Brook Taylor (1685–1731) and Colin Maclaurin.

NOTE   Be sure you understand Theorem 9.22. The theorem says that *if a power series converges to $f(x)$, the series must be a Taylor series.* The theorem does *not* say that every series formed with the Taylor coefficients $a_n = f^{(n)}(c)/n!$ will converge to $f(x)$.

> **Definitions of Taylor and Maclaurin Series**
>
> If a function f has derivatives of all orders at $x = c$, then the series
>
> $$\sum_{n=0}^{\infty} \frac{f^{(n)}(c)}{n!}(x - c)^n = f(c) + f'(c)(x - c) + \cdots + \frac{f^{(n)}(c)}{n!}(x - c)^n + \cdots$$
>
> is called the **Taylor series for $f(x)$ at c.** Moreover, if $c = 0$, then the series is the **Maclaurin series for f.**

If you know the pattern for the coefficients of the Taylor polynomials for a function, you can extend the pattern easily to form the corresponding Taylor series. For instance, in Example 4 in Section 9.7, you found the fourth Taylor polynomial for $\ln x$, centered at 1, to be

$$P_4(x) = (x - 1) - \frac{1}{2}(x - 1)^2 + \frac{1}{3}(x - 1)^3 - \frac{1}{4}(x - 1)^4.$$

From this pattern, you can obtain the Taylor series for $\ln x$ centered at $c = 1$,

$$(x - 1) - \frac{1}{2}(x - 1)^2 + \cdots + \frac{(-1)^{n+1}}{n}(x - 1)^n + \cdots.$$

EXAMPLE 1   Forming a Power Series

Use the function $f(x) = \sin x$ to form the Maclaurin series

$$\sum_{n=0}^{\infty} \frac{f^{(n)}(0)}{n!}x^n = f(0) + f'(0)x + \frac{f''(0)}{2!}x^2 + \frac{f^{(3)}(0)}{3!}x^3 + \frac{f^{(4)}(0)}{4!}x^4 + \cdots$$

and determine the interval of convergence.

Solution   Successive differentiation of $f(x)$ yields

$$
\begin{aligned}
f(x) &= \sin x & f(0) &= \sin 0 = 0 \\
f'(x) &= \cos x & f'(0) &= \cos 0 = 1 \\
f''(x) &= -\sin x & f''(0) &= -\sin 0 = 0 \\
f^{(3)}(x) &= -\cos x & f^{(3)}(0) &= -\cos 0 = -1 \\
f^{(4)}(x) &= \sin x & f^{(4)}(0) &= \sin 0 = 0 \\
f^{(5)}(x) &= \cos x & f^{(5)}(0) &= \cos 0 = 1
\end{aligned}
$$

and so on. The pattern repeats after the third derivative. So, the power series is as follows.

$$\sum_{n=0}^{\infty} \frac{f^{(n)}(0)}{n!}x^n = f(0) + f'(0)x + \frac{f''(0)}{2!}x^2 + \frac{f^{(3)}(0)}{3!}x^3 + \frac{f^{(4)}(0)}{4!}x^4 + \cdots$$

$$\sum_{n=0}^{\infty} \frac{(-1)^n x^{2n+1}}{(2n+1)!} = 0 + (1)x + \frac{0}{2!}x^2 + \frac{(-1)}{3!}x^3 + \frac{0}{4!}x^4 + \frac{1}{5!}x^5 + \frac{0}{6!}x^6$$

$$+ \frac{(-1)}{7!}x^7 + \cdots$$

$$= x - \frac{x^3}{3!} + \frac{x^5}{5!} - \frac{x^7}{7!} + \cdots$$

By the Ratio Test, you can conclude that this series converges for all x.

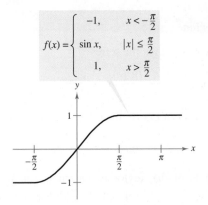

$$f(x) = \begin{cases} -1, & x < -\frac{\pi}{2} \\ \sin x, & |x| \le \frac{\pi}{2} \\ 1, & x > \frac{\pi}{2} \end{cases}$$

Figure 9.23

Notice that in Example 1 you cannot conclude that the power series converges to $\sin x$ for all x. You can simply conclude that the power series converges to some function, but you are not sure what function it is. This is a subtle, but important, point in dealing with Taylor or Maclaurin series. To persuade yourself that the series

$$f(c) + f'(c)(x - c) + \frac{f''(c)}{2!}(x - c)^2 + \cdots + \frac{f^{(n)}(c)}{n!}(x - c)^n + \cdots$$

might converge to a function other than f, remember that the derivatives are being evaluated at a single point. It can easily happen that another function will agree with the values of $f^{(n)}(x)$ when $x = c$ and disagree at other x-values. For instance, if you formed the power series (centered at 0) for the function shown in Figure 9.23, you would obtain the same series as in Example 1. You know that the series converges for all x, and yet it obviously cannot converge to both $f(x)$ and $\sin x$ for all x.

Let f have derivatives of all orders in an open interval I centered at c. The Taylor series for f may fail to converge for some x in I. Or, even if it is convergent, it may fail to have $f(x)$ as its sum. Nevertheless, Theorem 9.19 tells us that for each n,

$$f(x) = f(c) + f'(c)(x - c) + \frac{f''(c)}{2!}(x - c)^2 + \cdots + \frac{f^{(n)}(c)}{n!}(x - c)^n + R_n(x),$$

where

$$R_n(x) = \frac{f^{(n+1)}(z)}{(n + 1)!}(x - c)^{n+1}.$$

Note that in this remainder formula the particular value of z that makes the remainder formula true depends on the values of x and n. If $R_n \to 0$, then the following theorem tells us that the Taylor series for f actually converges to $f(x)$ for all x in I.

THEOREM 9.23   Convergence of Taylor Series

If $\lim\limits_{n \to \infty} R_n = 0$ for all x in the interval I, then the Taylor series for f converges and equals $f(x)$,

$$f(x) = \sum_{n=0}^{\infty} \frac{f^{(n)}(c)}{n!}(x - c)^n.$$

Proof   For a Taylor series, the nth partial sum coincides with the nth Taylor polynomial. That is, $S_n(x) = P_n(x)$. Moreover, because

$$P_n(x) = f(x) - R_n(x)$$

it follows that

$$\lim_{n \to \infty} S_n(x) = \lim_{n \to \infty} P_n(x)$$
$$= \lim_{n \to \infty} [f(x) - R_n(x)]$$
$$= f(x) - \lim_{n \to \infty} R_n(x).$$

So, for a given x, the Taylor series (the sequence of partial sums) converges to $f(x)$ if and only if $R_n(x) \to 0$ as $n \to \infty$.   ■

NOTE   Stated another way, Theorem 9.23 says that a power series formed with Taylor coefficients $a_n = f^{(n)}(c)/n!$ converges to the function from which it was derived at precisely those values for which the remainder approaches 0 as $n \to \infty$.

In Example 1, you derived the power series from the sine function and you also concluded that the series converges to some function on the entire real line. In Example 2, you will see that the series actually converges to $\sin x$. The key observation is that although the value of z is not known, it is possible to obtain an upper bound for $\left| f^{(n+1)}(z) \right|$.

EXAMPLE 2   A Convergent Maclaurin Series

Show that the Maclaurin series for $f(x) = \sin x$ converges to $\sin x$ for all x.

Solution   Using the result in Example 1, you need to show that

$$\sin x = x - \frac{x^3}{3!} + \frac{x^5}{5!} - \frac{x^7}{7!} + \cdots + \frac{(-1)^n x^{2n+1}}{(2n+1)!} + \cdots$$

is true for all x. Because

$$f^{(n+1)}(x) = \pm \sin x$$

or

$$f^{(n+1)}(x) = \pm \cos x$$

you know that $\left| f^{(n+1)}(z) \right| \le 1$ for every real number z. Therefore, for any fixed x, you can apply Taylor's Theorem (Theorem 9.19) to conclude that

$$0 \le \left| R_n(x) \right| = \left| \frac{f^{(n+1)}(z)}{(n+1)!} x^{n+1} \right| \le \frac{|x|^{n+1}}{(n+1)!}.$$

From the discussion in Section 9.1 regarding the relative rates of convergence of exponential and factorial sequences, it follows that for a fixed x

$$\lim_{n \to \infty} \frac{|x|^{n+1}}{(n+1)!} = 0.$$

Finally, by the Squeeze Theorem, it follows that for all x, $R_n(x) \to 0$ as $n \to \infty$. So, by Theorem 9.23, the Maclaurin series for $\sin x$ converges to $\sin x$ for all x.

Figure 9.24 visually illustrates the convergence of the Maclaurin series for $\sin x$ by comparing the graphs of the Maclaurin polynomials $P_1(x)$, $P_3(x)$, $P_5(x)$, and $P_7(x)$ with the graph of the sine function. Notice that as the degree of the polynomial increases, its graph more closely resembles that of the sine function.

$P_1(x) = x$

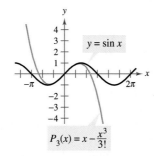

$P_3(x) = x - \dfrac{x^3}{3!}$

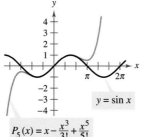

$P_5(x) = x - \dfrac{x^3}{3!} + \dfrac{x^5}{5!}$

$P_7(x) = x - \dfrac{x^3}{3!} + \dfrac{x^5}{5!} - \dfrac{x^7}{7!}$

As n increases, the graph of P_n more closely resembles the sine function.
Figure 9.24

The guidelines for finding a Taylor series for $f(x)$ at c are summarized below.

Guidelines for Finding a Taylor Series

1. Differentiate $f(x)$ several times and evaluate each derivative at c.

$$f(c), f'(c), f''(c), f'''(c), \cdots, f^{(n)}(c), \cdots$$

Try to recognize a pattern in these numbers.

2. Use the sequence developed in the first step to form the Taylor coefficients $a_n = f^{(n)}(c)/n!$, and determine the interval of convergence for the resulting power series

$$f(c) + f'(c)(x - c) + \frac{f''(c)}{2!}(x - c)^2 + \cdots + \frac{f^{(n)}(c)}{n!}(x - c)^n + \cdots.$$

3. Within this interval of convergence, determine whether or not the series converges to $f(x)$.

The direct determination of Taylor or Maclaurin coefficients using successive differentiation can be difficult, and the next example illustrates a shortcut for finding the coefficients indirectly—using the coefficients of a known Taylor or Maclaurin series.

EXAMPLE 3   Maclaurin Series for a Composite Function

Find the Maclaurin series for $f(x) = \sin x^2$.

Solution   To find the coefficients for this Maclaurin series directly, you must calculate successive derivatives of $f(x) = \sin x^2$. By calculating just the first two,

$$f'(x) = 2x \cos x^2 \quad \text{and} \quad f''(x) = -4x^2 \sin x^2 + 2 \cos x^2$$

you can see that this task would be quite cumbersome. Fortunately, there is an alternative. First consider the Maclaurin series for $\sin x$ found in Example 1.

$$g(x) = \sin x$$

$$= x - \frac{x^3}{3!} + \frac{x^5}{5!} - \frac{x^7}{7!} + \cdots$$

Now, because $\sin x^2 = g(x^2)$, you can substitute x^2 for x in the series for $\sin x$ to obtain

$$\sin x^2 = g(x^2)$$

$$= x^2 - \frac{x^6}{3!} + \frac{x^{10}}{5!} - \frac{x^{14}}{7!} + \cdots.$$

Be sure to understand the point illustrated in Example 3. Because direct computation of Taylor or Maclaurin coefficients can be tedious, the most practical way to find a Taylor or Maclaurin series is to develop power series for a *basic list* of elementary functions. From this list, you can determine power series for other functions by the operations of addition, subtraction, multiplication, division, differentiation, integration, or composition with known power series.

Binomial Series

Before presenting the basic list for elementary functions, you wll develop one more series—for a function of the form $f(x) = (1 + x)^k$. This produces the **binomial series.**

EXAMPLE 4   Binomial Series

Find the Maclaurin series for $f(x) = (1 + x)^k$ and determine its radius of convergence. Assume that k is not a positive integer.

Solution   By successive differentiation, you have

$$f(x) = (1 + x)^k \qquad\qquad f(0) = 1$$
$$f'(x) = k(1 + x)^{k-1} \qquad\qquad f'(0) = k$$
$$f''(x) = k(k - 1)(1 + x)^{k-2} \qquad\qquad f''(0) = k(k - 1)$$
$$f'''(x) = k(k - 1)(k - 2)(1 + x)^{k-3} \qquad f'''(0) = k(k - 1)(k - 2)$$
$$\vdots \qquad\qquad\qquad\qquad \vdots$$
$$f^{(n)}(x) = k \cdots (k - n + 1)(1 + x)^{k-n} \qquad f^{(n)}(0) = k(k - 1) \cdots (k - n + 1)$$

which produces the series

$$1 + kx + \frac{k(k - 1)x^2}{2} + \cdots + \frac{k(k - 1) \cdots (k - n + 1)x^n}{n!} + \cdots.$$

Because $a_{n+1}/a_n \to 1$, you can apply the Ratio Test to conclude that the radius of convergence is $R = 1$. So, the series converges to some function in the interval $(-1, 1)$.

Note that Example 4 shows that the Taylor series for $(1 + x)^k$ converges to some function in the interval $(-1, 1)$. However, the example does not show that the series actually converges to $(1 + x)^k$. To do this, you could show that the remainder $R_n(x)$ converges to 0, as illustrated in Example 2.

EXAMPLE 5   Finding a Binomial Series

Find the power series for $f(x) = \sqrt[3]{1 + x}$.

Solution   Using the binomial series

$$(1 + x)^k = 1 + kx + \frac{k(k - 1)x^2}{2!} + \frac{k(k - 1)(k - 2)x^3}{3!} + \cdots$$

let $k = \frac{1}{3}$ and write

$$(1 + x)^{1/3} = 1 + \frac{x}{3} - \frac{2x^2}{3^2 2!} + \frac{2 \cdot 5x^3}{3^3 3!} - \frac{2 \cdot 5 \cdot 8x^4}{3^4 4!} + \cdots$$

which converges for $-1 \le x \le 1$.

TECHNOLOGY   Use a graphing utility to confirm the result in Example 5. When you graph the functions

$$f(x) = (1 + x)^{1/3} \quad \text{and} \quad P_4(x) = 1 + \frac{x}{3} - \frac{x^2}{9} + \frac{5x^3}{81} - \frac{10x^4}{243}$$

in the same viewing window, you should obtain the result shown in Figure 9.25.

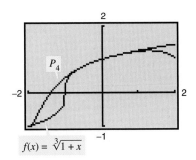

$f(x) = \sqrt[3]{1 + x}$

Figure 9.25

Deriving Taylor Series from a Basic List

The following list provides the power series for several elementary functions with the corresponding intervals of convergence.

Power Series for Elementary Functions

Function	Interval of Convergence
$\dfrac{1}{x} = 1 - (x - 1) + (x - 1)^2 - (x - 1)^3 + (x - 1)^4 - \cdots + (-1)^n (x - 1)^n + \cdots$	$0 < x < 2$
$\dfrac{1}{1 + x} = 1 - x + x^2 - x^3 + x^4 - x^5 + \cdots + (-1)^n x^n + \cdots$	$-1 < x < 1$
$\ln x = (x - 1) - \dfrac{(x - 1)^2}{2} + \dfrac{(x - 1)^3}{3} - \dfrac{(x - 1)^4}{4} + \cdots + \dfrac{(-1)^{n-1}(x - 1)^n}{n} + \cdots$	$0 < x \leq 2$
$e^x = 1 + x + \dfrac{x^2}{2!} + \dfrac{x^3}{3!} + \dfrac{x^4}{4!} + \dfrac{x^5}{5!} + \cdots + \dfrac{x^n}{n!} + \cdots$	$-\infty < x < \infty$
$\sin x = x - \dfrac{x^3}{3!} + \dfrac{x^5}{5!} - \dfrac{x^7}{7!} + \dfrac{x^9}{9!} - \cdots + \dfrac{(-1)^n x^{2n+1}}{(2n + 1)!} + \cdots$	$-\infty < x < \infty$
$\cos x = 1 - \dfrac{x^2}{2!} + \dfrac{x^4}{4!} - \dfrac{x^6}{6!} + \dfrac{x^8}{8!} - \cdots + \dfrac{(-1)^n x^{2n}}{(2n)!} + \cdots$	$-\infty < x < \infty$
$\arctan x = x - \dfrac{x^3}{3} + \dfrac{x^5}{5} - \dfrac{x^7}{7} + \dfrac{x^9}{9} - \cdots + \dfrac{(-1)^n x^{2n+1}}{2n + 1} + \cdots$	$-1 \leq x \leq 1$
$\arcsin x = x + \dfrac{x^3}{2 \cdot 3} + \dfrac{1 \cdot 3x^5}{2 \cdot 4 \cdot 5} + \dfrac{1 \cdot 3 \cdot 5x^7}{2 \cdot 4 \cdot 6 \cdot 7} + \cdots + \dfrac{(2n)!x^{2n+1}}{(2^n n!)^2(2n + 1)} + \cdots$	$-1 \leq x \leq 1$
$(1 + x)^k = 1 + kx + \dfrac{k(k - 1)x^2}{2!} + \dfrac{k(k - 1)(k - 2)x^3}{3!} + \dfrac{k(k - 1)(k - 2)(k - 3)x^4}{4!} + \cdots$	$-1 < x < 1*$

* The convergence at $x = \pm 1$ depends on the value of k.

NOTE The binomial series is valid for noninteger values of k. Moreover, if k happens to be a positive integer, the binomial series reduces to a simple binomial expansion.

EXAMPLE 6 Deriving a Power Series from a Basic List

Find the power series for $f(x) = \cos \sqrt{x}$.

Solution Using the power series

$$\cos x = 1 - \frac{x^2}{2!} + \frac{x^4}{4!} - \frac{x^6}{6!} + \frac{x^8}{8!} - \cdots$$

you can replace x by $\sqrt{x}$ to obtain the series

$$\cos \sqrt{x} = 1 - \frac{x}{2!} + \frac{x^2}{4!} - \frac{x^3}{6!} + \frac{x^4}{8!} - \cdots.$$

This series converges for all x in the domain of $\cos \sqrt{x}$—that is, for $x \geq 0$.

Power series can be multiplied and divided like polynomials. After finding the first few terms of the product (or quotient), you may be able to recognize a pattern.

EXAMPLE 7   Multiplication and Division of Power Series

Find the first three nonzero terms in each of the Maclaurin series.

a. $e^x \arctan x$ **b.** $\tan x$

Solution

a. Using the Maclaurin series for e^x and $\arctan x$ in the table, you have

$$e^x \arctan x = \left(1 + \frac{x}{1!} + \frac{x^2}{2!} + \frac{x^3}{3!} + \frac{x^4}{4!} + \cdots \right)\left(x - \frac{x^3}{3} + \frac{x^5}{5} - \cdots \right).$$

Multiply these expressions and collect like terms as you would for multiplying polynomials.

$$1 + x + \tfrac{1}{2}x^2 + \tfrac{1}{6}x^3 + \tfrac{1}{24}x^4 + \cdots$$

$$\underline{x \qquad\qquad - \tfrac{1}{3}x^3 \qquad\quad + \tfrac{1}{5}x^5 - \cdots}$$

$$x + \ x^2 + \tfrac{1}{2}x^3 + \ \tfrac{1}{6}x^4 + \tfrac{1}{24}x^5 + \cdots$$

$$- \tfrac{1}{3}x^3 - \tfrac{1}{3}x^4 - \tfrac{1}{6}x^5 - \cdots$$

$$\underline{\qquad\qquad\qquad\qquad\quad + \tfrac{1}{5}x^5 + \cdots}$$

$$x + \ x^2 + \tfrac{1}{6}x^3 - \tfrac{1}{6}x^4 + \tfrac{3}{40}x^5 + \cdots$$

So, $e^x \arctan x = x + x^2 + \tfrac{1}{6}x^3 + \cdots$.

b. Using the Maclaurin series for $\sin x$ and $\cos x$ in the table, you have

$$\tan x = \frac{\sin x}{\cos x} = \frac{x - \dfrac{x^3}{3!} + \dfrac{x^5}{5!} - \cdots}{1 - \dfrac{x^2}{2!} + \dfrac{x^4}{4!} - \cdots}.$$

Divide using long division.

$$
\begin{array}{r}
x + \dfrac{1}{3}x^3 + \dfrac{2}{15}x^5 + \cdots \\[4pt]
1 - \dfrac{1}{2}x^2 + \dfrac{1}{24}x^4 - \cdots \,\overline{\big)\, x - \dfrac{1}{6}x^3 + \dfrac{1}{120}x^5 - \cdots} \\[4pt]
\underline{x - \dfrac{1}{2}x^3 + \dfrac{1}{24}x^5 - \cdots} \\[4pt]
\dfrac{1}{3}x^3 - \dfrac{1}{30}x^5 + \cdots \\[4pt]
\underline{\dfrac{1}{3}x^3 - \dfrac{1}{6}x^5 + \cdots} \\[4pt]
\dfrac{2}{15}x^5 + \cdots
\end{array}
$$

So, $\tan x = x + \tfrac{1}{3}x^3 + \tfrac{2}{15}x^5 + \cdots$.

EXAMPLE 8 **A Power Series for sin² x**

Find the power series for $f(x) = \sin^2 x$.

Solution   Consider rewriting $\sin^2 x$ as follows.

$$\sin^2 x = \frac{1 - \cos 2x}{2} = \frac{1}{2} - \frac{\cos 2x}{2}$$

Now, use the series for $\cos x$.

$$\cos x = 1 - \frac{x^2}{2!} + \frac{x^4}{4!} - \frac{x^6}{6!} + \frac{x^8}{8!} - \cdots$$

$$\cos 2x = 1 - \frac{2^2}{2!}x^2 + \frac{2^4}{4!}x^4 - \frac{2^6}{6!}x^6 + \frac{2^8}{8!}x^8 - \cdots$$

$$-\frac{1}{2}\cos 2x = -\frac{1}{2} + \frac{2}{2!}x^2 - \frac{2^3}{4!}x^4 + \frac{2^5}{6!}x^6 - \frac{2^7}{8!}x^8 + \cdots$$

$$\sin^2 x = \frac{1}{2} - \frac{1}{2}\cos 2x = \frac{1}{2} - \frac{1}{2} + \frac{2}{2!}x^2 - \frac{2^3}{4!}x^4 + \frac{2^5}{6!}x^6 - \frac{2^7}{8!}x^8 + \cdots$$

$$= \frac{2}{2!}x^2 - \frac{2^3}{4!}x^4 + \frac{2^5}{6!}x^6 - \frac{2^7}{8!}x^8 + \cdots$$

This series converges for $-\infty < x < \infty$.

As mentioned in the preceding section, power series can be used to obtain tables of values of transcendental functions. They are also useful for estimating the values of definite integrals for which antiderivatives cannot be found. The next example demonstrates this use.

 EXAMPLE 9 **Power Series Approximation of a Definite Integral**

Use a power series to approximate

$$\int_0^1 e^{-x^2}\, dx$$

with an error of less than 0.01.

Solution   Replacing x with $-x^2$ in the series for e^x produces the following.

$$e^{-x^2} = 1 - x^2 + \frac{x^4}{2!} - \frac{x^6}{3!} + \frac{x^8}{4!} - \cdots$$

$$\int_0^1 e^{-x^2}\, dx = \left[x - \frac{x^3}{3} + \frac{x^5}{5 \cdot 2!} - \frac{x^7}{7 \cdot 3!} + \frac{x^9}{9 \cdot 4!} - \cdots \right]_0^1$$

$$= 1 - \frac{1}{3} + \frac{1}{10} - \frac{1}{42} + \frac{1}{216} - \cdots$$

Summing the first four terms, you have

$$\int_0^1 e^{-x^2}\, dx \approx 0.74$$

which, by the Alternating Series Test, has an error of less than $\frac{1}{216} \approx 0.005$.

In Exercises 1–10, use the definition to find the Taylor series (centered at c) for the function.

1. $f(x) = e^{2x}, \quad c = 0$

2. $f(x) = e^{3x}, \quad c = 0$

3. $f(x) = \cos x, \quad c = \dfrac{\pi}{4}$

4. $f(x) = \sin x, \quad c = \dfrac{\pi}{4}$

5. $f(x) = \ln x, \quad c = 1$

6. $f(x) = e^x, \quad c = 1$

7. $f(x) = \sin 2x, \quad c = 0$

8. $f(x) = \ln(x^2 + 1), \quad c = 0$

9. $f(x) = \sec x, \quad c = 0$ (first three nonzero terms)

10. $f(x) = \tan x, \quad c = 0$ (first three nonzero terms)

In Exercises 11–14, prove that the Maclaurin series for the function converges to the function for all x.

11. $f(x) = \cos x$

12. $f(x) = e^{-2x}$

13. $f(x) = \sinh x$

14. $f(x) = \cosh x$

In Exercises 15–20, use the binomial series to find the Maclaurin series for the function.

15. $f(x) = \dfrac{1}{(1 + x)^2}$

16. $f(x) = \dfrac{1}{\sqrt{1 - x}}$

17. $f(x) = \dfrac{1}{\sqrt{4 + x^2}}$

18. $f(x) = \sqrt[4]{1 + x}$

19. $f(x) = \sqrt{1 + x^2}$

20. $f(x) = \sqrt{1 + x^3}$

In Exercises 21–30, find the Maclaurin series for the function. (Use the table of power series for elementary functions.)

21. $f(x) = e^{x^2/2}$

22. $g(x) = e^{-3x}$

23. $g(x) = \sin 3x$

24. $f(x) = \cos 4x$

25. $f(x) = \cos x^{3/2}$

26. $g(x) = 2 \sin x^3$

27. $f(x) = \frac{1}{2}(e^x - e^{-x}) = \sinh x$

28. $f(x) = e^x + e^{-x} = 2 \cosh x$

29. $f(x) = \cos^2 x$

30. $f(x) = \sinh^{-1} x = \ln\!\left(x + \sqrt{x^2 + 1}\right)$

$\left(\textit{Hint: }\text{Integrate the series for } \dfrac{1}{\sqrt{x^2 + 1}}.\right)$

In Exercises 31–34, find the Maclaurin series for the function. (See Example 7.)

31. $f(x) = x \sin x$

32. $h(x) = x \cos x$

33. $g(x) = \begin{cases} \dfrac{\sin x}{x}, & x \ne 0 \\ 1, & x = 0 \end{cases}$

34. $f(x) = \begin{cases} \dfrac{\arcsin x}{x}, & x \ne 0 \\ 1, & x = 0 \end{cases}$

In Exercises 35 and 36, use a power series and the fact that $i^2 = -1$ to verify the formula.

35. $g(x) = \dfrac{1}{2i}(e^{ix} - e^{-ix}) = \sin x$

36. $g(x) = \frac{1}{2}(e^{ix} + e^{-ix}) = \cos x$

 In Exercises 37–42, find the first four nonzero terms of the Maclaurin series for the function by multiplying or dividing the appropriate power series. Use the table of power series for elementary functions on page 682. Use a graphing utility to graph the function and its corresponding polynomial approximation.

37. $f(x) = e^x \sin x$

38. $g(x) = e^x \cos x$

39. $h(x) = \cos x \ln(1 + x)$

40. $f(x) = e^x \ln(1 + x)$

41. $g(x) = \dfrac{\sin x}{1 + x}$

42. $f(x) = \dfrac{e^x}{1 + x}$

In Exercises 43–46, match the polynomial with its graph. [The graphs are labeled (a), (b), (c), and (d).] Factor a common factor from each polynomial and identify the function approximated by the remaining Taylor polynomial.

(a)

(b)

(c)

(d)

43. $y = x^2 - \dfrac{x^4}{3!}$

44. $y = x - \dfrac{x^3}{2!} + \dfrac{x^5}{4!}$

45. $y = x + x^2 + \dfrac{x^3}{2!}$

46. $y = x^2 - x^3 + x^4$

In Exercises 47 and 48, find a Maclaurin series for $f(x)$.

47. $f(x) = \int_0^x (e^{-t^2} - 1)\, dt$

48. $f(x) = \int_0^x \sqrt{1 + t^3}\, dt$

 In Exercises 49–52, verify the sum. Then use a graphing utility to approximate the sum with an error of less than 0.0001.

49. $\displaystyle\sum_{n=1}^{\infty} (-1)^{n+1} \frac{1}{n} = \ln 2$

50. $\displaystyle\sum_{n=0}^{\infty} (-1)^n \left[\frac{1}{(2n+1)!}\right] = \sin 1$

51. $\displaystyle\sum_{n=0}^{\infty} \frac{2^n}{n!} = e^2$

52. $\displaystyle\sum_{n=1}^{\infty} (-1)^{n-1} \left(\frac{1}{n!}\right) = \frac{e-1}{e}$

In Exercises 53 and 54, use the series representation of the function f to find $\lim_{x \to 0} f(x)$ (if it exists).

53. $f(x) = \dfrac{1 - \cos x}{x}$

54. $f(x) = \dfrac{\sin x}{x}$

In Exercises 55–58, use a power series to approximate the value of the integral with an error of less than 0.0001. (In Exercises 55 and 56, assume that the integrand is defined as 1 when $x = 0$.)

55. $\displaystyle\int_0^1 \frac{\sin x}{x}\, dx$

56. $\displaystyle\int_0^{1/2} \frac{\arctan x}{x}\, dx$

57. $\displaystyle\int_{0.1}^{0.3} \sqrt{1 + x^3}\, dx$

58. $\displaystyle\int_0^{1/4} x \ln(x+1)\, dx$

Area   **In Exercises 59 and 60, use a power series to approximate the area of the region. Use a graphing utility to verify the result.**

59. $\displaystyle\int_0^{\pi/2} \sqrt{x}\cos x\, dx$ **60.** $\displaystyle\int_{0.5}^{1} \cos\sqrt{x}\, dx$

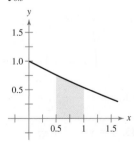

Probability   **In Exercises 61 and 62, approximate the normal probability with an error of less than 0.0001, where the probability is given by**

$$P(a < x < b) = \frac{1}{\sqrt{2\pi}} \int_a^b e^{-x^2/2}\, dx.$$

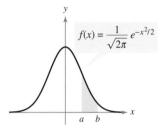

61. $P(0 < x < 1)$

62. $P(1 < x < 2)$

 In Exercises 63–66, use a computer algebra system to find the fifth-degree Taylor polynomial (centered at c) for the function. Graph the function and the polynomial. Use the graph to determine the largest interval on which the polynomial is a reasonable approximation of the function.

63. $f(x) = x \cos 2x, \quad c = 0$

64. $f(x) = \sin \dfrac{x}{2} \ln(1 + x), \quad c = 0$

65. $g(x) = \sqrt{x}\, \ln x, \quad c = 1$

66. $h(x) = \sqrt[3]{x}\, \arctan x, \quad c = 1$

Writing About Concepts

67. State the guidelines for finding a Taylor series.

68. If f is an even function, what must be true about the coefficients a_n in the Maclaurin series

$$f(x) = \sum_{n=0}^{\infty} a_n x^n?$$

Explain your reasoning.

69. Explain how to use the series

$$g(x) = e^x = \sum_{n=0}^{\infty} \frac{x^n}{n!}$$

to find the series for each function. Do not find the series.

(a) $f(x) = e^{-x}$

(b) $f(x) = e^{3x}$

(c) $f(x) = xe^x$

(d) $f(x) = e^{2x} + e^{-2x}$

70. Define the binomial series. What is its radius of convergence?

71. Projectile Motion   A projectile fired from the ground follows the trajectory given by

$$y = \left(\tan\theta - \frac{g}{kv_0\cos\theta}\right)x - \frac{g}{k^2}\ln\left(1 - \frac{kx}{v_0\cos\theta}\right)$$

where v_0 is the initial speed, θ is the angle of projection, g is the acceleration due to gravity, and k is the drag factor caused by air resistance. Using the power series representation

$$\ln(1 + x) = x - \frac{x^2}{2} + \frac{x^3}{3} - \frac{x^4}{4} + \cdots, \quad -1 < x < 1$$

verify that the trajectory can be rewritten as

$$y = (\tan\theta)x + \frac{gx^2}{2v_0^2\cos^2\theta} + \frac{kgx^3}{3v_0^3\cos^3\theta} + \frac{k^2gx^4}{4v_0^4\cos^4\theta} + \cdots.$$

72. Projectile Motion   Use the result of Exercise 71 to determine the series for the path of a projectile launched from ground level at an angle of $\theta = 60°$, with an initial speed of $v_0 = 64$ feet per second and a drag factor of $k = \frac{1}{16}$.

73. Investigation   Consider the function f defined by

$$f(x) = \begin{cases} e^{-1/x^2}, & x \neq 0 \\ 0, & x = 0. \end{cases}$$

(a) Sketch a graph of the function.

(b) Use the alternative form of the definition of the derivative (Section 3.1) and L'Hôpital's Rule to show that $f'(0) = 0$. [By continuing this process, it can be shown that $f^{(n)}(0) = 0$ for $n > 1$.]

(c) Using the result in part (b), find the Maclaurin series for f. Does the series converge to f?

74. Investigation

(a) Find the power series centered at 0 for the function

$$f(x) = \frac{\ln(x^2 + 1)}{x^2}.$$

(b) Use a graphing utility to graph f and the eighth-degree Taylor polynomial $P_8(x)$ for f.

(c) Complete the table, where

$$F(x) = \int_0^x \frac{\ln(t^2 + 1)}{t^2}\,dt \quad \text{and} \quad G(x) = \int_0^x P_8(t)\,dt.$$

x	0.25	0.50	0.75	1.00	1.50	2.00
$F(x)$						
$G(x)$						

(d) Describe the relationship between the graphs of f and P_8 and the results given in the table in part (c).

75. Prove that $\displaystyle\lim_{n\to\infty}\frac{x^n}{n!} = 0$ for any real x.

76. Find the Maclaurin series for

$$f(x) = \ln\frac{1 + x}{1 - x}$$

and determine its radius of convergence. Use the first four terms of the series to approximate $\ln 3$.

In Exercises 77–80, evaluate the binomial coefficient using the formula

$$\binom{k}{n} = \frac{k(k-1)(k-2)(k-3)\cdots(k-n+1)}{n!}$$

where k is a real number, n is a positive integer, and

$$\binom{k}{0} = 1.$$

77. $\dbinom{5}{3}$ **78.** $\dbinom{-2}{2}$

79. $\dbinom{0.5}{4}$ **80.** $\dbinom{-1/3}{5}$

81. Write the power series for $(1 + x)^k$ in terms of binomial coefficients.

82. Prove that e is irrational. $\Big[$Hint: Assume that $e = p/q$ is rational (p and q are integers) and consider

$$e = 1 + 1 + \frac{1}{2!} + \cdots + \frac{1}{n!} + \cdots.\Big]$$

83. Show that the Maclaurin series of the function

$$g(x) = \frac{x}{1 - x - x^2}$$

is

$$\sum_{n=1}^{\infty} F_n x^n$$

where F_n is the nth Fibonacci number with $F_1 = F_2 = 1$ and $F_n = F_{n-2} + F_{n-1}$, for $n \geq 3$.

(*Hint:* Write

$$\frac{x}{1 - x - x^2} = a_0 + a_1 x + a_2 x^2 + \cdots$$

and multiply each side of this equation by $1 - x - x^2$.)

Putnam Exam Challenge

84. Assume that $|f(x)| \leq 1$ and $|f''(x)| \leq 1$ for all x on an interval of length at least 2. Show that $|f'(x)| \leq 2$ on the interval.

This problem was composed by the Committee on the Putnam Prize Competition.
© The Mathematical Association of America. All rights reserved.

Review Exercises for Chapter 9

See www.CalcChat.com for worked-out solutions to odd-numbered exercises.

In Exercises 1 and 2, write an expression for the *n*th term of the sequence.

1. $1, \dfrac{1}{2}, \dfrac{1}{6}, \dfrac{1}{24}, \dfrac{1}{120}, \cdots$

2. $\dfrac{1}{2}, \dfrac{2}{5}, \dfrac{3}{10}, \dfrac{4}{17}, \cdots$

In Exercises 3–6, match the sequence with its graph. [The graphs are labeled (a), (b), (c), and (d).]

(a)

(b)

(c)

(d)
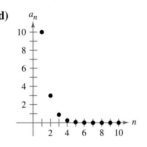

3. $a_n = 4 + \dfrac{2}{n}$

4. $a_n = 4 - \dfrac{1}{2}n$

5. $a_n = 10(0.3)^{n-1}$

6. $a_n = 6\left(-\dfrac{2}{3}\right)^{n-1}$

In Exercises 7 and 8, use a graphing utility to graph the first 10 terms of the sequence. Use the graph to make an inference about the convergence or divergence of the sequence. Verify your inference analytically and, if the sequence converges, find its limit.

7. $a_n = \dfrac{5n + 2}{n}$

8. $a_n = \sin \dfrac{n\pi}{2}$

In Exercises 9–16, determine the convergence or divergence of the sequence with the given *n*th term. If the sequence converges, find its limit. (*b* and *c* are positive real numbers.)

9. $a_n = \dfrac{n + 1}{n^2}$

10. $a_n = \dfrac{1}{\sqrt{n}}$

11. $a_n = \dfrac{n^3}{n^2 + 1}$

12. $a_n = \dfrac{n}{\ln n}$

13. $a_n = \sqrt{n + 1} - \sqrt{n}$

14. $a_n = \left(1 + \dfrac{1}{2n}\right)^n$

15. $a_n = \dfrac{\sin \sqrt{n}}{\sqrt{n}}$

16. $a_n = (b^n + c^n)^{1/n}$

17. *Compound Interest* A deposit of $5000 is made in an account that earns 5% interest compounded quarterly. The balance in the account after *n* quarters is

$$A_n = 5000\left(1 + \dfrac{0.05}{4}\right)^n, \quad n = 1, 2, 3, \cdots.$$

(a) Compute the first eight terms of the sequence $\{A_n\}$.

(b) Find the balance in the account after 10 years by computing the 40th term of the sequence.

18. *Depreciation* A company buys a machine for $120,000. During the next 5 years the machine will depreciate at a rate of 30% per year. (That is, at the end of each year, the depreciated value will be 70% of what it was at the beginning of the year.)

(a) Find a formula for the *n*th term of the sequence that gives the value *V* of the machine *t* full years after it was purchased.

(b) Find the depreciated value of the machine at the end of 5 full years.

Numerical, Graphical, and Analytic Analysis **In Exercises 19–22, (a) use a graphing utility to find the indicated partial sum S_k and complete the table, and (b) use a graphing utility to graph the first 10 terms of the sequence of partial sums.**

k	5	10	15	20	25
S_k					

19. $\displaystyle\sum_{n=1}^{\infty} \left(\dfrac{3}{2}\right)^{n-1}$

20. $\displaystyle\sum_{n=1}^{\infty} \dfrac{(-1)^{n+1}}{2n}$

21. $\displaystyle\sum_{n=1}^{\infty} \dfrac{(-1)^{n+1}}{(2n)!}$

22. $\displaystyle\sum_{n=1}^{\infty} \dfrac{1}{n(n + 1)}$

In Exercises 23–26, determine the convergence or divergence of the series.

23. $\displaystyle\sum_{n=0}^{\infty} (0.82)^n$

24. $\displaystyle\sum_{n=0}^{\infty} (1.82)^n$

25. $\displaystyle\sum_{n=1}^{\infty} \dfrac{(-1)^n n}{\ln n}$

26. $\displaystyle\sum_{n=0}^{\infty} \dfrac{2n + 1}{3n + 2}$

In Exercises 27–30, find the sum of the convergent series.

27. $\displaystyle\sum_{n=0}^{\infty} \left(\dfrac{2}{3}\right)^n$

28. $\displaystyle\sum_{n=0}^{\infty} \dfrac{2^{n+2}}{3^n}$

29. $\displaystyle\sum_{n=0}^{\infty} \left(\dfrac{1}{2^n} - \dfrac{1}{3^n}\right)$

30. $\displaystyle\sum_{n=0}^{\infty} \left[\left(\dfrac{2}{3}\right)^n - \dfrac{1}{(n + 1)(n + 2)}\right]$

In Exercises 31 and 32, (a) write the repeating decimal as a geometric series and (b) write its sum as the ratio of two integers.

31. $0.\overline{09}$ **32.** $0.\overline{923076}$

33. *Distance*   A ball is dropped from a height of 8 meters. Each time it drops h meters, it rebounds $0.7h$ meters. Find the total distance traveled by the ball.

34. *Salary*   You accept a job that pays a salary of $32,000 the first year. During the next 39 years, you will receive a 5.5% raise each year. What would be your total compensation over the 40-year period?

35. *Compound Interest*   A deposit of $200 is made at the end of each month for 2 years in an account that pays 6% interest, compounded continuously. Determine the balance in the account at the end of 2 years.

36. *Compound Interest*   A deposit of $100 is made at the end of each month for 10 years in an account that pays 3.5%, compounded monthly. Determine the balance in the account at the end of 10 years.

In Exercises 37–40, determine the convergence or divergence of the series.

37. $\displaystyle\sum_{n=1}^{\infty} \frac{\ln n}{n^4}$ **38.** $\displaystyle\sum_{n=1}^{\infty} \frac{1}{\sqrt[4]{n^3}}$

39. $\displaystyle\sum_{n=1}^{\infty} \left(\frac{1}{n^2} - \frac{1}{n}\right)$ **40.** $\displaystyle\sum_{n=1}^{\infty} \left(\frac{1}{n^2} - \frac{1}{2^n}\right)$

In Exercises 41–44, determine the convergence or divergence of the series.

41. $\displaystyle\sum_{n=1}^{\infty} \frac{1}{\sqrt{n^3 + 2n}}$ **42.** $\displaystyle\sum_{n=1}^{\infty} \frac{n+1}{n(n+2)}$

43. $\displaystyle\sum_{n=1}^{\infty} \frac{1 \cdot 3 \cdot 5 \cdots (2n-1)}{2 \cdot 4 \cdot 6 \cdots (2n)}$ **44.** $\displaystyle\sum_{n=1}^{\infty} \frac{1}{3^n - 5}$

In Exercises 45–48, determine the convergence or divergence of the series.

45. $\displaystyle\sum_{n=2}^{\infty} \frac{(-1)^n n}{n^2 - 3}$ **46.** $\displaystyle\sum_{n=1}^{\infty} \frac{(-1)^n \sqrt{n}}{n+1}$

47. $\displaystyle\sum_{n=4}^{\infty} \frac{(-1)^n n}{n-3}$ **48.** $\displaystyle\sum_{n=2}^{\infty} \frac{(-1)^n \ln n^3}{n}$

In Exercises 49–52, determine the convergence or divergence of the series.

49. $\displaystyle\sum_{n=1}^{\infty} \frac{n}{e^{n^2}}$

50. $\displaystyle\sum_{n=1}^{\infty} \frac{n!}{e^n}$

51. $\displaystyle\sum_{n=1}^{\infty} \frac{2^n}{n^3}$

52. $\displaystyle\sum_{n=1}^{\infty} \frac{1 \cdot 3 \cdot 5 \cdots (2n-1)}{2 \cdot 5 \cdot 8 \cdots (3n-1)}$

Numerical, Graphical, and Analytic Analysis   In Exercises 53 and 54, (a) verify that the series converges, (b) use a graphing utility to find the indicated partial sum S_n and complete the table, (c) use a graphing utility to graph the first 10 terms of the sequence of partial sums, and (d) use the table to estimate the sum of the series.

n	5	10	15	20	25
S_n					

53. $\displaystyle\sum_{n=1}^{\infty} n\left(\frac{3}{5}\right)^n$ **54.** $\displaystyle\sum_{n=1}^{\infty} \frac{(-1)^{n-1} n}{n^3 + 5}$

55. *Writing*   Use a graphing utility to complete the table for (a) $p = 2$ and (b) $p = 5$. Write a short paragraph describing and comparing the entries in the table.

N	5	10	20	30	40
$\displaystyle\sum_{n=1}^{N} \frac{1}{n^p}$					
$\displaystyle\int_{N}^{\infty} \frac{1}{x^p}\,dx$					

56. *Writing*   You are told that the terms of a positive series appear to approach zero very slowly as n approaches infinity. (In fact, $a_{75} = 0.7$.) If you are given no other information, can you conclude that the series diverges? Support your answer with an example.

In Exercises 57 and 58, find the third-degree Taylor polynomial centered at c.

57. $f(x) = e^{-x/2}$, $c = 0$

58. $f(x) = \tan x$, $c = -\dfrac{\pi}{4}$

In Exercises 59–62, use a Taylor polynomial to approximate the function with an error of less than 0.001.

59. $\sin 95°$ **60.** $\cos(0.75)$

61. $\ln(1.75)$ **62.** $e^{-0.25}$

63. A Taylor polynomial centered at 0 will be used to approximate the cosine function. Find the degree of the polynomial required to obtain the desired accuracy over each interval.

Maximum Error	Interval
(a) 0.001	$[-0.5, 0.5]$
(b) 0.001	$[-1, 1]$
(c) 0.0001	$[-0.5, 0.5]$
(d) 0.0001	$[-2, 2]$

64. Use a graphing utility to graph the cosine function and the Taylor polynomials in Exercise 63.

In Exercises 65–70, find the interval of convergence of the power series. (Be sure to include a check for convergence at the endpoints of the interval.)

65. $\displaystyle\sum_{n=0}^{\infty} \left(\frac{x}{10}\right)^n$

66. $\displaystyle\sum_{n=0}^{\infty} (2x)^n$

67. $\displaystyle\sum_{n=0}^{\infty} \frac{(-1)^n (x-2)^n}{(n+1)^2}$

68. $\displaystyle\sum_{n=1}^{\infty} \frac{3^n (x-2)^n}{n}$

69. $\displaystyle\sum_{n=0}^{\infty} n!(x-2)^n$

70. $\displaystyle\sum_{n=0}^{\infty} \frac{(x-2)^n}{2^n}$

In Exercises 71 and 72, show that the function represented by the power series is a solution of the differential equation.

71. $y = \displaystyle\sum_{n=0}^{\infty} (-1)^n \frac{x^{2n}}{4^n (n!)^2}$

$x^2 y'' + xy' + x^2 y = 0$

72. $y = \displaystyle\sum_{n=0}^{\infty} \frac{(-3)^n x^{2n}}{2^n n!}$

$y'' + 3xy' + 3y = 0$

In Exercises 73 and 74, find a geometric power series centered at 0 for the function.

73. $g(x) = \dfrac{2}{3-x}$

74. $h(x) = \dfrac{3}{2+x}$

75. Find a power series for the derivative of the function in Exercise 73.

76. Find a power series for the integral of the function in Exercise 74.

In Exercises 77 and 78, find a function represented by the series and give the domain of the function.

77. $1 + \dfrac{2}{3}x + \dfrac{4}{9}x^2 + \dfrac{8}{27}x^3 + \cdots$

78. $8 - 2(x-3) + \dfrac{1}{2}(x-3)^2 - \dfrac{1}{8}(x-3)^3 + \cdots$

In Exercises 79–86, find a power series for the function centered at c.

79. $f(x) = \sin x, \quad c = \dfrac{3\pi}{4}$

80. $f(x) = \cos x, \quad c = -\dfrac{\pi}{4}$

81. $f(x) = 3^x, \quad c = 0$

82. $f(x) = \csc x, \quad c = \dfrac{\pi}{2}$

(first three terms)

83. $f(x) = \dfrac{1}{x}, \quad c = -1$

84. $f(x) = \sqrt{x}, \quad c = 4$

85. $g(x) = \sqrt[5]{1+x}, \quad c = 0$

86. $h(x) = \dfrac{1}{(1+x)^3}, \quad c = 0$

In Exercises 87–92, find the sum of the convergent series by using a well-known function. Identify the function and explain how you obtained the sum.

87. $\displaystyle\sum_{n=1}^{\infty} (-1)^{n+1} \frac{1}{4^n n}$

88. $\displaystyle\sum_{n=1}^{\infty} (-1)^{n+1} \frac{1}{5^n n}$

89. $\displaystyle\sum_{n=0}^{\infty} \frac{1}{2^n n!}$

90. $\displaystyle\sum_{n=0}^{\infty} \frac{2^n}{3^n n!}$

91. $\displaystyle\sum_{n=0}^{\infty} (-1)^n \frac{2^{2n}}{3^{2n} (2n)!}$

92. $\displaystyle\sum_{n=0}^{\infty} (-1)^n \frac{1}{3^{2n+1} (2n+1)!}$

93. *Writing*   One of the series in Exercises 41 and 49 converges to its sum at a much lower rate than the other series. Which is it? Explain why this series converges so slowly. Use a graphing utility to illustrate the rate of convergence.

94. Use the binomial series to find the Maclaurin series for

$$f(x) = \frac{1}{\sqrt{1+x^3}}.$$

95. *Forming Maclaurin Series*   Determine the first four terms of the Maclaurin series for e^{2x}

(a) by using the definition of the Maclaurin series and the formula for the coefficient of the nth term, $a_n = f^{(n)}(0)/n!$.

(b) by replacing x by $2x$ in the series for e^x.

(c) by multiplying the series for e^x by itself, because $e^{2x} = e^x \cdot e^x$.

96. *Forming Maclaurin Series*   Follow the pattern of Exercise 95 to find the first four terms of the series for $\sin 2x$. (*Hint:* $\sin 2x = 2 \sin x \cos x$.)

In Exercises 97–100, find the series representation of the function defined by the integral.

97. $\displaystyle\int_0^x \frac{\sin t}{t} \, dt$

98. $\displaystyle\int_0^x \cos \frac{\sqrt{t}}{2} \, dt$

99. $\displaystyle\int_0^x \frac{\ln(t+1)}{t} \, dt$

100. $\displaystyle\int_0^x \frac{e^t - 1}{t} \, dt$

In Exercises 101 and 102, use a power series to find the limit (if it exists). Verify the result by using L'Hôpital's Rule.

101. $\displaystyle\lim_{x \to 0^+} \frac{\arctan x}{\sqrt{x}}$

102. $\displaystyle\lim_{x \to 0} \frac{\arcsin x}{x}$

1. The Cantor set (Georg Cantor, 1845–1918) is a subset of the unit interval $[0, 1]$. To construct the Cantor set, first remove the middle third $\left(\frac{1}{3}, \frac{2}{3}\right)$ of the interval, leaving two line segments. For the second step, remove the middle third of each of the two remaining segments, leaving four line segments. Continue this procedure indefinitely, as shown in the figure. The Cantor set consists of all numbers in the unit interval $[0, 1]$ that still remain.

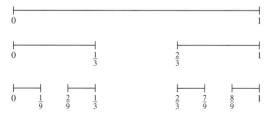

(a) Find the total length of all the line segments that are removed.

(b) Write down three numbers that are in the Cantor set.

(c) Let C_n denote the total length of the remaining line segments after n steps. Find $\lim_{n \to \infty} C_n$.

GEORG CANTOR (1845–1918)

Cantor was a German mathematician known for his work on the development of set theory, which is the basis of modern mathematical analysis. This theory extends to the concept of infinite (or transfinite) numbers.

The Granger Collection

2. It can be shown that

$$\sum_{n=1}^{\infty} \frac{1}{n^2} = \frac{\pi^2}{6} \text{ [see Example 3(b), Section 9.3].}$$

Use this fact to show that $\displaystyle\sum_{n=1}^{\infty} \frac{1}{(2n-1)^2} = \frac{\pi^2}{8}$.

3. Let T be an equilateral triangle with sides of length 1. Let a_n be the number of circles that can be packed tightly in n rows inside the triangle. For example, $a_1 = 1$, $a_2 = 3$, and $a_3 = 6$, as shown in the figure. Let A_n be the combined area of the a_n circles. Find $\lim_{n \to \infty} A_n$.

4. Identical blocks of unit length are stacked on top of each other at the edge of a table. The center of gravity of the top block must lie over the block below it, the center of gravity of the top two blocks must lie over the block below them, and so on (see figure).

(a) If there are three blocks, show that it is possible to stack them so that the left edge of the top block extends $\frac{11}{12}$ unit beyond the edge of the table.

(b) Is it possible to stack the blocks so that the right edge of the top block extends beyond the edge of the table?

(c) How far beyond the table can the blocks be stacked?

5. (a) Consider the power series

$$\sum_{n=0}^{\infty} a_n x^n = 1 + 2x + 3x^2 + x^3 + 2x^4 + 3x^5 + x^6 + \cdots$$

in which the coefficients $a_n = 1, 2, 3, 1, 2, 3, 1, \ldots$ are periodic of period $p = 3$. Find the radius of convergence and the sum of this power series.

(b) Consider a power series

$$\sum_{n=0}^{\infty} a_n x^n$$

in which the coefficients are periodic, $(a_{n+p} = a_p)$ and $a_n > 0$. Find the radius of convergence and the sum of this power series.

6. For what values of the positive constants a and b does the following series converge absolutely? For what values does it converge conditionally?

$$a - \frac{b}{2} + \frac{a}{3} - \frac{b}{4} + \frac{a}{5} - \frac{b}{6} + \frac{a}{7} - \frac{b}{8} + \cdots$$

7. (a) Find a power series for the function

$$f(x) = xe^x$$

centered at 0. Use this representation to find the sum of the infinite series

$$\sum_{n=1}^{\infty} \frac{1}{n!(n+2)}.$$

(b) Differentiate the power series for $f(x) = xe^x$. Use the result to find the sum of the infinite series

$$\sum_{n=0}^{\infty} \frac{n+1}{n!}.$$

8. Find $f^{(12)}(0)$ if $f(x) = e^{x^2}$. (*Hint:* Do not calculate 12 derivatives.)

9. The graph of the function

$$f(x) = \begin{cases} 1, & x = 0 \\ \dfrac{\sin x}{x}, & x > 0 \end{cases}$$

is shown below. Use the Alternating Series Test to show that the improper integral $\int_1^\infty f(x)\,dx$ converges.

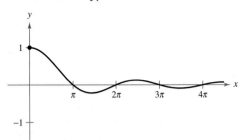

10. (a) Prove that $\displaystyle\int_2^\infty \frac{1}{x(\ln x)^p}\,dx$ converges if and only if $p > 1$.

   (b) Determine the convergence or divergence of the series

$$\sum_{n=4}^\infty \frac{1}{n \ln(n^2)}.$$

11. (a) Consider the following sequence of numbers defined recursively.

$$a_1 = 3$$
$$a_2 = \sqrt{3}$$
$$a_3 = \sqrt{3 + \sqrt{3}}$$
$$\vdots$$
$$a_{n+1} = \sqrt{3 + a_n}$$

Write the decimal approximations for the first six terms of this sequence. Prove that the sequence converges and find its limit.

   (b) Consider the following sequence defined recursively by $a_1 = \sqrt{a}$ and $a_{n+1} = \sqrt{a + a_n}$, where $a > 2$.

$$\sqrt{a},\ \sqrt{a + \sqrt{a}},\ \sqrt{a + \sqrt{a + \sqrt{a}}},\ \ldots$$

Prove that this sequence converges and find its limit.

12. Let $\{a_n\}$ be a sequence of positive numbers satisfying $\displaystyle\lim_{n\to\infty} (a_n)^{1/n} = L < \frac{1}{r}$, $r > 0$. Prove that the series $\displaystyle\sum_{n=1}^\infty a_n r^n$ converges.

13. Consider the infinite series $\displaystyle\sum_{n=1}^\infty \frac{1}{2^{n+(-1)^n}}$.

   (a) Find the first five terms of the sequence of partial sums.

   (b) Show that the Ratio Test is inconclusive for this series.

   (c) Use the Root Test to test for the convergence or divergence of this series.

14. Derive each identity using the appropriate geometric series.

   (a) $\dfrac{1}{0.99} = 1.01010101\ldots$ (b) $\dfrac{1}{0.98} = 1.0204081632\ldots$

15. Consider an idealized population with the characteristic that each member of the population produces one offspring at the end of every time period. Each member has a life span of three time periods and the population begins with 10 newborn members. The following table shows the population during the first five time periods.

		Time Period			
Age Bracket	**1**	**2**	**3**	**4**	**5**
0–1	10	10	20	40	70
1–2		10	10	20	40
2–3			10	10	20
Total	10	20	40	70	130

The sequence for the total population has the property that

$$S_n = S_{n-1} + S_{n-2} + S_{n-3}, \qquad n > 3.$$

Find the total population during each of the next five time periods.

16. Imagine you are stacking an infinite number of spheres of decreasing radii on top of each other, as shown in the figure. The radii of the spheres are 1 meter, $1/\sqrt{2}$ meter, $1/\sqrt{3}$ meter, etc. The spheres are made of a material that weighs 1 newton per cubic meter.

   (a) How high is this infinite stack of spheres?

   (b) What is the total surface area of all the spheres in the stack?

   (c) Show that the weight of the stack is finite.

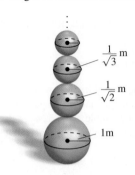

17. (a) Determine the convergence or divergence of the series

$$\sum_{n=1}^\infty \frac{1}{2n}.$$

   (b) Determine the convergence or divergence of the series

$$\sum_{n=1}^\infty \left(\sin \frac{1}{2n} - \sin \frac{1}{2n+1} \right).$$

10 Conics, Parametric Equations, and Polar Coordinates

During the 2002 Winter Olympic Games, the Olympic rings were lighted high on a mountainside in Salt Lake City. The volunteers who installed the lights for the display took care to minimize the environmental impact of the project. How can you calculate the area enclosed by the display? Explain.

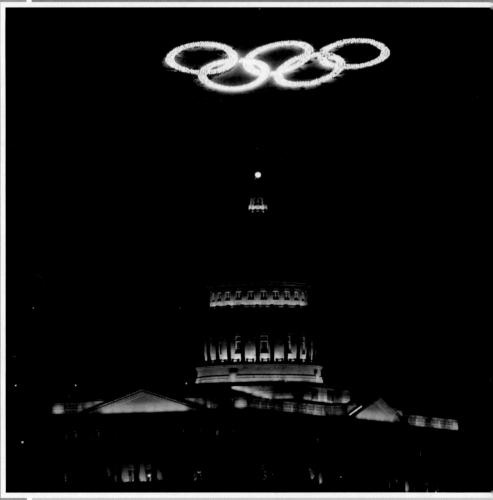

Harry Howe/Getty Images

In the polar coordinate system, graphing an equation involves tracing a curve about a fixed point called the pole. Consider a region bounded by a curve and by the rays that contain the endpoints of an interval on the curve. You can use sectors of circles to approximate the area of such a region. In Chapter 10, you will see how the limit process can be used to find this area.

693

Conics and Calculus

- Understand the definition of a conic section.
- Analyze and write equations of parabolas using properties of parabolas.
- Analyze and write equations of ellipses using properties of ellipses.
- Analyze and write equations of hyperbolas using properties of hyperbolas.

Conic Sections

Each **conic section** (or simply **conic**) can be described as the intersection of a plane and a double-napped cone. Notice in Figure 10.1 that for the four basic conics, the intersecting plane does not pass through the vertex of the cone. When the plane passes through the vertex, the resulting figure is a **degenerate conic,** as shown in Figure 10.2.

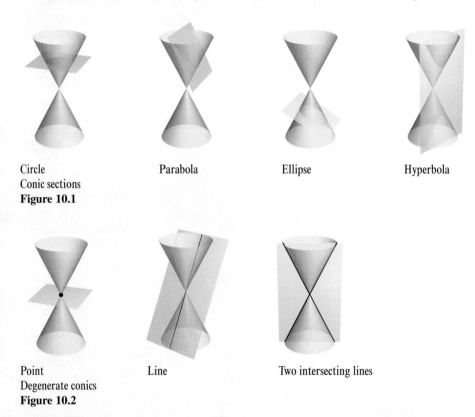

Circle
Conic sections
Figure 10.1

Parabola

Ellipse

Hyperbola

Point
Degenerate conics
Figure 10.2

Line

Two intersecting lines

HYPATIA (370–415 A.D.)

The Greeks discovered conic sections sometime between 600 and 300 B.C. By the beginning of the Alexandrian period, enough was known about conics for Apollonius (262–190 B.C.) to produce an eight-volume work on the subject. Later, toward the end of the Alexandrian period, Hypatia wrote a textbook entitled *On the Conics of Apollonius*. Her death marked the end of major mathematical discoveries in Europe for several hundred years.

The early Greeks were largely concerned with the geometric properties of conics. It was not until 1900 years later, in the early seventeenth century, that the broader applicability of conics became apparent. Conics then played a prominent role in the development of calculus.

There are several ways to study conics. You could begin as the Greeks did by defining the conics in terms of the intersections of planes and cones, or you could define them algebraically in terms of the general second-degree equation

$$Ax^2 + Bxy + Cy^2 + Dx + Ey + F = 0.$$ General second-degree equation

FOR FURTHER INFORMATION To learn more about the mathematical activities of Hypatia, see the article "Hypatia and Her Mathematics" by Michael A. B. Deakin in *The American Mathematical Monthly*. To view this article, go to the website *www.matharticles.com*.

However, a third approach, in which each of the conics is defined as a **locus** (collection) of points satisfying a certain geometric property, works best. For example, a circle can be defined as the collection of all points (x, y) that are equidistant from a fixed point (h, k). This locus definition easily produces the standard equation of a circle

$$(x - h)^2 + (y - k)^2 = r^2.$$ Standard equation of a circle

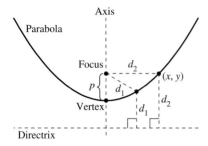

Parabola / Axis / Focus / d_2 / p / d_1 / (x, y) / Vertex / d_1 / d_2 / Directrix

Figure 10.3

Parabolas

A **parabola** is the set of all points (x, y) that are equidistant from a fixed line called the **directrix** and a fixed point called the **focus** not on the line. The midpoint between the focus and the directrix is the **vertex,** and the line passing through the focus and the vertex is the **axis** of the parabola. Note in Figure 10.3 that a parabola is symmetric with respect to its axis.

THEOREM 10.1   Standard Equation of a Parabola

The **standard form** of the equation of a parabola with vertex (h, k) and directrix $y = k - p$ is

$$(x - h)^2 = 4p(y - k). \qquad \text{Vertical axis}$$

For directrix $x = h - p$, the equation is

$$(y - k)^2 = 4p(x - h). \qquad \text{Horizontal axis}$$

The focus lies on the axis p units (*directed distance*) from the vertex. The coordinates of the focus are as follows.

$$(h, k + p) \qquad \text{Vertical axis}$$
$$(h + p, k) \qquad \text{Horizontal axis}$$

EXAMPLE 1   Finding the Focus of a Parabola

Find the focus of the parabola given by $y = -\frac{1}{2}x^2 - x + \frac{1}{2}$.

Solution   To find the focus, convert to standard form by completing the square.

$y = \frac{1}{2} - x - \frac{1}{2}x^2$	Rewrite original equation.
$y = \frac{1}{2}(1 - 2x - x^2)$	Factor out $\frac{1}{2}$.
$2y = 1 - 2x - x^2$	Multiply each side by 2.
$2y = 1 - (x^2 + 2x)$	Group terms.
$2y = 2 - (x^2 + 2x + 1)$	Add and subtract 1 on right side.
$x^2 + 2x + 1 = -2y + 2$	
$(x + 1)^2 = -2(y - 1)$	Write in standard form.

Comparing this equation with $(x - h)^2 = 4p(y - k)$, you can conclude that

$$h = -1, \qquad k = 1, \qquad \text{and} \qquad p = -\frac{1}{2}.$$

Because p is negative, the parabola opens downward, as shown in Figure 10.4. So, the focus of the parabola is p units from the vertex, or

$$(h, k + p) = \left(-1, \tfrac{1}{2}\right). \qquad \text{Focus}$$

$y = -\frac{1}{2}x^2 - x + \frac{1}{2}$

Vertex

$p = -\frac{1}{2}$ $\left(-1, \frac{1}{2}\right)$

Focus

Parabola with a vertical axis, $p < 0$
Figure 10.4

A line segment that passes through the focus of a parabola and has endpoints on the parabola is called a **focal chord.** The specific focal chord perpendicular to the axis of the parabola is the **latus rectum.** The next example shows how to determine the length of the latus rectum and the length of the corresponding intercepted arc.

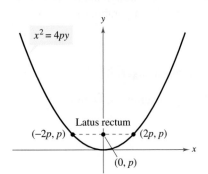

Length of latus rectum: $4p$
Figure 10.5

www **EXAMPLE 2**   **Focal Chord Length and Arc Length**

Find the length of the latus rectum of the parabola given by $x^2 = 4py$. Then find the length of the parabolic arc intercepted by the latus rectum.

Solution   Because the latus rectum passes through the focus $(0, p)$ and is perpendicular to the y-axis, the coordinates of its endpoints are $(-x, p)$ and (x, p). Substituting p for y in the equation of the parabola produces

$$x^2 = 4p(p) \quad \Longrightarrow \quad x = \pm 2p.$$

So, the endpoints of the latus rectum are $(-2p, p)$ and $(2p, p)$, and you can conclude that its length is $4p$, as shown in Figure 10.5. In contrast, the length of the intercepted arc is

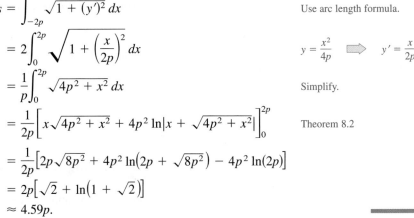

$$s = \int_{-2p}^{2p} \sqrt{1 + (y')^2}\, dx \qquad\qquad \text{Use arc length formula.}$$

$$= 2\int_{0}^{2p} \sqrt{1 + \left(\frac{x}{2p}\right)^2}\, dx \qquad\qquad y = \frac{x^2}{4p} \quad \Longrightarrow \quad y' = \frac{x}{2p}$$

$$= \frac{1}{p}\int_{0}^{2p} \sqrt{4p^2 + x^2}\, dx \qquad\qquad \text{Simplify.}$$

$$= \frac{1}{2p}\left[x\sqrt{4p^2 + x^2} + 4p^2 \ln\left|x + \sqrt{4p^2 + x^2}\right| \right]_{0}^{2p} \qquad \text{Theorem 8.2}$$

$$= \frac{1}{2p}\left[2p\sqrt{8p^2} + 4p^2 \ln\left(2p + \sqrt{8p^2}\right) - 4p^2 \ln(2p) \right]$$

$$= 2p\left[\sqrt{2} + \ln\left(1 + \sqrt{2}\right) \right]$$

$$\approx 4.59p.$$

One widely used property of a parabola is its reflective property. In physics, a surface is called **reflective** if the tangent line at any point on the surface makes equal angles with an incoming ray and the resulting outgoing ray. The angle corresponding to the incoming ray is the **angle of incidence,** and the angle corresponding to the outgoing ray is the **angle of reflection.** One example of a reflective surface is a flat mirror.

Another type of reflective surface is that formed by revolving a parabola about its axis. A special property of parabolic reflectors is that they allow us to direct all incoming rays parallel to the axis through the focus of the parabola—this is the principle behind the design of the parabolic mirrors used in reflecting telescopes. Conversely, all light rays emanating from the focus of a parabolic reflector used in a flashlight are parallel, as shown in Figure 10.6.

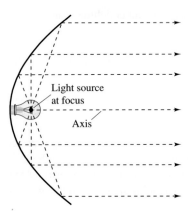

Parabolic reflector: light is reflected in parallel rays.
Figure 10.6

THEOREM 10.2   Reflective Property of a Parabola

Let P be a point on a parabola. The tangent line to the parabola at the point P makes equal angles with the following two lines.

1. The line passing through P and the focus
2. The line passing through P parallel to the axis of the parabola

---

www   *indicates that in the* HM mathSpace® *CD-ROM and the online* Eduspace® *system for this text, you will find an Open Exploration, which further explores this example using the computer algebra systems* Maple, Mathcad, Mathematica, *and* Derive.

Ellipses

NICOLAUS COPERNICUS (1473–1543)

Copernicus began to study planetary motion when asked to revise the calendar. At that time, the exact length of the year could not be accurately predicted using the theory that Earth was the center of the universe.

More than a thousand years after the close of the Alexandrian period of Greek mathematics, Western civilization finally began a Renaissance of mathematical and scientific discovery. One of the principal figures in this rebirth was the Polish astronomer Nicolaus Copernicus. In his work *On the Revolutions of the Heavenly Spheres,* Copernicus claimed that all of the planets, including Earth, revolved about the sun in circular orbits. Although some of Copernicus's claims were invalid, the controversy set off by his heliocentric theory motivated astronomers to search for a mathematical model to explain the observed movements of the sun and planets. The first to find an accurate model was the German astronomer Johannes Kepler (1571–1630). Kepler discovered that the planets move about the sun in elliptical orbits, with the sun not as the center but as a focal point of the orbit.

The use of ellipses to explain the movements of the planets is only one of many practical and aesthetic uses. As with parabolas, you will begin your study of this second type of conic by defining it as a locus of points. Now, however, *two* focal points are used rather than one.

An **ellipse** is the set of all points (x, y) the sum of whose distances from two distinct fixed points called **foci** is constant. (See Figure 10.7.) The line through the foci intersects the ellipse at two points, called the **vertices.** The chord joining the vertices is the **major axis,** and its midpoint is the **center** of the ellipse. The chord perpendicular to the major axis at the center is the **minor axis** of the ellipse. (See Figure 10.8.)

FOR FURTHER INFORMATION To learn about how an ellipse may be "exploded" into a parabola, see the article "Exploding the Ellipse" by Arnold Good in *Mathematics Teacher.* To view this article, go to the website *www.matharticles.com.*

Figure 10.7   **Figure 10.8**

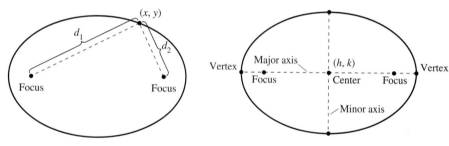

THEOREM 10.3   Standard Equation of an Ellipse

The standard form of the equation of an ellipse with center (h, k) and major and minor axes of lengths $2a$ and $2b$, where $a > b$, is

$$\frac{(x - h)^2}{a^2} + \frac{(y - k)^2}{b^2} = 1 \qquad \text{Major axis is horizontal.}$$

or

$$\frac{(x - h)^2}{b^2} + \frac{(y - k)^2}{a^2} = 1. \qquad \text{Major axis is vertical.}$$

The foci lie on the major axis, c units from the center, with $c^2 = a^2 - b^2$.

NOTE   You can visualize the definition of an ellipse by imagining two thumbtacks placed at the foci, as shown in Figure 10.9. If the ends of a fixed length of string are fastened to the thumbtacks and the string is drawn taut with a pencil, the path traced by the pencil will be an ellipse.

Figure 10.9

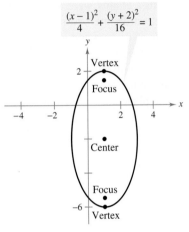

$$\frac{(x-1)^2}{4} + \frac{(y+2)^2}{16} = 1$$

Ellipse with a vertical major axis
Figure 10.10

EXAMPLE 3 **Completing the Square**

Find the center, vertices, and foci of the ellipse given by

$$4x^2 + y^2 - 8x + 4y - 8 = 0.$$

Solution By completing the square, you can write the original equation in standard form.

$$4x^2 + y^2 - 8x + 4y - 8 = 0 \qquad \text{Write original equation.}$$
$$4x^2 - 8x + y^2 + 4y = 8$$
$$4(x^2 - 2x + 1) + (y^2 + 4y + 4) = 8 + 4 + 4$$
$$4(x - 1)^2 + (y + 2)^2 = 16$$
$$\frac{(x-1)^2}{4} + \frac{(y+2)^2}{16} = 1 \qquad \text{Write in standard form.}$$

So, the major axis is parallel to the y-axis, where $h = 1$, $k = -2$, $a = 4$, $b = 2$, and $c = \sqrt{16 - 4} = 2\sqrt{3}$. So, you obtain the following.

Center:	$(1, -2)$	(h, k)
Vertices:	$(1, -6)$ and $(1, 2)$	$(h, k \pm a)$
Foci:	$\left(1, -2 - 2\sqrt{3}\right)$ and $\left(1, -2 + 2\sqrt{3}\right)$	$(h, k \pm c)$

The graph of the ellipse is shown in Figure 10.10.

NOTE If the constant term $F = -8$ in the equation in Example 3 had been greater than or equal to 8, you would have obtained one of the following degenerate cases.

1. $F = 8$, single point, $(1, -2)$: $\dfrac{(x-1)^2}{4} + \dfrac{(y+2)^2}{16} = 0$

2. $F > 8$, no solution points: $\dfrac{(x-1)^2}{4} + \dfrac{(y+2)^2}{16} < 0$

EXAMPLE 4 **The Orbit of the Moon**

The moon orbits Earth in an elliptical path with the center of Earth at one focus, as shown in Figure 10.11. The major and minor axes of the orbit have lengths of 768,800 kilometers and 767,640 kilometers, respectively. Find the greatest and least distances (the apogee and perigee) from Earth's center to the moon's center.

Solution Begin by solving for a and b.

$2a = 768,800$	Length of major axis
$a = 384,400$	Solve for a.
$2b = 767,640$	Length of minor axis
$b = 383,820$	Solve for b.

Now, using these values, you can solve for c as follows.

$$c = \sqrt{a^2 - b^2} \approx 21,108$$

The greatest distance between the center of Earth and the center of the moon is $a + c \approx 405,508$ kilometers, and the least distance is $a - c \approx 363,292$ kilometers.

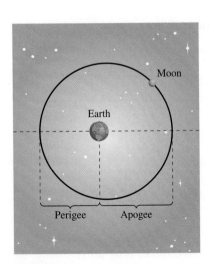

Figure 10.11

FOR FURTHER INFORMATION For more information on some uses of the reflective properties of conics, see the article "Parabolic Mirrors, Elliptic and Hyperbolic Lenses" by Mohsen Maesumi in *The American Mathematical Monthly.* Also see the article "The Geometry of Microwave Antennas" by William R. Parzynski in *Mathematics Teacher.*

Theorem 10.2 presented a reflective property of parabolas. Ellipses have a similar reflective property. You are asked to prove the following theorem in Exercise 110.

THEOREM 10.4   Reflective Property of an Ellipse

Let P be a point on an ellipse. The tangent line to the ellipse at point P makes equal angles with the lines through P and the foci.

One of the reasons that astronomers had difficulty in detecting that the orbits of the planets are ellipses is that the foci of the planetary orbits are relatively close to the center of the sun, making the orbits nearly circular. To measure the ovalness of an ellipse, you can use the concept of **eccentricity.**

Definition of Eccentricity of an Ellipse

The **eccentricity** e of an ellipse is given by the ratio

$$e = \frac{c}{a}.$$

To see how this ratio is used to describe the shape of an ellipse, note that because the foci of an ellipse are located along the major axis between the vertices and the center, it follows that

$$0 < c < a.$$

For an ellipse that is nearly circular, the foci are close to the center and the ratio c/a is small, and for an elongated ellipse, the foci are close to the vertices and the ratio is close to 1, as shown in Figure 10.12. Note that $0 < e < 1$ for every ellipse.

The orbit of the moon has an eccentricity of $e = 0.0549$, and the eccentricities of the nine planetary orbits are as follows.

Mercury:	$e = 0.2056$	Saturn:	$e = 0.0542$
Venus:	$e = 0.0068$	Uranus:	$e = 0.0472$
Earth:	$e = 0.0167$	Neptune:	$e = 0.0086$
Mars:	$e = 0.0934$	Pluto:	$e = 0.2488$
Jupiter:	$e = 0.0484$		

You can use integration to show that the area of an ellipse is $A = \pi ab$. For instance, the area of the ellipse

$$\frac{x^2}{a^2} + \frac{y^2}{b^2} = 1$$

is given by

$$A = 4 \int_0^a \frac{b}{a} \sqrt{a^2 - x^2}\, dx$$

$$= \frac{4b}{a} \int_0^{\pi/2} a^2 \cos^2 \theta\, d\theta. \qquad \text{Trigonometric substitution } x = a \sin \theta.$$

However, it is not so simple to find the *circumference* of an ellipse. The next example shows how to use eccentricity to set up an "elliptic integral" for the circumference of an ellipse.

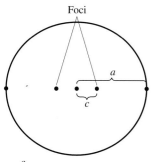

Foci

(a) $\dfrac{c}{a}$ is small.

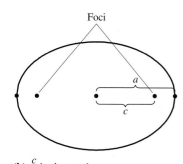

Foci

(b) $\dfrac{c}{a}$ is close to 1.

Eccentricity is the ratio $\dfrac{c}{a}$.

Figure 10.12

EXAMPLE 5   Finding the Circumference of an Ellipse

Show that the circumference of the ellipse $(x^2/a^2) + (y^2/b^2) = 1$ is

$$4a \int_0^{\pi/2} \sqrt{1 - e^2 \sin^2 \theta}\, d\theta. \qquad e = \frac{c}{a}$$

Solution   Because the given ellipse is symmetric with respect to both the *x*-axis and the *y*-axis, you know that its circumference *C* is four times the arc length of $y = (b/a)\sqrt{a^2 - x^2}$ in the first quadrant. The function *y* is differentiable for all *x* in the interval $[0, a]$ except at $x = a$. So, the circumference is given by the improper integral

$$C = \lim_{d \to a} 4 \int_0^d \sqrt{1 + (y')^2}\, dx = 4 \int_0^a \sqrt{1 + (y')^2}\, dx = 4 \int_0^a \sqrt{1 + \frac{b^2 x^2}{a^2(a^2 - x^2)}}\, dx.$$

Using the trigonometric substitution $x = a \sin \theta$, you obtain

$$\begin{aligned}
C &= 4 \int_0^{\pi/2} \sqrt{1 + \frac{b^2 \sin^2 \theta}{a^2 \cos^2 \theta}}\, (a \cos \theta)\, d\theta \\
&= 4 \int_0^{\pi/2} \sqrt{a^2 \cos^2 \theta + b^2 \sin^2 \theta}\, d\theta \\
&= 4 \int_0^{\pi/2} \sqrt{a^2(1 - \sin^2 \theta) + b^2 \sin^2 \theta}\, d\theta \\
&= 4 \int_0^{\pi/2} \sqrt{a^2 - (a^2 - b^2)\sin^2 \theta}\, d\theta.
\end{aligned}$$

Because $e^2 = c^2/a^2 = (a^2 - b^2)/a^2$, you can rewrite this integral as

$$C = 4a \int_0^{\pi/2} \sqrt{1 - e^2 \sin^2 \theta}\, d\theta.$$

A great deal of time has been devoted to the study of elliptic integrals. Such integrals generally do not have elementary antiderivatives. To find the circumference of an ellipse, you must usually resort to an approximation technique.

EXAMPLE 6   Approximating the Value of an Elliptic Integral

Use the elliptic integral in Example 5 to approximate the circumference of the ellipse

$$\frac{x^2}{25} + \frac{y^2}{16} = 1.$$

Solution   Because $e^2 = c^2/a^2 = (a^2 - b^2)/a^2 = 9/25$, you have

$$C = (4)(5) \int_0^{\pi/2} \sqrt{1 - \frac{9 \sin^2 \theta}{25}}\, d\theta.$$

Applying Simpson's Rule with $n = 4$ produces

$$\begin{aligned}
C &\approx 20\left(\frac{\pi}{6}\right)\left(\frac{1}{4}\right)[1 + 4(0.9733) + 2(0.9055) + 4(0.8323) + 0.8] \\
&\approx 28.36.
\end{aligned}$$

So, the ellipse has a circumference of about 28.36 units, as shown in Figure 10.13.

AREA AND CIRCUMFERENCE OF AN ELLIPSE

In his work with elliptic orbits in the early 1600's, Johannes Kepler successfully developed a formula for the area of an ellipse, $A = \pi ab$. He was less successful in developing a formula for the circumference of an ellipse, however; the best he could do was to give the approximate formula $C = \pi(a + b)$.

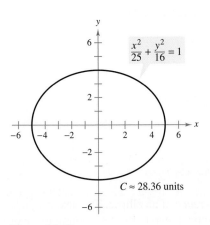

$$\frac{x^2}{25} + \frac{y^2}{16} = 1$$

$C \approx 28.36$ units

Figure 10.13

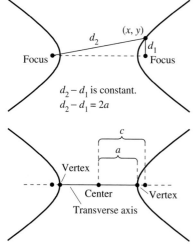

Figure 10.14

Hyperbolas

The definition of a hyperbola is similar to that of an ellipse. For an ellipse, the *sum* of the distances between the foci and a point on the ellipse is fixed, whereas for a hyperbola, the absolute value of the *difference* between these distances is fixed.

A **hyperbola** is the set of all points (x, y) for which the absolute value of the difference between the distances from two distinct fixed points called **foci** is constant. (See Figure 10.14.) The line through the two foci intersects a hyperbola at two points called the **vertices.** The line segment connecting the vertices is the **transverse axis,** and the midpoint of the transverse axis is the **center** of the hyperbola. One distinguishing feature of a hyperbola is that its graph has two separate *branches.*

THEOREM 10.5   Standard Equation of a Hyperbola

The standard form of the equation of a hyperbola with center at (h, k) is

$$\frac{(x - h)^2}{a^2} - \frac{(y - k)^2}{b^2} = 1 \qquad \text{Transverse axis is horizontal.}$$

or

$$\frac{(y - k)^2}{a^2} - \frac{(x - h)^2}{b^2} = 1. \qquad \text{Transverse axis is vertical.}$$

The vertices are a units from the center, and the foci are c units from the center, where, $c^2 = a^2 + b^2$.

NOTE   The constants a, b, and c do not have the same relationship for hyperbolas as they do for ellipses. For hyperbolas, $c^2 = a^2 + b^2$, but for ellipses, $c^2 = a^2 - b^2$.

An important aid in sketching the graph of a hyperbola is the determination of its **asymptotes,** as shown in Figure 10.15. Each hyperbola has two asymptotes that intersect at the center of the hyperbola. The asymptotes pass through the vertices of a rectangle of dimensions $2a$ by $2b$, with its center at (h, k). The line segment of length $2b$ joining $(h, k + b)$ and $(h, k - b)$ is referred to as the **conjugate axis** of the hyperbola.

THEOREM 10.6   Asymptotes of a Hyperbola

For a *horizontal* transverse axis, the equations of the asymptotes are

$$y = k + \frac{b}{a}(x - h) \qquad \text{and} \qquad y = k - \frac{b}{a}(x - h).$$

For a *vertical* transverse axis, the equations of the asymptotes are

$$y = k + \frac{a}{b}(x - h) \qquad \text{and} \qquad y = k - \frac{a}{b}(x - h).$$

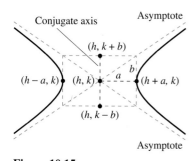

Figure 10.15

In Figure 10.15 you can see that the asymptotes coincide with the diagonals of the rectangle with dimensions $2a$ and $2b$, centered at (h, k). This provides you with a quick means of sketching the asymptotes, which in turn aids in sketching the hyperbola.

EXAMPLE 7   **Using Asymptotes to Sketch a Hyperbola**

Sketch the graph of the hyperbola whose equation is $4x^2 - y^2 = 16$.

TECHNOLOGY   You can use a graphing utility to verify the graph obtained in Example 7 by solving the original equation for y and graphing the following equations.

$$y_1 = \sqrt{4x^2 - 16}$$
$$y_2 = -\sqrt{4x^2 - 16}$$

Solution   Begin by rewriting the equation in standard form.

$$\frac{x^2}{4} - \frac{y^2}{16} = 1$$

The transverse axis is horizontal and the vertices occur at $(-2, 0)$ and $(2, 0)$. The ends of the conjugate axis occur at $(0, -4)$ and $(0, 4)$. Using these four points, you can sketch the rectangle shown in Figure 10.16(a). By drawing the asymptotes through the corners of this rectangle, you can complete the sketch as shown in Figure 10.16(b).

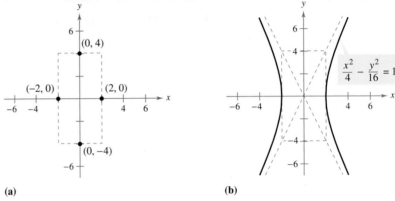

(a)

(b)

Figure 10.16

Definition of Eccentricity of a Hyperbola

The **eccentricity** e of a hyperbola is given by the ratio

$$e = \frac{c}{a}.$$

As with an ellipse, the **eccentricity** of a hyperbola is $e = c/a$. Because $c > a$ for hyperbolas, it follows that $e > 1$ for hyperbolas. If the eccentricity is large, the branches of the hyperbola are nearly flat. If the eccentricity is close to 1, the branches of the hyperbola are more pointed, as shown in Figure 10.17.

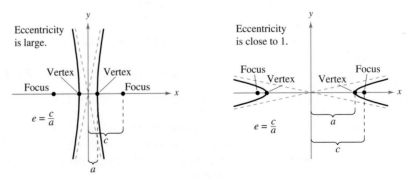

Figure 10.17

The following application was developed during World War II. It shows how the properties of hyperbolas can be used in radar and other detection systems.

EXAMPLE 8   A Hyperbolic Detection System

Two microphones, 1 mile apart, record an explosion. Microphone A receives the sound 2 seconds before microphone B. Where was the explosion?

Solution   Assuming that sound travels at 1100 feet per second, you know that the explosion took place 2200 feet farther from B than from A, as shown in Figure 10.18. The locus of all points that are 2200 feet closer to A than to B is one branch of the hyperbola $(x^2/a^2) - (y^2/b^2) = 1$, where

$$c = \frac{1 \text{ mile}}{2} = \frac{5280 \text{ ft}}{2} = 2640 \text{ feet}$$

and

$$a = \frac{2200 \text{ ft}}{2} = 1100 \text{ feet}.$$

Because $c^2 = a^2 + b^2$, it follows that

$$b^2 = c^2 - a^2$$
$$= 5,759,600$$

and you can conclude that the explosion occurred somewhere on the right branch of the hyperbola given by

$$\frac{x^2}{1,210,000} - \frac{y^2}{5,759,600} = 1.$$

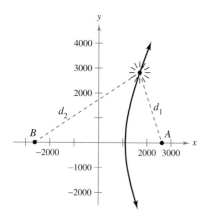

$2c = 5280$
$d_2 - d_1 = 2a = 2200$
Figure 10.18

In Example 8, you were able to determine only the hyperbola on which the explosion occurred, but not the exact location of the explosion. If, however, you had received the sound at a third position C, then two other hyperbolas would be determined. The exact location of the explosion would be the point at which these three hyperbolas intersect.

Another interesting application of conics involves the orbits of comets in our solar system. Of the 610 comets identified prior to 1970, 245 have elliptical orbits, 295 have parabolic orbits, and 70 have hyperbolic orbits. The center of the sun is a focus of each orbit, and each orbit has a vertex at the point at which the comet is closest to the sun. Undoubtedly, many comets with parabolic or hyperbolic orbits have not been identified—such comets pass through our solar system only once. Only comets with elliptical orbits such as Halley's comet remain in our solar system.

The type of orbit for a comet can be determined as follows.

1. Ellipse: $v < \sqrt{2GM/p}$
2. Parabola: $v = \sqrt{2GM/p}$
3. Hyperbola: $v > \sqrt{2GM/p}$

In these three formulas, p is the distance between one vertex and one focus of the comet's orbit (in meters), v is the velocity of the comet at the vertex (in meters per second), $M \approx 1.989 \times 10^{30}$ kilograms is the mass of the sun, and $G \approx 6.67 \times 10^{-8}$ cubic meters per kilogram-second squared is the gravitational constant.

CAROLINE HERSCHEL (1750–1848)

The first woman to be credited with detecting a new comet was the English astronomer Caroline Herschel. During her life, Caroline Herschel discovered a total of eight new comets.

Exercises for Section 10.1

In Exercises 1–8, match the equation with its graph. [The graphs are labeled (a), (b), (c), (d), (e), (f), (g), and (h).]

(a)

(b)

(c)

(d)

(e)

(f)

(g)

(h)

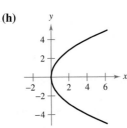

1. $y^2 = 4x$

2. $x^2 = 8y$

3. $(x + 3)^2 = -2(y - 2)$

4. $\dfrac{(x - 2)^2}{16} + \dfrac{(y + 1)^2}{4} = 1$

5. $\dfrac{x^2}{9} + \dfrac{y^2}{4} = 1$

6. $\dfrac{x^2}{9} + \dfrac{y^2}{9} = 1$

7. $\dfrac{y^2}{16} - \dfrac{x^2}{1} = 1$

8. $\dfrac{(x - 2)^2}{9} - \dfrac{y^2}{4} = 1$

In Exercises 9–16, find the vertex, focus, and directrix of the parabola, and sketch its graph.

9. $y^2 = -6x$

10. $x^2 + 8y = 0$

11. $(x + 3) + (y - 2)^2 = 0$

12. $(x - 1)^2 + 8(y + 2) = 0$

13. $y^2 - 4y - 4x = 0$

14. $y^2 + 6y + 8x + 25 = 0$

15. $x^2 + 4x + 4y - 4 = 0$

16. $y^2 + 4y + 8x - 12 = 0$

 In Exercises 17–20, find the vertex, focus, and directrix of the parabola. Then use a graphing utility to graph the parabola.

17. $y^2 + x + y = 0$

18. $y = -\frac{1}{6}(x^2 - 8x + 6)$

19. $y^2 - 4x - 4 = 0$

20. $x^2 - 2x + 8y + 9 = 0$

In Exercises 21–28, find an equation of the parabola.

21. Vertex: $(3, 2)$
Focus: $(1, 2)$

22. Vertex: $(-1, 2)$
Focus: $(-1, 0)$

23. Vertex: $(0, 4)$
Directrix: $y = -2$

24. Focus: $(2, 2)$
Directrix: $x = -2$

25.

26.

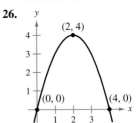

27. Axis is parallel to y-axis; graph passes through $(0, 3)$, $(3, 4)$, and $(4, 11)$.

28. Directrix: $y = -2$; endpoints of latus rectum are $(0, 2)$ and $(8, 2)$.

In Exercises 29–34, find the center, foci, vertices, and eccentricity of the ellipse, and sketch its graph.

29. $x^2 + 4y^2 = 4$

30. $5x^2 + 7y^2 = 70$

31. $\dfrac{(x - 1)^2}{9} + \dfrac{(y - 5)^2}{25} = 1$

32. $(x + 2)^2 + \dfrac{(y + 4)^2}{1/4} = 1$

33. $9x^2 + 4y^2 + 36x - 24y + 36 = 0$

34. $16x^2 + 25y^2 - 64x + 150y + 279 = 0$

 In Exercises 35–38, find the center, foci, and vertices of the ellipse. Use a graphing utility to graph the ellipse.

35. $12x^2 + 20y^2 - 12x + 40y - 37 = 0$

36. $36x^2 + 9y^2 + 48x - 36y + 43 = 0$

37. $x^2 + 2y^2 - 3x + 4y + 0.25 = 0$

38. $2x^2 + y^2 + 4.8x - 6.4y + 3.12 = 0$

In Exercises 39–44, find an equation of the ellipse.

39. Center: $(0, 0)$
Focus: $(2, 0)$
Vertex: $(3, 0)$

40. Vertices: $(0, 2)$, $(4, 2)$
Eccentricity: $\frac{1}{2}$

41. Vertices: $(3, 1)$, $(3, 9)$
Minor axis length: 6

42. Foci: $(0, \pm 5)$
Major axis length: 14

43. Center: $(0, 0)$

Major axis: horizontal

Points on the ellipse:

$(3, 1), (4, 0)$

44. Center: $(1, 2)$

Major axis: vertical

Points on the ellipse:

$(1, 6), (3, 2)$

In Exercises 45–52, find the center, foci, and vertices of the hyperbola, and sketch its graph using asymptotes as an aid.

45. $y^2 - \dfrac{x^2}{4} = 1$

46. $\dfrac{x^2}{25} - \dfrac{y^2}{9} = 1$

47. $\dfrac{(x - 1)^2}{4} - \dfrac{(y + 2)^2}{1} = 1$

48. $\dfrac{(y + 1)^2}{144} - \dfrac{(x - 4)^2}{25} = 1$

49. $9x^2 - y^2 - 36x - 6y + 18 = 0$

50. $y^2 - 9x^2 + 36x - 72 = 0$

51. $x^2 - 9y^2 + 2x - 54y - 80 = 0$

52. $9x^2 - 4y^2 + 54x + 8y + 78 = 0$

In Exercises 53–56, find the center, foci, and vertices of the hyperbola. Use a graphing utility to graph the hyperbola and its asymptotes.

53. $9y^2 - x^2 + 2x + 54y + 62 = 0$

54. $9x^2 - y^2 + 54x + 10y + 55 = 0$

55. $3x^2 - 2y^2 - 6x - 12y - 27 = 0$

56. $3y^2 - x^2 + 6x - 12y = 0$

In Exercises 57–64, find an equation of the hyperbola.

57. Vertices: $(\pm 1, 0)$

Asymptotes: $y = \pm 3x$

58. Vertices: $(0, \pm 3)$

Asymptotes: $y = \pm 3x$

59. Vertices: $(2, \pm 3)$

Point on graph: $(0, 5)$

60. Vertices: $(2, \pm 3)$

Foci: $(2, \pm 5)$

61. Center: $(0, 0)$

Vertex: $(0, 2)$

Focus: $(0, 4)$

62. Center: $(0, 0)$

Vertex: $(3, 0)$

Focus: $(5, 0)$

63. Vertices: $(0, 2), (6, 2)$

Asymptotes: $y = \frac{2}{3}x$

$y = 4 - \frac{2}{3}x$

64. Focus: $(10, 0)$

Asymptotes: $y = \pm \frac{3}{4}x$

In Exercises 65 and 66, find equations for (a) the tangent lines and (b) the normal lines to the hyperbola for the given value of x.

65. $\dfrac{x^2}{9} - y^2 = 1, \quad x = 6$

66. $\dfrac{y^2}{4} - \dfrac{x^2}{2} = 1, \quad x = 4$

In Exercises 67–76, classify the graph of the equation as a circle, a parabola, an ellipse, or a hyperbola.

67. $x^2 + 4y^2 - 6x + 16y + 21 = 0$

68. $4x^2 - y^2 - 4x - 3 = 0$

69. $y^2 - 4y - 4x = 0$

70. $25x^2 - 10x - 200y - 119 = 0$

71. $4x^2 + 4y^2 - 16y + 15 = 0$

72. $y^2 - 4y = x + 5$

73. $9x^2 + 9y^2 - 36x + 6y + 34 = 0$

74. $2x(x - y) = y(3 - y - 2x)$

75. $3(x - 1)^2 = 6 + 2(y + 1)^2$

76. $9(x + 3)^2 = 36 - 4(y - 2)^2$

Writing About Concepts

77. (a) Give the definition of a parabola.

(b) Give the standard forms of a parabola with vertex at (h, k).

(c) In your own words, state the reflective property of a parabola.

78. (a) Give the definition of an ellipse.

(b) Give the standard forms of an ellipse with center at (h, k).

79. (a) Give the definition of a hyperbola.

(b) Give the standard forms of a hyperbola with center at (h, k).

(c) Write equations for the asymptotes of a hyperbola.

80. Define the eccentricity of an ellipse. In your own words, describe how changes in the eccentricity affect the ellipse.

81. *Solar Collector*   A solar collector for heating water is constructed with a sheet of stainless steel that is formed into the shape of a parabola (see figure). The water will flow through a pipe that is located at the focus of the parabola. At what distance from the vertex is the pipe?

Not drawn to scale

Figure for 81   **Figure for 82**

82. *Beam Deflection*   A simply supported beam that is 16 meters long has a load concentrated at the center (see figure). The deflection of the beam at its center is 3 centimeters. Assume that the shape of the deflected beam is parabolic.

(a) Find an equation of the parabola. (Assume that the origin is at the center of the beam.)

(b) How far from the center of the beam is the deflection 1 centimeter?

83. Find an equation of the tangent line to the parabola $y = ax^2$ at $x = x_0$. Prove that the x-intercept of this tangent line is $(x_0/2, 0)$.

84. (a) Prove that any two distinct tangent lines to a parabola intersect.

(b) Demonstrate the result of part (a) by finding the point of intersection of the tangent lines to the parabola $x^2 - 4x - 4y = 0$ at the points $(0, 0)$ and $(6, 3)$.

85. (a) Prove that if any two tangent lines to a parabola intersect at right angles, their point of intersection must lie on the directrix.

(b) Demonstrate the result of part (a) by proving that the tangent lines to the parabola $x^2 - 4x - 4y + 8 = 0$ at the points $(-2, 5)$ and $\left(3, \frac{5}{4}\right)$ intersect at right angles, and that the point of intersection lies on the directrix.

86. Find the point on the graph of $x^2 = 8y$ that is closest to the focus of the parabola.

87. *Radio and Television Reception* In mountainous areas, reception of radio and television is sometimes poor. Consider an idealized case where a hill is represented by the graph of the parabola $y = x - x^2$, a transmitter is located at the point $(-1, 1)$, and a receiver is located on the other side of the hill at the point $(x_0, 0)$. What is the closest the receiver can be to the hill so that the reception is unobstructed?

88. *Modeling Data* The table shows the average amounts of time A (in minutes) women spent watching television each day for the years 1996 to 2002. *(Source: Nielsen Media Research)*

Year	1996	1997	1998	1999	2000	2001	2002
A	274	273	273	280	286	291	298

(a) Use the regression capabilities of a graphing utility to find a model of the form $A = at^2 + bt + c$ for the data. Let t represent the year, with $t = 6$ corresponding to 1996.

(b) Use a graphing utility to plot the data and graph the model.

(c) Find dA/dt and sketch its graph for $6 \le t \le 12$. What information about the average amount of time women spent watching television is given by the graph of the derivative?

89. *Architecture* A church window is bounded above by a parabola and below by the arc of a circle (see figure). Find the surface area of the window.

Figure for 89

Figure for 91

90. *Arc Length* Find the arc length of the parabola $4x - y^2 = 0$ over the interval $0 \le y \le 4$.

91. *Bridge Design* A cable of a suspension bridge is suspended (in the shape of a parabola) between two towers that are 120 meters apart and 20 meters above the roadway (see figure). The cables touch the roadway midway between the towers.

(a) Find an equation for the parabolic shape of each cable.

(b) Find the length of the parabolic supporting cable.

92. *Surface Area* A satellite-signal receiving dish is formed by revolving the parabola given by $x^2 = 20y$ about the y-axis. The radius of the dish is r feet. Verify that the surface area of the dish is given by

$$2\pi \int_0^r x \sqrt{1 + \left(\frac{x}{10}\right)^2}\, dx = \frac{\pi}{15}[(100 + r^2)^{3/2} - 1000].$$

93. *Investigation* Sketch the graphs of $x^2 = 4py$ for $p = \frac{1}{4}, \frac{1}{2}, 1, \frac{3}{2},$ and 2 on the same coordinate axes. Discuss the change in the graphs as p increases.

94. *Area* Find a formula for the area of the shaded region in the figure.

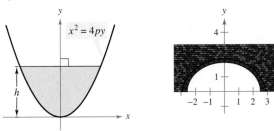

Figure for 94   Figure for 96

95. *Writing* On page 697, it was noted that an ellipse can be drawn using two thumbtacks, a string of fixed length (greater than the distance between the tacks), and a pencil. If the ends of the string are fastened at the tacks and the string is drawn taut with a pencil, the path traced by the pencil will be an ellipse.

(a) What is the length of the string in terms of a?

(b) Explain why the path is an ellipse.

96. *Construction of a Semielliptical Arch* A fireplace arch is to be constructed in the shape of a semiellipse. The opening is to have a height of 2 feet at the center and a width of 5 feet along the base (see figure). The contractor draws the outline of the ellipse by the method shown in Exercise 95. Where should the tacks be placed and what should be the length of the piece of string?

97. Sketch the ellipse that consists of all points (x, y) such that the sum of the distances between (x, y) and two fixed points is 16 units, and the foci are located at the centers of the two sets of concentric circles in the figure. To print an enlarged copy of the graph, go to the website *www.mathgraphs.com*.

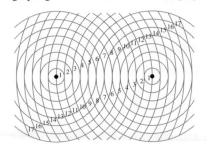

98. *Orbit of Earth*   Earth moves in an elliptical orbit with the sun at one of the foci. The length of half of the major axis is 149,598,000 kilometers, and the eccentricity is 0.0167. Find the minimum distance (*perihelion*) and the maximum distance (*aphelion*) of Earth from the sun.

99. *Satellite Orbit*   The *apogee* (the point in orbit farthest from Earth) and the *perigee* (the point in orbit closest to Earth) of an elliptical orbit of an Earth satellite are given by A and P. Show that the eccentricity of the orbit is

$$e = \frac{A - P}{A + P}.$$

100. *Explorer 18*   On November 27, 1963, the United States launched Explorer 18. Its low and high points above the surface of Earth were 119 miles and 123,000 miles. Find the eccentricity of its elliptical orbit.

101. *Halley's Comet*   Probably the most famous of all comets, Halley's comet, has an elliptical orbit with the sun at the focus. Its maximum distance from the sun is approximately 35.29 AU (astronomical unit $\approx 92.956 \times 10^6$ miles), and its minimum distance is approximately 0.59 AU. Find the eccentricity of the orbit.

102. The equation of an ellipse with its center at the origin can be written as

$$\frac{x^2}{a^2} + \frac{y^2}{a^2(1 - e^2)} = 1.$$

Show that as $e \to 0$, with a remaining fixed, the ellipse approaches a circle.

103. Consider a particle traveling clockwise on the elliptical path

$$\frac{x^2}{100} + \frac{y^2}{25} = 1.$$

The particle leaves the orbit at the point $(-8, 3)$ and travels in a straight line tangent to the ellipse. At what point will the particle cross the y-axis?

104. *Volume*   The water tank on a fire truck is 16 feet long, and its cross sections are ellipses. Find the volume of water in the partially filled tank as shown in the figure.

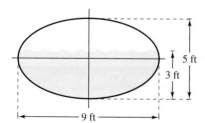

In Exercises 105 and 106, determine the points at which dy/dx is zero or does not exist to locate the endpoints of the major and minor axes of the ellipse.

105. $16x^2 + 9y^2 + 96x + 36y + 36 = 0$

106. $9x^2 + 4y^2 + 36x - 24y + 36 = 0$

Area and Volume   **In Exercises 107 and 108, find (a) the area of the region bounded by the ellipse, (b) the volume and surface area of the solid generated by revolving the region about its major axis (prolate spheroid), and (c) the volume and surface area of the solid generated by revolving the region about its minor axis (oblate spheroid).**

107. $\dfrac{x^2}{4} + \dfrac{y^2}{1} = 1$

108. $\dfrac{x^2}{16} + \dfrac{y^2}{9} = 1$

109. *Arc Length*   Use the integration capabilities of a graphing utility to approximate to two-decimal-place accuracy the elliptical integral representing the circumference of the ellipse

$$\frac{x^2}{25} + \frac{y^2}{49} = 1.$$

110. Prove that the tangent line to an ellipse at a point P makes equal angles with lines through P and the foci (see figure). [*Hint:* (1) Find the slope of the tangent line at P, (2) find the slopes of the lines through P and each focus, and (3) use the formula for the tangent of the angle between two lines.]

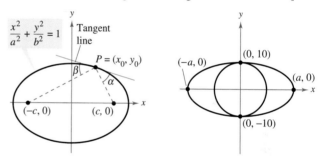

Figure for 110   **Figure for 111**

111. *Geometry*   The area of the ellipse in the figure is twice the area of the circle. What is the length of the major axis?

112. *Conjecture*

(a) Show that the equation of an ellipse can be written as

$$\frac{(x - h)^2}{a^2} + \frac{(y - k)^2}{a^2(1 - e^2)} = 1.$$

(b) Use a graphing utility to graph the ellipse

$$\frac{(x - 2)^2}{4} + \frac{(y - 3)^2}{4(1 - e^2)} = 1$$

for $e = 0.95$, $e = 0.75$, $e = 0.5$, $e = 0.25$, and $e = 0$.

(c) Use the results of part (b) to make a conjecture about the change in the shape of the ellipse as e approaches 0.

113. Find an equation of the hyperbola such that for any point on the hyperbola, the difference between its distances from the points $(2, 2)$ and $(10, 2)$ is 6.

114. Find an equation of the hyperbola such that for any point on the hyperbola, the difference between its distances from the points $(-3, 0)$ and $(-3, 3)$ is 2.

115. Sketch the hyperbola that consists of all points (x, y) such that the difference of the distances between (x, y) and two fixed points is 10 units, and the foci are located at the centers of the two sets of concentric circles in the figure. To print an enlarged copy of the graph, go to the website *www.mathgraphs.com*.

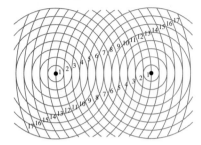

116. Consider a hyperbola centered at the origin with a horizontal transverse axis. Use the definition of a hyperbola to derive its standard form:

$$\frac{x^2}{a^2} - \frac{y^2}{b^2} = 1.$$

117. *Sound Location* A rifle positioned at point $(-c, 0)$ is fired at a target positioned at point $(c, 0)$. A person hears the sound of the rifle and the sound of the bullet hitting the target at the same time. Prove that the person is positioned on one branch of the hyperbola given by

$$\frac{x^2}{c^2 v_s^2 / v_m^2} - \frac{y^2}{c^2 (v_m^2 - v_s^2) / v_m^2} = 1$$

where v_m is the muzzle velocity of the rifle and v_s is the speed of sound, which is about 1100 feet per second.

118. *Navigation* LORAN (long distance radio navigation) for aircraft and ships uses synchronized pulses transmitted by widely separated transmitting stations. These pulses travel at the speed of light (186,000 miles per second). The difference in the times of arrival of these pulses at an aircraft or ship is constant on a hyperbola having the transmitting stations as foci. Assume that two stations, 300 miles apart, are positioned on the rectangular coordinate system at $(-150, 0)$ and $(150, 0)$ and that a ship is traveling on a path with coordinates $(x, 75)$ (see figure). Find the x-coordinate of the position of the ship if the time difference between the pulses from the transmitting stations is 1000 microseconds (0.001 second).

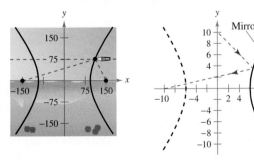

Figure for 118   **Figure for 119**

119. *Hyperbolic Mirror* A hyperbolic mirror (used in some telescopes) has the property that a light ray directed at the focus will be reflected to the other focus. The mirror in the figure has the equation $(x^2/36) - (y^2/64) = 1$. At which point on the mirror will light from the point $(0, 10)$ be reflected to the other focus?

120. Show that the equation of the tangent line to $\dfrac{x^2}{a^2} - \dfrac{y^2}{b^2} = 1$ at the point (x_0, y_0) is $(x_0/a^2)x - (y_0/b^2)y = 1$.

121. Show that the graphs of the equations intersect at right angles:

$$\frac{x^2}{a^2} + \frac{2y^2}{b^2} = 1 \quad \text{and} \quad \frac{x^2}{a^2 - b^2} - \frac{2y^2}{b^2} = 1.$$

122. Prove that the graph of the equation

$$Ax^2 + Cy^2 + Dx + Ey + F = 0$$

is one of the following (except in degenerate cases).

Conic	Condition
(a) Circle	$A = C$
(b) Parabola	$A = 0$ or $C = 0$ (but not both)
(c) Ellipse	$AC > 0$
(d) Hyperbola	$AC < 0$

True or False? **In Exercises 123–128, determine whether the statement is true or false. If it is false, explain why or give an example that shows it is false.**

123. It is possible for a parabola to intersect its directrix.

124. The point on a parabola closest to its focus is its vertex.

125. If C is the circumference of the ellipse

$$\frac{x^2}{a^2} + \frac{y^2}{b^2} = 1, \quad b < a$$

then $2\pi b \le C \le 2\pi a$.

126. If $D \ne 0$ or $E \ne 0$, then the graph of $y^2 - x^2 + Dx + Ey = 0$ is a hyperbola.

127. If the asymptotes of the hyperbola $(x^2/a^2) - (y^2/b^2) = 1$ intersect at right angles, then $a = b$.

128. Every tangent line to a hyperbola intersects the hyperbola only at the point of tangency.

Putnam Exam Challenge

129. For a point P on an ellipse, let d be the distance from the center of the ellipse to the line tangent to the ellipse at P. Prove that $(PF_1)(PF_2)d^2$ is constant as P varies on the ellipse, where PF_1 and PF_2 are the distances from P to the foci F_1 and F_2 of the ellipse.

130. Find the minimum value of $(u - v)^2 + \left(\sqrt{2 - u^2} - \dfrac{9}{v}\right)^2$ for $0 < u < \sqrt{2}$ and $v > 0$.

| Section 10.2 | **Plane Curves and Parametric Equations** |

- Sketch the graph of a curve given by a set of parametric equations.
- Eliminate the parameter in a set of parametric equations.
- Find a set of parametric equations to represent a curve.
- Understand two classic calculus problems, the tautochrone and brachistochrone problems.

Plane Curves and Parametric Equations

Until now, you have been representing a graph by a single equation involving *two* variables. In this section you will study situations in which *three* variables are used to represent a curve in the plane.

Consider the path followed by an object that is propelled into the air at an angle of 45°. If the initial velocity of the object is 48 feet per second, the object travels the parabolic path given by

$$y = -\frac{x^2}{72} + x \qquad \text{Rectangular equation}$$

Rectangular equation:
$$y = -\frac{x^2}{72} + x$$

Parametric equations:
$$x = 24\sqrt{2}\,t$$
$$y = -16t^2 + 24\sqrt{2}\,t$$

Curvilinear motion: two variables for position, one variable for time
Figure 10.19

as shown in Figure 10.19. However, this equation does not tell the whole story. Although it does tell you *where* the object has been, it doesn't tell you *when* the object was at a given point (x, y). To determine this time, you can introduce a third variable t, called a **parameter.** By writing both x and y as functions of t, you obtain the **parametric equations**

$$x = 24\sqrt{2}\,t \qquad \text{Parametric equation for } x$$

and

$$y = -16t^2 + 24\sqrt{2}\,t. \qquad \text{Parametric equation for } y$$

From this set of equations, you can determine that at time $t = 0$, the object is at the point $(0, 0)$. Similarly, at time $t = 1$, the object is at the point $\left(24\sqrt{2}, 24\sqrt{2} - 16\right)$, and so on. (You will learn a method for determining this particular set of parametric equations—the equations of motion—later, in Section 12.3.)

For this particular motion problem, x and y are continuous functions of t, and the resulting path is called a **plane curve.**

Definition of a Plane Curve

If f and g are continuous functions of t on an interval I, then the equations

$$x = f(t) \quad \text{and} \quad y = g(t)$$

are called **parametric equations** and t is called the **parameter.** The set of points (x, y) obtained as t varies over the interval I is called the **graph** of the parametric equations. Taken together, the parametric equations and the graph are called a **plane curve,** denoted by C.

NOTE   At times it is important to distinguish between a graph (the set of points) and a curve (the points together with their defining parametric equations). When it is important, we will make the distinction explicit. When it is not important, we will use C to represent the graph or the curve.

FOR FURTHER INFORMATION For more information on cycloids, see the article "The Geometry of Rolling Curves" by John Bloom and Lee Whitt in *The American Mathematical Monthly.* To view this article, go to the website *www.matharticles.com.*

EXAMPLE 5 Parametric Equations for a Cycloid

Determine the curve traced by a point P on the circumference of a circle of radius a rolling along a straight line in a plane. Such a curve is called a **cycloid.**

Solution Let the parameter θ be the measure of the circle's rotation, and let the point $P = (x, y)$ begin at the origin. When $\theta = 0$, P is at the origin. When $\theta = \pi$, P is at a maximum point $(\pi a, 2a)$. When $\theta = 2\pi$, P is back on the x-axis at $(2\pi a, 0)$. From Figure 10.25, you can see that $\angle APC = 180° - \theta$. So,

$$\sin \theta = \sin(180° - \theta) = \sin(\angle APC) = \frac{AC}{a} = \frac{BD}{a}$$

$$\cos \theta = -\cos(180° - \theta) = -\cos(\angle APC) = \frac{AP}{-a}$$

which implies that

$$AP = -a \cos \theta \quad \text{and} \quad BD = a \sin \theta.$$

Because the circle rolls along the x-axis, you know that $OD = \overset{\frown}{PD} = a\theta$. Furthermore, because $BA = DC = a$, you have

$$x = OD - BD = a\theta - a \sin \theta$$
$$y = BA + AP = a - a \cos \theta.$$

So, the parametric equations are

$$x = a(\theta - \sin \theta) \quad \text{and} \quad y = a(1 - \cos \theta).$$

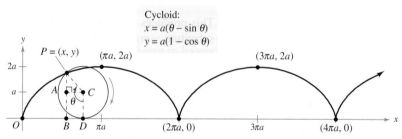

Cycloid:
$x = a(\theta - \sin \theta)$
$y = a(1 - \cos \theta)$

Figure 10.25

TECHNOLOGY Some graphing utilities allow you to simulate the motion of an object that is moving in the plane or in space. If you have access to such a utility, use it to trace out the path of the cycloid shown in Figure 10.25.

The cycloid in Figure 10.25 has sharp corners at the values $x = 2n\pi a$. Notice that the derivatives $x'(\theta)$ and $y'(\theta)$ are both zero at the points for which $\theta = 2n\pi$.

$$x(\theta) = a(\theta - \sin \theta) \qquad y(\theta) = a(1 - \cos \theta)$$
$$x'(\theta) = a - a \cos \theta \qquad y'(\theta) = a \sin \theta$$
$$x'(2n\pi) = 0 \qquad y'(2n\pi) = 0$$

Between these points, the cycloid is called **smooth.**

> **Definition of a Smooth Curve**
>
> A curve C represented by $x = f(t)$ and $y = g(t)$ on an interval I is called **smooth** if f' and g' are continuous on I and not simultaneously 0, except possibly at the endpoints of I. The curve C is called **piecewise smooth** if it is smooth on each subinterval of some partition of I.

The time required to complete a full swing of the pendulum when starting from point C is only approximately the same as when starting from point A.

Figure 10.26

JAMES BERNOULLI (1654–1705)

James Bernoulli, also called Jacques, was the older brother of John. He was one of several accomplished mathematicians of the Swiss Bernoulli family. James's mathematical accomplishments have given him a prominent place in the early development of calculus.

The Tautochrone and Brachistochrone Problems

The type of curve described in Example 5 is related to one of the most famous pairs of problems in the history of calculus. The first problem (called the **tautochrone problem**) began with Galileo's discovery that the time required to complete a full swing of a given pendulum is *approximately* the same whether it makes a large movement at high speed or a small movement at lower speed (see Figure 10.26). Late in his life, Galileo (1564–1642) realized that he could use this principle to construct a clock. However, he was not able to conquer the mechanics of actual construction. Christian Huygens (1629–1695) was the first to design and construct a working model. In his work with pendulums, Huygens realized that a pendulum does not take exactly the same time to complete swings of varying lengths. (This doesn't affect a pendulum clock, because the length of the circular arc is kept constant by giving the pendulum a slight boost each time it passes its lowest point.) But, in studying the problem, Huygens discovered that a ball rolling back and forth on an inverted cycloid does complete each cycle in exactly the same time.

An inverted cycloid is the path down which a ball will roll in the shortest time.
Figure 10.27

The second problem, which was posed by John Bernoulli in 1696, is called the **brachistochrone problem**—in Greek, *brachys* means short and chronos means time. The problem was to determine the path down which a particle will slide from point A to point B in the *shortest time*. Several mathematicians took up the challenge, and the following year the problem was solved by Newton, Leibniz, L'Hôpital, John Bernoulli, and James Bernoulli. As it turns out, the solution is not a straight line from A to B, but an inverted cycloid passing through the points A and B, as shown in Figure 10.27. The amazing part of the solution is that a particle starting at rest at *any* other point C of the cycloid between A and B will take exactly the same time to reach B, as shown in Figure 10.28.

A ball starting at point C takes the same time to reach point B as one that starts at point A.
Figure 10.28

FOR FURTHER INFORMATION To see a proof of the famous brachistochrone problem, see the article "A New Minimization Proof for the Brachistochrone" by Gary Lawlor in *The American Mathematical Monthly*. To view this article, go to the website *www.matharticles.com*.

Exercises for Section 10.2

See www.CalcChat.com for worked-out solutions to odd-numbered exercises.

1. Consider the parametric equations $x = \sqrt{t}$ and $y = 1 - t$.

(a) Complete the table.

t	0	1	2	3	4
x					
y					

(b) Plot the points (x, y) generated in the table, and sketch a graph of the parametric equations. Indicate the orientation of the graph.

(c) Use a graphing utility to confirm your graph in part (b).

(d) Find the rectangular equation by eliminating the parameter, and sketch its graph. Compare the graph in part (b) with the graph of the rectangular equation.

2. Consider the parametric equations $x = 4\cos^2\theta$ and $y = 2\sin\theta$.

(a) Complete the table.

θ	$-\dfrac{\pi}{2}$	$-\dfrac{\pi}{4}$	0	$\dfrac{\pi}{4}$	$\dfrac{\pi}{2}$
x					
y					

(b) Plot the points (x, y) generated in the table, and sketch a graph of the parametric equations. Indicate the orientation of the graph.

(c) Use a graphing utility to confirm your graph in part (b).

(d) Find the rectangular equation by eliminating the parameter, and sketch its graph. Compare the graph in part (b) with the graph of the rectangular equation.

(e) If values of θ were selected from the interval $[\pi/2, 3\pi/2]$ for the table in part (a), would the graph in part (b) be different? Explain.

In Exercises 3–20, sketch the curve represented by the parametric equations (indicate the orientation of the curve), and write the corresponding rectangular equation by eliminating the parameter.

3. $x = 3t - 1$, $\quad y = 2t + 1$

4. $x = 3 - 2t$, $\quad y = 2 + 3t$

5. $x = t + 1$, $\quad y = t^2$

6. $x = 2t^2$, $\quad y = t^4 + 1$

7. $x = t^3$, $\quad y = \dfrac{t^2}{2}$

8. $x = t^2 + t$, $\quad y = t^2 - t$

9. $x = \sqrt{t}$, $\quad y = t - 2$

10. $x = \sqrt[4]{t}$, $\quad y = 3 - t$

11. $x = t - 1$, $\quad y = \dfrac{t}{t - 1}$

12. $x = 1 + \dfrac{1}{t}$, $\quad y = t - 1$

13. $x = 2t$, $\quad y = |t - 2|$

14. $x = |t - 1|$, $\quad y = t + 2$

15. $x = e^t$, $\quad y = e^{3t} + 1$

16. $x = e^{-t}$, $\quad y = e^{2t} - 1$

17. $x = \sec\theta$, $\quad y = \cos\theta$, $\quad 0 \le \theta < \pi/2$, $\quad \pi/2 < \theta \le \pi$

18. $x = \tan^2\theta$, $\quad y = \sec^2\theta$

19. $x = 3\cos\theta$, $\quad y = 3\sin\theta$

20. $x = 2\cos\theta$, $\quad y = 6\sin\theta$

 In Exercises 21–32, use a graphing utility to graph the curve represented by the parametric equations (indicate the orientation of the curve). Eliminate the parameter and write the corresponding rectangular equation.

21. $x = 4\sin 2\theta, y = 2\cos 2\theta$

22. $x = \cos\theta, y = 2\sin 2\theta$

23. $x = 4 + 2\cos\theta$
$\quad y = -1 + \sin\theta$

24. $x = 4 + 2\cos\theta$
$\quad y = -1 + 2\sin\theta$

25. $x = 4 + 2\cos\theta$
$\quad y = -1 + 4\sin\theta$

26. $x = \sec\theta$
$\quad y = \tan\theta$

27. $x = 4\sec\theta$, $\quad y = 3\tan\theta$

28. $x = \cos^3\theta$, $\quad y = \sin^3\theta$

29. $x = t^3$, $\quad y = 3\ln t$

30. $x = \ln 2t$, $\quad y = t^2$

31. $x = e^{-t}$, $\quad y = e^{3t}$

32. $x = e^{2t}$, $\quad y = e^t$

Comparing Plane Curves **In Exercises 33–36, determine any differences between the curves of the parametric equations. Are the graphs the same? Are the orientations the same? Are the curves smooth?**

33. (a) $x = t$
$\quad\quad y = 2t + 1$

(b) $x = \cos\theta$
$\quad\quad y = 2\cos\theta + 1$

(c) $x = e^{-t}$
$\quad\quad y = 2e^{-t} + 1$

(d) $x = e^t$
$\quad\quad y = 2e^t + 1$

34. (a) $x = 2\cos\theta$
$\quad\quad y = 2\sin\theta$

(b) $x = \sqrt{4t^2 - 1}/|t|$
$\quad\quad y = 1/t$

(c) $x = \sqrt{t}$
$\quad\quad y = \sqrt{4 - t}$

(d) $x = -\sqrt{4 - e^{2t}}$
$\quad\quad y = e^t$

35. (a) $x = \cos\theta$
$\quad\quad y = 2\sin^2\theta$
$\quad\quad 0 < \theta < \pi$

(b) $x = \cos(-\theta)$
$\quad\quad y = 2\sin^2(-\theta)$
$\quad\quad 0 < \theta < \pi$

36. (a) $x = t + 1, y = t^3$

(b) $x = -t + 1, y = (-t)^3$

37. *Conjecture*

(a) Use a graphing utility to graph the curves represented by the two sets of parametric equations.

$\quad x = 4\cos t \quad\quad x = 4\cos(-t)$
$\quad y = 3\sin t \quad\quad y = 3\sin(-t)$

(b) Describe the change in the graph when the sign of the parameter is changed.

(c) Make a conjecture about the change in the graph of parametric equations when the sign of the parameter is changed.

(d) Test your conjecture with another set of parametric equations.

38. *Writing* Review Exercises 33–36 and write a short paragraph describing how the graphs of curves represented by different sets of parametric equations can differ even though eliminating the parameter from each yields the same rectangular equation.

In Exercises 39–42, eliminate the parameter and obtain the standard form of the rectangular equation.

39. Line through (x_1, y_1) and (x_2, y_2):
$$x = x_1 + t(x_2 - x_1), \quad y = y_1 + t(y_2 - y_1)$$
40. Circle: $x = h + r\cos\theta, \quad y = k + r\sin\theta$
41. Ellipse: $x = h + a\cos\theta, \quad y = k + b\sin\theta$
42. Hyperbola: $x = h + a\sec\theta, \quad y = k + b\tan\theta$

In Exercises 43–50, use the results of Exercises 39–42 to find a set of parametric equations for the line or conic.

43. Line: passes through $(0, 0)$ and $(5, -2)$
44. Line: passes through $(1, 4)$ and $(5, -2)$
45. Circle: center: $(2, 1)$; radius: 4
46. Circle: center: $(-3, 1)$; radius: 3
47. Ellipse: vertices: $(\pm 5, 0)$; foci: $(\pm 4, 0)$
48. Ellipse: vertices: $(4, 7), (4, -3)$; foci: $(4, 5), (4, -1)$
49. Hyperbola: vertices: $(\pm 4, 0)$; foci: $(\pm 5, 0)$
50. Hyperbola: vertices: $(0, \pm 1)$; foci: $(0, \pm 2)$

In Exercises 51–54, find two different sets of parametric equations for the rectangular equation.

51. $y = 3x - 2$ **52.** $y = \dfrac{2}{x - 1}$

53. $y = x^3$ **54.** $y = x^2$

 In Exercises 55–62, use a graphing utility to graph the curve represented by the parametric equations. Indicate the direction of the curve. Identify any points at which the curve is not smooth.

55. Cycloid: $x = 2(\theta - \sin\theta), \quad y = 2(1 - \cos\theta)$
56. Cycloid: $x = \theta + \sin\theta, \quad y = 1 - \cos\theta$
57. Prolate cycloid: $x = \theta - \frac{3}{2}\sin\theta, \quad y = 1 - \frac{3}{2}\cos\theta$
58. Prolate cycloid: $x = 2\theta - 4\sin\theta, \quad y = 2 - 4\cos\theta$
59. Hypocycloid: $x = 3\cos^3\theta, \quad y = 3\sin^3\theta$
60. Curtate cycloid: $x = 2\theta - \sin\theta, \quad y = 2 - \cos\theta$
61. Witch of Agnesi: $x = 2\cot\theta, \quad y = 2\sin^2\theta$
62. Folium of Descartes: $x = \dfrac{3t}{1 + t^3}, \quad y = \dfrac{3t^2}{1 + t^3}$

Writing About Concepts

63. State the definition of a plane curve given by parametric equations.
64. Explain the process of sketching a plane curve given by parametric equations. What is meant by the orientation of the curve?
65. State the definition of a smooth curve.

Writing About Concepts (continued)

66. Match each set of parametric equations with the correct graph. [The graphs are labeled (a), (b), (c), (d), (e), and (f).] Explain your reasoning.

(a)

(b)

(c)

(d)

(e)

(f)

(i) $x = t^2 - 1, \quad y = t + 2$
(ii) $x = \sin^2\theta - 1, \quad y = \sin\theta + 2$
(iii) Lissajous curve: $x = 4\cos\theta, \quad y = 2\sin 2\theta$
(iv) Evolute of ellipse: $x = \cos^3\theta, \quad y = 2\sin^3\theta$
(v) Involute of circle: $x = \cos\theta + \theta\sin\theta,$
$\quad y = \sin\theta - \theta\cos\theta$
(vi) Serpentine curve: $x = \cot\theta, \quad y = 4\sin\theta\cos\theta$

67. *Curtate Cycloid* A wheel of radius a rolls along a line without slipping. The curve traced by a point P that is b units from the center $(b < a)$ is called a **curtate cycloid** (see figure). Use the angle θ to find a set of parametric equations for this curve.

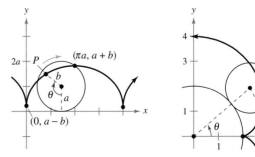

Figure for 67   Figure for 68

87. Hourglass: $(0 \le t \le 2\pi)$

$x = a \sin 2t$

$y = b \sin t$

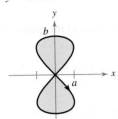

88. Teardrop: $(0 \le t \le 2\pi)$

$x = 2a \cos t - a \sin 2t$

$y = b \sin t$

Centroid **In Exercises 89 and 90, find the centroid of the region bounded by the graph of the parametric equations and the coordinate axes. (Use the result of Exercise 79.)**

89. $x = \sqrt{t}, y = 4 - t$

90. $x = \sqrt{4 - t}, y = \sqrt{t}$

Volume **In Exercises 91 and 92, find the volume of the solid formed by revolving the region bounded by the graphs of the given equations about the *x*-axis. (Use the result of Exercise 79.)**

91. $x = 3 \cos \theta, \quad y = 3 \sin \theta$

92. $x = \cos \theta, \quad y = 3 \sin \theta, \quad a > 0$

93. *Cycloid* Use the parametric equations

$$x = a(\theta - \sin \theta) \quad \text{and} \quad y = a(1 - \cos \theta), a > 0$$

to answer the following.

(a) Find dy/dx and d^2y/dx^2.

(b) Find the equations of the tangent line at the point where $\theta = \pi/6$.

(c) Find all points (if any) of horizontal tangency.

(d) Determine where the curve is concave upward or concave downward.

(e) Find the length of one arc of the curve.

94. Use the parametric equations

$$x = t^2\sqrt{3} \quad \text{and} \quad y = 3t - \frac{1}{3}t^3$$

to answer the following.

(a) Use a graphing utility to graph the curve on the interval $-3 \le t \le 3$.

(b) Find dy/dx and d^2y/dx^2.

(c) Find the equation of the tangent line at the point $\left(\sqrt{3}, \frac{8}{3}\right)$.

(d) Find the length of the curve.

(e) Find the surface area generated by revolving the curve about the *x*-axis.

95. *Involute of a Circle* The involute of a circle is described by the endpoint P of a string that is held taut as it is unwound from a spool that does not turn (see figure). Show that a parametric representation of the involute is

$$x = r(\cos \theta + \theta \sin \theta) \quad \text{and} \quad y = r(\sin \theta - \theta \cos \theta).$$

Figure for 95

96. *Involute of a Circle* The figure shows a piece of string tied to a circle with a radius of one unit. The string is just long enough to reach the opposite side of the circle. Find the area that is covered when the string is unwound counterclockwise.

97. (a) Use a graphing utility to graph the curve given by

$$x = \frac{1 - t^2}{1 + t^2}, \quad y = \frac{2t}{1 + t^2}, \quad -20 \le t \le 20.$$

(b) Describe the graph and confirm your result analytically.

(c) Discuss the speed at which the curve is traced as t increases from -20 to 20.

98. *Tractrix* A person moves from the origin along the positive *y*-axis pulling a weight at the end of a 12-meter rope. Initially, the weight is located at the point $(12, 0)$.

(a) In Exercise 86 of Section 8.7, it was shown that the path of the weight is modeled by the rectangular equation

$$y = -12 \ln\left(\frac{12 - \sqrt{144 - x^2}}{x}\right) - \sqrt{144 - x^2}$$

where $0 < x \le 12$. Use a graphing utility to graph the rectangular equation.

(b) Use a graphing utility to graph the parametric equations

$$x = 12 \operatorname{sech} \frac{t}{12} \quad \text{and} \quad y = t - 12 \tanh \frac{t}{12}$$

where $t \ge 0$. How does this graph compare with the graph in part (a)? Which graph (if either) do you think is a better representation of the path?

(c) Use the parametric equations for the tractrix to verify that the distance from the *y*-intercept of the tangent line to the point of tangency is independent of the location of the point of tangency.

True or False? **In Exercises 99 and 100, determine whether the statement is true or false. If it is false, explain why or give an example that shows it is false.**

99. If $x = f(t)$ and $y = g(t)$, then $d^2y/dx^2 = g''(t)/f''(t)$.

100. The curve given by $x = t^3$, $y = t^2$ has a horizontal tangent at the origin because $dy/dt = 0$ when $t = 0$.

Polar Coordinates and Polar Graphs

- Understand the polar coordinate system.
- Rewrite rectangular coordinates and equations in polar form and vice versa.
- Sketch the graph of an equation given in polar form.
- Find the slope of a tangent line to a polar graph.
- Identify several types of special polar graphs.

Polar Coordinates

So far, you have been representing graphs as collections of points (x, y) on the rectangular coordinate system. The corresponding equations for these graphs have been in either rectangular or parametric form. In this section you will study a coordinate system called the **polar coordinate system.**

To form the polar coordinate system in the plane, fix a point O, called the **pole** (or **origin**), and construct from O an initial ray called the **polar axis,** as shown in Figure 10.36. Then each point P in the plane can be assigned **polar coordinates** (r, θ), as follows.

$r = $ *directed distance* from O to P

$\theta = $ *directed angle*, counterclockwise from polar axis to segment $\overline{OP}$

Figure 10.37 shows three points on the polar coordinate system. Notice that in this system, it is convenient to locate points with respect to a grid of concentric circles intersected by **radial lines** through the pole.

Polar coordinates
Figure 10.36

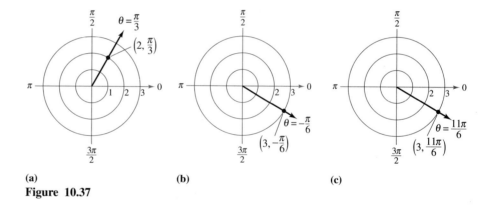

(a)   (b)   (c)
Figure 10.37

With rectangular coordinates, each point (x, y) has a unique representation. This is not true with polar coordinates. For instance, the coordinates (r, θ) and $(r, 2\pi + \theta)$ represent the same point [see parts (b) and (c) in Figure 10.37]. Also, because r is a *directed distance*, the coordinates (r, θ) and $(-r, \theta + \pi)$ represent the same point. In general, the point (r, θ) can be written as

$$(r, \theta) = (r, \theta + 2n\pi)$$

or

$$(r, \theta) = (-r, \theta + (2n + 1)\pi)$$

where n is any integer. Moreover, the pole is represented by $(0, \theta)$, where θ is any angle.

POLAR COORDINATES

The mathematician credited with first using polar coordinates was James Bernoulli, who introduced them in 1691. However, there is some evidence that it may have been Isaac Newton who first used them.

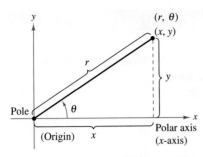

Relating polar and rectangular coordinates
Figure 10.38

Coordinate Conversion

To establish the relationship between polar and rectangular coordinates, let the polar axis coincide with the positive x-axis and the pole with the origin, as shown in Figure 10.38. Because (x, y) lies on a circle of radius r, it follows that $r^2 = x^2 + y^2$. Moreover, for $r > 0$, the definition of the trigonometric functions implies that

$$\tan \theta = \frac{y}{x}, \qquad \cos \theta = \frac{x}{r}, \qquad \text{and} \qquad \sin \theta = \frac{y}{r}.$$

If $r < 0$, you can show that the same relationships hold.

THEOREM 10.10   **Coordinate Conversion**

The polar coordinates (r, θ) of a point are related to the rectangular coordinates (x, y) of the point as follows.

1. $x = r \cos \theta$ **2.** $\tan \theta = \dfrac{y}{x}$

      $y = r \sin \theta$ $r^2 = x^2 + y^2$

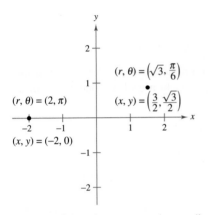

To convert from polar to rectangular coordinates, let $x = r \cos \theta$ and $y = r \sin \theta$.
Figure 10.39

EXAMPLE 1   Polar-to-Rectangular Conversion

a. For the point $(r, \theta) = (2, \pi)$,

$$x = r \cos \theta = 2 \cos \pi = -2 \qquad \text{and} \qquad y = r \sin \theta = 2 \sin \pi = 0.$$

So, the rectangular coordinates are $(x, y) = (-2, 0)$.

b. For the point $(r, \theta) = \left(\sqrt{3}, \pi/6\right)$,

$$x = \sqrt{3} \cos \frac{\pi}{6} = \frac{3}{2} \qquad \text{and} \qquad y = \sqrt{3} \sin \frac{\pi}{6} = \frac{\sqrt{3}}{2}.$$

So, the rectangular coordinates are $(x, y) = \left(3/2, \sqrt{3}/2\right)$. See Figure 10.39.

EXAMPLE 2   Rectangular-to-Polar Conversion

a. For the second quadrant point $(x, y) = (-1, 1)$,

$$\tan \theta = \frac{y}{x} = -1 \quad \Longrightarrow \quad \theta = \frac{3\pi}{4}.$$

Because θ was chosen to be in the same quadrant as (x, y), you should use a positive value of r.

$$\begin{aligned} r &= \sqrt{x^2 + y^2} \\ &= \sqrt{(-1)^2 + (1)^2} \\ &= \sqrt{2} \end{aligned}$$

This implies that *one* set of polar coordinates is $(r, \theta) = \left(\sqrt{2}, 3\pi/4\right)$.

b. Because the point $(x, y) = (0, 2)$ lies on the positive y-axis, choose $\theta = \pi/2$ and $r = 2$, and one set of polar coordinates is $(r, \theta) = (2, \pi/2)$.

See Figure 10.40.

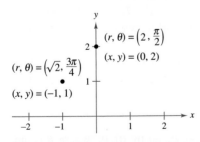

To convert from rectangular to polar coordinates, let $\tan \theta = y/x$ and $r = \sqrt{x^2 + y^2}$.
Figure 10.40

Polar Graphs

One way to sketch the graph of a polar equation is to convert to rectangular coordinates and then sketch the graph of the rectangular equation.

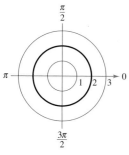

(a) Circle: $r = 2$

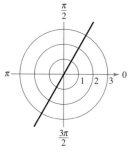

(b) Radial line: $\theta = \dfrac{\pi}{3}$

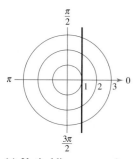

(c) Vertical line: $r = \sec\theta$
Figure 10.41

EXAMPLE 3   Graphing Polar Equations

Describe the graph of each polar equation. Confirm each description by converting to a rectangular equation.

a. $r = 2$ **b.** $\theta = \dfrac{\pi}{3}$ **c.** $r = \sec\theta$

Solution

a. The graph of the polar equation $r = 2$ consists of all points that are two units from the pole. In other words, this graph is a circle centered at the origin with a radius of 2. [See Figure 10.41(a).] You can confirm this by using the relationship $r^2 = x^2 + y^2$ to obtain the rectangular equation

$$x^2 + y^2 = 2^2. \qquad \text{Rectangular equation}$$

b. The graph of the polar equation $\theta = \pi/3$ consists of all points on the line that makes an angle of $\pi/3$ with the positive x-axis. [See Figure 10.41(b).] You can confirm this by using the relationship $\tan\theta = y/x$ to obtain the rectangular equation

$$y = \sqrt{3}\,x. \qquad \text{Rectangular equation}$$

c. The graph of the polar equation $r = \sec\theta$ is not evident by simple inspection, so you can begin by converting to rectangular form using the relationship $r\cos\theta = x$.

$$
\begin{aligned}
r &= \sec\theta && \text{Polar equation} \\
r\cos\theta &= 1 \\
x &= 1 && \text{Rectangular equation}
\end{aligned}
$$

From the rectangular equation, you can see that the graph is a vertical line. [See Figure 10.41(c.)]

TECHNOLOGY   Sketching the graphs of complicated polar equations *by hand* can be tedious. With technology, however, the task is not difficult. If your graphing utility has a *polar* mode, use it to graph the equations in the exercise set. If your graphing utility doesn't have a *polar* mode, but does have a *parametric* mode, you can graph $r = f(\theta)$ by writing the equation as

$$x = f(\theta)\cos\theta$$
$$y = f(\theta)\sin\theta.$$

For instance, the graph of $r = \frac{1}{2}\theta$ shown in Figure 10.42 was produced with a graphing calculator in *parametric* mode. This equation was graphed using the parametric equations

$$x = \frac{1}{2}\theta\cos\theta$$

$$y = \frac{1}{2}\theta\sin\theta$$

with the values of θ varying from -4π to 4π. This curve is of the form $r = a\theta$ and is called a **spiral of Archimedes.**

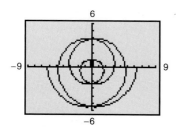

Spiral of Archimedes
Figure 10.42

EXAMPLE 4 **Sketching a Polar Graph**

Sketch the graph of $r = 2 \cos 3\theta$.

Solution Begin by writing the polar equation in parametric form.

$$x = 2 \cos 3\theta \cos \theta \qquad \text{and} \qquad y = 2 \cos 3\theta \sin \theta$$

After some experimentation, you will find that the entire curve, which is called a **rose curve,** can be sketched by letting θ vary from 0 to π, as shown in Figure 10.43. If you try duplicating this graph with a graphing utility, you will find that by letting θ vary from 0 to 2π, you will actually trace the entire curve *twice.*

NOTE One way to sketch the graph of $r = 2 \cos 3\theta$ by hand is to make a table of values.

θ	0	$\dfrac{\pi}{6}$	$\dfrac{\pi}{3}$	$\dfrac{\pi}{2}$	$\dfrac{2\pi}{3}$
r	2	0	-2	0	2

By extending the table and plotting the points, you will obtain the curve shown in Example 4.

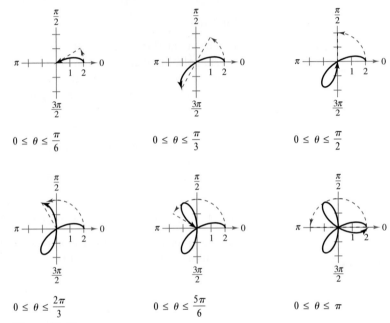

Figure 10.43

Use a graphing utility to experiment with other rose curves (they are of the form $r = a \cos n\theta$ or $r = a \sin n\theta$). For instance, Figure 10.44 shows the graphs of two other rose curves.

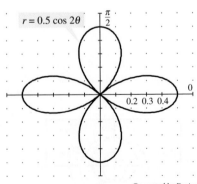

Generated by Derive

Rose curves
Figure 10.44

Slope and Tangent Lines

To find the slope of a tangent line to a polar graph, consider a differentiable function given by $r = f(\theta)$. To find the slope in polar form, use the parametric equations

$$x = r \cos \theta = f(\theta) \cos \theta \qquad \text{and} \qquad y = r \sin \theta = f(\theta) \sin \theta.$$

Using the parametric form of dy/dx given in Theorem 10.7, you have

$$\frac{dy}{dx} = \frac{dy/d\theta}{dx/d\theta}$$

$$= \frac{f(\theta) \cos \theta + f'(\theta) \sin \theta}{-f(\theta) \sin \theta + f'(\theta) \cos \theta}$$

which establishes the following theorem.

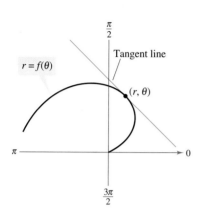

$r = f(\theta)$

Tangent line

(r, θ)

Tangent line to polar curve
Figure 10.45

> **THEOREM 10.11   Slope in Polar Form**
>
> If f is a differentiable function of θ, then the *slope* of the tangent line to the graph of $r = f(\theta)$ at the point (r, θ) is
>
> $$\frac{dy}{dx} = \frac{dy/d\theta}{dx/d\theta} = \frac{f(\theta) \cos \theta + f'(\theta) \sin \theta}{-f(\theta) \sin \theta + f'(\theta) \cos \theta}$$
>
> provided that $dx/d\theta \neq 0$ at (r, θ). (See Figure 10.45.)

From Theorem 10.11, you can make the following observations.

1. Solutions to $\dfrac{dy}{d\theta} = 0$ yield horizontal tangents, provided that $\dfrac{dx}{d\theta} \neq 0$.

2. Solutions to $\dfrac{dx}{d\theta} = 0$ yield vertical tangents, provided that $\dfrac{dy}{d\theta} \neq 0$.

If $dy/d\theta$ and $dx/d\theta$ are *simultaneously* 0, no conclusion can be drawn about tangent lines.

EXAMPLE 5   Finding Horizontal and Vertical Tangent Lines

Find the horizontal and vertical tangent lines of $r = \sin \theta, 0 \leq \theta \leq \pi$.

Solution   Begin by writing the equation in parametric form.

$$x = r \cos \theta = \sin \theta \cos \theta$$

and

$$y = r \sin \theta = \sin \theta \sin \theta = \sin^2 \theta$$

Next, differentiate x and y with respect to θ and set each derivative equal to 0.

$$\frac{dx}{d\theta} = \cos^2 \theta - \sin^2 \theta = \cos 2\theta = 0 \quad \Longrightarrow \quad \theta = \frac{\pi}{4}, \frac{3\pi}{4}$$

$$\frac{dy}{d\theta} = 2 \sin \theta \cos \theta = \sin 2\theta = 0 \quad \Longrightarrow \quad \theta = 0, \frac{\pi}{2}$$

So, the graph has vertical tangent lines at $\left(\sqrt{2}/2, \pi/4\right)$ and $\left(\sqrt{2}/2, 3\pi/4\right)$, and it has horizontal tangent lines at $(0, 0)$ and $(1, \pi/2)$, as shown in Figure 10.46.

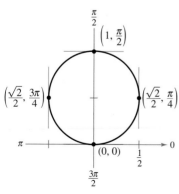

Horizontal and vertical tangent lines of $r = \sin \theta$
Figure 10.46

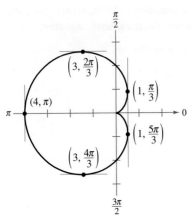

Horizontal and vertical tangent lines of
$r = 2(1 - \cos\theta)$
Figure 10.47

EXAMPLE 6   Finding Horizontal and Vertical Tangent Lines

Find the horizontal and vertical tangents to the graph of $r = 2(1 - \cos\theta)$.

Solution   Using $y = r\sin\theta$, differentiate and set $dy/d\theta$ equal to 0.

$$y = r\sin\theta = 2(1 - \cos\theta)\sin\theta$$

$$\frac{dy}{d\theta} = 2[(1 - \cos\theta)(\cos\theta) + \sin\theta(\sin\theta)]$$

$$= -2(2\cos\theta + 1)(\cos\theta - 1) = 0$$

So, $\cos\theta = -\frac{1}{2}$ and $\cos\theta = 1$, and you can conclude that $dy/d\theta = 0$ when $\theta = 2\pi/3, 4\pi/3$, and 0. Similarly, using $x = r\cos\theta$, you have

$$x = r\cos\theta = 2\cos\theta - 2\cos^2\theta$$

$$\frac{dx}{d\theta} = -2\sin\theta + 4\cos\theta\sin\theta = 2\sin\theta(2\cos\theta - 1) = 0.$$

So, $\sin\theta = 0$ or $\cos\theta = \frac{1}{2}$, and you can conclude that $dx/d\theta = 0$ when $\theta = 0$, π, $\pi/3$, and $5\pi/3$. From these results, and from the graph shown in Figure 10.47, you can conclude that the graph has horizontal tangents at $(3, 2\pi/3)$ and $(3, 4\pi/3)$, and has vertical tangents at $(1, \pi/3)$, $(1, 5\pi/3)$, and $(4, \pi)$. This graph is called a **cardioid**. Note that both derivatives $(dy/d\theta$ and $dx/d\theta)$ are 0 when $\theta = 0$. Using this information alone, you don't know whether the graph has a horizontal or vertical tangent line at the pole. From Figure 10.47, however, you can see that the graph has a cusp at the pole.

Theorem 10.11 has an important consequence. Suppose the graph of $r = f(\theta)$ passes through the pole when $\theta = \alpha$ and $f'(\alpha) \neq 0$. Then the formula for dy/dx simplifies as follows.

$$\frac{dy}{dx} = \frac{f'(\alpha)\sin\alpha + f(\alpha)\cos\alpha}{f'(\alpha)\cos\alpha - f(\alpha)\sin\alpha} = \frac{f'(\alpha)\sin\alpha + 0}{f'(\alpha)\cos\alpha - 0} = \frac{\sin\alpha}{\cos\alpha} = \tan\alpha$$

So, the line $\theta = \alpha$ is tangent to the graph at the pole, $(0, \alpha)$.

$f(\theta) = 2\cos 3\theta$

THEOREM 10.12   Tangent Lines at the Pole

If $f(\alpha) = 0$ and $f'(\alpha) \neq 0$, then the line $\theta = \alpha$ is tangent at the pole to the graph of $r = f(\theta)$.

Theorem 10.12 is useful because it states that the zeros of $r = f(\theta)$ can be used to find the tangent lines at the pole. Note that because a polar curve can cross the pole more than once, it can have more than one tangent line at the pole. For example, the rose curve

$$f(\theta) = 2\cos 3\theta$$

This rose curve has three tangent lines
$(\theta = \pi/6, \theta = \pi/2,$ and $\theta = 5\pi/6)$
at the pole.
Figure 10.48

has three tangent lines at the pole, as shown in Figure 10.48. For this curve, $f(\theta) = 2\cos 3\theta$ is 0 when θ is $\pi/6$, $\pi/2$, and $5\pi/6$. Moreover, the derivative $f'(\theta) = -6\sin 3\theta$ is not 0 for these values of θ.

Special Polar Graphs

Several important types of graphs have equations that are simpler in polar form than in rectangular form. For example, the polar equation of a circle having a radius of a and centered at the origin is simply $r = a$. Later in the text you will come to appreciate this benefit. For now, several other types of graphs that have simpler equations in polar form are shown below. (Conics are considered in Section 10.6.)

Limaçons

$r = a \pm b \cos \theta$

$r = a \pm b \sin \theta$

$(a > 0, b > 0)$

$\dfrac{a}{b} < 1$

Limaçon with
inner loop

$\dfrac{a}{b} = 1$

Cardioid
(heart-shaped)

$1 < \dfrac{a}{b} < 2$

Dimpled limaçon

$\dfrac{a}{b} \geq 2$

Convex limaçon

Rose Curves

n petals if n is odd

$2n$ petals if n is even

$(n \geq 2)$

$n = 3$

$r = a \cos n\theta$
Rose curve

$n = 4$

$r = a \cos n\theta$
Rose curve

$n = 5$

$r = a \sin n\theta$
Rose curve

$n = 2$

$r = a \sin n\theta$
Rose curve

Circles and Lemniscates

$r = a \cos \theta$
Circle

$r = a \sin \theta$
Circle

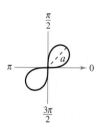

$r^2 = a^2 \sin 2\theta$
Lemniscate

$r^2 = a^2 \cos 2\theta$
Lemniscate

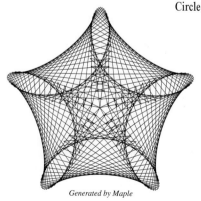

Generated by Maple

TECHNOLOGY   The rose curves described above are of the form $r = a \cos n\theta$ or $r = a \sin n\theta$, where n is a positive integer that is greater than or equal to 2. Use a graphing utility to graph $r = a \cos n\theta$ or $r = a \sin n\theta$ for some noninteger values of n. Are these graphs also rose curves? For example, try sketching the graph of $r = \cos \frac{2}{3}\theta$, $0 \leq \theta \leq 6\pi$.

FOR FURTHER INFORMATION For more information on rose curves and related curves, see the article "A Rose is a Rose . . ." by Peter M. Maurer in *The American Mathematical Monthly*. The computer-generated graph at the left is the result of an algorithm that Maurer calls "The Rose." To view this article, go to the website *www.matharticles.com*.

Exercises for Section 10.4

See www.CalcChat.com for worked-out solutions to odd-numbered exercises.

In Exercises 1–6, plot the point in polar coordinates and find the corresponding rectangular coordinates for the point.

1. $(4, 3\pi/6)$

2. $(-2, 7\pi/4)$

3. $(-4, -\pi/3)$

4. $(0, -7\pi/6)$

5. $\left(\sqrt{2}, 2.36\right)$

6. $(-3, -1.57)$

In Exercises 7–10, use the *angle* feature of a graphing utility to find the rectangular coordinates for the point given in polar coordinates. Plot the point.

7. $(5, 3\pi/4)$

8. $(-2, 11\pi/6)$

9. $(-3.5, 2.5)$

10. $(8.25, 1.3)$

In Exercises 11–16, the rectangular coordinates of a point are given. Plot the point and find *two* sets of polar coordinates for the point for $0 \le \theta < 2\pi$.

11. $(1, 1)$

12. $(0, -5)$

13. $(-3, 4)$

14. $(4, -2)$

15. $\left(\sqrt{3}, -1\right)$

16. $\left(3, -\sqrt{3}\right)$

In Exercises 17–20, use the *angle* feature of a graphing utility to find one set of polar coordinates for the point given in rectangular coordinates.

17. $(3, -2)$

18. $\left(3\sqrt{2}, 3\sqrt{2}\right)$

19. $\left(\frac{5}{2}, \frac{4}{3}\right)$

20. $(0, -5)$

21. Plot the point $(4, 3.5)$ if the point is given in (a) rectangular coordinates and (b) polar coordinates.

22. *Graphical Reasoning*

   (a) Set the window format of a graphing utility to rectangular coordinates and locate the cursor at any position off the axes. Move the cursor horizontally and vertically. Describe any changes in the displayed coordinates of the points.

   (b) Set the window format of a graphing utility to polar coordinates and locate the cursor at any position off the axes. Move the cursor horizontally and vertically. Describe any changes in the displayed coordinates of the points.

   (c) Why are the results in parts (a) and (b) different?

In Exercises 23–26, match the graph with its polar equation. [The graphs are labeled (a), (b), (c), and (d).]

(a)

(b)

(c)

(d)

23. $r = 2 \sin \theta$

24. $r = 4 \cos 2\theta$

25. $r = 3(1 + \cos \theta)$

26. $r = 2 \sec \theta$

In Exercises 27–34, convert the rectangular equation to polar form and sketch its graph.

27. $x^2 + y^2 = a^2$

28. $x^2 + y^2 - 2ax = 0$

29. $y = 4$

30. $x = 10$

31. $3x - y + 2 = 0$

32. $xy = 4$

33. $y^2 = 9x$

34. $(x^2 + y^2)^2 - 9(x^2 - y^2) = 0$

In Exercises 35–42, convert the polar equation to rectangular form and sketch its graph.

35. $r = 3$

36. $r = -2$

37. $r = \sin \theta$

38. $r = 5 \cos \theta$

39. $r = \theta$

40. $\theta = \dfrac{5\pi}{6}$

41. $r = 3 \sec \theta$

42. $r = 2 \csc \theta$

In Exercises 43–52, use a graphing utility to graph the polar equation. Find an interval for θ over which the graph is traced *only once.*

43. $r = 3 - 4 \cos \theta$

44. $r = 5(1 - 2 \sin \theta)$

45. $r = 2 + \sin \theta$

46. $r = 4 + 3 \cos \theta$

47. $r = \dfrac{2}{1 + \cos \theta}$

48. $r = \dfrac{2}{4 - 3 \sin \theta}$

49. $r = 2 \cos\left(\dfrac{3\theta}{2}\right)$

50. $r = 3 \sin\left(\dfrac{5\theta}{2}\right)$

51. $r^2 = 4 \sin 2\theta$

52. $r^2 = \dfrac{1}{\theta}$

53. Convert the equation

$$r = 2(h \cos \theta + k \sin \theta)$$

to rectangular form and verify that it is the equation of a circle. Find the radius and the rectangular coordinates of the center of the circle.

54. *Distance Formula*

(a) Verify that the Distance Formula for the distance between the two points (r_1, θ_1) and (r_2, θ_2) in polar coordinates is

$$d = \sqrt{r_1^2 + r_2^2 - 2r_1 r_2 \cos(\theta_1 - \theta_2)}.$$

(b) Describe the positions of the points relative to each other if $\theta_1 = \theta_2$. Simplify the Distance Formula for this case. Is the simplification what you expected? Explain.

(c) Simplify the Distance Formula if $\theta_1 - \theta_2 = 90°$. Is the simplification what you expected? Explain.

(d) Choose two points on the polar coordinate system and find the distance between them. Then choose different polar representations of the same two points and apply the Distance Formula again. Discuss the result.

In Exercises 55–58, use the result of Exercise 54 to approximate the distance between the two points in polar coordinates.

55. $\left(4, \dfrac{2\pi}{3}\right)$, $\left(2, \dfrac{\pi}{6}\right)$ **56.** $\left(10, \dfrac{7\pi}{6}\right)$, $(3, \pi)$

57. $(2, 0.5)$, $(7, 1.2)$ **58.** $(4, 2.5)$, $(12, 1)$

In Exercises 59 and 60, find dy/dx and the slopes of the tangent lines shown on the graph of the polar equation.

59. $r = 2 + 3\sin\theta$ **60.** $r = 2(1 - \sin\theta)$

 In Exercises 61–64, use a graphing utility to (a) graph the polar equation, (b) draw the tangent line at the given value of θ, and (c) find dy/dx at the given value of θ. (*Hint:* Let the increment between the values of θ equal $\pi/24$.)

61. $r = 3(1 - \cos\theta)$, $\theta = \dfrac{\pi}{2}$ **62.** $r = 3 - 2\cos\theta$, $\theta = 0$

63. $r = 3\sin\theta$, $\theta = \dfrac{\pi}{3}$ **64.** $r = 4$, $\theta = \dfrac{\pi}{4}$

In Exercises 65 and 66, find the points of horizontal and vertical tangency (if any) to the polar curve.

65. $r = 1 - \sin\theta$ **66.** $r = a\sin\theta$

In Exercises 67 and 68, find the points of horizontal tangency (if any) to the polar curve.

67. $r = 2\csc\theta + 3$ **68.** $r = a\sin\theta\cos^2\theta$

 In Exercises 69–72, use a graphing utility to graph the polar equation and find all points of horizontal tangency.

69. $r = 4\sin\theta\cos^2\theta$ **70.** $r = 3\cos 2\theta\sec\theta$

71. $r = 2\csc\theta + 5$ **72.** $r = 2\cos(3\theta - 2)$

In Exercises 73–80, sketch a graph of the polar equation and find the tangents at the pole.

73. $r = 3\sin\theta$ **74.** $r = 3\cos\theta$

75. $r = 2(1 - \sin\theta)$ **76.** $r = 3(1 - \cos\theta)$

77. $r = 2\cos 3\theta$ **78.** $r = -\sin 5\theta$

79. $r = 3\sin 2\theta$ **80.** $r = 3\cos 2\theta$

In Exercises 81–92, sketch a graph of the polar equation.

81. $r = 5$ **82.** $r = 2$

83. $r = 4(1 + \cos\theta)$ **84.** $r = 1 + \sin\theta$

85. $r = 3 - 2\cos\theta$ **86.** $r = 5 - 4\sin\theta$

87. $r = 3\csc\theta$ **88.** $r = \dfrac{6}{2\sin\theta - 3\cos\theta}$

89. $r = 2\theta$ **90.** $r = \dfrac{1}{\theta}$

91. $r^2 = 4\cos 2\theta$ **92.** $r^2 = 4\sin\theta$

 In Exercises 93–96, use a graphing utility to graph the equation and show that the given line is an asymptote of the graph.

	Name of Graph	Polar Equation	Asymptote
93.	Conchoid	$r = 2 - \sec\theta$	$x = -1$
94.	Conchoid	$r = 2 + \csc\theta$	$y = 1$
95.	Hyperbolic spiral	$r = 2/\theta$	$y = 2$
96.	Strophoid	$r = 2\cos 2\theta\sec\theta$	$x = -2$

Writing About Concepts

97. Describe the differences between the rectangular coordinate system and the polar coordinate system.

98. Give the equations for the coordinate conversion from rectangular to polar coordinates and vice versa.

99. For constants a and b, describe the graphs of the equations $r = a$ and $\theta = b$ in polar coordinates.

100. How are the slopes of tangent lines determined in polar coordinates? What are tangent lines at the pole and how are they determined?

101. Sketch the graph of $r = 4\sin\theta$ over each interval.

(a) $0 \le \theta \le \dfrac{\pi}{2}$ (b) $\dfrac{\pi}{2} \le \theta \le \pi$ (c) $-\dfrac{\pi}{2} \le \theta \le \dfrac{\pi}{2}$

 102. *Think About It*   Use a graphing utility to graph the polar equation $r = 6[1 + \cos(\theta - \phi)]$ for (a) $\phi = 0$, (b) $\phi = \pi/4$, and (c) $\phi = \pi/2$. Use the graphs to describe the effect of the angle ϕ. Write the equation as a function of $\sin\theta$ for part (c).

103. Verify that if the curve whose polar equation is $r = f(\theta)$ is rotated about the pole through an angle ϕ, then an equation for the rotated curve is $r = f(\theta - \phi)$.

104. The polar form of an equation for a curve is $r = f(\sin \theta)$. Show that the form becomes

(a) $r = f(-\cos \theta)$ if the curve is rotated counterclockwise $\pi/2$ radians about the pole.

(b) $r = f(-\sin \theta)$ if the curve is rotated counterclockwise π radians about the pole.

(c) $r = f(\cos \theta)$ if the curve is rotated counterclockwise $3\pi/2$ radians about the pole.

In Exercises 105–108, use the results of Exercises 103 and 104.

105. Write an equation for the limaçon $r = 2 - \sin \theta$ after it has been rotated by the given amount. Use a graphing utility to graph the rotated limaçon.

(a) $\dfrac{\pi}{4}$ (b) $\dfrac{\pi}{2}$ (c) π (d) $\dfrac{3\pi}{2}$

106. Write an equation for the rose curve $r = 2 \sin 2\theta$ after it has been rotated by the given amount. Verify the results by using a graphing utility to graph the rotated rose curve.

(a) $\dfrac{\pi}{6}$ (b) $\dfrac{\pi}{2}$ (c) $\dfrac{2\pi}{3}$ (d) π

107. Sketch the graph of each equation.

(a) $r = 1 - \sin \theta$ (b) $r = 1 - \sin\left(\theta - \dfrac{\pi}{4}\right)$

108. Prove that the tangent of the angle ψ ($0 \le \psi \le \pi/2$) between the radial line and the tangent line at the point (r, θ) on the graph of $r = f(\theta)$ (see figure) is given by $\tan \psi = |r/(dr/d\theta)|$.

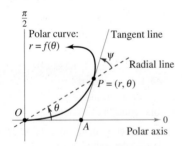

In Exercises 109–114, use the result of Exercise 108 to find the angle ψ between the radial and tangent lines to the graph for the indicated value of θ. Use a graphing utility to graph the polar equation, the radial line, and the tangent line for the indicated value of θ. Identify the angle ψ.

Polar Equation	Value of θ
109. $r = 2(1 - \cos \theta)$	$\theta = \pi$
110. $r = 3(1 - \cos \theta)$	$\theta = 3\pi/4$
111. $r = 2 \cos 3\theta$	$\theta = \pi/4$
112. $r = 4 \sin 2\theta$	$\theta = \pi/6$
113. $r = \dfrac{6}{1 - \cos \theta}$	$\theta = 2\pi/3$
114. $r = 5$	$\theta = \pi/6$

True or False? **In Exercises 115–118, determine whether the statement is true or false. If it is false, explain why or give an example that shows it is false.**

115. If (r_1, θ_1) and (r_2, θ_2) represent the same point on the polar coordinate system, then $|r_1| = |r_2|$.

116. If (r, θ_1) and (r, θ_2) represent the same point on the polar coordinate system, then $\theta_1 = \theta_2 + 2\pi n$ for some integer n.

117. If $x > 0$, then the point (x, y) on the rectangular coordinate system can be represented by (r, θ) on the polar coordinate system, where $r = \sqrt{x^2 + y^2}$ and $\theta = \arctan(y/x)$.

118. The polar equations $r = \sin 2\theta$ and $r = -\sin 2\theta$ have the same graph.

Section Project: Anamorphic Art

Use the anamorphic transformations

$$r = y + 16 \qquad \text{and} \qquad \theta = -\frac{\pi}{8}x, \qquad -\frac{3\pi}{4} \le \theta \le \frac{3\pi}{4}$$

to sketch the transformed polar image of the rectangular graph. When the reflection (in a cylindrical mirror centered at the pole) of each polar image is viewed from the polar axis, the viewer will see the original rectangular image.

(a) $y = 3$ (b) $x = 2$

(c) $y = x + 5$ (d) $x^2 + (y - 5)^2 = 5^2$

This example of anamorphic art is from the Museum of Science and Industry in Manchester, England. When the reflection of the transformed "polar painting" is viewed in the mirror, the viewer sees faces.

FOR FURTHER INFORMATION For more information on anamorphic art, see the article "Anamorphisms" by Philip Hickin in the *Mathematical Gazette*.

The area of a sector of a circle is $A = \frac{1}{2}\theta r^2$.
Figure 10.49

Area and Arc Length in Polar Coordinates

- Find the area of a region bounded by a polar graph.
- Find the points of intersection of two polar graphs.
- Find the arc length of a polar graph.
- Find the area of a surface of revolution (polar form).

Area of a Polar Region

The development of a formula for the area of a polar region parallels that for the area of a region on the rectangular coordinate system, but uses sectors of a circle instead of rectangles as the basic element of area. In Figure 10.49, note that the area of a circular sector of radius r is given by $\frac{1}{2}\theta r^2$, provided θ is measured in radians.

Consider the function given by $r = f(\theta)$, where f is continuous and nonnegative in the interval given by $\alpha \le \theta \le \beta$. The region bounded by the graph of f and the radial lines $\theta = \alpha$ and $\theta = \beta$ is shown in Figure 10.50(a). To find the area of this region, partition the interval $[\alpha, \beta]$ into n equal subintervals

$$\alpha = \theta_0 < \theta_1 < \theta_2 < \cdots < \theta_{n-1} < \theta_n = \beta.$$

Then, approximate the area of the region by the sum of the areas of the n sectors, as shown in Figure 10.50(b).

$$\text{Radius of } i\text{th sector} = f(\theta_i)$$

$$\text{Central angle of } i\text{th sector} = \frac{\beta - \alpha}{n} = \Delta\theta$$

$$A \approx \sum_{i=1}^{n} \left(\frac{1}{2}\right) \Delta\theta [f(\theta_i)]^2$$

Taking the limit as $n \to \infty$ produces

$$A = \lim_{n \to \infty} \frac{1}{2} \sum_{i=1}^{n} [f(\theta_i)]^2 \Delta\theta$$

$$= \frac{1}{2} \int_{\alpha}^{\beta} [f(\theta)]^2 \, d\theta$$

which leads to the following theorem.

(a)

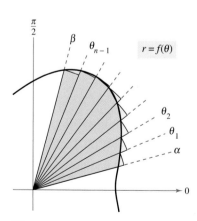

(b)
Figure 10.50

THEOREM 10.13   Area in Polar Coordinates

If f is continuous and nonnegative on the interval $[\alpha, \beta]$, $0 < \beta - \alpha \le 2\pi$, then the area of the region bounded by the graph of $r = f(\theta)$ between the radial lines $\theta = \alpha$ and $\theta = \beta$ is given by

$$A = \frac{1}{2} \int_{\alpha}^{\beta} [f(\theta)]^2 \, d\theta$$

$$= \frac{1}{2} \int_{\alpha}^{\beta} r^2 \, d\theta. \qquad 0 < \beta - \alpha \le 2\pi$$

NOTE   You can use the same formula to find the area of a region bounded by the graph of a continuous *nonpositive* function. However, the formula is not necessarily valid if f takes on both positive *and* negative values in the interval $[\alpha, \beta]$.

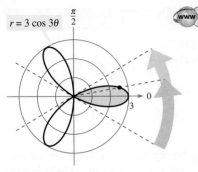

$r = 3 \cos 3\theta$

The area of one petal of the rose curve that lies between the radial lines $\theta = -\pi/6$ and $\theta = \pi/6$ is $3\pi/4$.
Figure 10.51

EXAMPLE 1 **Finding the Area of a Polar Region**

Find the area of one petal of the rose curve given by $r = 3 \cos 3\theta$.

Solution In Figure 10.51, you can see that the right petal is traced as θ increases from $-\pi/6$ to $\pi/6$. So, the area is

$$
\begin{aligned}
A = \frac{1}{2} \int_{\alpha}^{\beta} r^2 \, d\theta &= \frac{1}{2} \int_{-\pi/6}^{\pi/6} (3 \cos 3\theta)^2 \, d\theta && \text{Formula for area in polar coordinates} \\
&= \frac{9}{2} \int_{-\pi/6}^{\pi/6} \frac{1 + \cos 6\theta}{2} \, d\theta && \text{Trigonometric identity} \\
&= \frac{9}{4} \left[\theta + \frac{\sin 6\theta}{6} \right]_{-\pi/6}^{\pi/6} \\
&= \frac{9}{4} \left(\frac{\pi}{6} + \frac{\pi}{6} \right) \\
&= \frac{3\pi}{4}.
\end{aligned}
$$

NOTE To find the area of the region lying inside all three petals of the rose curve in Example 1, you could not simply integrate between 0 and 2π. In doing this you would obtain $9\pi/2$, which is twice the area of the three petals. The duplication occurs because the rose curve is traced twice as θ increases from 0 to 2π.

EXAMPLE 2 **Finding the Area Bounded by a Single Curve**

Find the area of the region lying between the inner and outer loops of the limaçon $r = 1 - 2 \sin \theta$.

Solution In Figure 10.52, note that the inner loop is traced as θ increases from $\pi/6$ to $5\pi/6$. So, the area inside the *inner loop* is

$$
\begin{aligned}
A_1 = \frac{1}{2} \int_{\alpha}^{\beta} r^2 \, d\theta &= \frac{1}{2} \int_{\pi/6}^{5\pi/6} (1 - 2 \sin \theta)^2 \, d\theta && \text{Formula for area in polar coordinates} \\
&= \frac{1}{2} \int_{\pi/6}^{5\pi/6} (1 - 4 \sin \theta + 4 \sin^2 \theta) \, d\theta \\
&= \frac{1}{2} \int_{\pi/6}^{5\pi/6} \left[1 - 4 \sin \theta + 4 \left(\frac{1 - \cos 2\theta}{2} \right) \right] d\theta && \text{Trigonometric identity} \\
&= \frac{1}{2} \int_{\pi/6}^{5\pi/6} (3 - 4 \sin \theta - 2 \cos 2\theta) \, d\theta && \text{Simplify.} \\
&= \frac{1}{2} \left[3\theta + 4 \cos \theta - \sin 2\theta \right]_{\pi/6}^{5\pi/6} \\
&= \frac{1}{2} \left(2\pi - 3\sqrt{3} \right) \\
&= \pi - \frac{3\sqrt{3}}{2}.
\end{aligned}
$$

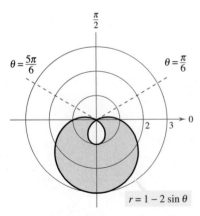

$\theta = \frac{5\pi}{6}$ $\theta = \frac{\pi}{6}$

$r = 1 - 2 \sin \theta$

The area between the inner and outer loops is approximately 8.34.
Figure 10.52

In a similar way, you can integrate from $5\pi/6$ to $13\pi/6$ to find that the area of the region lying inside the outer loop is $A_2 = 2\pi + (3\sqrt{3}/2)$. The area of the region lying between the two loops is the difference of A_2 and A_1.

$$
A = A_2 - A_1 = \left(2\pi + \frac{3\sqrt{3}}{2} \right) - \left(\pi - \frac{3\sqrt{3}}{2} \right) = \pi + 3\sqrt{3} \approx 8.34
$$

Points of Intersection of Polar Graphs

Because a point may be represented in different ways in polar coordinates, care must be taken in determining the points of intersection of two polar graphs. For example, consider the points of intersection of the graphs of

$$r = 1 - 2\cos\theta \quad \text{and} \quad r = 1$$

as shown in Figure 10.53. If, as with rectangular equations, you attempted to find the points of intersection by solving the two equations simultaneously, you would obtain

$r = 1 - 2\cos\theta$	First equation
$1 = 1 - 2\cos\theta$	Substitute $r = 1$ from 2nd equation into 1st equation.
$\cos\theta = 0$	Simplify.
$\theta = \dfrac{\pi}{2}, \dfrac{3\pi}{2}.$	Solve for θ.

FOR FURTHER INFORMATION For more information on using technology to find points of intersection, see the article "Finding Points of Intersection of Polar-Coordinate Graphs" by Warren W. Esty in *Mathematics Teacher*. To view this article, go to the website *www.matharticles.com*.

The corresponding points of intersection are $(1, \pi/2)$ and $(1, 3\pi/2)$. However, from Figure 10.53 you can see that there is a *third* point of intersection that did not show up when the two polar equations were solved simultaneously. (This is one reason why you should sketch a graph when finding the area of a polar region.) The reason the third point was not found is that it does not occur with the same coordinates in the two graphs. On the graph of $r = 1$, the point occurs with coordinates $(1, \pi)$, but on the graph of $r = 1 - 2\cos\theta$, the point occurs with coordinates $(-1, 0)$.

You can compare the problem of finding points of intersection of two polar graphs with that of finding collision points of two satellites in intersecting orbits about Earth, as shown in Figure 10.54. The satellites will not collide as long as they reach the points of intersection at different times (θ-values). Collisions will occur only at the points of intersection that are "simultaneous points"—those reached at the same time (θ-value).

NOTE   Because the pole can be represented by $(0, \theta)$, where θ is *any* angle, you should check separately for the pole when finding points of intersection.

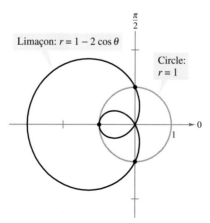

Three points of intersection: $(1, \pi/2)$, $(-1, 0), (1, 3\pi/2)$
Figure 10.53

The paths of satellites can cross without causing a collision.
Figure 10.54

EXAMPLE 3 Finding the Area of a Region Between Two Curves

Find the area of the region common to the two regions bounded by the following curves.

$$r = -6 \cos \theta \qquad \text{Circle}$$
$$r = 2 - 2 \cos \theta \qquad \text{Cardioid}$$

Solution Because both curves are symmetric with respect to the x-axis, you can work with the upper half-plane, as shown in Figure 10.55. The gray shaded region lies between the circle and the radial line $\theta = 2\pi/3$. Because the circle has coordinates $(0, \pi/2)$ at the pole, you can integrate between $\pi/2$ and $2\pi/3$ to obtain the area of this region. The region that is shaded red is bounded by the radial lines $\theta = 2\pi/3$ and $\theta = \pi$ and the cardioid. So, you can find the area of this second region by integrating between $2\pi/3$ and π. The sum of these two integrals gives the area of the common region lying *above* the radial line $\theta = \pi$.

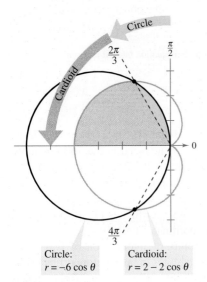

Circle: $r = -6 \cos \theta$ Cardioid: $r = 2 - 2 \cos \theta$

Figure 10.55

$$
\frac{A}{2} = \overbrace{\frac{1}{2} \int_{\pi/2}^{2\pi/3} (-6 \cos \theta)^2 \, d\theta}^{\substack{\text{Region between circle} \\ \text{and radial line } \theta = 2\pi/3}} + \overbrace{\frac{1}{2} \int_{2\pi/3}^{\pi} (2 - 2 \cos \theta)^2 \, d\theta}^{\substack{\text{Region between cardioid and} \\ \text{radial lines } \theta = 2\pi/3 \text{ and } \theta = \pi}}
$$

$$
= 18 \int_{\pi/2}^{2\pi/3} \cos^2 \theta \, d\theta + \frac{1}{2} \int_{2\pi/3}^{\pi} (4 - 8 \cos \theta + 4 \cos^2 \theta) \, d\theta
$$

$$
= 9 \int_{\pi/2}^{2\pi/3} (1 + \cos 2\theta) \, d\theta + \int_{2\pi/3}^{\pi} (3 - 4 \cos \theta + \cos 2\theta) \, d\theta
$$

$$
= 9 \left[\theta + \frac{\sin 2\theta}{2} \right]_{\pi/2}^{2\pi/3} + \left[3\theta - 4 \sin \theta + \frac{\sin 2\theta}{2} \right]_{2\pi/3}^{\pi}
$$

$$
= 9 \left(\frac{2\pi}{3} - \frac{\sqrt{3}}{4} - \frac{\pi}{2} \right) + \left(3\pi - 2\pi + 2\sqrt{3} + \frac{\sqrt{3}}{4} \right)
$$

$$
= \frac{5\pi}{2}
$$

$$
\approx 7.85
$$

Finally, multiplying by 2, you can conclude that the total area is 5π. ▬▬▬

NOTE To check the reasonableness of the result obtained in Example 3, note that the area of the circular region is $\pi r^2 = 9\pi$. So, it seems reasonable that the area of the region lying inside the circle and the cardioid is 5π.

To see the benefit of polar coordinates for finding the area in Example 3, consider the following integral, which gives the comparable area in rectangular coordinates.

$$
\frac{A}{2} = \int_{-4}^{-3/2} \sqrt{2\sqrt{1 - 2x} - x^2 - 2x + 2} \, dx + \int_{-3/2}^{0} \sqrt{-x^2 - 6x} \, dx
$$

Use the integration capabilities of a graphing utility to show that you obtain the same area as that found in Example 3.

Arc Length in Polar Form

NOTE   When applying the arc length formula to a polar curve, be sure that the curve is traced out only once on the interval of integration. For instance, the rose curve given by $r = \cos 3\theta$ is traced out once on the interval $0 \leq \theta \leq \pi$, but is traced out twice on the interval $0 \leq \theta \leq 2\pi$.

The formula for the length of a polar arc can be obtained from the arc length formula for a curve described by parametric equations. (See Exercise 77.)

THEOREM 10.14   Arc Length of a Polar Curve

Let f be a function whose derivative is continuous on an interval $\alpha \leq \theta \leq \beta$. The length of the graph of $r = f(\theta)$ from $\theta = \alpha$ to $\theta = \beta$ is

$$s = \int_{\alpha}^{\beta} \sqrt{[f(\theta)]^2 + [f'(\theta)]^2}\, d\theta = \int_{\alpha}^{\beta} \sqrt{r^2 + \left(\frac{dr}{d\theta}\right)^2}\, d\theta.$$

EXAMPLE 4   Finding the Length of a Polar Curve

Find the length of the arc from $\theta = 0$ to $\theta = 2\pi$ for the cardioid

$$r = f(\theta) = 2 - 2\cos\theta$$

as shown in Figure 10.56.

Solution   Because $f'(\theta) = 2\sin\theta$, you can find the arc length as follows.

$$s = \int_{\alpha}^{\beta} \sqrt{[f(\theta)]^2 + [f'(\theta)]^2}\, d\theta \qquad \text{Formula for arc length of a polar curve}$$

$$= \int_{0}^{2\pi} \sqrt{(2 - 2\cos\theta)^2 + (2\sin\theta)^2}\, d\theta$$

$$= 2\sqrt{2} \int_{0}^{2\pi} \sqrt{1 - \cos\theta}\, d\theta \qquad \text{Simplify.}$$

$$= 2\sqrt{2} \int_{0}^{2\pi} \sqrt{2\sin^2\frac{\theta}{2}}\, d\theta \qquad \text{Trigonometric identity}$$

$$= 4 \int_{0}^{2\pi} \sin\frac{\theta}{2}\, d\theta \qquad \sin\frac{\theta}{2} \geq 0 \text{ for } 0 \leq \theta \leq 2\pi$$

$$= 8\left[-\cos\frac{\theta}{2} \right]_{0}^{2\pi}$$

$$= 8(1 + 1)$$

$$= 16$$

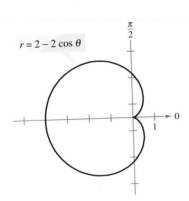

$r = 2 - 2\cos\theta$

Figure 10.56

In the fifth step of the solution, it is legitimate to write

$$\sqrt{2\sin^2(\theta/2)} = \sqrt{2}\sin(\theta/2)$$

rather than

$$\sqrt{2\sin^2(\theta/2)} = \sqrt{2}\,|\sin(\theta/2)|$$

because $\sin(\theta/2) \geq 0$ for $0 \leq \theta \leq 2\pi$.

NOTE   Using Figure 10.56, you can determine the reasonableness of this answer by comparing it with the circumference of a circle. For example, a circle of radius $\frac{5}{2}$ has a circumference of $5\pi \approx 15.7$.

In Exercises 65–72, convert the polar equation to rectangular form.

65. $r = 3 \cos \theta$

66. $r = 10$

67. $r = -2(1 + \cos \theta)$

68. $r = \dfrac{1}{2 - \cos \theta}$

69. $r^2 = \cos 2\theta$

70. $r = 4 \sec\left(\theta - \dfrac{\pi}{3}\right)$

71. $r = 4 \cos 2\theta \sec \theta$

72. $\theta = \dfrac{3\pi}{4}$

In Exercises 73–76, convert the rectangular equation to polar form.

73. $(x^2 + y^2)^2 = ax^2 y$

74. $x^2 + y^2 - 4x = 0$

75. $x^2 + y^2 = a^2 \left(\arctan \dfrac{y}{x}\right)^2$

76. $(x^2 + y^2)\left(\arctan \dfrac{y}{x}\right)^2 = a^2$

In Exercises 77–88, sketch a graph of the polar equation.

77. $r = 4$

78. $\theta = \dfrac{\pi}{12}$

79. $r = -\sec \theta$

80. $r = 3 \csc \theta$

81. $r = -2(1 + \cos \theta)$

82. $r = 3 - 4 \cos \theta$

83. $r = 4 - 3 \cos \theta$

84. $r = 2\theta$

85. $r = -3 \cos 2\theta$

86. $r = \cos 5\theta$

87. $r^2 = 4 \sin^2 2\theta$

88. $r^2 = \cos 2\theta$

 In Exercises 89–92, use a graphing utility to graph the polar equation.

89. $r = \dfrac{3}{\cos(\theta - \pi/4)}$

90. $r = 2 \sin \theta \cos^2 \theta$

91. $r = 4 \cos 2\theta \sec \theta$

92. $r = 4(\sec \theta - \cos \theta)$

In Exercises 93 and 94, (a) find the tangents at the pole, (b) find all points of vertical and horizontal tangency, and (c) use a graphing utility to graph the polar equation and draw a tangent line to the graph for $\theta = \pi/6$.

93. $r = 1 - 2 \cos \theta$

94. $r^2 = 4 \sin 2\theta$

95. Find the angle between the circle $r = 3 \sin \theta$ and the limaçon $r = 4 - 5 \sin \theta$ at the point of intersection $(3/2, \pi/6)$.

96. *True or False?*   There is a unique polar coordinate representation for each point in the plane. Explain.

In Exercises 97 and 98, show that the graphs of the polar equations are orthogonal at the points of intersection. Use a graphing utility to confirm your results graphically.

97. $r = 1 + \cos \theta$
    $r = 1 - \cos \theta$

98. $r = a \sin \theta$
    $r = a \cos \theta$

In Exercises 99–102, find the area of the region.

99. Interior of $r = 2 + \cos \theta$

100. Interior of $r = 5(1 - \sin \theta)$

101. Interior of $r^2 = 4 \sin 2\theta$

102. Common interior of $r = 4 \cos \theta$ and $r = 2$

In Exercises 103–106, use a graphing utility to graph the polar equation. Set up an integral for finding the area of the given region and use the integration capabilities of a graphing utility to approximate the integral accurate to two decimal places.

103. Interior of $r = \sin \theta \cos^2 \theta$

104. Interior of $r = 4 \sin 3\theta$

105. Common interior of $r = 3$ and $r^2 = 18 \sin 2\theta$

106. Region bounded by the polar axis and $r = e^\theta$ for $0 \le \theta \le \pi$

In Exercises 107 and 108, find the length of the curve over the given interval.

Polar Equation	*Interval*
107. $r = a(1 - \cos \theta)$	$0 \le \theta \le \pi$
108. $r = a \cos 2\theta$	$-\dfrac{\pi}{2} \le \theta \le \dfrac{\pi}{2}$

In Exercises 109 and 110, write an integral that represents the area of the surface formed by revolving the curve about the given line. Use a graphing utility to approximate the integral.

Polar Equation	*Interval*	*Axis of Revolution*
109. $r = 1 + 4 \cos \theta$	$0 \le \theta \le \dfrac{\pi}{2}$	Polar axis
110. $r = 2 \sin \theta$	$0 \le \theta \le \dfrac{\pi}{2}$	$\theta = \dfrac{\pi}{2}$

In Exercises 111–116, sketch and identify the graph. Use a graphing utility to confirm your results.

111. $r = \dfrac{2}{1 - \sin \theta}$

112. $r = \dfrac{2}{1 + \cos \theta}$

113. $r = \dfrac{6}{3 + 2 \cos \theta}$

114. $r = \dfrac{4}{5 - 3 \sin \theta}$

115. $r = \dfrac{4}{2 - 3 \sin \theta}$

116. $r = \dfrac{8}{2 - 5 \cos \theta}$

In Exercises 117–122, find a polar equation for the line or conic with its focus at the pole.

117. Circle
     Center: $(5, \pi/2)$
     Solution point: $(0, 0)$

118. Line
     Solution point: $(0, 0)$
     Slope: $\sqrt{3}$

119. Parabola
     Vertex: $(2, \pi)$

120. Parabola
     Vertex: $(2, \pi/2)$

121. Ellipse
     Vertices: $(5, 0), (1, \pi)$

122. Hyperbola
     Vertices: $(1, 0), (7, 0)$

1. Consider the parabola $x^2 = 4y$ and the focal chord $y = \frac{3}{4}x + 1$.

   (a) Sketch the graph of the parabola and the focal chord.

   (b) Show that the tangent lines to the parabola at the endpoints of the focal chord intersect at right angles.

   (c) Show that the tangent lines to the parabola at the endpoints of the focal chord intersect on the directrix of the parabola.

2. Consider the parabola $x^2 = 4py$ and one of its focal chords.

   (a) Show that the tangent lines to the parabola at the endpoints of the focal chord intersect at right angles.

   (b) Show that the tangent lines to the parabola at the endpoints of the focal chord intersect on the directrix of the parabola.

3. Prove Theorem 10.2, Reflective Property of a Parabola, as shown in the figure.

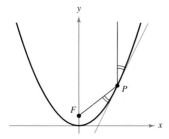

4. Consider the hyperbola

$$\frac{x^2}{a^2} - \frac{y^2}{b^2} = 1$$

with foci F_1 and F_2, as shown in the figure. Let T be the tangent line at a point M on the hyperbola. Show that incoming rays of light aimed at one focus are reflected by a hyperbolic mirror toward the other focus.

        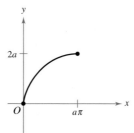

Figure for 4   **Figure for 5**

5. Consider a circle of radius a tangent to the y-axis and the line $x = 2a$, as shown in the figure. Let A be the point where the segment OB intersects the circle. The **cissoid of Diocles** consists of all points P such that $OP = AB$.

   (a) Find a polar equation of the cissoid.

   (b) Find a set of parametric equations for the cissoid that does not contain trigonometric functions.

   (c) Find a rectangular equation of the cissoid.

6. Consider the region bounded by the ellipse $x^2/a^2 + y^2/b^2 = 1$, with eccentricity $e = c/a$.

   (a) Show that the area of the region is πab.

   (b) Show that the solid (oblate spheroid) generated by revolving the region about the minor axis of the ellipse has a volume $V = 4\pi^2 b/3$ and a surface area of

   $$S = 2\pi a^2 + \pi\left(\frac{b^2}{e}\right)\ln\left(\frac{1+e}{1-e}\right).$$

   (c) Show that the solid (prolate spheroid) generated by revolving the region about the major axis of the ellipse has a volume of $V = 4\pi ab^2/3$ and a surface area of

   $$S = 2\pi b^2 + 2\pi\left(\frac{ab}{e}\right)\arcsin e.$$

7. The curve given by the parametric equations

$$x(t) = \frac{1 - t^2}{1 + t^2} \quad \text{and} \quad y(t) = \frac{t(1 - t^2)}{1 + t^2}$$

   is called a **strophoid.**

   (a) Find a rectangular equation of the strophoid.

   (b) Find a polar equation of the strophoid.

   (c) Sketch a graph of the strophoid.

   (d) Find the equations of the two tangent lines at the origin.

   (e) Find the points on the graph where the tangent lines are horizontal.

8. Find a rectangular equation of the portion of the cycloid given by the parametric equations $x = a(\theta - \sin \theta)$ and $y = a(1 - \cos \theta)$, $0 \le \theta \le \pi$, as shown in the figure.

9. Consider the **cornu spiral** given by

$$x(t) = \int_0^t \cos\left(\frac{\pi u^2}{2}\right) du \quad \text{and} \quad y(t) = \int_0^t \sin\left(\frac{\pi u^2}{2}\right) du.$$

   (a) Use a graphing utility to graph the spiral over the interval $-\pi \le t \le \pi$.

   (b) Show that the cornu spiral is symmetric with respect to the origin.

   (c) Find the length of the cornu spiral from $t = 0$ to $t = a$. What is the length of the spiral from $t = -\pi$ to $t = \pi$?

10. A particle is moving along the path described by the parametric equations $x = 1/t$ and $y = \sin t/t$, for $1 \le t < \infty$, as shown in the figure. Find the length of this path.

11. Let a and b be positive constants. Find the area of the region in the first quadrant bounded by the graph of the polar equation

$$r = \frac{ab}{(a \sin \theta + b \cos \theta)}, \quad 0 \le \theta \le \frac{\pi}{2}.$$

12. Consider the right triangle shown in the figure.

(a) Show that the area of the triangle is $A(\alpha) = \dfrac{1}{2}\displaystyle\int_0^\alpha \sec^2 \theta \, d\theta$.

(b) Show that $\tan \alpha = \displaystyle\int_0^\alpha \sec^2 \theta \, d\theta$.

(c) Use part (b) to derive the formula for the derivative of the tangent function.

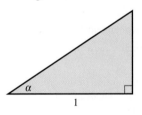

Figure for 12   **Figure for 13**

13. Determine the polar equation of the set of all points (r, θ), the product of whose distances from the points $(1, 0)$ and $(-1, 0)$ is equal to 1, as shown in the figure.

14. Four dogs are located at the corners of a square with sides of length d. The dogs all move counterclockwise at the same speed directly toward the next dog, as shown in the figure. Find the polar equation of a dog's path as it spirals toward the center of the square.

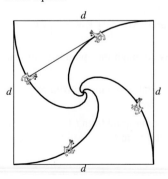

15. An air traffic controller spots two planes at the same altitude flying toward each other (see figure). Their flight paths are 20° and 315°. One plane is 150 miles from point P with a speed of 375 miles per hour. The other is 190 miles from point P with a speed of 450 miles per hour.

(a) Find parametric equations for the path of each plane where t is the time in hours, with $t = 0$ corresponding to the time at which the air traffic controller spots the planes.

(b) Use the result of part (a) to write the distance between the planes as a function of t.

 (c) Use a graphing utility to graph the function in part (b). When will the distance between the planes be minimum? If the planes must keep a separation of at least 3 miles, is the requirement met?

 16. Use a graphing utility to graph the curve shown below. The curve is given by

$$r = e^{\cos \theta} - 2 \cos 4\theta + \sin^5 \frac{\theta}{12}.$$

Over what interval must θ vary to produce the curve?

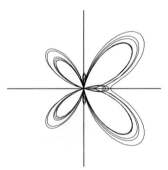

FOR FURTHER INFORMATION For more information on this curve, see the article "A Study in Step Size" by Temple H. Fay in *Mathematics Magazine*. To view this article, go to the website *www.matharticles.com*.

 17. Use a graphing utility to graph the polar equation $r = \cos 5\theta + n \cos \theta$, for $0 \le \theta < \pi$ and for the integers $n = -5$ to $n = 5$. What values of n produce the "heart" portion of the curve? What values of n produce the "bell" portion? (This curve, created by Michael W. Chamberlin, appeared in *The College Mathematics Journal*.)

Appendices

The remaining appendices are located on the website that accompanies this text at *college.hmco.com*.

A Proofs of Selected Theorems

> **THEOREM 2.2** **Properties of Limits (Properties 2, 3, 4, and 5)**
> **(page 79)**
>
> Let b and c be real numbers, let n be a positive integer, and let f and g be functions with the following limits.
>
> $$\lim_{x \to c} f(x) = L \quad \text{and} \quad \lim_{x \to c} g(x) = K$$
>
> **2.** Sum or difference: $\quad \lim_{x \to c} [f(x) \pm g(x)] = L \pm K$
>
> **3.** Product: $\quad \lim_{x \to c} [f(x)g(x)] = LK$
>
> **4.** Quotient: $\quad \lim_{x \to c} \dfrac{f(x)}{g(x)} = \dfrac{L}{K}, \quad$ provided $K \neq 0$
>
> **5.** Power: $\quad \lim_{x \to c} [f(x)]^n = L^n$

Proof To prove Property 2, choose $\varepsilon > 0$. Because $\varepsilon/2 > 0$, you know that there exists $\delta_1 > 0$ such that $0 < |x - c| < \delta_1$ implies $|f(x) - L| < \varepsilon/2$. You also know that there exists $\delta_2 > 0$ such that $0 < |x - c| < \delta_2$ implies $|g(x) - K| < \varepsilon/2$. Let δ be the smaller of δ_1 and δ_2; then $0 < |x - c| < \delta$ implies that

$$|f(x) - L| < \frac{\varepsilon}{2} \quad \text{and} \quad |g(x) - K| < \frac{\varepsilon}{2}.$$

So, you can apply the triangle inequality to conclude that

$$|[f(x) + g(x)] - (L + K)| \leq |f(x) - L| + |g(x) - K| < \frac{\varepsilon}{2} + \frac{\varepsilon}{2} = \varepsilon$$

which implies that

$$\lim_{x \to c} [f(x) + g(x)] = L + K = \lim_{x \to c} f(x) + \lim_{x \to c} g(x).$$

The proof that

$$\lim_{x \to c} [f(x) - g(x)] = L - K$$

is similar.

To prove Property 3, given that

$$\lim_{x \to c} f(x) = L \quad \text{and} \quad \lim_{x \to c} g(x) = K$$

you can write

$$f(x)g(x) = [f(x) - L][g(x) - K] + [Lg(x) + Kf(x)] - LK.$$

Because the limit of $f(x)$ is L, and the limit of $g(x)$ is K, you have

$$\lim_{x \to c} [f(x) - L] = 0 \quad \text{and} \quad \lim_{x \to c} [g(x) - K] = 0.$$

Let $0 < \varepsilon < 1$. Then there exists $\delta > 0$ such that if $0 < |x - c| < \delta$, then

$$|f(x) - L - 0| < \varepsilon \quad \text{and} \quad |g(x) - K - 0| < \varepsilon$$

which implies that

$$|[f(x) - L][g(x) - K] - 0| = |f(x) - L| \, |g(x) - K| < \varepsilon\varepsilon < \varepsilon.$$

So,

$$\lim_{x \to c} [f(x) - L][g(x) - K] = 0.$$

Furthermore, by Property 1, you have

$$\lim_{x \to c} Lg(x) = LK \quad \text{and} \quad \lim_{x \to c} Kf(x) = KL.$$

Finally, by Property 2, you obtain

$$\lim_{x \to c} f(x)g(x) = \lim_{x \to c} [f(x) - L][g(x) - K] + \lim_{x \to c} Lg(x) + \lim_{x \to c} Kf(x) - \lim_{x \to c} LK$$

$$= 0 + LK + KL - LK$$

$$= LK.$$

To prove Property 4, note that it is sufficient to prove that

$$\lim_{x \to c} \frac{1}{g(x)} = \frac{1}{K}.$$

Then you can use Property 3 to write

$$\lim_{x \to c} \frac{f(x)}{g(x)} = \lim_{x \to c} f(x) \frac{1}{g(x)} = \lim_{x \to c} f(x) \cdot \lim_{x \to c} \frac{1}{g(x)} = \frac{L}{K}.$$

Let $\varepsilon > 0$. Because $\lim_{x \to c} g(x) = K$, there exists $\delta_1 > 0$ such that if

$$0 < |x - c| < \delta_1, \text{ then } |g(x) - K| < \frac{|K|}{2}$$

which implies that

$$|K| = |g(x) + [|K| - g(x)]| \le |g(x)| + ||K| - g(x)| < |g(x)| + \frac{|K|}{2}.$$

That is, for $0 < |x - c| < \delta_1$,

$$\frac{|K|}{2} < |g(x)| \quad \text{or} \quad \frac{1}{|g(x)|} < \frac{2}{|K|}.$$

Similarly, there exists a $\delta_2 > 0$ such that if $0 < |x - c| < \delta_2$, then

$$|g(x) - K| < \frac{|K|^2}{2} \varepsilon.$$

Let δ be the smaller of δ_1 and δ_2. For $0 < |x - c| < \delta$, you have

$$\left| \frac{1}{g(x)} - \frac{1}{K} \right| = \left| \frac{K - g(x)}{g(x)K} \right| = \frac{1}{|K|} \cdot \frac{1}{|g(x)|} |K - g(x)| < \frac{1}{|K|} \cdot \frac{2}{|K|} \frac{|K|^2}{2} \varepsilon = \varepsilon.$$

So, $\lim_{x \to c} \dfrac{1}{g(x)} = \dfrac{1}{K}.$

Finally, the proof of Property 5 can be obtained by a straightforward application of mathematical induction coupled with Property 3.

> ### THEOREM 2.4   The Limit of a Function Involving a Radical (page 80)
>
> Let n be a positive integer. The following limit is valid for all c if n is odd, and is valid for $c > 0$ if n is even.
>
> $$\lim_{x \to c} \sqrt[n]{x} = \sqrt[n]{c}.$$

Proof   Consider the case for which $c > 0$ and n is any positive integer. For a given $\varepsilon > 0$, you need to find $\delta > 0$ such that

$$\left| \sqrt[n]{x} - \sqrt[n]{c} \right| < \varepsilon \quad \text{whenever} \quad 0 < |x - c| < \delta$$

which is the same as saying

$$-\varepsilon < \sqrt[n]{x} - \sqrt[n]{c} < \varepsilon \quad \text{whenever} \quad -\delta < x - c < \delta.$$

Assume $\varepsilon < \sqrt[n]{c}$, which implies that $0 < \sqrt[n]{c} - \varepsilon < \sqrt[n]{c}$. Now, let δ be the smaller of the two numbers

$$c - \left(\sqrt[n]{c} - \varepsilon \right)^n \quad \text{and} \quad \left(\sqrt[n]{c} + \varepsilon \right)^n - c.$$

Then you have

$$
\begin{array}{rcl}
-\delta < x - c & & < \delta \\
-\left[c - \left(\sqrt[n]{c} - \varepsilon \right)^n \right] < x - c & & < \left(\sqrt[n]{c} + \varepsilon \right)^n - c \\
\left(\sqrt[n]{c} - \varepsilon \right)^n - c < x - c & & < \left(\sqrt[n]{c} + \varepsilon \right)^n - c \\
\left(\sqrt[n]{c} - \varepsilon \right)^n < x & & < \left(\sqrt[n]{c} + \varepsilon \right)^n \\
\sqrt[n]{c} - \varepsilon < \sqrt[n]{x} & & < \sqrt[n]{c} + \varepsilon \\
-\varepsilon < \sqrt[n]{x} - \sqrt[n]{c} & & < \varepsilon.
\end{array}
$$

                                                              ∎

> ### THEOREM 2.5   The Limit of a Composite Function (page 81)
>
> If f and g are functions such that $\lim_{x \to c} g(x) = L$ and $\lim_{x \to L} f(x) = f(L)$, then
>
> $$\lim_{x \to c} f(g(x)) = f\left(\lim_{x \to c} g(x) \right) = f(L).$$

Proof   For a given $\varepsilon > 0$, you must find $\delta > 0$ such that

$$|f(g(x)) - f(L)| < \varepsilon \quad \text{whenever} \quad 0 < |x - c| < \delta.$$

Because the limit of $f(x)$ as $x \to L$ is $f(L)$, you know there exists $\delta_1 > 0$ such that

$$|f(u) - f(L)| < \varepsilon \quad \text{whenever} \quad |u - L| < \delta_1.$$

Moreover, because the limit of $g(x)$ as $x \to c$ is L, you know there exists $\delta > 0$ such that

$$|g(x) - L| < \delta_1 \quad \text{whenever} \quad 0 < |x - c| < \delta.$$

Finally, letting $u = g(x)$, you have

$$|f(g(x)) - f(L)| < \varepsilon \quad \text{whenever} \quad 0 < |x - c| < \delta. \qquad \blacksquare$$

THEOREM 2.7   Functions That Agree at All But One Point (page 82)

Let c be a real number and let $f(x) = g(x)$ for all $x \neq c$ in an open interval containing c. If the limit of $g(x)$ as x approaches c exists, then the limit of $f(x)$ also exists and

$$\lim_{x \to c} f(x) = \lim_{x \to c} g(x).$$

Proof   Let L be the limit of $g(x)$ as $x \to c$. Then, for each $\varepsilon > 0$ there exists a $\delta > 0$ such that $f(x) = g(x)$ in the open intervals $(c - \delta, c)$ and $(c, c + \delta)$, and

$$|g(x) - L| < \varepsilon \quad \text{whenever} \quad 0 < |x - c| < \delta.$$

Because $f(x) = g(x)$ for all x in the open interval other than $x = c$, it follows that

$$|f(x) - L| < \varepsilon \quad \text{whenever} \quad 0 < |x - c| < \delta.$$

So, the limit of $f(x)$ as $x \to c$ is also L.

THEOREM 2.8   The Squeeze Theorem (page 85)

If $h(x) \leq f(x) \leq g(x)$ for all x in an open interval containing c, except possibly at c itself, and if

$$\lim_{x \to c} h(x) = L = \lim_{x \to c} g(x)$$

then $\lim_{x \to c} f(x)$ exists and is equal to L.

Proof   For $\varepsilon > 0$ there exist $\delta_1 > 0$ and $\delta_2 > 0$ such that

$$|h(x) - L| < \varepsilon \quad \text{whenever} \quad 0 < |x - c| < \delta_1$$

and

$$|g(x) - L| < \varepsilon \quad \text{whenever} \quad 0 < |x - c| < \delta_2.$$

Because $h(x) \leq f(x) \leq g(x)$ for all x in an open interval containing c, except possibly at c itself, there exists $\delta_3 > 0$ such that $h(x) \leq f(x) \leq g(x)$ for $0 < |x - c| < \delta_3$. Let δ be the smallest of δ_1, δ_2, and δ_3. Then, if $0 < |x - c| < \delta$, it follows that $|h(x) - L| < \varepsilon$ and $|g(x) - L| < \varepsilon$, which implies that

$$-\varepsilon < h(x) - L < \varepsilon \quad \text{and} \quad -\varepsilon < g(x) - L < \varepsilon$$
$$L - \varepsilon < h(x) \quad \text{and} \quad g(x) < L + \varepsilon.$$

Now, because $h(x) \leq f(x) \leq g(x)$, it follows that $L - \varepsilon < f(x) < L + \varepsilon$, which implies that $|f(x) - L| < \varepsilon$. Therefore,

$$\lim_{x \to c} f(x) = L.$$

THEOREM 2.14   Vertical Asymptotes (page 105)

Let f and g be continuous on an open interval containing c. If $f(c) \neq 0$, $g(c) = 0$, and there exists an open interval containing c such that $g(x) \neq 0$ for all $x \neq c$ in the interval, then the graph of the function given by

$$h(x) = \frac{f(x)}{g(x)}$$

has a vertical asymptote at $x = c$.

Proof   Consider the case for which $f(c) > 0$, and there exists $b > c$ such that $c < x < b$ implies $g(x) > 0$. Then for $M > 0$, choose δ_1 such that

$$0 < x - c < \delta_1 \quad \text{implies that} \quad \frac{f(c)}{2} < f(x) < \frac{3f(c)}{2}$$

and δ_2 such that

$$0 < x - c < \delta_2 \quad \text{implies that} \quad 0 < g(x) < \frac{f(c)}{2M}.$$

Now let δ be the smaller of δ_1 and δ_2. Then it follows that

$$0 < x - c < \delta \quad \text{implies that} \quad \frac{f(x)}{g(x)} > \frac{f(c)}{2}\left[\frac{2M}{f(c)}\right] = M.$$

So, it follows that

$$\lim_{x \to c^+} \frac{f(x)}{g(x)} = \infty$$

and the line $x = c$ is a vertical asymptote of the graph of h.   ▬▬▬

Alternative Form of the Derivative (page 121)

The derivative of f at c is given by

$$f'(c) = \lim_{x \to c} \frac{f(x) - f(c)}{x - c}$$

provided this limit exists.

Proof   The derivative of f at c is given by

$$f'(c) = \lim_{\Delta x \to 0} \frac{f(c + \Delta x) - f(c)}{\Delta x}.$$

Let $x = c + \Delta x$. Then $x \to c$ as $\Delta x \to 0$. So, replacing $c + \Delta x$ by x, you have

$$f'(c) = \lim_{\Delta x \to 0} \frac{f(c + \Delta x) - f(c)}{\Delta x} = \lim_{x \to c} \frac{f(x) - f(c)}{x - c}.$$ ▬▬▬

THEOREM 3.11   The Chain Rule (page 152)

If $y = f(u)$ is a differentiable function of u, and $u = g(x)$ is a differentiable function of x, then $y = f(g(x))$ is a differentiable function of x and

$$\frac{dy}{dx} = \frac{dy}{du} \cdot \frac{du}{dx}$$

or, equivalently,

$$\frac{d}{dx}[f(g(x))] = f'(g(x))g'(x).$$

Proof   In Section 3.4, you let $h(x) = f(g(x))$ and used the alternative form of the derivative to show that $h'(c) = f'(g(c))g'(c)$, provided $g(x) \neq g(c)$ for values of x other than c. Now consider a more general proof. Begin by considering the derivative of f.

$$f'(x) = \lim_{\Delta x \to 0} \frac{f(x + \Delta x) - f(x)}{\Delta x} = \lim_{\Delta x \to 0} \frac{\Delta y}{\Delta x}$$

For a fixed value of x, define a function η such that

$$\eta(\Delta x) = \begin{cases} 0, & \Delta x = 0 \\ \dfrac{\Delta y}{\Delta x} - f'(x), & \Delta x \neq 0 \end{cases}.$$

Because the limit of $\eta(\Delta x)$ as $\Delta x \to 0$ doesn't depend on the value of $\eta(0)$, you have

$$\lim_{\Delta x \to 0} \eta(\Delta x) = \lim_{\Delta x \to 0} \left[\frac{\Delta y}{\Delta x} - f'(x) \right] = 0$$

and you can conclude that η is continuous at 0. Moreover, because $\Delta y = 0$ when $\Delta x = 0$, the equation

$$\Delta y = \Delta x \eta(\Delta x) + \Delta x f'(x)$$

is valid whether Δx is zero or not. Now, by letting $\Delta u = g(x + \Delta x) - g(x)$, you can use the continuity of g to conclude that

$$\lim_{\Delta x \to 0} \Delta u = \lim_{\Delta x \to 0} [g(x + \Delta x) - g(x)] = 0$$

which implies that

$$\lim_{\Delta x \to 0} \eta(\Delta u) = 0.$$

Finally,

$$\Delta y = \Delta u \eta(\Delta u) + \Delta u f'(u) \to \frac{\Delta y}{\Delta x} = \frac{\Delta u}{\Delta x} \eta(\Delta u) + \frac{\Delta u}{\Delta x} f'(u), \quad \Delta x \neq 0$$

and taking the limit as $\Delta x \to 0$, you have

$$\frac{dy}{dx} = \frac{du}{dx} \left[\lim_{\Delta x \to 0} \eta(\Delta u) \right] + \frac{du}{dx} f'(u) = \frac{dy}{dx}(0) + \frac{du}{dx} f'(u)$$

$$= \frac{du}{dx} f'(u)$$

$$= \frac{du}{dx} \cdot \frac{dy}{du}.$$

THEOREM 3.16 Continuity and Differentiability of Inverse Functions (page 175)

Let f be a function whose domain is an interval I. If f has an inverse, then the following statements are true.

1. If f is continuous on its domain, then f^{-1} is continuous on its domain.
2. If f is differentiable on an interval containing c and $f'(c) \neq 0$, then f^{-1} is differentiable at $f(c)$.

Proof To prove Property 1, you first need to define what is meant by a *strictly increasing* function or a *strictly decreasing* function. A function f is **strictly increasing** on an entire interval I if for any two numbers x_1 and x_2 in the interval, $x_1 < x_2$ implies $f(x_1) < f(x_2)$. The function f is **strictly decreasing** on the entire interval I if $x_1 < x_2$ implies $f(x_1) > f(x_2)$. The function f is **strictly monotonic** on the interval I if it is either strictly increasing or strictly decreasing. Now show that if f is continuous on I, and has an inverse, then f is strictly monotonic on I. Suppose that f were not strictly monotonic. Then there would exist numbers x_1, x_2, x_3 in I such that $x_1 < x_2 < x_3$, but $f(x_2)$ is not between $f(x_1)$ and $f(x_3)$. Without loss of generality, assume $f(x_1) < f(x_3) < f(x_2)$. By the Intermediate Value Theorem, there exists a number x_0 between x_1 and x_2 such that $f(x_0) = f(x_3)$. So, f is not one-to-one and cannot have an inverse. So, f must be strictly monotonic.

Because f is continuous, the Intermediate Value Theorem implies that the set of values of f, $\{f(x): x \in I\}$, forms an interval J. Assume that a is an interior point of J. From the previous argument, $f^{-1}(a)$ is an interior point of I. Let $\varepsilon > 0$. There exists $0 < \varepsilon_1 < \varepsilon$ such that

$$I_1 = (f^{-1}(a) - \varepsilon_1, f^{-1}(a) + \varepsilon_1) \subseteq I.$$

Because f is strictly monotonic on I_1, the set of values $\{f(x): x \in I_1\}$ forms an interval $J_1 \subseteq J$. Let $\delta > 0$ such that $(a - \delta, a + \delta) \subseteq J_1$. Finally, if $|y - a| < \delta$, then $|f^{-1}(y) - f^{-1}(a)| < \varepsilon_1 < \varepsilon$. So, f^{-1} is continuous at a. A similar proof can be given if a is an endpoint.

To prove Property 2, consider the limit

$$(f^{-1})'(a) = \lim_{y \to a} \frac{f^{-1}(y) - f^{-1}(a)}{y - a}$$

where a is in the domain of f^{-1} and $f^{-1}(a) = c$. Because f is differentiable on an interval containing c, f is continuous on that interval, and so is f^{-1} at a. So, $y \to a$ implies that $x \to c$, and you have

$$(f^{-1})'(a) = \lim_{x \to c} \frac{x - c}{f(x) - f(c)} = \lim_{x \to c} \frac{1}{\left(\dfrac{f(x) - f(c)}{x - c}\right)} = \frac{1}{\displaystyle\lim_{x \to c} \dfrac{f(x) - f(c)}{x - c}} = \frac{1}{f'(c)}.$$

So, $(f^{-1})'(a)$ exists, and f^{-1} is differentiable at $f(c)$. ▬▬▬

THEOREM 3.17 The Derivative of an Inverse Function (page 175)

Let f be a function that is differentiable on an interval I. If f has an inverse function g, then g is differentiable at any x for which $f'(g(x)) \neq 0$. Moreover,

$$g'(x) = \frac{1}{f'(g(x))}, \quad f'(g(x)) \neq 0.$$

Proof   From the proof of Theorem 3.16, letting $a = x$, you know that g is differentiable. Using the Chain Rule, differentiate both sides of the equation $x = f(g(x))$ to obtain

$$1 = f'(g(x)) \frac{d}{dx}[g(x)].$$

Because $f'(g(x)) \neq 0$, you can divide by this quantity to obtain

$$\frac{d}{dx}[g(x)] = \frac{1}{f'(g(x))}.$$

Concavity Interpretation (page 230)

1. Let f be differentiable on an open interval I. If the graph of f is concave *upward* on I, then the graph of f lies *above* all of its tangent lines on I.

2. Let f be differentiable on an open interval I. If the graph of f is concave *downward* on I, then the graph of f lies *below* all of its tangent lines on I.

Proof   Assume that f is concave upward on $I = (a, b)$. Then, f' is increasing on (a, b). Let c be a point in the interval $I = (a, b)$. The equation of the tangent line to the graph of f at c is given by

$$g(x) = f(c) + f'(c)(x - c).$$

If x is in the open interval (c, b), then the directed distance from the point $(x, f(x))$ (on the graph of f) to the point $(x, g(x))$ (on the tangent line) is given by

$$d = f(x) - [f(c) + f'(c)(x - c)]$$
$$= f(x) - f(c) - f'(c)(x - c).$$

Moreover, by the Mean Value Theorem, there exists a number z in (c, x) such that

$$f'(z) = \frac{f(x) - f(c)}{x - c}.$$

So, you have

$$d = f(x) - f(c) - f'(c)(x - c)$$
$$= f'(z)(x - c) - f'(c)(x - c)$$
$$= [f'(z) - f'(c)](x - c).$$

The second factor $(x - c)$ is positive because $c < x$. Moreover, because f' is increasing, it follows that the first factor $[f'(z) - f'(c)]$ is also positive. Therefore, $d > 0$ and you can conclude that the graph of f lies above the tangent line at x. If x is in the open interval (a, c), a similar argument can be given. This proves the first statement. The proof of the second statement is similar.

THEOREM 4.10   Limits at Infinity (page 239)

If r is a positive rational number and c is any real number, then

$$\lim_{x \to \infty} \frac{c}{x^r} = 0.$$

Furthermore, if x^r is defined when $x < 0$, then $\lim\limits_{x \to -\infty} \dfrac{c}{x^r} = 0.$

Proof   Begin by proving that

$$\lim_{x \to \infty} \frac{1}{x} = 0.$$

For $\varepsilon > 0$, let $M = 1/\varepsilon$. Then, for $x > M$, you have

$$x > M = \frac{1}{\varepsilon} \implies \frac{1}{x} < \varepsilon \implies \left| \frac{1}{x} - 0 \right| < \varepsilon.$$

So, by the definition of a limit at infinity, you can conclude that the limit of $1/x$ as $x \to \infty$ is 0. Now, using this result and letting $r = m/n$, you can write the following.

$$
\begin{aligned}
\lim_{x \to \infty} \frac{c}{x^r} &= \lim_{x \to \infty} \frac{c}{x^{m/n}} \\
&= c \left[\lim_{x \to \infty} \left(\frac{1}{\sqrt[n]{x}} \right)^m \right] \\
&= c \left(\lim_{x \to \infty} \sqrt[n]{\frac{1}{x}} \right)^m \\
&= c \left(\sqrt[n]{\lim_{x \to \infty} \frac{1}{x}} \right)^m \\
&= c \left(\sqrt[n]{0} \right)^m \\
&= 0
\end{aligned}
$$

The proof of the second part of the theorem is similar.   ▬▬▬

THEOREM 5.2   Summation Formulas (page 296)

1. $\displaystyle\sum_{i=1}^{n} c = cn$ **2.** $\displaystyle\sum_{i=1}^{n} i = \frac{n(n+1)}{2}$

3. $\displaystyle\sum_{i=1}^{n} i^2 = \frac{n(n+1)(2n+1)}{6}$ **4.** $\displaystyle\sum_{i=1}^{n} i^3 = \frac{n^2(n+1)^2}{4}$

Proof   The proof of Property 1 is straightforward. By adding c to itself n times, you obtain a sum of cn.

To prove Property 2, write the sum in increasing and decreasing order and add corresponding terms, as follows.

$$\sum_{i=1}^{n} i = \quad 1 \quad + \quad 2 \quad + \quad 3 \quad + \cdots + (n-1) + \quad n$$

$$\sum_{i=1}^{n} i = \quad n \quad + (n-1) + (n-2) + \cdots + \quad 2 \quad + \quad 1$$

$$2\sum_{i=1}^{n} i = (n+1) + (n+1) + (n+1) + \cdots + (n+1) + (n+1)$$

$$\underbrace{\qquad\qquad\qquad\qquad\qquad}_{n \text{ terms}}$$

So,

$$\sum_{i=1}^{n} i = \frac{n(n+1)}{2}.$$

To prove Property 3, use mathematical induction. First, if $n = 1$, the result is true because

$$\sum_{i=1}^{1} i^2 = 1^2 = 1 = \frac{1(1 + 1)(2 + 1)}{6}.$$

Now, assuming the result is true for $n = k$, you can show that it is true for $n = k + 1$, as follows.

$$\sum_{i=1}^{k+1} i^2 = \sum_{i=1}^{k} i^2 + (k + 1)^2$$

$$= \frac{k(k + 1)(2k + 1)}{6} + (k + 1)^2$$

$$= \frac{k + 1}{6}(2k^2 + k + 6k + 6)$$

$$= \frac{k + 1}{6}[(2k + 3)(k + 2)]$$

$$= \frac{(k + 1)(k + 2)[2(k + 1) + 1]}{6}$$

Property 4 can be proved using a similar argument with mathematical induction.

THEOREM 5.8   Preservation of Inequality (page 314)

1. If f is integrable and nonnegative on the closed interval $[a, b]$, then

$$0 \leq \int_{a}^{b} f(x)\, dx.$$

2. If f and g are integrable on the closed interval $[a, b]$ and $f(x) \leq g(x)$ for every x in $[a, b]$, then

$$\int_{a}^{b} f(x)\, dx \leq \int_{a}^{b} g(x)\, dx.$$

Proof   To prove Property 1, suppose, on the contrary, that

$$\int_{a}^{b} f(x)\, dx = I < 0.$$

Then, let $a = x_0 < x_1 < x_2 < \cdots < x_n = b$ be a partition of $[a, b]$, and let

$$R = \sum_{i=1}^{n} f(c_i)\, \Delta x_i$$

be a Riemann sum. Because $f(x) \geq 0$, it follows that $R \geq 0$. Now, for $\|\Delta\|$ sufficiently small, you have $|R - I| < -I/2$, which implies that

$$\sum_{i=1}^{n} f(c_i)\, \Delta x_i = R < I - \frac{I}{2} < 0$$

which is not possible. From this contradiction, you can conclude that

$$0 \leq \int_{a}^{b} f(x)\, dx.$$

To prove Property 2 of the theorem, note that $f(x) \leq g(x)$ implies that $g(x) - f(x) \geq 0$. So, you can apply the result of Property 1 to conclude that

$$0 \leq \int_a^b [g(x) - f(x)]\, dx$$

$$0 \leq \int_a^b g(x)\, dx - \int_a^b f(x)\, dx$$

$$\int_a^b f(x)\, dx \leq \int_a^b g(x)\, dx.$$

THEOREM 8.3   The Extended Mean Value Theorem (page 568)

If f and g are differentiable on an open interval (a, b) and continuous on $[a, b]$ such that $g'(x) \neq 0$ for any x in (a, b), then there exists a point c in (a, b) such that

$$\frac{f'(c)}{g'(c)} = \frac{f(b) - f(a)}{g(b) - g(a)}.$$

Proof You can assume that $g(a) \neq g(b)$, because otherwise, by Rolle's Theorem, it would follow that $g'(x) = 0$ for some x in (a, b). Now, define $h(x)$ to be

$$h(x) = f(x) - \left[\frac{f(b) - f(a)}{g(b) - g(a)}\right] g(x).$$

Then

$$h(a) = f(a) - \left[\frac{f(b) - f(a)}{g(b) - g(a)}\right] g(a) = \frac{f(a)g(b) - f(b)g(a)}{g(b) - g(a)}$$

and

$$h(b) = f(b) - \left[\frac{f(b) - f(a)}{g(b) - g(a)}\right] g(b) = \frac{f(a)g(b) - f(b)g(a)}{g(b) - g(a)}$$

and, by Rolle's Theorem, there exists a point c in (a, b) such that

$$h'(c) = f'(c) - \frac{f(b) - f(a)}{g(b) - g(a)} g'(c) = 0$$

which implies that

$$\frac{f'(c)}{g'(c)} = \frac{f(b) - f(a)}{g(b) - g(a)}.$$

> **THEOREM 8.4   L'Hôpital's Rule (page 568)**
>
> Let f and g be functions that are differentiable on an open interval (a, b) containing c, except possibly at c itself. Assume that $g'(x) \neq 0$ for all x in (a, b), except possibly at c itself. If the limit of $f(x)/g(x)$ as x approaches c produces the indeterminate form $0/0$, then
>
> $$\lim_{x \to c} \frac{f(x)}{g(x)} = \lim_{x \to c} \frac{f'(x)}{g'(x)}$$
>
> provided the limit on the right exists (or is infinite). This result also applies if the limit of $f(x)/g(x)$ as x approaches c produces any one of the indeterminate forms ∞/∞, $(-\infty)/\infty$, $\infty/(-\infty)$, or $(-\infty)/(-\infty)$.

You can use the Extended Mean Value Theorem to prove L'Hôpital's Rule. Of the several different cases of this rule, the proof of only one case is illustrated. The remaining cases where $x \to c^-$ and $x \to c$ are left for you to prove.

Proof   Consider the case for which

$$\lim_{x \to c^+} f(x) = 0 \quad \text{and} \quad \lim_{x \to c^+} g(x) = 0.$$

Define the following new functions:

$$F(x) = \begin{cases} f(x), & x \neq c \\ 0, & x = c \end{cases} \quad \text{and} \quad G(x) = \begin{cases} g(x), & x \neq c \\ 0, & x = c \end{cases}.$$

For any x, $c < x < b$, F and G are differentiable on $(c, x]$ and continuous on $[c, x]$. You can apply the Extended Mean Value Theorem to conclude that there exists a number z in (c, x) such that

$$\frac{F'(z)}{G'(z)} = \frac{F(x) - F(c)}{G(x) - G(c)}$$

$$= \frac{F(x)}{G(x)}$$

$$= \frac{f'(z)}{g'(z)}$$

$$= \frac{f(x)}{g(x)}.$$

Finally, by letting x approach c from the right, $x \to c^+$, you have $z \to c^+$ because $c < z < x$, and

$$\lim_{x \to c^+} \frac{f(x)}{g(x)} = \lim_{x \to c^+} \frac{f'(z)}{g'(z)}$$

$$= \lim_{z \to c^+} \frac{f'(z)}{g'(z)}$$

$$= \lim_{x \to c^+} \frac{f'(x)}{g'(x)}.$$

> **THEOREM 9.19   Taylor's Theorem (page 654)**
>
> If a function f is differentiable through order $n + 1$ in an interval I containing c, then, for each x in I, there exists z between x and c such that
>
> $$f(x) = f(c) + f'(c)(x - c) + \frac{f''(c)}{2!}(x - c)^2 + \cdots + \frac{f^{(n)}(c)}{n!}(x - c)^n + R_n(x)$$
>
> where
>
> $$R_n(x) = \frac{f^{(n+1)}(z)}{(n + 1)!}(x - c)^{n+1}.$$

Proof   To find $R_n(x)$, fix x in I ($x \neq c$) and write

$$R_n(x) = f(x) - P_n(x)$$

where $P_n(x)$ is the nth Taylor polynomial for $f(x)$. Then let g be a function of t defined by

$$g(t) = f(x) - f(t) - f'(t)(x - t) - \cdots - \frac{f^{(n)}(t)}{n!}(x - t)^n - R_n(x)\frac{(x - t)^{n+1}}{(x - c)^{n+1}}.$$

The reason for defining g in this way is that differentiation with respect to t has a telescoping effect. For example, you have

$$\frac{d}{dt}\left[-f(t) - f'(t)(x - t)\right] = -f'(t) + f'(t) - f''(t)(x - t)$$

$$= -f''(t)(x - t).$$

The result is that the derivative $g'(t)$ simplifies to

$$g'(t) = -\frac{f^{(n+1)}(t)}{n!}(x - t)^n + (n + 1)R_n(x)\frac{(x - t)^n}{(x - c)^{n+1}}$$

for all t between c and x. Moreover, for a fixed x,

$$g(c) = f(x) - [P_n(x) + R_n(x)] = f(x) - f(x) = 0$$

and

$$g(x) = f(x) - f(x) - 0 - \cdots - 0 = f(x) - f(x) = 0.$$

Therefore, g satisfies the conditions of Rolle's Theorem, and it follows that there is a number z between c and x such that $g'(z) = 0$. Substituting z for t in the equation for $g'(t)$ and then solving for $R_n(x)$, you obtain

$$g'(z) = -\frac{f^{(n+1)}(z)}{n!}(x - z)^n + (n + 1)R_n(x)\frac{(x - z)^n}{(x - c)^{n+1}} = 0$$

$$R_n(x) = \frac{f^{(n+1)}(z)}{(n + 1)!}(x - c)^{n+1}.$$

Finally, because $g(c) = 0$, you have

$$0 = f(x) - f(c) - f'(c)(x - c) - \cdots - \frac{f^{(n)}(c)}{n!}(x - c)^n - R_n(x)$$

$$f(x) = f(c) + f'(c)(x - c) + \cdots + \frac{f^{(n)}(c)}{n!}(x - c)^n + R_n(x).$$

> **THEOREM 9.20   Convergence of a Power Series (page 660)**
>
> For a power series centered at c, precisely one of the following is true.
>
> **1.** The series converges only at c.
> **2.** There exists a real number $R > 0$ such that the series converges absolutely for $|x - c| < R$, and diverges for $|x - c| > R$.
> **3.** The series converges absolutely for all x.
>
> The number R is the **radius of convergence** of the power series. If the series converges only at c, the radius of convergence is $R = 0$, and if the series converges for all x, the radius of convergence is $R = \infty$. The set of all values of x for which the power series converges is the **interval of convergence** of the power series.

Proof In order to simplify the notation, the theorem for the power series $\Sigma\, a_n x^n$ centered at $x = 0$ will be proved. The proof for a power series centered at $x = c$ follows easily. A key step in this proof uses the completeness property of the set of real numbers: If a nonempty set S of real numbers has an upper bound, then it must have a least upper bound (see page 601).

It must be shown that if a power series $\Sigma\, a_n x^n$ converges at $x = d$, $d \neq 0$, then it converges for all b satisfying $|b| < |d|$. Because $\Sigma\, a_n x^n$ converges, $\lim\limits_{x \to \infty} a_n d^n = 0$. So, there exists $N > 0$ such that $a_n d^n < 1$ for all $n \geq N$. Then for $n \geq N$,

$$\left| a_n b^n \right| = \left| a_n b^n \frac{d^n}{d^n} \right| = \left| a_n d^n \right| \left| \frac{b^n}{d^n} \right| < \left| \frac{b^n}{d^n} \right|.$$

So, for $|b| < |d|$, $\left| \dfrac{b}{d} \right| < 1$, which implies that

$$\Sigma \left| \frac{b^n}{d^n} \right|$$

is a convergent geometric series. By the Comparison Test, the series $\Sigma\, a_n b^n$ converges.

Similarly, if the power series $\Sigma\, a_n x^n$ diverges at $x = b$, where $b \neq 0$, then it diverges for all d satisfying $|d| > |b|$. If $\Sigma\, a_n d^n$ converged, then the argument above would imply that $\Sigma\, a_n b^n$ converged as well.

Finally, to prove the theorem, suppose that neither case 1 nor case 3 is true. Then there exist points b and d such that $\Sigma\, a_n x^n$ converges at b and diverges at d. Let $S = \{x : \Sigma\, a_n x^n \text{ converges}\}$. S is nonempty because $b \in S$. If $x \in S$, then $|x| \leq |d|$, which shows that $|d|$ is an upper bound for the nonempty set S. By the completeness property, S has a least upper bound, R.

Now, if $|x| > R$, then $x \notin S$ so $\Sigma\, a_n x^n$ diverges. And if $|x| < R$, then $|x|$ is not an upper bound for S, so there exists b in S satisfying $|b| > |x|$. Since $b \in S$, $\Sigma\, a_n b^n$ converges, which implies that $\Sigma\, a_n x^n$ converges. ▬▬▬

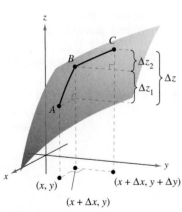

Figure A.1

> ### THEOREM 10.16   Classification of Conics by Eccentricity (page 748)
>
> Let F be a fixed point (*focus*) and let D be a fixed line (*directrix*) in the plane. Let P be another point in the plane and let e (*eccentricity*) be the ratio of the distance between P and F to the distance between P and D. The collection of all points P with a given eccentricity is a conic.
>
> **1.** The conic is an ellipse if $0 < e < 1$.
>
> **2.** The conic is a parabola if $e = 1$.
>
> **3.** The conic is a hyperbola if $e > 1$.

Proof   If $e = 1$, then, by definition, the conic must be a parabola. If $e \neq 1$, then you can consider the focus F to lie at the origin and the directrix $x = d$ to lie to the right of the origin, as shown in Figure A.1. For the point $P = (r, \theta) = (x, y)$, you have $|PF| = r$ and $|PQ| = d - r \cos \theta$. Given that $e = |PF|/|PQ|$, it follows that

$$|PF| = |PQ|e \quad \Longrightarrow \quad r = e(d - r \cos \theta).$$

By converting to rectangular coordinates and squaring each side, you obtain

$$x^2 + y^2 = e^2(d - x)^2 = e^2(d^2 - 2dx + x^2).$$

Completing the square produces

$$\left(x + \frac{e^2 d}{1 - e^2}\right)^2 + \frac{y^2}{1 - e^2} = \frac{e^2 d^2}{(1 - e^2)^2}.$$

If $e < 1$, this equation represents an ellipse. If $e > 1$, then $1 - e^2 < 0$, and the equation represents a hyperbola.

> ### THEOREM 13.4   Sufficient Condition for Differentiability (page 917)
>
> If f is a function of x and y, where f_x and f_y are continuous in an open region R, then f is differentiable on R.

$\Delta z = f(x + \Delta x, y + \Delta y) - f(x, y)$

Figure A.2

Proof   Let S be the surface defined by $z = f(x, y)$, where f, f_x, and f_y are continuous at (x, y). Let A, B, and C be points on surface S, as shown in Figure A.2. From this figure, you can see that the change in f from point A to point C is given by

$$\begin{aligned} \Delta z &= f(x + \Delta x, y + \Delta y) - f(x, y) \\ &= [f(x + \Delta x, y) - f(x, y)] + [f(x + \Delta x, y + \Delta y) - f(x + \Delta x, y)] \\ &= \Delta z_1 + \Delta z_2. \end{aligned}$$

Between A and B, y is fixed and x changes. So, by the Mean Value Theorem, there is a value x_1 between x and $x + \Delta x$ such that

$$\Delta z_1 = f(x + \Delta x, y) - f(x, y) = f_x(x_1, y) \, \Delta x.$$

Similarly, between B and C, x is fixed and y changes, and there is a value y_1 between y and $y + \Delta y$ such that

$$\Delta z_2 = f(x + \Delta x, y + \Delta y) - f(x + \Delta x, y) = f_y(x + \Delta x, y_1) \, \Delta y.$$

By combining these two results, you can write

$$\Delta z = \Delta z_1 + \Delta z_2 = f_x(x_1, y)\Delta x + f_y(x + \Delta x, y_1)\Delta y.$$

If you define ε_1 and ε_2 as

$$\varepsilon_1 = f_x(x_1, y) - f_x(x, y) \quad \text{and} \quad \varepsilon_2 = f_y(x + \Delta x, y_1) - f_y(x, y)$$

it follows that

$$\Delta z = \Delta z_1 + \Delta z_2 = \left[\varepsilon_1 + f_x(x, y)\right]\Delta x + \left[\varepsilon_2 + f_y(x, y)\right]\Delta y$$
$$= \left[f_x(x, y)\,\Delta x + f_y(x, y)\,\Delta y\right] + \varepsilon_1\Delta x + \varepsilon_2\Delta y.$$

By the continuity of f_x and f_y and the fact that $x \le x_1 \le x + \Delta x$ and $y \le y_1 \le y + \Delta y$, it follows that $\varepsilon_1 \to 0$ and $\varepsilon_2 \to 0$ as $\Delta x \to 0$ and $\Delta y \to 0$. Therefore, by definition, f is differentiable.

> **THEOREM 13.6   Chain Rule: One Independent Variable (page 923)**
>
> Let $w = f(x, y)$, where f is a differentiable function of x and y. If $x = g(t)$ and $y = h(t)$, where g and h are differentiable functions of t, then w is a differentiable function of t, and
>
> $$\frac{dw}{dt} = \frac{\partial w}{\partial x}\frac{dx}{dt} + \frac{\partial w}{\partial y}\frac{dy}{dt}.$$

Proof   Because g and h are differentiable functions of t, you know that both Δx and Δy approach zero as Δt approaches zero. Moreover, because f is a differentiable function of x and y, you know that

$$\Delta w = \frac{\partial w}{\partial x}\Delta x + \frac{\partial w}{\partial y}\Delta y + \varepsilon_1\Delta x + \varepsilon_2\Delta y$$

where both ε_1 and $\varepsilon_2 \to 0$ as $(\Delta x, \Delta y) \to (0, 0)$. So, for $\Delta t \ne 0$,

$$\frac{\Delta w}{\Delta t} = \frac{\partial w}{\partial x}\frac{\Delta x}{\Delta t} + \frac{\partial w}{\partial y}\frac{\Delta y}{\Delta t} + \varepsilon_1\frac{\Delta x}{\Delta t} + \varepsilon_2\frac{\Delta y}{\Delta t}$$

from which it follows that

$$\frac{dw}{dt} = \lim_{\Delta t \to 0}\frac{\Delta w}{\Delta t} = \frac{\partial w}{\partial x}\frac{dx}{dt} + \frac{\partial w}{\partial y}\frac{dy}{dt} + 0\left(\frac{dx}{dt}\right) + 0\left(\frac{dy}{dt}\right)$$
$$= \frac{\partial w}{\partial x}\frac{dx}{dt} + \frac{\partial w}{\partial y}\frac{dy}{dt}.$$

B Integration Tables

Forms Involving u^n

1. $\displaystyle\int u^n\,du = \frac{u^{n+1}}{n+1} + C,\ n \neq -1$

2. $\displaystyle\int \frac{1}{u}\,du = \ln|u| + C$

Forms Involving $a + bu$

3. $\displaystyle\int \frac{u}{a+bu}\,du = \frac{1}{b^2}\big(bu - a\ln|a+bu|\big) + C$

4. $\displaystyle\int \frac{u}{(a+bu)^2}\,du = \frac{1}{b^2}\left(\frac{a}{a+bu} + \ln|a+bu|\right) + C$

5. $\displaystyle\int \frac{u}{(a+bu)^n}\,du = \frac{1}{b^2}\left[\frac{-1}{(n-2)(a+bu)^{n-2}} + \frac{a}{(n-1)(a+bu)^{n-1}}\right] + C,\quad n \neq 1,2$

6. $\displaystyle\int \frac{u^2}{a+bu}\,du = \frac{1}{b^3}\left[-\frac{bu}{2}\,(2a-bu) + a^2\ln|a+bu|\right] + C$

7. $\displaystyle\int \frac{u^2}{(a+bu)^2}\,du = \frac{1}{b^3}\left(bu - \frac{a^2}{a+bu} - 2a\ln|a+bu|\right) + C$

8. $\displaystyle\int \frac{u^2}{(a+bu)^3}\,du = \frac{1}{b^3}\left[\frac{2a}{a+bu} - \frac{a^2}{2(a+bu)^2} + \ln|a+bu|\right] + C$

9. $\displaystyle\int \frac{u^2}{(a+bu)^n}\,du = \frac{1}{b^3}\left[\frac{-1}{(n-3)(a+bu)^{n-3}} + \frac{2a}{(n-2)(a+bu)^{n-2}} - \frac{a^2}{(n-1)(a+bu)^{n-1}}\right] + C,\quad n \neq 1,2,3$

10. $\displaystyle\int \frac{1}{u(a+bu)}\,du = \frac{1}{a}\ln\left|\frac{u}{a+bu}\right| + C$

11. $\displaystyle\int \frac{1}{u(a+bu)^2}\,du = \frac{1}{a}\left(\frac{1}{a+bu} + \frac{1}{a}\ln\left|\frac{u}{a+bu}\right|\right) + C$

12. $\displaystyle\int \frac{1}{u^2(a+bu)}\,du = -\frac{1}{a}\left(\frac{1}{u} + \frac{b}{a}\ln\left|\frac{u}{a+bu}\right|\right) + C$

13. $\displaystyle\int \frac{1}{u^2(a+bu)^2}\,du = -\frac{1}{a^2}\left[\frac{a+2bu}{u(a+bu)} + \frac{2b}{a}\ln\left|\frac{u}{a+bu}\right|\right] + C$

Forms Involving $a + bu + cu^2$, $b^2 \neq 4ac$

14. $\displaystyle \int \frac{1}{a + bu + cu^2}\, du = \begin{cases} \dfrac{2}{\sqrt{4ac - b^2}} \arctan \dfrac{2cu + b}{\sqrt{4ac - b^2}} + C, & b^2 < 4ac \\[4mm] \dfrac{1}{\sqrt{b^2 - 4ac}} \ln \left| \dfrac{2cu + b - \sqrt{b^2 - 4ac}}{2cu + b + \sqrt{b^2 - 4ac}} \right| + C, & b^2 > 4ac \end{cases}$

15. $\displaystyle \int \frac{u}{a + bu + cu^2}\, du = \frac{1}{2c}\left(\ln|a + bu + cu^2| - b \int \frac{1}{a + bu + cu^2}\, du \right)$

Forms Involving $\sqrt{a + bu}$

16. $\displaystyle \int u^n \sqrt{a + bu}\, du = \frac{2}{b(2n + 3)}\left[u^n(a + bu)^{3/2} - na \int u^{n-1}\sqrt{a + bu}\, du \right]$

17. $\displaystyle \int \frac{1}{u\sqrt{a + bu}}\, du = \begin{cases} \dfrac{1}{\sqrt{a}} \ln \left| \dfrac{\sqrt{a + bu} - \sqrt{a}}{\sqrt{a + bu} + \sqrt{a}} \right| + C, & a > 0 \\[4mm] \dfrac{2}{\sqrt{-a}} \arctan \sqrt{\dfrac{a + bu}{-a}} + C, & a < 0 \end{cases}$

18. $\displaystyle \int \frac{1}{u^n \sqrt{a + bu}}\, du = \frac{-1}{a(n - 1)}\left[\frac{\sqrt{a + bu}}{u^{n-1}} + \frac{(2n - 3)b}{2} \int \frac{1}{u^{n-1}\sqrt{a + bu}}\, du \right], \quad n \neq 1$

19. $\displaystyle \int \frac{\sqrt{a + bu}}{u}\, du = 2\sqrt{a + bu} + a \int \frac{1}{u\sqrt{a + bu}}\, du$

20. $\displaystyle \int \frac{\sqrt{a + bu}}{u^n}\, du = \frac{-1}{a(n - 1)}\left[\frac{(a + bu)^{3/2}}{u^{n-1}} + \frac{(2n - 5)b}{2} \int \frac{\sqrt{a + bu}}{u^{n-1}}\, du \right], \quad n \neq 1$

21. $\displaystyle \int \frac{u}{\sqrt{a + bu}}\, du = \frac{-2(2a - bu)}{3b^2}\sqrt{a + bu} + C$

22. $\displaystyle \int \frac{u^n}{\sqrt{a + bu}}\, du = \frac{2}{(2n + 1)b}\left(u^n \sqrt{a + bu} - na \int \frac{u^{n-1}}{\sqrt{a + bu}}\, du \right)$

Forms Involving $a^2 \pm u^2$, $a > 0$

23. $\displaystyle \int \frac{1}{a^2 + u^2}\, du = \frac{1}{a} \arctan \frac{u}{a} + C$

24. $\displaystyle \int \frac{1}{u^2 - a^2}\, du = - \int \frac{1}{a^2 - u^2}\, du = \frac{1}{2a} \ln \left| \frac{u - a}{u + a} \right| + C$

25. $\displaystyle \int \frac{1}{(a^2 \pm u^2)^n}\, du = \frac{1}{2a^2(n - 1)}\left[\frac{u}{(a^2 \pm u^2)^{n-1}} + (2n - 3) \int \frac{1}{(a^2 \pm u^2)^{n-1}}\, du \right], \quad n \neq 1$

Forms Involving $\sqrt{u^2 \pm a^2}$, $a > 0$

26. $\displaystyle \int \sqrt{u^2 \pm a^2}\, du = \frac{1}{2}\left(u\sqrt{u^2 \pm a^2} \pm a^2 \ln|u + \sqrt{u^2 \pm a^2}| \right) + C$

27. $\displaystyle \int u^2 \sqrt{u^2 \pm a^2}\, du = \frac{1}{8}\left[u(2u^2 \pm a^2)\sqrt{u^2 \pm a^2} - a^4 \ln|u + \sqrt{u^2 \pm a^2}| \right] + C$

28. $\displaystyle \int \frac{\sqrt{u^2 + a^2}}{u}\, du = \sqrt{u^2 + a^2} - a \ln \left| \frac{a + \sqrt{u^2 + a^2}}{u} \right| + C$

29. $\displaystyle\int \frac{\sqrt{u^2 - a^2}}{u}\, du = \sqrt{u^2 - a^2} - a \arcsec \frac{|u|}{a} + C$

30. $\displaystyle\int \frac{\sqrt{u^2 \pm a^2}}{u^2}\, du = \frac{-\sqrt{u^2 \pm a^2}}{u} + \ln\left|u + \sqrt{u^2 \pm a^2}\right| + C$

31. $\displaystyle\int \frac{1}{\sqrt{u^2 \pm a^2}}\, du = \ln\left|u + \sqrt{u^2 \pm a^2}\right| + C$

32. $\displaystyle\int \frac{1}{u\sqrt{u^2 + a^2}}\, du = \frac{-1}{a} \ln\left|\frac{a + \sqrt{u^2 + a^2}}{u}\right| + C$

33. $\displaystyle\int \frac{1}{u\sqrt{u^2 - a^2}}\, du = \frac{1}{a} \arcsec \frac{|u|}{a} + C$

34. $\displaystyle\int \frac{u^2}{\sqrt{u^2 \pm a^2}}\, du = \frac{1}{2}\left(u\sqrt{u^2 \pm a^2} \mp a^2 \ln\left|u + \sqrt{u^2 \pm a^2}\right|\right) + C$

35. $\displaystyle\int \frac{1}{u^2\sqrt{u^2 \pm a^2}}\, du = \mp \frac{\sqrt{u^2 \pm a^2}}{a^2 u} + C$

36. $\displaystyle\int \frac{1}{(u^2 \pm a^2)^{3/2}}\, du = \frac{\pm u}{a^2\sqrt{u^2 \pm a^2}} + C$

Forms Involving $\sqrt{a^2 - u^2}$, $a > 0$

37. $\displaystyle\int \sqrt{a^2 - u^2}\, du = \frac{1}{2}\left(u\sqrt{a^2 - u^2} + a^2 \arcsin \frac{u}{a}\right) + C$

38. $\displaystyle\int u^2\sqrt{a^2 - u^2}\, du = \frac{1}{8}\left[u(2u^2 - a^2)\sqrt{a^2 - u^2} + a^4 \arcsin \frac{u}{a}\right] + C$

39. $\displaystyle\int \frac{\sqrt{a^2 - u^2}}{u}\, du = \sqrt{a^2 - u^2} - a \ln\left|\frac{a + \sqrt{a^2 - u^2}}{u}\right| + C$

40. $\displaystyle\int \frac{\sqrt{a^2 - u^2}}{u^2}\, du = \frac{-\sqrt{a^2 - u^2}}{u} - \arcsin \frac{u}{a} + C$

41. $\displaystyle\int \frac{1}{\sqrt{a^2 - u^2}}\, du = \arcsin \frac{u}{a} + C$

42. $\displaystyle\int \frac{1}{u\sqrt{a^2 - u^2}}\, du = \frac{-1}{a} \ln\left|\frac{a + \sqrt{a^2 - u^2}}{u}\right| + C$

43. $\displaystyle\int \frac{u^2}{\sqrt{a^2 - u^2}}\, du = \frac{1}{2}\left(-u\sqrt{a^2 - u^2} + a^2 \arcsin \frac{u}{a}\right) + C$

44. $\displaystyle\int \frac{1}{u^2\sqrt{a^2 - u^2}}\, du = \frac{-\sqrt{a^2 - u^2}}{a^2 u} + C$

45. $\displaystyle\int \frac{1}{(a^2 - u^2)^{3/2}}\, du = \frac{u}{a^2\sqrt{a^2 - u^2}} + C$

Forms Involving $\sin u$ or $\cos u$

46. $\displaystyle\int \sin u \, du = -\cos u + C$

47. $\displaystyle\int \cos u \, du = \sin u + C$

48. $\displaystyle\int \sin^2 u \, du = \frac{1}{2}(u - \sin u \cos u) + C$

49. $\displaystyle\int \cos^2 u \, du = \frac{1}{2}(u + \sin u \cos u) + C$

50. $\displaystyle\int \sin^n u \, du = -\frac{\sin^{n-1} u \cos u}{n} + \frac{n-1}{n} \int \sin^{n-2} u \, du$

51. $\displaystyle\int \cos^n u \, du = \frac{\cos^{n-1} u \sin u}{n} + \frac{n-1}{n} \int \cos^{n-2} u \, du$

52. $\displaystyle\int u \sin u \, du = \sin u - u \cos u + C$

53. $\displaystyle\int u \cos u \, du = \cos u + u \sin u + C$

54. $\displaystyle\int u^n \sin u \, du = -u^n \cos u + n \int u^{n-1} \cos u \, du$

55. $\displaystyle\int u^n \cos u \, du = u^n \sin u - n \int u^{n-1} \sin u \, du$

56. $\displaystyle\int \frac{1}{1 \pm \sin u} \, du = \tan u \mp \sec u + C$

57. $\displaystyle\int \frac{1}{1 \pm \cos u} \, du = -\cot u \pm \csc u + C$

58. $\displaystyle\int \frac{1}{\sin u \cos u} \, du = \ln|\tan u| + C$

Forms Involving $\tan u$, $\cot u$, $\sec u$, $\csc u$

59. $\displaystyle\int \tan u \, du = -\ln|\cos u| + C$

60. $\displaystyle\int \cot u \, du = \ln|\sin u| + C$

61. $\displaystyle\int \sec u \, du = \ln|\sec u + \tan u| + C$

62. $\displaystyle\int \csc u \, du = \ln|\csc u - \cot u| + C$ or $\displaystyle\int \csc u \, du = -\ln|\csc u + \cot u| + C$

63. $\displaystyle\int \tan^2 u \, du = -u + \tan u + C$

64. $\displaystyle\int \cot^2 u \, du = -u - \cot u + C$

65. $\displaystyle\int \sec^2 u \, du = \tan u + C$

66. $\displaystyle\int \csc^2 u \, du = -\cot u + C$

67. $\displaystyle\int \tan^n u \, du = \frac{\tan^{n-1} u}{n-1} - \int \tan^{n-2} u \, du, \ n \neq 1$

68. $\displaystyle\int \cot^n u \, du = -\frac{\cot^{n-1} u}{n-1} - \int (\cot^{n-2} u) \, du, \ n \neq 1$

69. $\displaystyle\int \sec^n u \, du = \frac{\sec^{n-2} u \tan u}{n-1} + \frac{n-2}{n-1} \int \sec^{n-2} u \, du, \ n \neq 1$

70. $\displaystyle\int \csc^n u \, du = -\frac{\csc^{n-2} u \cot u}{n-1} + \frac{n-2}{n-1} \int \csc^{n-2} u \, du, \ n \neq 1$

71. $\displaystyle\int \frac{1}{1 \pm \tan u} \, du = \frac{1}{2}(u \pm \ln|\cos u \pm \sin u|) + C$

72. $\displaystyle\int \frac{1}{1 \pm \cot u} \, du = \frac{1}{2}(u \mp \ln|\sin u \pm \cos u|) + C$

73. $\displaystyle\int \frac{1}{1 \pm \sec u} \, du = u + \cot u \mp \csc u + C$

74. $\displaystyle\int \frac{1}{1 \pm \csc u} \, du = u - \tan u \pm \sec u + C$

Forms Involving Inverse Trigonometric Functions

75. $\displaystyle\int \arcsin u \, du = u \arcsin u + \sqrt{1 - u^2} + C$

76. $\displaystyle\int \arccos u \, du = u \arccos u - \sqrt{1 - u^2} + C$

77. $\displaystyle\int \arctan u \, du = u \arctan u - \ln\sqrt{1 + u^2} + C$

78. $\displaystyle\int \operatorname{arccot} u \, du = u \operatorname{arccot} u + \ln\sqrt{1 + u^2} + C$

79. $\displaystyle\int \operatorname{arcsec} u \, du = u \operatorname{arcsec} u - \ln\left|u + \sqrt{u^2 - 1}\right| + C$

80. $\displaystyle\int \operatorname{arccsc} u \, du = u \operatorname{arccsc} u + \ln\left|u + \sqrt{u^2 - 1}\right| + C$

Forms Involving e^u

81. $\displaystyle\int e^u \, du = e^u + C$

82. $\displaystyle\int u e^u \, du = (u - 1)e^u + C$

83. $\displaystyle\int u^n e^u \, du = u^n e^u - n \int u^{n-1} e^u \, du$

84. $\displaystyle\int \frac{1}{1 + e^u} \, du = u - \ln(1 + e^u) + C$

85. $\displaystyle\int e^{au} \sin bu \, du = \frac{e^{au}}{a^2 + b^2}(a \sin bu - b \cos bu) + C$

86. $\displaystyle\int e^{au} \cos bu \, du = \frac{e^{au}}{a^2 + b^2}(a \cos bu + b \sin bu) + C$

Forms Involving $\ln u$

87. $\displaystyle\int \ln u \, du = u(-1 + \ln u) + C$

88. $\displaystyle\int u \ln u \, du = \frac{u^2}{4}(-1 + 2 \ln u) + C$

89. $\displaystyle\int u^n \ln u \, du = \frac{u^{n+1}}{(n+1)^2}[-1 + (n+1)\ln u] + C, \; n \neq -1$

90. $\displaystyle\int (\ln u)^2 \, du = u\left[2 - 2\ln u + (\ln u)^2\right] + C$

91. $\displaystyle\int (\ln u)^n \, du = u(\ln u)^n - n \int (\ln u)^{n-1} \, du$

Forms Involving Hyperbolic Functions

92. $\displaystyle\int \cosh u \, du = \sinh u + C$

93. $\displaystyle\int \sinh u \, du = \cosh u + C$

94. $\displaystyle\int \operatorname{sech}^2 u \, du = \tanh u + C$

95. $\displaystyle\int \operatorname{csch}^2 u \, du = -\coth u + C$

96. $\displaystyle\int \operatorname{sech} u \tanh u \, du = -\operatorname{sech} u + C$

97. $\displaystyle\int \operatorname{csch} u \coth u \, du = -\operatorname{csch} u + C$

Forms Involving Inverse Hyperbolic Functions (in logarithmic form)

98. $\displaystyle\int \frac{du}{\sqrt{u^2 \pm a^2}} = \ln\left(u + \sqrt{u^2 \pm a^2}\right) + C$

99. $\displaystyle\int \frac{du}{a^2 - u^2} = \frac{1}{2a} \ln\left|\frac{a + u}{a - u}\right| + C$

100. $\displaystyle\int \frac{du}{u\sqrt{a^2 \pm u^2}} = -\frac{1}{a} \ln\frac{a + \sqrt{a^2 \pm u^2}}{|u|} + C$

C Business and Economic Applications

Previously, you learned that one of the most common ways to measure change is with respect to time. In this section, you will study some important rates of change in economics that are not measured with respect to time. For example, economists refer to **marginal profit, marginal revenue,** and **marginal cost** as the rates of change of the profit, revenue, and cost with respect to the number of units produced or sold.

Summary of Business Terms and Formulas

Basic Terms	*Basic Formulas*
x is the number of units produced (or sold).	
p is the price per unit.	
R is the total revenue from selling x units.	$R = xp$
C is the total cost of producing x units.	
$\overline{C}$ is the average cost per unit.	$\overline{C} = \dfrac{C}{x}$
P is the total profit from selling x units.	$P = R - C$

The **break-even point** is the number of units for which $R = C$.

Marginals

$\dfrac{dR}{dx}$ = Marginal revenue ≈ *extra* revenue from selling one additional unit

$\dfrac{dC}{dx}$ = Marginal cost ≈ *extra* cost of producing one additional unit

$\dfrac{dP}{dx}$ = Marginal profit ≈ *extra* profit from selling one additional unit

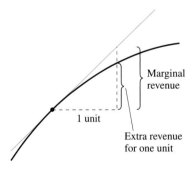

A revenue function
Figure C.1

In this summary, note that marginals can be used to approximate the *extra* revenue, cost, or profit associated with selling or producing one additional unit. This is illustrated graphically for marginal revenue in Figure C.1.

55. $y = 6 - |x|$
Symmetry: y-axis

57. $y_1 = \sqrt{x + 9}$
$y_2 = -\sqrt{x + 9}$
Symmetry: x-axis

21. (a)

Year ($6 \leftrightarrow 1996$)

(b) Population increased least rapidly: 2000–2001

23. $m = -\frac{1}{5}$, $(0, 4)$ **25.** m is undefined, no y-intercept
27. $3x - 4y + 12 = 0$ **29.** $2x - 3y = 0$

59. $y_1 = \sqrt{\dfrac{6 - x}{3}}$

$y_2 = -\sqrt{\dfrac{6 - x}{3}}$

Symmetry: x-axis

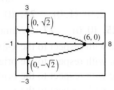

31. $3x - y - 11 = 0$ **33.** $3x - y = 0$

61. $(1, 1)$ **63.** $(-1, 5), (2, 2)$ **65.** $(-1, -2), (2, 1)$
67. $(-1, -1), (0, 0), (1, 1)$
69. $(-1, -5), (0, -1), (2, 1)$ **71.** $(-2, 2), \left(-3, \sqrt{3}\right)$
73. (a) $y = -0.007t^2 + 4.82t + 35.4$
(b)

(c) 217.0

35. $2x - y - 3 = 0$ **37.** $8x + 3y - 40 = 0$

75. $x \approx 3133$ units   **77.** $y = (x + 2)(x - 4)(x - 6)$
79. (i) b; $k = 2$ (ii) d; $k = -10$ (iii) a; $k = 3$ (iv) c; $k = 36$
81. False. $(-1, -2)$ is not a point on the graph of $x = \frac{1}{4}y^2$.
83. True   **85.** $x^2 + (y - 4)^2 = 4$

Section 1.2   (page 16)

1. $m = 1$ **3.** $m = 0$ **5.** $m = -12$
7.

9. $m = 3$

39. $x - 5 = 0$ **41.** $22x - 4y + 3 = 0$

11. m is undefined.

13. $m = 2$

43. $x - 3 = 0$ **45.** $3x + 2y - 6 = 0$ **47.** $x + y - 3 = 0$
49.

51.

15. $(0, 1), (1, 1), (3, 1)$ **17.** $(0, 10), (2, 4), (3, 1)$
19. (a) $\frac{1}{3}$ (b) $10\sqrt{10}$ ft

53.

55.

57. (a) (b)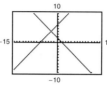

The lines in (a) do not appear perpendicular, but they do in (b) because a square setting is used. The lines are perpendicular.

59. (a) $2x - y - 3 = 0$ (b) $x + 2y - 4 = 0$

61. (a) $40x - 24y - 9 = 0$ (b) $24x + 40y - 53 = 0$

63. (a) $x - 2 = 0$ (b) $y - 5 = 0$

65. $V = 125t + 2040$ **67.** $V = -2000t + 28,400$

69. $y = 2x$ **71.** Not collinear, because $m_1 \neq m_2$

73. $\left(0, \dfrac{-a^2 + b^2 + c^2}{2c}\right)$ **75.** $\left(b, \dfrac{a^2 - b^2}{c}\right)$

77. $5F - 9C - 160 = 0$; $72°F \approx 22.2°C$

79. (a) $W_1 = 12.50 + 0.75x$; $W_2 = 9.20 + 1.30x$

(b) (c) When six units are produced, wages are $17.00 per hour with either option. Choose position 1 when less than six units are produced and position 2 otherwise.

81. (a) $x = (1330 - p)/15$

(b) (c) 49 units

$x(655) = 45$ units

83. $12y + 5x - 169 = 0$ **85.** 2 **87.** $(5\sqrt{2})/2$ **89.** $2\sqrt{2}$

91. Proof **93.** Proof **95.** Proof **97.** True

Section 1.3 (page 27)

1. (a) Domain of f: $[-4, 4]$; Range of f: $[-3, 5]$
Domain of g: $[-3, 3]$; Range of g: $[-4, 4]$

(b) $f(-2) = -1$; $g(3) = -4$

(c) $x = -1$ (d) $x \approx 1$ (e) $x \approx -1, x \approx 1$, and $x \approx 2$

3. (a) -3 (b) -9 (c) $2b - 3$ (d) $2x - 5$

5. (a) 3 (b) 0 (c) -1 (d) $2 + 2t - t^2$

7. (a) 1 (b) 0 (c) $-\frac{1}{2}$ **9.** $3x^2 + 3x\,\Delta x + (\Delta x)^2$, $\Delta x \neq 0$

11. $\left(\sqrt{x-1} - x + 1\right)/[(x-2)(x-1)]$
$= -1/\left[\sqrt{x-1}\left(1 + \sqrt{x-1}\right)\right], x \neq 2$

13. Domain: $[-3, \infty)$; Range: $(-\infty, 0]$

15. Domain: All real numbers t such that $t \neq 4n + 2$, where n is an integer; Range: $(-\infty, -1] \cup [1, \infty)$

17. Domain: $(-\infty, 0) \cup (0, \infty)$; Range: $(-\infty, 0) \cup (0, \infty)$

19. Domain: $[0, 1]$

21. Domain: All real numbers x such that $x \neq 2n\pi$, where n is an integer

23. Domain: $(-\infty, -3) \cup (-3, \infty)$

25. (a) -1 (b) 2 (c) 6 (d) $2t^2 + 4$
Domain: $(-\infty, \infty)$; Range: $(-\infty, 1) \cup [2, \infty)$

27. (a) 4 (b) 0 (c) -2 (d) $-b^2$
Domain: $(-\infty, \infty)$; Range: $(-\infty, 0] \cup [1, \infty)$

29. $f(x) = \begin{cases} x + 2, & -2 \leq x \leq -1 \\ -\frac{1}{2}x + \frac{1}{2}, & -1 < x \leq 3 \end{cases}$

31. $f(x) = 4 - x$ **33.** $h(x) = \sqrt{x - 1}$
Domain: $(-\infty, \infty)$ Domain: $[1, \infty)$
Range: $(-\infty, \infty)$ Range: $[0, \infty)$

 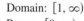

35. $f(x) = \sqrt{9 - x^2}$ **37.** $g(t) = 2\sin \pi t$
Domain: $[-3, 3]$ Domain: $(-\infty, \infty)$
Range: $[0, 3]$ Range: $[-2, 2]$

39. The student travels $\frac{1}{2}$ mi/min during the first 4 min, is stationary for the next 2 min, and travels 1 mi/min during the final 4 min.

41. y is not a function of x. **43.** y is a function of x.

45. y is not a function of x. **47.** y is not a function of x.

49. d **50.** b **51.** c **52.** a **53.** e **54.** g

55. (a) (b)

(c)

(d)

(e)

(f)

57. (a) Vertical translation   (b) Reflection about the x-axis

(c) Horizontal translation

59. (a) 0   (b) 0   (c) -1 (d) $\sqrt{15}$
    (e) $\sqrt{x^2 - 1}$ (f) $x - 1 \ (x \geq 0)$
61. $(f \circ g)(x) = x$; Domain: $[0, \infty)$
    $(g \circ f)(x) = |x|$; Domain: $(-\infty, \infty)$
    No, their domains are different.
63. $(f \circ g)(x) = 3/(x^2 - 1)$; Domain: $(-\infty, -1) \cup (-1, 1) \cup (1, \infty)$
    $(g \circ f)(x) = 9/x^2 - 1$; Domain: $(-\infty, 0) \cup (0, \infty)$
    No
65. (a) 4   (b) -2
    (c) Undefined. The graph of g does not exist at $x = -5$.
    (d) 3   (e) 2
    (f) Undefined. The graph of f does not exist at $x = -4$.
67. Answers will vary.
    Example: $f(x) = \sqrt{x}$; $g(x) = x - 2$; $h(x) = 2x$
69. Even   **71.** Odd   **73.** (a) $\left(\frac{3}{2}, 4\right)$ (b) $\left(\frac{3}{2}, -4\right)$
75. f is even. g is neither even nor odd. h is odd.
77. $f(x) = -2x - 5$ **79.** $y = -\sqrt{-x}$
81. ii, $c = -2$ **82.** i, $c = \frac{1}{4}$ **83.** iv, $c = 32$ **84.** iii, $c = 3$
85. (a) $T(4) = 16°$, $T(15) = 24°$
    (b) The changes in temperature will occur 1 hr later.
    (c) The temperatures are $1°$ lower.

87. (a)

    (b) $A(15) \approx 345$ acres/farm

89. $f(x) = |x| + |x - 2| = \begin{cases} 2x - 2, & \text{if } x \geq 2 \\ 2, & \text{if } 0 < x < 2 \\ -2x + 2, & \text{if } x \leq 0 \end{cases}$

91. Proof   **93.** Proof
95. (a) $V(x) = x(24 - 2x)^2$, $x > 0$ (b) 4 cm × 16 cm × 16 cm

(c)

Height, x	Length and Width	Volume, V
1	$24 - 2(1)$	$1[24 - 2(1)]^2 = 484$
2	$24 - 2(2)$	$2[24 - 2(2)]^2 = 800$
3	$24 - 2(3)$	$3[24 - 2(3)]^2 = 972$
4	$24 - 2(4)$	$4[24 - 2(4)]^2 = 1024$
5	$24 - 2(5)$	$5[24 - 2(5)]^2 = 980$
6	$24 - 2(6)$	$6[24 - 2(6)]^2 = 864$

    The dimensions of the box that yield a maximum volume are 4 cm × 16 cm × 16 cm.
97. False. For example, if $f(x) = x^2$, then $f(-1) = f(1)$.
99. True   **101.** Putnam Problem A1, 1988

Section 1.4   (page 34)

1. Quadratic   **3.** Linear
5. (a) and (b)

Approximately linear
    (c) 136
9. (a) $y = 0.124x + 0.82$
    $r \approx 0.838$
    (b)

    (d) $y = 0.134x + 0.28$
    $r \approx 0.968$

7. (a) $d = 0.066F$
    (b)

    The model is a "good fit".
    (c) 3.63 cm

    (c) Greater per capita electricity consumption by a country tends to relate to greater per capita gross national product of the country. Hong Kong, Venezuela, South Korea

11. (a) $y_1 = 0.03434t^3 - 0.3451t^2 + 0.884t + 5.61$
  $y_2 = 0.110t + 2.07$
  $y_3 = 0.092t + 0.79$
  (b) $y_1 + y_2 + y_3 = 0.03434t^3 - 0.3451t^2 + 1.086t + 8.47$

  31.1 cents/mi

13. (a) Linear: $y_1 = 4.83t + 28.6$
  Cubic: $y_2 = -0.1289t^3 + 2.235t^2 - 4.86t + 35.2$
  (b)   (c) Cubic

  (d) $y = -0.084t^2 + 5.84t + 26.7$

    (e) Linear:
     $N(14) \approx 96.2$ million people
     Cubic:
     $N(14) \approx 51.5$ million people
     (f) Answers will vary.

15. (a) $y = -1.806x^3 + 14.58x^2 + 16.4x + 10$
  (b)   (c) 214

17. (a) Yes. At time t there is one and only one displacement y.
  (b) Amplitude: 0.35; Period: 0.5   (c) $y = 0.35 \sin(4\pi t) + 2$
  (d)   The model appears to fit
       the data.

19. Answers will vary.

Section 1.5   (page 44)

1. (a) $f(g(x)) = 5[(x - 1)/5] + 1 = x$
  $g(f(x)) = [(5x + 1) - 1]/5 = x$
  (b)

3. (a) $f(g(x)) = \left(\sqrt[3]{x}\right)^3 = x$; $g(f(x)) = \sqrt[3]{x^3} = x$
  (b)

5. (a) $f(g(x)) = \sqrt{x^2 + 4 - 4} = x$;
  $g(f(x)) = \left(\sqrt{x - 4}\right)^2 + 4 = x$
  (b)

7. (a) $f(g(x)) = \dfrac{1}{1/x} = x$; $g(f(x)) = \dfrac{1}{1/x} = x$
  (b)

9. c   **10.** b   **11.** a   **12.** d
13. Inverse exists.   **15.** Inverse does not exist.
17. One-to-one   **19.** Not one-to-one; does not
                              have an inverse

21. One-to-one

23. One-to-one   **25.** Not one-to-one   **27.** One-to-one

49.

51. $\frac{1}{2}$

53.

55. $\frac{1}{5}[\ln(2x + 1) + \ln(2x - 1) - \ln(4x^2 + 1)]$

57. $\ln\left(\dfrac{3\sqrt[3]{4 - x^2}}{x}\right)$ **59.** $e^4 - 1 \approx 53.598$

61. (a) $f^{-1}(x) = e^{2x}$ **63.**

(b)

(c) Proof

P.S. Problem Solving (page 59)

1. (a) Center: $(3, 4)$; Radius: 5

(b) $y = -\frac{3}{4}x$ (c) $y = \frac{3}{4}x - \frac{9}{2}$ (d) $\left(3, -\frac{9}{4}\right)$

3.

(a) $H(x) - 2 = \begin{cases} -1, & x \ge 0 \\ -2, & x < 0 \end{cases}$ (b) $H(x - 2) = \begin{cases} 1, & x \ge 2 \\ 0, & x < 2 \end{cases}$

(c) $-H(x) = \begin{cases} -1, & x \ge 0 \\ 0, & x < 0 \end{cases}$ (d) $H(-x) = \begin{cases} 1, & x \le 0 \\ 0, & x > 0 \end{cases}$

(e) $\frac{1}{2}H(x) = \begin{cases} \frac{1}{2}, & x \ge 0 \\ 0, & x < 0 \end{cases}$ (f) $-H(x - 2) + 2 = \begin{cases} 1, & x \ge 2 \\ 2, & x < 2 \end{cases}$

5. (a) $A(x) = x[(100 - x)/2]$; Domain: $(0, 100)$

(b) Dimensions 50 m × 25 m yield maximum area of 1250 square meters.

(c) 50 m × 25 m; Area = 1250 square meters

7. $T(x) = \dfrac{2\sqrt{4 + x^2} + \sqrt{(3 - x)^2 + 1}}{4}$

9. (a) 5, less (b) 3, greater (c) 4.1, less (d) $4 + h$

(e) 4; Answers will vary.

11. (a) $x = -3 + \sqrt{18} \approx 1.2426$ **13.** Answers will vary.

$x = -3 - \sqrt{18} \approx -7.2426$

(b) $(x + 3)^2 + y^2 = 18$

Chapter 2

Section 2.1 (page 67)

1. Precalculus: 300 ft

3. Calculus: Slope of the tangent line at $x = 2$ is 0.16.

5. Precalculus: $\frac{15}{2}$ square units

7. (a)

(b) $1; \frac{3}{2}; \frac{5}{2}$

(c) 2. Use points closer to P.

9. (a) Area ≈ 10.417; Area ≈ 9.145 (b) Use more rectangles.

11. (a) 5.66 (b) 6.11 (c) Increase the number of line segments.

Section 2.2 (page 74)

1.

x	1.9	1.99	1.999	2.001	2.01	2.1
$f(x)$	0.3448	0.3344	0.3334	0.3332	0.3322	0.3226

$\lim\limits_{x \to 2} \dfrac{x-2}{x^2-x-2} \approx 0.3333 \left(\text{Actual limit is } \dfrac{1}{3}.\right)$

3.

x	2.9	2.99	2.999
$f(x)$	-0.0641	-0.0627	-0.0625

x	3.001	3.01	3.1
$f(x)$	-0.0625	-0.0623	-0.0610

$\lim\limits_{x \to 3} \dfrac{[1/(x+1)] - (1/4)}{x-3} \approx -0.0625 \left(\text{Actual limit is } -\dfrac{1}{16}.\right)$

5.

x	-0.1	-0.01	-0.001	0.001	0.01	0.1
$f(x)$	0.9983	0.99998	1.0000	1.0000	0.99998	0.9983

$\lim\limits_{x \to 0} \dfrac{\sin x}{x} \approx 1.0000$ (Actual limit is 1.)

7.

x	-0.1	-0.01	-0.001	0.001	0.01	0.1
$f(x)$	0.9516	0.9950	0.9995	1.0005	1.0050	1.0517

$\lim\limits_{x \to 0} \dfrac{e^x - 1}{x} \approx 1$ (Actual limit is 1.)

9.

x	-0.1	-0.01	-0.001	0.001	0.01	0.1
$f(x)$	1.0536	1.0050	1.0005	0.9995	0.9950	0.9531

$\lim\limits_{x \to 0} \dfrac{\ln(x+1)}{x} \approx 1$ (Actual limit is 1.)

11. 1

13. Limit does not exist. The function approaches 1 from the right side of 3 but it approaches -1 from the left side of 3.

15. 0

17. Limit does not exist. The function increases without bound as x approaches $\pi/2$ from the left and decreases without bound as x approaches $\pi/2$ from the right.

19. 1

21. (a) 2

(b) Limit does not exist. The function approaches 1 from the right side of 1 but it approaches 3.5 from the left side of 1.

(c) Value does not exist. The function is undefined at $x = 4$.

(d) 2

23. $\lim\limits_{x \to c} f(x)$ exists at all points on the graph except where $c = -3$.

25.   **27.**

$\lim\limits_{x \to c} f(x)$ exists at all points on the graph except where $c = 4$.

29. (a)

(b)

t	3	3.3	3.4	3.5	3.6	3.7	4
C	1.75	2.25	2.25	2.25	2.25	2.25	2.25

$\lim\limits_{t \to 3.5} C(t) = 2.25$

(c)

t	2	2.5	2.9	3	3.1	3.5	4
C	1.25	1.75	1.75	1.75	2.25	2.25	2.25

The limit does not exist, because the limits from the right and left are not equal.

31. $\delta = 0.4$ **33.** $\delta = \frac{1}{11} \approx 0.091$

35. $L = 8$. Let $\delta = 0.01/3 \approx 0.0033$.

37. $L = 1$. Let $\delta = 0.01/5 = 0.002$.

39. 5 **41.** -3 **43.** 3 **45.** 0 **47.** 4 **49.** 2

51. Answers will vary. **53.** Answers will vary.

$\lim\limits_{x \to 4} f(x) = \frac{1}{6}$ $\quad\lim\limits_{x \to 9} f(x) = 6$

Domain: $[-5, 4) \cup (4, \infty)$ Domain: $[0, 9) \cup (9, \infty)$

The graph has a hole The graph has a hole

at $x = 4$. at $x = 9$.

55. Answers will vary. Sample answer: As x approaches 8 from either side, $f(x)$ becomes arbitrarily close to 25.

57. No. The fact that $\lim\limits_{x \to 2} f(x) = 4$ has no bearing on the value of f at 2.

59. (a) $r = \dfrac{3}{\pi} \approx 0.9549$ cm

(b) $\dfrac{5.5}{2\pi} \leq r \leq \dfrac{6.5}{2\pi}$, or approximately $0.8754 < r < 1.0345$

(c) $\lim\limits_{r \to 3/\pi} 2\pi r = 6$; $\varepsilon = 0.5$; $\delta \approx 0.0796$

61.

x	-0.001	-0.0001	-0.00001
$f(x)$	2.7196	2.7184	2.7183

x	0.00001	0.0001	0.001
$f(x)$	2.7183	2.7181	2.7169

$\lim_{x \to 0} f(x) \approx 2.7183$

(0, 2.7183)

63.

(1.999, 0.001)
(2.001, 0.001)

$\delta \approx 0.001$
(1.999, 2.001)

65. False. The existence or nonexistence of $f(x)$ at $x = c$ has no bearing on the existence of the limit of $f(x)$ as $x \to c$.
67. False. See Exercise 23.
69. (a) Yes. As x approaches 0.25 from either side, $\sqrt{x}$ becomes arbitrarily close to 0.5.
(b) No. $\lim_{x \to 0} \sqrt{x}$ does not exist because for $x < 0$, $\sqrt{x}$ does not exist.
71–73. Proofs   **75.** Answers will vary.
77. Putnam Problem B1, 1986

Section 2.3   (page 87)

1.

3.

(a) 0   (b) 6   (a) 0   (b) ≈ 0.52 or $\pi/6$
5. 16   **7.** -1 **9.** 7   **11.** 1/2   **13.** $-2/5$
15. 35/3   **17.** 2   **19.** 1   **21.** $-1/2$ **23.** 1
25. 1/2   **27.** -1 **29.** 1   **31.** $\ln 3 + e$
33. (a) 4   (b) 64   (c) 64   **35.** (a) 3   (b) 2   (c) 2
37. (a) 15   (b) 5   (c) 6   (d) 2/3
39. (a) 64   (b) 2   (c) 12   (d) 8
41. (a) 1   (b) 3

$g(x) = \dfrac{-2x^2 + x}{x}$ and $f(x) = -2x + 1$ agree except at $x = 0$.

43. (a) 2   (b) 0

$g(x) = \dfrac{x^3 - x}{x - 1}$ and $f(x) = x^2 + x$ agree except at $x = 1$.

45. -2

$f(x) = \dfrac{x^2 - 1}{x + 1}$ and $g(x) = x - 1$ agree except at $x = -1$.

47. 12

$f(x) = \dfrac{x^3 - 8}{x - 2}$ and $g(x) = x^2 + 2x + 4$ agree except at $x = 2$.

49. $-\dfrac{\ln 2}{8} \approx -0.0866$

$f(x) = \dfrac{(x + 4)\ln(x + 6)}{x^2 - 16}$ and $g(x) = \dfrac{\ln(x + 6)}{x - 4}$ agree except at $x = -4$.

51. 1/10   **53.** 5/6   **55.** $\sqrt{5}/10$ **57.** 1/6   **59.** $-1/9$
61. 2   **63.** $2x - 2$
65.

The graph has a hole at $x = 0$.

Answers will vary. Example:

x	-0.1	-0.01	-0.001	0.001	0.01	0.1
$f(x)$	0.358	0.354	0.354	0.354	0.353	0.349

$\lim_{x \to 0} \dfrac{\sqrt{x + 2} - \sqrt{2}}{x} \approx 0.354$ $\left(\text{Actual limit is } \dfrac{1}{2\sqrt{2}} = \dfrac{\sqrt{2}}{4}.\right)$

67.

The graph has a hole at $x = 0$.

Answers will vary. Example:

x	-0.1	-0.01	-0.001
$f(x)$	-0.263	-0.251	-0.250

x	0.001	0.01	0.1
$f(x)$	-0.250	-0.249	-0.238

$\lim_{x \to 0} \dfrac{[1/(2 + x)] - (1/2)}{x} \approx -0.250$ $\left(\text{Actual limit is } -\dfrac{1}{4}.\right)$

69. 1/5   **71.** 0   **73.** 0   **75.** 0   **77.** 1   **79.** 1   **81.** 3/2
83.

The graph has a hole at $t = 0$.

Answers will vary. Example:

t	-0.1	-0.01	0	0.01	0.1
$f(t)$	2.96	2.9996	?	2.9996	2.96

$\lim_{t \to 0} \dfrac{\sin 3t}{t} = 3$

85. The graph has a hole at $x = 0$.

Answers will vary. Example:

x	-0.1	-0.01	-0.001	0	0.001	0.01	0.1
$f(x)$	-0.1	-0.01	-0.001	?	0.001	0.01	0.1

$\lim\limits_{x \to 0} \dfrac{\sin x^2}{x} = 0$

87. The graph has a hole at $x = 1$.

Answers will vary. Example:

x	0.5	0.9	0.99	1.01	1.1	1.5
$f(x)$	1.3863	1.0536	1.0050	0.9950	0.9531	0.8109

$\lim\limits_{x \to 1} \dfrac{\ln x}{x - 1} = 1$

89. 2   **91.** $-4/x^2$ **93.** 4

95. 0   **97.** 0

99. 0

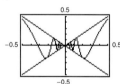 The graph has a hole at $x = 0$.

101. f and g agree at all but one point if c is a real number such that $f(x) = g(x)$ for all $x \neq c$.

103. An indeterminate form is obtained when the evaluation of a limit using direct substitution produces a meaningless fractional form, such as $\frac{0}{0}$.

105. The magnitudes of $f(x)$ and $g(x)$ are approximately equal when x is "close to" 0. Therefore, their ratio is approximately 1.

107. 160 ft/sec   **109.** -29.4 m/sec

111. Let $f(x) = 1/x$ and $g(x) = -1/x$.

$\lim\limits_{x \to 0} f(x)$ and $\lim\limits_{x \to 0} g(x)$ do not exist. However,

$\lim\limits_{x \to 0} [f(x) + g(x)] = \lim\limits_{x \to 0} \left[\dfrac{1}{x} + \left(-\dfrac{1}{x} \right) \right] = \lim\limits_{x \to 0} 0 = 0$

and therefore does exist.

113. Proof   **115.** Proof   **117.** Proof

119. False. The limit does not exist because the function approaches 1 from the right side of 0 and approaches -1 from the left side of 0. (See graph below.)

121. True. Theorem 2.7

123. False. The limit does not exist because $f(x)$ approaches 3 from the left side of 2 and approaches 0 from the right side of 2. (See graph below.)

125. Answers will vary. Example:

Let $f(x) = \begin{cases} 4, & \text{if } x \geq 0 \\ -4, & \text{if } x < 0 \end{cases}$

$\lim\limits_{x \to 0} |f(x)| = \lim\limits_{x \to 0} 4 = 4$

$\lim\limits_{x \to 0} f(x)$ does not exist because for $x < 0$, $f(x) = -4$ and for $x \geq 0$, $f(x) = 4$.

127. $\lim\limits_{x \to 0} f(x)$ does not exist because $f(x)$ oscillates between two fixed values as x approaches 0.

$\lim\limits_{x \to 0} g(x) = 0$ because, as x gets increasingly close to 0, the values of $g(x)$ become increasingly close to 0.

129. (a) $1/2$

(b) Because $\dfrac{1 - \cos x}{x^2} \approx \dfrac{1}{2}$, it follows that $1 - \cos x \approx \dfrac{1}{2}x^2$

$\cos x \approx 1 - \dfrac{1}{2}x^2$ when $x \approx 0$.

(c) 0.995   (d) Calculator: $\cos(0.1) \approx 0.9950$

Section 2.4   (page 98)

1. (a) 1   (b) 1   (c) 1; $f(x)$ is continuous on $(-\infty, \infty)$.

3. (a) 0   (b) 0   (c) 0; Discontinuity at $x = 3$

5. (a) 2   (b) -2 (c) Limit does not exist.
Discontinuity at $x = 4$

7. $\frac{1}{10}$

9. Limit does not exist. The function decreases without bound as x approaches -3 from the left.

11. -1 **13.** $-1/x^2$ **15.** 5/2   **17.** 2

19. Limit does not exist. The function decreases without bound as x approaches π from the left and increases without bound as x approaches π from the right.

21. 4

23. Limit does not exist. The function approaches 5 from the left side of 3 but approaches 6 from the right side of 3.

25. Does not exist.   **27.** $\ln 4$

29. Discontinuous at $x = -2$ and $x = 2$

31. Discontinuous at every integer

33. Continuous on $[-5, 5]$ **35.** Continuous on $[-1, 4]$

37. Continuous for all real x **39.** Continuous for all real x

41. Nonremovable discontinuity at $x = 1$
Removable discontinuity at $x = 0$

43. Continuous for all real x

45. Removable discontinuity at $x = -2$
Nonremovable discontinuity at $x = 5$

47. Nonremovable discontinuity at $x = -2$

49. Continuous for all real x

51. Nonremovable discontinuity at $x = 2$

53. Continuous for all real x

55. Nonremovable discontinuity at $x = 0$

57. Nonremovable discontinuities at integer multiples of $\pi/2$

59. Nonremovable discontinuity at each integer

61.

$$\lim_{x \to 0^+} f(x) = 0$$
$$\lim_{x \to 0^-} f(x) = 0$$
Discontinuity at $x = -2$

63. $a = 2$ **65.** $a = -1$, $b = 1$ **67.** Continuous for all real x

69. Nonremovable discontinuities at $x = 1$ and $x = -1$

71.

Nonremovable discontinuity at each integer

73.

Discontinuous at $x = 3$

75. Continuous on $(-\infty, \infty)$

77. Continuous on . . . , $(-6, -2)$, $(-2, 2)$, $(2, 6)$, . . .

79.

The graph has a hole at $x = 0$. The graph appears continuous but the function is not continuous on $[-4, 4]$. It is not obvious from the graph that the function has a discontinuity at $x = 0$.

81.

The graph has a hole at $x = 0$. The graph appears continuous but the function is not continuous on $[-4, 4]$. It is not obvious from the graph that the function has a discontinuity at $x = 0$.

83. $f(x)$ is continuous on $[2, 4]$.
$f(2) = -1$ and $f(4) = 3$
By the Intermediate Value Theorem, $f(c) = 0$ for at least one value of c between 2 and 4.

85. $h(x)$ is continuous on $[0, \pi/2]$.
$h(0) = -2 < 0$ and $h(\pi/2) \approx 0.9170 > 0$
By the Intermediate Value Theorem, $f(c) = 0$ for at least one value of c between 0 and $\pi/2$.

87. 0.68, 0.6823   **89.** 0.56, 0.5636

91. $f(3) = 11$ **93.** $f(2) = 4$

95. (a) The limit does not exist at $x = c$.
(b) The function is not defined at $x = c$.
(c) The limit exists, but it is not equal to the value of the function at $x = c$.
(d) The limit does not exist at $x = c$.

97.

Not continuous because $\lim\limits_{x \to 3} f(x)$ does not exist.

99. True

101. False. A rational function can be written as $P(x)/Q(x)$, where P and Q are polynomials of degree m and n, respectively. It can have at most n discontinuities.

103. $\lim\limits_{t \to 4^-} f(t) \approx 28$; $\lim\limits_{t \to 4^+} f(t) \approx 56$

At the end of day 3, the amount of chlorine in the pool is about 28 oz. At the beginning of day 4, the amount of chlorine in the pool is about 56 oz.

105. $C = \begin{cases} 1.04, & 0 < t \le 2 \\ 1.04 + 0.36[\![t - 1]\!], & t > 2, t \text{ is not an integer} \\ 1.04 + 0.36(t - 2), & t > 2, t \text{ is an integer} \end{cases}$

Nonremovable discontinuity at each integer greater than or equal to 2

107–109. Proofs   **111.** Answers will vary.

113. (a)

(b) There appears to be a limiting speed, and a possible cause is air resistance.

115. $c = \left(-1 \pm \sqrt{5}\right)/2$

117. Domain: $[-c^2, 0) \cup (0, \infty)$; Let $f(0) = 1/(2c)$.

119. $h(x)$ has a nonremovable discontinuity at every integer except 0.

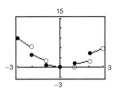

121. (a) Domain: $(-\infty, 0) \cup (0, \infty)$

(b)

(c) $\lim_{x \to 0^-} f(x) = 4$; $\lim_{x \to 0^+} f(x) = 0$ (d) Answers will vary.

123. Putnam Problem A2, 1971

Section 2.5 (page 108)

1. $\lim_{x \to -2^+} 2\left|\dfrac{x}{x^2 - 4}\right| = \infty$ $\lim_{x \to -2^-} 2\left|\dfrac{x}{x^2 - 4}\right| = \infty$

3. $\lim_{x \to -2^+} \tan((\pi x)/4) = -\infty$ $\lim_{x \to -2^-} \tan((\pi x)/4) = \infty$

5.

x	-3.5	-3.1	-3.01	-3.001
$f(x)$	0.31	1.64	16.6	167

x	-2.999	-2.99	-2.9	-2.5
$f(x)$	-167	-16.7	-1.69	-0.36

$\lim_{x \to -3^+} f(x) = -\infty$ $\lim_{x \to -3^-} f(x) = \infty$

7.

x	-3.5	-3.1	-3.01	-3.001
$f(x)$	3.8	16	151	1501

x	-2.999	-2.99	-2.9	-2.5
$f(x)$	-1499	-149	-14	-2.3

$\lim_{x \to -3^+} f(x) = -\infty$ $\lim_{x \to -3^-} f(x) = \infty$

9. $x = 0$ **11.** $x = 2$, $x = -1$ **13.** $x = \pm 2$

15. No vertical asymptote **17.** $x = \pi/4 + (n\pi)/2, n$ is an integer.

19. $t = 0$ **21.** $x = -2, x = 1$ **23.** No vertical asymptote

25. $x = 1$ **27.** $t = -2$ **29.** $x = 0$

31. $t = n\pi, n$ is a nonzero integer.

33. Removable discontinuity at $x = -1$

35. Vertical asymptote at $x = -1$

37. Removable discontinuity at $x = -1$

39. $-\infty$ **41.** ∞ **43.** $\frac{4}{5}$

45. $\frac{1}{2}$ **47.** $-\infty$ **49.** ∞ **51.** $-\infty$ **53.** Does not exist

55.

$\lim_{x \to 1^+} f(x) = \infty$

57.

$\lim_{x \to 5^-} f(x) = -\infty$

59. Answers will vary.

61. Answers will vary. Example: $f(x) = \dfrac{x - 3}{x^2 - 4x - 12}$

63.

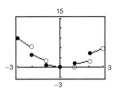

65. (a) $r = 200\pi/3$ (b) $r = 200\pi$ (c) ∞ **67.** ∞

69. (a) Proof; Domain: $x > 25$

(b)

x	30	40	50	60
y	150	66.667	50	42.857

Answers will vary.

(c) $\lim_{x \to 25^+} \dfrac{25x}{x - 25} = \infty$

As x gets close to 25 mph, y becomes larger and larger.

71. (a) $A = 50 \tan \theta - 50\theta$; Domain: $(0, \pi/2)$

(b)

θ	0.3	0.6	0.9	1.2	1.5
$f(\theta)$	0.47	4.21	18.01	68.61	630.07

(c) $\lim_{\theta \to (\pi/2)^-} A = \infty$

73. False; let $f(x) = (x^2 - 1)/(x - 1)$. **75.** True

77. Let $f(x) = \dfrac{1}{x^2}$, $g(x) = \dfrac{1}{x^4}$, and $c = 0$. $\lim_{x \to 0} \dfrac{1}{x^2} = \infty$ and $\lim_{x \to 0} \dfrac{1}{x^4} = \infty$, but $\lim_{x \to 0}\left(\dfrac{1}{x^2} - \dfrac{1}{x^4}\right) = \lim_{x \to 0}\left(\dfrac{x^2 - 1}{x^4}\right) = -\infty \neq 0$.

79. Given $\lim_{x \to c} f(x) = \infty$, let $g(x) = 1$. Then $\lim_{x \to c} \dfrac{g(x)}{f(x)} = 0$ by Theorem 2.15.

81. Answers will vary.

Review Exercises for Chapter 2 (page 111)

1. Calculus   Estimate: 8.268

3.

x	-0.1	-0.01	-0.001
$f(x)$	-1.0526	-1.0050	-1.0005

x	0.001	0.01	0.1
$f(x)$	-0.9995	-0.9950	-0.9524

$\lim_{x \to 0} f(x) = -1$

5.

x	-0.1	-0.01	-0.001
$f(x)$	0.8867	0.0988	0.0100

x	0.001	0.01	0.1
$f(x)$	-0.0100	-0.1013	-1.1394

$\lim\limits_{x \to 0} f(x) = 0$

7. (a) -2 (b) -3 **9.** (a) Limit does not exist. (b) 0
11. 2; Proof **13.** 1; Proof **15.** $\sqrt{6} \approx 2.45$ **17.** $-\frac{1}{4}$
19. $\frac{1}{4}$ **21.** -1 **23.** 75 **25.** 0 **27.** $\sqrt{3}/2$ **29.** 1
31. $-\frac{1}{2}$
33. (a)

x	1.1	1.01	1.001	1.0001
$f(x)$	0.5680	0.5764	0.5773	0.5773

$\lim\limits_{x \to 1^+} f(x) \approx 0.5773$

(b)
The graph has a hole at $x = 1$.
$\lim\limits_{x \to 1^+} f(x) \approx 0.5774$

(c) $\sqrt{3}/3$
35. -39.2 m/sec **37.** -1 **39.** 0
41. Limit does not exist. The limit as t approaches 1 from the left is 2, whereas the limit as t approaches 1 from the right is 1.
43. Nonremovable discontinuity at each integer
Continuous on $(k, k + 1)$ for all integers k
45. Removable discontinuity at $x = 1$
Continuous on $(-\infty, 1) \cup (1, \infty)$
47. Nonremovable discontinuity at $x = 2$
Continuous on $(-\infty, 2) \cup (2, \infty)$
49. Nonremovable discontinuity at $x = -1$
Continuous on $(-\infty, -1) \cup (-1, \infty)$
51. Nonremovable discontinuity at each even integer
Continuous on $(2k, 2k + 2)$ for all integers k
53. Nonremovable discontinuity at each integer
Continuous on $(k, k + 1)$ for all integers k
55. $c = -\frac{1}{2}$ **57.** Proof
59. Nonremovable discontinuity every 6 months

61. $x = 0$ **63.** $x = 10$ **65.** $x = -3, x = 3$ **67.** $-\infty$
69. $\frac{1}{3}$ **71.** $-\infty$ **73.** $\frac{4}{5}$ **75.** ∞ **77.** $-\infty$
79. (a) 2
(b) Yes, define as
$$f(x) = \begin{cases} \dfrac{\tan 2x}{x}, & x \neq 0 \\ 2, & x = 0 \end{cases}.$$

P.S. Problem Solving (page 113)

1. (a) Perimeter $\triangle PAO = 1 + \sqrt{(x^2 - 1)^2 + x^2} + \sqrt{x^4 + x^2}$
Perimeter $\triangle PBO = 1 + \sqrt{x^4 + (x - 1)^2} + \sqrt{x^4 + x^2}$

(b)

x	4	2	1
Perimeter $\triangle PAO$	33.0166	9.0777	3.4142
Perimeter $\triangle PBO$	33.7712	9.5952	3.4142
$r(x)$	0.9777	0.9461	1.0000

x	0.1	0.01
Perimeter $\triangle PAO$	2.0955	2.0100
Perimeter $\triangle PBO$	2.0006	2.0000
$r(x)$	1.0475	1.0050

(c) 1
3. (a) Area (hexagon) $= (3\sqrt{3})/2 \approx 2.5981$
Area (circle) $= \pi \approx 3.1416$
Area (circle) $-$ Area (hexagon) ≈ 0.5435
(b) $A_n = (n/2) \sin(2\pi/n)$
(c)

n	6	12	24	48	96
A_n	2.5981	3.0000	3.1058	3.1326	3.1394

(d) 3.1416 or π
5. (a) $m = -\frac{12}{5}$ (b) $y = \frac{5}{12}x - \frac{169}{12}$
(c) $m_x = \dfrac{-\sqrt{169 - x^2} + 12}{x - 5}$
(d) $\frac{5}{12}$; It is the same as the slope of the tangent line found in (b).
7. (a) Domain: $[-27, 1) \cup (1, \infty)$
(b)   (c) $\frac{1}{14}$ (d) $\frac{1}{12}$

The graph has a hole at $x = 1$.
9. (a) g_1, g_4 (b) g_1 (c) g_1, g_3, g_4
11.

The graph jumps at every integer.

(a) $f(1) = 0$, $f(0) = 0$, $f(\frac{1}{2}) = -1$, $f(-2.7) = -1$
(b) $\lim\limits_{x \to 1^-} f(x) = -1$, $\lim\limits_{x \to 1^+} f(x) = -1$, $\lim\limits_{x \to 1/2} f(x) = -1$
(c) There is a discontinuity at each integer.

13. (a)

(b) (i) $\lim\limits_{x \to a^+} P_{a,b}(x) = 1$

(ii) $\lim\limits_{x \to a^-} P_{a,b}(x) = 0$

(iii) $\lim\limits_{x \to b^+} P_{a,b}(x) = 0$

(iv) $\lim\limits_{x \to b^-} P_{a,b}(x) = 1$

(c) Continuous for all positive real numbers except a and b

(d) The area under the curve gives a value of 1.

Chapter 3

Section 3.1   (page 123)

1. (a) $m_1 = 0, m_2 = 5/2$ (b) $m_1 = -5/2, m_2 = 2$

3. $y = \dfrac{f(4) - f(1)}{4 - 1}(x - 1) + f(1) = x + 1$

5. $m = -2$ **7.** $m = 2$

9. $m = 3$ **11.** $f'(x) = 0$ **13.** $f'(x) = -5$ **15.** $h'(s) = \frac{2}{3}$

17. $f'(x) = 4x + 1$ **19.** $f'(x) = 3x^2 - 12$

21. $f'(x) = -1/(x - 1)^2$ **23.** $f'(x) = 1/2\sqrt{x + 1}$

25. (a) Tangent line:
$y = 4x - 3$

(b)

27. (a) Tangent line:
$y = 12x - 16$

(b)

29. (a) Tangent line:
$y = \frac{1}{2}x + \frac{1}{2}$

(b)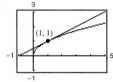

31. (a) Tangent line:
$y = \frac{3}{4}x + 2$

(b)

33. $y = 3x - 2; y = 3x + 2$ **35.** $y = -\frac{1}{2}x + \frac{3}{2}$

37. b   **38.** d   **39.** a   **40.** c   **41.** $g(5) = 2; g'(5) = -\frac{1}{2}$

43. $f'(x) = 1$

45. $f'(x) = 2x - 8$

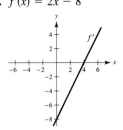

47. Answers will vary.
Sample answer: $y = -x$

49. $f(x) = 5 - 3x$
$c = 1$

51. $f(x) = -x^2$
$c = 6$

53. $f(x) = -3x + 2$

55. Answers will vary.
Sample answer: $f(x) = x^3$

57. $y = 2x + 1; y = -2x + 9$

59. (a) -3 (b) 0

(c) The graph is moving downward to the right when $x = 1$.

(d) The graph is moving upward to the right when $x = -4$.

(e) Positive. Because $g'(x) > 0$ on $[3, 6]$, the graph of g is moving upward to the right.

(f) No. Knowing only $g'(2)$ is not sufficient information. $g'(2)$ remains the same for any vertical translation of g.

61.

x	-2	-1.5	-1	-0.5	0	0.5	1	1.5	2
$f(x)$	-2	$-\frac{27}{32}$	$-\frac{1}{4}$	$-\frac{1}{32}$	0	$\frac{1}{32}$	$\frac{1}{4}$	$\frac{27}{32}$	2
$f'(x)$	3	$\frac{27}{16}$	$\frac{3}{4}$	$\frac{3}{16}$	0	$\frac{3}{16}$	$\frac{3}{4}$	$\frac{27}{16}$	3

63. $g(x) \approx f'(x)$

65. $f(2) = 4; f(2.1) = 3.99; f'(2) \approx -0.1$

67.   As x approaches infinity, the graph of f approaches a line of slope 0. Thus $f'(x)$ approaches 0.

69. (a)

(b) The graphs of S for decreasing values of Δx are secant lines approaching the tangent line to the graph of f at the point $(2, f(2))$.

71. 4   **73.** 4   **75.** $g(x)$ is not differentiable at $x = 0$.

77. $f(x)$ is not differentiable at $x = 6$.

79. $h(x)$ is not differentiable at $x = -5$.

81. $(-\infty, -1) \cup (-1, \infty)$ **83.** $(-\infty, 3) \cup (3, \infty)$

85. $(1, \infty)$

87. $(-\infty, -3) \cup (-3, \infty)$ **89.** $(-\infty, 0) \cup (0, \infty)$

91. The derivative from the left is -1 and the derivative from the right is 1, so f is not differentiable at $x = 1$.

93. The derivatives from both the right and the left are 0, so $f'(1) = 0$.

95. f is differentiable at $x = 2$.

97. (a) $d = (3|m + 1|)/\sqrt{m^2 + 1}$

(b)

Not differentiable at $m = -1$

99. False. The slope is $\lim\limits_{\Delta x \to 0} \dfrac{f(2 + \Delta x) - f(2)}{\Delta x}$.

101. False. For example: $f(x) = |x|$. The derivative from the left and the derivative from the right both exist but are not equal.

103. Proof

Section 3.2   (page 136)

1. (a) $\frac{1}{2}$ (b) 3   **3.** 0   **5.** $6x^5$ **7.** $1/(5x^{4/5})$ **9.** 1

11. $-4t + 3$ **13.** $2x + 12x^2$ **15.** $3t^2 - 2$ **17.** $6 - 5e^x$

19. $\dfrac{\pi}{2}\cos\theta + \sin\theta$ **21.** $2x + \dfrac{1}{2}\sin x$ **23.** $\dfrac{1}{2}e^x - 3\cos x$

	Function	Rewrite	Differentiate	Simplify
25.	$y = \dfrac{5}{2x^2}$	$y = \dfrac{5}{2}x^{-2}$	$y' = -5x^{-3}$	$y' = -\dfrac{5}{x^3}$
27.	$y = \dfrac{3}{(2x)^3}$	$y = \dfrac{3}{8}x^{-3}$	$y' = -\dfrac{9}{8}x^{-4}$	$y' = -\dfrac{9}{8x^4}$
29.	$y = \dfrac{\sqrt{x}}{x}$	$y = x^{-1/2}$	$y' = -\dfrac{1}{2}x^{-3/2}$	$y' = -\dfrac{1}{2x^{3/2}}$

31. -6 **33.** 0   **35.** 3   **37.** $\frac{3}{4}$ **39.** $2t + 12/t^4$

41. $(x^3 - 8)/x^3$ **43.** $3x^2 + 1$ **45.** $1/(2\sqrt{x}) - 2/x^{2/3}$

47. $4/5s^{1/5} - 2/3s^{1/3}$ **49.** $3/\sqrt{x} - 5\sin x$

51. $-2/x^3 - 2e^x$

53. (a) $5x + y + 3 = 0$

(b)

55. (a) $2x - y + 1 = 0$

(b)

57. $(0, 2), (-2, -14), (2, -14)$ **59.** (π, π)

61. $(\ln 4, 4 - 4\ln 4)$ **63.** $k = 2, k = -10$ **65.** $k = 3$

67. (a) A and B (b) Greater than

(c)

69. $g'(x) = f'(x)$

71.

The rate of change of f is constant and therefore f' is a constant function.

73. $y = 2x - 1$ $y = 4x - 4$

75. $f'(x) = 3 + \cos x \neq 0$ for all x.

77. $x - 4y + 4 = 0$

79.

$f'(1)$ appears to be close to -1.
$f'(1) = -1$

81. (a)

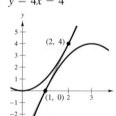

$(3.9, 7.7019)$; $S(x) = 2.981x - 3.924$

(b) $T(x) = 3(x - 4) + 8 = 3x - 4$

The slope (and equation) of the secant line approaches that of the tangent line at $(4, 8)$ as you choose points closer to $(4, 8)$.

(c) The approximation becomes less accurate.

(d)

Δx	-3	-2	-1	-0.5	-0.1	0
$f(4 + \Delta x)$	1	2.828	5.196	6.458	7.702	8
$T(4 + \Delta x)$	-1	2	5	6.5	7.7	8

Δx	0.1	0.5	1	2	3
$f(4 + \Delta x)$	8.302	9.546	11.180	14.697	18.520
$T(4 + \Delta x)$	8.3	9.5	11	14	17

83. False: let $f(x) = x$ and $g(x) = x + 1$.
85. False: $dy/dx = 0$.　　**87.** True
89. Average rate: $\frac{1}{2}$
　　Instantaneous rates:
　　$f'(1) = 1$
　　$f'(2) = \frac{1}{4}$
91. Average rate: $e \approx 2.718$
　　Instantaneous rates:
　　$g'(0) = 1$
　　$g'(1) = 2 + e \approx 4.718$
93. (a) $s(t) = -16t^2 + 1362$; $v(t) = -32t$　(b) -48 ft/sec
　　(c) $s'(1) = -32$ ft/sec; $s'(2) = -64$ ft/sec
　　(d) $t = \sqrt{1362}/4 \approx 9.226$ sec　(e) -295.242 ft/sec
95. $v(5) = 71$ m/sec; $v(10) = 22$ m/sec
97.　　　　　　　　　　**99.**

101. (a) $R(v) = 0.417v - 0.02$
　　(b) $B(v) = 0.0056v^2 + 0.001v + 0.04$
　　(c) $T(v) = 0.0056v^2 + 0.418v + 0.02$
　　(d)
　　(e) $T'(v) = 0.0112v + 0.418$
　　　　$T'(40) = 0.866$
　　　　$T'(80) = 1.314$
　　　　$T'(100) = 1.538$
　　(f) As speed increases, the total stopping distance increases.
103. $V'(4) = 48$ cm^2　　**105.** Proof
107. (a) The rate of change of the number of gallons of gasoline
　　sold when the price is $1.479
　　(b) In general, the rate of change when $p = 1.479$ should be
　　negative. As prices go up, sales go down.
109. $y = 2x^2 - 3x + 1$　　**111.** $y = -9x$, $y = -\frac{9}{4}x - \frac{27}{4}$
113. $a = \frac{1}{3}$, $b = -\frac{4}{3}$
115. $f_1(x) = |\sin x|$ is differentiable for all $x \neq n\pi$, n an integer.
　　$f_2(x) = \sin|x|$ is differentiable for all $x \neq 0$.

Section 3.3　(page 147)

1. $2(2x^3 - 3x^2 + x - 1)$　　**3.** $(7t^2 + 4)/(3t^{2/3})$
5. $x^2(3\cos x - x\sin x)$　　**7.** $(1 - x^2)/(x^2 + 1)^2$
9. $(1 - 8x^3)/[3x^{2/3}(x^3 + 1)^2]$　　**11.** $(x\cos x - 2\sin x)/x^3$
13. $f'(x) = (x^3 - 3x)(4x + 3) + (2x^2 + 3x + 5)(3x^2 - 3)$
　　　$= 10x^4 + 12x^3 - 3x^2 - 18x - 15$
　　$f'(0) = -15$
15. $f'(x) = \dfrac{x^2 - 6x + 4}{(x - 3)^2}$　　**17.** $f'(x) = \cos x - x\sin x$
　　$f'(1) = -\dfrac{1}{4}$　　　　　　　　　$f'\left(\dfrac{\pi}{4}\right) = \dfrac{\sqrt{2}}{8}(4 - \pi)$
19. $f'(x) = e^x(\cos x + \sin x)$
　　$f'(0) = 1$

Function	Rewrite	Differentiate	Simplify
21. $y = \dfrac{x^2 + 2x}{3}$	$y = \dfrac{1}{3}(x^2 + 2x)$	$y' = \dfrac{1}{3}(2x + 2)$	$y' = \dfrac{2(x + 1)}{3}$
23. $y = \dfrac{7}{3x^3}$	$y = \dfrac{7}{3}x^{-3}$	$y' = -7x^{-4}$	$y' = -\dfrac{7}{x^4}$
25. $y = \dfrac{4x^{3/2}}{x}$	$y = 4x^{1/2},$ $x > 0$	$y' = 2x^{-1/2}$	$y' = \dfrac{2}{\sqrt{x}},$ $x > 0$

27. $\dfrac{(x^2 - 1)(-2 - 2x) - (3 - 2x - x^2)(2x)}{(x^2 - 1)^2} = \dfrac{2}{(x + 1)^2}$, $x \neq 1$
29. $1 - 12/(x + 3)^2 = (x^2 + 6x - 3)/(x + 3)^2$
31. $[2\sqrt{x} - (2x + 5)/2\sqrt{x}]/x = (2x - 5)/2x^{3/2}$
33. $6s^2(s^3 - 2)$　　**35.** $-(2x^2 - 2x + 3)/[x^2(x - 3)^2]$
37. $(3x^3 + 4x)[(x - 5) \cdot 1 + (x + 1) \cdot 1]$
　　　$+ [(x - 5)(x + 1)](9x^2 + 4)$
　　　$= 15x^4 - 48x^3 - 33x^2 - 32x - 20$
39. $\dfrac{(x^2 - c^2)(2x) - (x^2 + c^2)(2x)}{(x^2 - c^2)^2} = -\dfrac{4xc^2}{(x^2 - c^2)^2}$
41. $t(t\cos t + 2\sin t)$　　**43.** $-(t\sin t + \cos t)/t^2$
45. $-e^x + \sec^2 x$　　**47.** $\dfrac{1}{4t^{3/4}} + 8\sec t\tan t$
49. $\dfrac{-6\cos^2 x + 6\sin x - 6\sin^2 x}{4\cos^2 x} = \dfrac{3}{2}(-1 + \tan x\sec x - \tan^2 x)$
　　　　　　　　　　　$= \dfrac{3}{2}\sec x(\tan x - \sec x)$
51. $\csc x\cot x - \cos x = \cos x\cot^2 x$　　**53.** $x(x\sec^2 x + 2\tan x)$
55. $2x\cos x + 2\sin x + x^2 e^x + 2xe^x$
57. $e^x/8x^{3/2}(2x - 1)$
59. $\left(\dfrac{x + 1}{x + 2}\right)(2) + (2x - 5)\left[\dfrac{(x + 2)(1) - (x + 1)(1)}{(x + 2)^2}\right]$
　　　$= (2x^2 + 8x - 1)/(x + 2)^2$
61. $\dfrac{1 - \sin\theta + \theta\cos\theta}{(1 - \sin\theta)^2}$　　**63.** $y' = \dfrac{-2\csc x\cot x}{(1 - \csc x)^2}$, $-4\sqrt{3}$
65. $h'(t) = \sec t(t\tan t - 1)/t^2$, $1/\pi^2$
67. (a) $y = -x - 2$
　　(b)

Section 4.7 (page 265)

1. (a) and (b)

First Number x	Second Number	Product P
10	$110 - 10$	$10(110 - 10) = 1000$
20	$110 - 20$	$20(110 - 20) = 1800$
30	$110 - 30$	$30(110 - 30) = 2400$
40	$110 - 40$	$40(110 - 40) = 2800$
50	$110 - 50$	$50(110 - 50) = 3000$
60	$110 - 60$	$60(110 - 60) = 3000$
70	$110 - 70$	$70(110 - 70) = 2800$
80	$110 - 80$	$80(110 - 80) = 2400$
90	$110 - 90$	$90(110 - 90) = 1800$
100	$110 - 100$	$100(110 - 100) = 1000$

(c) $P = x(110 - x)$

(d)

(e) 55 and 55

3. $S/2$ and $S/2$ **5.** 24 and 8 **7.** 50 and 25
9. $l = w = 25$ m **11.** $l = w = 8$ ft **13.** $\left(\frac{7}{2}, \sqrt{\frac{7}{2}}\right)$
15. $(1, 1)$ **17.** $x = Q_0/2$ **19.** 600 m × 300 m
21. (a) Proof (b) $V_1 = 99$ in.³, $V_2 = 125$ in.³, $V_3 = 117$ in.³
  (c) $5 \times 5 \times 5$ in.
23. Rectangular portion: $16/(\pi + 4)$ ft × $32/(\pi + 4)$ ft

25. (a) $L = \sqrt{x^2 + 4 + \dfrac{8}{x - 1} + \dfrac{4}{(x - 1)^2}}, \quad x > 1$

  (b)   Minimum when $x \approx 2.587$

  (c) $(0, 0), (2, 0), (0, 4)$
27. Width: $5\sqrt{2}/2$; Length: $5\sqrt{2}$
29. Dimensions of page: $\left(2 + \sqrt{30}\right)$ in. × $\left(2 + \sqrt{30}\right)$ in.
31. (a)

  (b)

Length x	Width y	Area xy
10	$2/\pi(100 - 10)$	$(10)(2/\pi)(100 - 10) \approx 573$
20	$2/\pi(100 - 20)$	$(20)(2/\pi)(100 - 20) \approx 1019$
30	$2/\pi(100 - 30)$	$(30)(2/\pi)(100 - 30) \approx 1337$
40	$2/\pi(100 - 40)$	$(40)(2/\pi)(100 - 40) \approx 1528$
50	$2/\pi(100 - 50)$	$(50)(2/\pi)(100 - 50) \approx 1592$
60	$2/\pi(100 - 60)$	$(60)(2/\pi)(100 - 60) \approx 1528$

The maximum area of the rectangle is approximately 1592 m².

(c) $A = 2/\pi(100x - x^2), \quad 0 < x < 100$
(d) $\dfrac{dA}{dx} = \dfrac{2}{\pi}(100 - 2x)$
  $= 0$ when $x = 50$
  The maximum value is
  approximately 1592
  when $x = 50$.

(e)

33. $18 \times 18 \times 36$ in. **35.** $32\pi r^3/81$
37. Answers will vary. If area is expressed as a function of either length or width, the feasible domain is the interval $(0, 10)$. No dimensions will yield a minimum area because the second derivative on this open interval is always negative.
39. $r = \sqrt[3]{9/\pi} \approx 1.42$ cm
41. Side of square: $\dfrac{10\sqrt{3}}{9 + 4\sqrt{3}}$; Side of triangle: $\dfrac{30}{9 + 4\sqrt{3}}$
43. $w = 8\sqrt{3}$ in., $h = 8\sqrt{6}$ in. **45.** $\theta = \pi/4$ **47.** $h = \sqrt{2}$ ft
49. One mile from the nearest point on the coast **51.** Proof
53.

(a) Origin to y-intercept: 2
  Origin to x-intercept: $\pi/2$
(b) $d = \sqrt{x^2 + (2 - 2\sin x)^2}$

(c) Minimum distance is 0.9795 when $x \approx 0.7967$.
55. $\theta = \pi/3\left(6 - 2\sqrt{6}\right) \approx 1.15°$
57. 4045 units **59.** $A = \sqrt{2}e^{-1/2}$ **61.** Proof
63. $y = \frac{64}{141}x$; $S_1 \approx 6.1$ mi **65.** $y = \frac{3}{10}x$; $S_3 \approx 4.50$ mi
67. Putnam Problem A1, 1986

Section 4.8 (page 276)

1. $T(x) = 4x - 4$

x	1.9	1.99	2	2.01	2.1
$f(x)$	3.610	3.960	4	4.040	4.410
$T(x)$	3.600	3.960	4	4.040	4.400

3. $T(x) = 80x - 128$

x	1.9	1.99	2	2.01	2.1
$f(x)$	24.761	31.208	32	32.808	40.841
$T(x)$	24.000	31.200	32	32.800	40.000

5. $T(x) = (\cos 2)(x - 2) + \sin 2$

x	1.9	1.99	2	2.01	2.1
$f(x)$	0.946	0.913	0.909	0.905	0.863
$T(x)$	0.951	0.913	0.909	0.905	0.868

7. $\Delta y = 0.6305$; $dy = 0.6000$ **9.** $\Delta y = -0.039$; $dy = -0.040$
11. $6x\, dx$ **13.** $-\dfrac{3}{(2x - 1)^2}\, dx$ **15.** $\dfrac{1 - 2x^2}{\sqrt{1 - x^2}}\, dx$

17. $x/(x^2 - 4)\,dx$ **19.** $(2 + 2\cot x + 2\cot^3 x)\,dx$

21. $-\pi\sin\!\left(\dfrac{6\pi x - 1}{2}\right)dx$ **23.** $\left(\arcsin x + \dfrac{x}{\sqrt{1 - x^2}}\right)dx$

25. (a) 0.9   (b) 1.04   **27.** (a) 1.05   (b) 0.98

29. (a) 8.035   (b) 7.95   **31.** $\pm\frac{3}{8}$ in.^{2} **33.** $\pm 7\pi$ in.^{2}

35. (a) $\frac{2}{3}$%   (b) 1.25%

37. (a) $\pm 2.88\pi$ in.^{3} (b) $\pm 0.96\pi$ in.^{2} (c) 1%, $\frac{2}{3}$%

39. 267.24, 3.1%   **41.** 80π cm^3

43. (a) 0.87%   (b) 2.16%   **45.** 4961 ft

47. $f(x) = \sqrt{x},\; dy = \dfrac{1}{2\sqrt{x}}\,dx$

$f(99.4) \approx \sqrt{100} + \dfrac{1}{2\sqrt{100}}(-0.6) = 9.97$

Calculator: 9.97

49. $f(x) = \sqrt[4]{x},\; dy = \dfrac{1}{4x^{3/4}}\,dx$

$f(624) \approx \sqrt[4]{625} + \dfrac{1}{4(625)^{3/4}}(-1) = 4.998$

Calculator: 4.998

51. $f(x) = \sqrt{x};\; dy = \dfrac{1}{2\sqrt{x}}\,dx$

$f(4.02) \approx \sqrt{4} + \dfrac{1}{2\sqrt{4}}(0.02) = 2 + \dfrac{1}{4}(0.02)$

53. $y - f(0) = f'(0)(x - 0)$

$y - 2 = \frac{1}{4}x$

$y = 2 + x/4$

55. $y - f(0) = f'(0)(x - 0)$

$y - 0 = 1(x)$

$y = x$

57. The value of dy becomes closer to the value of Δy as Δx decreases.

59. True   **61.** True

Review Exercises for Chapter 4   (page 278)

1. Let f be defined at c. If $f'(c) = 0$ or if f' is undefined at c, then c is a critical number of f.

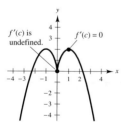

3. Maximum: $(2\pi, 17.57)$ **5.** $f'\!\left(\frac{1}{3}\right) = 0$

Minimum: $(2.73, 0.88)$

7. (a)

(b) f is not differentiable at $x = 4$.

9. $f'\!\left(\dfrac{2744}{729}\right) = \dfrac{3}{7}$ **11.** $f'(0) = 1$ **13.** $c = \dfrac{x_1 + x_2}{2}$

15. Critical numbers: $x = 1, \frac{7}{3}$

Increasing on $(-\infty, 1)$, $\left(\frac{7}{3}, \infty\right)$; Decreasing on $\left(1, \frac{7}{3}\right)$

17. Critical number: $x = 1$

Increasing on $(1, \infty)$; Decreasing on $(0, 1)$

19. Critical number: $t = 2 - 1/\ln 2$

Increasing on $(-\infty, 2 - 1/\ln 2)$

Decreasing on $(2 - 1/\ln 2, \infty)$

21. Minimum: $(2, -12)$

23. (a) $y = \frac{1}{4}$ in.; $v = 4$ in./sec   (b) Proof

(c) Period: $\pi/6$; Frequency: $6/\pi$

25. $(\pi/2, \pi/2)$, $(3\pi/2, 3\pi/2)$

27. Relative maxima: $\left(\sqrt{2}/2, 1/2\right), \left(-\sqrt{2}/2, 1/2\right)$

Relative minimum: $(0, 0)$

29.

31. Increasing and concave down

33. (a) $D = 0.00340t^4 - 0.2352t^3 + 4.942t^2 - 20.86t + 94.4$

(b)

(c) Maximum occurs in 1991; Minimum occurs in 1972.

(d) 1979

35. $\frac{2}{3}$ **37.** $-\infty$ **39.** 0   **41.** 6

43. Vertical asymptote: $x = 4$; Horizontal asymptote: $y = 2$

45. Vertical asymptote: $x = 0$; Horizontal asymptote: $y = -2$

47. Horizontal asymptotes: $y = 0$; $y = \frac{5}{3}$

49. Horizontal asymptote: $y = 0$

55. $y = 3e^{0.2x}$

57. $y = \frac{1}{2}e^{2x} + 2e^x + x + C$

59. $y = \frac{1}{2}\arctan(\tan x/2) + C$ **61.** $\frac{1}{2}$

63. $\frac{1}{2}(1 - e^{-1}) \approx 0.316$ **65.** 4   **67.** $\pi/18$

69. $5\sqrt{5} \approx 11.18$ **71.** $\frac{3}{2}\ln\left(\frac{34}{9}\right) + \frac{2}{3}\arctan\left(\frac{5}{3}\right) \approx 2.68$

73. $\frac{4}{3} \approx 1.333$

75. $\frac{1}{3}\arctan\left[\frac{1}{3}(x + 2)\right] + C$ **77.** $\tan\theta - \sec\theta + C$

Graphs will vary.   Graphs will vary.

Example: Example:

One graph is a vertical   One graph is a vertical

translation of the other.   translation of the other.

79. Power rule: $\int u^n \, du = \dfrac{u^{n+1}}{n+1} + C$; $u = x^2 + 1$, $du = 2x$, $n = 3$

81. Log rule: $\int \dfrac{du}{u} = \ln|u| + C$; $u = x^2 + 1$, $du = 2x$

83. Using laws of logarithms, $y_1 = e^{x+C_1} = e^x \cdot e^{C_1}$, where e^{C_1} is a constant. Therefore, e^{C_1} can be replaced by C, resulting in $y_2 = Ce^x$.

85. $a = \sqrt{2}$, $b = \dfrac{\pi}{4}$; $-\dfrac{1}{\sqrt{2}}\ln\left|\csc\left(x + \dfrac{\pi}{4}\right) + \cot\left(x + \dfrac{\pi}{4}\right)\right| + C$

87.

Negative; more area below the x-axis than above

89. a

91. (a)

(b)

$y = \sqrt{2x}$

(c)

$y = x$

93. (a) $\pi(1 - e^{-1}) \approx 1.986$ (b) $b = \sqrt{\ln\left(\dfrac{3\pi}{3\pi - 4}\right)} \approx 0.743$

95. $(8\pi/3)(10\sqrt{10} - 1) \approx 256.545$

97. $\frac{1}{3}\arctan 3 \approx 0.416$ **99.** ≈ 1.0320

101. (a) $\frac{1}{3}\sin x(\cos^2 x + 2)$

(b) $\frac{1}{15}\sin x(3\cos^4 x + 4\cos^2 x + 8)$

(c) $\frac{1}{35}\sin x(5\cos^6 x + 6\cos^4 x + 8\cos^2 x + 16)$

(d) Use the power reducing formula, $\int\cos^n x \, dx = \dfrac{\cos^{n-1} x \sin x}{n} + \dfrac{n-1}{n}\int\cos^{n-2} x \, dx$ until your integral is $\int\cos x \, dx$, which you can then integrate.

103. Proof

Section 8.2 (page 531)

1. b   **2.** d   **3.** c   **4.** a   **5.** $u = x$, $dv = e^{2x} \, dx$

7. $u = (\ln x)^2$, $dv = dx$ **9.** $u = x$, $dv = \sec^2 x \, dx$

11. $-(1/4e^{2x})(2x + 1) + C$ **13.** $e^x(x^3 - 3x^2 + 6x - 6) + C$

15. $\frac{1}{3}e^{x^3} + C$ **17.** $\frac{1}{4}[2(t^2 - 1)\ln|t + 1| - t^2 + 2t] + C$

19. $\frac{1}{3}(\ln x)^3 + C$ **21.** $e^{2x}/[4(2x + 1)] + C$

23. $(x - 1)^2 e^x + C$ **25.** $\frac{2}{15}(x - 1)^{3/2}(3x + 2) + C$

27. $x \sin x + \cos x + C$

29. $(6x - x^3)\cos x + (3x^2 - 6)\sin x + C$

31. $-t \csc t - \ln|\csc t + \cot t| + C$

33. $x\arctan x - \frac{1}{2}\ln(1 + x^2) + C$

35. $\frac{1}{5}e^{2x}(2\sin x - \cos x) + C$ **37.** $y = \frac{1}{2}e^{x^2} + C$

39. $y = \frac{2}{405}(27t^2 - 24t + 32)\sqrt{2 + 3t} + C$ **41.** $\sin y = x^2 + C$

43. (a)

(b) $2\sqrt{y} - \cos x - x\sin x = 3$

45.

47. $4 - 12/e^2$ **49.** $\pi/2 - 1$

51. $\left(\pi - 3\sqrt{3} + 6\right)/6 \approx 0.658$

53. $\frac{1}{2}[e(\sin 1 - \cos 1) + 1] \approx 0.909$

55. $(24\ln 2 - 7)/9 \approx 1.071$

57. $8\,\text{arcsec}\, 4 + \sqrt{3}/2 - \sqrt{15}/2 - 2\pi/3 \approx 7.380$

59. $(e^{2x}/4)(2x^2 - 2x + 1) + C$

61. $(3x^2 - 6)\sin x - (x^3 - 6x)\cos x + C$

63. $x\tan x + \ln|\cos x| + C$ **65.** $2\left(\sin\sqrt{x} - \sqrt{x}\cos\sqrt{x}\right) + C$

67. $\frac{128}{15}$ **69.** $\frac{1}{2}x[\cos(\ln x) + \sin(\ln x)] + C$ **71.** Product Rule

73. No   **75.** Yes. Let $u = x^2$ and $dv = e^{2x} \, dx$.

77. Yes. Let $u = x$, $dv = (1\sqrt{x + 1}) \, dx$. (Substitution also works. Let $u = \sqrt{x + 1}$).

79. (a) $-(e^{-4t}/128)(32t^3 + 24t^2 + 12t + 3) + C$

(b) Graphs will vary. Example: (c) One graph is a vertical translation of the other.

81. (a) $\frac{1}{13}(2e^{-\pi} + 3) \approx 0.2374$

(b) Graphs will vary. Example: (c) One graph is a vertical translation of the other.

83. $\frac{2}{5}(2x - 3)^{3/2}(x + 1) + C$ **85.** $\frac{1}{3}\sqrt{4 + x^2}(x^2 - 8) + C$

87. $n = 0$: $x(\ln x - 1) + C$

$n = 1$: $\frac{1}{4}x^2(2 \ln x - 1) + C$

$n = 2$: $\frac{1}{9}x^3(3 \ln x - 1) + C$

$n = 3$: $\frac{1}{16}x^4(4 \ln x - 1) + C$

$n = 4$: $\frac{1}{25}x^5(5 \ln x - 1) + C$

$\int x^n \ln x \, dx = \dfrac{x^{n+1}}{(n + 1)^2}[(n + 1) \ln x - 1] + C$

89–93. Proofs

95. $\frac{1}{16}x^4(4 \ln x - 1) + C$ **97.** $\frac{1}{13}e^{2x}(2 \cos 3x + 3 \sin 3x) + C$

99.

$1 - \dfrac{5}{e^4} \approx 0.908$

101.

$\dfrac{\pi}{1 + \pi^2}\left(\dfrac{1}{e} + 1\right) \approx 0.395$

103. (a) 1   (b) $\pi(e - 2) \approx 2.257$ (c) $\frac{1}{2}\pi(e^2 + 1) \approx 13.177$

(d) $\left(\dfrac{e^2 + 1}{4}, \dfrac{e - 2}{2}\right) \approx (2.097, 0.359)$

105. In Example 6, we showed that the centroid of an equivalent region was $(1, \pi/8)$. By symmetry, the centroid of this region is $(1, \pi/8)$.

107. $[7/(10\pi)](1 - e^{-4\pi}) \approx 0.223$ **109.** \$931,265

111. Proof   **113.** $b_n = [8h/(n\pi)^2] \sin(n\pi/2)$

115. Shell: $V = \pi\left[b^2 f(b) - a^2 f(a) - \displaystyle\int_a^b x^2 f'(x) \, dx\right]$

Disk: $V = \pi\left[b^2 f(b) - a^2 f(a) - \displaystyle\int_{f(a)}^{f(b)} [f^{-1}(y)]^2 \, dy\right]$

Both methods yield the same volume because $x = f^{-1}(y)$, $f'(x) \, dx = dy$, if $y = f(a)$ then $x = a$, and if $y = f(b)$ then $x = b$.

117. (a) $y = \frac{1}{4}(3 \sin 2x - 6x \cos 2x)$ (b)

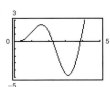

(c) You obtain the following points.

n	x^n	y^n
0	0	0
1	0.05	0.000250
2	0.10	0.001992
3	0.15	0.006689
4	0.20	0.015745
$\vdots$	$\vdots$	$\vdots$
80	4.00	1.615019

(d) You obtain the following points.

n	x^n	y^n
0	0	0
1	0.1	0.001992
2	0.2	0.015745
3	0.3	0.052081
4	0.4	0.119993
$\vdots$	$\vdots$	$\vdots$
40	4.0	1.615019

(e) Smaller step size, more data

119. The graph of $y = x \sin x$ is below the graph of $y = x$ on $[0, \pi/2]$.

Section 8.3   (page 540)

1. c   **2.** a   **3.** d   **4.** b

5. $-\frac{1}{4}\cos^4 x + C$ **7.** $\frac{1}{12}\sin^6 2x + C$

9. $-\frac{1}{3}\cos^3 x + \frac{2}{5}\cos^5 x - \frac{1}{7}\cos^7 x + C$

11. $\frac{2}{3}\sin^{3/2}\theta - \frac{2}{7}\sin^{7/2}\theta + C$ **13.** $\frac{1}{12}(6x + \sin 6x) + C$

15. $\frac{1}{8}(\alpha - \frac{1}{4}\sin 4\alpha) + C$ or $\frac{1}{8}\alpha - \frac{1}{32}\sin 4\alpha + C$

17. $\frac{1}{8}(2x^2 - 2x \sin 2x - \cos 2x) + C$ **19.** $\frac{2}{3}$ **21.** $\frac{16}{35}$

23. $5\pi/32$ **25.** $\frac{1}{3}\ln|\sec 3x + \tan 3x| + C$

27. $\frac{1}{15}\tan 5x(3 + \tan^2 5x) + C$

29. $\left(\sec \pi x \tan \pi x + \ln|\sec \pi x + \tan \pi x|\right)/(2\pi) + C$

31. $\tan^4(x/4) - 2\tan^2(x/4) - 4\ln|\cos(x/4)| + C$

33. $\frac{1}{2}\tan^2 x + C$ **35.** $\frac{1}{3}\tan^3 x + C$ **37.** $\frac{1}{24}\sec^6 4x + C$

39. $\frac{1}{3}\sec^3 x + C$ **41.** $\ln|\sec x + \tan x| - \sin x + C$

43. $(12\pi\theta - 8\sin 2\pi\theta + \sin 4\pi\theta)/(32\pi) + C$

45. $y = \frac{1}{9}\sec^3 3x - \frac{1}{3}\sec 3x + C$

47. (a)

(b) $y = \frac{1}{2}x - \frac{1}{4}\sin 2x$

49.

51. $-\frac{1}{10}(\cos 5x + 5 \cos x) + C$

53. $\frac{1}{8}(2 \sin 2\theta - \sin 4\theta) + C$ **55.** $\frac{1}{4}(\ln|\csc^2 2x| - \cot^2 2x) + C$

57. $-\cot \theta - \frac{1}{3}\cot^3 \theta + C$ **59.** $\ln|\csc t - \cot t| + \cos t + C$

61. $\ln|\csc x - \cot x| + \cos x + C$ **63.** $t - 2 \tan t + C$

65. π **67.** $\frac{1}{2}(1 - \ln 2)$ **69.** $\ln 2$ **71.** $\frac{4}{3}$

73. $\frac{1}{16}(6x + 8 \sin x + \sin 2x) + C$
Graphs will vary. Example:

75. $\Big[\sec^3 \pi x \tan \pi x +$
$\frac{3}{2}\big(\sec \pi x \tan \pi x + \ln|\sec \pi x + \tan \pi x|\big)\Big]/(4\pi) + C$
Graphs will vary. Example:

77. $(\sec^5 \pi x)/(5\pi) + C$ **79.** $3\sqrt{2}/10$ **81.** $3\pi/16$
Graphs will vary. Example:

83. (a) Save one sine factor and convert the remaining factors to cosines. Then, expand and integrate.
 (b) Save one cosine factor and convert the remaining factors to sines. Then, expand and integrate.
 (c) Make repeated use of the power reducing formulas to convert the integrand to odd powers of the cosine. Then, proceed as in part (b).

85. (a) $\frac{1}{18}\tan^6 3x + \frac{1}{12}\tan^4 3x + C_1,\ \frac{1}{18}\sec^6 3x - \frac{1}{12}\sec^4 3x + C_2$
 (b)   (c) Proof

87. $\frac{1}{3}$ **89.** 1 **91.** $2\pi(1 - \pi/4) \approx 1.348$

93. (a) $\pi^2/2$ (b) $(\bar{x}, \bar{y}) = (\pi/2, \pi/8)$ **95–97.** Proofs

99. $-\frac{1}{15}\cos x(3 \sin^4 x + 4 \sin^2 x + 8) + C$

101. $\frac{5}{6\pi}\tan \frac{2\pi x}{5}\left(\sec^2 \frac{2\pi x}{5} + 2\right) + C$

103. (a) $H(t) \approx 57.72 - 23.36 \cos(\pi t/6) - 2.75 \sin(\pi t/6)$
 (b) $L(t) \approx 42.04 - 20.91 \cos(\pi t/6) - 4.33 \sin(\pi t/6)$
 (c)   The maximum difference is at $t \approx 4.9$ or early summer.

105. Proof

Section 8.4 (page 549)

1. b **2.** d **3.** a **4.** c **5.** $x/\big(25\sqrt{25 - x^2}\big) + C$

7. $5 \ln\left|(5 - \sqrt{25 - x^2})/x\right| + \sqrt{25 - x^2} + C$

9. $\ln\left|x + \sqrt{x^2 - 4}\right| + C$ **11.** $\frac{1}{15}(x^2 - 4)^{3/2}(3x^2 + 8) + C$

13. $\frac{1}{3}(1 + x^2)^{3/2} + C$ **15.** $\frac{1}{2}[\arctan x + x/(1 + x^2)] + C$

17. $\frac{1}{2}x\sqrt{4 + 9x^2} + \frac{2}{3}\ln\left|3x + \sqrt{4 + 9x^2}\right| + C$

19. $\frac{25}{4}\arcsin(2x/5) + \frac{1}{2}x\sqrt{25 - 4x^2} + C$

21. $\sqrt{x^2 + 9} + C$ **23.** $\arcsin(x/4) + C$

25. $4\arcsin(x/2) + x\sqrt{4 - x^2} + C$ **27.** $\ln\left|x + \sqrt{x^2 - 9}\right| + C$

29. $-\dfrac{(1 - x^2)^{3/2}}{3x^3} + C$ **31.** $-\dfrac{1}{3}\ln\left|\dfrac{\sqrt{4x^2 + 9} + 3}{2x}\right| + C$

33. $5\sqrt{x^2 + 5}/(x + 5) + C$ **35.** $\frac{1}{3}(1 + e^{2x})^{3/2} + C$

37. $\frac{1}{2}\big(\arcsin e^x + e^x\sqrt{1 - e^{2x}}\big) + C$

39. $\frac{1}{4}\big(x/(x^2 + 2) + \big(1/\sqrt{2}\big)\arctan\big(x/\sqrt{2}\big)\big) + C$

41. $x \operatorname{arcsec} 2x - \frac{1}{2}\ln\left|2x + \sqrt{4x^2 - 1}\right| + C$

43. $\arcsin[(x - 2)/2] + C$

45. $\sqrt{x^2 + 4x + 8} - 2\ln\left|\sqrt{x^2 + 4x + 8} + (x + 2)\right| + C$

47. (a) and (b) $\sqrt{3} - \pi/3 \approx 0.685$

49. (a) and (b) $9\big(2 - \sqrt{2}\big) \approx 5.272$

51. (a) and (b) $-(9/2)\ln\big(2\sqrt{7}/3 - 4\sqrt{3}/3 - \sqrt{21}/3 + 8/3\big)$
$+ 9\sqrt{3} - 2\sqrt{7} \approx 12.644$

53. $\sqrt{x^2 - 9} - 3\arctan\big(\sqrt{x^2 - 9}/3\big) + 1$

55. $\frac{1}{2}(x - 15)\sqrt{x^2 + 10x + 9}$
$+ 33\ln\left|\sqrt{x^2 + 10x + 9} + (x + 5)\right| + C$

57. $\frac{1}{2}\big(x\sqrt{x^2 - 1} + \ln\left|x + \sqrt{x^2 - 1}\right|\big) + C$

59. (a) Let $u = a \sin \theta$, $\sqrt{a^2 - u^2} = a \cos \theta$, where $-\pi/2 \le \theta \le \pi/2$.
 (b) Let $u = a \tan \theta$, $\sqrt{a^2 + u^2} = a \sec \theta$, where $-\pi/2 < \theta < \pi/2$.
 (c) Let $u = a \sec \theta$, $\sqrt{u^2 - a^2} = \tan \theta$ if $u > a$ and $\sqrt{u^2 - a^2} = -\tan \theta$ if $u < -a$, where $0 \le \theta < \pi/2$ or $\pi/2 < \theta \le \pi$.

61. $\ln\sqrt{x^2 + 9} + C$ **63.** True

65. False: $\displaystyle\int_0^{\sqrt{3}} \frac{dx}{(1 + x^2)^{3/2}} = \int_0^{\pi/3} \cos \theta \, d\theta$

67. πab **69.** (a) $5\sqrt{2}$ (b) $25(1 - \pi/4)$ (c) $r^2(1 - \pi/4)$

71. $6\pi^2$ **73.** $\ln\left[\dfrac{5\big(\sqrt{2} + 1\big)}{\sqrt{26} + 1}\right] + \sqrt{26} - \sqrt{2} \approx 4.367$

75. Length of one arch of sine curve: $y = \sin x$, $y' = \cos x$

$$L_1 = \int_0^\pi \sqrt{1 + \cos^2 x}\,dx$$

Length of one arch of cosine curve: $y = \cos x$, $y' = -\sin x$

$$L_2 = \int_{-\pi/2}^{\pi/2} \sqrt{1 + \sin^2 x}\,dx$$

$$= \int_{-\pi/2}^{\pi/2} \sqrt{1 + \cos^2(x - \pi/2)}\,dx, \quad u = x - \pi/2, du = dx$$

$$= \int_{-\pi}^{0} \sqrt{1 + \cos^2 u}\,du = \int_0^\pi \sqrt{1 + \cos^2 u}\,du = L_1$$

77. (a) 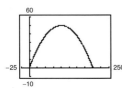 (b) 200

(c) $100\sqrt{2} + 50 \ln\left[(\sqrt{2} + 1)/(\sqrt{2} - 1)\right] \approx 229.559$

79. $(0, 0.422)$ **81.** $(\pi/32)\left[102\sqrt{2} - \ln(3 + 2\sqrt{2})\right] \approx 13.989$

83. (a) 187.2π lb   (b) $62.4\pi d$ lb

85. Proof   **87.** $12 + 9\pi/2 - 25 \arcsin(3/5) \approx 10.050$

Section 8.5   (page 559)

1. $\dfrac{A}{x} + \dfrac{B}{x - 10}$ **3.** $\dfrac{A}{x} + \dfrac{Bx + C}{x^2 + 10}$ **5.** $\dfrac{A}{x} + \dfrac{B}{x - 10}$

7. $\frac{1}{2} \ln|(x - 1)/(x + 1)| + C$ **9.** $\ln|(x - 1)/(x + 2)| + C$

11. $\frac{3}{2} \ln|2x - 1| - 2 \ln|x + 1| + C$

13. $5 \ln|x - 2| - \ln|x + 2| - 3 \ln|x| + C$

15. $x^2 + \frac{3}{2} \ln|x - 4| - \frac{1}{2} \ln|x + 2| + C$

17. $1/x + \ln|x^4 + x^3| + C$

19. $2 \ln|x - 2| - \ln|x| - 3/(x - 2) + C$

21. $\ln|(x^2 + 1)/x| + C$

23. $\frac{1}{6}\left[\ln|(x - 2)/(x + 2)| + \sqrt{2} \arctan(x/\sqrt{2})\right] + C$

25. $\frac{1}{16} \ln|(4x^2 - 1)/(4x^2 + 1)| + C$

27. $\ln|x + 1| + \sqrt{2} \arctan\left[(x - 1)/\sqrt{2}\right] + C$

29. $\ln 2$ **31.** $\frac{1}{2} \ln(8/5) - \pi/4 + \arctan 2 \approx 0.557$

33. $y = 3 \ln|x - 3| - 9/(x - 3) + 9$

35. $y = \left(\sqrt{2}/2\right) \arctan\left(x/\sqrt{2}\right) - 1/[2(x^2 + 2)] + 5/4$

37. $y = \ln|x - 2| + \frac{1}{2} \ln|x^2 + x + 1|$
$\qquad - \sqrt{3} \arctan\left[(2x + 1)/\sqrt{3}\right] - \frac{1}{2} \ln 13$
$\qquad + \sqrt{3} \arctan\left(7/\sqrt{3}\right) + 10$

39. $y = \frac{1}{4} \ln|(x - 2)/(x + 2)| + \frac{1}{4} \ln 2 + 4$

41. $\ln\left|\dfrac{\cos x}{\cos x - 1}\right| + C$ **43.** $\ln\left|\dfrac{-1 + \sin x}{2 + \sin x}\right| + C$

45. $\dfrac{1}{5} \ln\left|\dfrac{e^x - 1}{e^x + 4}\right| + C$ **47–49.** Proofs

51. $y = \dfrac{3}{2} \ln\left|\dfrac{2 + x}{2 - x}\right| + 3$ **53.** First divide x^3 by $(x - 5)$.

55. (a) Log Rule   (b) Partial fractions   (c) Inverse Tangent Rule

57. $12 \ln\left(\frac{9}{8}\right) \approx 1.4134$ **59.** 4.90 or \$490,000

61. $V = 2\pi\left(\arctan 3 - \frac{3}{10}\right) \approx 5.963$; $(\bar{x}, \bar{y}) \approx (1.521, 0.412)$

63. $x = n[e^{(n+1)kt} - 1]/[n + e^{(n+1)kt}]$ **65.** $\pi/8$

Section 8.6   (page 565)

1. $-\frac{1}{2}x(2 - x) + \ln|1 + x| + C$

3. $\frac{1}{2}\left[e^x\sqrt{e^{2x} + 1} + \ln\left(e^x + \sqrt{e^{2x} + 1}\right)\right] + C$

5. $-\sqrt{1 - x^2}/x + C$

7. $\frac{1}{16}(6x - 3 \sin 2x \cos 2x - 2 \sin^3 2x \cos 2x) + C$

9. $-2\left(\cot\sqrt{x} + \csc\sqrt{x}\right) + C$ **11.** $x - \frac{1}{2}\ln(1 + e^{2x}) + C$

13. $\frac{1}{16}x^4(4 \ln x - 1) + C$

15. (a) and (b) $e^x(x^2 - 2x + 2) + C$

17. (a) and (b) $\ln|(x + 1)/x| - 1/x + C$

19. $\frac{1}{2}\left[(x^2 + 1) \operatorname{arcsec}(x^2 + 1) - \ln\left(x^2 + 1 + \sqrt{x^4 + 2x^2}\right)\right] + C$

21. $\sqrt{x^2 - 4}/(4x) + C$ **23.** $\frac{2}{9}[\ln|1 - 3x| + 1/(1 - 3x)] + C$

25. $e^x \arccos(e^x) - \sqrt{1 - e^{2x}} + C$

27. $\frac{1}{2}(x^2 + \cot x^2 + \csc x^2) + C$

29. $\left(\sqrt{2}/2\right) \arctan\left[(1 + \sin \theta)/\sqrt{2}\right] + C$

31. $-\sqrt{2 + 9x^2}/(2x) + C$

33. $\frac{1}{4}\left(2 \ln|x| - 3 \ln|3 + 2 \ln|x||\right) + C$

35. $(3x - 10)/[2(x^2 - 6x + 10)] + \frac{3}{2} \arctan(x - 3) + C$

37. $\frac{1}{2} \ln\left|x^2 - 3 + \sqrt{x^4 - 6x^2 + 5}\right| + C$

39. $-\frac{1}{3}\sqrt{4 - x^2}(x^2 + 8) + C$

41. $2/(1 + e^x) - 1/[2(1 + e^x)^2] + \ln(1 + e^x) + C$

43. $\frac{1}{2}(e-1) \approx 0.8591$ **45.** $9\ln(3) - \frac{26}{9} \approx 6.9986$

47. $\pi/2$ **49.** $\pi^3/8 - 3\pi + 6 \approx 0.4510$ **51–55.** Proofs

57. $y = -2\sqrt{1-x}/\sqrt{x} + 7$

59. $y = \frac{1}{2}[(x-3)/(x^2 - 6x + 10) + \arctan(x-3)]$

61. $y = -\csc\theta + \sqrt{2} + 2$

63. $\frac{1}{\sqrt{5}}\ln\left|\dfrac{2\tan(\theta/2) - 3 - \sqrt{5}}{2\tan(\theta/2) - 3 + \sqrt{5}}\right| + C$ **65.** $\ln 2$

67. $\frac{1}{2}\ln(3 - 2\cos\theta) + C$ **69.** $2\sin\sqrt{\theta} + C$ **71.** $\frac{40}{3}$

73. Use Formula 23 and let $a = 1$, $u = e^x$, and $du = e^x\,dx$, because
it is in the form $\displaystyle\int \frac{1}{a^2 + u^2}\,du$.

75. Use Formula 81 and let $u = x^2$ and $du = 2x\,dx$, because it is in
the form $\displaystyle\int e^u\,du$.

77. Impossible; there is no integration formula that fits.

79. (a) $\displaystyle\int x\ln x\,dx = \frac{1}{2}x^2\ln x - \frac{1}{4}x^2 + C$

$\displaystyle\int x^2\ln x\,dx = \frac{1}{3}x^3\ln x - \frac{1}{9}x^3 + C$

$\displaystyle\int x^3\ln x\,dx = \frac{1}{4}x^4\ln x - \frac{1}{16}x^4 + C$

(b) $\displaystyle\int x^n\ln x\,dx = x^{n+1}\ln x/(n+1) - x^{n+1}/(n+1)^2 + C$

81. False. Substitutions may first have to be made to rewrite the
integral in a form that appears in the table.

83. 1919.145 ft-lb

85. (a) $V = 80\ln(\sqrt{10} + 3) \approx 145.5$ ft³
$W = 11,840\ln(\sqrt{10} + 3) \approx 21,530.4$ lb
(b) $(0, 1.19)$

87. (a) $k = 30/\ln 7 \approx 15.42$
(b)

89. Putnam problem A3, 1980

Section 8.7 (page 574)

1.

x	-0.1	-0.01	-0.001	0.001	0.01	0.1
$f(x)$	2.4132	2.4991	2.500	2.500	2.4991	2.4132

2.5

3.

x	1	10	10^2	10^3	10^4	10^5
$f(x)$	0.9900	90,483.7	3.7×10^9	4.5×10^{10}	0	0

0

5. $\frac{1}{3}$ **7.** $\frac{1}{4}$ **9.** $\frac{5}{3}$ **11.** 3 **13.** 0 **15.** 2

17. ∞ **19.** $\frac{2}{3}$ **21.** 1 **23.** $\frac{3}{2}$ **25.** ∞

27. 0 **29.** 1 **31.** 0 **33.** 0 **35.** ∞

37. (a) Not indeterminate **39.** (a) $0 \cdot \infty$
(b) ∞ (b) 1
(c) (c)

41. (a) Not indeterminate **43.** (a) ∞^0
(b) 0 (b) 1
(c) (c)

45. (a) 1^∞ (b) e **47.** (a) 0^0 (b) 3
(c) (c)

49. (a) 0^0 (b) 1 **51.** (a) $\infty - \infty$ (b) $-\frac{3}{2}$
(c) (c)

53. (a) $\infty - \infty$ (b) ∞ **55.** (a)
(c)

(b) $\frac{1}{2}$

57. (a)

(b) $\frac{5}{2}$

59. $\dfrac{0}{0}, \dfrac{\infty}{\infty}, 0 \cdot \infty, 1^\infty, 0^0, \infty - \infty$

61. Answers will vary. Examples:
(a) $f(x) = x^2 - 25$, $g(x) = x - 5$
(b) $f(x) = (x - 5)^2$, $g(x) = x^2 - 25$
(c) $f(x) = x^2 - 25$, $g(x) = (x - 5)^3$

63.

x	10	10^2	10^4	10^6	10^8	10^{10}
$\dfrac{(\ln x)^4}{x}$	2.811	4.498	0.720	0.036	0.001	0.000

65. 0   **67.** 0   **69.** 0

71. Horizontal asymptote: **73.** Horizontal asymptote:
$y = 1$ $y = 0$
Relative maximum: $(e, e^{1/e})$ Relative maximum: $(1, 2/e)$

75. Limit is not of the form $0/0$ or ∞/∞.

77. Limit is not of the form $0/0$ or ∞/∞.

79. (a) $\displaystyle \lim_{x\to\infty} \frac{x}{\sqrt{x^2 + 1}} = \lim_{x\to\infty} \frac{\sqrt{x^2 + 1}}{x} = \lim_{x\to\infty} \frac{x}{\sqrt{x^2 + 1}}$

Applying L'Hôpital's Rule twice results in the original limit, so L'Hôpital's Rule fails.

(b) 1

(c)

81.

As $x \to 0$, the graphs get closer together.

83. $v = 32t + v_0$ **85.** Proof   **87.** $c = \frac{2}{3}$ **89.** $c = \pi/4$

91. False: L'Hôpital's Rule does not apply, **93.** True
because $\displaystyle \lim_{x\to 0}(x^2 + x + 1) \neq 0$.

95. $\frac{3}{4}$ **97.** $\frac{4}{3}$ **99.** $a = 1, b = \pm 2$ **101.** Proof

103.

$g'(0) = 0$

105. (a) $0 \cdot \infty$ (b) 0
107–109. Proofs

111. (a)

(b) $\displaystyle \lim_{x\to\infty} h(x) = 1$
(c) No

Section 8.8   (page 585)

1. Improper; $0 < \frac{2}{3} < 1$ **3.** Not improper; continuous on $[0, 1]$
5. Infinite discontinuity at $x = 0$; 4
7. Infinite discontinuity at $x = 1$; diverges
9. Infinite limit of integration; 1
11. Infinite discontinuity at $x = 0$; diverges
13. Infinite limit of integration; converges to 1   **15.** 1
17. Diverges   **19.** Diverges   **21.** 2   **23.** $\frac{1}{2}$
25. $1/[2(\ln 4)^2]$ **27.** π **29.** $\pi/4$ **31.** Diverges
33. Diverges   **35.** 6   **37.** $-\frac{1}{4}$ **39.** Diverges   **41.** $\pi/3$
43. $\ln(2 + \sqrt{3})$ **45.** 0   **47.** $2\pi\sqrt{6}/3$ **49.** $p > 1$
51. Proof   **53.** Diverges   **55.** Converges   **57.** Converges
59. Diverges   **61.** Converges
63. An integral with infinite integration limits, an integral with an infinite discontinuity at or between the integration limits
65. The improper integral diverges.   **67.** e **69.** π
71. (a) 1   (b) $\pi/2$ (c) 2π
73.

Perimeter $= 48$

75. $8\pi^2$ **77.** (a) $W = 20,000$ mile-tons   (b) 4000 mi
79. (a) Proof   (b) $P = 43.53\%$ (c) $E(x) = 7$
81. (a) \$757,992.41   (b) \$837,995.15   (c) \$1,066,666.67
83. $P = \left[2\pi NI\left(\sqrt{r^2 + c^2} - c\right)\right]/\left(kr\sqrt{r^2 + c^2}\right)$
85. False. Let $f(x) = 1/(x + 1)$.   **87.** True
89. (a) $\displaystyle \int_1^\infty \frac{1}{x^n}\, dx$ will converge if $n > 1$ and diverge if $n \leq 1$.

(b)

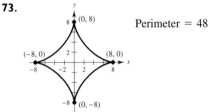

(c) Converges

91. (a) $\Gamma(1) = 1, \Gamma(2) = 1, \Gamma(3) = 2$ (b) Proof
(c) $\Gamma(n) = (n - 1)!$
93. $1/s, \ s > 0$ **95.** $2/s^3, \ s > 0$ **97.** $s/(s^2 + a^2), \ s > 0$
99. $s/(s^2 - a^2), \ s > |a|$
101. (a)

(b) ≈ 0.2525
(c) 0.2525; same by symmetry

103. $c = 1$; $\ln(2)$
105. $8\pi\left[(\ln 2)^{2/3} - (\ln 4)/9 + 2/27\right] \approx 2.01545$ **107.** 0.6278

35. $\frac{3}{4}$ **37.** 4   **39.** 2   **41.** $\frac{2}{3}$ **43.** $\frac{10}{9}$ **45.** $\frac{9}{4}$ **47.** $\frac{1}{2}$

49. $\dfrac{\sin(1)}{1 - \sin(1)}$ **51.** (a) $\displaystyle\sum_{n=0}^{\infty} \dfrac{4}{10}(0.1)^n$ (b) $\dfrac{4}{9}$

53. (a) $\displaystyle\sum_{n=0}^{\infty} \dfrac{81}{100}(0.01)^n$ (b) $\dfrac{9}{11}$

55. (a) $\displaystyle\sum_{n=0}^{\infty} \dfrac{3}{40}(0.01)^n$ (b) $\dfrac{5}{66}$ **57.** Diverges   **59.** Converges

61. Diverges   **63.** Converges   **65.** Diverges   **67.** Diverges
69. Diverges   **71.** Diverges   **73.** See definitions on page 606.
75. The series given by

$$\sum_{n=0}^{\infty} ar^n = a + ar + ar^2 + \cdots + ar^n + \cdots, a \neq 0$$

is a geometric series with ratio r. When $0 < |r| < 1$, the series

converges to the sum $\displaystyle\sum_{n=0}^{\infty} ar^n = \dfrac{a}{1 - r}$.

77. $\{a_n\}$ converges to 1 and $\displaystyle\sum_{n=1}^{\infty} a_n$ diverges. This fits Theorem 9.9,

which states that if $\displaystyle\lim_{n\to\infty} a_n \neq 0$, then $\displaystyle\sum_{n=1}^{\infty} a_n$ diverges.

79. $-2 < x < 2$; $x/(2 - x)$ **81.** $0 < x < 2$; $(x - 1)/(2 - x)$
83. $-1 < x < 1$; $1/(1 + x)$
85. x: $(-\infty, -1) \cup (1, \infty)$; $x/(x - 1)$
87. (a) Yes. Answers will vary.   (b) Yes. Answers will vary.
89. (a) x (b) $f(x) = 1/(1 - x)$, $|x| < 1$
(c)

91.

Horizontal asymptote: $y = 6$
The horizontal asymptote is the sum of the series.

93. The required terms for the two series are $n = 100$ and $n = 5$, respectively. The second series converges at a higher rate.
95. $80{,}000(1 - 0.9^n)$ units
97. $400(1 - 0.75^n)$ million dollars; Sum = \$400 million

99. 152.42 feet   **101.** $\dfrac{1}{8}$; $\displaystyle\sum_{n=0}^{\infty} \dfrac{1}{2}\left(\dfrac{1}{2}\right)^n = \dfrac{1/2}{1 - 1/2} = 1$

103. (a) $-1 + \displaystyle\sum_{n=0}^{\infty}\left(\dfrac{1}{2}\right)^n = -1 + \dfrac{a}{1 - r} = -1 + \dfrac{1}{1 - 1/2} = 1$
(b) No   (c) 2
105. (a) 126 in.^{2} (b) 128 in.^{2}
107. \$573,496.06; The \$1,000,000 sweepstakes has a present value of \$573,496.06. After accruing interest over the 20-year period, it attains its full value.
109. (a) \$5,368,709.11   (b) \$10,737,418.23   (c) \$21,474,836.47
111. (a) \$16,415.10   (b) \$16,421.83
113. (a) \$118,196.13   (b) \$118,393.43

115. (a) $a_n = 3484.1363e^{0.0490n}$ (b) \$50,809 million   (c) \$50,815 million

117. False. $\displaystyle\lim_{n\to\infty} \dfrac{1}{n} = 0$, but $\displaystyle\sum_{n=1}^{\infty} \dfrac{1}{n}$ diverges.

119. False. $\displaystyle\sum_{n=1}^{\infty} ar^n = \left(\dfrac{a}{1 - r}\right) - a$

The formula requires that the geometric series begins with $n = 0$.
121. True   **123.** Proof
125. Answers will vary. Example: $\displaystyle\sum_{n=0}^{\infty} 1$, $\displaystyle\sum_{n=0}^{\infty} (-1)$
127–131. Proofs
133. H = half-life of the drug
n = number of equal doses
P = number of units of the drug
t = equal time intervals
The total amount of the drug in the patient's system at the time the last dose is given is
$$T_n = P + Pe^{kt} + Pe^{2kt} + \cdots + Pe^{(n-1)kt}$$
where $k = -(\ln 2)/H$. One time interval after the last dose is given is
$$T_{n+1} = Pe^{kt} + Pe^{2kt} + Pe^{3kt} + \cdots + Pe^{nkt}$$
and so on. Because $k < 0$, $T_{n+s} \to 0$ as $s \to \infty$.
135. Putnam Problem A1, 1966

Section 9.3   (page 620)

1. Diverges   **3.** Converges   **5.** Converges   **7.** Diverges
9. Diverges   **11.** Diverges   **13.** Converges   **15.** Converges
17. Diverges   **19.** Diverges
21. $f(x)$ is not positive for $x \geq 1$.
23. $f(x)$ is not always decreasing.   **25.** Converges   **27.** Diverges
29. Diverges   **31.** Diverges   **33.** Converges   **35.** Converges
37. c; diverges   **38.** f; diverges   **39.** b; converges
40. a; diverges   **41.** d; converges   **42.** e; converges
43. (a)

n	5	10	20	50	100
S_n	3.7488	3.75	3.75	3.75	3.75

The partial sums approach the sum 3.75 very quickly.

(b)

n	5	10	20	50	100
S_n	1.4636	1.5498	1.5962	1.6251	1.635

The partial sums approach the sum $\pi^2/6 \approx 1.6449$ more slowly than the series in part (a).

45. See Theorem 9.10 on page 617. Answers will vary. For example, convergence or divergence can be determined for the series
$$\sum_{n=1}^{\infty} \frac{1}{n^2 + 1}.$$

47. No. Because $\sum_{n=1}^{\infty} \frac{1}{n}$ diverges, $\sum_{n=10,000}^{\infty} \frac{1}{n}$ also diverges. The convergence or divergence of a series is not determined by the first finite number of terms of the series.

49. The series diverges because the area under the rectangles is greater than the infinite area under the curve $y = 1/\sqrt{x}$ for $x \geq 1$.

51. $p > 1$ **53.** $p > 1$ **55.** Diverges
57. Converges   **59.** Proof
61. $S_6 \approx 1.0811$ **63.** $S_{10} \approx 0.9818$ **65.** $S_4 \approx 0.4049$
$R_6 \approx 0.0015$ $R_{10} \approx 0.0997$ $R_4 \approx 5.6 \times 10^{-8}$
67. $N \geq 7$ **69.** $N \geq 2$ **71.** $N \geq 1000$

73. (a) $\sum_{n=2}^{\infty} \frac{1}{n^{1.1}}$ converges by the p-Series Test since $1.1 > 1$.
$\sum_{n=2}^{\infty} \frac{1}{n \ln n}$ diverges by the Integral Test since $\int_2^{\infty} \frac{1}{x \ln x} dx$ diverges.

(b) $\sum_{n=2}^{\infty} \frac{1}{n^{1.1}} = 0.4665 + 0.2987 + 0.2176 + 0.1703$
$\qquad\qquad + 0.1393 + \cdots$
$\sum_{n=2}^{\infty} \frac{1}{n \ln n} = 0.7213 + 0.3034 + 0.1803 + 0.1243$
$\qquad\qquad + 0.0930 + \cdots$

(c) $n \geq 3.431 \times 10^{15}$

75. (a) Let $f(x) = 1/x$. f is positive, continuous, and decreasing on $[1, \infty)$.
$$S_n - 1 \leq \int_1^n \frac{1}{x} dx = \ln n$$
$$S_n \geq \int_1^{n+1} \frac{1}{x} dx = \ln(n + 1)$$
So, $\ln(n + 1) \leq S_n \leq 1 + \ln n$.

(b) $\ln(n + 1) - \ln n \leq S_n - \ln n \leq 1$.
Also, $\ln(n + 1) - \ln n > 0$ for $n \geq 1$. So, $0 \leq S_n - \ln n \leq 1$ and the sequence $\{a_n\}$ is bounded.

(c) $a_n - a_{n+1} = [S_n - \ln n] - [S_{n+1} - \ln(n + 1)]$
$\qquad\qquad = \int_n^{n+1} \frac{1}{x} dx - \frac{1}{n + 1} \geq 0$
So, $a_n \geq a_{n+1}$.

(d) Because the sequence is bounded and monotonic, it converges to a limit, γ.

(e) 0.5822

77. (a) Diverges   (b) Diverges

(c) $\sum_{n=2}^{\infty} x^{\ln n}$ converges for $x < 1/e$.

79. Diverges   **81.** Converges   **83.** Converges   **85.** Diverges
87. Diverges   **89.** Converges

Section 9.4   (page 628)

1. (a)

(b) $\sum_{n=1}^{\infty} \frac{6}{n^{3/2}}$; Converges

(c) The magnitudes of the terms are less than the magnitudes of the terms of the p-series. Therefore, the series converges.

(d) The smaller the magnitudes of the terms, the smaller the magnitudes of the terms of the sequence of partial sums.

3. Converges   **5.** Diverges   **7.** Converges   **9.** Diverges
11. Converges   **13.** Converges   **15.** Diverges   **17.** Diverges
19. Converges   **21.** Diverges   **23.** Converges   **25.** Diverges
27. Diverges   **29.** Diverges; p-Series Test

31. Converges; Direct Comparison Test with $\sum_{n=1}^{\infty} \left(\frac{1}{3}\right)^n$

33. Diverges; nth-Term Test   **35.** Converges; Integral Test

37. $\lim_{n \to \infty} \frac{a_n}{1/n} = \lim_{n \to \infty} n a_n$
$\lim_{n \to \infty} n a_n \neq 0$, but is finite.
The series diverges by the Limit Comparison Test.

39. Diverges   **41.** Converges

43. $\lim_{n \to \infty} n\left(\frac{n^3}{5n^4 + 3}\right) = \frac{1}{5} \neq 0$
So, $\sum_{n=1}^{\infty} \frac{n^3}{5n^4 + 3}$ diverges.

45. Diverges   **47.** Converges

49. Convergence or divergence is dependent on the form of the general term for the series and not necessarily on the magnitude of the terms.

51. See Theorem 9.13 on page 626. Answers will vary. For example,
$\sum_{n=2}^{\infty} \frac{1}{\sqrt{n - 1}}$ diverges because $\lim_{n \to \infty} \frac{1/\sqrt{n - 1}}{1/\sqrt{n}} = 1$ and
$\sum_{n=2}^{\infty} \frac{1}{\sqrt{n}}$ diverges (p-series).

53.
Because $0 < a_n < 1$, $0 < a_n^2 < a_n < 1$.

55. False. Let $a_n = 1/n^3$ and $b_n = 1/n^2$.

57. True   **59.** True   **61.** Proof   **63.** $\sum_{n=1}^{\infty} \frac{1}{n^2}, \sum_{n=1}^{\infty} \frac{1}{n^3}$

65–69. Proofs   **71.** Putnam Problem 1, afternoon session, 1953

Section 9.5   (page 636)

1. d   **2.** f   **3.** a   **4.** b   **5.** e   **6.** c

7. (a)

n	1	2	3	4	5
S_n	1.0000	0.6667	0.8667	0.7238	0.8349

n	6	7	8	9	10
S_n	0.7440	0.8209	0.7543	0.8131	0.7605

(b)

(c) The points alternate sides of the horizontal line $y = \pi/4$ that represents the sum of the series. The distances between the successive points and the line decrease.

(d) The distance in part (c) is always less than the magnitude of the next term of the series.

9. (a)

n	1	2	3	4	5
S_n	1.0000	0.7500	0.8611	0.7986	0.8386

n	6	7	8	9	10
S_n	0.8108	0.8312	0.8156	0.8280	0.8180

(b)

(c) The points alternate sides of the horizontal line $y = \pi^2/12$ that represents the sum of the series. The distances between the successive points and the line decrease.

(d) The distance in part (c) is always less than the magnitude of the next term of the series.

11. Converges   **13.** Converges   **15.** Diverges   **17.** Converges
19. Diverges   **21.** Diverges   **23.** Diverges   **25.** Converges
27. Converges   **29.** Converges   **31.** Converges
33. $2.3713 \le S \le 2.4937$ **35.** $0.7305 \le S \le 0.7361$
37. (a) 7 terms (Note that the sum begins with $N = 0$.)
  (b) 0.368
39. (a) 3 terms (Note that the sum begins with $N = 0$.)
  (b) 0.842
41. (a) 1000 terms   (b) 0.693   **43.** 10   **45.** 7
47. Converges absolutely   **49.** Converges conditionally
51. Diverges   **53.** Converges conditionally
55. Converges absolutely   **57.** Converges absolutely
59. Converges conditionally   **61.** Converges absolutely
63. An alternating series is a series whose terms alternate in sign. See Theorem 9.14 on page 631 for the Alternating Series Test.
65. A series $\Sigma\, a_n$ is absolutely convergent if $\Sigma\, |a_n|$ converges. A series $\Sigma\, a_n$ is conditionally convergent if $\Sigma\, a_n$ converges and $\Sigma\, |a_n|$ diverges.
67. False. Let $a_n = \dfrac{(-1)^n}{n}$.   **69.** True   **71.** $p > 0$
73. (a) Proof
  (b) The converse is false. For example: Let $a_n = 1/n$.

75. $\displaystyle\sum_{n=1}^{\infty} \frac{1}{n^2}$ converges, hence so does $\displaystyle\sum_{n=1}^{\infty} \frac{1}{n^4}$.
77. (a) No; $a_{n+1} \le a_n$ is not satisfied for all n. For example, $\frac{1}{9} < \frac{1}{8}$.   (b) Yes; 0.5
79. Converges; p-Series Test   **81.** Diverges; nth-Term Test
83. Converges; Geometric Series Test
85. Converges; Integral Test
87. Converges; Alternating Series Test
89. The first term of the series is 0, not 1. You cannot regroup series terms arbitrarily.
91. Putnam Problem 2, afternoon session, 1954

Section 9.6   (page 645)

1–3. Proofs   **5.** d   **6.** c   **7.** f   **8.** b   **9.** a   **10.** e
11. (a) Proof
(b)

n	5	10	15	20	25
S_n	9.2104	16.7598	18.8016	19.1878	19.2491

(c)

(d) 19.26

(e) The more rapidly the terms of the series approach 0, the more rapidly the sequence of partial sums approaches the sum of the series.

13. Diverges   **15.** Converges   **17.** Converges   **19.** Diverges
21. Converges   **23.** Diverges   **25.** Converges
27. Converges   **29.** Diverges   **31.** Converges
33–35. Proofs   **37.** Converges   **39.** Diverges   **41.** Converges
43. Diverges   **45.** Converges   **47.** Converges
49. Converges   **51.** Converges; Alternating Series Test
53. Converges; p-Series Test   **55.** Diverges; nth-Term Test
57. Diverges; Ratio Test
59. Converges; Limit Comparison Test with $b_n = 1/2^n$
61. Converges; Direct Comparison Test with $b_n = 1/2^n$
63. Converges; Ratio Test   **65.** Converges; Ratio Test
67. Converges; Ratio Test   **69.** a and c   **71.** a and b
73. $\displaystyle\sum_{n=0}^{\infty} \frac{n+1}{4^{n+1}}$ **75.** (a) 9   (b) -0.7769
77. Diverges; $\displaystyle\lim_{n\to\infty} \left|\frac{a_{n+1}}{a_n}\right| > 1$
79. Converges; $\displaystyle\lim_{n\to\infty} \left|\frac{a_{n+1}}{a_n}\right| < 1$
81. Diverges; $\text{Lim } a_n \ne 0$ **83.** Converges   **85.** Converges
87. $(-3, 3)$ **89.** $(-2, 0]$ **91.** $x = 0$
93. See Theorem 9.17 on page 639.
95. No; the series $\displaystyle\sum_{n=1}^{\infty} \frac{1}{n+10{,}000}$ diverges.
97. Absolutely; by definition   **99–103.** Proofs
105. Putnam Problem 3, morning session, 1942

Section 9.7   (page 656)

1. d   **2.** c   **3.** a   **4.** b

5. $P_1 = 6 - 2x$ **7.** $P_1 = \sqrt{2}x + \sqrt{2}(4 - \pi)/4$

P_1 is the tangent line to the curve $f(x) = 4/\sqrt{x}$ at the point $(1, 4)$.

P_1 is the tangent line to the curve $f(x) = \sec x$ at the point $(\pi/4, \sqrt{2})$.

9.

x	0	0.8	0.9	1	1.1
$f(x)$	Error	4.4721	4.2164	4.0000	3.8139
$P_2(x)$	7.5000	4.4600	4.2150	4.0000	3.8150

x	1.2	2
$f(x)$	3.6515	2.8284
$P_2(x)$	3.6600	3.5000

11. (a)

(b) $f^{(2)}(0) = -1$ $P_2^{(2)}(0) = -1$
$f^{(4)}(0) = 1$ $P_4^{(4)}(0) = 1$
$f^{(6)}(0) = -1$ $P_6^{(6)}(0) = -1$

(c) $f^{(n)}(0) = P_n^{(n)}(0)$

13. $1 - x + \frac{1}{2}x^2 - \frac{1}{6}x^3$ **15.** $1 + 2x + 2x^2 + \frac{4}{3}x^3 + \frac{2}{3}x^4$

17. $x - \frac{1}{6}x^3 + \frac{1}{120}x^5$ **19.** $x + x^2 + \frac{1}{2}x^3 + \frac{1}{6}x^4$

21. $1 - x + x^2 - x^3 + x^4$ **23.** $1 + \frac{1}{2}x^2$

25. $1 - (x - 1) + (x - 1)^2 - (x - 1)^3 + (x - 1)^4$

27. $1 + \frac{1}{2}(x - 1) - \frac{1}{8}(x - 1)^2 + \frac{1}{16}(x - 1)^3 - \frac{5}{128}(x - 1)^4$

29. $(x - 1) - \frac{1}{2}(x - 1)^2 + \frac{1}{3}(x - 1)^3 - \frac{1}{4}(x - 1)^4$

31. (a) $P_3(x) = x + (1/3)x^3$

(b) $Q_3(x) = 1 + 2(x - \pi/4) + 2(x - \pi/4)^2 + \frac{8}{3}(x - \pi/4)^3$

33. (a)

x	0	0.25	0.50	0.75	1.00
$\sin x$	0	0.2474	0.4794	0.6816	0.8415
$P_1(x)$	0	0.25	0.50	0.75	1.00
$P_3(x)$	0	0.2474	0.4792	0.6797	0.8333
$P_5(x)$	0	0.2474	0.4794	0.6817	0.8417

(b)

(c) As the distance increases, the polynomial approximation becomes less accurate.

35. (a) $P_3(x) = x + \frac{1}{6}x^3$

(b)

x	-0.75	-0.50	-0.25	0	0.25
$f(x)$	-0.848	-0.524	-0.253	0	0.253
$P_3(x)$	-0.820	-0.521	-0.253	0	0.253

x	0.50	0.75
$f(x)$	0.524	0.848
$P_3(x)$	0.521	0.820

(c)

37.

39.

41. 0.6042   **43.** 0.1823   **45.** $R_4 \le 2.03 \times 10^{-5}$; 0.000001

47. $R_3 \le 7.82 \times 10^{-3}$ **49.** 3   **51.** 5

53. $n = 9$; $\ln(1.5) \approx 0.4055$ **55.** $n = 16$; $e^{-\pi(1.3)} \approx 0.16838$

57. $-0.3936 < x < 0$ **59.** $-0.9467 < x < 0.9467$

61. The graph of the approximating polynomial P and the elementary function f both pass through the point $(c, f(c))$, and the slope of P is the same as the slope of the graph of f at the point $(c, f(c))$. If P is of degree n, then the first n derivatives of f and P agree at c. This allows for the graph of P to resemble the graph of f near the point $(c, f(c))$.

63. See "Definitions of nth Taylor Polynomial and nth Maclaurin Polynomial" on page 650.

65. As the degree of the polynomial increases, the graph of the Taylor polynomial becomes a better and better approximation of the function within the interval of convergence. Therefore, the accuracy is increased.

45. Center: $(0, 0)$
Foci: $(0, \pm\sqrt{5})$
Vertices: $(0, \pm 1)$

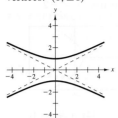

47. Center: $(1, -2)$
Foci: $(1 \pm \sqrt{5}, -2)$
Vertices: $(-1, -2), (3, -2)$

49. Center: $(2, -3)$
Foci: $(2 \pm \sqrt{10}, -3)$
Vertices: $(1, -3), (3, -3)$

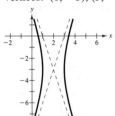

51. Degenerate hyperbola
Graph is two lines
$y = -3 \pm \frac{1}{3}(x + 1)$
intersecting at $(-1, -3)$.

53. Center: $(1, -3)$
Foci: $(1, -3 \pm 2\sqrt{5})$
Vertices: $(1, -3 \pm \sqrt{2})$
Asymptotes:
$y = \frac{1}{3}x - \frac{1}{3} - 3$;
$y = -\frac{1}{3}x + \frac{1}{3} - 3$

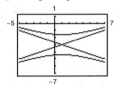

55. Center: $(1, -3)$
Foci: $(1 \pm \sqrt{10}, -3)$
Vertices: $(-1, -3), (3, -3)$
Asymptotes:
$y = \sqrt{6}x/2 - \sqrt{6}/2 - 3$;
$y = -\sqrt{6}x/2 + \sqrt{6}/2 - 3$

57. $x^2/1 - y^2/9 = 1$ **59.** $y^2/9 - (x - 2)^2/(9/4) = 1$
61. $y^2/4 - x^2/12 = 1$ **63.** $(x - 3)^2/9 - (y - 2)^2/4 = 1$
65. (a) $(6, \sqrt{3})$: $2x - 3\sqrt{3}y - 3 = 0$
$(6, -\sqrt{3})$: $2x + 3\sqrt{3}y - 3 = 0$
(b) $(6, \sqrt{3})$: $9x + 2\sqrt{3}y - 60 = 0$
$(6, -\sqrt{3})$: $9x - 2\sqrt{3}y - 60 = 0$
67. Ellipse **69.** Parabola **71.** Circle
73. Circle **75.** Hyperbola
77. (a) A parabola is the set of all points (x, y) that are equidistant from a fixed line and a fixed point not on the line.
(b) For directrix $y = k - p$: $(x - h)^2 = 4p(y - k)$
For directrix $x = h - p$: $(y - k)^2 = 4p(x - h)$
(c) If P is a point on a parabola, then the tangent line to the parabola at P makes equal angles with the line passing through P and the focus, and with the line passing through P parallel to the axis of the parabola.
79. (a) A hyperbola is the set of all points (x, y) for which the absolute value of the difference between the distances from two distinct fixed points is constant.

(b) Transverse axis is horizontal: $\dfrac{(x - h)^2}{a^2} - \dfrac{(y - k)^2}{b^2} = 1$
Transverse axis is vertical: $\dfrac{(y - k)^2}{a^2} - \dfrac{(x - h)^2}{b^2} = 1$
(c) Transverse axis is horizontal:
$y = k + (b/a)(x - h)$ and $y = k - (b/a)(x - h)$
Transverse axis is vertical:
$y = k + (a/b)(x - h)$ and $y = k - (a/b)(x - h)$
81. $\frac{9}{4}$ m **83.** $y = 2ax_0 x - ax_0^2$ **85.** (a) Proof (b) Proof
87. $x_0 = 2\sqrt{3}/3$; Distance from hill: $2\sqrt{3}/3 - 1$
89. $[16(4 + 3\sqrt{3} - 2\pi)]/3 \approx 15.536 \text{ ft}^2$
91. (a) $y = (1/180)x^2$
(b) $10\left[2\sqrt{13} + 9\ln\left(\dfrac{2 + \sqrt{13}}{3}\right)\right] \approx 128.4 \text{ m}$
93.

As p increases, the graph of $x^2 = 4py$ gets wider.

95. (a) $L = 2a$
(b) The thumbtacks are located at the foci and the length of string is the constant sum of distances from the foci.
97.

99. Proof

101. $e \approx 0.9672$ **103.** $\left(0, \frac{25}{3}\right)$
105. Minor-axis endpoints: $(-6, -2), (0, -2)$
Major-axis endpoints: $(-3, -6), (-3, 2)$
107. (a) Area $= 2\pi$
(b) Volume $= 8\pi/3$
Surface area $= [2\pi(9 + 4\sqrt{3}\pi)]/9 \approx 21.48$
(c) Volume $= 16\pi/3$
Surface area $= \dfrac{4\pi[6 + \sqrt{3}\ln(2 + \sqrt{3})]}{3} \approx 34.69$
109. 37.96 **111.** 40 **113.** $(x - 6)^2/9 - (y - 2)^2/7 = 1$
115.

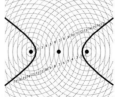

117. Proof

119. $x = (-90 + 96\sqrt{2})/7 \approx 6.538$
$y = (160 - 96\sqrt{2})/7 \approx 3.462$

121. There are four points of intersection.

At $\left(\dfrac{\sqrt{2}\,ac}{\sqrt{2a^2 - b^2}}, \dfrac{b^2}{\sqrt{2}\sqrt{2a^2 - b^2}} \right)$, the slopes of the tangent lines are $y'_e = -c/a$ and $y'_h = a/c$.

Since the slopes are negative reciprocals, the tangent lines are perpendicular. Similarly, the curves are perpendicular at the other three points of intersection.

123. False. See the definition of a parabola.　　**125.** True

127. True　　**129.** Putnam Problem B4, 1976

Section 10.2　(page 716)

1. (a)

t	0	1	2	3	4
x	0	1	$\sqrt{2}$	$\sqrt{3}$	2
y	1	0	-1	-2	-3

(b) and (c)　　　　(d) $y = 1 - x^2, \quad x \ge 0$

3. 　　**5.**

$2x - 3y + 5 = 0$　　$y = (x - 1)^2$

7.

$y = \frac{1}{2}x^{2/3}$

9. 　　**11.**

$y = x^2 - 2, \quad x \ge 0$　　$y = (x + 1)/x$

13.

$y = |x - 4|/2$

15.

$y = x^3 + 1, \quad x > 0$

17.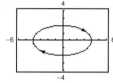

$y = 1/x, \quad |x| \ge 1$

19.

$x^2 + y^2 = 9$

21.

$\dfrac{x^2}{16} + \dfrac{y^2}{4} = 1$

23.

$\dfrac{(x - 4)^2}{4} + \dfrac{(y + 1)^2}{1} = 1$

25.

$\dfrac{(x - 4)^2}{4} + \dfrac{(y + 1)^2}{16} = 1$

27.

$\dfrac{x^2}{16} - \dfrac{y^2}{9} = 1$

29.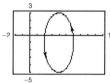

$y = \ln x$

31.

$y = 1/x^3, \quad x > 0$

33. Each curve represents a portion of the line $y = 2x + 1$.

Domain	Orientation	Smooth
(a) $-\infty < x < \infty$	Up	Yes
(b) $-1 \le x \le 1$	Oscillates	No, $\dfrac{dx}{d\theta} = \dfrac{dy}{d\theta} = 0$ when $\theta = 0, \pm\pi, \pm 2\pi, \ldots$
(c) $0 < x < \infty$	Down	Yes
(d) $0 < x < \infty$	Up	Yes

35. (a) and (b) represent the parabola $y = 2(1 - x^2)$ for $-1 \le x \le 1$. The curve is smooth. The orientation is from right to left in part (a) and in part (b).

37. (a)

(b) The orientation is reversed.   (c) The orientation is reversed.

(d) Answers will vary. For example,

$$x = 2 \sec t \quad\quad x = 2 \sec(-t)$$
$$y = 5 \sin t \quad\quad y = 5 \sin(-t)$$

have the same graphs, but their orientations are reversed.

39. $y - y_1 = \dfrac{y_2 - y_1}{x_2 - x_1}(x - x_1)$ **41.** $\dfrac{(x - h)^2}{a^2} + \dfrac{(y - k)^2}{b^2} = 1$

43. $x = 5t$
$y = -2t$
(Solution is not unique.)

45. $x = 2 + 4 \cos \theta$
$y = 1 + 4 \sin \theta$
(Solution is not unique.)

47. $x = 5 \cos \theta$
$y = 3 \sin \theta$
(Solution is not unique.)

49. $x = 4 \sec \theta$
$y = 3 \tan \theta$
(Solution is not unique.)

51. $x = t$
$y = 3t - 2;$
$x = t - 3$
$y = 3t - 11$
(Solution is not unique.)

53. $x = t$
$y = t^3;$
$x = \tan t$
$y = \tan^3 t$
(Solution is not unique.)

55.

57.

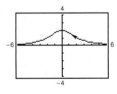

Not smooth when $\theta = 2n\pi$

59.

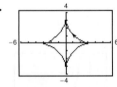

61.

Not smooth when $\theta = \frac{1}{2}n\pi$

63. See page 709.   **65.** See page 714.

67. $x = a\theta - b \sin \theta;\ y = a - b \cos \theta$

69. False. The graph of the parametric equations is the portion of the line $y = x$ when $x \geq 0$.

71. (a) $x = \left(\frac{440}{3} \cos \theta\right)t;\ y = 3 + \left(\frac{440}{3} \sin \theta\right)t - 16t^2$

(b)

Not a home run

(c)

Home run

(d) $19.4°$

Section 10.3   (page 725)

1. $-2/t$ **3.** -1

5. $\dfrac{dy}{dx} = \dfrac{3}{2}, \dfrac{d^2y}{dx^2} = 0$; Neither concave upward nor concave downward

7. $dy/dx = 2t + 3,\ d^2y/dx^2 = 2$
At $t = -1, dy/dx = 1,\ d^2y/dx^2 = 2$; Concave upward

9. $dy/dx = -\cot \theta,\ d^2y/dx^2 = -\csc^3 \theta/2$
At $\theta = \pi/4, dy/dx = -1,\ d^2y/dx^2 = -\sqrt{2}$;
Concave downward

11. $dy/dx = 2 \csc \theta,\ d^2y/dx^2 = -2 \cot^3 \theta$
At $\theta = \pi/6, dy/dx = 4,\ d^2y/dx^2 = -6\sqrt{3}$;
Concave downward

13. $dy/dx = -\tan \theta,\ d^2y/dx^2 = \sec^4 \theta \csc \theta/3$
At $\theta = \pi/4, dy/dx = -1,\ d^2y/dx^2 = 4\sqrt{2}/3$;
Concave upward

15. $\left(-2/\sqrt{3}, 3/2\right):\ 3\sqrt{3}x - 8y + 18 = 0$
$(0, 2):\ y - 2 = 0$
$\left(2\sqrt{3}, 1/2\right):\ \sqrt{3}x + 8y - 10 = 0$

17. (a) and (d)

(b) At $t = 2,\ dx/dt = 2,$
$dy/dt = 4$, and $dy/dx = 2$.
(c) $y = 2x - 5$

19. (a) and (d)

(b) At $t = -1,\ dx/dt = -3,$
$dy/dt = 0$, and $dy/dx = 0$.
(c) $y = 2$

21. $y = \pm\frac{3}{4}x$ **23.** $y = 3x - 5$ and $y = 1$

25. Horizontal: $(1, 0), (-1, \pi), (1, -2\pi)$
Vertical: $(\pi/2, 1), (-3\pi/2, -1), (5\pi/2, 1)$

27. Horizontal: $(1, 0)$
Vertical: None

29. Horizontal: $(0, -2), (2, 2)$
Vertical: None

31. Horizontal: $(0, 3), (0, -3)$
Vertical: $(3, 0), (-3, 0)$

33. Horizontal: $(4, 0), (4, -2)$
Vertical: $(2, -1), (6, -1)$

35. Horizontal: None
Vertical: $(1, 0), (-1, 0)$

37. Concave down: $-\infty < t < 0$
Concave up: $0 < t < \infty$

39. Concave up: $t > 0$

41. Concave down: $0 < t < \pi/2$
Concave up: $\pi/2 < t < \pi$

43. $\displaystyle\int_1^2 \sqrt{4t^2 + t + 4}\ dt$ **45.** $\displaystyle\int_{-2}^2 \sqrt{e^{2t} + 4}\ dt$

47. $2\sqrt{5} + \ln\left(2 + \sqrt{5}\right) \approx 5.916$ **49.** $\sqrt{2}\left(1 - e^{-\pi/2}\right) \approx 1.12$

51. $\frac{1}{12}\left[\ln\left(\sqrt{37} + 6\right) + 6\sqrt{37}\right] \approx 3.249$ **53.** $6a$ **55.** $8a$

57. (a)   (b) 219.2 ft

(c) 230.8 ft

59. (a)   (b) $(0, 0), \left(4\sqrt[3]{2}/3, 4\sqrt[3]{2}/3\right)$

(c) ≈ 6.557

61. (a)

(b) The average speed of the particle on the second path is twice the average speed of the particle on the first path.

(c) 4π

63. $S = 2\pi \int_0^2 \sqrt{17}(t + 1)\, dt = 8\pi\sqrt{17} \approx 103.625$

65. $S = 2\pi \int_0^{\pi/2} \left(\sin\theta\cos\theta\sqrt{4\cos^2\theta + 1}\right) d\theta = \dfrac{(5\sqrt{5} - 1)\pi}{6}$

≈ 5.330

67. (a) $32\pi\sqrt{5}$ (b) $16\pi\sqrt{5}$ **69.** 32π **71.** $12\pi a^2/5$

73. See Theorem 10.7, Parametric Form of the Derivative, on page 719.

75. Answers will vary. Example:

77. See Theorem 10.8, Arc Length in Parametric Form, on page 722.

79. Proof **81.** $3\pi/2$ **83.** d **84.** b **85.** f **86.** c

87. a **88.** e **89.** $\left(\frac{3}{4}, \frac{8}{5}\right)$ **91.** $V = 36\pi$

93. (a) $dy/dx = \sin\theta/(1 - \cos\theta)$; $d^2y/dx^2 = -1/[a(\cos\theta - 1)^2]$

(b) $y = \left(2 + \sqrt{3}\right)\left[x - a\left(\pi/6 - \frac{1}{2}\right)\right] + a\left(1 - \sqrt{3}/2\right)$

(c) $(a(2n + 1)\pi, 2a)$

(d) Concave down on $(0, 2\pi)$, $(2\pi, 4\pi)$, etc.

(e) $s = 8a$

95. Proof

97. (a)

(b) Circle of radius 1 and center at $(0, 0)$ except the point $(-1, 0)$

(c) As t increases from -20 to 0, the speed increases, and as t increases from 0 to 20, the speed decreases.

99. False: $\dfrac{d^2y}{dx^2} = \dfrac{\dfrac{d}{dt}\left[\dfrac{g'(t)}{f'(t)}\right]}{f'(t)} = \dfrac{f'(t)g''(t) - g'(t)f''(t)}{[f'(t)]^3}$.

Section 10.4 (page 736)

1.

$(0, 4)$

3.

$\left(-2, 2\sqrt{3}\right) \approx (-2, 3.464)$

5.

$(-1.004, 0.996)$

7.

9.

11.

$\left(\sqrt{2}, \pi/4\right), \left(-\sqrt{2}, 5\pi/4\right)$

13.

$(5, 2.214), (-5, 5.356)$

15.

$\left(-2, 5\pi/6\right), \left(2, 11\pi/6\right)$

17. $(3.606, -0.588)$ **19.** $(2.833, 0.490)$

21. (a)   (b)

23. c **24.** b **25.** a **26.** d

27. $r = a$

29. $r = 4 \csc \theta$

31. $r = \dfrac{-2}{3 \cos \theta - \sin \theta}$

33. $r = 9 \csc^2 \theta \cos \theta$

35. $x^2 + y^2 = 9$

37. $x^2 + y^2 - y = 0$

39. $\sqrt{x^2 + y^2} = \arctan (y/x)$

41. $x - 3 = 0$

43.

$0 \le \theta < 2\pi$

45.

$0 \le \theta < 2\pi$

47.

$-\pi < \theta < \pi$

49.

$0 \le \theta < 4\pi$

51.

$0 \le \theta < \pi/2$

53. $(x - h)^2 + (y - k)^2 = h^2 + k^2$

Radius: $\sqrt{h^2 + k^2}$

Center: (h, k)

55. $2\sqrt{5}$ **57.** ≈ 5.6

59. $\dfrac{dy}{dx} = \dfrac{2 \cos \theta (3 \sin \theta + 1)}{6 \cos^2 \theta - 2 \sin \theta - 3}$

$(5, \pi/2)$: $dy/dx = 0$

$(2, \pi)$: $dy/dx = -2/3$

$(-1, 3\pi/2)$: $dy/dx = 0$

61. (a) and (b)

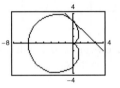

(c) $dy/dx = -1$

63. (a) and (b)

(c) $dy/dx = -\sqrt{3}$

65. Horizontal: $(2, 3\pi/2), \left(\frac{1}{2}, \pi/6\right), \left(\frac{1}{2}, 5\pi/6\right)$

Vertical: $\left(\frac{3}{2}, 7\pi/6\right), \left(\frac{3}{2}, 11\pi/6\right)$

67. $(5, \pi/2), (1, 3\pi/2)$

69.

$(0, 0), (1.4142, 0.7854),$
$(1.4142, 2.3562)$

71.

$(7, 1.5708), (3, 4.7124)$

73.

$\theta = 0$

75.

$\theta = \pi/2$

77.

$\theta = \pi/6, \pi/2, 5\pi/6$

79.

$\theta = 0, \pi/2$

81.

83.

(c)

85.

87.

103. Proof

105. (a) $r = 2 - \sin(\theta - \pi/4)$ (b) $r = 2 + \cos\theta$

$$= 2 - \frac{\sqrt{2}(\sin\theta - \cos\theta)}{2}$$

89.

91.

(c) $r = 2 + \sin\theta$ (d) $r = 2 - \cos\theta$

107. (a)   (b)

93.

95.

109.   **111.**

$\psi = \pi/2$ $\psi = \arctan\frac{1}{3} \approx 18.4°$

113.   **115.** True   **117.** True

97. The rectangular coordinate system is a collection of points of the form (x, y), where x is the directed distance from the y-axis to the point and y is the directed distance from the x-axis to the point. Every point has a unique representation.

The polar coordinate system is a collection of points of the form (r, θ), where r is the directed distance from the origin O to a point P and θ is the directed angle, measured counterclockwise, from the polar axis to the segment $\overline{OP}$. Polar coordinates do not have unique representations.

99. $r = a$: Circle of radius a centered at the pole
$\theta = b$: Line passing through the pole

101. (a)   (b)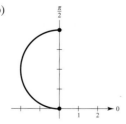

$\psi = \pi/3, \ 60°$

Section 10.5 (page 745)

1. $2\displaystyle\int_{\pi/2}^{\pi} \sin^2\theta \, d\theta$ **3.** $\dfrac{1}{2}\displaystyle\int_{\pi/2}^{3\pi/2} (1 - \sin\theta)^2 \, d\theta$

5. (a) and (b) 16π **7.** $\pi/3$ **9.** $\pi/8$ **11.** $3\pi/2$

115. Hyperbola

117. $r = 10 \sin \theta$

119. $r = 4/(1 - \cos \theta)$ **121.** $r = 5/(3 - 2 \cos \theta)$

P.S. Problem Solving   (page 759)

1. (a)

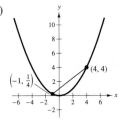

(b)–(c) Proofs

3. Proof

5. (a) $r = 2a \tan \theta \sin \theta$

(b) $x = 2at^2/(1 + t^2)$

$y = 2at^3/(1 + t^2)$

(c) $y^2 = x^3/(2a - x)$

7. (a) $y^2 = x^2[(1 - x)/(1 + x)]$

(b) $r = \cos 2\theta \cdot \sec \theta$

(c)

(d) $y = x, y = -x$

(e) $\left(\dfrac{\sqrt{5} - 1}{2}, \pm \dfrac{\sqrt{5} - 1}{2} \sqrt{-2 + \sqrt{5}} \right)$

9. (a)

Generated by Mathematica

(b) Proof

(c) $a, 2\pi$

11. $A = \frac{1}{2}ab$ **13.** $r^2 = 2 \cos 2\theta$

15. (a) First plane: $x_1 = \cos 70°(150 - 375t)$

$y_1 = \sin 70°(150 - 375t)$

Second plane: $x_2 = \cos 45°(450t - 190)$

$y_2 = \sin 45°(190 - 450t)$

(b) $\{[\cos 45°(450t - 190) - \cos 70°(150 - 375t)]^2$

$+ [\sin 45°(190 - 450t) - \sin 70°(150 - 375t)]^2\}^{1/2}$

(c)

0.4145 hr; Yes

17.

$n = 1, 2, 3, 4, 5$ produce "bells"; $n = -1, -2, -3, -4, -5$
produce "hearts."

APPENDIX C (page A27)

1. (a) Fixed cost

(b)

(c) Yes; the extremum occurs when production costs are increasing at their slowest rate.

3. 4500 **5.** 300 **7.** 200 **9.** 200

11. \$60 **13.** \$35 **15.** $x = 3$ **17.** Proof

19. (a)

Order size, x	Price	Profit, P
102	$90 - 2(0.15)$	$102[90 - 2(0.15)] - 102(60) = 3029.40$
104	$90 - 4(0.15)$	$104[90 - 4(0.15)] - 104(60) = 3057.60$
106	$90 - 6(0.15)$	$106[90 - 6(0.15)] - 106(60) = 3084.60$
108	$90 - 8(0.15)$	$108[90 - 8(0.15)] - 108(60) = 3110.40$
110	$90 - 10(0.15)$	$110[90 - 10(0.15)] - 110(60) = 3135.00$
112	$90 - 12(0.15)$	$112[90 - 12(0.15)] - 112(60) = 3158.40$

(b)

Order size, x	Price	Profit, P
.	.	.
.	.	.
.	.	.
146	$90 - 46(0.15)$	$146[90 - 46(0.15)] - 146(60) = 3372.60$
148	$90 - 48(0.15)$	$148[90 - 48(0.15)] - 148(60) = 3374.40$
150	$90 - 50(0.15)$	$150[90 - 50(0.15)] - 150(60) = 3375.00$
152	$90 - 52(0.15)$	$152[90 - 52(0.15)] - 152(60) = 3374.40$
154	$90 - 54(0.15)$	$154[90 - 54(0.15)] - 154(60) = 3372.60$
.	.	.
.	.	.
.	.	.

Maximum profit: \$3375.00

(c) $P = x[90 - (x - 100)(0.15)] - x(60) = 45x - 0.15x^2,$
$x \geq 100$

(d) 150 units

(e)

(150, 3375)

21. The line should run from the power station to a point across the river $3/(2\sqrt{7})$ mile downstream.

23. $x \approx 40$ units **25.** \$30,000

27. (a)

(b) July

(c) The cosine factor; 9.90

(d) The term $0.02t$ would mean a steady growth of sales over time. In this case, the maximum sales in 2008 (that is, on $49 \leq t \leq 60$) would be about 11.6 thousand gallons.

29. (a)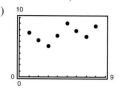

(b) $S = 6.2 + 0.25t + 1.5 \sin\left(\dfrac{\pi}{2}t\right)$

(c)   (d) \$12,000

31. $\eta = -\dfrac{17}{3}$; elastic **33.** $\eta = -\dfrac{1}{2}$; inelastic

Index of Applications

Engineering and Physical Sciences

Index

ALGEBRA

Factors and Zeros of Polynomials

Let $p(x) = a_n x^n + a_{n-1} x^{n-1} + \cdots + a_1 x + a_0$ be a polynomial. If $p(a) = 0$, then a is a *zero* of the polynomial and a solution of the equation $p(x) = 0$. Furthermore, $(x - a)$ is a *factor* of the polynomial.

Fundamental Theorem of Algebra

An nth degree polynomial has n (not necessarily distinct) zeros. Although all of these zeros may be imaginary, a real polynomial of odd degree must have at least one real zero.

Quadratic Formula

If $p(x) = ax^2 + bx + c$, and $0 \leq b^2 - 4ac$, then the real zeros of p are $x = \left(-b \pm \sqrt{b^2 - 4ac}\right)/2a$.

Special Factors

$$x^2 - a^2 = (x - a)(x + a) \qquad\qquad x^3 - a^3 = (x - a)(x^2 + ax + a^2)$$

$$x^3 + a^3 = (x + a)(x^2 - ax + a^2) \qquad\qquad x^4 - a^4 = (x^2 - a^2)(x^2 + a^2)$$

Binomial Theorem

$$(x + y)^2 = x^2 + 2xy + y^2 \qquad\qquad (x - y)^2 = x^2 - 2xy + y^2$$

$$(x + y)^3 = x^3 + 3x^2 y + 3xy^2 + y^3 \qquad\qquad (x - y)^3 = x^3 - 3x^2 y + 3xy^2 - y^3$$

$$(x + y)^4 = x^4 + 4x^3 y + 6x^2 y^2 + 4xy^3 + y^4 \qquad\qquad (x - y)^4 = x^4 - 4x^3 y + 6x^2 y^2 - 4xy^3 + y^4$$

$$(x + y)^n = x^n + nx^{n-1}y + \frac{n(n-1)}{2!}x^{n-2}y^2 + \cdots + nxy^{n-1} + y^n$$

$$(x - y)^n = x^n - nx^{n-1}y + \frac{n(n-1)}{2!}x^{n-2}y^2 - \cdots \pm nxy^{n-1} \mp y^n$$

Rational Zero Theorem .

If $p(x) = a_n x^n + a_{n-1} x^{n-1} + \cdots + a_1 x + a_0$ has integer coefficients, then every *rational zero* of p is of the form $x = r/s$, where r is a factor of a_0 and s is a factor of a_n.

Factoring by Grouping

$$acx^3 + adx^2 + bcx + bd = ax^2(cx + d) + b(cx + d) = (ax^2 + b)(cx + d)$$

Arithmetic Operations

$$ab + ac = a(b + c) \qquad \frac{a}{b} + \frac{c}{d} = \frac{ad + bc}{bd} \qquad \frac{a + b}{c} = \frac{a}{c} + \frac{b}{c}$$

$$\frac{\left(\dfrac{a}{b}\right)}{\left(\dfrac{c}{d}\right)} = \left(\frac{a}{b}\right)\left(\frac{d}{c}\right) = \frac{ad}{bc} \qquad \frac{\left(\dfrac{a}{b}\right)}{c} = \frac{a}{bc} \qquad \frac{a}{\left(\dfrac{b}{c}\right)} = \frac{ac}{b}$$

$$a\left(\frac{b}{c}\right) = \frac{ab}{c} \qquad \frac{a - b}{c - d} = \frac{b - a}{d - c} \qquad \frac{ab + ac}{a} = b + c$$

Exponents and Radicals

$$a^0 = 1, \quad a \neq 0 \qquad (ab)^x = a^x b^x \qquad a^x a^y = a^{x+y} \qquad \sqrt{a} = a^{1/2} \qquad \frac{a^x}{a^y} = a^{x-y} \qquad \sqrt[n]{a} = a^{1/n}$$

$$\left(\frac{a}{b}\right)^x = \frac{a^x}{b^x} \qquad \sqrt[n]{a^m} = a^{m/n} \qquad a^{-x} = \frac{1}{a^x} \qquad \sqrt[n]{ab} = \sqrt[n]{a}\,\sqrt[n]{b} \qquad (a^x)^y = a^{xy} \qquad \sqrt[n]{\frac{a}{b}} = \frac{\sqrt[n]{a}}{\sqrt[n]{b}}$$